Contemporary Physicists

Gian-Carlo Rota
David Sharp
Editors

Birkhäuser
Boston · Basel · Stuttgart

T.D. Lee in the early 1980s. (Photo taken by
Professor James Rainwater of the Columbia Physics Department).

T.D. Lee

Selected Papers
Volume 3

Random Lattices to Gravity

G. Feinberg
Editor

Birkhäuser
Boston · Basel · Stuttgart
1986

T.D. Lee
G. Feinberg *(Editor)*
Department of Physics
Columbia University
New York, NY 10027
USA

Library of Congress Cataloging in Publication Data
Lee, T.D., 1926–
T.D. Lee selected papers.
(Contemporary physicists)
Includes bibliographies.
1. Physics. 2. Nuclear physics. I. Feinberg, Gerald, 1933– . II. Title. III. Series.
QC71.L4325 1986 530 86-6169

CIP-Kurztitelaufnahme der Deutschen Bibliothek
Lee, Tsung D.:
Selected papers / T. D. Lee. G. Feinberg, ed. —
Boston : Basel : Stuttgart : Birkhäuser
(Contemporary physicists)
ISBN 3-7643-3344-8 (Stuttgart . . .);
ISBN 0-8176-3344-8 (Boston . . .)
NE: Lee, Tsung D.: [Sammlung]
Vol. 3. Random lattices to gravity. —1986.
ISBN 3-7643-3343-X (Stuttgart . . .)
ISBN 0-8176-3343-X (Boston . . .)

Printed in USA.

ISBN 0-8176-3344-8 (complete set)	3-7643-3344-8 (complete set)
ISBN 0-8176-3341-3 (Vol. 1)	3-7643-3341-3 (Vol. 1)
ISBN 0-8176-3342-1 (Vol. 2)	3-7643-3342-1 (Vol. 2)
ISBN 0-8176-3343-X (Vol. 3)	3-7643-3343-X (Vol. 3)

Preface

In these volumes, the most significant of the collected papers of the Chinese–American theoretical physicist Tsung-Dao Lee are printed. A complete list of his published papers, in order of publication, appears in the Bibliography of T.D. Lee. The papers have been arranged into ten categories, in most cases according to the subject matter. At the beginning of each of the first eight categories of papers, there is a commentary on the content and significance of all of the papers in the category. The two short final categories do not have any commentaries.

The editor would like to thank Dr. Richard Friedberg for his assistance in the early stages of the editorial work on this project, as well as for writing commentaries on the papers of Categories III and IV. I would also like to thank Dr. Norman Christ for writing the commentary on the papers of Category VII. The assistance of Irene Tramm was invaluable in many aspects of preparing this collection, including locating copies of Lee's papers.

GERALD FEINBERG

List of Categories of T.D. Lee's Papers

Volume 1

I. Weak Interactions
II. Early Papers on Astrophysics and Hydrodynamics
III. Statistical Mechanics
IV. Polarons and Solitons

Volume 2

V. Quantum Field Theory
VI. Symmetry Principles

Volume 3

VII. Discrete Physics
VIII. Strong Interaction Models
IX. Historical Papers
X. Gravity (Continuum Theory)

Contents (Volume 3)*

*The numbers in brackets preceding the titles of the papers are from the Bibliography of T.D. Lee (pages xiii–xxii in this volume).

Introduction

Tsung-Dao Lee was born in Shanghai, China, on November 25, 1926. After early education in China he came to the United States in 1946, through the help of Ta-You Wu, to do graduate study in physics at the University of Chicago. In that period, the presence of Enrico Fermi at Chicago had attracted many talented physics students, including J. Anderson, O. Chamberlain, G. Chew, R. Garwin, M. Goldberger, J. Orear, M. Rosenbluth, J. Steinberger R. Sternheimer, S. Treiman, L. Wolfenstein and C.N. Yang. Lee became a doctoral student of Fermi, and while a student already published the important paper, with Rosenbluth and Yang ([1]), on what came to be known as the universal Fermi interaction and the intermediate boson. Fermi directed Lee's research towards topics in astrophysics, which eventually resulted in his thesis, on white dwarf stars ([2]), which demonstrated conclusively that these stars contain no hydrogen and must therefore be the end points of stellar evolution.

Following the completion of his thesis, Lee spent eight months working at Chicago's Yerkes Observatory, with S. Chandrasekhar, on problems of turbulence ([3] and [5]). After a brief stay at the University of California at Berkeley, he went to the Institute for Advanced Study, at Princeton, where he spent the next two years. At the Institute, Lee, in collaboration with Yang, wrote two important papers on statistical mechanics ([7] and [8]). The first of these provides a rigorous definition of the thermodynamic limit for various phases, and shows that the thermodynamic functions can differ from phase to phase. In the second, the "unit circle theorem" on the roots of the grand partition function was proven.

In 1953, Lee went to Columbia University as an Assistant Professor in Physics. Except for a three-year period in the early 1960s, he has remained on the Columbia faculty, becoming Professor of Physics in 1956, and University Professor in 1983. Lee's first work at Columbia was on the renormalizable field theory model known as the Lee model ([13]). His attention soon turned to particle physics, and the developing puzzle of K meson decays. Lee and various collaborators suggested several models to account for the phenomena, but the observations eventually impelled him, with Yang, to make a careful study of the evidence about various symmetries in particle reactions. The result, in [21], was a precise criterion for what experiments need to be done to test the discrete symmetries C, P and T. The point was made that existing experiments on weak interactions did not test them, and experiments to do so, on beta decay and other weak interactions, were proposed. The paper was soon followed by another paper ([24]), proposing a specific model of two-component neutrinos, which made definite predictions for the amount of parity non-conservation expected. At the same time, Lee and collaborators also examined T and CP violations, with particular reference to the neutral kaon system [25]. The experiments were soon carried out by Wu and collaborators, by Lederman, Garwin and collaborators, and by Telegdi and collaborators, with the result that parity and charge conjugation invariance were not satisfied by weak interactions. This confirmation of their suggestion earned Lee and Yang the 1957 Nobel Prize for Physics (see [38] for the text of Lee's Nobel Prize Lecture).

In the following year Lee, with several collaborators, extended the new ideas about weak interactions to several other processes ([25], [27], [36] and [41]). The cumulation of this work on symmetry principles was the basis of the dominant theme in particle physics through the 1960s.

In the period 1957–1960 Lee, Yang and others studied the statistical mechanics of a Bose system of hard spheres ([26], [42] and [43]), finding a way of eliminating divergences in the theory by selective summation of series. They also, in this period, produced a general formalism for studying many-body problems in statistical mechanics ([45]–[49]).

Beginning in 1960, it became clear that experiments could be done with high-energy neutrinos. Lee, with several collaborators, provided much of the theoretical impetus for these early experiments ([50], [56], [58] and [60]). These experiments have proven to be the major source of information about weak interactions through the present.

The successful description of low-energy weak interactions led many physicists to revive the idea suggested in [1] that these interactions result from the exchange of intermediate vector bosons with universal coupling. Beginning in 1960, Lee wrote a number of papers on the properties and implications of these then hypothetical particles ([58], [61], [62], [68], and [115]). Of these, paper [61] was particularly important because it indicated the need for careful consideration of the gauge invariance questions involved in the interaction of spin-one intermediate vector bosons.

In 1964, Lee, with M. Nauenberg, analyzed the divergences connected with particles of zero rest mass, and described a general method for dealing with these divergences, which plays an important role in contemporary work in QCD, with its massless, self-interacting gluons.

The discovery in 1964 of *CP* non-invariance focussed the attention of many theorists on this problem. Lee, with several collaborators, suggested a number of models for *CP* violation ([77] and [80]), which later proved not to be in agreement with experiment. Some years later, in the context of spontaneously broken symmetry, he suggested another model ([146], [147] and [149]) which remains one of the viable explanations for *CP* violation.

In the period 1969–1971, Lee, together with G.C. Wick, proposed a method of eliminating ultraviolet divergences in quantum field theory, through the introduction of an indefinite metric in Hilbert space ([117], [118], [121] and [129]). They concluded that this type of theory is not in disagreement with any existing experiments.

In 1974, Lee and Wick began a study of the hypothesis that in some circumstances, such as inside a heavy nucleus, the broken symmetry that sometimes occurs through the vacuum expectation value of a scalar field could be restored ([150], [152], [155], [156] and [170]). This would show up as a change in the mass of nucleons in the heavy nucleus, with a resultant increase in nuclear density. This suggestion may eventually be testable through the collisions of beams of relativistic heavy ions.

In the late 1970s, theoretical physicists became interested in solutions to classical field equations (such as magnetic monopoles) that are stabilized by boundary conditions. Lee and collaborators showed that in several field theory models, stable configurations could be produced by other mechanisms than boundary conditions ([150], [159], [162], [163] and

[165]). These ideas were applied to models of hadrons in [167], with R. Friedberg, where it is shown that it can reproduce the results of the "bag" model of hadrons.

In 1982, Lee's interest turned to the lattice approach to field theory. In three papers with N.H. Christ and R. Friedberg ([179]–[181]), the concept of a random lattice was introduced, and it was shown how field theory calculations could be performed on such a lattice. Applications of these ideas to actual problems in QCD are presently under investigation.

The discrete space–time approach used in lattice theories led Lee to make the suggestion not previously considered by other workers with lattices, that space and time were really discrete. The general consequences of this idea for classical and quantum mechanics were explored in [182] and [183]. The possible application of discrete space time to general relativity is under current investigation by Lee and collaborators.

Bibliography of T.D. Lee

Note: **Boldface** numbers in parentheses at the end of each entry denote the volume that the entry appears in.

1949

[1] Interaction of Mesons with Nucleons and Light Particles (with M. Rosenbluth and C.N. Yang). *Phys. Rev.* **75,** 905. **(I)**

1950

[2] Hydrogen Content and Engery-Productive Mechanism of White Dwarfs. *Ap. J.* **111,** 625. **(I)**
[3] Note on the Coefficient of Eddy Viscosity in Isotropic Turbulence. *Phys. Rev.* **77,** 842. **(I)**
*[4] On the Proton–Proton Reaction in White Dwarf Stars. *Ap. J.* **112,** 561.

1951

[5] Difference Between Turbulence in a Two-Dimensional Fluid and in a Three-Dimensional Fluid. *J. Ap. Phys.* **22,** 524. **(I)**

1952

*[6] On Some Statistical Properties of Hydrodynamical and Magneto-Hydrodynamical Fields. *Quart. Ap. Math.* **10,** 69.
[7] Statistical Theory of Equations of State and Phase Transitions. I. Theory of Condensation (with C.N. Yang). *Phys. Rev.* **87,** 404. **(I)**
[8] Statistical Theory of Equations of State and Phase Transitions. II. Lattice Gas and Ising Model (with C.N. Yang). *Phys. Rev.* **87,** 410. **(I)**
[9] Motion of Slow Electrons in Polar Crystals (with David Pines). *Phys. Rev.* **88,** 960. **(I)**

1953

[10] The Motion of Slow Electrons in Polar Crystals (with F. Low and D. Pines). *Phys. Rev.* **90,** 297. **(I)**
[11] Interaction of a Nonrelativistic Particle and a Scalar Field with Applications to Slow Electrons in Polar Crystal (with D. Pines). *Phys. Rev.* **92,** 883. **(I)**

1954

[12] Intermediate Coupling Theory of Meson-Nucleon Scattering (with R. Christian). *Phys. Rev.* **94,** 1760. **(II)**
[13] Some Special Examples in Renormalizable Field Theory. *Phys. Rev.* **95,** 1329. **(II)**

*Does not appear in these volumes.

1955

[14] Conservation of Heavy Particles and Generalized Gauge Transformations (with C.N. Yang). *Phys. Rev.* **98,** 1501. **(II)**
*[15] Absorption Experiments Involving Heavy Mesons. *Phys. Rev.* **99,** 337.
[16] Speculations on Heavy Mesons (with J. Orear). *Phys. Rev.* **100,** 932. **(II)**
*[17] Scattering of Pions by Nucleons in Intermediate Coupling (with M. Friedman and R. Christian). *Phys. Rev.* **100,** 1494.

1956

*[18] Model for Multiple Meson Production (with E.M. Henley). *Phys. Rev.* **101,** 1536.
*[19] Mass Degeneracy of the Heavy Mesons (with C.N. Yang). *Phys. Rev.* **102,** 290.
[20] Charge Conjugation, a New Quantum Number G, and Selection Rules Concerning a Nucleon-Antinucleon System (with C.N. Yang). *Nuovo Cimento* **3,** 749. **(II)**
[21] Question of Parity Conservation in Weak Interactions (with C.N. Yang). *Phys. Rev.* **104,** 254. **(II)**
[21a] Errata to "Question of Parity Conservation in Weak Interactions" (with C.N. Yang). *Phys. Rev.* **106,** 1371 (1957). **(II)**
*[22] Possible Interference Phenomena Between Parity Doublets (with C.N. Yang). *Phys. Rev.* **104,** 822.

1957

[23] Many-Body Problem in Quantum Mechanics and Quantum Statistical Mechanics (with C.N. Yang). *Phys. Rev.* **105,** 1119. **(I)**
[24] Parity Nonconservation and a Two-Component Theory of the Neutrino (with C.N. Yang). *Phys. Rev.* **105,** 1671. **(II)**
[25] Remarks on Possible Noninvariance Under Time Reversal and Charge Conjugation (with R. Oehme and C.N. Yang). *Phys. Rev.* **106,** 340. **(II)**
[26] Eigenvalues and Eigenfunctions of a Bose System of Hard Spheres and Its Low-Temperature Properties (with K. Huang and C.N. Yang). *Phys. Rev.* **106,** 1135. **(I)**
[27] Introductory Talk at the 1957 Rochester Conference, Session on Weak Interactions. Interscience Publishers, Section VII, pp. 1–12. **(II)**
[28] Possible Detection of Parity Nonconservation in Hyperon Decay (with J. Steinberger, G. Feinberg, P.K. Kabir and C.N. Yang). *Phys. Rev.* **106,** 1367. **(II)**
[29] Capture of μ^- Mesons by Protons (with K. Huang and C.N. Yang). *Phys. Rev.* **108,** 1340. **(I)**
[30] Possible Nonlocal Effects in μ Decay (with C.N. Yang). *Phys. Rev.* **108,** 1611. **(I)**
[31] General Partial Wave Analysis of the Decay of a Hyperon of Spin ½ (with C.N. Yang). *Phys. Rev.* **108,** 1645. **(I)**
[32] Elementary Particles and Weak Interactions (with C.N. Yang). BNL 443 (T-91). **(II)**
[33] Theoretical Implications of Parity Violations in Beta Interactions. *Proc. of Rehovoth Conference on Nuclear Structure,* North-Holland Publishing Co. (1958), p. 336. **(I, II)**
*[34] Theoretical Implications of Parity Violations in Beta Interactions. *Proc. of Padua–Venice Conference on Mesons and Newly Discovered Particles,* September.

1958

[35] Possible Determination of the Spin of Λ^0 from Its Large Decay Angular Asymmetry (with C.N. Yang). *Phys. Rev.* **109,** 1755. **(I, II)**
[36] Effect of the Hyperfine Splitting of a μ-Mesic Atom on Its Lifetime (with J. Bernstein, H. Primakoff and C.N. Yang). *Phys. Rev.* **111,** 313. **(I)**
*[37] Weak Interactions and Nonconservation of Parity (Nobel lecture). *Science* **127,** 569.

[38] Weak Interactions and Nonconservation of Parity, Le Prix Nobel Lecture, December 11, 1957. Stockholm. **(I, II)**

*[39] The Low Temperature Properties of a Bose System of Hard Spheres (with C.N. Yang). *Proc. of Midwest Conference on Theoretical Physics, St. Louis, March 14–15*, p. 149.

*[40] A Review of Weak Interactions. *Proc. of Midwest Conference on Theoretical Physics, St. Louis, March 14–15*, p. 167.

[41] Decay Modes of a $(\theta + \bar{\theta})$ System (with M. Goldhaber and C.N. Yang). *Phys. Rev.* **112,** 1796. **(II)**

[42] Low Temperature Behavior of a Dilute Bose System of Hard Spheres. I. Equilibrium Properties (with C.N. Yang). *Phys. Rev.* **112,** 1419. **(I)**

1959

[43] Low Temperature Behavior of a Dilute Bose System of Hard Spheres. II. Non-Equilibrium Properties (with C.N. Yang). *Phys. Rev.* **113,** 1406. **(I)**

[44] Possible Determination of the Helicity of Elementary Excitations in Liquid He II (with F. Mohling). *Phys. Rev. Letters* **2,** 284. **(I)**

[45] Many-Body Problem in Quantum Statistical Mechanics. I. General Formulation (with C.N. Yang). *Phys. Rev.* **113,** 1165. **(I)**

[46] Many-Body Problem in Quantum Statistical Mechanics. II. Virial Expansion for Hard-Sphere Gas (with C.N. Yang). *Phys. Rev.* **116,** 25. **(I)**

1960

[47] Many-Body Problem in Quantum Statistical Mechanics. III. Zero Temperature Limit for Dilute Hard Spheres (with C.N. Yang). *Phys. Rev.* **117,** 12. **(I)**

[48] Many-Body Problem in Quantum Statistical Mechanics. IV. Formulation in Terms of Average Occupation Number in Momentum Space (with C.N. Yang). *Phys. Rev.* **117,** 22. **(I)**

[49] Many-Body Problem in Quantum Statistical Mechanics. V. Degenerate Phase in Bose–Einstein Condensation (with C.N. Yang). *Phys. Rev.* **117,** 897. **(I)**

[50] Theoretical Discussions on Possible High Energy Neutrino Experiments (with C.N. Yang). *Phys. Rev. Letters* **4,** 307. **(I)**

[51] Implications of the Intermediate Boson Basis of the Weak Interactions: Existence of a Quartet of Intermediate Bosons and Their Dual Isotopic Spin Transformation Properties (with C.N. Yang). *Phys. Rev.* **119,** 1410. **(I)**

*[52] Intermediate Boson Hypothesis of Weak Interactions. 1960 Rochester Conference on High Energy Physics, p. 567.

[53] Elementary Particles. *Physics Today* **13,** 30. **(I, II)**

1961

[54] Some Considerations on Global Symmetry (with C.N. Yang). *Phys. Rev.* **122,** 1954. **(II)**

*[55] Some Theoretical Considerations on the Desirability of a 300 to 1000 Bev Proton Accelerator (with R. Serber, G.C. Wick and C.N. Yang). BNL 772 (T-290), p. 15.

[56] Production Cross Section of Intermediate Bosons by Neutrinos in the Coulomb Field of Protons and Iron (with P. Markstein and C.N. Yang). *Phys. Rev. Letters* **7,** 429. **(I)**

*[57] Theoretical Considerations on *W*-Production by "High"-Energy Neutrinos. CERN Conference on Theoretical Aspects of Very High Energy Phenomena, Session on Weak Interactions, June 1961, p. 44.

[58] High-Energy Neutrino Experiments. CERN 61-31: Particle Physics—CERN Seminars 1961, p. 65. **(I)**

*[59] Lepton Physics. Aix-en-Provence International Conference on Elementary Particles, September 1961, Vol. II, p. 15.

1962

[60] High Energy Neutrino Reactions Without Production of Intermediate Bosons (with C.N. Yang). *Phys. Rev.* **126,** 2239. **(I)**
[61] A Theory of Charged Vector Mesons Interacting with Electromagnetic Field (with C.N. Yang). *Phys. Rev.* **128,** 885. **(I)**
[62] Application of ξ-Limiting Process to Intermediate Bosons. *Phys. Rev.* **128,** 899. **(I)**
[62a] Errata to "Application of ξ-Limiting Processes to Intermediate Bosons." *Phys. Rev.* **173,** 1806 (1968). **(I)**
*[63] Speculations Concerning the $\Delta S = -\Delta Q$ Reactions and the Intermediate Boson Structures in Weak Interactions. *Phys. Rev. Letters* **9,** 319.

1963

*[64] Quantum Mechanical Many-Body Problem and the Low Temperature Properties of a Bose System of Hard Spheres (with K. Huang and C.N. Yang). *The Many-Body Problem,* Interscience Publishers, p. 165.
[65] Spin ½ Wave Equation in De-Sitter Space (with F. Gürsey). *Proc. of National Academy of Sciences* **49,** 179. **(II).**
*[66] Remarks Concerning the Recent High-Energy Neutrino Experiment (with G. Danby, J.M. Gaillard, K. Goulianos, L. Lederman, N. Mistry, M. Schwartz and J. Steinberger). *Phys. Rev. Letters* **10,** 260.
[67] Intensity of Upward Muon Flux Due to Cosmic-Ray Neutrinos Produced in the Atmosphere (with H. Robinson, M. Schwartz and R. Cool). *Phys. Rev.* **132,** 1297. **(I)**
[68] Electromagnetic Form Factor of the Neutrinos (with J. Bernstein). *Phys. Rev. Letters* **11,** 512. **(I)**

1964

[69] Degenerate Systems and Mass Singularities (with M. Nauenberg). *Phys. Rev.* **133,** B1549. **(II)**
[70] A Possible Method of Determining the Moment of Charge of the ν_e (with A. Sirlin). *Reviews of Modern Physics* **36,** 666. **(I)**
[71] Radiative Corrections to Electromagnetic Moments and Leptonic Decays of the Intermediate Boson *W*. *Phys. Rev. Letters* **12,** 569. **(I)**
[72] Implications of Approximate SU_3 Symmetry and Mass Formulas for the Mesons (with F. Gürsey and M. Nauenberg). *Phys. Rev.* **135,** B467. **(III)**
[73] CP Invariance and the 2π Decay Mode of the K_2^0 (with J. Bernstein and N. Cabibbo). *Physics Letters* **12,** 146. **(II)**

1965

*[74] Remarks on Triplet Models in SU_3 Symmetry. *Nuovo Cimento* **35,** 933.
[75] Symmetry Properties of Leptons in the Zero-Mass Limit. *Nuovo Cimento* **35,** 945. **(II)**
*[76] Comments on Elementary Particle Physics. *Nature of Matter, Purposes of High Energy Physics,* edited by L.C.L. Yuan, Brookhaven National Laboratory #888 (T-360), p. 50.
[77] Hypercharge Conservation, CP Invariance and the Possible Existence of a Zero-Mass Zero-Spin Field. *Phys. Rev.* **137,** B1621. **(II)**
[78] An Analysis of CP Noninvariant Interactions and the K_1^0, K_2^0 System (with L. Wolfenstein). *Phys. Rev.* **138,** B1490. **(II)**
*[79] Possible *C* Noninvariant Effects in the 3π Decay Modes of η^0 and ω^0. *Phys. Rev.* **139,** B1415.
[80] Possible *C, T* Noninvariance in the Electromagnetic Interaction (with J. Bernstein and G. Feinberg). *Phys. Rev.* **139,** B1650. **(II)**
*[81] Minimal Electromagnetic Interaction and *C, T* Noninvariance. *Phys. Rev.* **140,** B967.

*[82] Classification of All *C* Noninvariant Electromagnetic Interactions and the Possible Existence of a Charged, But $C = 1$, Particle. *Phys. Rev.* **140,** B959.

*[83] Neutrino Physics. *Informal Conference on Experimental Neutrino Physics,* CERN 1965, edited by C. Franzinetti, CERN #65-32, p. 275.

[84] Weak Interactions (with C.S. Wu). *Ann. Rev. Nuclear Science* **15,** 381. **(I)**

[84a] Corrections to "Weak Interactions" (with C.S. Wu). *Ann. Rev. Nuclear Science* **17,** 513 (1967). **(I)**

1966

[85] An Elementary Discussion of Possible Noninvariance Under *T, CP* and *CPT* in Hyperon Decays. *Preludes in Theoretical Physics* (in Honor of V.F. Weisskopf), edited by De-Shalit, Feshbach and Van Hove, North-Holland Publishing Co., p. 5. **(II)**

*[86] Theoretical Survey of High-Energy Neutrino Interactions. *Weak Interactions and High Energy Neutrino Physics* (Rendiconti della Scuola Internazionale di Fisica "E. Fermi"-XXXII Corso), Academic Press, p. 142.

*[87] Mass Formulae and SU_3 Symmetry. *Ibid,* p. 311.

*[88] Possible *C* Noninvariance in the Electromagnetic Interaction. *Particle Symmetries* (Vol. II of 1965 Brandeis University Summer Institute in Theoretical Physics), edited by M. Chretien and S. Deser, Gordon and Breach, p. 453.

[89] Weak Interactions and Questions of C, P, T Noninvariances. *Proc. Oxford International Conference on Elementary Particles, 19/25 September 1965,* Rutherford High Energy Laboratory, p. 225. **(II)**

*[90] Theoretical Remarks on CP Noninvariance. *Proc. of International Conference on Weak Interactions, October 25–27, 1965,* Argonne National Laboratory #7130, p. 431.

[91] Space Inversion, Time Reversal and Particle-Antiparticle Conjugation (based on a lecture delivered at the Fourth Annual Science Conference of the Belfer Graduate School of Science, Yeshiva University), *Physics Today* **19,** 23. **(II)**

*[92] Possible Tests of C_{st} and T_{st} Invariances in $l^{\pm} + N \rightarrow l^{\pm} + \Gamma$ and $A \rightarrow B + e^{+} + e^{-}$ (with N.H. Christ). *Phys. Rev.* **143,** 1310.

[93] Space Inversion, Time Reversal and Other Discrete Symmetries in Local Field Theories (with G.C. Wick). *Phys. Rev.* **148,** 1385. **(II)**

*[94] Reciprocity Relations in Photo-Pion Reactions (with N.H. Christ). *Phys. Rev.* **148,** 1520.

[95] Weak Interactions: Chapter 8—Decays of Charged K Mesons, and Chapter 9—Decays of Neutral K Mesons (with C.S. Wu). *Ann. Rev. Nuclear Science* **16,** 471. **(I)**

*[95a] Corrections to "Weak Interactions" (Chapters 8 and 9) (with C.S. Wu). *Ann. Rev. Nuclear Science* **17,** 513 (1967). **(I)**

1967

*[96] Space Inversion, Time Reversal and Charge Conjugation. *Cargese Lectures, 1966,* Gordon and Breach, p. 55.

[97] Other Aspects of Weak Interactions: High Energy Neutrino Physics and Questions of C, P, T Noninvariance. *Proc. of XIIIth International Conference on High Energy Physics, Berkeley, 1966,* Univ. California Press, p. 75. **(II)**

*[98] Time Reversal Symmetry. *Plenarvorträge,* Teil II (Physikertagung 1966 München), B.G. Teubner, p. 243.

*[99] Hadronic Electromagnetic Current and Neutral Vector Mesons. *Proc. of Fifth Eastern Theoretical Physics Conference, Brown University, 1966,* W.A. Benjamin, p. 101.

[100] Neutral Vector Mesons and the Hadronic Electromagnetic Current (with N.M. Kroll and Bruno Zumino). *Phys. Rev.* **157,** 1376. **(II)**

[101] Algebra of Fields (with S. Weinberg and Bruno Zumino). *Phys. Rev. Letters* **18,** 1029. **(II)**

*[102] High Energy Weak Interactions. *Proc. of Argonne High Energy Physics Meeting.*

[103] Field-Current Identities and the Algebra of Fields (with Bruno Zumino). *Phys. Rev.* **163,** 1667. **(II)**

*[104] CP Noninvariance, Electromagnetic Interaction and Field-Current Identities. *Proc. of International Symposium on Electron and Photon Interactions at High Energies, SLAC, 1967,* p. 389.

1968

*[105] Field Current Identities and Intermediate Bosons. *Phys. Rev.* **168,** 1714.

*[106] Hadronic Contributions to the Photon Propagator (with Bruno Zumino). *Old and New Problems in Elementary Particles* (a volume dedicated to G. Bernardini), edited by G. Puppi, Academic Press, p. 167.

*[107] Algebra of the Observed Current Operators. *Proc. of Second Hawaii Topical Conference in Particle Physics, Honolulu, 1967,* edited by S. Pakvasa and S.F. Tuan, Univ. Hawaii Press, p. 13.

*[108] CPT Symmetries. *Ibid.*, p. 67.

*[109] Algebra of the Observed Current Operators. *Fundamental Particle Physics* (1967 Tokyo Summer Lectures in Theoretical Physics), edited by G. Takeda and Y. Hara, W.A. Benjamin, Inc., p. 74.

[110] CPT Symmetries. *Ibid.*, p. 102. **(II)**

*[111] Field Current Identities. The Fermi Lectures, a series given at the Scuola Normale Superiore, Pisa, January 1968.

*[112] CPT Symmetries. *Ibid.*

*[113] Intermediate Bosons and the Electromagnetic Field. *Phys. Rev.* **171,** 1731.

[114] Weak Interactions and CPT Asymmetries. *Contemporary Physics,* Volume II, Trieste Symposium, 1968, International Atomic Energy Agency, p. 283. **(II)**

1969

[115] Analysis of Divergences in a Neutral Spin 1 Meson Theory with Parity Non-Conserving Interactions. *Nuovo Cimento* **59A,** 579. **(I)**

*[116] A Possible Way to Remove Divergence Difficulties in Physics. *Topical Conference on Weak Interactions, CERN, January 1969,* edited by J.S. Bell, CERN 69-7, p. 427.

[117] Negative Metric and the Unitarity of the S-Matrix (with G.C. Wick). *Nuclear Physics* **B9,** 209. **(II)**

[118] Unitarity in the $N\theta\theta$ Sector of a Soluble Model with Indefinite Metric (with G.C. Wick). *Nuclear Physics* **B10,** 1. **(II)**

*[119] Current-Field Identities. *Theory and Phenomenology in Particle Physics,* Part A (1968 International School of Physics "Ettore Majorana"), edited by A. Zichichi, Academic Press, p. 262.

1970

*[120] A Relativistic Complex Pole Model with Indefinite Metric. *Quanta* (Essays in Theoretical Physics dedicated to Gregor Wentzel), Univ. Chicago Press, p. 260.

[121] Finite Theory of Quantum Electrodynamics (with G.C. Wick). *Phys. Rev.* **D 2,** 1033. **(II)**

[122] Symmetry Principles in Physics. *Evolution of Particle Physics* (a volume dedicated to Edoardo Amaldi), edited by M. Conversi, Academic Press, p. 176. **(II)**

*[123] On the Algebra of Fields. *Methods and Problems of Theoretical Physics* (in honour of R.E. Peierls), edited by J.E. Bowcock, North-Holland Publishing Co., p. 415; also "Concluding Lecture," p. 421.

[124] Remarks on the $|\Delta\vec{I}| = \frac{1}{2}$ Rule in Nonleptonic Weak Decays. Brookhaven National Laboratory Report #50260 (T-590), printed by the National Technical Information Service, U.S. Department of Commerce. **(I)**

*[125] Symmetry Principles in Physics. Brookhaven National Laboratory Report #50261 (T-591), printed by the National Technical Information Service, U.S. Department of Commerce.

*[126] Possible Existence of Spin 0 $W^{\pm}$ and Some of Its Experimental Consequences. *Phys. Rev. Letters* **25,** 1144.

*[127] Recent Progress in Constructing Finite, Relativistic and Unitary Field Theories. *Comments on Nuclear and Particle Physics,* **IV** (6), 189.

1971

*[128] Scaling Properties in *e-p* Scattering. Lecture given at the National Accelerator Laboratory, January, Columbia University Report #NYO-1932(2)-189.

[129] Questions of Lorentz Invariance in Field Theories with Indefinite Metric (with G.C. Wick). *Phys. Rev.* **D3,** 1046. **(II)**

*[130] Speculative Determination of the Intermediate-Boson Mass. *Phys. Rev. Letters* **26,** 801.

*[131] History of Weak Interactions. Talk given at Columbia University, March, Report #NYO-1932(2)-185.

*[132] Scaling Properties in Inelastic Electron Scattering with a Fixed Final Multiplicity. *Annals of Physics* **66.**

*[133] A Finite Theory of Quantum Electrodynamics. Lectures given at the "Ettore Majorana" International School of Physics in Erice, Sicily during July 1970, in *Elementary Processes at High Energy*, edited by A. Zichichi, Academic Press.

*[134] Remarks on the $|\Delta\vec{I}| = \frac{1}{2}$ Rule in Nonleptonic Weak Decays. *Ibid.*

*[135] Symmetry Principles in Physics. *Ibid.*

[136] History of Weak Interactions. *Ibid.* **(III)**

[137] *CP* Nonconservation and Inequalities between $\mu^+\mu^-$ and 2γ Decay Rates of K_S^0 and K_L^0 (with N.H. Christ). *Phys. Rev.* **D4,** 209. **(II)**

[138] High Energy Electromagnetic and Weak Interaction Processes, Brookhaven National Laboratory Report CRISP 71-57, Isabelle Project, August. **(I)**

1972

*[139] Scaling Properties in Inelastic Electron Scattering with a Fixed Final Multiplicity, in *Nuclear, Particle and Many-Body Physics,* Volume II, Academic Press, p. 857.

*[140] Some New Applications of Indefinite Metric to Quantum Electrodynamics and Weak Interactions, in *Elementary Particles* (Proc. of Amsterdam International Conference), edited by A.G. Tenner and M. Veltman, North-Holland Publishing Co., p. 399.

*[141] High Energy Physics and What Lies Beyond, a talk before the Joint Committee on Atomic Energy, Washington, D.C., March 1.

*[142] A New High-Energy Scale? *Physics Today* **25,** 23.

[143] Scaling Properties and the Bound-State Nature of the Physical Nucleon (with S.D. Drell). *Phys. Rev.* **D5,** 1738. **(III)**

*[143a] Scaling Properties in Weak and Electromagnetic Processes, in *Neutrino '72* (Proc. of Conference held at Balatonfüred, Hungary, 11–17 June), edited by A. Frenkel and G. Marx, OMKDK Technoinform, Budapest, Volume II, p. 1.

[144] A Simple Bound-State Model of the Physical Nucleon. *Phys. Rev.* **D6,** 1110. **(III)**

[145] Multiplicity Distribution in the Multiperipheral Model and a One-Dimensional Gas. *Phys. Rev.* **D6,** 3617. **(III)**

1973

[146] A Theory of Spontaneous *T* Violation. *Phys. Rev.* **D8,** 1226. **(II)**
[147] Spontaneous *T* Violation, a talk presented at the Topical Seminar on Weak Interactions, Trieste, Italy, June 29, 1973. **(II)**
*[148] Scaling Properties and the Bound-State Model of Physical Baryons, CERN Report 73-15.

1974

[149] *CP* Nonconservation and Spontaneous Symmetry Breaking. *Physics Reports* **9C,** No. 2. **(II)**
[150] Vacuum Stability and Vacuum Excitation in a Spin 0 Field Theory (with G.C. Wick). *Phys. Rev.* **D9,** 2291. **(II)**
*[151] Abnormal Nuclear States, *Proc. of XVII International Conference on High Energy Physics,* London, Science Research Council, III, 81.

1975

[152] Abnormal Nuclear States and Vacuum Excitations, an invited talk given at the Annual Bevatron Users' Meeting, Lawrence Berkeley Laboratory, 1974; *Revs. Mod. Phys.* **47,** 267. **(III)**
*[153] Abnormal Nuclear States and Vacuum Excitations, *Lepton and Hadron Structure,* ed. A. Zichichi, Academic Press, pp. 697–715.
*[154] Spontaneous *CP* Nonconservation, *Lepton and Hadron Structure,* ed. A. Zichichi, Academic Press, pp. 721–731.
[155] A Possible New Form of Matter at High Density, *Report of the Workshop on BeV/Nucleon Collisions of Heavy Ions—How and Why,* Bear Mountain, 1974 (BNL No. 50445), 1. **(III)**
[156] Interactions of a Dense Fermion Medium with a Scalar Meson Field (with M. Margulies). *Phys. Rev.* **D11,** 1591. **(III)**
[156a] Errata to "Interactions of a Dense Fermion Medium with a Scalar Meson Field" (with M. Margulies). *Phys. Rev.* **D12,** 4008 (1976). **(III)**
[157] Quantum Expansion of Soliton Solutions (with N.H. Christ). *Phys. Rev.* **D12,** 1606. **(I)**

1976

*[158] A Possible New Form of Matter at High Density, *Physics and Our World: Symposium in Honor of V.F. Weisskopf,* M.I.T. 1974, ed. K. Huang (AIP Conference Proc. No. 28), pp. 65–81.
[159] A Class of Scalar-Field Soliton Solutions in Three Space Dimensions (with R. Friedberg and A. Sirlin). *Phys. Rev.* **D13,** 2739. **(I)**
*[160] Unconfined, Heavy Fermion Model of Psions (with G. Feinberg). *Phys. Rev.* **D13,** 3071.
*[161] Examples of Four-Dimensional Soliton Solutions and Abnormal Nuclear States. *Phys. Reports* **23C,** 254.
[162] Gauge-Field Nontopological Solitons in Three Space Dimensions (I) (with R. Friedberg and A. Sirlin). *Nucl. Phys.* **B115,** 1. **(I)**
[163] Gauge-Field Nontopological Solitons in Three Space Dimensions (II) (with R. Friedberg and A. Sirlin). *Nucl. Phys.* **B115,** 32. **(I)**
[164] Nontopological Solitons, in *Proceedings of the Symposium on Frontier Problems in High Energy Physics* (in honour of G. Bernardini), edited by L. Foa and L.A. Radicati, Scuola Normale Superiore, Pisa, p. 47. **(I)**

1977

[165] Fermion-Field Nontopological Solitons I (with R. Friedberg). *Phys. Rev.* **D15,** 1694. **(I)**

*[166] Solitons and Hadrons, in *Problèmes Théorique Liés Aux Nouvelles Particules,* Centre National de la Recherche Scientifique, Montpellier, p. 413.

[167] Fermion-Field Nontopological Solitons II. Models for Hadrons (with R. Friedberg). *Phys. Rev.* **D16,** 1096. **(I)**

1978

[168] Nontopological Solitons and Applications to Hadrons. *Comments on Nuclear and Particle Physics* **7,** 165. **(I)**

*[169] QCD and the Soliton Model of Hadrons (with R. Friedberg). *Phys. Rev.* **D18,** 2623.

1979

[170] Abnormal Nuclear States (with G.C. Wick). *Mesons in Nuclei,* Vol. III, edited by M. Rho and D. Wilkinson, North-Holland Publishing Co., p. 869. **(III)**

*[171] Recent Results on Nontopological Soliton Solutions, *Science of Matter* (Festschrift in honor of T.Y. Wu), ed. S. Fujita, Gordon and Breach, p. 89.

*[172] Nontopological Solitons and Applications to Hadrons (Chalmers Symposium on Solitons, Göteborg), edited by H. Wilhelmsson, *Physica Scripta* **20,** 440.

[173] Feynman Rules of QCD inside a Hadron. *Phys. Rev.* **D19,** 1802. **(II)**

1980

[174] Operator Ordering and Feynman Rules in Gauge Theories (with N.H. Christ). *Phys. Rev.* **D22,** 939. **(II)**

[175] Is the Physical Vacuum a Medium?, in *A Festschrift for Maurice Goldhaber,* edited by G. Feinberg, A.W. Sunyar and J. Weneser, *Transactions of the NY Academy of Sciences,* Series II, **40,** 111. **(II)**

[176] QCD and the Bag Model of Hadrons, *Proceedings of the 1980 Guangzhou Conference on Theoretical Particle Physics,* edited by Ning Hu and Hung-yuan Tzu, Science Press and Van Nostrand Reinhold Co., p. 32. **(III)**

1981

*[177] Is the Vacuum a Physical Medium?, in *Statistical Mechanics of Quarks and Hadrons,* edited by H. Satz, North-Holland Publishing Co., 3.

*[178] *Particle Physics and Introduction to Field Theory,* Harwood Academic Publishers GmbH.

1982

[179] Random Lattice Field Theory: General Formulation (with N.H. Christ and R. Friedberg). *Nucl. Phys.* **B202,** 89. **(III)**

[180] Gauge Theory on a Random Lattice (with N.H. Christ and R. Friedberg). *Nucl. Phys.* **B210** [FS6], 310. **(III)**

[181] Weights of Links and Plaquettes in a Random Lattice (with N.H. Christ and R. Friedberg). *Nucl. Phys.* **B210** [FS6], 337. **(III)**

1983

[182] Can Time Be a Discrete Dynamical Variable? *Phys. Letters* **122B,** 217. **(III)**

[183] Discrete Quantum Mechanics (with R. Friedberg). *Nucl. Phys.* **B225** [FS9], 1. **(III)**

*[184] Time As a Dynamical Variable, in *Shelter Island II,* edited by R. Jackiw, N.N. Khuri, S. Weinberg and E. Witten, The MIT Press (1985).

[185] Discrete Mechanics, to be published in the *Proceedings of the International School of Subnuclear Physics,* Erice, Plenum Press. **(III)**

[186] Broken Parity. **(III)**

1984

[187] Relativistic Heavy Ion Collisions and Future Physics, in *Symmetries in Particle Physics,* Plenum Press, 93.

[188] Derivation of Regge's Action from Einstein's Theory of General Relativity (with R. Friedberg). *Nucl. Phys.* **B242,** 145. **(III)**

[189] Lattice Gravity Near the Continuum Limit (with G. Feinberg, R. Friedberg and H.C. Ren). *Nucl. Phys.* **B245,** 343. **(III)**

*[190] High Energy Theory, in *Pief-Fest,* edited by P. McDonough, Stanford Linear Accelerator Center, 1985, p. 664.

1985

*[191] Difference Equations As the Basis of Fundamental Physical Theories, to be published in the *Proceedings of the Symposium "Old and New Problems in Fundamental Physics."*

[192] Are Black Holes Black Bodies? *Nucl. Phys.* **B264,** (1986), 437. **(III)**

*[193] A Theorem on Black Hole Radiation, to be published in the *Proceedings of Meson 50.*

*[194] Difference Equations As the Basis of Fundamental Physical Theories, to be published in the *Proceedings of the APS Division of Particles and Fields,* Eugene, Oregon meeting.

[195] Physics in Terms of Difference Equations, in *The Lesson of Quantum Theory*, edited by J. de Boer, E. Dal, and O. Ulfbeck, North-Holland Physics Publishing (1986), p. 181.

VII

Discrete Physics

Commentary

A novel, discrete approximation to continuous field theory is developed in paper [179]. In this approach the space-time continuum is represented by an array of points which are distributed at random. Central to the paper is the natural definition of the derivative of a function defined on such a random lattice and an analysis of its properties. By averaging quantities of interest over many such lattices or over a large region of one lattice, one obtains a discrete approximation which is nevertheless Lorentz invariant. This approach is extended to field theories of spin ½ and spin 1 particles in papers [180] and [181]. In particular, in the latter paper an appealing scheme is proposed for assigning interaction strengths to the various geometrical entities appearing in the field-theory action.

This discrete formulation of field theory was developed in the previous papers as a starting point for a variety of non-perturbative approximations to an underlying theory defined on the conventional space-time continuum. However, paper [182] takes a radically different approach. By postulating that time is a discrete dynamical variable, one recovers the random lattice as the exact description of a possible quantum field theory history. It is shown that this new interpretation of time can be consistently applied through all phases of mechanics: from non-relativistic motion of a single particle to quantum field theory. These ideas are examined and developed further in papers [183] and [185].

One of the most interesting applications of this new perspective (from both theoretical and observational viewpoints) is to the theory of gravity. Papers [188] and [189] begin studying this application. In the first it is shown how the discrete action invented by Regge twenty-three years earlier can be precisely related to the corresponding action in Einstein's theory of general relativity. The second paper examines this relationship in detail, showing how the discrete formulation gives a systematic approximation to the usual continuous general relativity.

Nuclear Physics B202 (1982) 89–125

RANDOM LATTICE FIELD THEORY: General formulation*

N.H. CHRIST, R. FRIEDBERG and T.D. LEE

Columbia University, New York, N.Y. 10027, USA

Received 4 February 1982

A new type of lattice field theory is formulated in which the sites are chosen randomly in space. An algorithm is given for linking nearby sites; these links form the edges of non-overlapping simplices which fill the entire volume. All physical quantities averaged over such a lattice can therefore be translationally and rotationally symmetric.

1. Introduction

Among the more promising techniques currently being used to investigate the properties of quantum field theory is the replacement of the space-time continuum on which the theory is defined by a discrete lattice**. Such a lattice approximation to a continuum field theory makes finite the number of degrees of freedom per unit volume, allowing the application of numerical methods to the approximate theory. Furthermore, the ultraviolet divergences of the theory are regulated in a manner not requiring the use of perturbation theory. Applications of this method to date have used a lattice based on a regular, hypercubic array of points in space or space-time. Although such a regular lattice is the simplest choice, its introduction explicitly violates the Poincaré symmetry possessed by the original theory. Both the translation and Lorentz invariance are lost.

In this paper we will develop an alternative, discrete approximation to a continuum field theory which is based on the introduction of a random lattice. The sites of this lattice are points distributed at random in the system volume. An algorithm is presented for linking nearby sites in a quite natural fashion. The links introduced form the edges of non-overlapping simplices which fill the entire volume. In four dimensions these 4-simplices have sides which are 3-dimensional tetrahedra. In turn, the sides of these tetrahedra are elementary triangles. Just as in the case of a regular

* This research was supported in part by the US Department of Energy.

** The present interest in this technique has been motivated largely by the work of Wilson [1]. For recent reviews containing further references, see Parisi [2] and Kogut [3].

lattice, derivatives of the continuum field variables are replaced by differences of the field variables on linked sites. Similarly, a lattice gauge theory can be defined just as for a regular lattice by using as fundamental plaquettes the elementary triangles referred to above.

Although a discretized field theory built on such a random lattice is necessarily more complex than a field theory defined on a conventional regular lattice, the translation and Lorentz symmetries can be preserved. This can be accomplished in two ways. First consider a particular lattice in a finite volume Ω with a relatively large number $\mathcal{N}$ of random points. In contrast to a regular lattice, such a random array of points specifies no preferred direction or origin in space-time. Consequently one would expect a theory defined on such a lattice to be, to a large extent, translationally and Lorentz symmetric. As we shall see, these symmetries are exact for any intensive or intrinsic physical quantity, such as Wilson string tension or large-distance behavior of correlation functions, evaluated at a constant site density ρ in the limit $\Omega \to \infty$. The same result can also be derived by the following alternative approach.

In this second approach, we begin with a Feynman path integral description of the quantum field theory. For a lattice theory, such a path integral reduces to an ordinary multiple integral over the field variables corresponding to the discrete lattice points in space-time. One can then add to this integration over field variables an average over the positions of the points in the random lattice, but keeping again the site density ρ constant. We may define this combined integral over field variables and lattice points as our discrete approximation to the original continuum field theory. Such a definition provides an approximation to the continuum theory which consists of a finite number of degrees of freedom per unit volume, contains no ultraviolet singularities and is manifestly Lorentz and translationally symmetric.

This regularization of continuum field theory provided by the introduction of a random lattice will be developed in a series of papers. In the present paper we define our random lattice, derive some of its geometrical properties, and examine some features of a massless scalar field theory defined on the lattice. In subsequent papers, we will: (i) Apply this approximation to the Yang-Mills theory of the strong interactions, first defining a non-abelian gauge theory on a random lattice and then systematically investigating in particular the strong coupling expansion for such a lattice gauge theory. (ii) Describe numerical methods and results for both scalar and gauge theories defined on a random lattice.

The present paper is arranged as follows. In sect. 2 the introduction of links and the resulting division of space into simplices is described. Euler's theorem and related geometrical arguments are employed in sect. 3 to derive relationships between the average numbers of simplices of different dimensions which occur in the random lattice. In sect. 4 and the appendix, we compute various distributions and average values for link length, triangle area, coordination number, etc. A brief description of the dual lattice is given in sect. 5. Finally, in sect. 6, we consider the theory of a massless scalar field defined on a random lattice. In particular we study the long-distance limit of the propagator for the scalar field. Although the scalar propagator agrees in this limit with the form of the continuum propagator, the normalization is different even for a free theory. This anomalous normalization constant Z is related to the bulk resistivity of a random array of resistors and reflects the fluctuations inherent in the random lattice.

2. Division into simplices

Consider a D-dimensional euclidean space. Suppose that a random distribution of $\mathcal{N}$ points (called lattice sites, or simply *sites*) has already been given in a finite volume Ω. The density of lattice sites is

$$\rho = \mathcal{N}/\Omega . \tag{2.1}$$

Our first task is to divide Ω into simplices whose vertices are those random sites. To avoid complications due to the boundary, we may assume Ω to be a D-dimensional rectangular volume with the standard periodic boundary condition. [Alternatively, Ω may instead be a D-dimensional spherical surface.] For $D=2$, the simplices are triangles; for $D=3$, they are tetrahedra. In general, a D-simplex consists of $D+1$ vertices. Between any pair of vertices, we draw a straight line which forms a link. Hence, there are altogether

$$C_2^{D+1} = \tfrac{1}{2}(D+1)D$$

different links per simplex. Regarding the links as sides of triangles, we can form

$$C_3^{D+1} = \frac{1}{3!}(D+1)D(D-1)$$

different triangles per simplex. Likewise, since the triangles can be viewed as surfaces of tetrahedra, there are

$$C_4^{D+1} = \frac{1}{4!}(D+1)D(D-1)(D-2)$$

different tetrahedra per simplex, etc.

For physical applications, we wish to draw links only between nearby sites. This together with the problem of partitioning Ω into non-overlapping simplices leads us to introduce the following concept of clusters of neighboring sites: Consider an arbitrary group of $D+1$ random lattice sites. They lie on the surface of a hypersphere in D dimensions, which will be referred to as their circumscribed sphere. If the inside of that sphere is free of all lattice sites, then this group of $D+1$ sites is called a *cluster*. We form a D-simplex for every cluster, using its sites as vertices.

We note that for any $D+1$ infinitesimal volumes $d^D r_1, d^D r_2, \ldots, d^D r_{D+1}$ in Ω, the probability that each volume element should contain a lattice site is

$$\rho^{D+1} \prod_{i=1}^{D+1} d^D r_i ;$$

the probability that there is no lattice point inside their circumscribed sphere is given by the Poisson formula

$$\exp(-v\rho),$$

where v is the volume of the sphere, which is related to its radius R by

$$v = \begin{cases} (2\pi)^{D/2} R^D / D!!, & \text{if } D \text{ is even}, \\ 2(2\pi)^{(D-1)/2} R^D / D!!, & \text{if } D \text{ is odd}. \end{cases} \tag{2.2}$$

Hence, the probability that $d^D r_1, d^D r_2, \ldots, d^D r_{D+1}$ should form a cluster and therefore be the vertices of a D-simplex is

$$\rho^{D+1} \prod_{i=1}^{D+1} d^D r_i \exp(-v\rho). \tag{2.3}$$

The factor $\exp(-v\rho)$ insures that the vertices of the simplex cannot be too far apart.

Let us examine, out of the $\mathcal{N}$ given sites, all $C^{\mathcal{N}}_{D+1}$ combinations of possible groupings of $D+1$ sites. Whenever a group fulfills the above condition of a cluster, a D-simplex is formed. The simplices thus constructed define our basic random lattice, which has the following properties:

Theorem 1. (i) There is no overlap between the volumes of any two simplices, and (ii) the volume-sum of all simplices is Ω.

Proof. Consider two clusters $\{a_i\}$ and $\{b_i\}$ where a_i and b_i denote the sites, with $i = 1, 2, \ldots, D+1$. Let their circumscribed spheres be respectively O_a and O_b. By definition, the insides of both O_a and O_b are free of any lattice sites. Hence, O_a cannot be completely inside O_b, nor vice versa. Consequently, O_a and O_b must either be totally separated from each other, which makes (i) an obvious conclusion, or intersect each other*. In the latter case, the intersection of these two surfaces O_a and O_b defines a sphere o in $(D-1)$ dimensions. As shown in fig. 1, o cuts O_a into two separate caps C_a and C'_a with C'_a lying entirely inside O_b. For definiteness, we exclude the border o from the caps C_a and C'_a so that the sets C_a, C'_a and o are disjoint. The sites in $\{a_i\}$ can be on C_a and o, but not on C'_a; otherwise, some of the

* The possibility that two different clusters share the same circumsphere (i.e., O_a and O_b are identical) is of zero measure, and is therefore ignored.

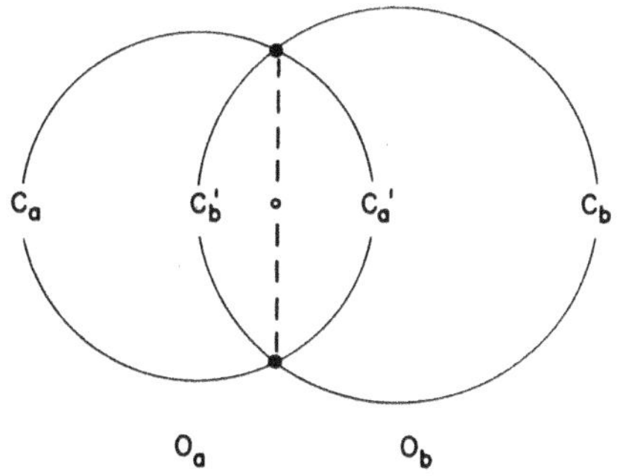

Fig. 1. The intersection o of two hyperspheres O_a and O_b cuts each sphere O_i into two caps C_i and C_i' where $i = a$ or b.

a_i would be inside the other sphere O_b, in violation of the assumption. Likewise, o separates O_b into two caps C_b and C_b' with C_b' inside O_a. [Again o is not included in C_b and C_b'.] Hence, none of the b_i can be on C_b'. From $\{a_i\}$ and $\{b_i\}$ we now construct two D-simplices, S_a and S_b. These two simplices must lie on the opposite sides of the $(D-1)$-dimensional subspace that contains o. Proposition (i) now follows.

We note that S_a and S_b may share some common boundary. If the sets $\{a_i\}$ and $\{b_i\}$ have a single site in common, then there is a common vertex between S_a and S_b; if there are two sites in common, then they share a common line, etc. The maximum contact between these two simplices is for $\{a_i\}$ and $\{b_i\}$ to have D sites in common; in that case, S_a and S_b share a common $(D-1)$-simplex, say s, whose vertices are all on o.

We now turn to the proof of (ii). Every D-simplex is bounded by $(D-1)$-simplices, each of which has two sides. We shall show that every such $(D-1)$-simplex is shared by two D-simplices, one on each side. [In two dimensions, this means that every link is shared by two triangles; in three dimensions, every triangle is shared by two tetrahedra, etc.] Assuming that this is not true, then there must be a D-simplex, say S_a, which has on its surface a $(D-1)$-simplex, say s, that is not shared by any other D-simplices. Let O_a be the circumscribed sphere of S_a, and o that of s. We denote by O_b any hypersphere that intersects O_a at o but, as shown in fig. 1, we require the centers of O_a and O_b to lie on opposite sides of s. Consider now the family of all such O_b. By assumption, none of the O_b can contain any lattice points except those on o (i.e., on s). This is clearly false, since O_b can be arbitrarily large and Ω is finite with a periodic boundary condition. [A convenient way to visualize this situation is to consider an array of identical volumes Ω, each containing exactly the same site distribution. These volumes are placed side by side to form an infinite periodic rectangular lattice whose unit cell is Ω. It is then easy to see that the radius of O_b can be arbitrarily large.]

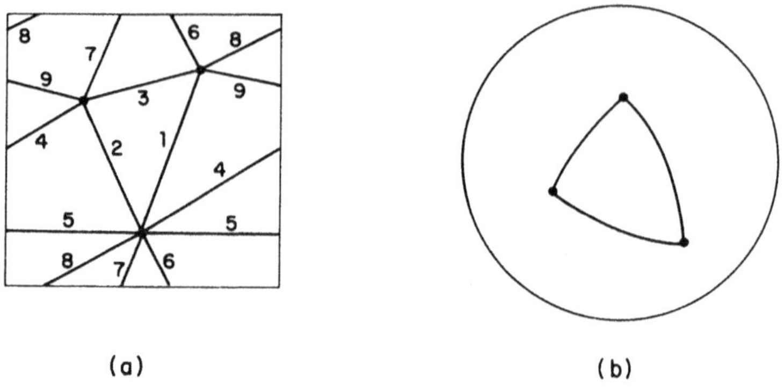

Fig. 2. (a) Ω is a two-dimensional square with a periodic boundary condition. There are three sites, nine links and six triangles. The links are marked 1 through 9. (b) Ω is a two-dimensional spherical surface which contains three sites, three links (arcs) and two (spherical) triangles. The triangles are shown here with different solid angles whose sum equals 4π.

From any D-simplex, we can go through its boundary and reach other D-simplices. By repeating this process, we must eventually cover a volume Ω' which is within Ω and also has no boundary. Consequently, $\Omega' = \Omega$, and the proof of theorem 1 is complete.

By following the same argument, one can also establish the theorem for the case that Ω is a D-dimensional spherical surface. In fig. 2, we illustrate the simplest configuration when the dimension $D = 2$ and the total number of sites $\mathcal{N} = 3$.

3. Some general relations

Consider a large random lattice in D dimensions. Let

$$\begin{aligned} \mathcal{N}_n &= \text{total number of } n\text{-simplices in } \Omega\,, \\ \rho_n &= \text{density of } n\text{-simplices} = \mathcal{N}_n/\Omega\,, \\ N_{n/m} &= \text{average number of } n\text{-simplices per } m\text{-simplex} \\ &\quad \text{(i.e., each } m\text{-simplex is on the average shared} \\ &\quad \text{by } N_{n/m}\ n\text{-simplices)}\,, \end{aligned} \tag{3.1}$$

where

$$0 \leqq m < n \leqq D$$

Thus, in the notations of the previous section

$$\mathcal{N}_0 = \mathcal{N} = \text{total number of lattice sites in } \Omega ,$$

$$\rho_0 = \rho = \mathcal{N}/\Omega . \tag{3.2}$$

Since each n-simplex has on its surface a total of

$$C_{m+1}^{n+1} = \frac{(n+1)!}{(m+1)!(n-m)!}$$

m-simplices, we have the identity

$$N_{n/m}\mathcal{N}_m = C_{m+1}^{n+1}\mathcal{N}_n . \tag{3.3}$$

The Euler formula gives for Ω with a periodic boundary condition

$$\sum_{n=0}^{D} (-1)^n \mathcal{N}_n = 0 . \tag{3.4a}$$

On the other hand, if Ω were a spherical surface, then (3.4a) would be modified to

$$\sum_{n=0}^{D} (-1)^n \mathcal{N}_n = \begin{cases} 2, & \text{if } D \text{ is even},\\ 0, & \text{if } D \text{ is odd}. \end{cases} \tag{3.4b}$$

As we shall see, there are additional linear relations between these $\mathcal{N}_n$.

Theorem 2. For any $m \leqslant D-1$,

$$\sum_{n=m+1}^{D} (-1)^{n-m-1} N_{n/m} = \begin{cases} 2, & \text{if } D-m \text{ is odd},\\ 0, & \text{if } D-m \text{ is even}. \end{cases} \tag{3.5}$$

Proof. Take any lattice site, say p. Using p as the center, draw a very tiny hypersphere around it. Let $\mathbb{S}$ be the $(D-1)$-dimensional surface of the hypersphere. The intersection between $\mathbb{S}$ and any n-simplex having p as one of its vertices is an $(n-1)$-simplex; e.g., any link (1-simplex) that has p as one of its ends intersects $\mathbb{S}$ at a point (0-simplex), etc. Apply the Euler formula (3.4b) to the lattice structure (formed by these intersections) on $\mathbb{S}$, and then average over p. The result is

$$N_{1/0} - N_{2/0} + \cdots + (-1)^{D-1} N_{D/0} = \begin{cases} 2, & \text{if } D-1 \text{ is even},\\ 0, & \text{if } D-1 \text{ is odd}, \end{cases} \tag{3.6}$$

which is the special case of (3.5) for $m = 0$.

In a similar way, we may select any m-simplex, say s, in the random lattice. Take any point p_s inside s. Consider the $(D-m)$-dimensional subspace which contains p_s and is perpendicular to the m-simplex s. Using p_s as the center, draw a very tiny hypersphere in this $(D-m)$-dimensional subspace. Let $\mathbb{S}$ now be the $(D-m-1)$-dimensional surface of the hypersphere, and $N_{n/m}(s)$ be the number of n-simplices that share s; i.e., the same m-simplex s appears on the surfaces of all these n-simplices $(n>m)$. The intersection of $\mathbb{S}$ and each of these n-simplices is an $(n-m-1)$-simplex. Theorem 2 is proved by applying the Euler formula (3.4b) to the lattice structure (formed by these intersections) on the surface $\mathbb{S}$, and then averaging over s. For $m=D-1$, (3.5) becomes $N_{D/D-1}=2$, which is simply the statement that each $(D-1)$-simplex is shared by two D-simplices.

We can rewrite the result of theorem 2 in terms of the $D+1$ densities $\mathfrak{N}_n$ if we multiply (3.5) by $\mathfrak{N}_m$ and use (3.3):

$$\sum_{n=m+1}^{D} (-1)^{n-m-1} C_{m+1}^{n+1} \mathfrak{N}_n = \begin{cases} 2\mathfrak{N}_m, & \text{if } D-m \text{ is odd}, \\ 0, & \text{if } D-m \text{ is even}, \end{cases} \tag{3.7}$$

for $0 \leqslant m \leqslant D-1$. If we define

$$\mathfrak{N}_{-1} = \begin{cases} 0, & \text{if } \Omega = \text{euclidean volume with periodic boundary condition}, \\ 1, & \text{if } \Omega = \text{spherical surface}, \end{cases} \tag{3.8}$$

then (3.7) also holds for $m=-1$, being simply the original Euler formula (3.4a) or (3.4b).

An alternative form of (3.7) is

$$\sum_{n=m}^{D} (-1)^{D-n} C_{m+1}^{n+1} \mathfrak{N}_n = \mathfrak{N}_m . \tag{3.9}$$

To recover (3.7) from (3.9), transpose the term for $n=m$ from the left- to the right-hand side.

We remark that (3.9) holds as well for $m=D$, but trivially so:

$$\mathfrak{N}_D = \mathfrak{N}_D . \tag{3.10}$$

Also, for $m=D-2$ (3.9) becomes

$$-D\mathfrak{N}_{D-1} + \tfrac{1}{2}(D+1)D\mathfrak{N}_D = 0 , \tag{3.11}$$

which is identical to the equation for $m=D-1$. It turns out that a similar redundancy applies to every other equation, in accord with the following theorem.

Theorem 3. For each integer s, $0 \leqslant 2s \leqslant D+1$, there exists a linear combination (with non-zero coefficients) of (3.9) for $m=D-2s,\ldots,m=D$ which is an identity in all the $\mathcal{N}_n$.

Proof. Multiply (3.9) by $(-1)^m(m+1)!/(m-D+2s+1)!$ and sum over all m from $D-2s$ to D. The result is

$$\sum_{m=D-2s}^{D} \sum_{n=m}^{D} (-1)^{D-n+m} \frac{(n+1)!}{(n-m)!(m-D+2s+1)!} \mathcal{N}_n$$

$$= \sum_{m=D-2s}^{D} (-1)^m \frac{(m+1)!}{(m-D+2s+1)!} \mathcal{N}_m . \tag{3.12}$$

Transposing the right-hand side of (3.12) to the left, with m replaced by n, we obtain

$$\sum_{m=D-2s-1}^{D} \sum_{n=m}^{D} (-1)^{D-n+m} \frac{(n+1)!}{(n-m)!(m-D+2s+1)!} \mathcal{N}_n$$

$$-(-1)^D (D-2s)! \mathcal{N}_{D-2s-1} = 0 , \tag{3.13}$$

in which the last term is included simply to cancel the $m=n=D-2s-1$ term in the summation so that the rest of the $m=D-2s-1$ terms are the ones transposed from (3.12).

The coefficient of $\mathcal{N}_n$, for $n \geqslant D-2s$, in (3.13) is

$$\sum_{m=D-2s-1}^{n} (-1)^{D-n+m} \frac{(n+1)!}{(n-m)!(m-D+2s+1)!} , \tag{3.14}$$

which can be written as

$$\frac{(n+1)!}{R!} \sum_{r=0}^{R} (-1)^{D-r} C_r^R = \frac{(n+1)!}{R!} (-1)^D (1-1)^R = 0 \tag{3.15}$$

where $r=n-m$, $R=n-D+2s+1$. Hence (3.13) is an identity and therefore so is (3.12). The theorem is then proved.

From theorem 3 it follows that all eqs. (3.7) for even $D-m$ are redundant, being linear combinations of equations of higher m. But all the equations for odd $D-m$ ($-1 \leqslant m \leqslant D-1$) are independent, because each contains two variables $\mathcal{N}_m$ and $\mathcal{N}_{m+1}$, that do not appear in (3.7) for larger m. Therefore the number of independent equations obtained from (3.7) is $\frac{1}{2}D+1$ if D is even, or $\frac{1}{2}(D+1)$ if D is odd. Among the D ratios $\mathcal{N}_1/\mathcal{N}_0, \mathcal{N}_2/\mathcal{N}_0,\ldots,$ and $\mathcal{N}_D/\mathcal{N}_0$, this leaves u_D undetermined quantities, where

$$u_D = \tfrac{1}{2}D - 1\,, \qquad \text{if } D \text{ is even}\,,$$

$$u_D = \tfrac{1}{2}(D-1)\,, \qquad \text{if } D \text{ is odd}\,. \tag{3.16}$$

Example 1. $D=2$: For $m=1$, (3.7) becomes

$$3\mathfrak{N}_2 = 2\mathfrak{N}_1\,. \tag{3.17}$$

This expresses the fact that when $D=2$ each link is shared by two triangles, and each triangle has three sides. Hence, 2 times the number of links $\mathfrak{N}_1$ equals 3 times the number of triangles $\mathfrak{N}_2$. If (3.7) is evaluated for $m=0$ only the quantities $\mathfrak{N}_2$ and $\mathfrak{N}_1$ enter and in fact (3.17) is reproduced. The case $m=-1$ involves in addition $\mathfrak{N}_0$ and gives the Euler formula (3.4a) or (3.4b). For a euclidean volume Ω with periodic boundary conditions, we have

$$\mathfrak{N}_0 - \mathfrak{N}_1 + \mathfrak{N}_2 = 0\,. \tag{3.18}$$

Combining (3.16) and (3.18) we find

$$\mathfrak{N}_1 = 3\mathfrak{N}_0\,, \qquad \mathfrak{N}_2 = 2\mathfrak{N}_0 \tag{3.19}$$

or referring to (3.3),

$$N_{1/0} = C_1^2 \mathfrak{N}_1/\mathfrak{N}_0 = 6\,,$$

$$N_{2/0} = C_1^3 \mathfrak{N}_2/\mathfrak{N}_0 = 6\,. \tag{3.20}$$

In two dimensions, the average number of links/site is 6, which is also the average number of triangles/site. There are no undetermined ratios, in agreement with (3.16).

Example 2. $D=3$: In three dimensions, (3.7) gives for $m=2$

$$\mathfrak{N}_2 = 2\mathfrak{N}_3\,, \tag{3.21}$$

which expresses the fact that when $D=3$ each triangle is shared by two tetrahedra and that each tetrahedron contains four triangles. For $m=1$, (3.7) reproduces (3.21). For $m=0$, (3.7) becomes

$$2\mathfrak{N}_0 - 2\mathfrak{N}_1 + 3\mathfrak{N}_2 - 4\mathfrak{N}_3 = 0\,, \tag{3.22}$$

while for $m=-1$ (Euler's formula)

$$\mathfrak{N}_0 - \mathfrak{N}_1 + \mathfrak{N}_2 - \mathfrak{N}_3 = 0\,. \tag{3.23}$$

As we expect from theorem 3, (3.23) is redundant with (3.21) and (3.22). Hence there are two independent relations, say (3.21) and (3.22) between the three ratios $\mathcal{N}_1/\mathcal{N}_0$, $\mathcal{N}_2/\mathcal{N}_0$ and $\mathcal{N}_3/\mathcal{N}_0$. Using (3.3), we can write these relations as

$$N_{2/0} = \tfrac{3}{2} N_{3/0}, \qquad N_{1/0} = 2 + \tfrac{1}{2} N_{3/0}. \tag{3.24}$$

In accordance with (3.16) there is a single undetermined ratio $N_{3/0}$, which we shall calculate in the next section.

Example 3. $D = 4$: The pattern established in $D = 2$ and 3 repeats itself in four dimensions. For $m = 3$ and 2, (3.7) gives the single equation

$$5\mathcal{N}_4 = 2\mathcal{N}_3. \tag{3.25}$$

For $m = 1$ and 0, (3.7) becomes

$$2\mathcal{N}_1 - 3\mathcal{N}_2 + 6\mathcal{N}_3 - 10\mathcal{N}_4 = 0, \tag{3.26}$$

and

$$2\mathcal{N}_1 - 3\mathcal{N}_2 + 4\mathcal{N}_3 - 5\mathcal{N}_4 = 0. \tag{3.27}$$

Again only a single combination of the variables $\mathcal{N}_1$ and $\mathcal{N}_2$ enters into these two equations, and in fact the difference of (3.26) and (3.27) is simply (3.25). The final $m = -1$ equation (Euler's formula) introduces $\mathcal{N}_0$. In the case of a euclidean volume with periodic boundary conditions, we have

$$\mathcal{N}_0 - \mathcal{N}_1 + \mathcal{N}_2 - \mathcal{N}_3 + \mathcal{N}_4 = 0,$$

which together with, say, (3.25) and (3.26) provides three linearly independent relations among the four ratios $\mathcal{N}_i/\mathcal{N}_0$ for $i = 1, 2, 3$ and 4. Using (3.3) we can write them as

$$\begin{aligned} N_{3/0} &= 2N_{4/0}, \\ N_{2/0} &= 6 + \tfrac{6}{5} N_{4/0}, \\ N_{1/0} &= 6 + \tfrac{1}{5} N_{4/0}. \end{aligned} \tag{3.28}$$

In accordance with (3.16), there is a single undetermined ratio $N_{4/0}$, which we shall calculate in the next section.

4. Distribution functions and average values

In D dimensions we see from (2.1)–(2.3) that the probability of having in our random lattice a D-simplex $S(1, 2, \ldots, D+1)$ with vertices at the positions

$\boldsymbol{r}_1, \boldsymbol{r}_2, \ldots, \boldsymbol{r}_{D+1}$ within the infinitesimal volumes $\mathrm{d}^D r_1, \mathrm{d}^D r_2, \ldots, \mathrm{d}^D r_{D+1}$ is

$$(D+1)!\mathcal{D}(1,2,\ldots,D+1)\prod_{i=1}^{D+1}\mathrm{d}^D r_i, \tag{4.1}$$

where

$$\mathcal{D}(1,2,\ldots,D+1) \equiv \frac{1}{(D+1)!}\rho^{D+1}\mathrm{e}^{-v\rho} \tag{4.2}$$

is called the *distribution function* of D-simplices and satisfies

$$\int \mathcal{D}(1,2,\ldots,D+1)\prod_{i=1}^{D+1}\mathrm{d}^D r_i = \mathcal{N}_D, \tag{4.3}$$

with each $\mathrm{d}^D r_i$ varying independently over the entire volume Ω. Here $\mathcal{N}_D$ is the total number of D-simplices in Ω as before. We can use (3.2) and the relationship (3.3) between $\mathcal{N}_D$ and the average number of D-simplices per site, $N_{D/0}$,

$$(D+1)\mathcal{N}_D = N_{D/0}\mathcal{N}_0 = \rho\Omega N_{D/0}, \tag{4.4}$$

to rewrite (4.3) as an equation for $N_{D/0}$:

$$\int \mathcal{D}(1,2,\ldots,D+1)\prod_{i=2}^{D+1}\mathrm{d}^D r_i = \frac{1}{D+1}\rho N_{D/0}. \tag{4.5}$$

If we integrate the probability function (4.1) for D-simplices over all positions of one of the coordinates, $\boldsymbol{r}_{D+1}$, we obtain twice the probability of finding a $(D-1)$-simplex in our lattice whose vertices lie at the positions $\boldsymbol{r}_1, \ldots, \boldsymbol{r}_D$ within the infinitesimal volumes $\mathrm{d}^D r_1, \ldots, \mathrm{d}^D r_D$. The factor of two arises because each $(D-1)$-simplex appears in precisely two D-simplices. Thus, in analogy with (4.1), we define

$$\mathcal{D}'(1,2,\ldots,D)\prod_{i=1}^{D}\mathrm{d}^D r_i \tag{4.6}$$

as $(D!)^{-1}$ times this probability. Consequently,

$$\mathcal{D}'(1,2,\ldots,D) = \tfrac{1}{2}(D+1)\int \mathcal{D}(1,2,\ldots,D+1)\,\mathrm{d}^D r_{D+1} \tag{4.7}$$

is the distribution function for $(D-1)$-simplices and satisfies the normalization condition

$$\int \mathcal{D}'(1,2,\ldots,D)\prod_{i=1}^{D}\mathrm{d}^D r_i = \mathcal{N}_{D-1}. \tag{4.8}$$

Combining (4.3) and (4.8) reproduces the previous result

$$(D+1)\mathcal{N}_D = 2\mathcal{N}_{D-1}. \tag{4.9}$$

We shall now calculate the average values of, say, the link length $\bar{l}$, the triangle area $\bar{\Delta}$ and the tetrahedron volume $\bar{\tau}$, per D-simplex; i.e.,

$$C_2^{D+1}\bar{l} = \text{average of total link length in a } D\text{-simplex}, \tag{4.10}$$

$$C_3^{D+1}\bar{\Delta} = \text{average of total triangle area in a } D\text{-simplex}, \tag{4.11}$$

$$C_4^{D+1}\bar{\tau} = \text{average of total tetrahedron volume in a } D\text{-simplex}. \tag{4.12}$$

Consider a specific function $f(1,2,\ldots,D+1)$ defined for a D-simplex. Its average over our random lattice is

$$\bar{f} \equiv \frac{1}{\mathcal{N}_D}\int f(1,2,\ldots,D+1)\mathcal{D}(1,2,\ldots,D+1)\prod_{i=1}^{D+1} \mathrm{d}^D r_i. \tag{4.13}$$

We denote by c the center of the circumscribed sphere of the D-simplex $\mathrm{S}(1,2,\ldots, D+1)$; its position vector $\boldsymbol{r}_{\mathrm{c}}$ satisfies

$$|\boldsymbol{r}_i - \boldsymbol{r}_{\mathrm{c}}| = R, \tag{4.14}$$

for all $i = 1,2,\ldots,D+1$, where, as before, R is the radius of the sphere. It is useful to introduce the unit vector

$$\hat{\boldsymbol{\sigma}}_i \equiv \frac{1}{R}(\boldsymbol{r}_i - \boldsymbol{r}_{\mathrm{c}}), \tag{4.15}$$

whose differential is the solid angle $\mathrm{d}^{D-1}\hat{\boldsymbol{\sigma}}_i$. Since most of the functions $f(1,2,\ldots, D+1)$ that we are interested in depend only on relative coordinates, the integration (4.13) can be facilitated by a coordinate transformation

$$\boldsymbol{r}_1, \boldsymbol{r}_2, \ldots, \boldsymbol{r}_{D+1} \to \boldsymbol{r}_{\mathrm{c}}, R, \hat{\boldsymbol{\sigma}}_1, \hat{\boldsymbol{\sigma}}_2, \ldots, \hat{\boldsymbol{\sigma}}_{D+1}. \tag{4.16}$$

Theorem 4.

$$\prod_{i=1}^{D+1} \mathrm{d}^D r_i = D!\,V R^{D^2-D-1}\,\mathrm{d}R\,\mathrm{d}^D r_{\mathrm{c}} \prod_{i=1}^{D+1} \mathrm{d}^{D-1}\hat{\boldsymbol{\sigma}}_i, \tag{4.17}$$

where V is the volume of the D-simplex.

Proof. Let $\delta^D(\boldsymbol{r}-\boldsymbol{r}_c)$ be a D-dimensional δ-function. Multiplying the integral

$$I(f)\equiv\int f(1,2,\dots,D+1)\prod_{i=1}^{D+1}\mathrm{d}^D r_i \tag{4.18}$$

by the identity

$$1=\int\delta^D(\boldsymbol{r}-\boldsymbol{r}_c)\,\mathrm{d}^D r,$$

we obtain for any function $f(1,2,\dots,D+1)$

$$I(f)=\int f\prod_{i=1}^{D+1}\mathrm{d}^D r_i\int\delta^D(\boldsymbol{r}-\boldsymbol{r}_c)\,\mathrm{d}^D r, \tag{4.19}$$

in which $\boldsymbol{r}_c$ is regarded as a function of $\boldsymbol{r}_1,\boldsymbol{r}_2,\dots,\boldsymbol{r}_{D+1}$ through (4.14). Define

$$\boldsymbol{\sigma}_i\equiv\boldsymbol{r}_i-\boldsymbol{r},\qquad \sigma_i\equiv|\boldsymbol{\sigma}_i|. \tag{4.20}$$

At fixed $\boldsymbol{r}_i$ $(i=1,2,\dots,D+1)$, we have $\mathrm{d}\boldsymbol{r}_c=0$, $\mathrm{d}\boldsymbol{\sigma}_i=-\mathrm{d}\boldsymbol{r}$,

$$\mathrm{d}\sigma_i=-\frac{1}{\sigma_i}\boldsymbol{\sigma}_i\cdot\mathrm{d}\boldsymbol{r},$$

$$\delta^D(\boldsymbol{r}-\boldsymbol{r}_c)=|M|\delta(\sigma_{D+1}-\sigma_1)\delta(\sigma_{D+1}-\sigma_2)\cdots\delta(\sigma_{D+1}-\sigma_D), \tag{4.21}$$

and $|M|$ is the determinant of a matrix whose (m,n)th element is

$$M_{mn}=\frac{\partial}{\partial r_n}(\sigma_{D+1}-\sigma_m)=(\hat{\boldsymbol{\sigma}}_m-\hat{\boldsymbol{\sigma}}_{D+1})_n,$$

where m and n vary from 1 to D, r_n is the nth component of $\boldsymbol{r}$ and $\hat{\boldsymbol{\sigma}}_i$ is a unit vector along $\boldsymbol{\sigma}_i$. When $\boldsymbol{r}=\boldsymbol{r}_c$, $\hat{\boldsymbol{\sigma}}_i$ is given by (4.15); M_{mn} becomes

$$M_{mn}=\frac{1}{R}(\boldsymbol{r}_m-\boldsymbol{r}_{D+1})_n,$$

and therefore its determinant is related to V, the volume of the D-simplex, by

$$|M|=D!R^{-D}V. \tag{4.22}$$

Interchanging the order of integration in (4.19) we write

$$I(f)=\int\mathrm{d}^D r\int\delta^D(\boldsymbol{r}-\boldsymbol{r}_c)f\prod_{i=1}^{D+1}\mathrm{d}^D r_i. \tag{4.23}$$

Because of (4.20), at fixed $\boldsymbol{r}$

$$d^D r_i = d^D \sigma_i = \sigma_i^{D-1} d\sigma_i d^{D-1}\hat{\boldsymbol{\sigma}}_i .$$

Substituting the above expression and (4.21) into (4.23), we find

$$I(f) = \int d^D r \int |M| f \delta(\sigma_{D+1} - \sigma_1)\delta(\sigma_{D+1} - \sigma_2) \cdots$$
$$\times \delta(\sigma_{D+1} - \sigma_D) \prod_{i=1}^{D+1} \sigma_i^{D-1} d\sigma_i d^{D-1}\hat{\boldsymbol{\sigma}}_i . \tag{4.24}$$

Carry out the integrations over $d\sigma_1, d\sigma_2, \ldots, d\sigma_D$ and note that when $\boldsymbol{r} = \boldsymbol{r}_c$

$$\sigma_1 = \sigma_2 = \cdots = \sigma_{D+1} = R ,$$

$$d\sigma_{D+1} = dR , \qquad d^D r = d^D r_c .$$

Hence, (4.24) becomes

$$I(f) = \int |M| R^{D^2-1} f d^D r_c dR \prod_{i=1}^{D+1} d^{D-1}\hat{\boldsymbol{\sigma}}_i .$$

Theorem 4 is then proved because of (4.22).

Example 1. $D=2$: A D-simplex is now a triangle; in (4.2), v is the area of its circumscribed circle,

$$v = \pi R^2 .$$

The distribution function $\mathcal{D}$ is therefore given by

$$\mathcal{D}(1,2,3) = \frac{1}{3!}\rho^3 e^{-\rho\pi R^2} . \tag{4.25}$$

Because of (4.13), the average of any function $f(1,2,3)$ is given by

$$\bar{f} = \frac{1}{3!\mathcal{N}_2} \int f(1,2,3)\rho^3 e^{-\pi R^2 \rho} \prod_1^3 d^2 r_i , \tag{4.26}$$

where, as before, $\mathcal{N}_2$ is the number of 2-simplices (triangles).

Let $\Delta(1,2,3)$ be the triangle whose vertices are sites 1, 2 and 3. As in fig. 3a, the polar coordinates of $\boldsymbol{\sigma}_i = \boldsymbol{r}_i - \boldsymbol{r}_c$ are R and ϕ_i; hence, (4.17) becomes

$$\prod_1^3 d^2 r_i = 2\Delta R\, dR\, d^2 r_c \prod_1^3 d\phi_i , \tag{4.27}$$

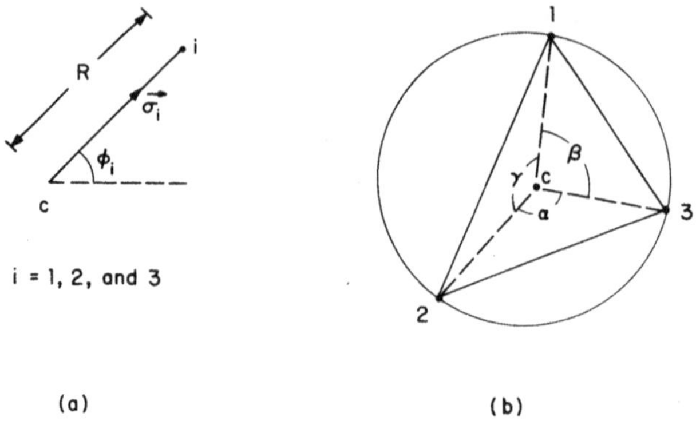

Fig. 3. In two dimensions, let c be the center of the circumscribed circle of $\Delta(1,2,3)$. (a) The polar coordinates of $\boldsymbol{\sigma}_i=\boldsymbol{r}_i-\boldsymbol{r}_c$ are R and ϕ_i, where R is the radius of the circumscribed circle. (b) The relative polar angles of 1, 2, 3 are γ, α and β; they are chosen to be always positive and satisfy $\alpha+\beta+\gamma=2\pi$.

where Δ is the area of the triangle. [$\Delta=V$ of (4.17).] As in fig. 3b, we may arrange 1, 2 and 3 in a definite order, say counter-clockwise, on their circumscribed circle, with their relative polar angles $=\alpha$, β and γ; these angles are chosen to be positive and satisfy

$$\alpha+\beta+\gamma=2\pi. \tag{4.28}$$

Thus,

$$\int\prod_1^3 \mathrm{d}\phi_i=2(2\pi)\int\delta(\alpha+\beta+\gamma-2\pi)\,\mathrm{d}\alpha\,\mathrm{d}\beta\,\mathrm{d}\gamma, \tag{4.29}$$

where the first factor two is due to the definite counter-clockwise order chosen for the sites 1, 2 and 3. In terms of α, β, γ and R, we have for the area of the triangle

$$\Delta=\tfrac{1}{2}R^2(\sin\alpha+\sin\beta+\sin\gamma), \tag{4.30}$$

and for the average length of its sides

$$l=\tfrac{1}{3}2R(\sin\tfrac{1}{2}\alpha+\sin\tfrac{1}{2}\beta+\sin\tfrac{1}{2}\gamma). \tag{4.31}$$

By substituting (4.27) and (4.29) into (4.26), and assuming that $f(1,2,3)$ depends only on relative coordinates, we find

$$\bar{f}=\frac{4\pi\rho^3}{3\rho_2}\int f(1,2,3)R\,\mathrm{d}R\,\mathrm{e}^{-\rho\pi R^2}\Delta\delta(\alpha+\beta+\gamma-2\pi)\,\mathrm{d}\alpha\,\mathrm{d}\beta\,\mathrm{d}\gamma, \tag{4.32}$$

where, as in (3.1), (3.2), ρ_2 is the density of triangles and ρ is the density of sites.

Setting f equal to 1, we can carry out the integration (4.32). The result gives

$$\rho_2 = 2\rho\,; \tag{4.33}$$

likewise, by setting $f = \Delta$ and l we obtain

$$\bar{\Delta} = 1/2\rho\,, \tag{4.34}$$

$$\bar{l} = \frac{32}{9\pi}\rho^{-1/2} \cong 1.1318\rho^{-1/2}\,. \tag{4.35}$$

Thus, (4.33) confirms the second equation in (3.19), and (4.34) can then be directly derived by using

$$\rho_2 \bar{\Delta} = 1\,.$$

When $D = 2$, each link is shared by two triangles and the total numbers $\mathcal{N}_1$ and $\mathcal{N}_2$ of links and triangles are related by

$$\mathcal{N}_1 = \tfrac{3}{2}\mathcal{N}_2\,.$$

Hence, from (4.10) we see that the total link length in Ω is equal to

$$\tfrac{3}{2}\bar{l}\mathcal{N}_2 = \bar{l}\mathcal{N}_1\,,$$

and thus $\bar{l}$ is also the average link length of the entire lattice*.

Example 2. $D = 3$ and 4: In three dimensions, the D-simplex is a tetrahedron, say $\tau(1,2,3,4)$, whose vertices are sites 1, 2, 3 and 4. Let the spherical coordinates of $\boldsymbol{\sigma}_i = \boldsymbol{r}_i - \boldsymbol{r}_c$ be R, θ_i and ϕ_i, where θ_i is the polar angle and ϕ_i the azimuthal angle. In accordance with (4.2), (4.13) and (4.17), the average of any function $f(1,2,3,4)$ that depends only on relative coordinates is given by

$$\bar{f} = \frac{\rho^4}{4\rho_3}\int f R^5\,\mathrm{d}R\,\mathrm{e}^{-(4\pi R^3\rho)/3}\tau \prod_{i=1}^{4} \sin\theta_i\,\mathrm{d}\theta_i\,\mathrm{d}\phi_i\,, \tag{4.36}$$

where τ is the volume of the tetrahedron and, as before, ρ_3 is the density of tetrahedra and ρ the density of sites.

In four dimensions, let $S(1,2,3,4,5)$ be the D-simplex whose vertices are sites $1,2,\ldots,5$. The corresponding average of any function $f(1,2,\ldots,5)$ that depends only

* This is to be contrasted with the situation in other dimensions. For example, when $D = 3$, similar reasoning leads to $\bar{\Delta}$ and $\bar{\tau}$, defined by (4.11) and (4.12), as the triangle area and the tetrahedron volume averaged over the entire Ω. But $\bar{l}$ is only the average link length per tetrahedron; it is not the average link length of the whole lattice.

on relative coordinates is given by

$$\bar{f}=\frac{\rho^5}{5\rho_4}\int fR^{11}\,\mathrm{d}R\,\mathrm{e}^{-(\pi^2R^4\rho)/2}V\prod_{i=1}^{5}\sin^2\theta_i\sin\theta_i'\,\mathrm{d}\theta_i\,\mathrm{d}\theta_i'\,\mathrm{d}\phi_i, \tag{4.37}$$

where V is the volume of the 4-simplex, R, θ_i, θ_i' and ϕ_i are the spherical coordinates of $\boldsymbol{\sigma}_i=\boldsymbol{r}_i-\boldsymbol{r}_c$, with the polar angles θ_i, θ_i' varying independently from 0 to π, and the azimuthal angle ϕ_i from 0 to 2π.

For most of the functions of physical interest, such as the link length, the triangle area, $N_{n/m}$, etc., it turns out that these multidimensional integrals (4.36) and (4.37) can all be reduced to elementary integrations. The details are given in the appendix and the results are listed in tables 1 and 2.

From table 2 and (3.3), we see that when $D=3$, the ratio

$$\frac{\mathcal{N}_i}{\mathcal{N}_0}=\begin{cases}1+\frac{24}{35}\pi^2, & \text{for } i=1,\\ \frac{48}{35}\pi^2, & \text{for } i=2,\\ \frac{24}{35}\pi^2, & \text{for } i=3,\end{cases} \tag{4.38}$$

Table 1

Average values of l (the length of a link), Δ (the area of a triangle), τ (the volume of a tetrahedron) and V (the volume of a 4-simplex) per D-simplex

$D=2$	$D=3$	$D=4$
$\bar{l}=\frac{32}{9\pi}$ $\cong 1.1318$	$\left(\frac{3}{4\pi}\right)^{1/3}\frac{5\cdot 7^3}{3^2\cdot 4^4}\Gamma(\frac{1}{3})$ $\cong 1.2371$	$\left(\frac{2}{\pi^2}\right)^{1/4}\frac{27}{77\pi}\frac{16!!}{15!!}\Gamma(\frac{1}{4})$ $\cong 1.3825$
$\bar{\Delta}=\frac{1}{2}$	$\left(\frac{3}{4\pi}\right)^{2/3}\frac{7\cdot 5^3}{3^5\pi}\Gamma(\frac{2}{3})$ $\cong 0.5973$	$\sqrt{2}\,\frac{2}{13}\left(\frac{8}{11}\right)^2\frac{17!!}{16!!}\Gamma(\frac{1}{2})$ $\cong 0.6809$
$\bar{\tau}=$	$\frac{35}{24\pi^2}$ $\cong 0.1478$	$\left(\frac{2}{\pi^2}\right)^{3/4}\frac{32}{385}\left(\frac{14}{13}\right)^2\frac{18!!}{17!!}\Gamma(\frac{3}{4})$ $\cong 0.1923$
$\bar{V}=$		$\frac{9}{286}$

See (4.10)–(4.13) for their definitions. The site density ρ is set to be unity.

TABLE 2
Average number of n-simplies/m-simplex

		$D=2$	$D=3$	$D=4$
$N_{1/0}$	$=$	6	$2(1+\frac{24}{35}\pi^2)$	$\frac{340}{9}$
$N_{2/0}$	$=$	6	$\frac{144}{35}\pi^2$	$\frac{590}{3}$
$N_{3/0}$	$=$		$\frac{96}{35}\pi^2$	$\frac{2860}{9}$
$N_{4/0}$	$=$			$\frac{1430}{9}$
$N_{2/1}$	$=$	2	$\dfrac{144\pi^2}{35+24\pi^2}$	$\frac{177}{17}$
$N_{3/1}$	$=$		$\dfrac{144\pi^2}{35+24\pi^2}$	$\frac{429}{17}$
$N_{4/1}$	$=$			$\frac{286}{17}$
$N_{3/2}$	$=$		2	$\frac{286}{59}$
$N_{4/2}$	$=$			$\frac{286}{59}$
$N_{4/3}$	$=$			2

E.g., $N_{1/0}$, $N_{2/0}$ and $N_{3/0}$ are the average numbers of links, triangles and tetrahedra per site, $N_{2/1}$ and $N_{3/1}$ are the average numbers of triangles and tetrahedra per link, etc.

and when $D=4$

$$\frac{\mathcal{N}_i}{\mathcal{N}_0}=\begin{cases}\frac{170}{9}, & \text{for } i=1,\\ \frac{590}{9}, & \text{for } i=2,\\ \frac{715}{9}, & \text{for } i=3,\\ \frac{286}{9}, & \text{for } i=4.\end{cases} \tag{4.39}$$

Also, we note that

$$N_{D/(D-1)}=2, \tag{4.40}$$

which is the consequence of the fact that in D dimensions, each $(D-1)$-simplex is shared by two D-simplices. In addition, from the same table we have

$$N_{D/(D-2)}=N_{(D-1)/(D-2)}=\begin{cases}6, & \text{when } D=2,\\ 5.2276, & \text{when } D=3,\\ 4.8475, & \text{when } D=4.\end{cases} \tag{4.41}$$

We note that the first equality $N_{D/(D-2)}=N_{(D-1)/(D-2)}$ follows from (3.5) with $m=D-2$.

Some insight into these numbers may be obtained by examining a regular D-simplex, one in which all links are of identical length. The surface of a D-simplex consists of a number of $(D-1)$-simplices. It can be readily verified that the angle between two such neighboring $(D-1)$-simplices in a regular D-simplex is

$$\cos^{-1}\frac{1}{D}. \tag{4.42}$$

For example, when $D=2$, the angle between two neighboring sides in a regular triangle is $\cos^{-1}\frac{1}{2}$; when $D=3$, the angle between two neighboring triangular surfaces in a regular tetrahedron is $\cos^{-1}\frac{1}{3}$; etc. Now, consider any $(D-2)$-simplex in a D-dimensional random lattice. Assuming that it is shared only by regular D-simplices, we derive a rather good approximate formula

$$N_{D/(D-2)} = N_{(D-1)/(D-2)} \cong \frac{2\pi}{\cos^{-1}(1/D)} \tag{4.43}$$

which is 6 for $D=2$, 5.104 for $D=3$, 4.767 for $D=4$, and approaches 4 when $D\to\infty$.

5. Dual lattice

Let i be an arbitrary site of a D-dimensional random lattice ($i=1,2,\ldots,\mathcal{N}$). Take any point p in Ω. We say p belongs to i if i is the nearest site to p. The volume ω_i consists of all p belonging to i. Each ω_i is a convex polyhedron (called a cell) in D dimensions. The surface of a cell consists of a number of $(D-1)$-dimensional faces, formed by points that belong to two or more sites. The intersections of these $(D-1)$-dimensional faces are $(D-2)$-dimensional faces, which consist of points each belonging to three or more sites, etc. The corners (i.e., vertices) of these cells are single points, each belonging to, and only to, $D+1$ sites. Let c be one of these corners. From our construction we see that c is equidistant from all the $D+1$ sites to which it belongs. Using c as the center, draw a hypersphere through these $D+1$ sites. Since no site is nearer to c than the aforementioned $D+1$ sites, there cannot be any lattice site inside the hypersphere. Hence, these $D+1$ sites satisfy the definition of a cluster, and form a D-simplex.

The division of Ω into cells gives the dual of our random lattice. Each link in the random lattice is perpendicular to (but need not intersect) a $(D-1)$-dimensional face in its dual; each triangle is perpendicular to a $(D-2)$-dimensional face, etc. This gives a one-to-one correspondence between an n-simplex in the random lattice and a $(D-n)$-dimensional face in its dual. While each D-simplex has a fixed structure (e.g., fixed number of vertices and faces, etc.), this is not so for the cells. For this reason, as we shall discuss, it is often more convenient to formulate the action in terms of simplices, instead of cells. This is particularly true in applications to gauge field theories.

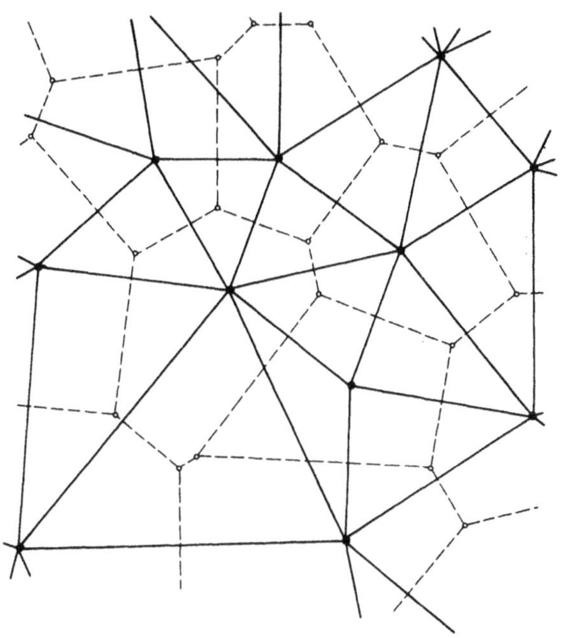

Fig. 4. An example of a two-dimensional random lattice (solid lines) and its dual (dashed lines). Dots are the lattice sites, and open circles are the corners of the dual cells. Inside each cell there is always a dot; but inside each 2-simplex (triangle) there may or may not be an open circle.

The cells ω_i are called Voronoi polyhedra in the literature [4]. Their construction is a generalization of the Wigner-Seitz procedure [5] for a regular lattice.

In fig. 4, we give a two-dimensional example of a random lattice and its dual. Each cell is now a polygon, whose sides will be referred to as borders. Because of the one-to-one correspondence between cells and sites, borders and links, and corners and triangles, the total numbers of cells, borders and corners are, respectively, $\mathcal{N}_0$, $\mathcal{N}_1$ and $\mathcal{N}_2$ defined by (3.1). The average number of borders per cell is 6, since it is the same as that of links per site. Likewise, the average number of corners per cell is also 6. The average area of a cell is ρ^{-1}, where ρ is the site-density. By using (4.32), one can show that the average length of a border is

$$\bar{b} = \frac{2}{3\sqrt{\rho}} \,. \tag{5.1}$$

The properties of the cells in the dual lattice have been studied in the literature [6, 7]; hence, the details will be omitted here.

6. Massless scalar field

As an application of the preceding ideas let us consider a discrete approximation to a continuum free field theory in D dimensions using our random lattice. We begin with the usual generating functional for a free massless scalar field in D euclidean dimensions expressed as a path integral

$$\mathcal{G}_C(J) = \int e^{-A_C(\Omega)} d[\phi]. \tag{6.1}$$

Here $A_C(\Omega)$ is the continuum action

$$A_C(\Omega) = \int_\Omega \left\{ \tfrac{1}{2}[\nabla\phi(\boldsymbol{r})]^2 - J(\boldsymbol{r})\phi(\boldsymbol{r}) \right\} d^D r, \tag{6.2}$$

$\boldsymbol{r}$ is a D-dimensional euclidean vector, ∇ the corresponding gradient operator, $J(\boldsymbol{r})$ the external current density and Ω the D-dimensional euclidean volume of the system.

6.1. LATTICE ACTION

In the same volume Ω, consider now a random lattice of site-density $\rho = \mathcal{N}/\Omega$. On each site i, we assign a value ϕ_i for the scalar field and a value J_i for the external current. The lattice action is

$$A_L(\Omega) = \tfrac{1}{4}\sum_{i,j} \lambda_{ij}[\phi_i - \phi_j]^2 - \sum_i \phi_i J_i, \tag{6.3}$$

where the second sum is over all sites i and the first extends over all sites i and j connected by a link. If i and j are not linked we choose the coefficient $\lambda_{ij} = 0$; otherwise it is a function λ_l of the link length $l = |\boldsymbol{r}_i - \boldsymbol{r}_j|$. The function λ_l will be chosen to satisfy the following requirement: If ϕ_i is obtained from a continuum function $\phi(\boldsymbol{r})$ by

$$\phi_i = \phi(\boldsymbol{r}_i), \tag{6.4}$$

where $\boldsymbol{r}_i$ is the position vector of i, then the lattice action should approach the continuum action when we increase ρ but keep Ω fixed; i.e.

$$\lim_{\rho\to\infty} A_L(\Omega) = A_C(\Omega). \tag{6.5}$$

As we shall see, this does not determine λ_l but imposes a constraint

$$\frac{1}{D}\int_0^\infty f(l) l^2 \lambda_l \, dl = 1, \tag{6.6}$$

where $f(l)\,dl$ is the density of links of length between l and $l+dl$ so that

$$\int_0^\infty f(l)\,dl = \rho_1, \tag{6.7}$$

the link density in accordance with (3.1).

To show this, consider the limit of site density $\rho \to \infty$ and make the replacement

$$\phi_i - \phi_j \to \boldsymbol{l}_{ij} \cdot \nabla\phi, \tag{6.8}$$

where $\boldsymbol{l}_{ij} = \boldsymbol{r}_i - \boldsymbol{r}_j$ in the lattice action (6.5). One obtains

$$\tfrac{1}{2}\sum_{i,j} \lambda_{ij}[\phi_i - \phi_j]^2 \to \int_\Omega d^D r \int_0^\infty f(l)\lambda_l \langle (\boldsymbol{l}\cdot\nabla\phi)^2\rangle_{av}\,dl, \tag{6.9}$$

where $\langle\ \rangle_{av}$ denotes the average over the angle θ between $\boldsymbol{l}$ and $\nabla\phi$. For a random lattice, we have

$$\langle(\boldsymbol{l}\cdot\nabla\phi)^2\rangle_{av} = l^2(\nabla\phi)^2\langle\cos^2\theta\rangle_{av}, \tag{6.10}$$

and

$$\langle\cos^2\theta\rangle_{av} = 1/D,$$

since $\cos^2\theta$ is the square of one component of a D-dimensional unit vector. In order that $A_L \to A_C$, λ_l must satisfy (6.6), at least in the limit.

Thus we may approximate the continuum generating functional $\mathcal{G}_C(J)$ by the lattice function

$$\mathcal{G}_L(J_i) = \int e^{-A_L(\Omega)} \prod_{i=1}^{\mathfrak{N}} d\phi_i. \tag{6.11}$$

6.2. KIRCHOFF'S LAWS

It is instructive to examine the classical Euler-Lagrange equations determined by the actions A_C and A_L of (6.2) and (6.3). For the continuum case we have

$$-\nabla^2\phi(\boldsymbol{r}) = J(\boldsymbol{r}), \tag{6.12}$$

while for our lattice formulation

$$\sum_j \lambda_{ij}(\phi_i - \phi_j) = J_i. \tag{6.13}$$

Eq. (6.13) may be read as one of Kirchoff's laws for an electric circuit. We may regard ϕ_i as the electric potential at i and λ_{ij}^{-1} as the resistance of the link connecting i and j. The current I_{ij} flowing along a link l_{ij} from j to i is then given by

$$I_{ij}=\lambda_{ij}(\phi_j-\phi_i). \tag{6.14}$$

Thus (6.13) expresses current conservation at the site i if J_i is interpreted as the external current entering i. A similar interpretation can be made of the continuum equation (6.12) if space is viewed as a uniform medium with unit resistivity.

6.3. LATTICE PROPAGATOR

In an infinite volume, the continuum free-field propagator $G_C(\boldsymbol{r})$ determined by the generating functional $\mathcal{G}_C(J)$,

$$G_C(\boldsymbol{r})\equiv\mathcal{G}_C(J)^{-1}\frac{\delta}{\delta J(\boldsymbol{r}_a)}\frac{\delta}{\delta J(\boldsymbol{r}_b)}\mathcal{G}_C(J)\bigg|_{J=0},$$

where $\boldsymbol{r}=\boldsymbol{r}_a-\boldsymbol{r}_b$, solves the classical field equation (6.12) when J is a δ-function:

$$-\nabla^2 G_C=\delta^D(\boldsymbol{r}). \tag{6.15}$$

Because of spherical symmetry, G_C depends only on $r=|\boldsymbol{r}|$. If (6.15) is integrated over a spherical region of radius r and Gauss' law employed, we find

$$-\frac{dG_C}{dr}=\frac{1}{\mathcal{S}_r}, \tag{6.16}$$

where

$$\mathcal{S}_r=\begin{cases}\dfrac{(2\pi)^{D/2}}{(D-2)!!}r^{D-1}, & \text{if } D \text{ is even},\\[2ex] \dfrac{(2\pi)^{(D-1)/2}}{(D-2)!!}2r^{D-1}, & \text{if } D \text{ is odd},\end{cases} \tag{6.17}$$

is the "area" of a hypersphere of radius r in D dimensions.

Likewise in the lattice theory (again for infinite volume), the propagator G_{ab} determined by the generating function $\mathcal{G}_L(J_i)$,

$$G_{ab}\equiv\mathcal{G}_L(J_i)^{-1}\frac{\partial}{\partial J_a}\frac{\partial}{\partial J_b}\mathcal{G}_L(J_i)\bigg|_{J_i=0}, \tag{6.18}$$

solves the classical field equation (6.13) when the current J_i is a Kronecker δ-function δ_{ib}:

$$\sum_j \lambda_{ij}(G_{ib}-G_{jb}) = \delta_{ib} = \begin{cases} 1, & \text{if } i=b, \\ 0, & \text{otherwise}. \end{cases} \tag{6.19}$$

This equation determines the propagator G_{ab} for a massless field on the lattice up to an arbitrary, additive constant. (For $D>2$ we will fix this constant by requiring $G_{ab}\to 0$ as $|\boldsymbol{r}_a-\boldsymbol{r}_b|\to\infty$.)

Now let us determine the limiting form of G_{ab} when the separation between a and b becomes large. For simplicity we choose the point b to lie at the origin $\boldsymbol{r}_b=0$ and define an average function $\bar{G}(r)$ obtained by averaging G_{ab} over a sphere of radius r centered at the origin. To that end define a potential $\phi_{ij}(\xi)$ for a point on a link connecting i and j a distance ξ from the end j:

$$\phi_{ij}(\xi) \equiv G_{jb} + \frac{\xi}{l_{ij}}(G_{ib}-G_{jb}), \tag{6.20}$$

where $l_{ij}=|\boldsymbol{l}_{ij}|$. We can then define $\bar{G}(r)$ as the weighted average

$$\bar{G}(r) = \frac{\sum_{(i,j)_r} w_{ij}\phi_{ij}(\xi_{ij}(r))}{\sum_{(i,j)_r} w_{ij}}, \tag{6.21}$$

in which both sums extend over pairs $(i,j)_r$ of points i and j connected by a link cut by our sphere of radius r at distance $\xi=\xi_{ij}(r)$ from the end j. The weights $w_{ij}>0$ can be chosen with some freedom without affecting the large-distance behavior of $\bar{G}(r)$. Our specification, given below in (6.31), is made in such a way as to simplify the analysis. For large r, $\bar{G}(r)$ will obey an equation analogous to the continuum equation (6.16):

$$-\frac{d\bar{G}(r)}{dr} = \frac{Z}{\mathbb{S}_r}, \tag{6.22}$$

where Z is the bulk resistivity of the lattice. For example if we integrate (6.22) between r and $r+\Delta r$ with $\Delta r \ll r$, we find

$$\bar{G}(r) - \bar{G}(r+\Delta r) = \frac{\Delta r}{\mathbb{S}_r} Z. \tag{6.23}$$

The left-hand side of this equation can be viewed as the potential difference between spheres of radius r and $r+\Delta r$ divided by the unit current flowing between them. The right-hand side is then the resistance of the "medium" between them, or $\Delta r/\mathbb{S}_r$ times the bulk resistivity Z. Thus we expect $G_C(r)$ and $\bar{G}(r)$ to have a similar r^{-D+2} large-r behavior with a possible difference in scale. We will now show that even for a lattice action with a properly normalized continuum limit, eqs. (6.5) and (6.6), the constant Z differs from 1. (In a forthcoming paper, we shall show that by not restricting λ_{ij} to be a function only of the link length l_{ij}, it is possible to have $Z=1$ rigorously.)

6.4. BULK RESISTIVITY

The bulk resistivity Z can be most easily determined in a simpler, planar geometry. Begin with a random lattice constructed in a large cube of volume $\Omega = L^D$ and satisfying periodic boundary conditions. We can view such a lattice as defined over all of space but invariant under the translation $x_\mu \to x_\mu + n_\mu L$ for integers n_μ, where x_μ are the components of $\boldsymbol{r}$ with $\mu = 1,2,\ldots,D$. As above we can introduce a potential ϕ_i on each site i of the lattice. We can impose a potential difference V between the $x_D=0$ face and the $x_D=L$ face if we require

$$\phi_j = \phi_i + \delta_{\mu D} V, \qquad \text{for } \boldsymbol{r}_j = \boldsymbol{r}_i + L\hat{\boldsymbol{e}}_\mu, \tag{6.24}$$

where the unit vector $\hat{\boldsymbol{e}}_\mu$ has coordinates $x_\nu = \delta_{\nu\mu}$. Thus ϕ_i is periodic in the directions 1 through $D-1$, but increases by V when translated in the D-direction.

Just as in continuum electrostatics there is a unique potential function ϕ_i which satisfies the boundary conditions (6.24) and the homogeneous Kirchoff equations

$$\sum_j \lambda_{ij}(\phi_j - \phi_i) = 0, \qquad \text{for all } i.$$

Henceforth, let ϕ_i denote this solution. If we can find the total current I flowing within our cube in the negative D-direction,

$$I = \sum_{(i,j)_y} \lambda_{ij}(\phi_i - \phi_j), \tag{6.25}$$

then the bulk resistivity Z can also be obtained: since the current density is I/L^{D-1} and the potential change V extends over a length L, Z is given by

$$Z = L^{D-2} V/I. \tag{6.26}$$

The sum in (6.25) runs over all pairs $(i, j)_y$ of points i and j which are connected by a link cut by the plane $x_D = y$. The intersection of the plane and the link must lie inside the cube with

$$(\boldsymbol{r}_i)_D > y > (\boldsymbol{r}_j)_D. \tag{6.27}$$

In order to compute I, we define an average potential $\bar{\phi}(y)$ for the plane $x_D = y$ in analogy with (6.21):

$$\bar{\phi}(y) = \frac{\sum_{(i,j)_y} w_{ij}\phi_{ij}(\xi_{ij}(y))}{\sum_{(i,j)_y} w_{ij}}, \tag{6.28a}$$

with

$$\phi_{ij}(\xi)=\phi_j+\frac{\xi}{l_{ij}}(\phi_i-\phi_j), \tag{6.28b}$$

in the notation of (6.20) and (6.21). Next we evaluate $\mathrm{d}\bar{\phi}(y)/\mathrm{d}y$. We now choose the weights w_{ij} so that

$$w_{ij}\frac{\mathrm{d}\phi_{ij}(\xi_{ij}(y))}{\mathrm{d}y}=\lambda_{ij}(\phi_i-\phi_j), \tag{6.29}$$

the current flowing through the link from i to j. From (6.28b)

$$\frac{\mathrm{d}}{\mathrm{d}y}\phi_{ij}(\xi_{ij}(y))=(\phi_i-\phi_j)\frac{1}{l_{ij}}\frac{\mathrm{d}\xi_{ij}(y)}{\mathrm{d}y}. \tag{6.30}$$

Using $\mathrm{d}\xi_{ij}(y)/\mathrm{d}y=(\cos\theta_{ij})^{-1}$, we find

$$w_{ij}=l_{ij}\lambda_{ij}\cos\theta_{ij}, \tag{6.31}$$

where $\cos\theta_{ij}\equiv \boldsymbol{l}_{ij}\cdot\hat{\boldsymbol{e}}_D/l_{ij}$ which, because of (6.27), is $\geqslant 0$. With this choice of w_{ij}, the sum in the denominator of (6.28a) can be evaluated:

$$\begin{aligned}\sum_{(i,j)_y} w_{ij} &= \sum_{(i,j)_y} l_{ij}\lambda_{ij}\cos\theta_{ij} \\ &= L^{D-1}\int_0^\infty f(l)l^2\lambda_l\langle\cos^2\theta\rangle \\ &= L^{D-1}.\end{aligned} \tag{6.32}$$

The third equality in (6.32) follows from the normalization condition (6.6) obeyed by $f(l)$. The second equality can be obtained if we recall that $f(l)\,\mathrm{d}l$ is the density of links with length between l and $l+\mathrm{d}l$ and that such a link with $\cos\theta_{ij}$ will be cut by our plane if, e.g., one of its ends lies within a volume $l\cos(\theta_{ij})L^{D-1}$.

Thus (6.28a) for $\bar{\phi}(y)$ becomes

$$\bar{\phi}(y)=L^{-D+1}\sum_{(i,j)_y} w_{ij}\phi_{ij}(\xi_{ij}(y)). \tag{6.33}$$

There are two separate contributions to the y-derivative of (6.33):

$$\frac{\mathrm{d}\bar{\phi}(y)}{\mathrm{d}y}=\left[\frac{\mathrm{d}\bar{\phi}(y)}{\mathrm{d}y}\right]_{\text{link}}+\left[\frac{\mathrm{d}\bar{\phi}(y)}{\mathrm{d}y}\right]_{\text{site}}. \tag{6.34}$$

$[\mathrm{d}\bar{\phi}/\mathrm{d}y]_{\text{link}}$ comes solely from the y dependence of $\phi_{ij}(\xi_{ij}(y))$ as $\xi_{ij}(y)$ moves along a link and is therefore completely determined by (6.29) and (6.25):

$$\left[\frac{\mathrm{d}\bar{\phi}}{\mathrm{d}y}\right]_{\text{link}} = L^{-D+1} \sum_{(i,j)_y} \lambda_{ij}(\phi_i - \phi_j)$$

$$= L^{-D+1} I. \tag{6.35}$$

If this were the only term, then

$$V = \int_0^L \left[\frac{\mathrm{d}\bar{\phi}}{\mathrm{d}y}\right]_{\text{link}} \mathrm{d}y = L^{-D+2} I \tag{6.36}$$

and, from (6.26), Z would be 1.

The second term $[\mathrm{d}\bar{\phi}(y)/\mathrm{d}y]_{\text{site}}$ arises because the $\Sigma_{(i,j)_y}$ changes discontinuously as the plane $x_D = y$ crosses the location of a site:

$$\left[\frac{\mathrm{d}\bar{\phi}(y)}{\mathrm{d}y}\right]_{\text{site}} = L^{-D+1} \sum_i \delta(y - (\boldsymbol{r}_i)_D)\phi_i \left[\sum_{j(+)} w_{ji} - \sum_{j(-)} w_{ij}\right]. \tag{6.37}$$

Here ϕ_i times the quantity in square brackets is the discontinuity in the sum (6.33) when y equals the x_D coordinate $(\boldsymbol{r}_i)_D$ of the ith site. The sum $\Sigma_{j(+)}$ extends over those sites j linked to i with $(\boldsymbol{r}_j)_D > y$ and $\Sigma_{j(-)}$ over those with $(\boldsymbol{r}_j)_D < y$. (See fig. 5.) Substituting the formula (6.31) for w_{ij} and recognizing that $\cos\theta_{ij} = -\cos\theta_{ji}$, we find

$$\left[\frac{\mathrm{d}\bar{\phi}(y)}{\mathrm{d}y}\right]_{\text{site}} = L^{-D+1} \sum_i \delta(y - (\boldsymbol{r}_i)_D)\phi_i \left[\sum_j \lambda_{ji} \boldsymbol{l}_{ji} \cdot \hat{\boldsymbol{e}}_D\right], \tag{6.38}$$

where Σ_j now sums over all sites j connected to i. In the limit that the dimensions of the cube transverse to $\hat{\boldsymbol{e}}_D$ become large,

$$\sum_i \delta(y - (\boldsymbol{r}_i)_D) \to \rho L^{D-1}, \tag{6.39}$$

so that

$$\left[\frac{\mathrm{d}\bar{\phi}(y)}{\mathrm{d}y}\right]_{\text{site}} = \rho \left\langle \phi_i \sum_j \lambda_{ji} \boldsymbol{l}_{ji} \cdot \hat{\boldsymbol{e}}_D \right\rangle_i, \tag{6.40}$$

where ρ is the site density and $\langle\ \rangle$ represents an average of sites i in the plane $x_D = y$.

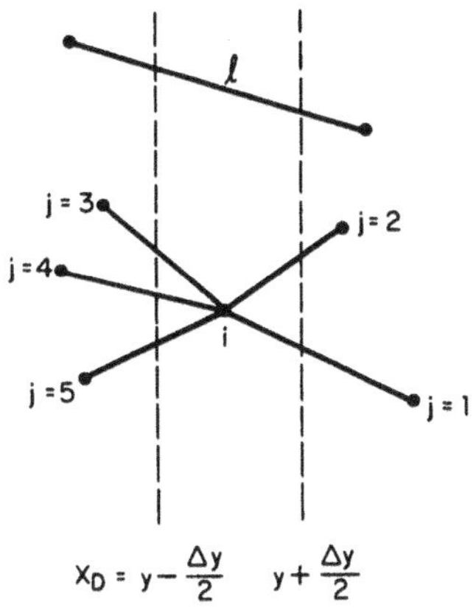

Fig. 5. Link l contributes to $[\mathrm{d}\bar{\phi}(y)/\mathrm{d}y]_{\text{link}}$ and site i to $[\mathrm{d}\bar{\phi}(y)/\mathrm{d}y]_{\text{site}}$. Sites $j = 1$ and 2 participate in the sum $\Sigma_{j(+)}$ in (6.37) and $j = 3$, 4 and 5 in $\Sigma_{j(-)}$.

In order to evaluate the right-hand side of (6.40) more explicitly, we replace the factor ϕ_i by $\phi_i - \bar{\phi}(y)$; the extra constant piece, $-\bar{\phi}(y)$, drops out of the average because of the random link direction:

$$\left[\frac{\mathrm{d}\bar{\phi}(y)}{\mathrm{d}y}\right]_{\text{site}} = \rho\left\langle \delta\phi_i \sum_j \lambda_{ji} \boldsymbol{l}_{ji} \cdot \hat{\boldsymbol{e}}_D \right\rangle_i , \tag{6.41}$$

where $\delta\phi_i = \phi_i - \bar{\phi}((\boldsymbol{r}_i)_D)$ is the variation of ϕ_i from the average $\bar{\phi}(y)$ at that point. Next we use the fact that, unlike ϕ_i, $\delta\phi_i$ is periodic in all D directions and obeys the inhomogeneous Kirchoff equation

$$\begin{aligned}\sum_j \lambda_{ji}(\delta\phi_i - \delta\phi_j) &= \sum_j \lambda_{ji}\left[\bar{\phi}((r_j)_D) - \bar{\phi}((r_i)_D)\right] \\ &= \sum_j \lambda_{ji} \boldsymbol{l}_{ji} \cdot \hat{\boldsymbol{e}}_D \frac{\mathrm{d}\bar{\phi}}{\mathrm{d}y} ,\end{aligned} \tag{6.42}$$

where the last equality is true for a large volume Ω, in which case $\bar{\phi}(y)$ is linear in y. If we introduce a Green function G_{ij}, identical to that given by (6.18) and (6.19) but for a finite volume satisfying periodic boundary conditions, then we can solve (6.42) for $\delta\phi_i$:

$$\delta\phi_i = \sum_k G_{ik}\left\{\sum_j \lambda_{jk} \boldsymbol{l}_{jk} \cdot \hat{\boldsymbol{e}}_D \frac{\mathrm{d}\bar{\phi}}{\mathrm{d}y}\right\} . \tag{6.43}$$

Finally, combining (6.34), (6.35), (6.41) and (6.43), and recognizing that $\mathrm{d}\bar{\phi}/\mathrm{d}y =$

V/L is independent of y, we find

$$\frac{\mathrm{d}\bar{\phi}}{\mathrm{d}y} = L^{-D+1}I + \rho\left\langle \sum_k G_{ik}\left(\sum_j \lambda_{jk} \boldsymbol{l}_{jk} \cdot \hat{\boldsymbol{e}}_D\right)\right.$$

$$\left. \times \left(\sum_{j'} \lambda_{j'i} \boldsymbol{l}_{j'i} \cdot \hat{\boldsymbol{e}}_D\right)\right\rangle_i \frac{\mathrm{d}\bar{\phi}}{\mathrm{d}y}. \tag{6.44}$$

For very large Ω, the quantity averaged in (6.44) should be rotationally symmetric so the product $(\hat{\boldsymbol{e}}_D)_a(\hat{\boldsymbol{e}}_D)_b$ can be replaced by δ_{ab}/D. Solving (6.44) for $\mathrm{d}\bar{\phi}(y)/\mathrm{d}y$, we find

$$\frac{\mathrm{d}\bar{\phi}}{\mathrm{d}y} = ZL^{-D+1}I, \tag{6.45}$$

with

$$Z = \left[1 - \frac{\rho}{D}\left\langle \sum_k G_{ik}\left(\sum_j \lambda_{jk} \boldsymbol{l}_{jk}\right) \cdot \left(\sum_{j'} \lambda_{j'i} \boldsymbol{l}_{j'i}\right)\right\rangle_i\right]^{-1}. \tag{6.46}$$

In a regular lattice $(\Sigma_j \lambda_{jk} \boldsymbol{l}_{jk}) = 0$ and therefore $Z = 1$. In a random lattice the inherent fluctuations cause Z to differ from 1. We expect that these fluctuations, and hence the difference $Z - 1$, approaches zero as the space-time dimension D becomes large. Let us now estimate the deviation of Z from 1.

6.5. MEAN FIELD APPROXIMATION

We can make an estimate of the bulk resistivity Z by returning to (6.40) and expanding the right-hand side in the deviation of ϕ_i from its mean value $\bar{\phi}((\boldsymbol{r}_i)_D)$ in the following way.

Zeroth approximation. Here we simply replace ϕ_i by $\bar{\phi}((\boldsymbol{r}_i)_D)$ in (6.40). The resulting expression then averages to zero and

$$\frac{\mathrm{d}\bar{\phi}}{\mathrm{d}y} = \left[\frac{\mathrm{d}\bar{\phi}}{\mathrm{d}y}\right]_{\text{link}} = L^{-D+1}I, \tag{6.47}$$

so that $Z = 1$ as observed above.

First approximation. As a next approximation we might assume that ϕ_j on the sites linked to i is given by the mean field $\bar{\phi}((\boldsymbol{r}_j)_D)$ and compute the resulting value of ϕ_i from Kirchoff's law:

$$\sum_j \lambda_{ji}\left[\bar{\phi}((\boldsymbol{r}_j)_D) - \phi_i\right] = 0, \tag{6.48}$$

or

$$\phi_i = \sum_j \lambda_{ji} \bar{\phi}((r_j)_D) / \sum_j \lambda_{ji}$$

$$\cong \bar{\phi}((\boldsymbol{r}_i)_D) + \frac{\Sigma_j \lambda_{ji} \boldsymbol{l}_{ji} \cdot \nabla \bar{\phi}((\boldsymbol{r}_i)_D)}{\Sigma_j \lambda_{ji}}. \tag{6.49}$$

By substituting this expression for ϕ_i into (6.40) we find

$$\left[\frac{\mathrm{d}\bar{\phi}}{\mathrm{d}y}\right]_{\text{site}} = \rho \left\langle \frac{(\Sigma_k \lambda_{ki} \boldsymbol{l}_{ki} \cdot \hat{\boldsymbol{e}}_D)^2}{\Sigma_j \lambda_{ji}} \right\rangle_i \frac{\mathrm{d}\bar{\phi}}{\mathrm{d}y}, \tag{6.50}$$

which, using rotational invariance as before for the large-volume limit, and (6.34), (6.35) and (6.45), yields

$$Z = \left[1 - \frac{\rho}{D} \left\langle \frac{(\Sigma_k \lambda_{ki} \boldsymbol{l}_{ki})^2}{\Sigma_j \lambda_{ji}} \right\rangle_i \right]^{-1}. \tag{6.51}$$

Eq. (6.51) can be easily evaluated if we make two further approximations:
(i) Assuming uncorrelated link directions, i.e.

$$\langle \boldsymbol{l}_{ji} \cdot \boldsymbol{l}_{ki} \rangle_{\text{av}} = \langle l^2 \rangle_{\text{av}} \delta_{jk}, \tag{6.52}$$

we replace $\langle\ \rangle_i$ in (6.51) by

$$\left\langle \frac{\Sigma_k \lambda_{ki}^2 l_{ki}^2}{\Sigma_j \lambda_{ji}} \right\rangle_i. \tag{6.53}$$

(ii) From table 2 we see that the number of links/site, $N_{1/0}$, is a relatively large number, which $\to \infty$ when $D \to \infty$. Thus for large dimension D, we may replace the sum $\Sigma_j g(l_{ji})$ for any function $g(l)$ by the i-independent average

$$N_{1/0} \overline{g(l)}, \tag{6.54}$$

where

$$\overline{g(l)} \equiv \frac{\int_0^\infty f(l) g(l)\, \mathrm{d}l}{\int_0^\infty f(l)\, \mathrm{d}l}, \tag{6.55}$$

and $f(l)$ is given by (6.7). Therefore, (6.53) becomes $\overline{\lambda_l^2 l^2} / \overline{\lambda_l}$ and the bulk resistivity is

$$Z \cong \left[1 - \frac{\rho}{D} \frac{\overline{\lambda_l^2 l^2}}{\overline{\lambda_l}}\right]^{-1}. \tag{6.56}$$

From (3.1) and (3.3), we have

$$\rho N_{1/0} = 2\int_0^\infty f(l)\,\mathrm{d}l,$$

which together with (6.6) gives

$$\frac{\rho}{D} = \frac{2}{N_{1/0}} \frac{1}{\overline{\lambda_l l^2}}. \tag{6.57}$$

Thus, (6.56) can also be written as

$$Z \cong \left[1 - \frac{2}{N_{1/0}} \frac{\overline{\lambda_l^2 l^2}}{\overline{\lambda_l}\,\overline{\lambda_l l^2}}\right]^{-1}. \tag{6.58}$$

So far, except for the constraint (6.6) the dependence of λ_l on l is entirely arbitrary. If we assume

$$\lambda_l \propto l^{-n},$$

then (6.58) becomes

$$Z \cong \left[1 - \frac{2}{N_{1/0}} \frac{\overline{l^{2-2n}}}{\overline{l^{2-n}}\,\overline{l^{-n}}}\right]^{-1}. \tag{6.59}$$

Since $N_{1/0} \to \infty$ as $D \to \infty$, we have $Z \to 1$ as D increases. For $n = 0$ or 2, since $\overline{l^0} = 1$, we have

$$Z \cong \frac{N_{1/0}}{N_{1/0} - 2} \tag{6.60}$$

which, according to table 2, takes on the values

$$Z \cong \begin{cases} 1.50 & D = 2 \\ 1.15 & \text{when } D = 3\,. \\ 1.06 & D = 4 \end{cases}$$

For $n=1$, the corresponding values of Z are uniformly smaller:

$$Z \cong \begin{cases} 1.28 & D=2 \\ 1.12 & \text{when } D=3\,. \\ 1.05 & D=4 \end{cases}$$

Therefore, Z is fairly close to 1 when $D=4$.

We wish to thank Dr. Bernard Grossman for a discussion of the Wigner-Seitz construction which was instrumental in our arriving at the particular algorithm for linking sites given in sect. 2. One of us (T.D.L.) wishes to thank the theoretical physics groups at Scuola Normale Superiore in Pisa and CERN in Geneva for their very kind hospitality during the summer of 1981 when part of the present work was done.

Note added in proof

A different approach to rotational invariance in lattice field theory has been discussed in ref. [8]. In 3-space, for a different construction based on tetrahedra, see ref. [9].

Appendix

In this appendix we indicate how the multi-dimensional integrals such as (4.36) and (4.37) can be carried out. An essential step is the reduction from integrals in D to $(D-1)$ dimensions. This may best be illustrated by considering the three-dimensional case. When $D=3$, the simplex is a tetrahedron, say $\tau(1,2,3,4)$. Let the position vector of its ith vertex (site) be $\boldsymbol{r}_i$ $(i=1,\dots,4)$ and that of its circumcenter C be $\boldsymbol{r}_{\mathrm{C}}$. As in (4.15), we define

$$\hat{\boldsymbol{\sigma}}_i = \frac{1}{R}(\boldsymbol{r}_i - \boldsymbol{r}_{\mathrm{C}}), \tag{A.1}$$

where R is the radius of the circumsphere; (4.17) then becomes

$$\prod_{i=1}^{4} \mathrm{d}^3 r_i = 3!\tau R^5 \,\mathrm{d}R\, \mathrm{d}^3 r_{\mathrm{C}} \prod_{i=1}^{4} \mathrm{d}^2 \hat{\sigma}_i, \tag{A.2}$$

where τ is the volume of the tetrahedron [$\tau = V$ of (4.17)].

The sites 1, 2 and 3 define a triangle $\Delta(1,2,3)$. As shown in fig. 6, we denote c to be its circumcenter and $\hat{\boldsymbol{n}}$ its unit normal vector whose direction is determined by

$$\hat{\boldsymbol{n}} \,\|\, (\boldsymbol{r}_2 - \boldsymbol{r}_1) \times (\boldsymbol{r}_3 - \boldsymbol{r}_1). \tag{A.3}$$

On the plane defined by 1, 2 and 3 (which also contains c), we can introduce, as in fig. 3a, three azimuthal angles ϕ_1, ϕ_2 and ϕ_3, or their relative angles α, β and γ, as in

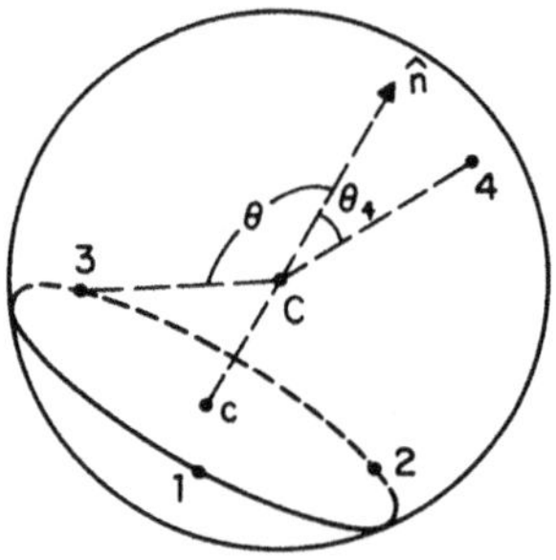

Fig. 6. When $D=3$, let the center of the circumscribed sphere of the tetrahedron $\tau(1,2,3,4)$ be C, and the circumcenter of the triangle $\Delta(1,2,3)$ be c (which lies on the same plane). See (A.3), (A.5) and (A.19) for the definitions of $\hat{\boldsymbol{n}}$, θ and θ_4.

fig. 3b. The area of $\Delta(1,2,3)$ is then given by, similar to (4.30),

$$\Delta = \tfrac{1}{2}R^2\sin^2\theta(\sin\alpha + \sin\beta + \sin\gamma)\,, \tag{A.4}$$

where θ is defined by

$$\cos\theta = \hat{\boldsymbol{n}}\cdot\hat{\boldsymbol{\sigma}}_1 = \hat{\boldsymbol{n}}\cdot\hat{\boldsymbol{\sigma}}_2 = \hat{\boldsymbol{n}}\cdot\hat{\boldsymbol{\sigma}}_3\,. \tag{A.5}$$

Let the spherical coordinates of $\boldsymbol{r}_i - \boldsymbol{r}_{\mathrm{C}} = R\hat{\boldsymbol{\sigma}}_i$ be

$$R\,,\qquad \theta_i\,,\qquad \phi_i\,. \tag{A.6}$$

If the z-axis is chosen to be parallel to $\hat{\boldsymbol{n}}$, then θ is also given by

$$\theta = \theta_1 = \theta_2 = \theta_3\,. \tag{A.7}$$

Consider now the transformation

$$\hat{\boldsymbol{\sigma}}_1, \hat{\boldsymbol{\sigma}}_2, \hat{\boldsymbol{\sigma}}_3 \to \hat{\boldsymbol{n}}, \theta, \phi_1, \phi_2, \phi_3$$

in which both sides consist of six variables. We shall show that

$$\prod_{i=1}^{3} \mathrm{d}^2\hat{\boldsymbol{\sigma}}_i = \frac{2\Delta}{R^2}\mathrm{d}^2\hat{\boldsymbol{n}}\sin\theta\,\mathrm{d}\theta \prod_{i=1}^{3}\mathrm{d}\phi_i\,. \tag{A.8}$$

To establish (A.8), let us take any function F of $\hat{\boldsymbol{\sigma}}_1$, $\hat{\boldsymbol{\sigma}}_2$ and $\hat{\boldsymbol{\sigma}}_3$. We may multiply the integral

$$\mathcal{I}(F) \equiv \int F \prod_{i=1}^{3} \mathrm{d}^2\hat{\boldsymbol{\sigma}}_i \tag{A.9}$$

by the identity

$$1 = \int \mathrm{d}^2\hat{\xi}\,\delta^2(\hat{\xi} - \hat{\boldsymbol{n}})\,, \tag{A.10}$$

where the integration extends over the surface of a unit sphere $|\hat{\xi}| = 1$. The result is

$$\mathcal{G}(F) = \int F \prod_{i=1}^{3} \mathrm{d}^2\hat{\sigma}_i \int \mathrm{d}^2\hat{\xi}\, \delta^2(\hat{\xi} - \hat{n}) \tag{A.11}$$

in which $\hat{n}$ is regarded as a function of $\hat{\sigma}_1$, $\hat{\sigma}_2$ and $\hat{\sigma}_3$ through (A.1) and (A.3). We note that when $\hat{\xi} - \hat{n} = 0$,

$$a \equiv \hat{\xi} \cdot (\hat{\sigma}_2 - \hat{\sigma}_1) = 0,$$

$$b \equiv \hat{\xi} \cdot (\hat{\sigma}_3 - \hat{\sigma}_1) = 0. \tag{A.12}$$

Hence,

$$\delta^2(\hat{\xi} - \hat{n}) = \mathcal{J}\delta(a)\delta(b), \tag{A.13}$$

where the jacobian is

$$\mathcal{J} = \hat{n} \cdot [(\hat{\sigma}_2 - \hat{\sigma}_1) \times (\hat{\sigma}_3 - \hat{\sigma}_1)] = 2\Delta/R^2. \tag{A.14}$$

As a result, (A.11) can be written as

$$\mathcal{G}(F) = \int F \mathrm{d}^2\hat{\xi} \int \frac{2\Delta}{R^2} \delta(a)\delta(b) \prod_{i=1}^{3} \mathrm{d}^2\hat{\sigma}_i, \tag{A.15}$$

in which we have exchanged the order of integrations, so that the integration over $\prod \mathrm{d}^2\hat{\sigma}_i$ is to be carried out at a fixed $\hat{\xi}$. Regarding $\hat{\xi}$ now as the unit vector along the z-axis we have, in accordance with (A.6) and (A.12)

$$\prod_{i=1}^{3} \mathrm{d}^2\hat{\sigma}_i = \prod_{i=1}^{3} \mathrm{d}\cos\theta_i\, \mathrm{d}\phi_i,$$

$$a = \cos\theta_2 - \cos\theta_1,$$

$$b = \cos\theta_3 - \cos\theta_1.$$

We first integrate with respect to $\cos\theta_2$ and $\cos\theta_3$; (A.15) then becomes

$$\mathcal{G}(F) = \int \frac{2F\Delta}{R^2} \mathrm{d}^2\hat{\xi}\, \mathrm{d}\cos\theta_1 \prod_{i=1}^{3} \mathrm{d}\phi, \tag{A.16}$$

in which because of the original factor $\delta^2(\hat{\xi} - \hat{n})$ in (A.11), we may set $\hat{\xi} = \hat{n}$ and therefore, on account of (A.7), $\theta_1 = \theta$. Consequently, (A.9) equals

$$\mathcal{I}(F) = \int \frac{2F\Delta}{R^2} \mathrm{d}^2\hat{n}\, \mathrm{d}\cos\theta \prod_{i=1}^{3} \mathrm{d}\phi_i \tag{A.17}$$

for any F, and that leads to (A.8).

As shown in fig. 6, for definiteness let us arrange the order of sites 1, 2, 3 and 4 such that

$$(\boldsymbol{r}_4 - \boldsymbol{r}_1)\cdot[(\boldsymbol{r}_2 - \boldsymbol{r}_1)\times(\boldsymbol{r}_3 - \boldsymbol{r}_1)] > 0\,; \tag{A.18}$$

therefore,

$$\hat{\boldsymbol{n}}\cdot\hat{\boldsymbol{\sigma}}_4 \equiv \cos\theta_4 > \cos\theta\,. \tag{A.19}$$

According to (4.2), when $D = 3$ the distribution function is given by

$$\mathfrak{D}(1,2,3,4) = \frac{1}{4!}\rho^4 \mathrm{e}^{-4\pi R^3\rho/3} \prod_{i=1}^{4} \mathrm{d}^3 r_i\,. \tag{A.20}$$

By using (A.2) and (A.8) we find for any function $f(1,2,3,4)$ that depends only on the relative coordinates

$$\begin{aligned} I(f) &\equiv \int f(1,2,3,4)\mathfrak{D}(1,2,3,4) \prod_{i=1}^{4} \mathrm{d}^3 r_i \\ &= \tfrac{1}{4}\rho^4 \Omega \int_0^\infty \mathrm{e}^{-4\pi R^3\rho/3} R^5\, \mathrm{d}R \int \prod_{i=1}^{4} \mathrm{d}^2\hat{\sigma}_i\, f\tau\,, \end{aligned} \tag{A.21}$$

in which the integration over $\mathrm{d}^2\hat{\sigma}_i$ can also be written as

$$\int \prod_{i=1}^{4} \mathrm{d}^2\hat{\sigma}_i\, f\tau = 2\int \mathrm{d}^2\hat{n} \int_0^\pi \sin\theta\, \mathrm{d}\theta \int_0^\theta \sin\theta_4\, \mathrm{d}\theta_4 \int \frac{2f\tau\Delta}{R^2} \prod_{i=1}^{4} \mathrm{d}\phi_i\,, \tag{A.22}$$

where the first factor 2 is due to (A.18) which allows only even permutations of 1, 2, 3 and 4 (and that changes the factor 1/4! in (A.20) to 2/4!). As an example, we may consider the simplest case, $f = 1$:

$$I(f) = I(1) = \mathfrak{N}_3 = \text{total number of tetrahedra}\,.$$

Hence,

$$\begin{aligned} \rho_3 = \frac{\mathfrak{N}_3}{\Omega} &= 16\pi^3\rho^4 \int_0^\infty R^3 \mathrm{e}^{-4\pi R^3\rho/3}\, \mathrm{d}R \int_0^{2\pi} \sin\theta\, \mathrm{d}\theta \\ &\quad \times \int_0^\theta \sin\theta_4\, \mathrm{d}\theta_4 \int \mathrm{d}\alpha\, \mathrm{d}\beta\, \mathrm{d}\gamma\, \delta(\alpha+\beta+\gamma-2\pi)\tau\Delta\,, \end{aligned} \tag{A.23}$$

where $16\pi^3$ is the product of $4\pi = \int d^2\hat{n}$ and $(2\pi)^2 = \int d\phi_4 d\phi_1$,

$$\tau = \tfrac{1}{3}R(\cos\theta_4 - \cos\theta)\Delta, \tag{A.24}$$

and Δ is given by (A.4). The integration (A.23) is completely elementary; it is not more complicated than the two-dimensional case discussed in (4.25)–(4.35). We find that when $D=3$ the density of tetrahedra is related to the site density ρ by

$$\rho_3 = \tfrac{24}{35}\pi^2\rho. \tag{A.25}$$

The density of triangles is given by

$$\rho_2 = 2\rho_3 = \tfrac{48}{35}\pi^2\rho. \tag{A.26}$$

Likewise, other average values can be found. The results are listed in tables 1 and 2.

References

[1] K. Wilson, Phys. Rev. D10 (1974) 2455
[2] G. Parisi, High-energy physics 1980, ed. L. Durand and L. Pondrom (AIP Conf. Proc. no. 68), p. 1530
[3] J. Kogut, Revs. Mod. Phys. 51 (1979) 659
[4] J. M. Ziman, Models of disorder (Cambridge University Press, New York, 1979)
[5] E. Wigner and F. Seitz, Phys. Rev. 43 (1933) 804;
F. Seitz, Modern theory of solids (McGraw-Hill, New York, 1940)
[6] J.L. Meijering, Philips research report 8 (1953) 270
[7] L.A. Santaló, Integral geometry and geometrical probability (Addison-Wesley, Reading, Mass., 1976)
[8] K.I. Macrae, Phys. Rev. D23 (1981) 886, 893, 900
[9] B. Hasslacher and M.J. Perry, Phys. Lett. 103B (1981) 21

Nuclear Physics B210[FS6] (1982) 310–336

GAUGE THEORY ON A RANDOM LATTICE*

N.H. CHRIST, R. FRIEDBERG and T.D. LEE

Columbia University, New York, N.Y. 10027, USA

Received 7 June 1982

A general formulation of gauge theory on a random lattice is developed and the strong coupling limit of the Wilson string tension worked out. The confining force found in this strong coupling limit is identical to that predicted by the relativistic string model. In particular, the force between two color-triplet charges is a constant for large separation and the tube of electric flux joining the charges fluctuates, giving it a net thickness proportional to the logarithm of its length.

1. Introduction

In an earlier paper** [1] we developed a discrete approximation to continuum field theory based on the introduction of a random lattice. A particularly attractive definition of a random lattice was given and some properties of a scalar field theory defined on such a lattice were investigated. In this paper we carry these ideas further, considering a gauge theory defined on our random lattice. The construction follows closely the regular lattice gauge theory defined by Wilson [2] with the triangles which naturally occur in the random lattice taking the place of the elementary square plaquettes of the regular lattice. Just as for the theory defined on a regular lattice, this random lattice gauge theory has a strong coupling limit which can be analyzed analytically and which confines charges in the fundamental representation of the gauge group. However, in contrast to the strong coupling limit on the regular lattice, the confining potential is rotationally symmetric and the tube of electric flux connecting the separated charges acquires a thickness which grows as the logarithm of its length. The properties of this flux tube are identical to those of a relativistic string.

In order to obtain these analytical results, it is necessary to perform an average over an ensemble of random lattices. This averaging process can be viewed in two ways: The first perspective approximates the space-time continuum by a particular random lattice chosen to satisfy some conditions of randomness and isotropy. Quantum mechanical expectation values are then to be computed for this single lattice. For example, a numerical Monte Carlo calculation might be performed with a single random lattice. In order to determine such an expectation value analytically, one might attempt to compute an average of that expectation value over an ensemble

* This research was supported in part by the US Department of Energy.

** Hereafter referred to as I.

of such lattices, which corresponds to a "quenched" average in the usual terminology.

The second point of view treats the positions of the lattice sites themselves as dynamical variables, so that we may add to the usual path integration an additional integration over the positions of random lattice sites. This can be called an "annealed" average. As we shall see, these two perspectives agree in the strong coupling limit, Outside the strong coupling region these two averages may differ, each having its own advantages.

In sect. 2 we define a gauge theory on our random lattice, and in sect. 3 we discuss the issues associated with these alternative averages in the context of computing a Wilson loop expectation value. The strong coupling limit of this Wilson loop expectation value is evaluated in sect. 4 and the string tension deduced. Finally, in sect. 5 we compute the thickness of the confining string implied by this strong coupling calculation and compare it with the calculation of a similar quantity in the relativistic string model.

We note that the approach to the strong coupling limit developed here can be applied in a straightforward way to the calculation of other quantities, such as the glueball mass and that the analogy with the string model persists. This will be discussed in detail in a separate paper.

2. Lattice gauge theory action

As in the conventional treatment of lattice gauge theory, we introduce gauge variables by assigning a group element $U(i,j)$ to a link connecting the points (i.e., sites) i and j in the lattice. We will consider an SU(N) gauge theory so that $U(i,j)$ can be viewed as an arbitrary $N\times N$ unitary matrix with determinant 1. It is convenient to define both $U(i,j)$ and $U(j,i)$ with $U(j,i)=U(i,j)^{-1}$. Next, with each elementary triangle of vertices i, j and k we associate a group element $U_{ijk}=U(i,j)U(j,k)U(k,i)$. Recall that in our construction of the random lattice we identify a "cluster" of $D+1$ vertices with the property that the $D-1$ dimensional sphere, on whose surface the cluster of $D+1$ points lies, has no lattice points in its interior. Each pair of points in such a cluster is joined by a link. An elementary triangle Δ_{ijk} is then formed of any three points i, j and k in the same cluster and the corresponding links.

The lattice action is defined using these group variables U_{ijk}:

$$\mathcal{A}_{\mathrm{L}}=\frac{1}{g^2}\sum_{\Delta_{ijk}}\kappa_{ijk}f(U_{ijk}), \tag{2.1}$$

where g is the coupling constant, and the sum extends over all triangles Δ_{ijk}. The coefficients κ_{ijk} obey a condition, specified below, which assures the proper continuum limit*. The action (2.1) will be invariant under the gauge transformation

$$U(i,j)\to u_i U(i,j) u_j^{-1}, \tag{2.2}$$

for an assignment of group elements u_i to the ith lattice site if the function $f(U)$ obeys

$$f(uUu^{-1}) = f(U) \tag{2.3}$$

for all elements u and U. Finally we require that when the SU(N) matrix U is near the identity matrix I,

$$U = I + i\lambda^a \varepsilon^a + \mathrm{O}(\varepsilon^2), \tag{2.4}$$

then

$$f(U) \sim \sum_a (\varepsilon^a)^2 + \mathrm{O}(\varepsilon^3), \tag{2.5}$$

where ε^a are small, the repeated index a is summed over from 1 to N^2-1, and the $N\times N$ hermitian generators λ^a obey the usual normalization condition

$$\mathrm{tr}\,(\lambda^a\lambda^b) = 2\delta^{ab}. \tag{2.6}$$

The simple choice for the function $f(U)$,

$$f(U) = \tfrac{1}{2}\,\mathrm{tr}\,(I-U) + \mathrm{c.c.} = \mathrm{Re}\,[\mathrm{tr}\,(I-U)], \tag{2.7}$$

corresponds to the conventional Wilson action.

Next let us determine the condition that the coefficient κ_{ijk} must satisfy in order that our lattice action approach the usual continuum gauge theory action

$$\mathcal{A}_\mathrm{C} = \tfrac{1}{4}\int_\Omega F^a_{\mu\nu} F^a_{\mu\nu}\, \mathrm{d}^D x \tag{2.8}$$

in the continuum limit. Here the Yang-Mills field strength is given by

$$F^a_{\mu\nu} = \partial_\mu A^a_\nu - \partial_\nu A^a_\mu + g f^{abc} A^b_\mu A^c_\nu, \tag{2.9}$$

where A^a_μ is the gauge potential and f^{abc} the SU(N) structure constants with

$$[\lambda^a, \lambda^b] = 2i\lambda^c f^{abc}. \tag{2.10}$$

Begin with a continuum gauge field $A^a_\mu(x)$ defined in the volume Ω. We can introduce a group element $U(i,j)$ for linked sites i and j by

$$U(i,j) = \mathrm{P}\left[\exp\left(-ig\int_{x_j}^{x_i} A^a_\mu \lambda^a\, \mathrm{d}x_\mu\right)\right], \tag{2.11}$$

where the symbol P indicates the usual "path ordering" of the exponential. If $\mathcal{A}_\mathrm{L}$

* A particularly attractive choice of κ_{ijk} will be given in the companion paper.

is then evaluated for this assignment of $U(i, j)$, we require that $\mathscr{A}_L$ should approach the continuum action $\mathscr{A}_C$ of the gauge field A_μ^a in the limit that the density ρ of lattice sites becomes large for a fixed volume Ω:

$$\lim_{\rho\to\infty} \mathscr{A}_L = \mathscr{A}_C . \tag{2.12}$$

In the limit of large density, the range of integration in the exponent of (2.11) becomes small and we can approximate $U(i, j)$ by the power series

$$U(i, j) = I - ig\int_{x_j}^{x_i} A_\mu^a \lambda^a \, dx_\mu + \tfrac{1}{2}\mathrm{P}\left[-ig\int_{x_j}^{x_i} A_\mu^a \lambda^a \, \mathrm{d}x_\mu\right]^2 + \mathrm{O}(\bar{l}^3), \tag{2.13}$$

where $\bar{l}$ is the average length of a link. Thus the large density limit in (2.12) is also a "weak field" limit in the sense that our lattice variables $U(i, j)$ approach I. If the approximation (2.13) is used for the group elements defining the product U_{ijk}, then we find

$$U_{ijk} = I - \tfrac{1}{2}ig\lambda^a F_{\mu\nu}^a \Delta_{\mu\nu}(ijk) + \mathrm{O}(\bar{l}^3) . \tag{2.14}$$

Throughout, the subscripts μ and ν denote the euclidean-space index which can be 1, 2, . . . , D, the superscripts a, b and c are internal symmetry indices, each varying from 1 to $N^2 - 1$ and i, j and k (whether as subscripts or arguments) refer to the lattice sites 1, 2, . . . , $\mathscr{N}$. Here $\Delta_{\mu\nu}(ijk)$ is an antisymmetric tensor in μ and ν, parallel to the plane of the i, j, k triangle with a magnitude related to the triangle's area, A_{ijk}:

$$\Delta_{\mu\nu}(ijk)\Delta_{\mu\nu}(ijk) = 2(A_{ijk})^2 ; \tag{2.15}$$

more specifically,

$$\Delta_{\mu\nu}(ijk) = \tfrac{1}{6}[l_\mu(ij)l_\nu(jk) + l_\mu(jk)l_\nu(ki) + l_\mu(ki)l_\nu(ij) - (\mu \rightleftarrows \nu)] , \tag{2.16}$$

where $l_\mu(ij)$ is the μth component of the relative coordinate vector from j to i. Those terms linear in g in (2.14) follow from an application of Stokes' theorem while the order g^2 terms can be determined by gauge invariance; both are $\mathrm{O}(\bar{l}^2)$. If this form for U_{ijk} is substituted into the lattice action (2.1) and the condition (2.5) used to evaluate $f(U_{ijk})$, we find in the limit of $\rho \to \infty$

$$\mathscr{A}_L \to \tfrac{1}{4} \sum_{\Delta_{ijk}} \kappa_{ijk}\Delta_{\mu\nu}(ijk)\Delta_{\rho\sigma}(ijk)F_{\mu\nu}^a F_{\rho\sigma}^a . \tag{2.17}$$

If we assume that κ_{ijk} is a rotationally invariant function of the triangle i, j, k (i.e., κ_{ijk} depends only on the length of the three links ij, jk and ki), then (2.17) becomes

$$\mathscr{A}_L \to \int \mathrm{d}^D x F_{\mu\nu}^a F_{\mu\nu}^a \frac{\rho_2}{D(D-1)} \langle \kappa_{ijk}(A_{ijk})^2 \rangle . \tag{2.18}$$

Here ρ_2 is the number of triangles (2-simplices) per unit volume and $\langle\ \rangle$ represents an average over the distribution of triangles appearing in the random lattice. In obtaining the result (2.18) we have used the following formula for the angular average of the product $\Delta_{\mu\nu}(ijk)\Delta_{\rho\sigma}(ijk)$:

$$\langle \Delta_{\mu\nu}(ijk)\Delta_{\rho\sigma}(ijk)\rangle_{\rm ang} = \frac{2(A_{ijk})^2}{D(D-1)}[\delta_{\mu\rho}\delta_{\nu\sigma} - \delta_{\mu\sigma}\delta_{\nu\rho}], \tag{2.19}$$

which follows from the antisymmetry of $\Delta_{\mu\nu}(ijk)$ in μ and ν and the normalization condition (2.15). Thus, if the coefficients κ_{ijk} are related to the area A_{ijk} of the triangle by the condition

$$\langle \kappa_{ijk}(A_{ijk})^2\rangle \frac{\rho_2}{D(D-1)} = \frac{1}{4}, \tag{2.20}$$

then (2.1) defines a lattice gauge theory action with the proper weak-field, continuum limit.

If we add to the lattice action (2.1) a linear coupling between the group element $U(i,j)$ and an external $N \times N$ matrix source $J(i,j)$,

$$\sum_{i,j} \operatorname{tr}[U(i,j)J(i,j) + \text{h.c.}], \tag{2.21}$$

where the sum extends over all pairs of sites i and j connected by a link, we can define the generating functional for a quantum gauge theory on the random lattice as the Feynman path integral

$$\mathcal{G}_{\rm L}(J) = \prod_{i,j} \int d[U(i,j)] \exp\left\{-\mathcal{A}_{\rm L} - \sum_{i,j} \operatorname{tr}[U(i,j)J(i,j) + \text{h.c.}]\right\}. \tag{2.22}$$

This equation defines the quantum gauge theory for a particular random lattice. Later we will find it convenient to generalize (2.22) by adding to the functional integral an average over an ensemble of random lattices.

3. Wilson loop and ensemble averages

The question of confinement is closely connected with the expectation value of the Wilson loop operator, which in a continuum theory is given by

$$\langle 0|\mathrm{P}\left[\exp\left(-ig\oint_{\rm C} A_\mu^a \lambda^a \, dx_\mu\right)\right]|0\rangle, \tag{3.1}$$

where C denotes a smooth closed loop lying in a two-dimensional plane. For a given random lattice L we can find a second continuous curve $\mathrm{C_L}$ made up of links belonging to the lattice and approximating the smooth curve C as follows: Recall (see sect. 5 of I) that in the dual lattice we divide space into cells of convex polyhedra ω_i composed of those points in space nearest to the site i. A lattice link then joins the sites i and i' if the corresponding polyhedra ω_i and $\omega_{i'}$ share a common face. Thus, by following the smooth curve C as it passes through a sequence of bordering polyhedra ω_i, ω_j, $\omega_k, \ldots$, we can map out a corresponding sequence of sites $i, j, k, \ldots$. The chain of links connecting these sites forms our curve C_L. We define the random lattice approximation to (3.1) as

$$W_L(C) = \frac{1}{Q_L} \mathscr{W}_L(C), \tag{3.2}$$

where

$$Q_L \equiv \int \prod_{k,l} d[U(k,l)]\, e^{-\mathscr{A}_L}, \tag{3.3}$$

$$\mathscr{W}_L(C) \equiv \int \prod_{k,l} d[U(k,l)]\, e^{-\mathscr{A}_L} \operatorname{tr}\Big[\prod_{i,j \in C_L} U(i,j)\Big], \tag{3.4}$$

in which the first product extends over all linked pairs of sites k and l, and the second product over $i, j \in C_L$ is understood to be path-ordered with the point i following j as one moves around the curve.

In order to give $W_L(C)$ a unique definition and to render it manifestly invariant under both translations and rotations, it is necessary to perform an average such as one of the following:

(i) Consider a particular random lattice L in a large volume Ω which contains $\mathscr{N} = \rho\Omega$ random sites. Draw a Wilson loop C along any smooth closed path. Evaluate $W_L(C)$ according to (3.2). The loop C can of course be placed anywhere in Ω. The average of $W_L(C)$ over all translations of C (called C-average and indicated by a subscript C),

$$\langle W_L(C) \rangle_C = \Big\langle \frac{1}{Q_L} \mathscr{W}_L(C) \Big\rangle_C, \tag{3.5}$$

in the limit $\Omega \to \infty$ at a constant density ρ then gives the physical result. Since Q_L is independent of C, it can be taken outside the average; hence (3.5) can also be written as

$$\langle W_L(C) \rangle_C = \frac{1}{Q_L} \langle \mathscr{W}_L(C) \rangle_C. \tag{3.6}$$

(ii) Alternatively, we may perform an average over an ensemble of lattices, each having $\mathscr{N}$ points randomly distributed in a volume Ω. If we use (2.22) to define a generating functional $\mathscr{G}_L(J)$ for a particular random lattice L with sites at $\boldsymbol{r}_1, \ldots, \boldsymbol{r}_{\mathscr{N}}$ in Ω, then the corresponding ensemble average over different L is*

$$\langle \mathscr{G}_L(J) \rangle_L = \Omega^{-\mathscr{N}} \int \prod_{i=1}^{\mathscr{N}} d^D r_i\, \mathscr{G}_L(J), \tag{3.7}$$

* In order that $\mathscr{G}_L(J)$ can be averaged for fixed J, it is necessary that the matrix source $J(i,j)$ be obtained from a continuous function $J(\boldsymbol{r}, \boldsymbol{r}')$ with $J(i,j) = J(\boldsymbol{r}_i, \boldsymbol{r}_j)$.

where the subscript L outside the bracket indicates that the average is over an ensemble of lattices.

The corresponding ensemble average of the Wilson loop functional may be done in two different ways:

(a) the quenched average

$$\langle W_{\mathrm{L}}(\mathrm{C})\rangle_{\mathrm{L}} \equiv \Omega^{-\mathcal{N}} \int \prod_{i=1}^{\mathcal{N}} \mathrm{d}^D r_i W_{\mathrm{L}}(\mathrm{C})$$

$$= \Omega^{-\mathcal{N}} \int \prod_{i=1}^{\mathcal{N}} \mathrm{d}^D r_i \frac{1}{Q_{\mathrm{L}}} \mathcal{W}_{\mathrm{L}}(\mathrm{C}) , \tag{3.8}$$

and (b) the annealed average

$$W(\mathrm{C}) \equiv \frac{\langle \mathcal{W}_{\mathrm{L}}(\mathrm{C})\rangle_{\mathrm{L}}}{\langle Q_{\mathrm{L}}\rangle_{\mathrm{L}}} , \tag{3.9}$$

where

$$\langle \mathcal{W}_{\mathrm{L}}(\mathrm{C})\rangle_{\mathrm{L}} = \Omega^{-\mathcal{N}} \int \prod_{i=1}^{\mathcal{N}} \mathrm{d}^D r_i \mathcal{W}_{\mathrm{L}}(\mathrm{C}) , \tag{3.10}$$

$$\langle Q_{\mathrm{L}}\rangle_{\mathrm{L}} = \Omega^{-\mathcal{N}} \int \prod_{i=1}^{\mathcal{N}} \mathrm{d}^D r_i Q_{\mathrm{L}} . \tag{3.11}$$

In the limit of infinite Ω, these averages are all manifestly invariant under translation and rotation.

We shall now show that the C-average in (i) is equivalent to the quenched average in (ii). Let us label the lattice and its volume used in (i) by L and Ω, those used in (ii) by L$'$ and Ω'. We take

$$\Omega = N\Omega' , \tag{3.12}$$

where N is a large number. The much larger random lattice L can be viewed as a composite of N different random lattices L$'(1)$, L$'(2), \ldots,$ L$'(N)$, each in a region of volume Ω' which is assumed to be of a sufficient size so that effects due to surfaces between different L$'(n)$ can be neglected. Hence,

$$Q_{\mathrm{L}} = \prod_{n=1}^{N} Q_{\mathrm{L}'(n)} . \tag{3.13}$$

Consider now a specific term $\mathcal{W}_{\mathrm{L}}(\mathrm{C})$ in the C-average (3.6). When $\Omega \to \infty$ with N held fixed, we may assume that the Wilson loop C lies within one of the smaller lattices, say L$'(a)$. From (3.3) and (3.4), we see that

$$\mathcal{W}_{\mathrm{L}}(\mathrm{C}) = \mathcal{W}_{\mathrm{L}'(a)}(\mathrm{C}) \prod_{n \neq a} Q_{\mathrm{L}'(n)} , \tag{3.14}$$

and therefore

$$W_L(C) = \frac{1}{Q_L}\mathcal{W}_L(C) = \frac{1}{Q_{L'(a)}}\mathcal{W}_{L'(a)}(C) = W_{L'(a)}(C)\,. \tag{3.15}$$

The average over the position of C is then the same as the average over different L′(a); i.e.,

$$\langle W_L(C)\rangle_C = \langle W_L(C)\rangle_L\,. \tag{3.16}$$

Consequently, the C-average in (i) is equivalent to the quenched average in (ii).

In the sections that follow, we are mostly concerned with the evaluation of the Wilson loop functional in the strong coupling limit. As we shall see (eq. (4.6) in the next section), in that limit $Q_L \to$ constant, which makes the quenched average (3.8) the *same* as the annealed average (3.9). Away from the strong coupling limit, these two averages are in general different. We view them as alternative possibilities. In the quenched average, the lattice sites r_i appear as purely kinematic variables; each random lattice L is weighted entirely according to the kinematic factor $\prod d^D r_i$. However, in the annealed average, the lattice sites are treated on a par with other dynamic variables. In (3.9)–(3.11), the degrees of freedom describing our random lattice take on an added physical significance; therefore, this particular approach appears to us to be of a deeper nature. It generalizes the usual rigid lattice field theory in a fundamental way, and gives a formulation of the random lattice field theory that is perhaps more far-reaching.

Remark. There exist also other possibilities which we do *not* adopt. For example, we may consider a sequence of larger and larger loops C of a given shape, all passing through the same arbitrarily selected site in a particular random lattice L in a volume Ω. The string tension T can be obtained by taking the logarithm of $W_L(C)$ in the limits first $\Omega \to \infty$ and then the loop-size $\to\infty$. The result can be quite different from that obtained from the logarithm of either (3.8) or (3.9). [This difference remains even if we eventually also carry out an average of T; i.e., an average of $\ln W_L(C)$, not the logarithm of an average of $W_L(C)$.]

As an illustration of this difference, let us consider an example of simplified roulette games, each consisting of spinning the wheel N times. The croupier always wins, and his take per game is $M = e^{n\lambda}$ where λ is a constant and n is the number of times that, say, 00 appears. Since the probability is $C^N_n p^n q^{N-n}$ where $p = \frac{1}{38}$ and $q = \frac{37}{38}$, the averages of M and $\ln M$ are $\langle M\rangle = (pe^\lambda + q)^N$ and $\langle \ln M\rangle = Np\lambda$. For a single game of very large N, $\lim_{N\to\infty}(1/N)\ln M$ is $p\lambda = (1/N)\langle \ln M\rangle$ which is *different* from $(1/N)\ln\langle M\rangle = \ln(pe^\lambda + q)$. This difference disappears when $\lambda \to 0$ (weak coupling), but is largest when $\lambda\to\infty$ (strong coupling). If we regard M as the analog of $W_L(C)$ and N as that of the loop area, then $(1/N)\ln\langle M\rangle$ corresponds to the string tension evaluated according to either the quenched average (3.8) or the annealed average (3.9). In this example since the analog of Q_L is

$$Q = \sum_{n=1}^{N} C^N_n p^n q^{N-n} = (p+q)^N = 1\,,$$

these two averages (quenched and annealed) are the same. Notice that when $\lambda \to \infty$, $(1/N) \ln \langle M \rangle \to \lambda + \ln p$ which seeks out the single configuration $n = N$, similar to the strong coupling calculation to be discussed in the next section.

To avoid ambiguity, throughout this paper we shall adhere only to the approaches described in (i) or (ii); i.e., either the quenched average (3.8) or the annealed average (3.9) of $W_L(C)$ is always taken at the beginning, and then we let the volume $\Omega \to \infty$. In the evaluation of the string tension, we take the limit of an infinite loop-size only in the very end.

4. String tension in the strong coupling limit

One of the attractive features of the conventional regular lattice gauge theory is the possibility of an analytic analysis of the strong coupling limit. Of particular interest is the Wilson loop functional, defined by (3.1). If the closed loop C lies in a lattice plane and is the border of an area made up of n elementary plaquettes, then, in the strong coupling limit, the Wilson loop functional behaves as

$$\sim (1/g^2)^n . \tag{4.1}$$

This result follows if we use the Wilson action, based on the function (2.7), and indicates a constant force between color triplet charges with an energy per unit separation or "string tension" proportional to $\ln g^2$. For a regular lattice, the Wilson loop functional varies when we change the orientation of the loop. In this section, we shall carry out a similar strong coupling calculation for the random lattice gauge theory, and shall obtain a rotationally symmetric result.

Let us take an arbitrary random lattice and begin with the expression $W_L(C)$ given by (3.2). For simplicity, we will study the strong coupling limit of $W_L(C)$ using the action

$$\mathcal{A}_L = \frac{1}{g^2} \sum_{\Delta_{ijk}} \kappa_{ijk} \, \mathrm{Re}\,[\mathrm{tr}\,(1 - U_{ijk})] \tag{4.2}$$

in accordance with (2.1) and (2.7). For large coupling we can expand the exponential of the action, keeping only the zeroth and first order terms for each of the fundamental triangles:

$$W_L(C) = \frac{1}{Q_L} \mathcal{W}_L(C) ,$$

where

$$Q_L = \int \prod_{k,l} \mathrm{d}[U(k,l)] \prod_{\Delta_{ijk}} \left\{ 1 - \frac{1}{g^2} \kappa_{ijk} \, \mathrm{Re}\,[\mathrm{tr}\,(1 - U_{ijk})] \right\} , \tag{4.3}$$

$$W_L(C) = \int \prod_{k,l} d[U(k,l)] \, \mathrm{tr}\left[\prod_{i,j \in C_L} U(i,j)\right]$$
$$\times \prod_{\Delta_{ijk}} \left\{1 - \frac{1}{g^2} \kappa_{ijk} \, \mathrm{Re}\,[\mathrm{tr}\,(1 - U_{ijk})]\right\}. \tag{4.4}$$

Since the differential $d[U]$ and the matrix elements of U and U^{-1} satisfy the identities

$$\int d[U] = N,$$
$$\int d[U] U_{\alpha\beta} = 0, \tag{4.5}$$
$$\int d[U] U_{\alpha\beta} (U^{-1})_{\gamma\rho} = \delta_{\alpha\rho}\delta_{\beta\gamma},$$

the leading contribution to Q_L when $g^2 \to \infty$ is

$$Q_L = N^{\mathcal{N}_1} = \text{constant}, \tag{4.6}$$

where $\mathcal{N}_1$ is the total number of links (1-simplices) in the lattice. Therefore,

$$W_L = N^{-\mathcal{N}_1} \mathcal{W}_L(C) \tag{4.7}$$

in the strong coupling limit. The limiting value can be obtained by taking only the smallest number of $1/g^2$ terms in the third product in (4.4) that is necessary for a non-zero result. We observe that for each $U(i,j)$ appearing in the second product in (4.4), its inverse $U(j,i)$ must be present in the expansion of $\exp(-\mathcal{A}_L)$, i.e., in the last product in (4.4). Similarly for each factor tr U_{ijk} there must appear elsewhere in the integrand the variables $U(j,i)$, $U(k,j)$ and $U(i,k)$.

Thus the leading order term in the strong coupling expansion of $W_L(C)$ corresponds to a continuous surface $\mathcal{S}_C$ formed of triangles Δ_{ijk} and bounded by the links in C_L. In fact, it must correspond to that surface containing the smallest number of triangles for the given curve C. If we assign each triangle an orientation given by the ordering of the Wilson loop which forms the boundary of the surface $\mathcal{S}_C$, then for SU(N) with $N \neq 2$, each triangle Δ_{ijk} corresponds to that term in the expression

$$\mathrm{Re}\,\mathrm{tr}\,(U_{ijk}) = \tfrac{1}{2}[\mathrm{tr}\,U_{ijk} + \mathrm{tr}\,U_{kji}] \tag{4.8}$$

in which the order of the vertices i, j, k is opposite to the orientation of the loop product $\prod_{i,j\in C_L} U(i,j)$ in (4.4). For SU(2) these two terms in (4.8) are equal and we obtain an extra factor of 2. Hence, substituting (4.5) into (4.4) we obtain the strong coupling limit of the Wilson loop expectation value for a particular random lattice L:

$$W_L(C) \cong g^{-2n} N^{1-\mathcal{N}_e} \prod_{\Delta_{ijk} \in \mathcal{S}_C} [\tfrac{1}{2}(1+\delta_{N2})\kappa_{ijk}], \tag{4.9}$$

where

$$n = \text{number of triangles contained in } \mathcal{S}_C \tag{4.10}$$

and $\mathcal{N}_\ell$ is the number of links contained in these n triangles. Next, we evaluate the ensemble average of $W_L(C)$. From (4.9) and (3.8), or (3.9), it follows that in the strong coupling limit either the quenched or the annealed average gives (omitting the subscript L outside the bracket in (3.8) or (3.9))

$$\langle W_L(C)\rangle = \Omega^{-\mathcal{N}} \int \prod_{i=1}^{\mathcal{N}} d^D r_i N^{1-\mathcal{N}_\ell} g^{-2n} \prod_{\Delta_{ijk}\in\mathcal{S}_C} [\tfrac{1}{2}(1+\delta_{N2})\kappa_{ijk}] . \tag{4.11}$$

Both n and $\mathcal{S}_C$ are functions of the sites $1, 2, \ldots, \mathcal{N}$ which define the lattice. In this calculation the factor g^{-2n} weights very strongly those lattice configurations which contain a minimal surface $\mathcal{S}_C$ made up of a small number n of triangles.

To evaluate the lattice average (4.9), we first relabel the lattice sites so that

$$1, 2, \ldots, \nu \text{ are the sites defining the surface } \mathcal{S}_C , \tag{4.12}$$

and $\nu+1, \ldots, \mathcal{N}$ the remaining ones. In order that a particular triangle, say Δ_{ijk}, that is present in $\mathcal{S}_C$ actually be realized as a plaquette in the lattice, there must be a cluster of $D+1$ sites including the three sites i, j, k. We recall from sect. 2 of I that $D+1$ sites form a cluster if and only if the $(D-1)$-dimensional sphere passing through them has an empty interior. Thus it is necessary, in order that Δ_{ijk} be a plaquette, that *some* $(D-1)$-dimensional sphere passing through i, j, k have an empty interior. But this condition is also sufficient since the largest such sphere must pass through $D+1$ sites, otherwise it could be made larger. It follows that for any plaquette Δ_{ijk} there is a *smallest* $(D-1)$-dimensional sphere S_{ijk} that passes through i, j, k and has an empty interior. This sphere may pass through any number

$$m_{ijk} = 0, 1, \ldots, D-2 \tag{4.13}$$

of additional sites besides i, j and k. [We can ignore the case when S_{ijk} passes through more than $D+1$ sites (i.e., $m_{ijk} > D-2$), since it is of zero measure.]

For notational clarity, we shall refer to those Δ_{ijk} on the surface $\mathcal{S}_C$ and their associated smallest empty spheres S_{ijk} by Δ_α and S_α where $\alpha = 1, 2, \ldots, n$. Define

$$V \equiv \text{union of volumes contained inside } S_1, S_2, \ldots, S_n . \tag{4.14}$$

By definition, there is no lattice site inside V. Given the positions of sites $1,2,\ldots,\nu$ which determine $\mathcal{S}_C$, the volume V has a minimum value $V_{\min}(1,\ldots,\nu)$.

In the limit $\Omega\to\infty$ and $\mathcal{N}\to\infty$ at a constant site density $\rho = \mathcal{N}/\Omega$, the integration over the positions of the lattice sites in (4.11) can be written as a sum of integrals over ν-sites given by (4.12) for various values of ν:

$$\lim_{\Omega\to\infty} \Omega^{-\mathcal{N}} \int \prod_{i=1}^{\mathcal{N}} d^D r_i = \sum_{\nu=1}^{\mathcal{N}} \int \prod_{j=1}^{\nu} d^D r_j \mathcal{P}(1, \ldots, \nu) . \tag{4.15}$$

The weighting factor $\mathscr{P}(1,\ldots,\nu)$ thus defined is due to the integration over the remaining $\nu+1, \nu+2, \ldots, \mathscr{N}$ sites. In terms of $\mathscr{P}(1,\ldots,\nu)$, (4.11) becomes

$$\langle W_L(C)\rangle = \sum_{\nu=1}^{\mathscr{N}} \int \prod_{i=1}^{\nu} d^D r_i N^{1-\mathscr{N}_\ell} g^{-2n} \times \prod_{\Delta_{ijk}\in\mathscr{S}_C} [\tfrac{1}{2}(1+\delta_{N2})\kappa_{ijk}]\mathscr{P}(1,\ldots,\nu) . \tag{4.16}$$

It is useful to express $\mathscr{P}(1,\ldots,\nu)$ as an integral over the volume V defined by (4.14):

$$\mathscr{P}(1,\ldots,\nu) = \int_{V_{min}}^{\infty} J(V)\,dV\,e^{-\rho V} , \tag{4.17}$$

where $e^{-\rho V}$ is the probability that the inside of V is empty of lattice sites and $J(V)$ is the jacobian.

To evaluate $\langle W_L(C)\rangle$ in the limit of large C and large g^2, we shall use the saddle-point approximation. The leading large g^2 behavior of $\langle W_L(C)\rangle$ is determined by the factors g^{-2n} in (4.16) and $e^{-\rho V}$ in (4.17). As g^2 becomes large, the factor g^{-2n} enhances those configurations with a small number n of triangles. However, as n decreases, the volume V, defined by (4.14), necessarily grows, making $e^{-\rho V}$ small. Hence at given site-positions $1, 2, \ldots, \nu$ (therefore, also at fixed n), the largest contribution to $\langle W_L(C)\rangle$ comes from regions near the lower limit $V = V_{min}$ in the integral (4.17), i.e. the maximum of $e^{-\rho V}$; in particular, as we shall see, the deviation of $\mathscr{P}(1,\ldots,\nu)$ from its saddle-point value $e^{-\rho V_{min}}$ can be estimated to be

$$\mathscr{P}(1,\ldots,\nu) \sim \frac{\rho^\nu}{\nu!} \prod_{\alpha=1}^{n} (\rho^{2/D} A_\alpha)^{D(D-2)/4}\, e^{-\rho V_{min}} , \tag{4.18}$$

where A_α is the area of the plaquette Δ_α contained in $\mathscr{S}_C$. In the strong coupling limit it will turn out that* V_{min} is $\sim A(\ln g^2)^{(D-2)/D}$, A_α is $\sim(\ln g^2)^{2/D}$, and ν and n are $\sim A(\ln g^2)^{-2/D}$, where

$$A = \sum_{\alpha=1}^{n} A_\alpha \tag{4.19}$$

is the area of the Wilson loop. Thus, if we neglect $\ln\ln g^2$ as compared to $\ln g^2$, then the approximation

$$\mathscr{P}(1,\ldots,\nu) \cong e^{-\rho V_{min}} \tag{4.20}$$

suffices for the strong coupling limit calculation. In what follows we shall first discuss how the estimate (4.18) is obtained, and then evaluate the string tension by using the approximation (4.20).

* See (4.47)–(4.49) below.

To understand the estimate (4.18) it is useful to discuss further the previously introduced notion of the smallest empty sphere S_{ijk} that is associated with each plaquette Δ_{ijk}. Consider the simple case when the space-dimension D is 3. From (4.13), we see that m_{ijk} can only be 0 or 1. The probability $P(i, j, k)$ that sites i, j and k are linked to form a plaquette may be written as a sum according to $m_{ijk} = 0$ or 1:

$$P(i, j, k) = P_0(i, j, k) + P_1(i, j, k) . \tag{4.21}$$

By $P_0(i, j, k)$ we mean the probability that the smallest empty sphere S_{ijk} passing through i, j and k passes through *no* other point (i.e., $m_{ijk} = 0$). But this would mean that i, j and k must lie on a great circle of S_{ijk}. [Otherwise, we could make the sphere smaller by moving its center towards the plaquette Δ_{ijk}, keeping of course i, j and k on its surface. Because we are considering the case $m_{ijk} = 0$, the surface S_{ijk} cannot encounter any other site during this move; thus eventually i, j and k must lie on a great circle of S_{ijk}.] In the plane that contains Δ_{ijk}, let r be the radius of the circumscribed circle of Δ_{ijk}. $P_0(i, j, k)$ is then simply the probability that a sphere of radius $R = r$ has empty interior, or

$$P_0(i, j, k) = e^{-4\pi r^3 \rho/3} , \tag{4.22}$$

where, as before, ρ is the site density. Likewise, by $P_1(i, j, k)$ we mean the probability that the smallest empty sphere passing through i, j and k also passes through one other site, say p. [Hence, it corresponds to $m_{ijk} = 1$.] The plane that contains Δ_{ijk} cuts the sphere S_{ijk} into two caps; p must lie on the smaller cap. Let x be the perpendicular distance from the center of the sphere to the plaquette Δ_{ijk}. As x increases by dx, the smaller cap sweeps out an infinitesimal volume $d\tau(x)$, which is the volume available to the site p. The whole sphere, whose interior we require to be empty, has a radius $R = (r^2 + x^2)^{1/2}$. Therefore

$$P_1(i, j, k) = 2 \int_0^\infty e^{-4\pi (r^2+x^2)^{3/2} \rho/3} \rho \, d\tau(x) , \tag{4.23}$$

where the factor 2 allows for displacement in either direction.

In (4.23), the exponent for small x is

$$-\tfrac{1}{3} 4\pi (r^2 + x^2)^{3/2} \rho = -\tfrac{1}{3} 4\pi r^3 \rho - 2\pi \rho r x^2 + O(x^4) ,$$

which implies that only those x for which $\rho r x^2 = O(1)$ are significant. In the strong coupling limit, as we shall see, $\rho r^3 \sim \ln g^2 \gg 1$; therefore we need only consider $x \ll r$. In this approximation one has $d\tau(x) \cong \pi r^2 \, dx$, and so (4.23) becomes

$$\begin{aligned} P_1(i, j, k) &\cong 2 \, e^{-4\pi r^3 \rho/3} \int_0^\infty e^{-2\pi \rho r x^2} \pi r^2 \rho \, dx \\ &= \pi (\tfrac{1}{2} \rho r^3)^{1/2} \, e^{-4\pi r^3 \rho/3} . \end{aligned} \tag{4.24}$$

In terms of the area A_{ijk} of the triangle Δ_{ijk}, we see that (4.24) can be written as

$$P_1(i,j,k) = \text{constant} \times (\rho A_{ijk}^{3/2})^{1/2}\, e^{-4\pi r^3 \rho/3}\,, \tag{4.25}$$

where the constant depends on the shape of Δ_{ijk}, but is O(1), while A_{ijk} is $O(r^2) = O(\ln^{2/3} g^2)$. Note that from (4.22) and (4.24)

$$P_1(i,j,k) \gg P_0(i,j,k) \tag{4.26}$$

in the strong coupling limit.

The above discussion can be generalized to any dimension D. Just as in (4.21), the probability $P(i,j,k)$ that sites i, j and k are linked to form a plaquette can be written as

$$P(i,j,k) = \sum_{m=0}^{D-2} P_m(i,j,k)\,, \tag{4.27}$$

where $P_m(i,j,k)$ corresponds to the case $m = m_{ijk}$ of (4.13). The Poisson formula gives

$$P_0(i,j,k) = \exp(-\rho v_{ijk}) \tag{4.28}$$

where v_{ijk} is the volume of a sphere of radius r in D-dimensions. Similarly to (4.24), we have (shown in the appendix)

$$P_m(i,j,k) = \text{constant} \times (\rho A_{ijk}^{D/2})^{m/2} \exp(-\rho v_{ijk})\,. \tag{4.29}$$

When $D = 3$ and $m = 1$, (4.29) reduces to (4.25).

Next, we examine the more complicated probability that the ν sites $1, 2, \ldots, \nu$ of (4.12) should be linked to form the n plaquettes $\Delta_1, \Delta_2, \ldots, \Delta_n$ which define the surface $\mathcal{S}_C$. To each Δ_α ($\alpha = 1, 2, \ldots, n$) there is a smallest $(D-1)$-dimensional sphere S_α with empty interior. The surface S_α passes through i, j, k and m_α additional sites, where $m_\alpha = 0, 1, \ldots, D-2$. In accord with (4.12), we may label these additional sites as

$$\nu+1, \ldots, \nu+m_1; \quad \nu+m_1+1, \ldots, \nu+m_1+m_2\,; \quad \ldots, \nu + \sum_{\alpha=1}^{n} m_\alpha\,, \tag{4.30}$$

so that $\nu+1, \ldots, \nu+m_1$ refer to sites on S_1, $\nu+m_1+1, \ldots, \nu+m_1+m_2$ to sites on S_2, etc. (The possibility that two or more of these $\nu+\sum_{\alpha=1}^{n} m_\alpha$ sites are identical is discussed in the appendix.)

As we shall see, this probability is closely related to the evaluation of $\mathcal{P}(1, \ldots, \nu)$ defined earlier. From the definition (4.15), $\mathcal{P}(1, \ldots, \nu)$ is determined by the integration over the positions of all the sites $\nu+1, \nu+2, \ldots, \mathcal{N}$ not on $\mathcal{S}_C$. We have

$$\mathcal{P}(1,2,\ldots,\nu)=\Omega^{-\mathcal{N}}\sum_{m_\alpha}\frac{\mathcal{N}!}{(\mathcal{N}-\nu-M)!\nu!\prod_\alpha m_\alpha!}$$

$$\times\int\prod_{i=1}^{M}\mathrm{d}^D r_{\nu+i}\int\prod_{j=\nu+M+1}^{\mathcal{N}}\mathrm{d}^D r_j, \tag{4.31}$$

where

$$M=\sum_{\alpha=1}^{n} m_\alpha, \tag{4.32}$$

and as before each m_α can vary from 0 to $D-2$. The sites $1, 2, \ldots, \nu$ are on $\mathcal{S}_C$, $\nu+1, \ldots, \nu+M$ are on the surface of V given by (4.14), but $\nu+M+1, \ldots, \mathcal{N}$ are outside V. Therefore, the last integration in (4.31) can be readily carried out:

$$\int\prod_{j=\nu+M+1}^{\mathcal{N}}\mathrm{d}^D r_j=(\Omega-V)^{\mathcal{N}-\nu-M}. \tag{4.33}$$

Substituting this expression into (4.31), we find in the limit $\Omega\to\infty$ and $\mathcal{N}\to\infty$ at a constant $\rho=\mathcal{N}/\infty$:

$$\mathcal{P}(1,\ldots,\nu)=\frac{\rho^\nu}{\nu!}\sum_{m_\alpha}\frac{\rho^M}{\prod_\alpha m_\alpha!}\int \mathrm{e}^{-\rho V}\prod_{i=1}^{M}\mathrm{d}^D r_{\nu+i}. \tag{4.34}$$

The integration over $\prod_{i=1}^{M}\mathrm{d}^D r_{\nu+i}$ can be carried out by following arguments similar to those used in the derivation of (4.24) and (4.29). The details are given in the appendix; for each distribution $\{m_\alpha\}$, the result is, similar to (4.29),

$$\frac{\rho^M}{\prod_\alpha m_\alpha!}\int \mathrm{e}^{-\rho V}\prod_{i=1}^{M}\mathrm{d}^D r_{\nu+i}\cong \mathrm{e}^{-\rho V_{\min}}\prod_{\alpha=1}^{n}\left[\sum_{m=0}^{D-2}C_\alpha(m)\rho^{m/2}A_\alpha^{mD/4}\right], \tag{4.35}$$

where $V_{\min}$ is the minimum of V and $C_\alpha(m)$, although dependent on $1, 2, \ldots, \nu$, remains O(1) when $g\to\infty$. On the other hand, as we shall see, the area A_α of the triangle Δ_α is

$$A_\alpha\sim(\ln g^2)^{2/D}. \tag{4.36}$$

Substituting (4.35) into (4.34) we have

$$\mathcal{P}(1,\ldots,\nu)\cong\frac{\rho^\nu}{\nu!}\mathrm{e}^{-\rho V_{\min}}\prod_{\alpha=1}^{n}\left[\sum_{m=0}^{D-2}C_\alpha(m)\rho^{m/2}A_\alpha^{mD/4}\right]. \tag{4.37}$$

Because of (4.36), the dominant term inside the square bracket is when $m=D-2$; (4.37) then leads to (4.18). For the leading strong coupling calculation, approximation (4.20) is sufficient; i.e., we can set

$$\mathcal{P}(1,\ldots,\nu) \cong e^{-\rho V_{\min}}.$$

To derive $V_{\min}$, we need only consider the case when the smallest empty sphere S_{ijk} of Δ_{ijk} is the one that has sites i, j and k on its great circle. Since the surface $\mathcal{S}_C$ is made up of n triangles Δ_{ijk}, the volume $V_{\min}$ is the union of the corresponding n spheres S_{ijk} determined in this way. If we consider a term in the expression (4.16) for $\langle W_L(C)\rangle$ with a fixed number of triangles n, the largest contribution to it will come from that arrangement of the n triangles with the maximum value of $\mathcal{P}$ or, referring to (4.20), with the smallest minimum empty volume $V_{\min}$. Clearly $V_{\min}$ will be smallest if the n triangles and their ν vertices $1,\ldots,\nu$ all lie in the two-dimensional plane containing the original curve C. The dependence of $V_{\min}$ on the size and shape of a particular triangle, say Δ_{ijk}, is complicated by the overlap between the $D-1$ dimensional sphere S_{ijk} determined by that triangle and the spheres determined by neighboring triangles. However, we can uniquely associate with a triangle Δ_{ijk} a D-dimensional volume V_{ijk} consisting of that portion of the interior of the sphere S_{ijk} which is swept out by the triangle if it is displaced upward and downward in directions perpendicular to the surface $\mathcal{S}_C$ (see fig. 1).

For a given triangle this corresponding swept-out volume V_{ijk} is straightforward to compute. If the triangle has sides with lengths l_1, l_2 and l_3, the radius of the circumscribing circle is given by

$$r = \frac{l_1 l_2 l_3}{[(l_1^2+l_2^2+l_3^2)^2 - 2(l_1^4+l_2^4+l_3^4)]^{1/2}}\,. \qquad (4.38)$$

The volume V_{ijk} for such a triangle is then the volume of the $D-1$ dimensional

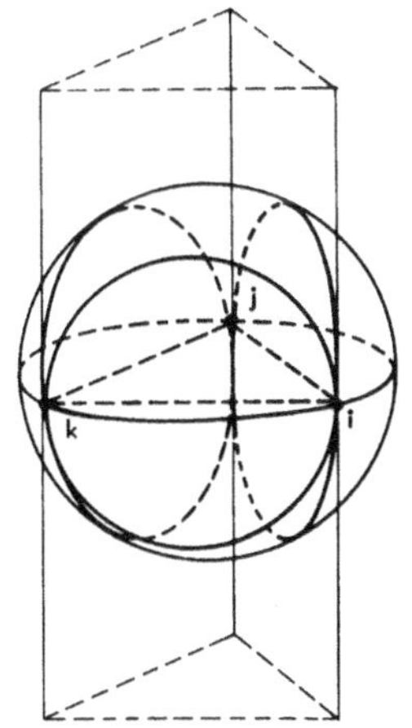

Fig. 1. The sphere S_{ijk} of minimum volume that must be empty if the triangle Δ_{ijk} is to occur in our lattice. That portion of the sphere lying inside the vertical, right triangular prism contains the volume V_{ijk}, uniquely associated with that triangle.

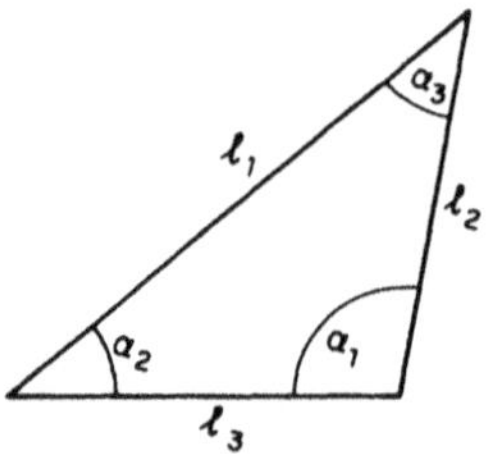

Fig. 2. One of the triangles making up the surface bounded by the Wilson loop C. Here we show the sides l_i and the opposite angles α_i $(i = 1, 2, 3)$.

sphere with radius r minus the three spherical caps visible in fig. 1. The cap corresponding to the side l_i has a base which is a $D-2$ dimensional sphere of diameter l_i. The volume for such a cap is

$$\tfrac{1}{3}\pi r^3(2-3\cos\alpha_i+\cos^3\alpha_i)\,, \qquad \text{for } D=3\,, \tag{4.39}$$

$$\tfrac{1}{6}\pi r^4(3\alpha_i-5\sin\alpha_i\cos\alpha_i+2\sin\alpha_i\cos^3\alpha_i)\,, \qquad \text{for } D=4\,. \tag{4.40}$$

Here α_i is the interior angle opposite the side l_i (see fig. 2). Thus for such a triangle Δ_{ijk} in three or four dimensions, the corresponding volume V_{ijk} is

$$V_{ijk}=\tfrac{1}{3}\pi r^3\left[-2+\sum_{i=1}^{3}(3\cos\alpha_i-\cos^3\alpha_i)\right], \qquad \text{for } D=3\,, \tag{4.41}$$

$$V_{ijk}=\tfrac{1}{6}\pi r^4\left[3\pi-\sum_{i=1}^{3}(3\alpha_i-5\sin\alpha_i\cos\alpha_i+2\sin\alpha_i\cos^3\alpha_i)\right], \qquad \text{for } D=4\,. \tag{4.42}$$

The sum of V_{ijk} over all triangles Δ_{ijk} in $\mathscr{S}_C$ gives the corresponding minimum empty volume V_{min} in (4.20).

Next we will show that the arrangement of n triangles which covers a given large area A and corresponds to the smallest V_{min} is a regular array of equilateral triangles. Consider a triangle Δ_{ijk} of a particular shape specified by the angles α_1, α_2, α_3. We can relate the triangle's area A_{ijk} and the corresponding volume V_{ijk} by

$$V_{ijk}=\gamma_{ijk}[A_{ijk}]^{D/2}\,. \tag{4.43}$$

Thus $\gamma_{ijk}=\gamma_{ijk}(\alpha_1,\alpha_2,\alpha_3)$ serves as a scale factor; it is determined by the ratio of the volume V_{ijk}, given by (4.41) or (4.42), divided by the $\tfrac{1}{2}D$ power of the triangle's area

$$A_{ijk}=r^2\sum_{i=1}^{3}\sin\alpha_i\cos\alpha_i\,.$$

We observe that the function γ_{ijk} is minimum for an equilateral triangle, $\alpha_1=\alpha_2=$

$\alpha_3 = \frac{1}{3}\pi$, and this minimum is

$$\gamma_{\min} = \begin{cases} \dfrac{17\pi}{27(3)^{1/4}}, & D=3, \\ \dfrac{\pi}{\sqrt{3}}, & D=4. \end{cases} \qquad (4.44)$$

Hence if we consider a particular arrangement of n triangles covering our large area A, each with individual area A_α and scale factor γ_α, where α (as before, replacing the subscripts ijk) varies from 1 to n, then the minimum empty volume that appears in (4.20) for this arrangement is

$$V_{\min} = \sum_{\alpha=1}^{n} \gamma_\alpha (A_\alpha)^{D/2} \geqslant \gamma_{\min} \sum_\alpha (A_\alpha)^{D/2} \geqslant \gamma_{\min} n (A/n)^{D/2}. \qquad (4.45)$$

Here the last inequality states that if an area A is separated into a sum of n pieces $A = \sum_{\alpha=1}^{n} A_\alpha$, it is the separation into equal areas $A_\alpha = A/n$ which makes the sum $\sum_{\alpha=1}^{n} (A_\alpha)^{D/2}$ minimum for $D > 2$. The inequality (4.45) implies that the smallest value of $V_{\min}$ is obtained for an arrangement of equilateral triangles of equal area.

With this result and the coefficient $\gamma_{\min}$ given above, we can write down the leading contributions to $\langle W_L(C)\rangle$ coming from those lattices in our average which contain a planar surface $\mathscr{S}_C$ bounded by the curve C, and consisting of n equilateral triangles of equal area $A_\alpha = A/n$:

$$\langle W_L(C)\rangle \sim \sum_n \exp\left[-n \ln g^2 - n\rho\gamma_{\min}(A/n)^{D/2}\right]. \qquad (4.46)$$

The largest term in the sum corresponds to that value of n for which the exponent is minimum:

$$n = A\left[\frac{(D-2)\rho\gamma_{\min}}{2 \ln g^2}\right]^{2/D}, \qquad (4.47)$$

which gives for each triangle an area

$$\bar{A}_\Delta = \left[\frac{2 \ln g^2}{(D-2)\rho\gamma_{\min}}\right]^{2/D} \qquad (4.48)$$

and a side-length

$$\bar{l} = 2(\tfrac{1}{3})^{1/4}\left[\frac{2 \ln g^2}{(D-2)\rho\gamma_{\min}}\right]^{1/D}. \qquad (4.49)$$

Hence,

$$\langle W_L(C)\rangle \sim \exp\left(-AD[\tfrac{1}{2}\rho\gamma_{\mathrm{mim}}]^{2/D}[\ln g^2/(D-2)]^{(D-2)/D}\right). \qquad (4.50)$$

Thus the Wilson-loop expectation value $\langle W_L(C)\rangle$ has a computable area-law behavior in the strong-coupling limit with the string tension $T \equiv -(1/A)\ln\langle W_L(C)\rangle$ given by

$$T = \begin{cases} \dfrac{1}{\sqrt{3}}\left(\dfrac{17\pi}{6}\right)^{2/3}(\rho^2 \ln g^2)^{1/3} = 2.4797(\rho^2 \ln g^2)^{1/3}, & D=3, \\ 2\left(\dfrac{\pi}{\sqrt{3}}\right)^{1/2}(\rho \ln g^2)^{1/2} = 2.6936(\rho \ln g^2)^{1/2}, & D=4. \end{cases} \tag{4.51}$$

5. String thickness

In the preceding section we analyzed the strong coupling limit of the Wilson-loop expectation value. That limit was determined in a saddle-point approximation with the saddle point corresponding to a regular array of equilateral triangles lying in the plane of the Wilson loop. The area of each of these triangles was proportional to $\ln^{2/D} g^2$ as implied by (4.48). Of course, in this saddle-point approximation the major contribution to the integral comes from a continuum of such triangular arrays corresponding to fluctuations about the saddle point. Because of these fluctuations, the typical surface of triangles does not lie precisely in the plane of the Wilson loop but fluctuates about that plane with an amplitude which grows logarithmically with the loop area. As we shall see, the average square of the distance between the fluctuating surface of triangles and the plane of the Wilson loop is given by

$$d^2 = \frac{1}{2\pi T}\ln A, \tag{5.1}$$

where T is the strong coupling string tension found in the previous section and the distance d is measured near the center of the Wilson loop. We will refer to d as the "string thickness". The relation (5.1) between string thickness and string tension is identical to that for a relativistic string [3].

This rather simple result is to be contrasted with the behavior of the string thickness obtained in the strong coupling limit for a gauge theory on a regular, hypercubic lattice [2]. For such a regular lattice we should distinguish two cases: (i) If the Wilson loop lies in a lattice plane, then the strong coupling limit comes from a single set of plaquettes, those lying in the plane of the Wilson loop and enclosed by it. Nearby configurations involving plaquettes outside this plane require an extra factor $(1/g^2)^4$ and thus have an action separated by a non-zero gap from the action of the unique leading order term. The strong-coupling string thickness for such a loop is zero. As g^2 is decreased from a "strong-coupling" value, it has been argued [4, 5] that the string thickness changes discontinuously, the "roughening transition". Below the roughening transition the string thickness is expected to increase with area as in (5.1). (ii) If the Wilson loop does not lie in a lattice plane, then there will be a number of different "staircase"-like surfaces of plaquettes that make equal contributions to the Wilson-loop expectation value. The thickness corresponding to such a configuration grows as the square root of the area of the Wilson loop in the limit of large loop area provided that the shape and orientation of the loop are held fixed. Thus in the strong-coupling limit for a regular lattice, both the string tension and thickness depend strongly on the orientation of the Wilson loop from which they were derived*.

In order to discuss the string thickness for our random lattice, we begin with a large, fixed Wilson loop C, chosen to lie in the x, y plane. Let the vector $\boldsymbol{r}_{l,m}$ locate the vertices of that regular array of triangles which is the saddle point of the expectation-value integration. The two integers l, m are the coordinates $\boldsymbol{r}_{l,m}$ with respect to axes meeting at a 60° angle:

$$\boldsymbol{r}_{l,m} = l\hat{\alpha} + m\hat{\beta} . \tag{5.2}$$

Here $\hat{\alpha}$ and $\hat{\beta}$ are unit vectors which lie in the x, y plane wth $\hat{\alpha} \cdot \hat{\beta} = \frac{1}{2}$. We will choose the vertex $\boldsymbol{r}_{0,0}$ to lie near the center of the Wilson loop. Next consider a second random lattice in our ensemble which lies near the saddle point referred to above. It also contains a surface of triangles whose vertices $\boldsymbol{r}'_{l,m}$ lie near the positions $\boldsymbol{r}_{l,m}$:

$$\boldsymbol{r}'_{l,m} = \boldsymbol{r}_{l,m} + \delta\boldsymbol{r}_{l,m} . \tag{5.3}$$

As will be shown below, the major contribution to the expectation-value integration comes from the region $|\delta\boldsymbol{r}_{l,m}|^2 \sim 1/\ln^{2/D} g^2$ so the fluctuations $\delta\boldsymbol{r}_{l,m}$ about the saddle point can be treated as small.

To leading order the $\delta\boldsymbol{r}_{l,m}$ dependence in the integrand of (4.15) appears only in the excluded volume V. Again to leading order, we can replace V by $V_{\min}(\delta\boldsymbol{r}_{l,m})$, the smallest volume which must be free of all lattice sites if the vertices given by (5.3) are to be linked into the specified triangular array. Let us decompose the fluctuations $\delta\boldsymbol{r}_{l,m}$ into two components, $\boldsymbol{\xi}_{l,m}$ parallel to the x, y plane and $\boldsymbol{\zeta}_{l,m}$ perpendicular to that plane:

$$\delta\boldsymbol{r}_{l,m} = \boldsymbol{\xi}_{l,m} + \boldsymbol{\zeta}_{l,m} . \tag{5.4}$$

Since $V_{\min}$ is symmetric under reflection in the x, y plane, $\boldsymbol{\zeta}_{l,m} \to -\boldsymbol{\zeta}_{l,m}$, the variables $\boldsymbol{\xi}_{l,m}$ and $\boldsymbol{\zeta}_{l,m}$ decouple in $V_{\min}$ through the quadratic approximation. Thus we can compute the string thickness, which involves only the fluctuations $\boldsymbol{\zeta}_{l,m}$, by considering only configurations with $\boldsymbol{\xi}_{l,m} = 0$. The string thickness d is then given by

$$d^2 = \frac{\int \prod_{l,m} \mathrm{d}^2\zeta_{l,m} [\boldsymbol{\zeta}_{0,0}]^2 \, \mathrm{e}^{-\rho V_{\min}(\boldsymbol{\zeta}_{l,m})}}{\int \prod_{l,m} \mathrm{d}^2\zeta_{l,m} \, \mathrm{e}^{-V_{\min}(\boldsymbol{\zeta}_{l,m})}} , \tag{5.5}$$

where in $V_{\min}$ we can neglect all powers of $\boldsymbol{\zeta}_{l,m}$ higher than quadratic.

* Kogut, Sinclair, Pearson, Richardson and Shigemitsu [6] have shown that higher-order terms in the strong-coupling expansion tend to restore the rotational symmetry at least for the string tension.

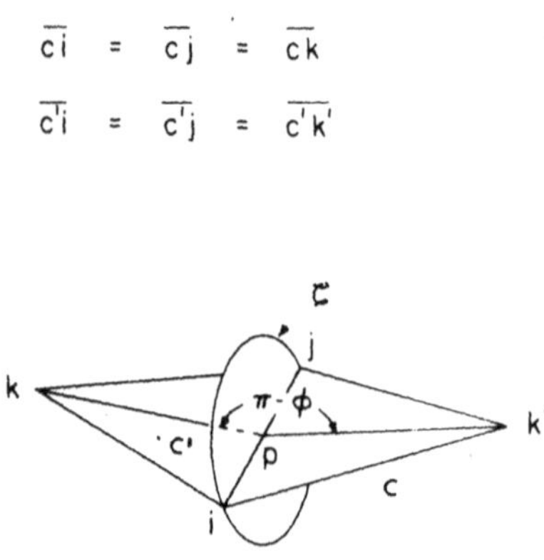

Fig. 3. Circle $\mathscr{C}$ is the intersection of the circumspheres S and S′, whose centers are c and c′. Link $\overline{ij}$ is perpendicular to $\overline{kp}$ and $\overline{k'p}$. When $\xi_{l,m}=0$, p is the midpoint of $\overline{ij}$.

We will now compute $V_{\min}(\zeta_{l,m})$. Consider two adjacent triangles, say Δ_{ijk} and $\Delta_{ijk'}$. Let S and S′ be the corresponding circumspheres which contribute to $V_{\min}(\zeta_{l,m})$ and which must be empty if the triangles Δ_{ijk} and $\Delta_{ijk'}$ are to occur in the lattice. The vertices of Δ_{ijk} lie on the surface of S and those of $\Delta_{ijk'}$ on S′. If all the $\zeta_{l,m}$ are zero, the triangle Δ_{ijk} lies on a great circle of S, and $\Delta_{ijk'}$ on that of S′; however for non-zero $\zeta_{l,m}$ this is no longer true. For example, if the angle ϕ between the planes defined by the triangles Δ_{ijk} and $\Delta_{ijk'}$ is non-zero, the centers c and c′ of the spheres S and S′ will be drawn out of the planes of the respective triangles as shown in fig. 3. We can divide the overlapping volumes of S and S′ into two pieces, each associated with one of the two triangles, if we erect a plane passing through their common edge $\overline{ij}$ which contains the $D-2$ dimensional sphere (the circle $\mathscr{C}$ in fig. 3) of points common to the surfaces of S and S′. Thus to the triangle Δ_{ijk} we associate the volume of the sphere S minus the volume of the cap defined by the plane just described. If this process is continued to eliminate overlaps with the other two triangles which border Δ_{ijk}, the resulting volume V_{ijk} will be that of S minus the volume of three caps, defined by the three planes passing through the three sides of Δ_{ijk}. In contrast to the planar array of triangles discussed in sect. 4, these caps may overlap, leaving a volume V_{ijk} with faces of various sorts: segments of spheres, disks and cylinders (fig. 4). The complete volume $V_{\min}(\zeta_{l,m})$ is the union of the V_{ijk} corresponding to all the triangles Δ_{ijk} in the surface.

For a particular triangle Δ_{ijk}, the volume V_{ijk} depends on the lengths of the three sides of Δ_{ijk} and on the locations c, c′, c″ and c‴ of the centers of the sphere S and the spheres S′, S″ and S‴ for the three triangles adjacent to Δ_{ijk}. Since each of the lengths of the sides of Δ_{ijk} depends quadratically on $\zeta_{l,m}$, we can separate these two influences on V_{ijk} through second order

$$\delta V_{ijk} = \delta V^{\mathrm{I}}_{ijk} + \delta V^{\mathrm{II}}_{ijk} . \qquad (5.6)$$

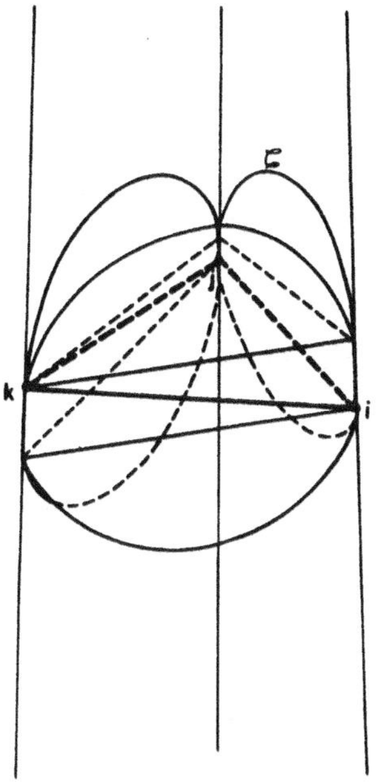

Fig. 4. Points i, j, k and the circle $\mathscr{C}$ are the same as those shown in fig. 3. Likewise the three vertical lines form a distorted version of the right triangular prism of fig. 1. The three circles represent the intersection of the sphere determined by the triangle ijk with the sphere defined by each of the neighboring triangles. The three nearly vertical lines are formed by the intersections of the planes defined by these circles.

Here δV_{ijk} is the change in V_{ijk} from the value of (4.41) and (4.42) for the planar, equilateral array. $\delta V^{\mathrm{I}}_{ijk}$ is the change in V_{ijk} that comes from the variation in the lengths of the triangle's sides, treating the centers c, c′, c″ and c‴ as each lying in the plane of Δ_{ijk}. Likewise, $\delta V^{\mathrm{II}}_{ijk}$ is the variation in V_{ijk} arising from the displacement of the centers c, c′, c″ and c‴ away from the plane of Δ_{ijk}. The corresponding change in V_{min} from its saddle-point value is

$$\delta V_{\mathrm{min}} = \sum_{\Delta_{ijk}} \delta V_{ijk} . \tag{5.7}$$

If we denote by δl_1, δl_2 and δl_3 the deviation of the length of each of the sides of the triangle Δ_{ijk} from the saddle-point values, then dimensional argument and the threefold symmetry of the equilateral triangle imply

$$\delta V^{\mathrm{I}}_{ijk} = \tfrac{1}{3}(\delta l_1 + \delta l_2 + \delta l_3) D \bar{V}_{ijk} / \bar{l} , \tag{5.8}$$

where $\bar{V}_{ijk} = \gamma_{\mathrm{min}}(\bar{A}_{\Delta})^{D/2}$ and $\bar{l}$ are the saddle-point values of the excluded volume and triangle-side length, given by (4.44), (4.48) and (4.49). The sum of all the $\delta V^{\mathrm{I}}_{ijk}$

in terms of our orignal variables $\zeta_{l,m}$ is

$$\delta V^{\mathrm{I}}_{\min} = \sum_{\Delta_{ijk}} \delta V^{\mathrm{I}}_{ijk}$$
$$= \tfrac{1}{3}[D\gamma_{\min}(\bar{A}_\Delta)^{D/2}/\bar{l}^2] \sum_{l,m} \{(\zeta_{l+1,m} - \zeta_{l,m})^2$$
$$+ (\zeta_{l,m+1} - \zeta_{l,m})^2 + (\zeta_{l-1,m+1} - \zeta_{l,m})^2\}\,. \tag{5.9}$$

In deriving this result we have used the expression

$$\tfrac{1}{2}(\zeta_{l,m} - \zeta_{l',m'})^2/\bar{l} \tag{5.10}$$

for the change in the length of the link connecting the vertices l, m and l', m', and also the fact that each link appears in the $\delta V^{\mathrm{I}}_{ijk}$ of two triangles.

We may regard $\zeta_{l,m} = \zeta(\boldsymbol{r}_{l,m})$ where $\boldsymbol{r}_{l,m}$ is given by (5.2). In the long-wavelength limit, $\zeta_{l+1,m} - \zeta_{l,m}$ can be approximated by $(\hat{\alpha} \cdot \boldsymbol{\nabla})\zeta$; likewise, the sum inside the curly bracket in (5.9) becomes $\frac{3}{2}\bar{l}^2[(\partial\zeta/\partial x)^2 + (\partial\zeta/\partial y)^2]$. Since the number of lattice sites equals $\frac{1}{2}$ times that of triangles, we have

$$\delta V^{\mathrm{I}}_{\min} = \tfrac{1}{4}D\gamma_{\min}(\bar{A}_\Delta)^{D/2-1} \int_A \mathrm{d}x\,\mathrm{d}y\left[\left(\frac{\partial\zeta}{\partial x}\right)^2 + \left(\frac{\partial\zeta}{\partial y}\right)^2\right] + \mathrm{O}((\boldsymbol{\nabla}\zeta)^4)\,, \tag{5.11}$$

where the integral extends over the total area A of the Wilson loop.

The second contribution $\delta V^{\mathrm{II}}_{ijk}$ in (5.6) is in general considerably more complicated to evaluate. However, by definition, it vanishes when the angle ϕ between the planes of the neighboring triangles Δ_{ijk} and $\Delta_{ijk'}$ in fig. 3 is zero. Hence, $\delta V^{\mathrm{II}}_{ijk}$ is proportional to the curvature of the surface of triangles. As will be shown below, this makes it quartic in the gradient of ζ. Therefore, in the long-wavelength limit, we can neglect $\delta V^{\mathrm{II}}_{ijk}$ and set $\delta V_{\min} = \delta V^{\mathrm{I}}_{\min}$. By using (4.48)–(4.51) and neglecting $\mathrm{O}(\boldsymbol{\nabla}\zeta)^4$, we see that

$$\rho\delta V_{\min} = \tfrac{1}{2}T \int_A \mathrm{d}x\,\mathrm{d}y\left[\left(\frac{\partial\zeta}{\partial x}\right)^2 + \left(\frac{\partial\zeta}{\partial y}\right)^2\right], \tag{5.12}$$

which is identical to the action in the string model [7, 8], with T as the string tension.

To exhibit more explicitly the behavior of $\delta V^{\mathrm{II}}_{ijk}$, we introduce the Fourier transform

$$\zeta(x, y) = \sum_{q_1,q_2} \frac{1}{\sqrt{A}} \mathrm{e}^{i(q_1x+q_2y)}\tilde{\zeta}(q_1q_2)\,. \tag{5.13}$$

The q-dependence of $\delta V^{\mathrm{II}}_{ijk}$ can be derived if we first express ϕ in terms of the displacements ζ_i, ζ_j, ζ_k, $\zeta_{k'}$ of the vertices of Δ_{ijk} and $\Delta_{ijk'}$. From fig. 3, we see that

$$\cos(\pi - \phi) = \frac{\boldsymbol{k'p} \cdot \boldsymbol{kp}}{|\boldsymbol{k'p}||\boldsymbol{kp}|}, \tag{5.14}$$

where

$$\begin{aligned} \boldsymbol{kp} &= -\boldsymbol{h} + \boldsymbol{\zeta}_k - \tfrac{1}{2}(\boldsymbol{\zeta}_i + \boldsymbol{\zeta}_j)\,, \\ \boldsymbol{k'p} &= +\boldsymbol{h} + \boldsymbol{\zeta}_{k'} - \tfrac{1}{2}(\boldsymbol{\zeta}_i + \boldsymbol{\zeta}_j)\,, \end{aligned} \tag{5.15}$$

with $\boldsymbol{h}$ the vector connecting the midpoint p of $\overline{ij}$ with k' for the saddle-point configuration. If (5.15) is substituted into (5.14), one obtains

$$\phi^2 = \frac{1}{h^2}(\boldsymbol{\zeta}_k + \boldsymbol{\zeta}_{k'} - \boldsymbol{\zeta}_i - \boldsymbol{\zeta}_j)^2\,, \tag{5.16}$$

where $h = \frac{1}{2}\sqrt{3}\ \bar{l}$. Next, we replace the function $\boldsymbol{\zeta}_{l,m}$ in (5.16) with a single Fourier amplitude $\tilde{\boldsymbol{\zeta}}(q_1, q_2)\exp(i\boldsymbol{q}\cdot\boldsymbol{r}_{l,m})$ from (5.13). Let us choose the line $\overline{ij}$ to be along the y-axis and parallel to $\hat{\beta}$ in (5.2); ϕ^2 and consequently also $\delta V^{\mathrm{II}}_{ijk}$ are proportional to

$$[\mathrm{e}^{i(q_\beta - q_\alpha)} + \mathrm{e}^{iq_\alpha} - \mathrm{e}^{iq_\beta} - 1]^2 = \mathrm{O}(\boldsymbol{q}^4)\,, \tag{5.17}$$

where $q_\beta = \boldsymbol{q}\cdot\hat{\beta} = q_2$ and $q_\alpha = \boldsymbol{q}\cdot\hat{\alpha} = \frac{1}{2}\sqrt{3}q_1 + \frac{1}{2}q_2$. Thus, in the long-wavelength limit (5.12) is valid. In terms of the Fourier components of $\boldsymbol{\zeta}$, it becomes, when the area A of the Wilson loop is large,

$$\rho\delta V_{\min} = \int \frac{T\,\mathrm{d}^2 q}{2(2\pi)^2}|\tilde{\boldsymbol{\zeta}}(q_1, q_2)|^2[q_1^2 + q_2^2]\,. \tag{5.18}$$

Substituting this expression into (5.5) and recalling that $\boldsymbol{\zeta}$ is a $(D-2)$-dimensional vector, we obtain

$$d^2 = (D-2)\int \frac{\mathrm{d}^2 q}{(2\pi)^2}\frac{1}{Tq^2} = \frac{D-2}{4\pi T}\ln A\,, \tag{5.19}$$

since the lower limit of the integral is $q^2 \sim A^{-1}$. When $D = 4$, this becomes (5.1).

We thank Professor J.M. Luttinger for a discussion of the quenched and annealed averages as used in condensed matter physics and Professor A.H. Mueller for a very helpful explanation of the physics of the relativistic string.

Appendix

To pass from (4.34) to (4.37), we recall from the text that each triangle (indexed by $\alpha = 1, \ldots, n$) is associated with m_α additional sites lying on the smallest empty sphere that passes through the three vertices of the triangle. We shall first ignore the fact that neighboring spheres may overlap. In that case the integral in (4.34) yields a product of independent factors similar to (4.29), one for each triangle. The result is of the form (4.37), provided that we can show that the integral over m sites associated with a triangle Δ of area A_Δ (where $\rho A_\Delta^{D/2} \gg 1$) yields a result proportional to $A_\Delta^{mD/4}$. This was done in the text for the case $D = 3$, in which m can take only the values 0 and 1. For $m = 1$, the extra factor is given by (4.24):

$$\pi(\tfrac{1}{2}\rho r^3)^{1/2} \sim A_\Delta^{3/4} = A_\Delta^{mD/4}\,, \tag{A.1}$$

since $A_\Delta \sim r^2$.

To show how a similar result arises for higher D, we shall first consider a simpler problem, in which the triangle Δ_{ijk} is replaced by a link l_{ij} of length $2r$. The number m now ranges from 0 to $D-1$. For $D=2$, the reasoning parallels the derivation of (4.22) and (4.24) in the text, and gives

$$P_0(i,j)=e^{-\pi r^2\rho}, \tag{A.2}$$

$$\begin{aligned} P_1(i,j) &\cong 2\,e^{-\pi r^2}\int_0^\infty e^{-\pi\rho x^2}2r\rho\,dx \\ &= 2(\rho r^2)^{1/2}\,e^{-\pi r^2\rho}. \end{aligned} \tag{A.3}$$

For $D=3$, there are three possibilities, $m=0, 1, 2$. To see how they arise, imagine that we have found an empty sphere passing through i and j. We move the center of the sphere toward the midpoint of $\overline{ij}$, shrinking the radius so that the sphere always passes through i and j. If we can reach the midpoint with the sphere still empty, then $m=0$. If not, we must have encountered a third site p_1 that prevented us from moving the sphere further in the desired direction. Now the sites i, j, p_1 lie on a small circle of the sphere. We move the center of the sphere toward the center of this circle, shrinking the radius so that the sphere always passes through i, j and p_1. If we can reach our goal, then $m=1$. If not, we encountered a fourth point p_2. Once the sphere passes through i, j, p_1, p_2 it cannot be moved any more, and $m=2$.

For the case $m=0$ the probability is given by an argument like that leading to (A.2):

$$P_0=e^{-4\pi r^3\rho/3}. \tag{A.4}$$

To find P_1, we let x denote the displacement of the center of the sphere (in any direction normal to l_{ij}) from the midpoint of l_{ij}. Then (4.23) holds, but without the factor 2, and with a different meaning for $d\tau(x)$. For a fixed x, the site p_1 lies on the surface of revolution generated by the smaller arc through i and j with radius $\sqrt{r^2+x^2}$. As x increases by dx, this surface sweeps out a volume

$$\begin{aligned} d\tau(x) &= \int_0^\pi (2\pi r\sin\theta) r\sin\theta\,d\theta\cdot dx \\ &= \pi^2 r^2\,dx, \end{aligned} \tag{A.5}$$

where $r\sin\theta$ is the perpendicular distance from p to $\overline{ij}$ in the approximation $x \ll r$, as explained before (4.24). Therefore

$$\begin{aligned} P_1(i,j) &\cong e^{-4\pi r^3\rho/3}\int_0^\infty e^{-2\pi\rho r x^2}\pi^2 r^2\rho\,dx \\ &= \tfrac{1}{2}\pi^2(\tfrac{1}{2}\rho r^3)^{1/2}\,e^{-4\pi r^3\rho/3}. \end{aligned} \tag{A.6}$$

Comparing (A.6) with (A.3), we see that in each case we obtained a factor $(\rho r^D)^{1/2} = \rho r^{Dm/2}$ for $m = 1$.

To evaluate P_2, we remember that i, j, p_1 lie on a small circle of the sphere passing through i, j, p_1, p_2. We denote by x_1 the displacement of the center of this circle from the midpoint of l_{ij}, and by x_2 the displacement of the center of the sphere from the center of the circle. The integral over x_2 is exactly like the one evaluated in (4.24), with x replaced by x_2, r by $\sqrt{r^2 + x_1^2}$, and p_1 playing the role of k. Thus

$$P_2(i, j) = \tfrac{1}{2} \int_0^{\infty} F(x_1)\, d\tau(x_1)\,, \tag{A.7}$$

where

$$\begin{aligned} F(x_1) &\cong 2\, e^{-\pi(r^2+x_1^2)^{3/2}\rho/3} \int_0^{\infty} e^{-2\pi\rho\sqrt{r^2+x_1^2}x_2^2}\pi(r^2 + x_1^2)\rho\, dx_2 \\ &= \pi[\tfrac{1}{2}\rho(r^2 + x_1^2)^{3/2}]^{1/2}\, e^{-4\pi(r^2+x_1^2)^{3/2}\rho/3}\,, \end{aligned} \tag{A.8}$$

and $d\tau(x_1)$ is given by exactly the same reasoning that led to (A.5). The factor $\frac{1}{2}$ in (A.7) compensates for overcounting $p_1 \leftrightarrow p_2$; it corresponds to the factor $1/m_\alpha!$ in (4.34).

Substituting (A.5) and (A.8) into (A.7), and making the approximation $x_1 \ll r$ because of the exponential in (A.8), we find

$$\begin{aligned} P_2(i, j) &\cong \tfrac{1}{2}\pi(\tfrac{1}{2}\rho r^3)^{1/2}\, e^{-4\pi r^3\rho/3} \int_0^{\infty} e^{-2\pi\rho x_1^2} \cdot \pi^2 r^2 \rho\, dx_1 \\ &= \tfrac{1}{2}\pi(\tfrac{1}{2}\rho r^3)^{1/2}\, e^{-4\pi r^3\rho/3} \tfrac{1}{2}\pi^2 (\tfrac{1}{2}\rho r^3)^{1/2} \\ &= \tfrac{1}{4}\pi^3(\tfrac{1}{2}\rho r^3)\, e^{-4\pi r^3\rho/3}\,. \end{aligned} \tag{A.9}$$

The significant factor is now $\rho r^3 = \rho r^D = \rho r^{Dm/2}$.

The preceding examples illustrate the pattern which holds for higher D as well. Each new site restricting the sphere brings in a factor $(\rho r^D)^{1/2}$. Therefore, in general we have

$$P_m(i, j) = b_m(\rho r^D)^{m/2}\, e^{-\rho v_{ij}}\,, \tag{A.10}$$

where b_m is O(1) and v_{ij} is the volume of the sphere of radius r, in D dimensions.

For a triangle Δ_{ijk} instead of a link l_{ij}, the calculations are similar, only everything is raised by one dimension. Again one obtains

$$P_m(i, j, k) = b'_m(\rho r^D)^{m/2}\, e^{-\rho v_{ijk}}\,, \tag{A.11}$$

where b'_m is O(1), r is the radius of the circle drawn through i, j, k, and v_{ijk} is the volume in D dimensions of the sphere of radius r. Eq. (A.11) is identical to (4.29) in the text. Since the area $A_{ijk} \sim r^2$, we can write (A.11) as

$$P_m(i, j, k) = c'_m (\rho A_{ijk}^{D/2})^{m/2} e^{-\rho v_{ijk}}, \tag{A.12}$$

where c'_m depends on m, D and the shape of Δ_{ijk}, but is O(1).

To pass from (A.12) to (4.37), we must consider the overlap of neighboring spheres. The main effect of this overlap is that the final formula contains not $e^{-\rho \Sigma v_{ijk}}$ but $e^{-\rho V_{min}}$ where V_{min} is the volume of the union of spheres as discussed in the text. In addition, the factors c'_m are replaced by smaller factors c_m, also of O(1), since the extra m_α sites for Δ_α are not allowed to penetrate neighboring spheres.

There is also the possibility that one additional site may do double duty, lying on the intersection of two neighboring spheres. Letting $\tilde{x}$ stand for a typical displacement and $\tilde{r}$ for a typical radius, such a point would have an available volume $\sim \tilde{r}^{D-2}\tilde{x}^2$ (as compared to a single site on one sphere which has available volume $\sim \tilde{r}^{D-1}x$). Since $\rho\tilde{r}^{D-2}\tilde{x}^2 \sim 1$, the order of magnitude of the corresponding term of (4.37) is the same as if the double-duty site were absent. (Triple-duty sites actually reduce the order of magnitude.) The effect is that the leading term of (4.37), $m_\alpha = D-2$ for all α, is unaffected by multiple-duty sites, although the lower-order terms are not quite correctly represented by the product form given there. In any case, the estimate (4.18) is established.

References

[1] N.H. Christ, R. Friedberg and T.D. Lee, Nucl. Phys. B202 (1982) 89
[2] K. Wilson, Phys. Rev. D10 (1974) 2455
[3] M. Luscher, Nucl. Phys. B180 [FS2] (1981) 1
[4] A. Hasenfratz, E. Hasenfratz and P. Hasenfratz, Nucl. Phys. B180 [FS2] (1981) 353
[5] C. Itzykson, M.E. Peskin and J.B. Zuber, Phys. Lett. 95B (1980) 259
[6] J.B. Kogut, D.R. Sinclair, R.B. Pearson, I.L. Richardson and J. Shigemitsu, Phys. Rev. D23 (1981) 2945
[7] Y. Nambu, Proc. Conf. on Symmetries and quark models, Detroit, 1969 (Gordon and Breach, New York, 1970), p. 269
[8] C. Rebbi, Phys. Reports 12 (1974) 1

Nuclear Physics B210[FS6] (1982) 337–346

WEIGHTS OF LINKS AND PLAQUETTES IN A RANDOM LATTICE*

N.H. CHRIST, R. FRIEDBERG and T.D. LEE

Columbia University, New York, N.Y. 10027, USA

Received 7 June 1982

In a random lattice, links and plaquettes have varying sizes and orientations. Therefore, they should enter the action function with different weights. A particularly attractive choice of these weights is presented; its consequences for spin-0, $\frac{1}{2}$ and 1 fields on such a lattice are discussed.

1. Introduction

In two preceding papers** [1, 2], we gave the construction of a random lattice and a discussion of the spin-0 and spin-1 fields on such a lattice. The usual gradient or curl operator in a continuum theory is replaced by the appropriate difference of operators on linked points or their product around a plaquette. These links and plaquettes have varying sizes and orientations, and can carry different weights in the lattice action. A single constraint is placed on these weights by the requirement that the lattice action should approach the continuum limit when the site density approaches infinity. In I and II, the weights are left arbitrary except for this constraint. In this paper, we focus on a particularly natural choice of these weights which, as we shall see, has several very attractive consequences. With this choice of weights, the spin-0 and spin-1 field equations are solved exactly by potentials which depend linearly on the position of the lattice sites (provided the weak coupling limit is taken for the spin-1 case). Likewise, the action for such fields can be evaluated analytically.

2. Spin-0 field

In the notation of I, the lattice action for a zero-mass free field is

$$A_{\mathrm{L}}(\Omega)=\tfrac{1}{4}\sum_{i,j}\lambda_{ij}(\phi_i-\phi_j)^2\,, \tag{1}$$

where i and j denote the random lattice sites 1, 2, ..., $\mathcal{N}$ and λ_{ij} are the weights:

$$\lambda_{ij}=0\,,\qquad \text{if } i \text{ and } j \text{ are not linked}\,. \tag{2}$$

* This research was supported in part by the US Department of Energy.

** Hereafter, they will be referred to as I and II.

The whole system is contained in a D-dimensional euclidean volume Ω. We recall that to each site i in the random lattice there corresponds a convex D-dimensional polyhedron ω_i in the dual lattice. To each link l_{ij} connecting sites i and j there corresponds a $(D-1)$-dimensional polyhedral surface s_{ij} of ω_i, and to each plaquette Δ_{ijk} with vertices at i, j and k there corresponds a $(D-2)$-dimensional polyhedral surface τ_{ijk} of s_{ij}, etc. We choose the weight λ_{ij} between any linked pair i and j to be

$$\lambda_{ij} = s_{ij}/l_{ij}\,, \tag{3}$$

where l_{ij} is the length of the link and s_{ij} the "volume" of the corresponding $(D-1)$-dimensional surface in the dual lattice. [If $D=2$, then s_{ij} denotes the border length in the dual lattice. See sect. 5 of I.]

Theorem 1.

$$\sum_j l_{ij}^{\mu}\lambda_{ij} = 0 \tag{4}$$

of all lattice sites i, and

$$L^{\mu\nu} \equiv \sum_{i,j} l_{ij}^{\mu} l_{ij}^{\nu}\lambda_{ij} = 2\Omega\delta^{\mu\nu}\,, \tag{5}$$

where λ_{ij} is given by (2) and (3), and l_{ij}^{μ} is the μ-component of $\boldsymbol{l}_{ij} = \boldsymbol{r}_j - \boldsymbol{r}_i$ with $\mu = 1, 2, \ldots, D$. Throughout this paper, the position vector of any point p is denoted by $\boldsymbol{r}_{\mathrm{p}}$.

Proof. Take any constant vector $\boldsymbol{V}$. From Gauss' theorem we have

$$0 = \int_{\omega_i} \boldsymbol{\nabla}\cdot\boldsymbol{V}\,\mathrm{d}^D r = \int \boldsymbol{V}\cdot \mathrm{d}\boldsymbol{S}\,. \tag{6}$$

Because the $(D-1)$-dimensional polyhedra s_{ij} (with i fixed but j varying) form the boundary of ω_i, and because $\hat{l}_{ij} \equiv \boldsymbol{l}_{ij}/l_{ij}$ is the normal vector of s_{ij}, (6) becomes

$$0 = \boldsymbol{V}\cdot\sum_j \hat{l}_{ij} s_{ij}$$

which implies the first part, (4), of the theorem.

To prove (5), we again apply Gauss' theorem, but to the following integral:

$$\omega_i\delta^{\mu\nu} = \int_{\omega_i} \frac{\partial}{\partial x^{\nu}}(\boldsymbol{r}-\boldsymbol{r}_i)^{\mu}\,\mathrm{d}^D r = \int (\boldsymbol{r}-\boldsymbol{r}_i)^{\mu}\,\mathrm{d}S^{\nu}$$

$$= \sum_j r_{ic}^{\mu}\hat{l}_{ij}^{\nu} s_{ij}\,, \tag{7}$$

where x^{ν} is the ν-component of $\boldsymbol{r}$ and r_{ic}^{μ} is the μ-component of $\boldsymbol{r}_{ic} \equiv \boldsymbol{r}_c - \boldsymbol{r}_i$. The vector $\boldsymbol{r}_c$ is given by

$$\boldsymbol{r}_c \equiv \frac{1}{s_{ij}}\int_{s_{ij}} \boldsymbol{r}\,\mathrm{d}^{D-1}r\,, \tag{8}$$

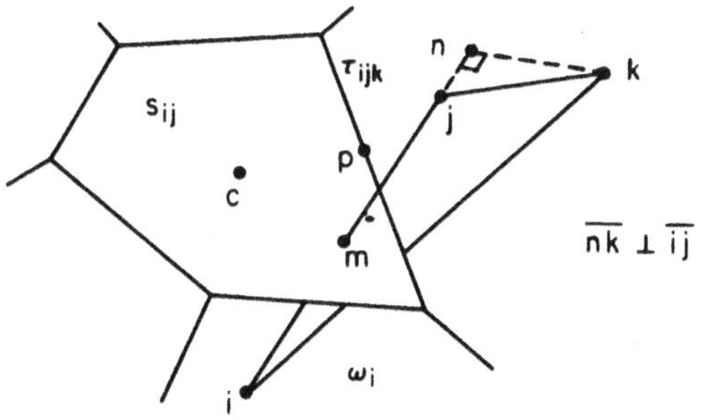

Fig. 1. The duals of i, l_{ij} and Δ_{ijk} are ω_i, s_{ij} and τ_{ijk}. The center of mass of s_{ij} is c, and midpoint of l_{ij} is m. Hence, $\boldsymbol{im} = \boldsymbol{mj} = \frac{1}{2}\boldsymbol{l}_{ij}$ and $\mathbf{mc} = \frac{1}{2}\boldsymbol{d}_{ij}$ where $\boldsymbol{d}_{ij}$ is defined by (10).

and is the position vector of the center of mass c of s_{ij}. (See fig. 1.) The link-vector l_{ij} is related to $\boldsymbol{r}_{ic}$ and $\boldsymbol{r}_{jc}$ by

$$\boldsymbol{l}_{ij} = \boldsymbol{r}_{ic} - \boldsymbol{r}_{jc}\,. \tag{9}$$

Define

$$\boldsymbol{d}_{ij} \equiv \boldsymbol{r}_{ic} + \boldsymbol{r}_{jc}\,. \tag{10}$$

Eq. (7) can then be written as

$$\omega_i \delta^{\mu\nu} = \tfrac{1}{2} \sum_j (\boldsymbol{l}_{ij} + \boldsymbol{d}_{ij})^\mu \hat{l}^\nu_{ij} s_{ij}\,.$$

Summing over all dual cells ω_i and noting that $\boldsymbol{d}_{ij}$ and s_{ij} are even under the exchange of i and j but $\boldsymbol{l}_{ij}$ and $\hat{l}_{ij}$ are odd, we derive

$$\Omega \delta^{\mu\nu} = \sum_i \omega_i \delta^{\mu\nu} = \tfrac{1}{2} \sum_{i,j} l^\mu_{ij} \hat{l}^\nu_{ij} s_{ij} \tag{11}$$

which establishes (5) and completes the proof of theorem 1.

By using the action (1), we see that the massless free field equation is

$$\sum_j \lambda_{ij} (\phi_i - \phi_j) = 0\,. \tag{12}$$

From theorem 1 it follows that a field ϕ_i which depends linearly on the site position $\boldsymbol{r}_i$ solves this field equation. Let

$$\phi_i = \phi_0 + \boldsymbol{V} \cdot \boldsymbol{r}_i\,, \tag{13}$$

where ϕ_0 and $\boldsymbol{V}$ are constants. Substituting (13) into (12) and applying (4), we find

$$\sum_j \lambda_{ij} (\phi_i - \phi_j) = -\boldsymbol{V} \cdot \sum_j \lambda_{ij} \boldsymbol{l}_{ij} = 0\,, \tag{14}$$

at all sites i. Furthermore, because of (5) the corresponding action is

$$A_{\mathrm{L}}(\Omega)=\tfrac{1}{4}\sum_{i,j}\lambda_{ij}(\boldsymbol{V}\cdot\boldsymbol{l}_{ij})^2=\tfrac{1}{2}\boldsymbol{V}^2\Omega\,. \tag{15}$$

Another consequence of (4) is that the bulk resistivity Z, given by eq. (6.46) of I, is unity; this is, of course, related to the statement that (13) is a solution of the massless free field equation (12).

For $D=2$, a stronger form of (15) exists (which, however, does not generalize to higher D or to spin 1). This is discussed in the appendix.

3. Spin-$\frac{1}{2}$ field

The continuum action in a D-dimensional euclidean volume Ω is a hermitian bilinear form given by*

$$A_{\mathrm{C}}(\Omega)=\int_{\Omega}\left(\psi^{\dagger}\gamma^0\gamma^{\mu}\frac{\partial}{\partial x^{\mu}}\psi+m\psi^{\dagger}\gamma^0\psi\right)\mathrm{d}^D r\,, \tag{16}$$

where the dagger denotes hermitian conjugation and $\gamma^0, \gamma^1, \dots, \gamma^D$ are anticommuting hermitian matrices whose squares are the unit matrix. The corresponding lattice action is

$$A_{\mathrm{L}}(\Omega)-\tfrac{1}{2}\sum_{i,j}\psi_i^{\dagger}\gamma^0\gamma^{\mu}l_{ij}^{\mu}\lambda_{ij}\psi_j+m\sum_{i}\omega_i\psi_i^{\dagger}\gamma^0\psi_i\,, \tag{17}$$

where λ_{ij} is given by (2) and (3) and ω_i is the volume of the D-dimensional polyhedron dual to i. Because of (4), it follows that

$$\psi_j=\text{a constant spinor} \tag{18}$$

satisfies the massless field equation:

$$\sum_{j}\gamma^{\mu}l_{ij}^{\mu}\lambda_{ij}\psi_j=0 \tag{19}$$

at all sites i.

Next, we consider the function

$$\psi_j=\chi\,\mathrm{e}^{i\boldsymbol{p}\cdot\boldsymbol{r}_j} \tag{20}$$

where χ is a constant spinor; the corresponding lattice action is

$$A_{\mathrm{L}}(\Omega)=\tfrac{1}{2}(\chi^{\dagger}\gamma^0\gamma^{\mu}\chi)\sum_{i,j}l_{ij}^{\mu}\lambda_{ij}\,\mathrm{e}^{i\boldsymbol{p}\cdot\boldsymbol{l}_{ij}}+m\chi^{\dagger}\gamma^0\chi\Omega\,.$$

* We exclude other hermitian choices such as $A_{\mathrm{C}}(\Omega)=\int_{\Omega}(i\psi^{\dagger}\gamma^{\mu}(\partial/\partial x^{\mu})\psi+m\psi^{\dagger}\psi)\,\mathrm{d}^D r$ by requiring the field equation to be $(\gamma^{\mu}(\partial/\partial x^{\mu})+m)\psi=0$. The index μ is summed from 1 to D.

For small $\boldsymbol{p}$, $e^{i\boldsymbol{p}\cdot\boldsymbol{l}_{ij}} = 1 + i\boldsymbol{p}\cdot\boldsymbol{l}_{ij} - \frac{1}{2}(\boldsymbol{p}\cdot\boldsymbol{l}_{ij})^2 + \cdots$. By using (5) and neglecting $O(\boldsymbol{p}^3)$, we see that

$$A_L(\Omega) = [i(\chi^\dagger\gamma^0\gamma^\mu\chi)p^\mu + m\chi^\dagger\gamma^0\chi]\Omega\,,$$

which has the same form as in the continuum case.

The lattice action (17) may be resolved in terms of its eigenvectors $\psi_i(n)$ and eigenvalues $E(n)$:

$$\tfrac{1}{2}\sum_j \gamma^\mu l^\mu_{ij}\lambda_{ij}\psi_j(n) + m\omega_i\psi_i(n) = E(n)\omega_i\gamma^0\psi_i(n)\,. \tag{21}$$

We write

$$A_L(\Omega) = \sum_{n,i} E(n)\omega_i\psi_i(n)^\dagger\psi_i(n)\,.$$

It is easy to verify that (20) is not a solution of the eigenvector equation (21). Hence, the low-lying spectrum $E(n)$ of the lattice action is more complicated than the continuum case; its structure will be analyzed in a later paper.

4. Spin-1 field

In the notation of II, the action of a lattice gauge theory is

$$A_L(\Omega) = \frac{1}{g^2}\sum_{\Delta_{ijk}} \kappa_{ijk} f(U_{ijk})\,, \tag{22}$$

where κ_{ijk} is the weight of the plaquette Δ_{ijk}, g is the coupling constant and in the weak field approximation

$$f(U_{ijk}) = \tfrac{1}{4}g^2(\Delta^{\mu\nu}_{ijk}F^{\mu\nu})(\Delta^{\rho\sigma}_{ijk}F^{\rho\sigma})\,, \tag{23}$$

where*

$$\Delta^{\mu\nu}_{ijk} = \tfrac{1}{6}[l^\mu_{ij}l^\nu_{jk} + l^\mu_{jk}l^\nu_{ki} + l^\mu_{ki}l^\nu_{ij} - (\mu \rightleftarrows \nu)]\,, \tag{24}$$

which makes

$$A_L(\Omega) = \tfrac{1}{4}\sum_{\Delta_{ijk}} \kappa_{ijk}\Delta^{\mu\nu}_{ijk}\Delta^{\rho\sigma}_{ijk}F^{\mu\nu}F^{\rho\sigma}\,, \tag{25}$$

where, as in (22), the sum extends over all plaquettes Δ_{ijk}. [See (2.17) of II.]

As mentioned earlier, each plaquette Δ_{ijk} in the random lattice has a dual which is a $(D-2)$-dimensional convex polyhedron τ_{ijk}. These τ_{ijk}, with i and j fixed but k varying, form the surface of the $(D-1)$-dimensional polyhedron s_{ij}, which in

* $\Delta^{\mu\nu}_{ijk}$, $F^{\mu\nu}$ here are the same tensors as $\Delta_{\mu\nu}(ijk)$ and $F^a_{\mu\nu}$ of II, but with the internal symmetry index a suppressed.

turn is the dual of the link l_{ij}. We choose the weight κ_{ijk} of the plaquette Δ_{ijk} to be

$$\kappa_{ijk} = \frac{\tau_{ijk}}{2\Delta_{ijk}} , \tag{26}$$

where Δ_{ijk} is the area of the plaquette and τ_{ijk} the $(D-2)$-dimensional volume of its dual.

Theorem 2.

$$\sum_k \kappa_{ijk}\Delta_{ijk}^{\mu\nu} = 0 \tag{27}$$

for all sites i and j that are linked, and

$$T^{\mu\nu\rho\sigma} \equiv \tfrac{1}{4} \sum_{\Delta_{ijk}} \kappa_{ijk}\Delta_{ijk}^{\mu\nu}\Delta_{ijk}^{\rho\sigma} = \tfrac{1}{8}(\delta^{\mu\rho}\delta^{\nu\sigma} - \delta^{\mu\sigma}\delta^{\nu\rho})\Omega . \tag{28}$$

Proof. Consider the triangle Δ_{ijk}. Choose the point n lying on $\overline{ij}$ (or its extension) so that $\overline{nk}$ is perpendicular to $\overline{ij}$, as shown in fig. 1. Let $\hat{r}_{nk}$ be the unit vector along $\boldsymbol{r}_{nk} \equiv \boldsymbol{r}_k - \boldsymbol{r}_n$. The tensor $\Delta_{ijk}^{\mu\nu}$, (24), can also be expressed in terms of the components of $\hat{l}_{ij}$ and $\hat{r}_{nk}$:

$$\Delta_{ijk}^{\mu\nu} = (\hat{l}_{ij}^{\mu}\hat{r}_{nk}^{\nu} - \hat{l}_{ij}^{\nu}\hat{r}_{nk}^{\mu})\Delta_{ijk} . \tag{29}$$

To derive (27), take any constant vector $\boldsymbol{V}$ lying in the $(D-1)$-dimensional space that contains s_{ij}. Similarly to (6), we apply Gauss' theorem to the volume integral of the derivative of the μ-component of $\boldsymbol{V}$ over s_{ij}:

$$0 = \int_{s_{ij}} \frac{\partial}{\partial x^{\nu}} V^{\mu}\, \mathrm{d}^{D-1}r = \sum_k \int_{\tau_{ijk}} \hat{e}_{ijk} V^{\mu}\, \mathrm{d}^{D-2}r , \tag{30}$$

where as before x^{ν} is the ν-component of $\boldsymbol{r}$; $\hat{e}_{ijk}$ denotes the unit vector normal to τ_{ijk} and lying in s_{ij}, the dual of l_{ij}. Hence $\hat{e}_{ijk}$ is also normal to $\boldsymbol{l}_{ij}$. Recall that $\hat{r}_{nk} \perp \boldsymbol{l}_{ij}$ and is also normal to τ_{ijk} (since $\hat{r}_{nk}$ is on Δ_{ijk} which is normal to its dual τ_{ijk}). Thus, $\hat{e}_{ijk} = \hat{r}_{nk}$ and (30) becomes

$$0 = \sum_k \hat{r}_{nk}^{\nu}\tau_{ijk}V^{\mu} . \tag{31}$$

Therefore we have

$$\sum_k \tau_{ijk}\hat{r}_{nk}^{\nu} = 0 , \tag{32}$$

which, when multiplied by $\hat{l}_{ij}^{\mu}$, gives

$$\sum_k \tau_{ijk}\hat{l}_{ij}^{\mu}\hat{r}_{nk}^{\nu} = 0 . \tag{33}$$

Combining (26) and (29) with (33), we see that (27) holds.

Next, we turn to $T^{\mu\nu\rho\sigma}$, defined by (28). We may write

$$T^{\mu\nu\rho\sigma} = \tfrac{1}{48} \sum_{i,j,k} \frac{1}{\Delta_{ijk}} \Delta_{ijk}^{\mu\nu} \Delta_{ijk}^{\rho\sigma} \tau_{ijk}$$

$$= \tfrac{1}{48} \sum_{i,j,k} \frac{1}{\Delta_{ijk}} \Delta_{ijk}^{\mu\nu} \int_{\tau_{ijk}} \Delta_{ijk}^{\rho\sigma} \, \mathrm{d}^{D-2} r_{\mathrm{p}} \,, \tag{34}$$

where p is a point inside τ_{ijk} as shown in fig. 1, and the sum extends over all i, j, k which counts each Δ_{ijk} 3! times. Draw lines $\overline{i\mathrm{p}}$, $\overline{j\mathrm{p}}$ and $\overline{k\mathrm{p}}$. By considering the line integrals $\oint (x^\rho \, \mathrm{d}x^\sigma - x^\sigma \, \mathrm{d}x^\rho)$ around the triangles ijk, ijp, jkp and kip, we have

$$\Delta_{ijk}^{\rho\sigma} = \Delta_{ij\mathrm{p}}^{\rho\sigma} + \Delta_{jk\mathrm{p}}^{\rho\sigma} + \Delta_{ki\mathrm{p}}^{\rho\sigma} \,. \tag{35}$$

Because $\Delta_{ijk}^{\mu\nu}$ is symmetric under a cyclic permutation of ijk, the three terms on the right-hand side of (35) give equal contributions to (34), which can then be written as

$$T^{\mu\nu\rho\sigma} = \tfrac{1}{16} \sum_{i,j,k} \frac{1}{\Delta_{ijk}} \Delta_{ijk}^{\mu\nu} \int_{\tau_{ijk}} \Delta_{ij\mathrm{p}}^{\rho\sigma} \, \mathrm{d}^{D-2} r_{\mathrm{p}} \,. \tag{36}$$

Let m be the midpoint of l_{ij} and $\boldsymbol{r}_{\mathrm{mp}} = \boldsymbol{r}_{\mathrm{p}} - \boldsymbol{r}_{\mathrm{m}}$. Because s_{ij} is the dual of l_{ij}, the $(D-1)$-dimensional space that contains s_{ij} bisects l_{ij}; therefore m is inside that space and $\boldsymbol{r}_{\mathrm{mp}} \perp \boldsymbol{l}_{ij}$. In terms of the components of $\boldsymbol{l}_{ij}$ and $\boldsymbol{r}_{\mathrm{mp}}$, we write

$$\Delta_{ij\mathrm{p}}^{\rho\sigma} = \tfrac{1}{2}(l_{ij}^{\rho} r_{\mathrm{mp}}^{\sigma} - l_{ij}^{\sigma} r_{\mathrm{mp}}^{\rho}) \,. \tag{37}$$

Substituting this expression and (29) into (36), we have

$$T^{\mu\nu\rho\sigma} = \tfrac{1}{32} \sum_{i,j,k} (\hat{l}_{ij}^{\mu} \hat{r}_{nk}^{\nu} - \hat{l}_{ij}^{\nu} \hat{r}_{nk}^{\mu}) \int_{\tau_{ijk}} (l_{ij}^{\rho} r_{\mathrm{mp}}^{\sigma} - l_{ij}^{\sigma} r_{\mathrm{mp}}^{\rho}) \, \mathrm{d}^{D-2} r_{\mathrm{p}}$$

$$= t^{\mu\nu\rho\sigma} - t^{\nu\mu\rho\sigma} - t^{\mu\nu\sigma\rho} + t^{\nu\mu\sigma\rho} \,, \tag{38}$$

where

$$t^{\mu\nu\rho\sigma} = \tfrac{1}{32} \sum_{i,j} \hat{l}_{ij}^{\mu} l_{ij}^{\rho} \sum_{k} \int_{\tau_{ijk}} \hat{r}_{nk}^{\nu} r_{\mathrm{mp}}^{\sigma} \, \mathrm{d}^{D-2} r_{\mathrm{p}} \,. \tag{39}$$

Apply Gauss' theorem to the following integral over the $(D-1)$-dimensional polyhedron s_{ij}. Remembering that the surface of s_{ij} is τ_{ijk} and $\hat{r}_{nk}$ is the normal to τ_{ijk} that lies in s_{ij}, we obtain

$$\delta^{\nu\sigma} s_{ij} = \int_{\mathrm{s}_{ij}} \frac{\partial}{\partial x^{\nu}} (\boldsymbol{r} - \boldsymbol{r}_{\mathrm{m}})^{\sigma} \, \mathrm{d}^{D-1} r = \sum_{k} \int_{\tau_{ijk}} \hat{r}_{nk}^{\nu} r_{\mathrm{mp}}^{\sigma} \, \mathrm{d}^{D-2} r_{\mathrm{p}} \,, \tag{40}$$

which, when substituted into (39), leads to

$$t^{\mu\nu\rho\sigma} = \tfrac{1}{32} \sum_{i,j} \hat{l}_{ij}^{\mu} l_{ij}^{\rho} s_{ij} \delta^{\nu\sigma}$$

$$= \tfrac{1}{16} \Omega \delta^{\mu\rho} \delta^{\nu\sigma} \,, \tag{41}$$

where the last equality follows from (5). Combining (41) with (38), we establish (28). Theorem 2 is then proved.

Consider now the case when the field $F^{\mu\nu}$ is a constant; i.e., in (23), it is independent of the plaquette Δ_{ijk}. On account of (28), the action $A_L(\Omega)$ in the weak field approximation (25) becomes simply

$$A_L(\Omega) = \tfrac{1}{4}F^{\mu\nu}F^{\mu\nu}\Omega\,. \tag{42}$$

In addition, because of (27), the corresponding field equation

$$\sum_k \kappa_{ijk}\Delta^{\mu\nu}_{ijk}F^{\mu\nu} = 0 \tag{43}$$

is valid for any linked pair of sites i and j.

We thank Mr. H.-C. Ren for his considerable help in recognizing the properties of this definition of the action for a scalar field on the random lattice.

Appendix

We present here a special result for the spin-zero field in two dimensions. Consider any triangle Δ_{ijk}. The border* b_{ij} (dual to the link l_{ij}) can be cut into an internal and an external part:

$$b_{ij} = b_{ij}^{\text{int}\,(ijk)} + b_{ij}^{\text{ext}\,(ijk)}\,. \tag{A.1}$$

If the border does not cut the link, (A.1) still holds with one of the terms on the right negative.

Since λ_{ij} is proportional to b_{ij}, it can be similarly decomposed:

$$\lambda_{ij} = \lambda_{ij}^{\text{int}\,(ijk)} + \lambda_{ij}^{\text{ext}\,(ijk)}\,. \tag{A.2}$$

Then the action (1) can be written

$$A_L(\Omega) = \tfrac{1}{6}\sum_{i,j,k} A_L(ijk)\,, \tag{A.3}$$

where

$$\begin{aligned} A_L(ijk) &= \tfrac{1}{2}\lambda_{ij}^{\text{int}\,(ijk)}(\phi_i - \phi_j)^2 + \text{cyclic permutations} \\ &= \tfrac{1}{2}\frac{b_{ij}^{\text{int}\,(ijk)}}{l_{ij}}(\phi_i - \phi_j)^2 + \text{cyclic permutations} \end{aligned} \tag{A.4}$$

and the sum in (A.3) is over triples i, j, k forming a plaquette.

The foregoing can be generalized to any dimension, but the following theorem holds only for $D = 2$.

* The s_{ij} in the text becomes b_{ij} for $D = 2$.

Theorem 3. Suppose sites i, j, k form a plaquette. Let ϕ_i, ϕ_j, ϕ_k be given any values whatever, and let $\boldsymbol{E}$ be the unique vector satisfying

$$\boldsymbol{E}\cdot\boldsymbol{r}_i-\phi_i=\boldsymbol{E}\cdot\boldsymbol{r}_j-\phi_j=\boldsymbol{E}\cdot\boldsymbol{r}_k-\phi_k\,. \tag{A.5}$$

Then

$$A_{\mathrm{L}}(ijk)=\tfrac{1}{2}E^2\Delta_{ijk}\,, \tag{A.6}$$

where Δ_{ijk} is the area of the plaquette.

Proof. Without loss of generality we may take the circumcenter of the triangle as the origin of coordinates. Moreover, we may rotate the coordinate system so that $\boldsymbol{E}$ points in the y-direction. Then (writing β_{ij} for $b_{ij}^{\mathrm{int}\,(ijk)}$)

$$\begin{aligned} A_{\mathrm{L}}&=\frac{1}{2}\frac{\beta_{ij}}{l_{ij}}(l_{ij}^y)^2E^2+\text{cyclic permutations}\\ &=\tfrac{1}{2}\beta_{ij}^x l_{ij}^y E^2+\text{cyclic permutations}\,. \end{aligned} \tag{A.7}$$

Here we regard $\boldsymbol{\beta}_{ij}$ as a vector pointing from the circumcenter to the midpoint of l_{ij}, and assume that the vertices i, j, k appear in counterclockwise order.

Since the origin is at the circumcenter, we have

$$\boldsymbol{\beta}_{ij}=\tfrac{1}{2}(\boldsymbol{r}_j+\boldsymbol{r}_i)\,, \tag{A.8}$$

and so

$$\begin{aligned} A_{\mathrm{L}}&=\tfrac{1}{4}(r_j^x+r_i^x)(r_j^y-r_i^y)E^2+\text{cyclic permutations}\\ &=\tfrac{1}{4}E^2(r_j^xr_j^y-r_i^xr_i^y+r_i^xr_j^y-r_j^xr_i^y)+\text{cyclic permutations}\\ &=\tfrac{1}{4}E^2(r_i^xr_j^y-r_j^xr_i^y)+\text{cyclic permutations}\,, \end{aligned} \tag{A.9}$$

since the first two terms of the second line cancel in the sum over permutations. But $\frac{1}{2}(r_i^xr_j^y-r_j^xr_i^y)$ is the area of the triangle made by i, j and the circumcenter; therefore (A.9) is identical to (A.6). The same result is obtained if the circumcenter lies outside the triangle, when due allowance is made for negative signs in the intermediate steps.

In three dimensions we have the following counter-example: let ijk form an equilateral triangle, with $\phi_i=\phi_j=\phi_k$. Let p be equidistant from i, j, k, but a distance h from their plane. Let $\phi_{\mathrm{p}}-\phi_i=Eh$. Let s be the distance from p to the midpoint of l_{ij}, and $s\gamma$ the distance from p to the circumcenter of i, j, p. Then the analog to (A.4) gives

$$A_{\mathrm{L}}(ijkl)=\tfrac{1}{4}\gamma h\Delta_{ijk}E^2\,, \tag{A.10}$$

whereas the volume of the tetrahedron is

$$\Omega_{ijkl}=\tfrac{1}{3}h\Delta_{ijk}\,. \tag{A.11}$$

The analog to (A.6) holds only if $\gamma = \frac{2}{3}$, which makes the circumcenter of i, j, l coincide with the intersection of medians. This happens only if the tetrahedron is regular.

References

[1] N.H. Christ, R. Friedberg and T.D. Lee, Nucl. Phys. B202 (1982) 89
[2] N.H. Christ, R. Friedberg and T.D. Lee, Nucl. Phys. B210[FS6] (1982) 310

CAN TIME BE A DISCRETE DYNAMICAL VARIABLE? ☆

T.D. LEE
Columbia University, New York, NY 10027, USA

Received 19 November 1982

The possibility that time can be regarded as a discrete dynamical variable is examined through all phases of mechanics: from classical mechanics to nonrelativistic quantum mechanics, and to relativistic quantum field theories.

1. Introduction. Throughout the development of physics, "time" always appears as a continuous parameter. Take the example of a nonrelativistic particle in quantum mechanics. In Feynman's path integration formulation, the probability amplitude for the particle to be at the position $\boldsymbol{r}_0$ at an initial time t_0 and at $\boldsymbol{r}_f$ at a final time t_f is given by the amplitude sum over all paths $\boldsymbol{r}(t)$ connecting $\boldsymbol{r}(t_0) = \boldsymbol{r}_0$ and $\boldsymbol{r}(t_f) = \boldsymbol{r}_f$. Apart from a normalization constant, it is equal to

$$\int \prod_t \mathrm{d}^3 r(t) \exp(\mathrm{i}A_\mathrm{c}), \qquad (1)$$

where the action A_c is a functional of $\boldsymbol{r}(t)$, related to the lagrangian L by

$$A_\mathrm{c} = \int_{t_0}^{t_f} L(\boldsymbol{r}(t), \dot{\boldsymbol{r}}(t))\, \mathrm{d}t, \qquad (2)$$

with the subscript c denoting the familiar case that t is continuous. We note that the position of the particle $\boldsymbol{r}$ is not treated on the same basis as the time t. At a given time the integration $\mathrm{d}^3 r(t)$ in eq. (1) can be viewed as that over the whole range of eigenvalues of the operator $\boldsymbol{r}_{\mathrm{op}}(t)$. This then underlies the familiar difference between $\boldsymbol{r}$ as an operator and t as a parameter.

Actually this asymmetry can be traced back to classical mechanics. The classical trajectory of a particle is determined by the extremity of the action A_c, which is a functional of $\boldsymbol{r}(t)$. While $\boldsymbol{r}$ is the dynamical variable, t appears only as a continuous parameter. By setting the variational derivative $\delta A_\mathrm{c}/\delta \boldsymbol{r}(t) = 0$, we obtain the usual Lagrange equation of motion, whose solution gives the classical path $\boldsymbol{r}(t)$.

In the relativistic quantum field theory, space $\boldsymbol{r}$ and time t have to be treated symmetrically due to Lorentz invariance. Our usual approach is to regard $\boldsymbol{r}$ and t all as parameters; the operators are now the field variables, say the scalar field $\phi(\boldsymbol{r},t)$. Expression (1) is then replaced by

$$\int \prod_{\boldsymbol{r}} \prod_t \mathrm{d}\phi(\boldsymbol{r}, t) \exp(\mathrm{i}A_\mathrm{c}), \qquad (3)$$

where the action A_c is related to the lagrangian density $\mathcal{L}$ by

$$A_\mathrm{c} = \int \mathcal{L}(\phi, \nabla\phi, \dot{\phi})\, \mathrm{d}^3 r\, \mathrm{d}t. \qquad (4)$$

Again, the integration over $\mathrm{d}\phi(\boldsymbol{r},t)$ can be viewed as that over the eigenvalues of the operator $\phi_{\mathrm{op}}(\boldsymbol{r},t)$.

In the following we wish to explore some alternative possibilities. In place of treating time as a continuous parameter, we may ask:

(1) Can time be a discrete parameter (discrete time formulation)?

(2) Can time be discrete and treated as a bona fide dynamical variable (discrete mechanics)?

As we shall see, both possibilities can be realized in all stages of mechanics. In a relativistic field theory, the discrete time formulation becomes the discrete space–

☆ This research was supported in part by the US Department of Energy.

0 031-9163/83/0000–0000/\$ 03.00 © 1983 North-Holland

time formulation. In the corresponding discrete mechanics both space and time, in addition to being discrete, are treated as genuine dynamical variables. Because this approach involves a fundamental change in our concept of space–time, we will develop our ideas gradually: first in classical mechanics, then in nonrelativistic quantum mechanics and finally in relativistic quantum field theory. The result is that in this new formalism our usual idea of continuous time or continuous space–time structure will appear only as an approximation.

2. Classical mechanics. Consider the simple example of a nonrelativistic particle of unit mass moving in a potential $V(\boldsymbol{r})$. The usual lagrangian is (regarding time as continuous)

$$L = \tfrac{1}{2}\dot{\boldsymbol{r}}^2 - V(\boldsymbol{r}). \tag{5}$$

The action is then given by the integral (2) from the initial position $\boldsymbol{r}(t_0) = \boldsymbol{r}_0$ to the final $\boldsymbol{r}(t_f) = \boldsymbol{r}_f$.

In the discrete time formalism we replace the continuous function $\boldsymbol{r}(t)$ by a sequence of discrete values: $(\boldsymbol{r}_0, t_0), (\boldsymbol{r}_1, t_1), ..., (\boldsymbol{r}_{N+1}, t_{N+1})$, where $(\boldsymbol{r}_{N+1}, t_{N+1}) = (\boldsymbol{r}_f, t_f)$. The action (2) is then replaced by

$$A = \sum_{n=1}^{N+1} \left(\frac{1}{2} \frac{(\boldsymbol{r}_n - \boldsymbol{r}_{n-1})^2}{t_n - t_{n-1}} - \tfrac{1}{2}(t_n - t_{n-1})[V(\boldsymbol{r}_n) + V(\boldsymbol{r}_{n-1})] \right), \tag{6}$$

with $t_n > t_{n-1}$. Newton's equation of motion is derived by setting the derivative

$$\partial A / \partial \boldsymbol{r}_n = 0. \tag{7}$$

Keeping the initial and final configurations fixed, we have altogether N such equations:

$$\boldsymbol{v}_{n+1} - \boldsymbol{v}_n = -\tfrac{1}{2}(t_{n+1} - t_{n-1}) \nabla V(\boldsymbol{r}_n) \tag{8}$$

where $\boldsymbol{v}_n = (\boldsymbol{r}_n - \boldsymbol{r}_{n-1})/(t_n - t_{n-1})$ is the velocity. For any given time distribution $t_1, ..., t_N$, the positions, $\boldsymbol{r}_1, ..., \boldsymbol{r}_N$ can then be determined.

An ad hoc discrete time distribution $t_1, t_2, ..., t_N$ is clearly not satisfactory for a closed mechanical system. Thus, in discrete mechanics we require that the time distribution should also be determined by the same action (6). This can be achieved by treating t_n as dynamical variables, on the same basis as $\boldsymbol{r}_n$. By setting

$$\partial A / \partial t_n = 0, \tag{9}$$

we derive N additional equations:

$$E_n \equiv \tfrac{1}{2}\boldsymbol{v}_n^2 + \tfrac{1}{2}[V(\boldsymbol{r}_n) + V(\boldsymbol{r}_{n-1})] = E_{n+1}. \tag{10}$$

Thus, the entire set $(\boldsymbol{r}_1, t_1), ..., (\boldsymbol{r}_N, t_N)$ can be determined from (8) and (10).

In the continuum case, the energy conservation law is a consequence of Newton's equation of motion. In the discrete case, these two are independent. By treating t_n and $\boldsymbol{r}_n$ both as dynamical variables, we manage to employ the same action principle for their determination. In our formulation of discrete mechanics, there is a *fundamental length* or time l (in natural units). Given any time interval $T = t_f - t_0$, the total number N of discrete points that define the trajectory is given by the integer nearest T/l.

As illustrations, we may consider some special examples:

(i) $V = 0$. In this case (8) gives $\boldsymbol{v}_n$ = constant, which also satisfies (10). The trajectory of the particle is always a straight line, independent of the time distribution.

(ii) ∇V = constant. It can be readily verified that (8) and (10) yield equal spacings for the time intervals. The trajectory consists of N discrete points $(\boldsymbol{r}_1, t_1), ..., (\boldsymbol{r}_N, t_N)$ all lying on a parabola, similar to the continuum case. (For more complicated potentials, the time intervals are in general of varying lengths.)

3. Nonrelativistic quantum mechanics. As above, let the initial and final configurations be $(\boldsymbol{r}_0, t_0)$ and $(\boldsymbol{r}_f, t_f)$. In the discrete quantum mechanics, we assume that within the given time interval $T = t_f - t_0$, there can be maximally N measurements of the particle's configurations possible, where the ratio $N/T = l$ is the fundamental constant mentioned before. Each measurement yields a set of values $(\boldsymbol{r}_n, t_n)$ where $n = 1,2,...,N$. [The initial $(\boldsymbol{r}_0, t_0)$ can be regarded as the result of a previous experiment, and the final $(\boldsymbol{r}_f, t_f) = (\boldsymbol{r}_{N+1}, t_{N+1})$ as that of the $(N + 1)$th measurement.] Without any loss of generality, we may arrange $t_1, ..., t_N$ into an ordered sequence: $t_0 < t_1 < ... < t_N < t_f$. The probability amplitude of observing such a set $(\boldsymbol{r}_1, t_1), ..., (\boldsymbol{r}_N, t_N)$ is proportional to $\exp(\mathrm{i}A)$ with A given by eq. (6).

Summing over all possible $(\boldsymbol{r}_1, t_1), ..., (\boldsymbol{r}_N, t_N)$ we obtain the overall quantum mechanical amplitude $\langle \mathrm{f} | G | 0 \rangle$ leading from $(\boldsymbol{r}_0, t_0)$ to $(\boldsymbol{r}_f, t_f)$:

$$\langle \mathrm{f} | G | 0 \rangle = \int J \exp(\mathrm{i}A) \prod_{n=1}^{N} \mathrm{d}t_n \, \mathrm{d}^3 \boldsymbol{r}_n, \tag{11}$$

where J is a jacobian, which can be either a simple normalization constant independent of r_n and t_n, or some other slowly varying functions. The classical trajectory is determined by the stationary phase condition $\delta A = 0$. In the path integration (11), the integrations d^3r_n and dt_n may be viewed as those over the eigenvalues of the operators $(r_n)_{op}$ and $(t_n)_{op}$. The discreteness in this new mechanics enters through the finite number of measurements allowed in any given time interval. The results of each measurement can be any of the continuous eigenvalues of $(r_n)_{op}$ and $(t_n)_{op}$. In the usual continuum case, as the time-parameter t varies continuously there are an infinite number of operators $r_{op}(t)$. Now, by replacing these with a finite number of operators we have elevated "time" to the status of an operator, the same as "space".

As an illustration, let us consider the special case that the jacobian is

$$J = \frac{N!}{T^N} \prod_{n=1}^{N+1} \left(\frac{1}{2\pi\epsilon_n}\right)^{3/2} \tag{12}$$

where $\epsilon_n \equiv t_n - t_{n-1} > 0$; G is then related to the operator

$$U(\epsilon) \equiv \exp(-i\tfrac{1}{2}\epsilon V)\exp(i\tfrac{1}{2}\epsilon\nabla^2)\exp(-i\tfrac{1}{2}\epsilon V) \tag{13}$$

by

$$G = \frac{N!}{T^N}\int U(\epsilon_{N+1})\, U(\epsilon_N) ... U(\epsilon_1) \prod_1^n d\epsilon_n, \tag{14}$$

where ∇^2 is the laplacian operator and $T = \Sigma_1^{N+1}\epsilon_n$. Thus, for example, for a free particle ($V = 0$), $G = \exp(i\frac{1}{2}T\nabla^2)$, identical to the usual continuum case. In a subsequent paper with R. Friedberg, we shall prove that for arbitrary V when T becomes very large $G \sim \exp(-iTH)$ (after averaging over an appropriate grand canonical ensemble), where H is a hermitian whose power series expansion in l can be written as

$$H = H_0 + l^2 H_1 + O(l^3), \tag{15}$$

in which $H_0 = -\frac{1}{2}\nabla^2 + V$ is the usual continuum hamiltonian and

$$H_1 = \tfrac{1}{8}[\nabla^2 - V, [\nabla^2, V]] \tag{16}$$

gives the first correction, We may call H an effective hamiltonian. Its hermiticity insures the unitarity of the S-matrix.

There is quite a latitude in the choice of the jacobian. For example, if instead of (12), J is simply a normalization constant independent of t_n and r_n, then to $O(l^0)$ the effective hamiltonian remains the same H_0; the $O(l^2)$ correction differs from (16) only by a simple numerical factor. Different J correspond to different discrete mechanical systems that have the same continuum limit.

4. Analytical continuation. As in the continuum case, it is often useful to introduce $\tau \equiv it$ and consider the analytic continuation of the region of real t to that of real τ. For a nonrelativistic particle, the action A in (6) is replaced by

$$\mathcal{A} \equiv -iA = \sum_{n=1}^{N+1}\left(\frac{1}{2}\frac{(r_n - r_{n-1})^2}{\tau_n - \tau_{n-1}} + \tfrac{1}{2}(\tau_n - \tau_{n-1})[V(r_n) + V(r_{n-1})]\right). \tag{17}$$

The minimum of $\mathcal{A}$ determines the classical path $(r_1, \tau_1), ..., (r_N, \tau_N)$. As is well known, the resulting trajectory is one in which the particle is moving as if in a "potential" $-V$. The corresponding Green's function G in quantum mechanics is now given by, instead of (11),

$$\langle f|G|0\rangle = \int J \exp(-\mathcal{A}) \prod_1^N d\tau_n d^3r_n. \tag{18}$$

In the following extension of the discrete mechanics to a relativistic theory, we shall stay in the region of real τ. The usual Lorentz transformation becomes a rotation in the four-dimensional euclidean space. For convenience, we shall now refer to τ as "time".

5. Relativistic field theory. The relativistic generalization is identical to the annealed version of the random lattice field theory [1–3]. Consider the example of a scalar field $\phi(r,\tau)$. For any given finite four-dimensional volume Ω, we assume that maximally there can only be N measurements, each yielding a set of values (r_i, τ_i, ϕ_i), where $i = 1,...,N$. The ratio $\Omega/N = l^4$ is a fixed constant. The probability amplitude of observing such a sequence $(r_1, \tau_1, \phi_1), ..., (r_N, \tau_N, \phi_N)$ is proportional to $\exp(-\mathcal{A})$ where $\mathcal{A}$ is the action; its integral

$$J \exp(-\mathcal{A}) \prod_{i=1}^N d\phi_i\, d\tau_i\, d^3r_i \tag{19}$$

gives the overall quantum mechanical amplitude where J is the jacobian which again can be taken as a simple normalization constant, independent of $\boldsymbol{r}_i$ and τ_i. By integrating over the entire domain $d\tau_i d^3 r_i$, we see that rotational invariance is maintained, provided $\mathcal{A}$ is invariant. To construct $\mathcal{A}$, let us consider an arbitrary distribution of N sites of coordinates $(\boldsymbol{r}_1, \tau_1), ..., (\boldsymbol{r}_N, \tau_N)$ in Ω. To each site we assign a value ϕ_i for the scalar field. The sites are linked according to the algorithm given in ref. [1]. The action $\mathcal{A}$ is

$$\mathcal{A} = \tfrac{1}{2} \sum_{l_{ij}} (\phi_i - \phi_j)^2 \lambda_{ij} + \sum_i \omega_i V(\phi_i), \tag{20}$$

where the first sum extends over all links l_{ij} and the second over all sites i. The weights λ_{ij} and ω_i are functions of the sites, given explicitly in ref. [3]. This discrete formalism can, of course, be applied to fields of higher spins.

In the usual continuum relativistic quantum field theory, the space–time coordinates are not regarded as dynamical variables; e.g., in eq. (3) the integration is over $\Pi_r \Pi_\tau d\phi(\boldsymbol{r},\tau)$, which corresponds to an infinite number of operators $\phi_{op}(\boldsymbol{r},\tau)$, with $\boldsymbol{r}$ and τ as parameters. Here, there is a finite number of operators $(\phi_i)_{op}$, $(\boldsymbol{r}_i)_{op}$ and $(\tau_i)_{op}$, each of which can have a continuous range of eigenvalues that can be observed. The discreteness appears, as before, only in the maximal number N of measurements allowed in any finite space–time volume Ω. Each measurement determines the physical field as well as the space–time location of the measurement itself. In this sense, this new approach seems to be nearer to our actual experimental experience. Our usual concept that all physical fields should be embedded in a continuous space–time manifold may well be only an approximation, far from fundamental.

In this new formulation, any finite number of observations may be made very close to each other in space and in time. However, within an infinitesimal space–time volume it is not possible to perform an infinite number of measurements. In the usual continuum quantum field theory, such unphysical possibilities are in principle allowed, and that leads to the well-known complications in the ultraviolet region. If one wishes, one may also regard the fundamental length l of the discrete mechanics as a covariant regularization parameter for the elimination of ultraviolet divergence.

I wish to thank R. Friedberg for discussions. I am grateful for the very kind hospitality of the theoretical group at CERN, especially during the summer of 1982 when most of the present idea was formed.

References

[1] N.H. Christ, R. Friedberg and T.D. Lee, Nucl. Phys. B202 (1982) 89.
[2] N.H. Christ, R. Friedberg and T.D. Lee, Nucl. Phys. B210 (1982) 310.
[3] N.H. Christ, R. Friedberg and T.D. Lee, Nucl. Phys. B210 (1982) 337.

Nuclear Physics B225 [FS9] (1983) 1–52

DISCRETE QUANTUM MECHANICS*

R. FRIEDBERG and T.D. LEE

Columbia University, New York, NY 10027, USA

Received 31 January 1983

Recently the possibility was raised that time can be regarded as a dynamical variable. This leads to the formulation of discrete mechanics, with the usual continuum mechanics appearing as an approximation. The difference between these two is examined in this paper.

1. Introduction

At present, with QCD for the strong interaction, a unifying gauge theory for the electromagnetic and weak interactions, and the Einstein theory of gravitation (plus possible further unification of some or all of these interactions), we may feel that we have arrived at a closed system of physical laws. Nevertheless, the fact that any of the unifying gauge theoretical models requires about twenty parameters suggests that our understanding of these forces most likely remains phenomenological, far from fundamental. This uneasy feeling is reinforced by the enormous technical difficulty of quantizing gravity. Attempts to improve this situation have so far invariably led back to the original unsatisfactory state, with only a change in scale. The underlying cause may be traced to the foundation of our relativistic quantum local field theory. In such a theory only fields are observables, represented by operator functions embedded in a space–time continuum $x = (\boldsymbol{r}, t)$. Hence we can conceive of performing an infinite sequence of observations, at $x_1, x_2, \ldots,$ all lying infinitesimally close but with the result of the field measurements, say $\phi(x_1)$, $\phi(x_2), \ldots,$ fluctuating violently. The probability amplitude of such an unlikely event is proportional to e^{iA} with A as the corresponding action. Yet only the part of A that involves the gradient of the fields is effective in making such rapid fluctuations improbable. This then severely limits the candidates for an acceptable physical theory, and results in the present small set of renormalizable field theories: QED, QCD, Higgs, ... All of these have been intensively investigated. Except for the variation in group algebra, we have virtually exhausted the possible theories within this limited repertoire.

In order to find a way out, it has been suggested recently [1]** that a limit should exist on the density of measurements per unit space–time domain. Such a limit

* This research was supported in part by the US Department of Energy.
** Hereafter referred to as (I).

would automatically remove all ultraviolet divergences. It turns out that this can be achieved by treating time as a discrete dynamical variable, instead of as a continuous parameter. This change can be introduced into all stages of physics: from classical theory to non-relativistic quantum mechanics, and to relativistic quantum field theories; it results in a new formulation, called discrete mechanics, with the usual continuum mechanics appearing as an approximation. The difference between these two will be examined in detail in this paper.

Take the example of a non-relativistic particle of unit mass moving in a potential $V(\boldsymbol{r})$. The usual continuum hamiltonian is

$$H_0 = -\tfrac{1}{2}\nabla^2 + V(\boldsymbol{r}), \tag{1.1}$$

where ∇^2 is the laplacian. The matrix element $\langle \boldsymbol{r}_f | e^{-iTH_0} | \boldsymbol{r}_0 \rangle$ can be written in terms of Feynman's path integration, as a sum over all paths $\boldsymbol{r}(t)$ connecting the initial position $\boldsymbol{r} = \boldsymbol{r}_0$ at time $t = 0$ to the final $\boldsymbol{r} = \boldsymbol{r}_f$ at $t = T$; each path carries a probability amplitude proportional to

$$\exp\left(i \int_0^T [\tfrac{1}{2}\dot{\boldsymbol{r}}^2 - V(\boldsymbol{r})]\, dt\right). \tag{1.2}$$

In the discrete quantum mechanics the continuous path is replaced by N discrete points in the $(\boldsymbol{r}, t)$ space, with their time-coordinates $t_1, t_2, \ldots, t_N$ varying independently within 0 and T, but their space-coordinates anywhere in $\boldsymbol{r}$. It is convenient to denote the time-spacings between these points in sequential order as $\varepsilon_1, \varepsilon_2, \ldots, \varepsilon_N$ so that there is a discrete point at each of

$$(\boldsymbol{r}_1, \varepsilon_1), (\boldsymbol{r}_2, \varepsilon_1 + \varepsilon_2), \ldots, (\boldsymbol{r}_N, \varepsilon_1 + \varepsilon_2 + \cdots + \varepsilon_N), \tag{1.3}$$

where $\varepsilon_n > 0$ and

$$\sum_{n=1}^{N+1} \varepsilon_n = T. \tag{1.4}$$

The two sequences $t_1, t_2, \ldots$ and $\varepsilon_1, \varepsilon_1 + \varepsilon_2, \ldots$ are related by a permutation.

As in (I), in place of (1.2), the amplitude $e^{i\mathscr{A}}$ carried by each such path of discrete points is given by

$$\mathscr{A} = \sum_{n=1}^{N+1} \left\{ \frac{(\boldsymbol{r}_n - \boldsymbol{r}_{n-1})^2}{2\varepsilon_n} - \tfrac{1}{2}\varepsilon_n [V(\boldsymbol{r}_n) + V(\boldsymbol{r}_{n-1})] \right\}, \tag{1.5}$$

in which the initial point is at $\boldsymbol{r} = \boldsymbol{r}_0$, $t = 0$ and the final one at $\boldsymbol{r} = \boldsymbol{r}_{N+1} = \boldsymbol{r}_f$, $t = T$. Each set $(\boldsymbol{r}_i, t_i)$ denotes the result of an observation, which measures both the position and the time of the particle. The average density of measurements per unit time is the inverse of the average spacing $l = T/N$, which is a fundamental constant of the theory. In the usual continuum formalism, the sum over paths is given by an integration over an infinite number of variables: $\int \prod_t d^3 r(t)$. As the time-parameter t varies continuously, there is an infinite number of operators $\boldsymbol{r}_{\rm op}(t)$;

the integration $d^3r(t)$ is over the eigenvalues of the operator $r_{op}(t)$. These operators at different t do not commute, and the integrand is simply a product of $\langle r(t+dt)|r(t)\rangle$ where $|r(t)\rangle$ is the eigenvector of $r_{op}(t)$. Here, in the discrete mechanics, it is a finite-dimensional integration $\int\prod_{n=1}^{N} dt_n\, d^3r_n$ which gives the time t_n, appearing as a dynamical variable, the same status as the position r_n. In like manner, we can regard the integrations d^3r_n and dt_n as those over the eigenvalues of the operators $(r_n)_{op}$ and $(t_n)_{op}$. The discreteness in this new mechanics enters through the finite number of measurements allowed in any given time interval.

If l were zero, then $e^{i\mathscr{A}}$ would approach expression (1.2) and the result would become the familiar continuum limit. For $l\neq 0$ the predictions of the discrete quantum mechanics are different from the usual continuum theory. In sects. 2 and 3, these differences will be calculated systematically when l is small. In particular, it will be proved that the S-matrix, suitably defined, remains unitary.

As we shall see, the final S-matrix can be generated by an effective hamiltonian H (in terms of the usual continuum language). The difference between H in the discrete mechanics and its continuum limit H_0 can be expanded in powers of l. As an illustration of the general method, this power series is calculated explicitly up to $O(l^4)$ and the result is given in (2.22) and (3.20).

Extension to a non-relativistic multiparticle system is straightforward and given in sect. 4. Generalization to a relativistic field theory is somewhat more complicated. By using a scalar field as an example, we give the systematic corrections to the usual continuum theory when l is small but non-zero. These corrections can be carried out by introducing techniques similar to Mayer's cluster expansion for an imperfect gas. The details are given in sect. 5.

2. A non-relativistic particle

2.1. THE GREEN FUNCTION

In a continuum theory, it is often more convenient to deal with the operator e^{-TH_0} instead of e^{-iTH_0}. Likewise, we find it useful to introduce $\tau\equiv it$ and analytically continue from t real to τ real*. The discrete action $\mathscr{A}$ is then replaced by

$$A\equiv -i\mathscr{A}=\sum_{n=1}^{N+1}\left\{\frac{(r_n-r_{n-1})^2}{2\varepsilon_n}+\tfrac{1}{2}\varepsilon_n[V(r_n)+V(r_{n-1})]\right\}, \tag{2.1}$$

where, as in (1.5), $r_{N+1}=r_f$, ε_n are real >0, and satisfy (1.4). In the (r,τ) space, the N discrete points that define the path remain given by (1.3). Let the τ coordinates of these N points be $\tau_1,\tau_2,\ldots,\tau_N$, each of which varies independently between 0 and T. Thus, the two sequences $\varepsilon_1,\varepsilon_1+\varepsilon_2,\ldots$ and $\tau_1,\tau_2,\ldots$ are now related by a permutation.

* Our discussions can also be carried out in the real time domain, without the analytic continuation. These will be given in appendix A.

In place of the matrix element $\langle r_f|e^{-TH_0}|r_0\rangle$ in the continuum theory, the Green function in the discrete mechanics is given by the following sum over all discrete paths leading from r_0 at $\tau=0$ to $r_{N+1}=r_f$ at $\tau=T$:

$$\langle r_f|\mathscr{G}(T,l)|r_0\rangle \equiv \sum_{N=0}^{\infty}\frac{1}{N!}\left(\frac{1}{l}\right)^N \int J\prod_{n=1}^{N} d\tau_n\, d^3r_n\, e^{-A}\,, \tag{2.2}$$

where l^{-1} plays the same role as the fugacity in a grand partition function in statistical mechanics, and J is the jacobian which can be either a simple normalization constant independent of τ_n and r_n, or some other slowly varying function (compared to e^{-A}). As we shall see, different J correspond to different discrete mechanical systems that have the same continuum limit.

2.2. A SIMPLE CHOICE OF JACOBIAN

In this section we consider the simplest case*

$$J=J_0\equiv e^{-T/l}\prod_{n=1}^{N+1}\left(\frac{1}{2\pi\varepsilon_n}\right)^{3/2}. \tag{2.3}$$

(Other choices of J will be analyzed in sect. 3.) Recalling that

$$\langle r'|e^{\frac{1}{2}\varepsilon\nabla^2}|r\rangle=\left(\frac{1}{2\pi\varepsilon}\right)^{3/2}\exp\left(-\frac{1}{2\varepsilon}(r'-r)^2\right), \tag{2.4}$$

we see that (2.2) becomes

$$\langle r_f|\mathscr{G}(T,l)|r_0\rangle=\sum_{N=0}^{\infty}\frac{1}{N!}\left(\frac{1}{l}\right)^N e^{-T/l}\int\prod_{n=1}^{N} d\tau_n \langle r_f|G_N|r_0\rangle\,, \tag{2.5}$$

where G_N can be written as an operator product

$$G_N=U(\varepsilon_{N+1})U(\varepsilon_N)\cdots U(\varepsilon_1)\,, \tag{2.6}$$

with

$$U(\varepsilon)\equiv e^{-\frac{1}{2}\varepsilon V}e^{\frac{1}{2}\varepsilon\nabla^2}e^{-\frac{1}{2}\varepsilon V}, \tag{2.7}$$

so that its matrix element is

$$\langle r'|U(\varepsilon)|r\rangle=\left(\frac{1}{2\pi\varepsilon}\right)^{3/2}\exp\left(-\frac{1}{2\varepsilon}(r'-r)^2-\tfrac{1}{2}\varepsilon[V(r')+V(r)]\right).$$

In (2.3), we include an overall normalization factor $e^{-T/l}$. Since $\int\prod_1^N d\tau_n=T^N$, when $U=1$ we have $\mathscr{G}(T,l)=1$, in which case the average of $N=T/l$. (In general, the average of N is given by $\partial \ln\langle r_f|\mathscr{G}|r_0\rangle/\partial\ln l^{-1}$ for a given r_0 and r_f.) Because

$$\int\prod_1^N d\tau_n\cdots=N!\int\prod_1^N d\varepsilon_n\cdots, \tag{2.8}$$

* See appendix B for a supersymmetric formulation, which simplifies the jacobian.

(2.5) reduces to

$$\mathscr{G}(T,l)=\sum_{N=0}^{\infty}\left(\frac{1}{l}\right)^{N}\mathrm{e}^{-T/l}\int\prod_{n=1}^{N}\mathrm{d}\varepsilon_n\,G_N\,. \tag{2.9}$$

We note that for a free particle $V=0$, $H_0=-\frac{1}{2}\nabla^2$; hence, $U(\varepsilon)=\mathrm{e}^{\frac{1}{2}\varepsilon\nabla^2}$ and therefore $\mathscr{G}(T,l)=G_N=\mathrm{e}^{-TH_0}$, the same as the continuum case.

For a general V, we multiply the right-hand side of (2.9) by

$$1=\int\mathrm{d}\varepsilon_{N+1}\delta\left(\sum_{1}^{N+1}\varepsilon_n-T\right)=\int_{-i\infty}^{i\infty}\frac{\mathrm{d}z}{2\pi i}\int_0^{\infty}\mathrm{d}\varepsilon_{N+1}\exp\left(Tz-\sum_{1}^{N+1}\varepsilon_n z\right), \tag{2.10}$$

which takes care of the constraint (1.4), and write $\exp(-T/l)$ as $\exp(-\sum_1^{N+1}\varepsilon_n/l)$. Hence,

$$\mathscr{G}(T,l)=\int_{-i\infty}^{i\infty}\frac{\mathrm{d}z}{2\pi i}\mathrm{e}^{Tz}\sum_{N=0}^{\infty}\left(\frac{M(z)}{l}\right)^{N}M(z)\,, \tag{2.11}$$

where

$$M(z)=\int_0^{\infty}\mathrm{d}\varepsilon\;\mathrm{e}^{-\varepsilon(1/l+z)}U(\varepsilon)\,. \tag{2.12}$$

Defining

$$\bar{H}(z)\equiv M^{-1}(z)-z-\frac{1}{l}\,, \tag{2.13}$$

we have

$$\mathscr{G}(T,l)=\int_{-i\infty}^{i\infty}\frac{\mathrm{d}z}{2\pi i}\,\frac{\mathrm{e}^{Tz}}{z+\bar{H}(z)}\,. \tag{2.14}$$

Anticipating that the average ε_n is $O(l)$ which is small, we expand the deviation of $U(\varepsilon_n)$ from $\mathrm{e}^{-\varepsilon_n H_0}$ in powers of ε_n. Using (1.1) and (2.7), and writing

$$K\equiv-\tfrac{1}{2}\nabla^2\,,\qquad H_0=K+V\,,$$
$$U(\varepsilon)=\mathrm{e}^{-\frac{1}{2}\varepsilon V}\,\mathrm{e}^{-\varepsilon K}\,\mathrm{e}^{-\frac{1}{2}\varepsilon V}\,, \tag{2.15}$$

we find that the expansion

$$U(\varepsilon)-\mathrm{e}^{-\varepsilon H_0}=\sum_{1}^{\infty}\frac{\varepsilon^n}{n!}h_n\,, \tag{2.16}$$

is given by

$$h_1=h_2=0\,,$$
$$h_3=-\tfrac{1}{2}[K+\tfrac{1}{2}V,[K,V]],$$
$$h_4=\tfrac{1}{2}[K^2+(K+V)^2,[K,V]]+\tfrac{1}{2}[K,[K,V^2]]-[K,V]^2\,, \tag{2.17}$$

etc. Combining (2.12), (2.13) and (2.16), we obtain the power series of $\bar{H}(z)$

$$\bar{H}(z)=H_0+l^2\bar{H}_2(z)+l^3\bar{H}_3(z)+l^4\bar{H}_4(z)+\cdots, \tag{2.18}$$

where

$$\bar{H}_2(z)=-h_3,$$

$$\bar{H}_3(z)=2zh_3-h_4-h_3H_0-H_0h_3,$$

$$\bar{H}_4(z)=-3z^2h_3+3z(h_4+h_3H_0+H_0h_3)-(h_5+H_0h_4+h_4H_0+H_0h_3H_0). \tag{2.19}$$

2.3. THE EFFECTIVE HAMILTONIAN

Substituting (2.18) into (2.14), we shall establish the following asymptotic expression for $\mathcal{G}(T, l)$ when T is large.

Theorem.

$$\mathcal{G}(T,l)=\xi^\dagger\,\mathrm{e}^{-TH}\xi+\mathrm{O}(\mathrm{e}^{-T/l}), \tag{2.20}$$

where in powers of l

$$\xi=1-l^3h_3-3l^4(H_0h_3+h_3H_0+\tfrac{1}{2}h_4)+\mathrm{O}(l^5)=\xi^\dagger, \tag{2.21}$$

$$H=H_0-l^2h_3-l^3(h_4+2h_3H_0+2H_0h_3)$$
$$-l^4(h_5+\tfrac{5}{2}h_4H_0+\tfrac{5}{2}H_0h_4+3h_3H_0^2+3H_0^2h_3+4H_0h_3H_0)+\mathrm{O}(l^5), \tag{2.22}$$

with $H_0=-\frac{1}{2}\nabla^2+V$ as before, and $h_3, h_4, \ldots$ given by (2.16) and (2.17).

Proof. Let us assume that the potential $V(\boldsymbol{r})$ is bounded from below. Without any loss of generality, we may set* $V(\boldsymbol{r})\geqslant 0$.

Lemma. In the complex z-plane, if $\mathrm{Re}\,z>0$ the determinant $|z+\bar{H}(z)|\neq 0$.

Take any normalized state vector ψ. We have $0<\psi^\dagger\,\mathrm{e}^{-\varepsilon K}\psi\leqslant 1$ where $K=-\frac{1}{2}\nabla^2$, as before. Thus, the expectation value of $U(\varepsilon)=\mathrm{e}^{-\frac{1}{2}\varepsilon V}\,\mathrm{e}^{-\varepsilon K}\,\mathrm{e}^{-\frac{1}{2}\varepsilon V}$ has the same bounds: $0<\psi^\dagger U(\varepsilon)\psi\leqslant 1$. From (2.12) we see that

$$\psi^\dagger M(z)\psi=\int_0^\infty \mathrm{d}\varepsilon\;\mathrm{e}^{-\varepsilon(1/l+z)}\psi^\dagger U(\varepsilon)\psi\,;$$

its magnitude satisfies

$$|\psi^\dagger M(z)\psi|\leqslant\int_0^\infty \mathrm{d}\varepsilon\,|\mathrm{e}^{-\varepsilon(1/l+z)}\psi^\dagger U(\varepsilon)\psi|$$
$$=\int_0^\infty \mathrm{d}\varepsilon\;\mathrm{e}^{-\varepsilon(1/l+\mathrm{Re}\,z)}|\psi^\dagger U(\varepsilon)\psi|\leqslant\frac{l}{1+l\,\mathrm{Re}\,z}. \tag{2.23}$$

* A shift $V(\boldsymbol{r})\to V(\boldsymbol{r})+v$ where v is a constant leads, in accordance with (2.5)–(2.7), to $U(\varepsilon)\to U(\varepsilon)\,\mathrm{e}^{-\varepsilon v}$, $G_N\to G_N\,\mathrm{e}^{-Tv}$, and therefore $\mathcal{G}(T,l)\to\mathcal{G}(T,l)\,\mathrm{e}^{-Tv}$.

Hence when Re $z > 0$, we find

$$|\psi^\dagger M(z)\psi| < l ,$$

which implies that the magnitude of every eigenvalue of $M(z)$ is less than l, and therefore that of $M^{-1}(z)$ is greater than l^{-1}; consequently the determinant $|M^{-1}(z) - l^{-1}| \neq 0$. From (2.13) we have $z + \bar{H}(z) = M^{-1}(z) - l^{-1}$ and the lemma is proved.

We now turn to the proof of the theorem.

In the complex z-plane, the roots of $|z + \bar{H}(z)| = 0$ all lie on the half-plane Re $z \leqslant 0$. For small l, those roots that are near $|z| = \mathrm{O}(l^{-1})$ or larger give an $\leqslant \mathrm{O}(\mathrm{e}^{-T/l})$ contribution to $\mathscr{G}(T, l)$; they can be ignored when $T \to \infty$. To find the roots that are O(1), we can expand $\bar{H}(z)$ in powers of l.

(i) Neglecting $\mathrm{O}(l^3)$, (2.18) becomes

$$\bar{H}(z) = H_0 - l^2 h_3 , \tag{2.24}$$

which is independent of z. Hence, in accordance with (2.14),

$$\mathscr{G}(T, l) = \int_{-i\infty}^{i\infty} \frac{\mathrm{d}z}{2\pi i} \frac{\mathrm{e}^{Tz}}{z + H_0 - l^2 h_3} = \mathrm{e}^{-TH} ,$$

where

$$H = H_0 - l^2 h_3 + \mathrm{O}(l^3) ,$$

in agreement with (2.22).

(ii) Neglecting $\mathrm{O}(l^4)$, $\bar{H}(z)$ is a linear function of z:

$$\bar{H}(z) = \alpha + \beta z , \tag{2.25}$$

where

$$\alpha = H_0 - l^2 h_3 - l^3 (h_4 + h_3 H_0 + H_0 h_3) + \mathrm{O}(l^4) , \tag{2.26}$$

$$\beta = 2 l^3 h_3 + \mathrm{O}(l^4) . \tag{2.27}$$

Thus,

$$z + \bar{H}(z) = (1+\beta)^{1/2} (\bar{\alpha} + z)(1+\beta)^{1/2} + \mathrm{O}(l^4) , \tag{2.28}$$

where

$$\bar{\alpha} = (1+\beta)^{-1/2} \alpha (1+\beta)^{-1/2} . \tag{2.29}$$

Substituting (2.28) into (2.14), we obtain

$$\mathscr{G}(T, l) = \xi^\dagger \, \mathrm{e}^{-TH} \xi ,$$

where

$$\xi = (1+\beta)^{-1/2} = 1 - l^3 h_3 + \mathrm{O}(l^4) ,$$

$$H = \bar{\alpha} + \mathrm{O}(l^4) = H_0 - l^2 h_3 - l^3 (h_4 + 2 h_3 H_0 + 2 H_0 h_3) + \mathrm{O}(l^4) ,$$

in agreement with (2.21) and (2.22).

(iii) Neglecting $O(l^5)$, we have

$$\bar{H}(z) = \alpha + \beta z + \gamma z^2 ,$$

where, according to (2.18) and (2.19)

$$\alpha = H_0 - l^2 h_3 - l^3(h_4 + h_3 H_0 + H_0 h_3) - l^4(h_5 + h_4 H_0 + H_0 h_4 + H_0 h_3 H_0) + O(l^5) , \tag{2.30}$$

$$\beta = 2l^3 h_3 + 3l^4(h_4 + h_3 H_0 + H_0 h_3) + O(l^5) , \tag{2.31}$$

$$\gamma = -3l^4 h_3 + O(l^5) . \tag{2.32}$$

Thus,

$$z + \bar{H}(z) = (1+\beta)^{1/2}[\bar{\alpha} + z + \gamma z^2](1+\beta)^{1/2} + O(l^5) , \tag{2.33}$$

in which $\bar{\alpha}$ is given by (2.29). Since $\bar{\alpha}$ is hermitian we can transform it into a diagonal matrix:

$$u\ \bar{\alpha} u = \alpha' = \begin{pmatrix} \alpha'_1 & 0 & 0 & 0 & \cdots & 0 \\ 0 & \alpha'_2 & 0 & 0 & \cdots & 0 \\ 0 & 0 & \alpha'_3 & 0 & \cdots & 0 \\ & & \cdots & & & \\ 0 & 0 & 0 & 0 & \cdots & \alpha'_n \end{pmatrix} , \tag{2.34}$$

with $u^\dagger u = 1$. Eq. (2.33) can be written as

$$z + \bar{H}(z) = (1+\beta)^{1/2} u (\alpha' + z + \gamma' z^2) u^\dagger (1+\beta)^{1/2} , \tag{2.35}$$

where

$$\gamma' = u^\dagger \gamma u . \tag{2.36}$$

Let

$$\hat{e}_1 = \begin{pmatrix} 1 \\ 0 \\ 0 \\ \vdots \\ 0 \end{pmatrix} , \quad \hat{e}_2 = \begin{pmatrix} 0 \\ 1 \\ 0 \\ \vdots \\ 0 \end{pmatrix} , \ldots , \tag{2.37}$$

be the normalized eigenvectors of α' satisfying

$$\alpha' \hat{e}_i = \alpha'_i \hat{e}_i , \tag{2.38}$$

and $\psi_i(z)$ be the corresponding normalized eigenvector of the matrix $\alpha' + \gamma' z^2$,

$$(\alpha' + \gamma' z^2)\psi_i = \lambda_i(z)\psi_i . \tag{2.39}$$

Since γ' is $O(l^4)$, we may apply the standard first-order perturbation formula:

$$\lambda_i(z) = \alpha'_i + \gamma'_{ii} z^2 ,$$

$$\psi_i(z) = \hat{e}_i + z^2 \sum_{j \neq i} \frac{\gamma'_{ji}}{\alpha'_i - \alpha'_j} \hat{e}_j , \tag{2.40}$$

where γ'_{ji} is the (ji)th matrix element of γ'. Let $(\psi_i)_j$ be the jth component of the ith eigenvector ψ_i. Define $U_{ji} \equiv (\psi_i)_j$. For z real, U is unitary. We have

$$U(z)^\dagger(\alpha' + z + \gamma' z^2)U(z) = z + \lambda(z), \tag{2.41}$$

where $\lambda(z)$ is the diagonal matrix

$$\lambda(z) = \begin{pmatrix} \lambda_1(z) & 0 & 0 & \cdots & 0 \\ 0 & \lambda_2(z) & 0 & \cdots & 0 \\ & \cdots & & & \\ 0 & 0 & 0 & \cdots & \lambda_n(z) \end{pmatrix}. \tag{2.42}$$

From (2.35) and (2.41), we find

$$\frac{1}{z + \bar{H}(z)} = (1+\beta)^{-1/2} u U(z) \frac{1}{z + \lambda(z)} U^\dagger(z) u^\dagger (1+\beta)^{-1/2}. \tag{2.43}$$

The roots of $z + \lambda(z) = 0$ are determined by

$$z + \lambda_i(z) = \alpha'_i + z + \gamma'_{ii} z^2 = 0.$$

Because γ'_{ii} is $O(l^4)$, one of the roots is $O(1)$ given by

$$z \equiv z_i = -\alpha'_i - \gamma'_{ii}(\alpha'_i)^2 + O(l^8), \tag{2.44}$$

while the other is $\cong -1/\gamma'_{ii} = O(l^{-4})$ which will be ignored in accordance with the discussion given before. We see that z_i is real. When $z \to z_i$, $\psi_i(z) \to \psi_i(z_i)$,

$$\lambda_i(z) \to \lambda_i(z_i) = \alpha'_i + \gamma'_{ii}(\alpha'_i)^2 + O(l^8),$$

$$z + \lambda_i(z) \to (z - z_i)[1 - 2\gamma'_{ii}\alpha'_i + O(l^8)].$$

Thus, neglecting all higher-order terms

$$\int_{-i\infty}^{i\infty} \frac{dz}{2\pi i} U(z) \frac{e^{Tz}}{z + \lambda(z)} U(z)^\dagger = \sum_i \psi_i(z_i) \frac{e^{-T\lambda_i(z_i)}}{1 - 2\gamma'_{ii}\alpha'_i} \psi_i(z_i)^\dagger$$

$$= \sum_i \frac{\psi_i(z_i)}{1 - \gamma'_{ii}\alpha'_i} e^{-T\lambda_i(z_i)} \frac{\psi_i(z_i)^\dagger}{1 - \gamma'_{ii}\alpha'_i}. \tag{2.45}$$

Introduce

$$\Lambda \equiv \alpha' + \tfrac{1}{2}(\alpha'^2\gamma' + \gamma'\alpha'^2);$$

its diagonal matrix elements are

$$\Lambda_{ii} = \lambda_i(z_i),$$

and its normalized eigenvectors are χ_i. Since α' is a diagonal matrix given by (2.34) and $\gamma' = O(l^4)$, neglecting $O(l^8)$ we have

$$\Lambda \chi_i = \Lambda_{ii} \chi_i, \tag{2.46}$$

$$\chi_i = \hat{e}_i + \sum_{j \neq i} \frac{1}{2} \frac{\alpha_i'^2 + \alpha_j'^2}{\alpha'_i - \alpha'_j} \gamma'_{ji} \hat{e}_j, \tag{2.47}$$

where $\hat{e}_i$ are given by (2.37). Thus, to this order Λ_{ii} are the eigenvalues of Λ. From (2.40), (2.44) and (2.47) it follows that the column matrices $\psi_i(z_i)$ and χ_i are related by

$$\frac{\psi_i(z_i)}{1-\gamma'_{ii}\alpha'_i} = [1+\tfrac{1}{2}(\alpha'\gamma'+\gamma'\alpha')]\chi_i + O(l^8)\,,$$

since the scalar products of both sides with $\hat{e}_i$ are $1+\gamma'_{ii}\alpha'_i$, and those with $\hat{e}_j$ $(j \neq i)$ are $\gamma'_{ji}(\alpha'_i)^2/(\alpha'_i-\alpha'_j)$. Consequently, (2.45) becomes

$$\int_{-i\infty}^{i\infty} \frac{dz}{2\pi i} U(z) \frac{e^{Tz}}{z+\lambda(z)} U(z)^\dagger = [1+\tfrac{1}{2}(\alpha'\gamma'+\gamma'\alpha')]\, e^{-T\Lambda}[1+\tfrac{1}{2}(\alpha'\gamma'+\gamma'\alpha')]\,. \quad (2.48)$$

Recalling

$$u\alpha' u^\dagger = \bar{\alpha} = (1+\beta)^{-1/2}\alpha(1+\beta)^{-1/2} = H_0 + O(l^2)\,, \qquad u\gamma' u^\dagger = \gamma\,,$$

we find

$$\int_{-i\infty}^{i\infty} \frac{dz}{2\pi i} uU(z) \frac{e^{Tz}}{z+\lambda(z)} U^\dagger(z) u^\dagger$$

$$= (1+\tfrac{1}{2}H_0\gamma+\tfrac{1}{2}\gamma H_0)\exp[-T(\bar{\alpha}+\tfrac{1}{2}H_0^2\gamma+\tfrac{1}{2}\gamma H_0^2)](1+\tfrac{1}{2}H_0\gamma+\tfrac{1}{2}\gamma H_0)\,.$$

Combining with (2.43), we derive

$$\mathcal{G}(T, l) = \xi\, e^{-TH}\xi\,, \quad (2.49)$$

where

$$H = (1+\beta)^{-1/2}\alpha(1+\beta)^{-1/2} + \tfrac{1}{2}H_0^2\gamma + \tfrac{1}{2}\gamma H_0^2\,, \quad (2.50)$$

$$\xi = 1 - \tfrac{1}{2}\beta + \tfrac{1}{2}H_0\gamma + \tfrac{1}{2}\gamma H_0 = \xi^\dagger\,, \quad (2.51)$$

which, on account of (2.30)–(2.32), are the same as (2.21) and (2.22). It is not difficult to extend the above series expansion to still higher powers of l. The theorem is established.

Remark. When T is large, we can neglect $O(e^{-T/l})$. The Green function $\mathcal{G}(T, l)$ reduces to e^{-TH}, provided we choose the metric g in the Hilbert space to be

$$g = \xi^\dagger \xi\,. \quad (2.52)$$

The eigenvalues of the hermitian operator H give the energy spectrum in the discrete quantum mechanics, and the unitary operator e^{-iTH} is the S-matrix. Thus, H can be regarded as the effective hamiltonian. Further discussions are given in subsect. 2.4 below.

2.4. METRIC IN THE HILBERT SPACE

In the usual continuum quantum mechanics, we may introduce a *positive* metric

$$g \equiv \xi^\dagger \xi = g^\dagger\,, \quad (2.53)$$

in the Hilbert space. As before, † denotes hermitian conjugation. Let ϕ be a state vector. The scalar product between two state vectors ϕ_a and ϕ_b is

$$\phi_b^{\dagger} g \phi_a .$$

We define the adjoint of any operator O by

$$\bar{O} \equiv g^{-1} O^{\dagger} g .$$

A physical observable corresponds to a self-adjoint operator

$$O = \bar{O} ; \qquad (2.54)$$

its expectation value over any state vector ϕ is

$$\langle O \rangle \equiv \phi^{\dagger} g O \phi ,$$

which is real.

The Schroedinger equation in such a Hilbert space is

$$i \frac{\mathrm{d}\phi}{\mathrm{d}t} = h\phi , \qquad (2.55)$$

where $h = \bar{h}$ is the hamiltonian. From (2.55) one sees that

$$\frac{\mathrm{d}}{\mathrm{d}t} (\phi^{\dagger} g \phi) = 0 , \qquad (2.56)$$

which preserves the unitarity. Let $\phi = \phi_0$ at $t = 0$; then the transition amplitude $T_{0\to n}$ for the state to be in ϕ_n at t is

$$T_{0\to n} = \phi_n^{\dagger} g \, \mathrm{e}^{-iht} \phi_0 . \qquad (2.57)$$

Define

$$H \equiv \xi h \xi^{-1} . \qquad (2.58)$$

Since $h^{\dagger} = g h g^{-1}$, we have

$$H^{\dagger} = (\xi^{\dagger})^{-1} h^{\dagger} \xi^{\dagger} = \xi h \xi^{-1} = H ,$$
$$g \, \mathrm{e}^{-iht} = \xi^{\dagger} \xi \xi^{-1} \, \mathrm{e}^{-iHt} \xi = \xi^{\dagger} \, \mathrm{e}^{-iHt} \xi ,$$

and therefore the transition amplitude (2.57) becomes

$$T_{0\to n} = \phi_n^{\dagger} \xi^{\dagger} \, \mathrm{e}^{-iHt} \xi \phi_0 . \qquad (2.59)$$

The hermitian operator H given by (2.22) acts as an effective hamiltonian. The unusual feature is that the metric g, given by (2.53) and (2.21), depends on $V(\boldsymbol{r})$. In the asymptotic region $r \to \infty$, we have $g \to 1$, provided that $V \to 0$.

2.5. ENERGY LEVELS AND UNITARITY

As mentioned before, the hermiticity of H ensures that the S-matrix, defined by

$$S = \lim_{T\to\infty} e^{-iTH}, \tag{2.60}$$

is unitary. The eigenvalues of H give the energy spectrum.

It is of interest to study the lowest-order correction to the energy levels in the usual continuum theory. Let $|\,\rangle$ be an eigenstate of the continuum hamiltonian H_0:

$$H_0|\,\rangle = E_0|\,\rangle\,. \tag{2.61}$$

Because $K+\frac{1}{2}V = H_0 - \frac{1}{2}V$ and

$$\langle|[H_0, [K, V]]|\rangle = 0\,,$$

to $O(l^2)$ the corresponding energy in the discrete quantum mechanics for a non-relativistic particle of unit mass moving in a potential V is given by, on account of (2.17) and (2.22),

$$\begin{aligned} E &= \langle|H|\rangle + O(l^3) \\ &= E_0 - \tfrac{1}{4}l^2\langle|(\nabla V)^2|\rangle + O(l^3)\,. \end{aligned} \tag{2.62}$$

Examples. (i) In the case of a one-dimensional harmonic oscillator $V=\frac{1}{2}\omega^2x^2$, $\nabla V = \omega^2 x$ we have $E_0 = (n+\frac{1}{2})\omega$ and

$$E = (1-\tfrac{1}{4}l^2\omega^2)(n+\tfrac{1}{2})\omega + O(l^3)\,, \tag{2.63}$$

where $n = 0, 1, 2, \ldots$.

(ii) Let V be the Coulomb potential experienced by an electron of mass m_e at a distance r from a nucleus of charge Z distributed uniformly within a sphere of radius R:

$$V = \begin{cases} -Z\alpha/r, & \text{for } r \geq R \\ -\frac{1}{2}Z\alpha\left[3-\left(\dfrac{r}{R}\right)^2\right]\Big/R, & \text{for } r \leq R\,, \end{cases} \tag{2.64}$$

where $\alpha \simeq \frac{1}{137}$ is the fine-structure constant. The additional energy shift ΔE in the discrete mechanics may be estimated by using the non-relativistic wave function $\psi(\boldsymbol{r}) = \langle \boldsymbol{r}|\rangle$. We find

$$\Delta E = -\frac{l^2}{4m_e}\langle|(\nabla V)^2|\rangle \simeq -\frac{6\pi}{5m_eR}|Z\alpha l\psi(0)|^2\,. \tag{2.65}$$

For the 2s orbit in hydrogen

$$|\psi(0)|^2 = (8\pi a^3)^{-1}\,, \tag{2.66}$$

where $a=(m_e\alpha)^{-1}$ is the Bohr radius. Setting the agreement between the experimental observation and the theoretical QED calculation of this level to an accuracy ε Rydberg, we obtain

$$lm_e<[\tfrac{10}{3}(137)^3\varepsilon m_e R]^{1/2}, \tag{2.67}$$

which, for $R\sim 0.8$ fm and [2] $\varepsilon\sim 10^{-11}$, gives

$$l<1.6\times 10^{-14}\text{ cm}. \tag{2.68}$$

3. Theories with different J

We now turn to questions concerning different choices of the jacobian. Let us consider in place of (2.3) a more general form

$$J\propto\prod_{n=1}^{N+1}\left(\frac{1}{2\pi\varepsilon_n}\right)^{3/2} f(\varepsilon_n), \tag{3.1}$$

where the proportionality constant is independent of ε_n. When $f(\varepsilon)=1$, the above expression reduces to (2.3), and if $f(\varepsilon)\propto\varepsilon^{3/2}$ then J becomes simply an overall normalization constant, independent of $\boldsymbol{r}_n$ and τ_n. Similar to (2.9), we see that the Green function $\mathscr{G}(T,l)$, defined by (2.1), (2.2) and (3.1), can be written as

$$\mathscr{G}(T,l)=\frac{\sum\limits_{N=0}^{\infty}\left(\frac{1}{l}\right)^N\int\delta\left(\sum\limits_{1}^{N+1}\varepsilon_n-T\right)G_N\prod\limits_{1}^{N+1}f(\varepsilon_n)\,\mathrm{d}\varepsilon_n}{\sum\limits_{N=0}^{\infty}\left(\frac{1}{l}\right)^N\int\delta\left(\sum\limits_{1}^{N+1}\varepsilon_n-T\right)\prod\limits_{1}^{N+1}f(\varepsilon_n)\,\mathrm{d}\varepsilon_n}, \tag{3.2}$$

where G_N is the same operator given by (2.6) and (2.7), and the denominator is simply the inverse of the overall proportionality constant in (3.1). Thus, when $G_N=1$, we have $\mathscr{G}(T,l)=1$, and if $f(\varepsilon)=1$ then the denominator $=\mathrm{e}^{T/l}$ and the above formula reduces to (2.9).

In order to examine more explicitly the effect of different $f(\varepsilon)$, we assume

$$f(\varepsilon)=\left(\frac{\varepsilon}{a}\right)^m, \tag{3.3}$$

where a is a constant having the same dimension as l. We discuss first the normalization of $f(\varepsilon)$. From (3.2) we see that under the transformation

$$f(\varepsilon)\to\lambda f(\varepsilon), \tag{3.4}$$

we have

$$\frac{1}{l}f(\varepsilon)\to\frac{\lambda}{l}f(\varepsilon),$$

and therefore $\mathscr{G}(T,l)\to\mathscr{G}(T,l/\lambda)$ provided a is independent of l. On the other hand, if $a = \text{constant} \times l$, then under the same transformation (3.4)

$$\frac{1}{l} f(\varepsilon) \propto \frac{\varepsilon^m}{l^{m+1}} \to \frac{\lambda}{l} f(\varepsilon) \propto \frac{\lambda \varepsilon^m}{l^{m+1}},$$

with the same proportionality constant on both sides; hence

$$\mathcal{G}(T, l) \to \mathcal{G}(T, l/\lambda^{1/(m+1)}). \tag{3.5}$$

In any case, different normalization conditions satisfied by $f(\varepsilon)$ correspond to simple scale changes in l. Thus, without any loss of generality we can set

$$f(\varepsilon) = \frac{1}{\Gamma(m+1)} \left(\frac{\varepsilon}{l}\right)^m. \tag{3.6}$$

Consequently,

$$\int_0^\infty f(\varepsilon)\, \mathrm{e}^{-\varepsilon/l}\, \mathrm{d}\varepsilon = l.$$

Let us define the average of $F(N)$ to be

$$\langle F(N) \rangle \equiv \frac{\sum_{N=0}^{\infty} \left(\frac{1}{l}\right)^N \int \delta\left(\sum_1^{N+1} \varepsilon_n - T\right) F(N) \prod_1^{N+1} f(\varepsilon_n)\, \mathrm{d}\varepsilon_n}{\sum_{N=0}^{\infty} \left(\frac{1}{l}\right)^N \int \delta\left(\sum_1^{N+1} \varepsilon_n - T\right) \prod_1^{N+1} f(\varepsilon_n)\, \mathrm{d}\varepsilon_n}. \tag{3.7}$$

Hence,

$$\mathcal{G}(T, l) = \langle G_N \rangle, \tag{3.8}$$

in accordance with (3.2). By using (3.6)–(3.7) we can show that when $T \to \infty$ the average density of discrete points is

$$\rho \equiv \frac{\langle N \rangle}{T} = \frac{1}{(m+1)l}, \tag{3.9}$$

and its fluctuation is

$$\frac{\langle N^2 \rangle - \langle N \rangle^2}{\langle N \rangle} = \frac{1}{m+1}, \tag{3.10}$$

which decreases to 0 as m increases to ∞.

Multiplying the numerator and the denominator in (3.2) by $\mathrm{e}^{-T/l}$, we may follow steps identical to those from (2.9) to (2.14). This leads to

$$\mathcal{G}(T, l) = \frac{Q(T, l)}{Q_0(T, l)}, \tag{3.11}$$

where

$$Q(T, l) = \int_{-i\infty}^{i\infty} \frac{\mathrm{d}z}{2\pi i} \frac{\mathrm{e}^{Tz}}{z + \mathcal{H}(z)}, \tag{3.12}$$

$$Q_0(T, l) = \int_{-i\infty}^{i\infty} \frac{dz}{2\pi i} \frac{e^{Tz}}{z + \bar{\mathcal{H}}_0(z)}, \tag{3.13}$$

$$\bar{\mathcal{H}}(z) = \mathcal{M}^{-1}(z) - z - \frac{1}{l}, \tag{3.14}$$

$$\bar{\mathcal{H}}_0(z) = \mathcal{M}_0^{-1}(z) - z - \frac{1}{l}, \tag{3.15}$$

$$\mathcal{M}(z) = \int_0^\infty d\varepsilon \; e^{-\varepsilon(1/l+z)} f(\varepsilon) \, e^{-\frac{1}{2}\varepsilon V} e^{\frac{1}{2}\varepsilon \nabla^2} e^{-\frac{1}{2}\varepsilon V}, \tag{3.16}$$

$$\mathcal{M}_0(z) = \int_0^\infty d\varepsilon \; e^{-\varepsilon(1/l+z)} f(\varepsilon) \, . \tag{3.17}$$

By using (2.16) and (2.17), we find similar to (2.20)–(2.22)

$$\mathcal{G}(T, l) = \xi^\dagger \, e^{-TH} \xi + O(e^{-T/l}) \, , \tag{3.18}$$

where

$$\xi = 1 + O(l^3) = 1 + O(\rho^{-3}) \, , \tag{3.19}$$

and the effective hamiltonian is

$$H = K + V - \frac{l^2}{3!}(m+3)(m+2)h_3 + O(l^3) \, , \tag{3.20}$$

where, as before, $K = -\frac{1}{2}\nabla^2$ and

$$h_3 = -\tfrac{1}{2}[K + \tfrac{1}{2}V, [K, V]] \, .$$

When $m = 0$, (3.18)–(3.20) reduce to (2.20)–(2.22). In terms of ρ given by (3.9), H can be written as

$$H = K + V - \frac{1}{3!} \frac{(m+3)(m+2)}{(m+1)^2 \rho^2} h_3 + O(\rho^{-3}) \, .$$

The operator H is hermitian. As before, we may choose the metric g in the Hilbert space to be $g = \xi^\dagger \xi$, the S-matrix is then given by the unitary matrix e^{-iTH} with $T \to \infty$. Thus, different $f(\varepsilon)$ give different discrete mechanical systems that have the same continuum limit.

Let $|\;\rangle$ be the eigenstate of the continuum hamiltonian $H_0 = K + V$ of eigenvalue E_0, as in (2.61). The corresponding energy in discrete quantum mechanics is given by

$$E = E_0 - \frac{1}{24} \frac{(m+3)(m+2)}{(m+1)^2 \rho^2} \langle |(\nabla V)^2| \rangle + O(\rho^{-3}) \, . \tag{3.21}$$

In the case of a one-dimensional harmonic oscillator $V = \frac{1}{2}\omega^2 x^2$, we have $E_0 = (n+\frac{1}{2})\omega$ and

$$E = \left[1 - \frac{1}{24}\frac{(m+3)(m+2)}{(m+1)^2}\frac{\omega^2}{\rho^2}\right](n+\tfrac{1}{2})\omega + \mathrm{O}(\rho^{-3})\,, \tag{3.22}$$

where $n = 0, 1, 2, \ldots$. As shown in appendix C, when $m \to \infty$, (3.22) reduces to that for a regular spacing discrete time distribution, in agreement with (3.10).

4. Multiparticle system

The generalization to a system of $\mathcal{N}$ non-relativistic particles is straightforward. For example, if they are all Boltzmann particles of unit mass, we can simply regard $\boldsymbol{r}$ as a $3\mathcal{N}$-dimensional euclidean vector and ∇ the corresponding gradient. The usual continuum hamiltonian and action are given by the same expressions (1.1) and (1.2). Likewise, the discrete action is given by the same (1.5). Except for the replacements of $\mathrm{d}^3 r$ by $\mathrm{d}^{3\mathcal{N}} r$ and (2.3), (2.4) and (3.1) by

$$J = J_0 \equiv \mathrm{e}^{-T/l} \prod_{n=1}^{N+1} \left(\frac{1}{2\pi\varepsilon_n}\right)^{\frac{1}{2}3\mathcal{N}},$$

$$\langle \boldsymbol{r}' | \mathrm{e}^{\frac{1}{2}\varepsilon\nabla^2} | \boldsymbol{r} \rangle = \left(\frac{1}{2\pi\varepsilon}\right)^{\frac{1}{2}3\mathcal{N}} \exp\left(-\frac{1}{2\varepsilon}(\boldsymbol{r}-\boldsymbol{r}')^2\right),$$

$$J \propto \prod_{n=1}^{N+1} \left(\frac{1}{2\pi\varepsilon_n}\right)^{\frac{1}{2}3\mathcal{N}} f(\varepsilon_n)\,,$$

all previous formulas remain valid.

The Green function for particles obeying Bose or Fermi statistics can be obtained from that of Boltzmann statistics through the usual symmetrization or antisymmetrization, either of the initial or the final configuration.

5. Relativistic field theory

The extension of discrete quantum mechanics to the relativistic field theory is identical to the annealed version of the random lattice field theory [3–5]. There is, however, a difference in philosophy. In refs. [3–5], the lattice theory is regarded as an approximation to the continuum theory; therefore only the zero lattice-spacing limit is of physical interest. Here the opposite view is maintained: the continuum theory is considered to be an approximation. In this section we use the massless scalar field as an example to illustrate how techniques very similar to Mayer's cluster expansion [6] can be developed, which enable us to carry out a systematic calculation of corrections to the continuum theory when the average distance l between lattice sites is small but non-zero.

5.1. MASSLESS SCALAR THEORY (CONTINUUM THEORY)

We first review the trivial problem of a massless scalar field $\phi(\boldsymbol{r}, t)$ in the usual continuum theory, where t is the time and $\boldsymbol{r}$ is a $(D-1)$-dimensional euclidean vector. Let $\Pi(\boldsymbol{r}, t)$ be the conjugate momentum, and

$$H_0 = \tfrac{1}{2}\int \left[\Pi_{\rm op}^2 + \left(\frac{\partial}{\partial \boldsymbol{r}}\phi_{\rm op}\right)^2\right] {\rm d}^{D-1}r \tag{5.1}$$

be the hamiltonian operator, in which $\partial/\partial\boldsymbol{r}$ is the $(D-1)$-dimensional gradient vector and the integration is over a $(D-1)$-dimensional volume

$$V = L^{D-1}, \tag{5.2}$$

assumed to have the periodic boundary conditions. The subscript op indicates operators, so the usual commutation relation

$$[\Pi_{\rm op}(\boldsymbol{r}, t), \phi_{\rm op}(\boldsymbol{r}', t)] = -i\delta^{D-1}(\boldsymbol{r}-\boldsymbol{r}'),$$

holds. The matrix element of e^{-TH_0} between the initial configuration $\phi_0(\boldsymbol{r})$ and the final $\phi_f(\boldsymbol{r})$ can be expressed, in Feynman's path integration language, as a sum over all paths $\phi(\boldsymbol{r}, \tau)$ from $\phi = \phi_0(\boldsymbol{r})$ at $\tau = 0$ to $\phi = \phi_f(\boldsymbol{r})$ at $\tau = T$:

$$\langle \phi_f(\boldsymbol{r})|e^{-TH_0}|\phi_0(\boldsymbol{r})\rangle = \int e^{-A_C(\phi)}\, {\rm d}[\phi], \tag{5.3}$$

where τ is real (as before, $\tau = it$ is the analytic continuation of t) and

$$A_C(\phi) = \int_\Omega \frac{1}{2}\left(\frac{\partial\phi}{\partial x^\mu}\right)^2 {\rm d}^D x$$

is the continuum action; $x^\mu = (\boldsymbol{r}, \tau)$ is the component of a D-dimensional euclidean vector $\boldsymbol{x}$; the repeated index μ is summed over 1 to D and

$$\Omega = VT, \tag{5.4}$$

is a D-dimensional volume. For convenience, τ is sometimes simply referred to as "time".

Let $\phi^c(x)$ be the classical solution of the Laplace equation

$$\frac{\partial^2\phi^c}{\partial x^\mu\,\partial x^\mu} = 0, \tag{5.5}$$

which satisfies the periodic boundary conditions in the space dimension (i.e. along the $(D-1)$ $\boldsymbol{r}$ directions); in time it conforms to the boundary conditions

$$\phi^c = \phi_0(\boldsymbol{r}), \qquad \text{at } \tau = 0,$$
$$\phi^c = \phi_f(\boldsymbol{r}), \qquad \text{at } \tau = T. \tag{5.6}$$

By making the shift $\phi = \phi^c + \delta\phi$ and integrating over the fluctuation $\delta\phi$, we obtain

$$\langle \phi_f(\boldsymbol{r})| e^{-TH_0} |\phi_0(\boldsymbol{r})\rangle = \langle 0| e^{-TH_0}|0\rangle \, e^{-A_C(\phi^c)} , \tag{5.7}$$

where

$$A_C(\phi^c) = \int_\Omega \frac{1}{2}\left(\frac{\partial \phi^c}{\partial x^\mu}\right)^2 d^D x , \tag{5.8}$$

is the classical action. The normalization constant $\langle 0|e^{-TH_0}|0\rangle$ is the same matrix element when $\phi_0 = \phi_f = 0$; it is singular, as is well known. (More complicated cases when interactions are included will be discussed briefly in subsect. 5.9.)

5.2. MASSLESS SCALAR THEORY (DISCRETE THEORY)

The corresponding discrete problem is however a non-trivial one (except in its continuum limit), as we shall see. Let us consider a random lattice of N sites in the same D-dimensional euclidean volume Ω, given by (5.4). We first assume the periodic boundary conditions in τ as well as in $\boldsymbol{r}$. These N sites are linked according to the algorithm given in ref. [3]. The entire volume Ω is then divided into non-overlapping simplices S. Let

$$i_S = 1_S, 2_S, \ldots, (D+1)_S , \tag{5.9}$$

be the vertices of S. The sphere whose surface passes through these $D+1$ sites is called the circumscribed sphere. By construction, its interior is free of lattice sites.

We recall that to each site i in the random lattice there corresponds a convex D-dimensional polyhedron ω_i in the dual lattice. To each link l_{ij} connecting sites i and j, there corresponds a $(D-1)$-dimensional polyhedral surface s_{ij} of ω_i. According to ref. [5], the lattice action for a massless scalar field that satisfies periodic boundary conditions in all D directions is

$$A_{\mathscr{L}}(\phi) = \tfrac{1}{4} \sum_{i,j} \lambda_{ij} (\phi_i - \phi_j)^2 , \tag{5.10}$$

where the weight $\lambda_{ij} = \lambda_{ji}$ is given by

$$\begin{aligned} \lambda_{ij} &= 0 , \qquad \text{if sites } i \text{ and } j \text{ are not linked} ,\\ \lambda_{ij} &= s_{ij}/l_{ij} , \qquad \text{otherwise} , \end{aligned} \tag{5.11}$$

with l_{ij} denoting the length of the link and s_{ij} the $(D-1)$-dimensional "surface area" of its dual. As proved in ref. [5], the λ_{ij} thus chosen satisfies the identities

$$\sum_j l_{ij}^\mu \lambda_{ij} = 0 , \tag{5.12}$$

at all sites i, and

$$\sum_{i,j} l_{ij}^\mu l_{ij}^\nu \lambda_{ij} = 2\Omega\, \delta^{\mu\nu} . \tag{5.13}$$

Next, we cut the volume Ω open along a $(D-1)$-dimensional "plane" P perpendicular to the τ axis. Let

$$\{l_{ab}\} \text{ be the set of links cut by P}, \tag{5.14}$$

with all a sites lying on one side of P, and all b on the other side; each link l_{ab} is severed at, say, point p. Through the cut, p is turned into two end-points: p_- and p_+, with p_- connected to a and p_+ to b. We choose the set

$$\begin{aligned} &\{p_-\}, \qquad \text{to define } \tau = 0, \\ &\{p_+\}, \qquad \text{to define } \tau = T. \end{aligned} \tag{5.15}$$

The volume Ω is now open. Although it still has no boundary in the $\boldsymbol{r}$ direction, in the τ direction it is bounded by these two "planes": $\tau = 0$ and $\tau = T$. Denote ϕ_{p_-} and ϕ_{p_+} as the fields at these end-points. The corresponding lattice action is changed from $A_{\mathscr{L}}(\phi)$, given by (5.10) for the closed volume, to

$$A_{\mathrm{L}}(\phi) \equiv \langle \phi_{p_+} | A_{\mathrm{L}} | \phi_{p_-} \rangle = \tfrac{1}{4} \sum_{\alpha,\beta} \lambda_{\alpha\beta} (\phi_\alpha - \phi_\beta)^2 , \tag{5.16}$$

where the subscript L denotes the lattice whose volume is open (whereas $\mathscr{L}$ refers to the original lattice with periodic boundary conditions in all D directions), α and β can be either the original sites $1, 2, \dots, N$ or the new end-points $p_\pm$. For the former $\lambda_{\alpha\beta}$ is given by (5.11), and for the latter the only non-vanishing weights are

$$\lambda_{ap_-} = s_{ab}/l_{ap_-} , \qquad \lambda_{bp_+} = s_{ab}/l_{bp_+} , \tag{5.17}$$

where

$$l_{ap_-} + l_{bp_+} = l_{ab} . \tag{5.18}$$

(λ_{ab} is now missing in (5.16), since l_{ab} is cut.) All $\lambda_{\alpha\beta}$ are symmetric with respect to their indices, as before. Because of (5.17), the two actions for the closed and open volumes, $A_{\mathscr{L}}$ of (5.10) and A_{L} of (5.16), are related by setting $\phi_{p_-} = \phi_{p_+} = \phi_p$ and integrating over $\mathrm{d}\phi_p$:

$$\int \exp[-\langle \phi_p | A_{\mathrm{L}} | \phi_p \rangle] \prod_p \left(\frac{\lambda_{ap_-} \lambda_{bp_+}}{4\pi^2} \right)^{1/2} \mathrm{d}\phi_p = \prod_{l_{ab}} \left(\frac{\lambda_{ab}}{2\pi} \right)^{1/2} \exp[-A_{\mathscr{L}}] . \tag{5.19}$$

Since each link l_{ab} is cut only once at p, the product over p is identical to that over l_{ab}.

The discrete version of the continuum matrix element $\langle \phi_{\mathrm{f}}(\boldsymbol{r}) | \mathrm{e}^{-TH_0} | \phi_0(\boldsymbol{r}) \rangle$, given by (5.7), is

$$\langle \phi_{\mathrm{f}} | \mathscr{G}_N | \phi_0 \rangle \equiv \int J \prod_{i=1}^{N} \mathrm{d}^D x_i \, \mathrm{d}\phi_i \exp[-\langle \phi_{p_+} | A_{\mathrm{L}} | \phi_{p_-} \rangle] , \tag{5.20}$$

where

$$\phi_{p_-} = \phi_0(\boldsymbol{r}_{p_-}) , \qquad \phi_{p_+} = \phi_{\mathrm{f}}(\boldsymbol{r}_{p_+}) , \tag{5.21}$$

and J is the jacobian which is assumed to be a slowly varying function of $\boldsymbol{x}_i = (\boldsymbol{r}_i, \tau_i)$ compared to e^{-A_L}. As in the non-relativistic case, there is a certain latitude in the choice of J.

5.3. CLASSICAL LATTICE ACTION

For a given lattice L and the boundary condition (5.21), let ϕ^c_α be the classical solution which satisfies

$$\sum_\beta \lambda_{\alpha\beta}(\phi^c_\alpha - \phi^c_\beta) = 0\,, \tag{5.22}$$

at all sites $\alpha = 1, 2, \ldots, N$ in the interior of Ω, and

$$\phi^c_{p_-} = \phi_0(\boldsymbol{r}_{p_-})\,, \qquad \phi^c_{p_+} = \phi_f(\boldsymbol{r}_{p_+})\,, \tag{5.23}$$

at the boundary. The corresponding action is

$$A_L(\phi^c) = \tfrac{1}{4} \sum_{\alpha,\beta} \lambda_{\alpha\beta}(\phi^c_\alpha - \phi^c_\lambda)^2\,, \tag{5.24}$$

where α and β can be either $1, 2, \ldots, N$ or $p_\pm$.

Example. Consider first the boundary condition that $\phi_0(\boldsymbol{r})$ and $\phi_f(\boldsymbol{r})$ are both linear functions of $\boldsymbol{r}$ with the same gradient:

$$\phi_0(\boldsymbol{r}) = \bar{\phi}_0 + \boldsymbol{E}\cdot\boldsymbol{r}\,, \qquad \phi_f(\boldsymbol{r}) = \bar{\phi}_f + \boldsymbol{E}\cdot\boldsymbol{r} \tag{5.25}$$

where $\bar{\phi}_0$ and $\bar{\phi}_f$ are constants and $\boldsymbol{E}$ is a constant $(D-1)$-dimensional vector. In this case, as we shall see, the classical lattice solution at site α depends linearly on its position vector $\boldsymbol{x}_\alpha = (\boldsymbol{r}_\alpha, \tau_\alpha)$:

$$\phi^c_\alpha = \bar{\phi}_0 + \boldsymbol{\mathscr{E}}\cdot x_\alpha\,, \tag{5.26}$$

where $\boldsymbol{\mathscr{E}} = (\boldsymbol{E}_x, \mathscr{E}_\tau)$ is a D-dimensional vector whose space-component is $\boldsymbol{E}$ and time-component is

$$\mathscr{E}_\tau = (\bar{\phi}_f - \bar{\phi}_0)/T\,. \tag{5.27}$$

We observe that before the volume Ω is opened, λ_{ij} (with λ_{ab} present since, in accordance with the notation of (5.14), l_{ab} is still uncut) satisfies (5.12) and (5.13). After the cut, a is linked to p_-, instead of b; likewise b is linked to p_+. By using (5.26) we see that the left-hand side of (5.22) becomes

$$\boldsymbol{\mathscr{E}}\cdot\sum_\beta \lambda_{\alpha\beta}\boldsymbol{l}_{\alpha\beta}\,, \tag{5.28}$$

which is zero for $\alpha \neq a$ or b, on account of (5.12). When $\alpha = a$, in the above sum β now goes over p_-, instead of b. But from (5.11) and (5.17), we have

$$\lambda_{ap_-}\boldsymbol{l}_{ap_-} = \lambda_{ab}\boldsymbol{l}_{ab}\,.$$

Likewise, when $\alpha = b$,

$$\lambda_{bp_+} \boldsymbol{l}_{bp_+} = \lambda_{ba} \boldsymbol{l}_{ba} .$$

In either case (5.28) vanishes because of (5.12), and that establishes the solution (5.26).

The corresponding continuum classical solution with the same boundary condition (5.25) is

$$\phi^c(x) = \bar{\phi}_0 + \boldsymbol{\mathcal{E}} \cdot \boldsymbol{x} , \tag{5.29}$$

which is related to the lattice solution (5.26) by

$$\phi^c_\alpha = \phi^c(x_\alpha) . \tag{5.30}$$

By using (5.8), (5.13) and (5.16) we find the classical action for the continuum and the discrete cases to be identical in this simple example:

$$A_C(\phi^c) = A_L(\phi^c) = \tfrac{1}{2}\boldsymbol{\mathcal{E}}^2 \Omega . \tag{5.31}$$

From (5.7), we see that in the continuum case the matrix element of e^{-TH_0}, when $\phi_0(\boldsymbol{r})$ and $\phi_f(\boldsymbol{r})$ are given by (5.25), is

$$\langle \phi_f | e^{-TH_0} | \phi_0 \rangle = \langle 0 | e^{-TH_0} | 0 \rangle \exp(-\tfrac{1}{2}\boldsymbol{\mathcal{E}}^2 \Omega) . \tag{5.32}$$

By the same reasoning, the corresponding Green function $\mathcal{G}_N$, defined by (5.20), is given by

$$\langle \phi_f | \mathcal{G}_N | \phi_0 \rangle = \langle 0 | \mathcal{G}_N | 0 \rangle \exp(-\tfrac{1}{2}\boldsymbol{\mathcal{E}}^2 \Omega) , \tag{5.33}$$

where $\langle 0 | \mathcal{G}_N | 0 \rangle$ is the same Green function, but with the boundary condition $\phi_0 = \phi_f = 0$.

In this example, $\phi_0(\boldsymbol{r})$ and $\phi_f(\boldsymbol{r})$ are functions of $\bar{\phi}_0$, $\bar{\phi}_f$ and $\boldsymbol{E}$ appearing in the boundary condition (5.25), which in turn determines $\boldsymbol{\mathcal{E}}$. Apart from a multiplicative constant, the matrix elements of $\mathcal{G}_N$ and e^{-TH_0} have the *same* dependence on these variables. The constant $\langle 0 | e^{-TH_0} | 0 \rangle$ in the continuum theory is singular, but the corresponding $\langle 0 | \mathcal{G}_N | 0 \rangle$ in the discrete case is well-defined.

A particularly simple jacobian that corresponds to the choice (2.3) for a non-relativistic system is*

$$J = J_0 \equiv \left\{ \int \prod_{i=1}^{N} d\phi_i \exp[-\langle 0 | A_L | 0 \rangle] \right\}^{-1} , \tag{5.34}$$

where $\langle 0 | A_L | 0 \rangle$ is the same $\langle \phi_{p_+} | A_L | \phi_{p_-} \rangle$ given by (5.16), but with the boundary condition $\phi_{p_-} = \phi_{p_+} = 0$. Representing ϕ as an $N \times 1$ column matrix with ϕ_i as its elements, we may write

$$\langle 0 | A_L | 0 \rangle = \tfrac{1}{2} \tilde{\phi} M \phi , \tag{5.35}$$

* See appendix B for a supersymmetric formulation, which simplifies the jacobian.

where M is an $N \times N$ symmetric matrix, and $\tilde{\phi}$ the transpose of ϕ. Thus, J_0 is

$$J_0 = \left(\frac{1}{2\pi}\right)^{N/2} \sqrt{|M|}\,.$$

In this case, in accordance with (5.20), we have

$$\langle 0|\mathcal{G}_N|0\rangle = \Omega^N\,; \tag{5.36}$$

(5.33) then becomes

$$\langle \phi_f|\mathcal{G}_N|\phi_0\rangle = \Omega^N \exp\left(-\tfrac{1}{2}\mathcal{E}^2\Omega\right). \tag{5.37}$$

Remark. In establishing the equality between the matrix elements of $\mathcal{G}_N$ in the discrete mechanics and those of e^{-TH_0} in the usual continuum theory (apart from a proportionality constant) in this simple example, it is crucial to have (5.11) for the weight λ_{ij} so that (5.12) and (5.13) hold. Another consequence of this particular choice of λ_{ij} is that the classical lattice solution ϕ_α^c, determined by (5.22)–(5.23) for an arbitrary given set of boundary values ϕ_0 and ϕ_f, is a continuous function of the lattice sites $\boldsymbol{x}_i$ ($i = 1, 2, \ldots, N$). The details are given in appendix D.

5.4. ENSEMBLE AVERAGE AND CLUSTER EXPANSION

We now turn to the general problem of an arbitrary boundary condition (5.21). Let ϕ_α^c be the classical lattice solution and $A_L(\phi^c)$ its action, as before. Because $A_L(\phi)$, eq. (5.16), is a quadratic function of ϕ_α, and its minimum is at $\phi_\alpha = \phi_\alpha^c$, we find

$$\langle \phi_f|\mathcal{G}_N|\phi_0\rangle = \int J \prod_1^N d^D x_i\, d\phi_i \exp\left[-\langle 0|A_L|0\rangle - A_L(\phi^c)\right].$$

For convenience, we shall set $J = J_0$ given by (5.34). Hence,

$$\langle \phi_f|\mathcal{G}_N|\phi_0\rangle = \int \prod_1^N d^D x_i\, e^{-A_L(\phi^c)}, \tag{5.38}$$

where $A_L(\phi^c)$ depends on the lattice sites $\boldsymbol{x}_1, \ldots, \boldsymbol{x}_N$ and the boundary-value functions $\phi_0(\boldsymbol{r})$ and $\phi_f(\boldsymbol{r})$.

As in the non-relativistic case, in analogy to (2.2) and (2.5), we introduce the ensemble average

$$\begin{aligned}\mathcal{G} &\equiv e^{-\sigma\Omega} \sum_{N=0}^{\infty} \frac{1}{N!} \sigma^N \mathcal{G}_N \\ &= e^{-\sigma\Omega} \sum_{N=0}^{\infty} \frac{1}{N!} \sigma^N \int \prod_1^N d^D x_i\, e^{-A_L(\phi^c)},\end{aligned} \tag{5.39}$$

where $\mathscr{G}_N$ stands for the matrix whose elements are $\langle\phi_f|\mathscr{G}_N|\phi_0\rangle$,

$$\sigma \equiv l^{-D}, \tag{5.40}$$

is the fugacity and l the fundamental length of the discrete theory. Because of (5.36), it follows that when $\phi_0 = \phi_f = 0$ the matrix element of $\mathscr{G}$ is

$$\langle 0|\mathscr{G}|0\rangle = 1 .$$

We shall now show that for arbitrary ϕ_0 and ϕ_f, the following relation between $\mathscr{G}$ and the classical continuum action $A_C(\phi^c)$ holds:

$$\langle\phi_f|\mathscr{G}|\phi_0\rangle = e^{-\mathscr{A}},$$

where

$$\mathscr{A} = A_C(\phi^c) - \Omega \sum_{m=1}^{\infty} b_m(\sigma), \tag{5.41}$$

and $b_m(\sigma)$ are cluster integrals obtainable through integrations over the positions of m simplices. Like Mayer's cluster integrals of the virial expansion, $b_m(\sigma)$ has a limit when $\Omega \to \infty$. But unlike the usual virial coefficients, $b_m(\sigma) \to 0$ when $\sigma \to \infty$ (i.e. $l \to 0$) as will be seen. Consequently, when $l \to 0$, $\langle\phi_f|\mathscr{G}|\phi_0\rangle$ approaches its continuum limit

$$e^{-A_C(\phi^c)} = \frac{\langle\phi_f|e^{-TH_0}|\phi_0\rangle}{\langle 0|e^{-TH_0}|0\rangle} . \tag{5.42}$$

To establish (5.41), we rewrite (5.39) as

$$\mathscr{G} = e^{-A_C(\phi^c)} \sum_{N=0}^{\infty} \frac{\sigma^N}{N!} e^{-\sigma\Omega} \int \prod_1^N d^D x_i \, e^{-A_L(\phi^c) + A_C(\phi^c)} , \tag{5.43}$$

and decompose the classical actions $A_C(\phi^c)$ and $A_L(\phi^c)$ for the continuum and discrete cases as sums over simplices S (for given lattice sites $\boldsymbol{x}_1, \ldots, \boldsymbol{x}_N$):

$$A_C(\phi^c) = \sum_S A_C(S) , \qquad A_L(\phi^c) = \sum_S A_L(S) \tag{5.44}$$

where

$$A_C(S) = \int_S \frac{1}{2}\left(\frac{\partial\phi^c}{\partial x^\mu}\right)^2 d^D x ,$$

$$A_L(S) \equiv \tfrac{1}{4} \sum_{i,j} \lambda_{ij}^{\text{int}} (\phi_i^c - \phi_j^c)^2 , \tag{5.45}$$

in which i and j run over all vertices, 1 to $D+1$, of the simplex S. To derive $\lambda_{ij}^{\text{int}}$, we first separate the "area" s_{ij} of the dual to the link l_{ij} into two parts:

$$s_{ij} = s_{ij}^{\text{int}} + s_{ij}^{\text{ext}} , \tag{5.46}$$

where s_{ij}^{int} is the section inside S, and s_{ij}^{ext} that outside. If l_{ij} does not intersect s_{ij}, then s_{ij}^{int} may be negative, given by the analytic continuation of the case when l_{ij} does go through s_{ij}. From (5.11), we see that λ_{ij} is proportional to s_{ij}, and therefore can be similarly decomposed:

$$\lambda_{ij} = \lambda_{ij}^{\text{int}} + \lambda_{ij}^{\text{ext}}, \tag{5.47}$$

with

$$\lambda_{ij}^{\text{int}} = s_{ij}^{\text{int}}/l_{ij}, \qquad \lambda_{ij}^{\text{ext}} = s_{ij}^{\text{ext}}/l_{ij}.$$

Defining f_{S} by

$$\exp[-A_{\text{L}}(\text{S}) + A_{\text{C}}(\text{S})] = 1 + f_{\text{S}}, \tag{5.48}$$

we have

$$\begin{aligned}\int \prod_1^N \mathrm{d}^D x_i \, \mathrm{e}^{-A_{\text{L}}(\phi^c) + A_{\text{C}}(\phi^c)} &= \int \prod_1^N \mathrm{d}^D x_i \prod_{\text{S}} (1 + f_{\text{S}}) \\ &= \int \prod_1^N \mathrm{d}^D x_i \Big[1 + \sum_{\text{S}} f_{\text{S}} + \sum_{\text{S,S}'} f_{\text{S}} f_{\text{S}'} + \cdots \Big].\end{aligned} \tag{5.49}$$

By introducing

$$f(1, 2, \ldots, D+1) \equiv \begin{cases} f_{\text{S}} \text{ if } 1, \ldots, D+1 \text{ are vertices of S} \\ 0, \quad \text{otherwise}, \end{cases} \tag{5.50}$$

we shall establish

$$\int \prod_1^N \mathrm{d}^D x_i \sum_{\text{S}} f_{\text{S}} = \frac{N!}{(N-n)!\,n!} \int (\Omega - \mathcal{S})^{N-n} f(1, 2, \ldots, n) \prod_1^n \mathrm{d}^D x_i, \tag{5.51}$$

where

$$n = D + 1$$

and $\mathcal{S}$ is the volume of the circumscribed sphere of S. The proof is as follows. Consider a specific term f_{S} on the left-hand side. Suppose $1, 2, \ldots, n$ are the vertices of S, then by our construction of the random lattice [3], the remaining sites $n+1, \ldots, N$ must be outside $\mathcal{S}$; this gives the factor $(\Omega - \mathcal{S})^{N-n}$ after the integration. Eq. (5.51) now follows from the usual combinatorial consideration and the definition (5.50).

We then multiply (5.51) by $\sigma^N \mathrm{e}^{-\sigma\Omega}/N!$ and sum over N:

$$\begin{aligned}\sum_{N=0}^{\infty} \frac{\sigma^N}{N!} \mathrm{e}^{-\sigma\Omega} \int \prod_1^N \mathrm{d}^D x_i \sum_{\text{S}} f_{\text{S}} &= \sum_{N=n}^{\infty} \frac{\sigma^N \mathrm{e}^{-\sigma\Omega}}{(N-n)!\,n!} \int (\Omega - \mathcal{S})^{N-n} f(1, 2, \ldots, n) \prod_1^n \mathrm{d}^D x_i \\ &\equiv \Omega b_1 .\end{aligned} \tag{5.52}$$

Carrying out the sum from $N-n=0$ to ∞, we derive

$$b_1=\frac{1}{\Omega}\int\frac{\sigma^{D+1}}{(D+1)!}\,\mathrm{e}^{-\sigma\mathscr{S}}f(1,2,\ldots,D+1)\prod_{1}^{D+1}\mathrm{d}^D x_i\,. \tag{5.53}$$

The factor $\mathrm{e}^{-\sigma\mathscr{S}}$ restricts the integration region of all relative coordinates between $1,2,\ldots,D+1$. Hence b_1 has a limit when $\Omega\to\infty$ (assuming that all derivatives of ϕ^c are O(1)). Next, we replace $\sum_S f_S$ on the left-hand side of (5.52) by the double sum $\sum_{S,S'} f_S f_{S'}$. If the integrations over these two simplices S and S′ were independent, then the result would be simply $\frac{1}{2}(\Omega b_1)^2$. The deviation from $\frac{1}{2}(\Omega b_1)^2$ can be easily seen to be proportional to Ω, which gives rise to Ωb_2.

To derive b_2 explicitly, let us write

$$\sum_{S,S'} f_S f_{S'}=\Big\{\sum_{[SS']}+\sum_{(SS')}+\sum_{(S)(S')}\Big\}f_S f_{S'}\,, \tag{5.54}$$

where [SS′] means that S and S′ have at least one vertex in common, (SS′) means that they have no vertex in common but their circumspheres overlap, and (S)(S′) refers to the case of non-overlapping circumspheres. (By construction, the interiors of all circumspheres of simplices are free of lattice sites.) Similar to (5.52), we define b_2 by

$$\sum_{N=0}^{\infty}\frac{\sigma^N}{N!}\,\mathrm{e}^{-\sigma\Omega}\int\prod_1^N\mathrm{d}^D x_i\sum_{S,S'} f_S f_{S'}\equiv\tfrac{1}{2}(\Omega b_1)^2+\Omega b_2\,.$$

Hence, b_2 is given by

$$\begin{aligned}
b_2=&\frac{1}{\Omega}\sum_{n=D+2}^{2D+1}\frac{\sigma^n}{n!}\int_{[S_1S_2]}\mathrm{e}^{-\sigma\mathscr{S}}f_{S_1}f_{S_2}\prod_1^n\mathrm{d}^D x_i\\
&+\frac{1}{\Omega}\frac{\sigma^{2D+2}}{(2D+2)!}\int_{(S_1S_2)}[\mathrm{e}^{-\sigma\mathscr{S}}-\mathrm{e}^{-\sigma(\mathscr{S}_1+\mathscr{S}_2)}]f_{S_1}f_{S_2}\prod_1^{2D+2}\mathrm{d}^D x_i\\
&-\frac{1}{\Omega}\frac{\sigma^{2D+2}}{(2D+2)!}\int_{R}\mathrm{e}^{-\sigma(\mathscr{S}_1+\mathscr{S}_2)}f_{S_1}f_{S_2}\prod_1^{2D+2}\mathrm{d}^D x_i\,,
\end{aligned} \tag{5.55}$$

where $\mathscr{S}_1$ and $\mathscr{S}_2$ are the volumes of the circumspheres of S_1 and S_2, $\mathscr{S}$ is their union. The integration $\int_{[S_1S_2]}$ extends over the region when the n vertices ($2D+1\geqslant n\geqslant D+2$) form two connected simplicies, and $\int_{(S_1S_2)}$ over the region when the $2D+2$ vertices form two disconnected simplices but with overlapping circumspheres. The last integral is over a region R in which the $2D+2$ sites are placed in such a way that half of the sites lie on the surface of one sphere and half on one other sphere; the volumes of these two spheres, say $\mathscr{S}_1$ and $\mathscr{S}_2$, overlap and at least one of their interiors is not empty. (Hence, R is not included in the sum (5.54), but it does contribute to $\frac{1}{2}(\Omega b_1)^2$.)

5.5. MEAN-FIELD APPROXIMATION

The dependence of the cluster integral b_m on the average lattice spacing l as $l \to 0$ can be most easily arrived at by using the mean-field approximation, even though our main conclusions are independent of this approximation (as will be discussed later in subsect. 5.7).

Zeroth approximation. Here we simply approximate the classical discrete solution ϕ_α^c by the corresponding classical continuum solution $\phi^c(x)$ at α:

$$\phi_\alpha^c = \phi^c(x_\alpha) . \tag{5.56}$$

First approximation. As an improvement, in (5.22) we apply the zeroth approximation to all ϕ_β^c on the sites β linked to α; i.e. ϕ_α^c is determined by (5.22) in which all ϕ_β^c are approximated by $\phi^c(x_\beta)$, the continuum solution at β. Hence

$$\phi_\alpha^c = \sum_\beta \lambda_{\alpha\beta}\phi^c(x_\beta) \Big/ \sum_\beta \lambda_{\alpha\beta} . \tag{5.57}$$

Expand the classical continuum solution $\phi^c(x)$ in powers of $x - x_\alpha$:

$$\phi^c(x) = \phi^c(x_\alpha) + (x - x_\alpha)^\mu \left(\frac{\partial \phi^c}{\partial x^\mu}\right)_\alpha + \tfrac{1}{2}(x - x_\alpha)^\mu (x - x_\alpha)^\nu \left(\frac{\partial^2 \phi^c}{\partial x^\mu\, \partial x^\nu}\right)_\alpha + \cdots ,$$

where, as before, the subscript α denotes the lattice site, the superscript c indicates the classical solution and μ, ν refer to the space-time indices which, because of their repetition, are summed over from 1 to D. When $x = x_\beta$, we have

$$(x_\beta - x_\alpha)^\mu = l_{\alpha\beta}^\mu ,$$

the μth component of the link vector $\boldsymbol{l}_{\alpha\beta}$. Write

$$\phi_\alpha^c = \phi^c(x_\alpha) + \delta\phi_\alpha^c . \tag{5.58}$$

Because of the identity (5.12), we find the correction $\delta\phi_\alpha^c$ to be $O(l^2)$. More explicitly,

$$\delta\phi_\alpha^c = \frac{1}{2}\left(\frac{\partial^2 \phi^c}{\partial x^\mu\, \partial x^\nu}\right)_\alpha \sum_\beta \lambda_{\alpha\beta} l_{\alpha\beta}^\mu l_{\alpha\beta}^\nu \Big/ \sum_\beta \lambda_{\alpha\beta} + O(l^3) . \tag{5.59}$$

Furthermore, for a random lattice if $l_{\alpha\beta}^\mu l_{\alpha\beta}^\nu$ is replaced by its average (over the link direction)

$$\langle l_{\alpha\beta}^\mu l_{\alpha\beta}^\nu \rangle = \frac{1}{D} l_{\alpha\beta}^2 \delta^{\mu\nu} ,$$

then the corresponding correction becomes

$$\langle \delta\phi_\alpha^c \rangle = \frac{\sum_\beta \lambda_{\alpha\beta} l_{\alpha\beta}^2}{2D \sum_\beta \lambda_{\alpha\beta}} \frac{\partial^2 \phi^c}{\partial x^\mu\, \partial x^\mu} + O(l^3) .$$

Since the classical continuum solution $\phi^c(x)$ satisfies $\partial^2\phi^c/\partial x^\mu\,\partial x^\mu = 0$, we have

$$\langle \delta\phi^c_\alpha \rangle = O(l^3)\,. \tag{5.60}$$

(Note that for a constant $\partial^2\phi^c/\partial x^\mu\,\partial x^\nu$, because of (5.13) the average of $\delta\phi^c_\alpha$ over all lattice sites α is zero.) In any case, we see that the correction to the zeroth approximation (5.56) is $O(l^2)$; because of (5.60), its magnitude is down by an additional factor

$$\sim (N_{1/0})^{-1/2} \tag{5.61}$$

where $N_{1/0}$ is the average number of links per site. These numbers have been computed in ref. [3], and $N_{1/0}$ is found to be 6 for $D=2$, 15.54 for $D=3$, 37.78 for $D=4$, etc. Consequently, when the dimension $D\to\infty$, the first-order correction $\delta\phi^c_\alpha$ is even smaller, only $O(l^3)$.

Remark. We may wonder whether the lattice solution ϕ^c_α should be expanded around a mean field $\bar\phi(x)$, different from the classical continuous solution $\phi^c(x)$. For example, for a regular square (or cubic) lattice, to $O(l^2)$ the best mean field $\bar\phi(x)$ is not a solution of the laplace equation; instead, it satisfies eq. (C.41) given in appendix C (with $l=\varepsilon$ and $\bar\phi(x)=\phi^c_L$). The situation is, however, different for a random lattice: $\bar\phi(x)$ should be rotationally and reflectionally symmetric; hence, for dimensional reasons its differential equation must be of the form

$$\nabla^2(1+l^2c_1\nabla^2+l^4c_2(\nabla^2)^2+\cdots)\bar\phi(x)=0\,, \tag{5.62}$$

where $c_1, c_2, \ldots$ are dimensionless constants. Since any solution of $\nabla^2\bar\phi=0$ is also a solution of (5.62), we may set $\bar\phi(x)=\phi^c(x)$ without any loss of generality. (The other factor in (5.62) would lead to $(1+l^2c_1\nabla^2+\cdots)\bar\phi(x)=0$ which gives a rapidly varying solution, clearly unsuitable for a mean field.)

5.6. EVALUATION OF b_1

Using M, the center of mass of a simplex S, as the origin, let us denote $\boldsymbol{x}$ as the position vector of any point in S. Hence, the integral of $\boldsymbol{x}$ over the simplex is zero,

$$\int_S \boldsymbol{x}\, d^Dx = 0\,.$$

The classical continuum solution $\phi^c(x)$, determined by (5.5) and (5.6), can be expanded in a power series in x:

$$\phi^c(x)=\phi^c(M)+E^\mu x^\mu+\tfrac{1}{2}T^{\mu\nu}x^\mu x^\nu+\cdots\,, \tag{5.63}$$

where $\phi^c(M)$ is $\phi^c(x)$ at M,

$$E^\mu=\frac{\partial\phi^c}{\partial x^\mu}\,,\quad \text{at } M\,,$$
$$T^{\mu\nu}=\frac{\partial^2\phi^c}{\partial x^\mu\,\partial x^\mu}\,,\quad \text{at } M\,. \tag{5.64}$$

The classical continuum action over the simplex S is, in accordance with (5.45),

$$A_C(S) = \tfrac{1}{2}vE^2 + \tfrac{1}{2}\,\text{trace}\, IT^2 + O(vl^3)\,, \tag{5.65}$$

where

$$v = \text{volume of S}\,, \tag{5.66}$$

I is the matrix of the moment of inertia tensor

$$I^{\mu\nu} \equiv \int_S x^\mu x^\nu\, d^D x\,, \tag{5.67}$$

and T^2 is the square of

$$T = \text{matrix } T^{\mu\nu}\,. \tag{5.68}$$

In (5.65) we regard the linear size of the simplex to be $O(l)$ where $l = \sigma^{-1/D}$, as in (5.40).

To illustrate how we may evaluate the corresponding lattice action $A_L(S)$, let us consider the special case of $D = 2$. The simplex is then a triangle; its "volume" v is simply the area Δ of the triangle, related to the sides a, b, c by the familiar formula

$$v = \Delta = \tfrac{1}{4}[(a+b+c)(a+b-c)(a-b+c)(-a+b+c)]^{1/2}\,. \tag{5.69}$$

From theorem 3 established in ref. [5], the lattice action $A_L(S)$, defined by (5.44)–(5.45) is

$$A_L(S) = \tfrac{1}{2}\boldsymbol{E}_L^2\Delta\,, \tag{5.70}$$

where*

$$\boldsymbol{E}_L = \frac{1}{2\Delta}(\boldsymbol{a}\phi_1^c + \boldsymbol{b}\phi_2^c + \boldsymbol{c}\phi_3^c), \tag{5.71}$$

with ϕ_i^c as the classical lattice solution at the vertex located at $\boldsymbol{x}_i$,

$$\boldsymbol{a} = \boldsymbol{x}_3 - \boldsymbol{x}_2\,, \qquad \boldsymbol{b} = \boldsymbol{x}_1 - \boldsymbol{x}_3\,,$$

$$\boldsymbol{c} = \boldsymbol{x}_2 - \boldsymbol{x}_1\,,$$

as shown in fig. 1.

Using the zeroth order of the mean-field approximation, we set $\phi_i^c = \phi^c(x_i)$ in accordance with (5.56). Substituting (5.63) into (5.71), we see that, neglecting higher-order corrections in l,

$$\boldsymbol{E}_L = \boldsymbol{E} \times \hat{z} + \boldsymbol{E}_L^T\,, \tag{5.72}$$

where $\hat{z}$ is a unit vector normal to the two-dimensional plane that contains the

* The field $\boldsymbol{E}_L$ is related to the vector field "$\boldsymbol{E}$", defined by eq. (A.5) in ref. [5], by a 90° rotation; their magnitudes are identical.

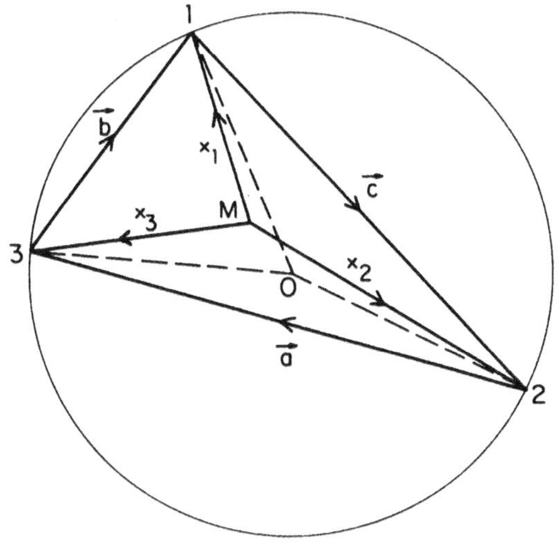

Fig. 1. The vertices of the triangle $\triangle$ are 1, 2, 3; its c.m. is M and the center of its circumscribed circle is O. The angles α, β, γ, used in appendix E, are $\angle 203$, $\angle 301$ and $\angle 102$.

triangle, and

$$\boldsymbol{E}_{\mathrm{L}}^{T}=\frac{1}{4\triangle}(\boldsymbol{a}x_1^{\mu}x_1^{\nu}+\boldsymbol{b}x_2^{\mu}x_2^{\nu}+\boldsymbol{c}x_3^{\mu}x_3^{\nu})T^{\mu\nu}\,, \tag{5.73}$$

with $\boldsymbol{E}$ and T defined by (5.64). Thus,

$$A_{\mathrm{L}}(\mathrm{S})=\tfrac{1}{2}\boldsymbol{E}^2\triangle+\hat{z}\cdot(\boldsymbol{E}_{\mathrm{L}}^{T}\times\boldsymbol{E})\triangle+\tfrac{1}{2}(\boldsymbol{E}_{\mathrm{L}}^{T})^2\triangle+\mathrm{O}(l^3\triangle)\,. \tag{5.74}$$

The difference between (5.65) and the above expression gives

$$-A_{\mathrm{L}}(\mathrm{S})+A_{\mathrm{C}}(\mathrm{S})=-\hat{z}\cdot(\boldsymbol{E}_{\mathrm{L}}^{T}\times\boldsymbol{E})\triangle-\tfrac{1}{2}(\boldsymbol{E}_{\mathrm{L}}^{T})^2\triangle+\tfrac{1}{2}\,\mathrm{trace}\,IT^2+\mathrm{O}(l^3\triangle)\,. \tag{5.75}$$

From (5.48), we see that

$$f_{\mathrm{S}}=-\hat{z}\cdot(\boldsymbol{E}_{\mathrm{L}}^{T}\times\boldsymbol{E})\triangle-\tfrac{1}{2}(\boldsymbol{E}_{\mathrm{L}}^{T})^2\triangle+\tfrac{1}{2}\,\mathrm{trace}\,IT^2+\mathrm{O}(l^3\triangle)\,, \tag{5.76}$$

in which $\boldsymbol{E}$ and T are counted as $\mathrm{O}(l^0)$, $\boldsymbol{E}_{\mathrm{L}}^{T}$ as $\mathrm{O}(l)$, and therefore

$$(\boldsymbol{E}_{\mathrm{L}}^{T}\times\boldsymbol{E})\triangle=\mathrm{O}(l^3)\,,$$

whose square, being $\mathrm{O}(l^6)=\mathrm{O}(l^4\triangle)$, is not exhibited explicitly. According to (5.53), when $D=2$

$$b_1=\frac{1}{\Omega}\int\frac{\sigma^3}{3!}\,\mathrm{e}^{-\sigma\pi R^2}f_{\mathrm{S}}\prod_1^3\mathrm{d}^2x_i\,, \tag{5.77}$$

where

R = radius of the circumscribed circle.

Keeping M and the shape and size of the triangle fixed, but averaging over the azimuthal angle (around $\hat{z}$), we have, since $\boldsymbol{E}_{\mathrm{L}}^{T}$ is odd in $\boldsymbol{a}$, $\boldsymbol{b}$ and $\boldsymbol{c}$,

$$\langle \boldsymbol{E}_{\mathrm{L}}^{T} \times \boldsymbol{E} \rangle = 0 .$$

Therefore,

$$\langle f_{\mathrm{S}} \rangle = \mathrm{O}(l^2 \triangle) . \tag{5.78}$$

The integration (5.77) is carried out in appendix E, and the result is (neglecting higher orders in l)

$$b_1 = -\frac{l^2}{24\pi} \frac{1}{\Omega} \int \left(\frac{\partial^2 \phi^{\mathrm{c}}}{\partial x^{\mu}\, \partial x^{\nu}} \right) \left(\frac{\partial^2 \phi^{\mathrm{c}}}{\partial x^{\mu}\, \partial x^{\nu}} \right) \mathrm{d}^2 x , \tag{5.79}$$

where $l^2 = 1/\sigma$ since $D = 2$. Because so far we have used only the zeroth order mean-field approximation, the above b_1 is not exact, even to $\mathrm{O}(l^2)$.

5.7. l-DEPENDENCE OF b_m

The cluster expansion (5.41) is established without any approximation. Given the classical continuum solution $\phi^{\mathrm{c}}(x)$ and the positions of all lattice sites, the classical lattice solution $\phi_{\alpha}^{\mathrm{c}}$ is determined (since $\phi^{\mathrm{c}}(x)$ contains the information of the boundary). When $l \to 0$, $\phi_{\alpha}^{\mathrm{c}}$ becomes $\phi^{\mathrm{c}}(x_{\alpha})$. For $l \neq 0$, but $\phi^{\mathrm{c}}(x)$ a linear function of x, we also have $\phi_{\alpha}^{\mathrm{c}} = \phi^{\mathrm{c}}(x_{\alpha})$, in accordance with (5.30). In the mean-field approach, but without restriction to zeroth-order approximation, when l is small $\phi_{\alpha}^{\mathrm{c}}$ can be written as (e.g. in (5.58)–(5.59))

$$\phi_{\alpha}^{\mathrm{c}} = \phi^{\mathrm{c}}(x_{\alpha}) + L^{\mu\nu}(\alpha) \left(\frac{\partial^2 \phi^{\mathrm{c}}}{\partial x^{\mu}\, \partial x^{\nu}} \right)_{\alpha} + \cdots ,$$

where $L^{\mu\nu}(\alpha)$ is $\mathrm{O}(l^2)$ and depends on the lattice. As discussed in subsect. 5.5, $L^{\mu\nu}(\alpha)$ depends on the sites β linked to α in the first approximation, and also on the sites γ linked to β in the second approximation, etc. For a random lattice, because of (5.61), the corrections due to each successive approximation are, on the average, down by a factor $(N_{1/0})^{-1/2}$. Hence, we expect $L^{\mu\nu}(\alpha)$, and likewise $\phi_{\alpha}^{\mathrm{c}}$, to depend only on the lattice structure in a region near α. From (5.45) and (5.48), a similar conclusion can also be reached for f_{S}; furthermore, to the lowest powers of l, f_{S} depends quadratically on ϕ^{c}. By using (5.53) and dimensional considerations and rotational symmetry, we see that b_1 can be written as

$$b_1 = -\frac{1}{2\mu^2} \frac{1}{\Omega} \int \frac{\partial^2 \phi^{\mathrm{c}}}{\partial x^{\lambda}\, \partial x^{\nu}} \frac{\partial^2 \phi^{\mathrm{c}}}{\partial x^{\lambda}\, \partial x^{\nu}} \mathrm{d}^D x + \mathrm{O}(l^4) , \tag{5.80}$$

where $\mu^2 = \mathrm{O}(l^{-2})$. (This follows because $\phi^{\mathrm{c}}(x)$ satisfies the laplace equation; all other invariants, quadratic in ϕ^{c} and having four derivatives such as $(\partial^2 \phi^{\mathrm{c}}/\partial x^{\mu}\, \partial x^{\mu})^2$, $(\partial \phi^{\mathrm{c}}/\partial x^{\nu})(\partial^3 \phi^{\mathrm{c}}/\partial x^{\mu}\, \partial x^{\mu}\, \partial x^{\nu})$ and $\phi^{\mathrm{c}}(\partial^4 \phi^{\mathrm{c}}/\partial x^{\mu}\, \partial x^{\mu}\, \partial x^{\nu}\, \partial x^{\nu})$, are all zero.) When the dimension $D=2$, and if we use only the zeroth-order mean-field approximation then $\mu^2 \simeq 12\pi/l^2$, as given by (5.79).

Next, we shall show that in general as $l \to 0$

$$b_m \sim \begin{cases} l^{D(m-1)+m}, & \text{if } m \text{ is even} \\ l^{D(m-1)+m+1}, & \text{if } m \text{ is odd}. \end{cases} \tag{5.81}$$

Let us recall, once again, that $A_L(S)$ and $A_C(S)$, given by (5.45), are both quadratic functions of ϕ^c; their difference vanishes when either $l \to 0$, or ϕ^c is a linear function of x, in accordance with (5.29)–(5.31). Thus, as in (5.75) we may write

$$-A_L(S)+A_C(S)=O(lvET)+O(l^2vT^2)+O(l^3v),$$

where $v=O(l^D)$ is the volume of the simplex, E and T represent $\partial\phi^c/\partial x^\mu$ and $\partial^2\phi^c/\partial x^\mu\,\partial x^\nu$ at M, the c.m. of S. Hence,

$$f_S=O(lvET)+O(l^2vT^2)+O(l^3v). \tag{5.82}$$

Both E and T are $O(l^0)$. Substituting (5.82) into (5.55) we note that the integration

$$\int \prod_1^n d^D x_i\, e^{-\sigma\mathscr{S}} \sim \sigma^{-n+1}\Omega, \tag{5.83}$$

and (in the integrand) the factor $f_{S_1}f_{S_2} \sim (lvET)^2$. Thus

$$b_2 \sim (lvET)^2\sigma \sim l^{D+2},$$

since $v \sim \sigma^{-1} \sim l^D$.

To give an estimate of b_m, we may write, similar to (5.55),

$$b_m = \frac{1}{\Omega}\sum_{n=D+m}^{m(D+1)-1}\frac{\sigma^n}{n!}\int_{[S_1\cdots S_m]} e^{-\sigma\mathscr{S}} f_{S_1}f_{S_2}\cdots f_{S_m}\prod_1^n d^D x_i+\cdots, \tag{5.84}$$

where the integration $[S_1\cdots S_m]$ extends over the region where these n vertices form m connected simplices with $\mathscr{S}$ as the union of their circumspheres. The remaining $\cdots$ is the generalization of the second and third terms in (5.55). We note that the factor $f_{S_1}f_{S_2}\cdots f_{S_m}$ is

$$\sim (lvET)^m, \qquad \text{if } m \text{ is even},$$

but

$$\sim (lvET)^{m-1}(l^2vT^2), \qquad \text{if } m \text{ is odd}. \tag{5.85}$$

The difference occurs because for an odd m, $(ET)^m$ is odd in $(\partial\phi^c/\partial x^\lambda)\partial^2\phi^c/\partial x^\mu\,\partial x^\nu$ which gives zero after averaging over the overall angular orientation of the simplices. By using (5.83)–(5.85) and $v \sim \sigma^{-1} \sim l^D$, we prove (5.81).

From (5.85) we see that each b_m depends on at least $2m$ powers of the field. Hence, the discrete version of a continuum "free" field is not a free theory.

5.8. THE EFFECTIVE HAMILTONIAN

From (5.41) we see that the discrete version of $\langle\phi_f|e^{-TH_0}|\phi_0\rangle$, where H_0 is the free continuum hamiltonian operator (5.1), is

$$\langle\phi_f|\mathcal{G}|\phi_0\rangle = e^{-\mathcal{A}} = \exp\left(-A_C(\phi^c) + \Omega \sum_m b_m\right), \tag{5.86}$$

apart from a normalization constant independent of ϕ_0 and ϕ_f. To $O(l^2)$, on account of (5.8) and (5.80), $\mathcal{A}$ is given by

$$\mathcal{A} = \int \frac{1}{2}\left[\left(\frac{\partial\phi^c}{\partial x^\lambda}\right)^2 + \frac{1}{\mu^2}\left(\frac{\partial^2\phi^c}{\partial x^\lambda\,\partial x^\nu}\right)^2\right] d^D x, \tag{5.87}$$

where $\mu^2 = \text{constant}/l^2$ and ϕ^c is the solution of the laplace equation (5.5) with the boundary condition (5.6). In this section we shall show that to this order the matrix elements of $\mathcal{G}$ can be derived by using an effective hamiltonian H (in the continuum theory):

$$\langle\phi_f|\mathcal{G}|\phi_0\rangle = \frac{\langle\phi_f|e^{-TH}|\phi_0\rangle}{\langle 0|e^{-TH}|0\rangle}, \tag{5.88}$$

where

$$H = \tfrac{1}{2}\int\left[M^{-1}\Pi_{\mathrm{op}}^2 + M\left(\frac{\partial}{\partial \boldsymbol{r}}\phi_{\mathrm{op}}\right)^2\right] d^D x, \tag{5.89}$$

$$M = 1 - \frac{2}{\mu^2}\left(\frac{\partial}{\partial \boldsymbol{r}}\right)^2 .$$

Proof. Consider the standard Fourier expansion:

$$\begin{aligned}\phi_{\mathrm{op}}(\boldsymbol{r},t) &= \frac{1}{\sqrt{V}}\left\{q_0(t) + \sum_{\boldsymbol{k}}{}' \sqrt{2}[q_{\boldsymbol{k},c}(t)\cos(\boldsymbol{k}\cdot\boldsymbol{r}) + q_{\boldsymbol{k},s}(t)\sin(\boldsymbol{k}\cdot\boldsymbol{r})]\right\},\\ \Pi_{\mathrm{op}}(\boldsymbol{r},t) &= \frac{1}{\sqrt{V}}\left\{p_0(t) + \sum_{\boldsymbol{k}}{}' \sqrt{2}\,[p_{\boldsymbol{k},c}(t)\cos(\boldsymbol{k}\cdot\boldsymbol{r}) + p_{\boldsymbol{k},s}(t)\sin(\boldsymbol{k}\cdot\boldsymbol{r})]\right\},\end{aligned} \tag{5.90}$$

where V is given by (5.2) and $\sum_{\boldsymbol{k}}'$ sums over the half-$\boldsymbol{k}$ space (excluding $\boldsymbol{k}=0$). In terms of these Fourier components, the hamiltonian of a massless scalar field in the usual continuum theory is

$$H_0 = \sum_\alpha \tfrac{1}{2}[p_\alpha^2 + k^2 q_\alpha^2],$$

and the effective hamiltonian (5.89) becomes

$$H = \sum_\alpha \tfrac{1}{2}\left[\left(1 + \frac{2k^2}{\mu^2}\right)^{-1} p_\alpha^2 + \left(1 + \frac{2k^2}{\mu^2}\right)k^2 q_\alpha^2\right], \tag{5.91}$$

where $k=|\boldsymbol{k}|$ and the sum extends over all modes

$$\alpha=0;\ \boldsymbol{k},c \quad \text{and} \quad \boldsymbol{k},s \tag{5.92}$$

with the convention that $\alpha=0$ means $\boldsymbol{k}=0$.

Let $q_\alpha^c(\tau)$ be the corresponding Fourier component of the classical free continuum solution $\phi^c(x)$, which satisfies, as before,

$$\frac{\partial^2}{\partial x^\mu\,\partial x^\mu}\phi^c(x)=0\,,$$

with $x^\mu=(\boldsymbol{r},\tau)$ and $\tau=it$ (the analytic continuation of t). Hence, $q_\alpha^c(\tau)$ is the solution of

$$\frac{d^2}{d\tau^2}q_\alpha^c=k^2q_\alpha^c\,,$$

and can be written as

$$q_\alpha^c(\tau)=\frac{1}{\sinh kT}\left[q_\alpha(T)\sinh k\tau+q_\alpha(0)\sinh\left[k(T-\tau)\right]\right], \tag{5.93}$$

with $q_\alpha(0)$ and $q_\alpha(T)$ the Fourier components of $\phi_0(\boldsymbol{r})$ and $\phi_f(\boldsymbol{r})$ at the boundary $\tau=0$ and $\tau=T$, introduced in (5.6). Substituting this solution into (5.87), we obtain

$$\mathscr{A}=\sum_\alpha\frac{k}{2\sinh kT}\left(1+\frac{2k^2}{\mu^2}\right)\{[q_\alpha(T)^2+q_\alpha(0)^2]\cosh kT-2q_\alpha(T)q_\alpha(0)\}\,. \tag{5.94}$$

In the usual continuum theory, the matrix elements of e^{-TH_0} are

$$\langle\phi_f|e^{-TH_0}|\phi_0\rangle=\langle 0|e^{-TH_0}|0\rangle e^{-A_C(\phi^c)}\,,$$

where

$$A_C(\phi^c)=\sum_\alpha\frac{k}{2\sinh kT}\{[q_\alpha(T)^2+q_\alpha(0)^2]\cosh kT-2q_\alpha(T)q_\alpha(0)\}\,. \tag{5.95}$$

By making the canonical transformations

$$Q_\alpha=\left(1+\frac{2k^2}{\mu^2}\right)^{1/2}q_\alpha\,,$$
$$P_\alpha=\left(1+\frac{2k^2}{\mu^2}\right)^{-1/2}p_\alpha\,, \tag{5.96}$$

we see that the effective hamiltonian (5.91) becomes

$$H=\sum_\alpha\tfrac{1}{2}[P_\alpha^2+k^2Q_\alpha^2]\,. \tag{5.97}$$

Hence, (5.88) follows from (5.94)–(5.97).

Remarks. (i) The energy spectra of H and H_0 are exactly the same. If we have a regular lattice, then there would be an energy shift to $O(l^2)$. This difference can be traced to the different symmetry properties in these two cases. From eq. (C.41) in appendix C we see that the $O(l^2)$ frequency shift in a regular lattice is the consequence of a mean-field equation which has only 90° rotational symmetry. On the other hand, in our discrete theory the corresponding equation must be fully rotationally symmetric, which reduces to the Laplace equation as commented on at the end of subsect. 5.5.

(ii) The continuum propagator for a free massless scalar field between two points 1 and 2 in a D-dimensional euclidean space is (for $D>2$).

$$\mathscr{D}_C = \frac{1}{D-2}\begin{cases} \dfrac{(D-2)!!}{(2\pi)^{D/2}} R^{2-D}, & \text{if } D \text{ is even} \\[2ex] \dfrac{(D-2)!!}{2(2\pi)^{(D-1)/2}} R^{2-D}, & \text{if } D \text{ is odd}, \end{cases} \tag{5.98}$$

where

$$R = |\boldsymbol{x}_1 - \boldsymbol{x}_2|,$$

is the distance between these two points. Because to $O(l^2)$ the effective hamiltonian in the discrete theory is related to the continuum free hamiltonian by a canonical transformation, the corresponding propagator $\mathscr{D}_L$ has the same dependence on $\boldsymbol{R}$ as $\mathscr{D}_C$ at large distances. We have

$$\mathscr{D}_L = \mathscr{D}_C[1 + O(e^{-R/l})]; \tag{5.99}$$

i.e. there is no $O(l^2/R^2)$ correction.

(iii) To higher order, as remarked before, b_m depends on at least $2m$ powers of the field. Hence, the effective hamiltonian H is quite different from that in a free continuum theory.

(iv) The hermiticity of H insures the unitarity of the S-matrix, as in the non-relativistic case.

5.9. DISCUSSIONS

(i) Instead of J_0, given by (5.34), we assume the jacobian J to be of the form

$$J = J_0 \prod_S (1 + g_S),$$

where g_S depends only on the lattice structure near a simplex S. Replacing f_S, defined by (5.48), by $f_S + g_S + f_S g_S$ we can use the same argument given in subsect. 5.4, and establish the cluster expansion (5.41).

(ii) For an interacting field, the usual continuum hamiltonian becomes, instead of (5.1),

$$H_0 = \int \left[\tfrac{1}{2}\Pi_{op}^2 + \tfrac{1}{2}\left(\frac{\partial}{\partial \boldsymbol{r}}\phi_{op}\right)^2 + U(\phi_{op})\right] d^{D-1}r.$$

The corresponding action in the discrete theory is, in place of (5.10),

$$A_{\mathscr{L}}(\phi)=\tfrac{1}{4}\sum_{i,j}\lambda_{ij}(\phi_i-\phi_j)^2+\sum_i \omega_i U(\phi_i)\,,$$

where λ_{ij} is given by (5.11) and ω_i is the volume of the Voronoi polyhedron in the dual lattice. By following the same steps from (5.14)–(5.21), we define the discrete version of $\langle\phi_f|e^{-TH_0}|\phi_0\rangle$ to be

$$\langle\phi_f|\mathscr{G}_N|\phi_0\rangle=\int J\prod_{i=1}^{N}d^Dx_i\,d\phi_i\,\exp[-\langle\phi_{p_+}|A_L|\phi_{p_-}\rangle]\,,$$

with

$$\langle\phi_{p_+}|A_L|\phi_{p_-}\rangle=\tfrac{1}{4}\sum_{\alpha,\beta}\lambda_{\alpha\beta}(\phi_\alpha-\phi_\beta)^2+\sum_i\omega_i U(\phi_i)\,,$$

where α and β can be either the interior sites $i=1, 2,\ldots, N$ or the end-points $p_\pm$, as before. The ensemble average $\mathscr{G}$ is, similar to (5.39),

$$\mathscr{G}=e^{-\sigma\Omega}\sum_{N=0}^{\infty}\frac{1}{N!}\sigma^N\mathscr{G}_N\,. \tag{5.100}$$

For a very large volume Ω, the volume dependence of $\mathscr{G}$ can be analyzed by expanding $U(\phi_i)$ in a power series and using the lattice propagator (5.99). The diagrams are more complicated because of the additional non-linearity in b_m. Nevertheless, by exponentiating the contributions of all connected diagrams, we should find a similar volume-dependence of $\mathscr{G}$ to that given by (5.41).

(iii) When $U(\phi)$ is a quadratic function the analysis becomes much simpler, since the continuum theory is a free massive scalar field. An identical cluster expansion formula (5.41) can be established for $\mathscr{G}$, with b_1 given by the same (5.53), and b_2 by (5.55) etc. provided that (5.45) is properly modified to include $U(\phi)$.

(iv) The present high energy e^+e^- experiments indicate that [7] there is no deviation from the standard electro-weak model up to a c.m. energy ~36 GeV. This implies that l should be less than $\approx 10^{-15}$ cm. If we expect that the quantization of gravity is to be carried out in the context of discrete mechanics, then a lower bound of l can be set in terms of Newton's constant G:

$$l\gtrsim G^{1/2}\simeq 1.6\times10^{-33}\ \text{cm}.$$

The wide difference between the upper and lower bounds of l indicates the paucity of our knowledge of physics at close distances. Our present continuum equations are based mainly on experience at large distances, being essentially generalizations of Maxwell's theory. While the analysis given in this and the preceding [1] paper is exploratory in spirit, it suggests the possibility that underneath the present continuum edifice, there may lie an entirely different layer, hitherto unexplored but more fundamental in concept and equally rich in structure.

We wish to thank G. Feinberg and J.M. Luttinger for discussions.

Appendix A

If we stay within the region t real, then (2.2) is replaced by

$$\langle r_f|G(T,l)|r_0\rangle = \sum_{N=0}^{\infty} \frac{1}{N!}\left(\frac{i}{l}\right)^N \int J \prod_{n=1}^{N} dt_n\, d^3r_n\, e^{i\mathscr{A}}, \tag{A.1}$$

where $\mathscr{A}$ is given by (1.5) and l remains real. Instead of (2.3), the simplest choice of J is

$$J = J_0 \equiv e^{-iT/l} \prod_{n=1}^{N+1} \left(\frac{1}{2\pi i \varepsilon_n}\right)^{3/2}, \tag{A.2}$$

where ε_n is real and positive, as before. Thus, (2.9) becomes

$$G(T,l) = \sum_{N=0}^{\infty} \left(\frac{i}{l}\right)^N e^{-iT/l} \int \prod_1^N d\varepsilon_n U(\varepsilon_{N+1}) \cdots U(\varepsilon_1), \tag{A.3}$$

where instead of (2.7)

$$U(\varepsilon) = e^{-i\frac{1}{2}\varepsilon V}\, e^{i\frac{1}{2}\varepsilon \nabla^2}\, e^{-i\frac{1}{2}\varepsilon V}. \tag{A.4}$$

By following the same arguments given in sect. 2, we derive for large T

$$G(T,l) = \xi^\dagger e^{-iTH} \xi + O(e^{-iT/l}), \tag{A.5}$$

where ξ and H are given by the *same* (2.21) and (2.22). As $T \to \infty$, the $O(e^{-iT/l})$ term can be neglected because of the very rapid change of phase.

We note that $U(\varepsilon)$ is unitary, and therefore so is the product $U(\varepsilon_{N+1}) \cdots U(\varepsilon_1)$. Since unitarity is not preserved under addition of matrices, it is reassuring that e^{-iTH} is unitary.

Appendix B

Consider the discrete version of either a non-relativistic free particle or a relativistic free massless scalar field (in the usual continuum theory). The actions are, respectively,

$$\begin{aligned} \langle r_f|A|r_0\rangle &= \sum_{n=1}^{N+1} \frac{1}{2\varepsilon_n}(r_n - r_{n-1})^2, \\ \langle \phi_f|A|\phi_0\rangle &= \sum_{\alpha,\beta} \tfrac{1}{4}\lambda_{\alpha\beta}(\phi_\alpha - \phi_\beta)^2, \end{aligned} \tag{B.1}$$

in accordance with (2.1) and (5.16). The jacobians (2.3) and (5.34) are related to the integrals of e^{-A}. Define

$$\mathscr{J}_0 \equiv \left\{ \int \prod_1^N d^3r_n\, e^{-\langle 0|A|0\rangle} \right\}^{-1}, \tag{B.2}$$

for the non-relativistic particle, and

$$\mathcal{J}_0 \equiv \left\{ \int \prod_1^N \mathrm{d}\phi_i \, \mathrm{e}^{-\langle 0|A|0\rangle} \right\}^{-1}, \tag{B.3}$$

for the relativistic field, where $\langle 0|A|0\rangle$ is the same action (B.1) when the boundary conditions are $\boldsymbol{r}_0 = \boldsymbol{r}_f = 0$ and $\phi_0 = \phi_f = 0$. In the non-relativistic case, the jacobian J_0, given by (2.3), is related to $\mathcal{J}_0$ by

$$J_0 = \left(\frac{1}{2\pi T}\right)^{3/2} \mathrm{e}^{-T/l} \mathcal{J}_0 ; \tag{B.4}$$

i.e. apart from an overall normalization constant (introduced for the notational convenience of some of the later formulas in sect. 2), the detailed dependence of J_0 on ε_n is entirely contained in $\mathcal{J}_0$. In the relativistic case, according to (5.34) we have

$$J_0 = \mathcal{J}_0 . \tag{B.5}$$

The corresponding Green functions are

$$\int \prod_{n=1}^N \mathrm{d}\varepsilon_n \mathcal{J}_0 \prod_{n=1}^N \mathrm{d}^3 r_n \, \mathrm{e}^{-\langle r_f|A|r_0\rangle} , \tag{B.6}$$

$$\int \prod_{i=1}^N \mathrm{d}^D x_i \mathcal{J}_0 \prod_{i=1}^N \mathrm{d}\phi_i \, \mathrm{e}^{-\langle \phi_f|A|\phi_0\rangle} . \tag{B.7}$$

In both cases, we can write

$$\langle 0|A|0\rangle = \tfrac{1}{2}\tilde{\phi} M \phi , \tag{B.8}$$

where ϕ is an $\mathcal{N} \times 1$ column matrix, $\tilde{\phi}$ its transpose and M an $\mathcal{N} \times \mathcal{N}$ real symmetric matrix. In the non-relativistic case, the components of ϕ are the cartesian components of $\boldsymbol{r}_1, \boldsymbol{r}_2, \ldots, \boldsymbol{r}_N$; hence, $\mathcal{N} = 3N$. In the relativistic case, the components of ϕ are simply ϕ_i, the scalar field on the ith site; therefore, $\mathcal{N} = N$. We introduce $2\mathcal{N}$ elements of a Grassmann algebra: $z_1, \ldots, z_{\mathcal{N}}$ and $\bar{z}_1, \ldots, \bar{z}_{\mathcal{N}}$ which satisfy

$$\{z_i, z_j\} = \{z_i, \bar{z}_j\} = \{\bar{z}_i, \bar{z}_j\} = 0 ,$$

for $i, j = 1, \ldots, \mathcal{N}$. Define

$$\begin{aligned} \langle 0|\bar{A}|0\rangle &\equiv \tfrac{1}{2}\tilde{\phi} M \phi + \bar{\Psi} M^{1/2} \Psi \\ &= \langle 0|A|0\rangle + \bar{\Psi} M^{1/2} \Psi , \end{aligned} \tag{B.9}$$

where

$$\Psi = \begin{pmatrix} z_1 \\ \vdots \\ z_{\mathcal{N}} \end{pmatrix}, \qquad \bar{\Psi} = (\bar{z}_1 \cdots \bar{z}_{\mathcal{N}}) . \tag{B.10}$$

The function $\bar{A}$ belongs to a general class of supersymmetrical functions considered by Parisi and Sourlas [8]. It is invariant under the supersymmetric transformation

$$\delta\phi_i = \bar{\Psi}_i\varepsilon + \bar{\varepsilon}\Psi_1\,,$$
$$\delta\Psi = -\varepsilon M^{1/2}\phi\,, \tag{B.11}$$
$$\delta\bar{\Psi} = -\tilde{\phi}M^{1/2}\bar{\varepsilon}\,,$$

where ε and $\bar{\varepsilon}$ are anticommuting parameters.

One can readily verify that

$$\int \prod_1^{\mathcal{N}} \mathrm{d}\phi_i\, \mathrm{d}\bar{z}_i\, \mathrm{d}z_i\, \mathrm{e}^{-\langle 0|\bar{A}|0\rangle} = (2\pi)^{\mathcal{N}/2}\,. \tag{B.12}$$

Thus, apart from the factor $(2\pi)^{\mathcal{N}/2}$, (B.6) is identical to

$$\int \prod_\nu \mathrm{d}\varepsilon_n\, \mathrm{d}^3 r_n \prod_i \mathrm{d}\bar{z}_i\, \mathrm{d}z_i\, \mathrm{e}^{-\langle r_f|\bar{A}|r_i\rangle}\,, \tag{B.13}$$

and (B.7) to

$$\int \prod_i \mathrm{d}^D x_i\, \mathrm{d}\phi_i\, \mathrm{d}\bar{z}_i\, dz_i\, \mathrm{e}^{-\langle \phi_f|\bar{A}|\phi_0\rangle}\,, \tag{B.14}$$

where, similar to (B.9)

$$\langle \boldsymbol{r}_f|\bar{A}|\boldsymbol{r}_0\rangle = \langle \boldsymbol{r}_f|A|\boldsymbol{r}_0\rangle + \bar{\Psi}M^{1/2}\Psi\,, \tag{B.15}$$

in the non-relativistic case, and

$$\langle \phi_f|\bar{A}|\phi_0\rangle = \langle \phi_f|A|\phi_0\rangle + \bar{\Psi}M^{1/2}\Psi\,, \tag{B.16}$$

for the relativistic theory. The conversion of (B.6)–(B.7) to a more natural expression (B.13)–(B.14) is reminiscent of a similar circumstance in statistical mechanics: the transition from integrals in the coordinate space $\int\prod_i dq_i \cdots$ to those in the phase space $\int\prod_i \mathrm{d}p_i\, \mathrm{d}q_i \cdots$.

Appendix C

In this appendix, we review some well-known and some less familiar properties of harmonic oscillators on a regular lattice.

C.1. A SINGLE OSCILLATOR

Consider a non-relativistic particle of unit mass whose potential energy is

$$V = \tfrac{1}{2}\omega^2 x^2\,. \tag{C.1}$$

For a regular lattice (in "time" τ), we set the spacing between sites

$$\varepsilon_1 = \varepsilon_2 = \cdots = \varepsilon_{N+1} = \varepsilon\,, \tag{C.2}$$

with

$$\varepsilon = T/(N+1)\,,$$

in agreement with (1.4). The matrix elements of G_N, defined by (2.6), are given by

$$\langle x'|G_N|x\rangle = \left(\frac{1}{2\pi\varepsilon}\right)^{(N+1)/2} \int \prod_{n=1}^{N} \mathrm{d}x_n\, \mathrm{e}^{-A}\,, \tag{C.3}$$

where, in accordance with (2.1),

$$A = \sum_{n=1}^{N+1} \frac{(x_n - x_{n-1})^2}{2\varepsilon} + \tfrac{1}{4}\varepsilon\omega^2 \sum_{n=1}^{N+1} (x_n^2 + x_{n-1}^2)\,, \tag{C.4}$$

and the boundary condition is

$$x_0 = x\,, \qquad x_{N+1} = x'\,. \tag{C.5}$$

As we shall see, without any approximation, the following relations hold:

$$\langle x'|G_N|x\rangle = \left[\frac{\omega(1+\frac{1}{4}\varepsilon^2\omega^2)^{1/2}}{2\pi \sinh \nu T}\right]^{1/2} \mathrm{e}^{-A_\mathrm{L}^\mathrm{c}}\,, \tag{C.6}$$

where A_L^c is the classical lattice action (with the subscript L indicating the lattice and the superscript c for classical), given by

$$A_\mathrm{L}^\mathrm{c} = \frac{\omega(1+\frac{1}{4}\varepsilon^2\omega^2)^{1/2}}{2 \sinh \nu T}[(x^2 + x'^2)\cosh \nu T - 2xx']\,, \tag{C.7}$$

$$\nu = \frac{2}{\varepsilon} \sinh^{-1}(\tfrac{1}{2}\varepsilon\omega)\,. \tag{C.8}$$

The trace of G_N is

$$\text{trace } G_N \equiv \int \mathrm{d}x \langle x|G_N|x\rangle = [2 \sinh(\tfrac{1}{2}\nu T)]^{-1}\,. \tag{C.9}$$

Proof. From (C.4), it can be readily verified that the classical solution of

$$\frac{\partial A}{\partial x_n} = 0\,, \tag{C.10}$$

with the boundary condition (C.5), is

$$x_n = x_n^\mathrm{c} \equiv \frac{1}{\sinh \nu T}[x' \sinh \nu\tau_n + x \sinh\{\nu(T-\tau_n)\}]\,, \tag{C.11}$$

where $\tau_n = n\varepsilon$, and therefore $\tau_{N+1} = T$. Multiplying (C.10) by x_n and summing from $n = 1$ to N, we obtain (C.7) for the classical lattice action A_L^c. By making the shift $x_n = x_n^\mathrm{c} + \delta x_n$ and integrating over the fluctuation δx_n, we convert (C.3) into

$$\langle x'|G_N|x\rangle = \langle 0|G_N|0\rangle\, \mathrm{e}^{-A_\mathrm{L}^\mathrm{c}}\,, \tag{C.12}$$

where $\langle 0|G_N|0\rangle$ is the same $\langle x'|G_N|x\rangle$, but with $x = x' = 0$.

It is useful to set $x_0 = x_{N+1}$ in (C.4), and to write

$$A = \tfrac{1}{2}\phi^{\dagger}\bar{A}\phi = \tfrac{1}{2}\phi'^{\dagger}\Lambda\phi' , \tag{C.13}$$

where $\bar{A}$ is an $(N+1)\times(N+1)$ cyclic matrix,

$$\bar{A} = \begin{pmatrix} \varepsilon\omega^2+2\varepsilon^{-1} & -\varepsilon^{-1} & 0 & 0 & \cdots & 0 & -\varepsilon^{-1} \\ -\varepsilon^{-1} & \varepsilon\omega^2+2\varepsilon^{-1} & -\varepsilon^{-1} & 0 & \cdots & 0 & 0 \\ & & \cdots & & & & \\ -\varepsilon^{-1} & 0 & 0 & \cdots & 0 & -\varepsilon^{-1} & \varepsilon\omega^2+2\varepsilon^{-1} \end{pmatrix}, \tag{C.14}$$

Λ is a diagonal matrix

$$\Lambda = \begin{pmatrix} \lambda_0 & 0 & \cdots & \cdots & 0 \\ 0 & \lambda_1 & 0 & \cdots & 0 \\ & \cdots & \cdots & \cdots & \\ 0 & 0 & \cdots & 0 & \lambda_N \end{pmatrix}, \tag{C.15}$$

with

$$\lambda_n = \varepsilon\omega^2 + \frac{4}{\varepsilon}\sin^2\left(\frac{\pi\varepsilon}{T}n\right), \tag{C.16}$$

and ϕ, ϕ' are both $(N+1)\times 1$ column matrices

$$\phi = \begin{pmatrix} x_0 \\ x_1 \\ \vdots \\ x_N \end{pmatrix}, \qquad \phi' = \begin{pmatrix} x'_0 \\ x'_1 \\ \vdots \\ x'_N \end{pmatrix} = U\phi . \tag{C.17}$$

The transformation matrix U is unitary; its elements are

$$U_{nn'} = (N+1)^{-1/2}\, e^{i2\pi nn'\varepsilon/T} , \tag{C.18}$$

where n and $n' = 0, 1, \ldots, N$.

Substituting $A = \tfrac{1}{2}\phi'^{\dagger}\Lambda\phi'$ into (C.3) and carrying out the integration (C.9) over dx'_n, we find

$$\text{trace } G_N = \prod_{n=0}^{N} (\varepsilon\lambda_n)^{-1/2} . \tag{C.19}$$

To evaluate $\langle 0|G_N|0\rangle$, we note that, from (C.17)–(C.18),

$$x_0 = (N+1)^{-1/2} \sum_{n=0}^{N} x'_n .$$

Hence,

$$\langle 0|G_N|0\rangle = \int \mathrm{d}x_0\, \delta(x_0)\langle x_0|G_N|x_0\rangle$$

$$= (N+1)^{1/2}\left(\frac{1}{2\pi\varepsilon}\right)^{(N+1)/2} \int \prod_0^N \mathrm{d}x'_n\, \delta\left(\sum_0^N x'_n\right) \exp\left(-\tfrac{1}{2}\sum_0^N \lambda_n x'^2_n\right)$$

$$= \left[\frac{(N+1)\lambda}{2\pi}\right]^{1/2} \text{trace } G_N, \tag{C.20}$$

where

$$\lambda^{-1} = \sum_0^N \lambda_n^{-1}. \tag{C.21}$$

By multiplying

$$z^{N+1} - 1 = \prod_{n=0}^N [z - \mathrm{e}^{i2\pi n/(N+1)}],$$

and

$$1 - z^{-(N+1)} = \prod_{n=0}^N [1 - z^{-1}\,\mathrm{e}^{-i2\pi n/(N+1)}],$$

and setting $z = \mathrm{e}^{2x}$, we have

$$4\sinh^2[(N+1)x] = \prod_{n=0}^N \left(4\sinh^2 x + 4\sin^2\frac{\pi n}{N+1}\right), \tag{C.22}$$

which, for $\varepsilon\omega = 2\sinh x$, gives

$$2\sinh[(N+1)\sinh^{-1}\tfrac{1}{2}\varepsilon\omega] = \left[\prod_{n=0}^N \left(\varepsilon^2\omega^2 + 4\sin^2\frac{\pi n}{N+1}\right)\right]^{1/2}. \tag{C.23}$$

According to (C.16) and (C.19), the right-hand side is $(\text{trace } G_N)^{-1}$. By differentiating the logarithm of (C.23) with respect to ω, we obtain $\sum_0^N \lambda_n^{-1}$ which, together with (C.20)–(C.21), leads to

$$\langle 0|G_N|0\rangle = \left\{\frac{\omega(1+\frac{1}{4}\varepsilon^2\omega^2)^{1/2}}{2\pi\sinh[2(N+1)\sinh^{-1}(\frac{1}{2}\varepsilon\omega)]}\right\}^{1/2}. \tag{C.24}$$

This completes the proof of (C.6)–(C.9).

From (C.8) it follows that when ε is small but non-zero, the effect of a regular lattice (in τ) is to change the frequency of the oscillator from ω to

$$\nu = \frac{2}{\varepsilon}\sinh^{-1}(\tfrac{1}{2}\varepsilon\omega) = \omega - \tfrac{1}{24}\varepsilon^2\omega^2 + \mathrm{O}(\varepsilon^4). \tag{C.25}$$

Returning to the spectrum (3.22) in the discrete quantum mechanics we see that to the same order, when $m \to \infty$ and for $\rho = \varepsilon^{-1}$, the energy levels are

$$E = (1 - \tfrac{1}{24}\varepsilon^2\omega^2)(n + \tfrac{1}{2})\omega ,$$

in agreement with (C.25). This is expected because (3.10) shows that the spacing ε_n becomes regular in the limit $m \to \infty$.

C.2. MASSLESS SCALAR FIELD IN TWO DIMENSIONS

Consider a regular two-dimensional array of lattice sites whose coordinates are (τ_n, x_m) where

$$\tau_n = n\varepsilon , \qquad x_m = m\varepsilon , \tag{C.26}$$

with

$$n = 0, 1, 2, \ldots, N+1 , \qquad m = 1, 2, \ldots, M+1,$$
$$T = (N+1)\varepsilon , \qquad L = (M+1)\varepsilon . \tag{C.27}$$

On each site (τ_n, x_m), we assign a field strength $\phi_{n,m}$. The lattice action is, in accordance with (5.16),

$$A_{\mathrm{L}}(\phi) = \tfrac{1}{2} \sum_{n=1}^{N+1} \sum_{m=1}^{M+1} [(\phi_{n,m} - \phi_{n-1,m})^2 + (\phi_{n,m} - \phi_{n,m-1})^2] , \tag{C.28}$$

where

$$\phi_{n,0} = \phi_{n,M+1} , \tag{C.29}$$

so that it is periodic in the x direction. Because of (C.29), we may write

$$\phi_{n,m} = L^{-1/2}\left\{q_n(0) + \sqrt{2} \sum_{k>0} [q_n(k, c) \cos kx_m + q_n(k, s) \sin kx_m]\right\} , \tag{C.30}$$

where the sum extends over (assuming M even)

$$\frac{kL}{2\pi} = l = 1, 2, \ldots, \tfrac{1}{2}M . \tag{C.31}$$

Substituting (C.30) into (C.28), we find

$$A_{\mathrm{L}} = \sum_{\alpha,n} \left\{\frac{1}{2\varepsilon} [q_n(\alpha) - q_{n-1}(\alpha)]^2 + \tfrac{1}{4}\varepsilon\omega^2[q_n^2(\alpha) + q_{n-1}^2(\alpha)]\right\} , \tag{C.32}$$

where

$$\alpha = 0 , \qquad k, c \text{ and } k, s , \tag{C.33}$$

$$\omega = \frac{2}{\varepsilon} \sin (\tfrac{1}{2}k\varepsilon) \tag{C.34}$$

with the convention that $\alpha = 0$ means $k = \omega = 0$.

By using (C.11), we see that the classical solution of $\partial A_L/\partial\phi_{n,m}=0$ with the boundary condition

$$q_0(\alpha)=q(\alpha)\,,\qquad q_{N+1}(\alpha)=q'(\alpha)\,, \tag{C.35}$$

is

$$\phi^c_{n,m}\equiv\phi^c_L(\tau_n,x_m)=L^{-1/2}\Big\{q^c_n(0)+\sqrt{2}\sum_{k>0}[q^c_n(k,c)\cos kx_m+q^c_n(k,s)\sin kx_m]\Big\}\,, \tag{C.36}$$

where

$$q^c_n(\alpha)=\frac{1}{\sinh\nu T}[q'(\alpha)\sinh\nu\tau_n+q(\alpha)\sinh\{\nu(T-\tau_n)\}]\,, \tag{C.37}$$

$$\nu=\frac{2}{\varepsilon}\sinh^{-1}(\tfrac{1}{2}\varepsilon\omega)\,, \tag{C.38}$$

as in (C.8). When $\varepsilon\to0$, $\phi^c_L(\tau_n,x_m)$ approaches the classical continuum solution ϕ^c_C at (τ_n,x_m), where ϕ^c_C satisfies

$$\left(\frac{\partial^2}{\partial\tau^2}+\frac{\partial^2}{\partial x^2}\right)\phi^c_C(\tau,x)=0\,. \tag{C.39}$$

For $\varepsilon\neq0$ but small, we find from (C.34) and (C.38) that

$$\nu=k-\tfrac{1}{12}\varepsilon^2k^3+O(\varepsilon^4)\,. \tag{C.40}$$

Therefore, the classical lattice solution (C.36) satisfies

$$\left[\frac{\partial^2}{\partial\tau^2}+\frac{\partial^2}{\partial x^2}+\tfrac{1}{12}\varepsilon^2\left(\frac{\partial^4}{\partial\tau^4}+\frac{\partial^4}{\partial x^4}\right)+O(\varepsilon^4)\right]\phi^c_L(\tau,x)=0\,, \tag{C.41}$$

in which τ and x refer to any point inside the entire area TL. Eq. (C.41) can also be more easily derived by changing the boundary condition (C.35) to a periodic one in τ, as well as in x.

Appendix D

We prove here that the classical lattice solution ϕ_α and the related classical lattice action A_L are continuous functions of the lattice sites $\boldsymbol{x}_\alpha(\alpha=1,2,\ldots,N)$, when the boundary conditions (5.23) are kept fixed. Our discussion will be divided into three parts: (i) when the sites do not coalesce, (ii) when only two sites coalesce, and (iii) when more than two coalesce. Throughout our discussions, we denote V as the ND-dimensional space in which each point P is

$$P=(\boldsymbol{x}_1,\boldsymbol{x}_2,\ldots,\boldsymbol{x}_N)\,. \tag{D.1}$$

(i) Pick any pair of sites i, j. Take V_{ij} to be the subset of V for which $\lambda_{ij} = s_{ij}/l_{ij}$ is >0; its complement V'_{ij} then consists only of points for which $\lambda_{ij} = 0$. When $P \in V_{ij}$, we have in the dual lattice the Voronoi cells ω_i and ω_j (dual to sites i and j) sharing a common boundary s_{ij} (dual to l_{ij}). Excluding those configurations for which $\boldsymbol{x}_i = \boldsymbol{x}_j$, it is obvious that λ_{ij} is continuous inside either V_{ij} or V'_{ij}. It remains to show that λ_{ij} is continuous at P, when P lies on the boundary between V_{ij} and V'_{ij}.

If $\lambda_{ij}(P) > 0$, then $\lambda_{ij} > 0$ in the neighborhood of P, so that P is not on the boundary. Therefore, when P is on the boundary

$$\lambda_{ij}(P) = 0\,.$$

Since

$$l_{ij}(P) > 0\,, \tag{D.2}$$

and is finite, we have

$$s_{ij}(P) = 0\,. \tag{D.3}$$

Let P' be close to P. If $P' \notin V_{ij}$ then $\lambda_{ij}(P') = 0 = \lambda_{ij}(P)$. If $P' \in V_{ij}$, then as $P' \to P$, $s_{ij}(P') \to 0$ by (D.3), and $l_{ij}(P') \to l_{ij}(P) \neq 0$ by (D.2). Hence, $\lambda_{ij}(P') \to 0 = \lambda_{ij}(P)$. Therefore λ_{ij} is continuous at P.

The continuity of λ_{ij} insures that the same property holds for the classical lattice solution and also for the classical lattice action.

(ii) When two sites, say 1 and 2, approach each other, $l_{12} \to 0$ and therefore

$$\lambda_{12} \to \infty\,, \tag{D.4}$$

provided all other sites are kept separate and fixed (which makes s_{12} finite and non-zero). The lattice field equation at 1 is

$$\sum_k \lambda_{1k}(\phi_1 - \phi_k) = \lambda_{12}(\phi_1 - \phi_2) + \sum_{k \neq 2} \lambda_{1k}(\phi_1 - \phi_k) = 0\,. \tag{D.5}$$

Hence, because of (D.4),

$$\phi_1 - \phi_2 \to 0\,. \tag{D.6}$$

At any site k other than 1 and 2, we have

$$\lambda_{k1}(\phi_k - \phi_1) + \lambda_{k2}(\phi_k - \phi_2) + \sum_{1 \neq l \neq 2} \lambda_{kl}(\phi_k - \phi_l) = 0\,. \tag{D.7}$$

Consider first the case that k is linked to both 1 and 2. The links $\boldsymbol{l}_{k1}$ and $\boldsymbol{l}_{k2}$ are normal to their dual surfaces s_{k1} and s_{k2} which are $(D-1)$-dimensional polyhedra in contact with each other and forming part of the boundary of the Voronoi cell ω_k. When $1 \to 2$, $\boldsymbol{l}_{k1} \to \boldsymbol{l}_{k2}$; therefore the surfaces s_{k1} and s_{k2} become parallel and form a single $(D-1)$-dimensional polyhedron of "area"

$$s_{k0} \equiv s_{k1} + s_{k2}\,. \tag{D.8}$$

We may view the two coalescing sites 1 and 2 as a single point, labeled 0, in which case the lattice consists of only $N-1$ distinct sites (as before, not counting the sites on the boundary):

$$0, 3, \ldots, N. \tag{D.9}$$

Eq. (D.7) becomes then

$$\lambda_{k0}(\phi_k - \phi_0) + \sum_{l \neq 0} \lambda_{kl}(\phi_k - \phi_l) = 0\,, \tag{D.10}$$

where

$$\phi_0 = \phi_1 = \phi_2\,, \tag{D.11}$$

on account of (D.6) and

$$\lambda_{k0} = s_{k0}/l_{k0} = \lambda_{k1} + \lambda_{k2}\,, \tag{D.12}$$

with

$$l_{k0} = l_{k1} = l_{k2}\,.$$

Hence we see that ϕ_k is continuous when $1 \to 2$. In this limit, although $\lambda_{12} \to \infty$, the field remains continuous; because of (D.5)–(D.6), $\lambda_{12}(\phi_1 - \phi_2)^2 \to 0$ and therefore the lattice action is also continuous. If k is linked neither to 1 nor 2, then $\lambda_{k1} = \lambda_{k2} = \lambda_{k0} = 0$, and (D.10) is trivially true. If k is linked only to 2, but not to 1 (before the limit), then the corresponding $\lambda_{k1} = 0$; as $1 \to 2$, $\lambda_{k2} = \lambda_{k0}$ and therefore (D.10) remains valid. It follows then that both the field and the lattice action are continuous.

It is of interest to examine the evolution of Voronoi cells during the coalescence. When $\boldsymbol{x}_1$ and $\boldsymbol{x}_2$ coalesce in configuration P at the position $\boldsymbol{x}_0$, we may denote ω_0 to be the Voronoi cell dual to 0 (regarding 0 as a single site, as in (D.9)). Before coalescence, let P' be close to P. As $P' \to P$, the union of the Voronoi cells ω_1 and ω_2 (dual to sites 1 and 2) approaches ω_0. The boundary, say $s(1)$, of ω_1 is composed of two connected parts:

$$s(1) = s_{\text{in}}(1) + s_{\text{out}}(1)\,, \tag{D.13}$$

where $s_{\text{out}}(1)$ lies on the boundary of ω_0, but $s_{\text{in}}(1)$ is inside ω_0; hence,

$$s_{\text{in}}(1) = s_{12}\,, \qquad s_{\text{out}}(1) = \sum_{k>2} s_{1k} \tag{D.14}$$

where the sum extends over those $k > 2$ linked to 1, and s_{1k} is the dual to link l_{1k} (consequently, s_{1k} is the boundary shared by ω_1 and ω_k). As $P' \to P$, $s_{\text{in}}(1)$ remains finite and non-zero. At P, we can simply remove $s_{\text{in}}(1)$ and view cells ω_1 and ω_2 as forming a single cell ω_0.

(iii) When more than two sites, say $1, 2, \ldots, n$, coalesce, new complications can occur: for example, some of these sites may have no link connected to any distant sites (i.e. they are linked only to the n sites that are approaching each other); the corresponding $\lambda_{ij}=s_{ij}/l_{ij}$ depend on how the n sites coalesce, since both s_{ij} and $l_{ij}\to 0$ in the limit. Nevertheless, as we shall see, both the classical field and the classical action remain continuous.

Let sites $\boldsymbol{x}_1, \boldsymbol{x}_2, \ldots, \boldsymbol{x}_n$ coalesce in configuration P at the position $\boldsymbol{x}_0$. As before, take P' to be a configuration before the coalescence, but close to P. Suppose that as $P'\to P$, the classical field values $\phi_1, \ldots, \phi_n$ all approach the same value ϕ_0. Then by summing (5.22) over $\alpha=1\cdots n$, we find that ϕ_0 is independent of the trajectory of P', since $\sum_{i\leqslant n,k>n}\lambda_{ik}$ is so independent by arguments similar to those given in (ii). Therefore if any of $\phi_1, \ldots, \phi_n$ is *not* continuous at P, it can only mean that for some $\varepsilon>0$ there exist P' arbitrarily close to P such that

$$(\exists i, j\leqslant n)(|\phi_i-\phi_j|>\varepsilon \text{ in } P')\,. \tag{D.15}$$

We shall show that (D.15) leads to contradictions.

Arrange ϕ_i in an ascending order, say, $\phi_{a_1}\leqslant\phi_{a_2}\leqslant\cdots\leqslant\phi_{a_n}$. At P' close to P and satisfying (D.15), the average of $\phi_{a_i}-\phi_{a_{i-1}}$ must be $>\varepsilon/(n-1)$. Hence, we can find a set $\mathrm{I}\subset(1\cdots n)$ such that

$$\phi_i-\phi_k>\frac{\varepsilon}{n-1}\,, \qquad (i\in \mathrm{I}, k\leqslant n, k\notin \mathrm{I})\,; \tag{D.16}$$

furthermore, some of the λ_{ik} must be non-zero. The set I may consist of disconnected subsets (i.e. with no direct links between them). Let J be a maximally connected subset of I; therefore, by construction each site j in J is either linked to other sites in the same set J or to sites $k\notin$ I. As in (D.16) we have

$$\phi_j-\phi_k>\frac{\varepsilon}{n-1}\,, \qquad (j\in \mathrm{J}, k\leqslant n, k\notin \mathrm{I})\,, \tag{D.17}$$

and with

$$\text{some } \lambda_{jk}>0\,. \tag{D.18}$$

Let ω_j be the union of all Voronoi cells ω_j $(j\in \mathrm{J})$:

$$\omega_{\mathrm{J}}=\bigcup_{j\in \mathrm{J}}\omega_j\,. \tag{D.19}$$

By summing (5.22) over the set J, we obtain

$$\lambda_{\mathrm{A}}\frac{\varepsilon}{n-1}<\lambda_{\mathrm{B}}(\phi_{\max}-\phi_{\min})\,, \tag{D.20}$$

where

$$\lambda_{\mathrm{A}}=\sum_{j\in \mathrm{J}}\sum_{k\notin \mathrm{J}}^{k\leqslant n}\lambda_{jk}\,, \tag{D.21}$$

$$\lambda_{\mathrm{B}}=\sum_{j\in \mathrm{J}}\sum_{k>n}\lambda_{jk}\,, \tag{D.22}$$

and $\phi_{\max}$, $\phi_{\min}$ are the extreme values of the classical field in the entire lattice for the given boundary conditions. Because of (D.18) we have, at P',

$$\lambda_A > 0 . \tag{D.23}$$

At P, sites $\boldsymbol{x}_1, \ldots, \boldsymbol{x}_n$ coalesce at $\boldsymbol{x}_0$. We may, as in (ii), view these n sites as a single site 0, the corresponding lattice then consists of only $N-n+1$ sites:

$$0, n+1, \ldots, N \tag{D.24}$$

with

$$\omega_0 = \text{Voronoi cell dual to } 0 . \tag{D.25}$$

Draw two spheres of radii R and R' with 0 as the center and $R < R'$, so that the boundary of ω_0 lies entirely within these two $(D-1)$-dimensional spherical surfaces. Clearly:

$$\begin{gathered} \text{the sphere of radius } R \text{ is inside } \omega_0 \text{ and ;} \\ R < \text{distance between 0 and its nearest neighboring sites} . \end{gathered} \tag{D.26}$$

As in (D.13), we may decompose the boundary $s(\mathrm{J})$ of ω_J, defined by (D.19), into two connected parts:

$$s(\mathrm{J}) = s_{\mathrm{in}}(\mathrm{J}) + s_{\mathrm{out}}(\mathrm{J}) , \tag{D.27}$$

where $s_{\mathrm{out}}(\mathrm{J})$ lies on the boundary of ω_0 and $s_{\mathrm{in}}(\mathrm{J})$ lies inside ω_0. Similar to (D.14), we have

$$\begin{aligned} s_{\mathrm{in}}(\mathrm{J}) &= \sum_{j\in \mathrm{J}} \sum_{k \notin \mathrm{J}}^{k\leq n} s_{jk} , \\ s_{\mathrm{out}}(\mathrm{J}) &= \sum_{j\in \mathrm{J}} \sum_{k>n} s_{jk} , \end{aligned} \tag{D.28}$$

where the sum only extends over links $l_{jk} \neq 0$, with $s_{jk} =$ dual of l_{jk}, as before.

At P' sufficiently close to P, $s(\mathrm{J})$ can be similarly decomposed in the form (D.27) by continuity. Take 0 again as the origin and enclose the n sites within a small sphere of radius $\frac{1}{2}r \ll R$ (hence $|\boldsymbol{x}_i - \boldsymbol{x}_j| < r$ for all $i, j = 1, \ldots, n$). From (D.21)–(D.22) and (D.26)–(D.28), it follows that

$$\lambda_A > s_{\mathrm{in}}(\mathrm{J})/r , \tag{D.29}$$

$$\lambda_B < s_{\mathrm{out}}(\mathrm{J})/R . \tag{D.30}$$

There are three different possibilities for the limit $P' \to P$:

(i) If every site j in J is linked only to $k \leq n$, then $s_{\mathrm{out}}(\mathrm{J}) = 0$, and therefore $\lambda_B = 0$, at P' before coalescence; this makes (D.20) impossible. (In this case, $s_{\mathrm{in}}(\mathrm{J}) \neq 0$ at P', but $\to 0$ as $P' \to P$.) In cases (ii) and (iii) below, we assume that there exist some sites j in J linked to some $k > n$.

(ii) At P, $s_{out}(J) \neq 0$. In this case, since $s_{out}(J)$ lies on the boundary of ω_0 and $s_{in}(J)$ is connected to $s_{out}(J)$, we must also have $s_{in}(J) \neq 0$ at P. By choosing at P'

$$r < \frac{\varepsilon}{n-1} \frac{s_{in}(J)}{s_{out}(J)} \frac{R}{\phi_{max} - \phi_{min}}, \tag{D.31}$$

and using (D.29)–(D.30), we arrive at a result contradictory to (D.20).

(iii) At P', $s_{out}(J) \neq 0$; but as $P' \to P$

$$s_{out}(J) \to 0 . \tag{D.32}$$

Consequently, the entire volume $\omega_J \to 0$ in the limit. To understand this rather special situation, let $b(J)$ be the $(D-2)$-dimensional border of $s_{in}(J)$ that lies on $s_{out}(J)$; i.e. $b(J) = s_{in}(J) \cap s_{out}(J)$, in which $s_{in}(J)$ is understood to include its boundary. When $D = 2$, $b(J)$ consists of two points, say p_1 and p_2 before the coalescence, at P'. From (D.26) and the construction preceding it, we see that since $\overline{0p_1}$ and $\overline{0p_2}$ are both $>R$, $s_{in}(J) > 2R$; in addition if we denote the angle $\angle p_1 0 p_2 = 2\theta$, then (because of convexity) $s_{out}(J) < 2R'\theta$. To arrive at (D.32), we must have $p_1 = p_2$ at P. Thus,

$$\frac{s_{out}(J)}{s_{in}(J)} < \frac{R'}{R} \theta \to 0 , \qquad \text{as } P' \to P . \tag{D.33}$$

When $D = 3$, $b(J)$ consists of a closed figure made of straight lines. Let α be the solid angle sustained by $b(J)$ at 0, and 2θ its opening angle. When α is sufficiently small, we have $\alpha \simeq \pi\theta^2$, $s_{in}(J) > 2\pi R^2\theta$ and $s_{out}(J) < R'^2\alpha$. This gives

$$\frac{s_{out}(J)}{s_{in}(J)} < \left(\frac{R'}{R}\right)^2 \tfrac{1}{2}\theta \to 0 , \qquad \text{as } P' \to P .$$

At any dimension D, the same reasoning leads to

$$\frac{s_{out}(J)}{s_{in}(J)} < \left(\frac{R'}{R}\right)^{D-1} \frac{\theta}{D-1} \to 0 , \qquad \text{as } P' \to P , \tag{D.34}$$

and that makes the right-hand side of (D.31) approach ∞. Consequently, at P' we can always choose r to satisfy (D.31), which makes (D.20), and therefore also (D.15), impossible. Hence, the classical field is indeed continuous at P.

By using (5.22) and the fact that (D.15) is untenable, we see that

$$\sum_{i,j \leq n} \lambda_{ij} (\phi_i - \phi_j)^2 \to 0 , \qquad \text{as } P' \to P .$$

Hence, the action is also continuous at P.

Appendix E

In this appendix we give further details on the evaluation of b_1 and some of the related properties of the classical lattice action A_L in a two-dimensional volume

$\Omega = LT$. As in (5.44), we may decompose

$$A_L = \sum_S A_L(S)\,, \tag{E.1}$$

in which the classical lattice solution ϕ_i^c is assumed to be

$$\phi_i^c = \bar{\phi}(x_i)\,, \tag{E.2}$$

with $\bar{\phi}(x)$ an arbitrary given function of x. Using the same formulas (5.70)–(5.71), we write

$$A_L(S) = \tfrac{1}{2}\boldsymbol{E}_L^2 \Delta\,, \tag{E.3}$$

$$\boldsymbol{E}_L = \frac{1}{2\Delta}(\boldsymbol{a}\phi_1^c + \boldsymbol{b}\phi_2^c + \boldsymbol{c}\phi_3^c)\,. \tag{E.4}$$

Similar to (5.48) and (5.53), let us define

$$\exp[-A_L(S) + \bar{A}_C(S)] \equiv 1 + \bar{f}_S\,, \tag{E.5}$$

$$\bar{b}_1 \equiv \frac{1}{\Omega}\int \frac{\sigma^3}{3!} e^{-\sigma\pi R^2} \bar{f}_S \prod_1^3 d^2x_i\,, \tag{E.6}$$

where

$$\bar{A}_C(S) \equiv \int_S \frac{1}{2}\left(\frac{\partial\bar{\phi}}{\partial x^\mu}\right)^2 d^2x\,, \tag{E.7}$$

and R is the radius of the circumscribed circle of the simplex S (which is a triangle with area Δ and vertices 1, 2 and 3). If $\bar{\phi}(x) = \phi^c(x)$ the classical continuum solution, then (E.2) becomes the zeroth-order mean-field approximation (5.56), and the above $\bar{A}_C(S)$, $\bar{f}_S$ and $\bar{b}_1$ reduce to the corresponding $A_C(S)$, f_S and b_1 (given by (5.45), (5.48) and (5.53)) under the same approximation.

As in (5.63)–(5.64), we expand $\bar{\phi}(x)$ in $\boldsymbol{x}$, the coordinate measured from the c.m. M of the triangle:

$$\bar{\phi}(x) = \bar{\phi}(M) + E^\mu x^\mu + \tfrac{1}{2}T^{\mu\nu}x^\mu x^\nu\,, \tag{E.8}$$

where $\bar{\phi}(M)$ is $\bar{\phi}(x)$ at M,

$$E^\mu = \frac{\partial\bar{\phi}}{\partial x^\mu}\,, \quad \text{at } M\,,$$
$$T^{\mu\nu} = \frac{\partial^2\bar{\phi}}{\partial x^\mu\,\partial x^\nu}\,, \quad \text{at } M\,. \tag{E.9}$$

By following exactly the same steps given in subsect. 5.6 from (5.65) to (5.75), we derive, as in (5.76),

$$\bar{f}_S = -\hat{z}\cdot(\boldsymbol{E}_L^T \times \boldsymbol{E})\Delta - \tfrac{1}{2}(\boldsymbol{E}_L^T)^2\Delta + \tfrac{1}{2}\,\mathrm{trace}\, IT^2 + O(l^3\Delta) \tag{E.10}$$

where $\boldsymbol{E}_{\rm L}^T$, I and T are defined by (5.73), (5.67) and (5.68), and $\hat{z}$ is, as before, a unit vector normal to the plane that contains the triangle.

Let $\langle F\rangle$ denote the average of F over the azimuthal angle around $\hat{z}$, while the c.m. $\boldsymbol{M}$ and the shape and size of the triangle are kept fixed. We have

$$\langle \boldsymbol{E}_{\rm L}^T \times \boldsymbol{E}\rangle = 0 ,$$

and for any given sites i and j

$$\left\langle \sum_{\mu,\nu} x_i^\mu x_i^\nu T^{\mu\nu} \sum_{\rho,\lambda} x_j^\rho x_j^\lambda T^{\rho\lambda}\right\rangle$$
$$= \tfrac{1}{8}[3x_i^2x_j^2 - 2(\boldsymbol{x}_i\cdot\boldsymbol{x}_j)^2](\text{trace } T)^2 + \tfrac{1}{4}[-x_i^2x_j^2 + 2(\boldsymbol{x}_i\cdot\boldsymbol{x}_j)^2](\text{trace } T^2) , \tag{E.11}$$

where $x_i^2 = \boldsymbol{x}_i^2$. The vectors $\boldsymbol{x}_i$ satisfy

$$\boldsymbol{x}_1 + \boldsymbol{x}_2 + \boldsymbol{x}_3 = 0 ,$$

as shown in fig. 1. It is useful to introduce

$$u_2 = \sum_1^3 x_i^2 , \qquad u_4 = \sum_1^3 x_i^4 ,$$
$$u_6 = \sum_1^3 x_i^6 , \qquad u_{22} = \sum_{i>j} x_i^2 x_j^2 , \tag{E.12}$$
$$u_{42} = \sum_{i\neq j} x_i^4 x_j^2 , \qquad u_{222} = x_1^2 x_2^2 x_3^2 .$$

The area of the triangle can be written as

$$\triangle = \tfrac{3}{4}(2u_{22} - u_4)^{1/2} . \tag{E.13}$$

In terms of the u's the averages of $\bar{A}_{\rm C}({\rm S})$ and $A_{\rm L}({\rm S})$ are, neglecting $O(l^3\triangle)$,

$$\langle \bar{A}_{\rm C}({\rm S})\rangle = \tfrac{1}{2}\boldsymbol{E}^2\triangle + \tfrac{1}{48}u_2(\text{trace } T^2)\triangle , \tag{E.14}$$

$$\langle A_{\rm L}({\rm S})\rangle = \tfrac{1}{2}\boldsymbol{E}^2\triangle + \frac{1}{256\triangle}[(\text{trace } T^2)(-5u_6 + 9u_{42} - 12u_{222})$$
$$+ (\text{trace } T)^2(\tfrac{1}{2}u_6 + \tfrac{3}{2}u_{42} - 24u_{222})] . \tag{E.15}$$

Taking their difference and using (E.5) we find, to the same order,

$$\langle \bar{f}_{\rm S}\rangle = -\frac{1}{256\triangle}[(\text{trace } T^2)(-2u_6 + 6u_{42} - 30u_{222})$$
$$+ (\text{trace } T)^2(\tfrac{1}{2}u_6 + \tfrac{3}{2}u_{42} - 24u_{222})] . \tag{E.16}$$

Define

$$v_6 = \tfrac{1}{27}(a^6 + b^6 + c^6), \qquad v_{222} = \tfrac{1}{27}a^2b^2c^2,$$

$$v_{42} = \tfrac{1}{27}(a^4b^2 + b^4c^2 + c^4a^2 + b^4a^2 + c^4b^2 + a^4c^2), \tag{E.17}$$

$$v = \begin{pmatrix} v_6 \\ v_{42} \\ v_{222} \end{pmatrix}, \qquad u = \begin{pmatrix} u_6 \\ u_{42} \\ u_{222} \end{pmatrix}$$

where a, b and c are the three sides of the triangle, as before. One can verify

$$u = Mv, \qquad v = Mu,$$

where

$$M = \frac{1}{9}\begin{pmatrix} 5 & 6 & -24 \\ 4 & 3 & 24 \\ -\frac{4}{3} & 2 & 1 \end{pmatrix}, \tag{E.18}$$

and $M^2 = 1$. In terms of the v's, (E.16) becomes

$$\langle \bar{f}_s \rangle = -\frac{3}{128\triangle}[(\text{trace } T^2)(v_6 - v_{42} + 3v_{222}) + \tfrac{3}{4}(\text{trace } T)^2(v_6 - v_{42})]. \tag{E.19}$$

Let $\boldsymbol{x}_0$ be the position vector of the center O of the circumscribed circle, measured from any fixed origin unrelated to the triangle. As in the appendix of ref. [3], we have

$$\int \prod_1^3 d^2x_i = 8\pi R \triangle \, d^2x_0 \, dR \, d\alpha \, d\beta \, d\gamma \, \delta(\alpha + \beta + \gamma - 2\pi), \tag{E.20}$$

where the integration is only over the azimuthal angle around $\hat{z}$, and α, β, γ are the angles: $\angle 203$, $\angle 301$ and $\angle 102$, as shown in fig. 1. The integral (E.6) becomes

$$\bar{b}_1 = \frac{\sigma^3}{3!\Omega}\int e^{-\sigma\pi R^2}\langle \bar{f}_s \rangle \, d^2x_0 (8\pi R\triangle) \, dR \, d\alpha \, d\beta \, d\gamma \, \delta(\alpha + \beta + \gamma - 2\pi). \tag{E.21}$$

By using (E.19) and carrying out the integration we obtain, to $O(\sigma^{-1}) = O(l^2)$,

$$\bar{b}_1 = \frac{1}{24\sigma\pi}\frac{1}{\Omega}\int [-(\text{trace } T^2) + \tfrac{3}{2}(\text{trace } T)^2] \, d^2x. \tag{E.22}$$

When $\bar{\phi} = \phi^c$, because $\partial^2\phi^c/\partial x^\mu \, \partial x^\mu = 0$ we have trace $T = 0$; hence, (E.22) reduces to (5.79).

References

[1] T.D. Lee, Phys. Lett. 122B (1983) 217
[2] S.R. Lundeen and F.M. Pipkin, Phys. Rev. Lett. 46 (1981) 232

[3] N.H. Christ, R. Friedberg and T.D. Lee, Nucl. Phys. B202 (1982) 89
[4] N.H. Christ, R. Friedberg and T.D. Lee, Nucl. Phys. B210 [FS6] (1982) 310
[5] N.H. Christ, R. Friedberg and T.D. Lee, Nucl. Phys. B210 [FS6] (1982) 337
[6] J.E. Mayer and M.G. Mayer, Statistical mechanics, chapter 13 (New York, John Wiley and Sons, Inc., 1940)
[7] B. Adeva et al., Phys. Rev. Lett. 48 (1982) 967
[8] G. Parisi and N. Sourlas, Nucl. Phys. B206 (1982) 321

CU - TP - 267

DISCRETE MECHANICS

T. D. Lee

Columbia University, New York, N. Y. 10027

A series of four lectures given at the International School of Subnuclear Physics, Erice, August 1983.

This research was supported in part by the U.S. Department of Energy.

CONTENTS

I. INTRODUCTION

At present, with QCD for the strong interaction, a unifying gauge theory for the electromagnetic and weak interactions, and the Einstein theory of gravitation (plus possible further unification of some or all of these interactions), we may feel that we have arrived at a closed system of physical laws. Nevertheless, the fact that any of the unifying gauge theoretical models requires about twenty parameters suggests that our understanding of these forces most likely remains phenomenological, far from fundamental. This uneasy feeling is reinforced by the enormous technical difficulty of quantizing gravity. Attempts to improve this situation have so far invariably led back to the original unsatisfactory state, with only a change in scale. The underlying cause may be traced to the foundation of our relativistic quantum local field theory. In such a theory only fields are observables, represented by operator functions embedded in a space-time continuum $x = (\vec{r}, t)$. Hence we can conceive of performing an infinite sequence of observations, at $x_1, x_2, \cdots$, all lying infinitesimally close but with the result of the field measurements, say $\phi(x_1)$, $\phi(x_2), \cdots$, fluctuating violently. The probability amplitude of such an unlikely event is proportional to e^{iA} with A as the corresponding action. Yet only the part of A that involves the gradient of the fields is effective in making such rapid fluctuations improbable. This then severely limits the candidates for an acceptable physical theory, and results in the present small set of renormalizable field theories: QED, QCD, Higgs, $\cdots$. All of these have been intensively investigated. Except for the variation in group algebra, we have virtually exhausted the possible theories within this limited repertoire.

Let us suppose that the concept of locality is only an approximate one. After all, it is not difficult to imagine that the present quantum local field theory should break down at a distance ℓ of the order of Planck's length. Once the notion of a fundamental length ℓ is accepted, one could contemplate that ℓ might even be much bigger (though about 10^{-16} cm in order to be consistent with present experimental observations). In any case, the existence of a fundamental length would break the present confinement of renormalizable field theories; it could naturally inhibit the possibility of making an infinite number of measurements that are infinitesimally close to each other in space-time.

Assuming that ℓ is the only scale in such a theory, at distances $r \gg \ell$ we may set $\ell/r \cong 0$. As a result, the theory becomes approximately scale-invariant. Scale-invariant theories are very few in number. We have the chiral symmetric theories for spin $\frac{1}{2}$, the gauge theories for spin 1, etc. These are precisely the type of theories that are in use at present. They are almost all generalizations of the (classical) Maxwell equation which was originally derived from large-distance phenomena.

From this point of view, our present theories may well be approximations valid only at large distances. When we extrapolate these theories to small distances, new phenomenological constants are needed. This then accounts for the very large number of parameters that are currently required.

In this series of lectures, I wish to explore a new direction which may provide a fundamental length. I shall begin the discussion by regarding time as a dynamical variable. This will lead to the following two inter-related topics:

1. discrete quantum mechanics

and

2. random lattice field theories.

These subjects have recently been developed at Columbia[1-4] with the collaboration of N. H. Christ, R. Friedberg and H. C. Ren. In addition, there are also the recent papers by C. Itzykson and others.[5]

II. AN OVERVIEW: TIME AS A DYNAMICAL VARIABLE

As we all know, time has always been regarded as a continuous parameter in physics. Even in general relativity, although the metric is a dynamical variable, the continuous four-dimensional space that the metric is embedded in is not. This concept can be traced to Newtonian mechanics. The following quotation was written by James Clerk Maxwell[6] in 1877:

> Absolute, true and mathematical Time is conceived by Newton as flowing at a constant rate, unaffected by the speed or slowness of material things.

There is no evidence that this long-held view is in any kind of trouble. Nevertheless, we wish to explore an alternative: that time is a dynamical variable. Just like any other dynamical variable, time is affected by material things, and in fact it is part of matter.

Before introducing this new concept, let us review the role of time throughout all phases of mechanics. This is illustrated in the following table.

TABLE 1

	Usual Continuum Theory	
Classical Mechanics	$\vec{r}(t)$	dynamical variable
	t	parameter
Non-relativistic Quantum Mechanics	$\vec{r}(t)$	operator (observable)
	t	parameter
Relativistic Quantum Theory	field $\phi(\vec{r}, t)$	operator (observable)
	$\vec{r}$, t	parameters

6.

In the usual continuum theory the position $\vec{r}(t)$ of a particle is a dynamical variable in classical mechanics, but the time t is a parameter. When we go over to the non-relativistic quantum mechanics, the observable $\vec{r}(t)$ becomes an operator while t remains a parameter. In the relativistic theory, $\vec{r}$ and t have to be treated on an equal basis. Two choices are open. Either regard t as an operator or $\vec{r}$ as a parameter. Our traditional course is to opt for the latter: only the fields are operators or observables. The space-time coordinates are merely parameters. An alternative route is to see whether we can regard t as an operator; this is then the essence of this new approach, which I call discrete mechanics.

Table 2 summarizes the fundamental concepts of the discrete mechanics.

TABLE 2

	Discrete Theory	
Classical Mechanics	$\vec{r}$, t	both dynamical variables
Non-relativistic Quantum Mechanics	$\vec{r}$, t	both operators (observables)
Relativistic Quantum Theory	$\vec{r}$, t, ϕ	all operators (observables)

Thus, in the discrete version of relativistic quantum theory, the space-time position, as well as the field, is considered a dynamical variable. For example, in a collision experiment of, say, $e^+e^- \to \mu^+\mu^-$ at LEP region P4 in 1988, the precise location and time of the collision should be regarded as part of the measurement, on the same footing as the electric field, magnetic field, $\cdots$. In order to incorporate such a view, let us start with the discrete theory in its classical form.

III. CLASSICAL MECHANICS

Consider the example of a non-relativistic point particle of unit mass moving in a potential $V(\vec{r})$. In Table 3, which appears on the next page, we give the familiar formulation of classical continuum mechanics in the left column with A_c = action. The corresponding discrete version is given in the right column.

Let us consider a large time interval T. A fundamental postulate of discrete mechanics is that within such a time interval a particle can only assume N space-time positions

$$(\vec{r}_n , t_n)$$

with $n = 1, 2, \cdots , N$. The ratio

$$\frac{N}{T} \equiv \rho$$

is a fundamental constant of the theory. For convenience and without loss of generality we have arranged $t_1 , t_2 , \cdots , t_N$ in ascending order. The discrete action A_d is then given in Table 3. Unlike the continuum case (where only $\vec{r}(t)$ is the dynamical variable) we regard $\vec{r}_n$ and t_n both as dynamical variables. Consequently there are two sets of equations:

$$\frac{\partial A_d}{\partial \vec{r}_n} = 0 \tag{3.1}$$

which gives the discrete version of Newton's law, and in addition

$$\frac{\partial A_d}{\partial t_n} = 0 \tag{3.2}$$

which yields the energy conservation. There are altogether $4N$ unknowns:

$$\vec{r}_n = (x_n , y_n , z_n) \quad \text{and} \quad t_n .$$

8.

TABLE 3

Continuum Mechanics	Discrete Mechanics
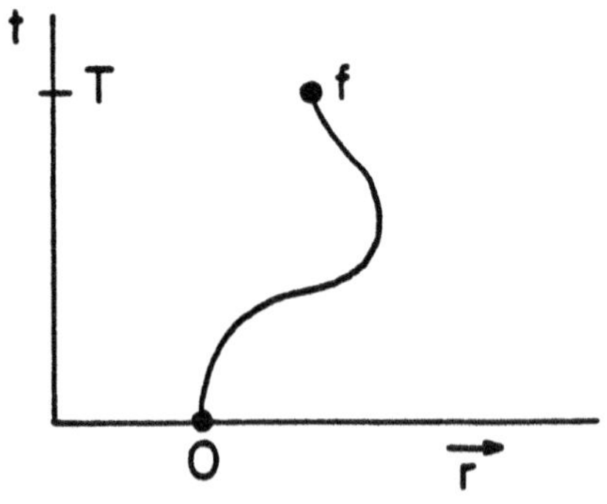	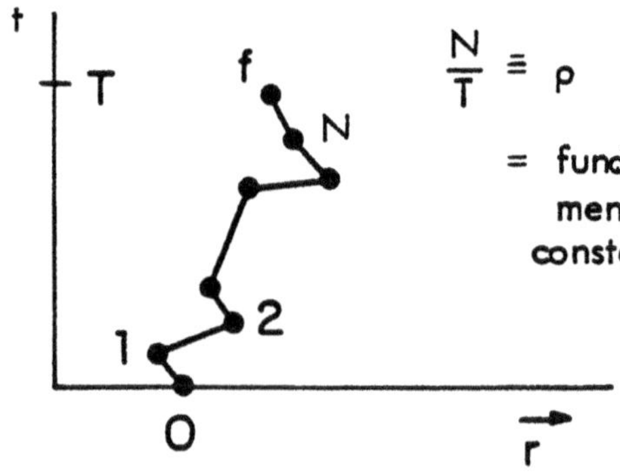 $\frac{N}{T} \equiv \rho$ = fundamental constant
$A_c = \int_0^T (\frac{1}{2}\dot{\vec{r}}^2 - V)\, dt$	$A_d = \sum_n \left\{\frac{1}{2}\vec{v}_n^{\,2} - \frac{1}{2}\left[V(\vec{r}_n) + V(\vec{r}_{n-1})\right]\right\} \cdot (t_n - t_{n-1})$ $\vec{v}_n = \frac{\vec{r}_n - \vec{r}_{n-1}}{t_n - t_{n-1}}$
fix $\vec{r}(0) = \vec{r}_0$ $\vec{r}(T) = \vec{r}_f$	fix $(\vec{r}_n, t_n) = \begin{cases} (\vec{r}_0, 0) & \text{when } n = 0 \\ (\vec{r}_f, T) & \text{when } n = N+1 \end{cases}$
$\frac{\delta A_c}{\delta \vec{r}(t)} = 0$ gives $\ddot{\vec{r}} = -\nabla V$	$\frac{\partial A_d}{\partial \vec{r}_n} = 0$ gives $\frac{\vec{v}_{n+1} - \vec{v}_n}{\frac{1}{2}(t_{n+1} - t_{n-1})} = -\vec{\nabla} V(\vec{r}_n)$
$\vec{r}(t)$ = dynamical variable t = parameter	$\frac{\partial A_d}{\partial t_n} = 0$ gives $E_n \equiv \frac{1}{2}v_n^2 + \frac{1}{2}\left[V(\vec{r}_n) + V(\vec{r}_{n-1})\right] = E_{n+1}$

Exactly the same number of equations is supplied by (3.1) and (3.2). In continuum mechanics conservation of energy is a consequence of Newton's equation. This is not so in discrete mechanics.

Had we treated time merely as a predetermined discrete parameter, the system would then violate time-translational invariance, and lead to energy non-conservation. Historically this has always been the difficulty encountered in any attempt to treat time as a discrete parameter. Here, by viewing $\vec{r}_n$ and t_n as dynamical variables we bypass this problem.

Example 1. For a free particle $V = 0$, Eq. (3.1) gives $\vec{v}_n =$ constant, which also satisfies (3.2). The trajectory of the particle is always a straight line, the same as in the continuum version. Therefore Newton's first law remains unaltered.

Example 2. In the case that the particle is subject to a constant force,

$$\vec{\nabla} V \quad = \quad \text{constant,}$$

it is easy to see that the solution of (3.1) and (3.2) determines that $(\vec{r}_1, t_1)$, $(\vec{r}_2, t_2), \cdots, (\vec{r}_N, t_N)$ all lie on a parabola, similar to the continuum case. In addition, the time intervals $t_2 - t_1$, $t_3 - t_2$, $\cdots$ are all equal.

Example 3. For a harmonic oscillator $V = \frac{1}{2}\omega^2 x^2$, Eqs (3.1) and (3.2) can be solved numerically. In Figure 1 on the next page we give the result for $\omega = 1$, $N = 40$ and $T = 6$ (which is near 2π). The ordinate is x and the abscissa t for $n = 1, 2, \cdots, 40$. At each point (x_n, t_n) the discrete action A_d is stationary with respect to both x_n and t_n. The former gives the equality between momentum-change and impulse, whereas the latter gives energy conservation. Thus both the position x_n and the time t_n respond to the potential. Both depend on the dynamics.

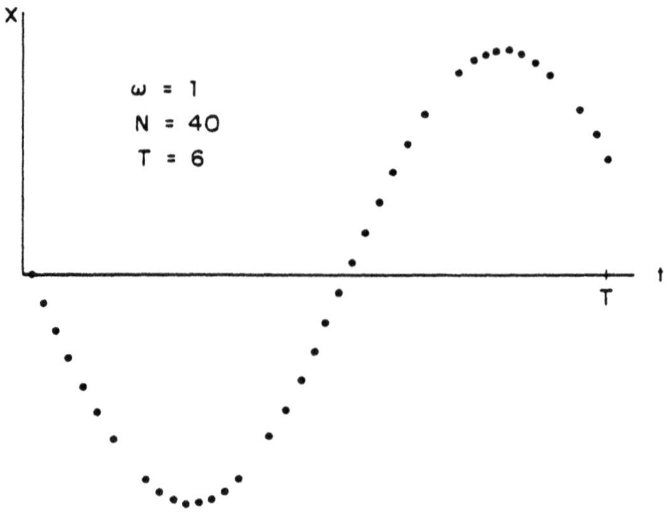

Figure 1. Numerical solution for a harmonic oscillator potential $V = \frac{1}{2}\omega x^2$ in discrete mechanics.

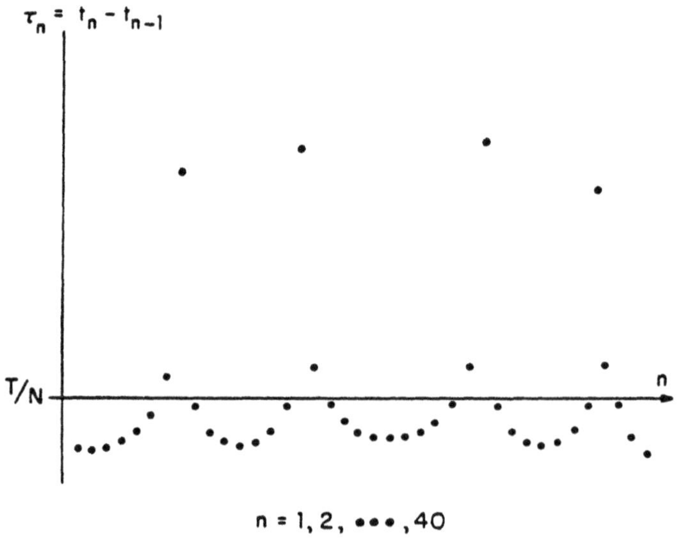

Figure 2. Plot of the consecutive time-spacings $\tau_n = t_n - t_{n-1}$ for n running from 1 to 40. The horizontal line indicates the average τ_n (= T/N = 6/40).

The consecutive time spacing $\tau_n \equiv t_n - t_{n-1}$ is not uniform and is plotted in Figure 2 for n running from 1 to 40. The horizontal line indicates the average τ_n (= T/N = 6/40). Note that the τ_n for n = 8, 17, 28 and 36 are much larger than the average. As we can see, τ_n is almost a periodic function in n, but with a frequency $\simeq 4\omega$, which can also be established analytically when N/T is very large.

IV. NON-RELATIVISTIC QUANTUM MECHANICS

4.1. General Formulation

Extension of the above classical system of a point particle to quantum mechanics is straightforward, as shown in the following table.

TABLE 4

Non-relativistic Continuum Quantum Mechanics	Non-relativistic Discrete Quantum Mechanics
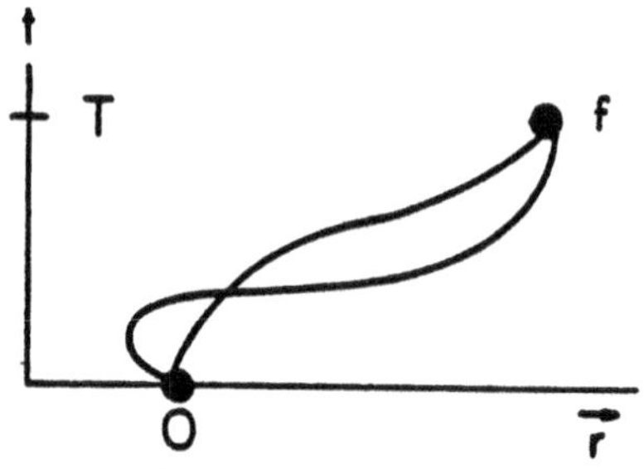	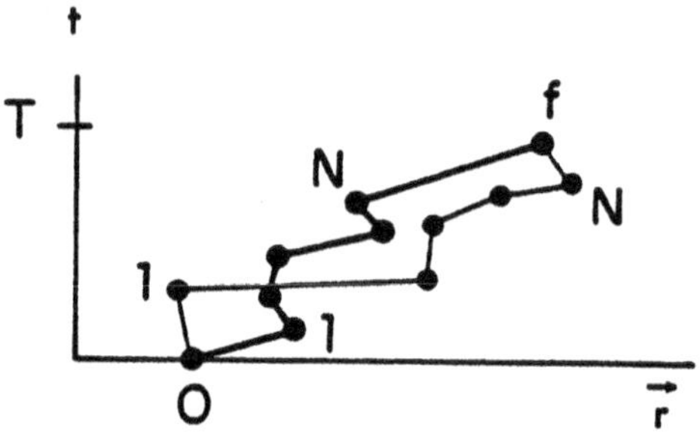
$H_0 = -\frac{1}{2}\nabla^2 + V \qquad (\hbar = m = 1)$	$\frac{N}{T} = \rho \equiv \frac{1}{\ell}$ = fundamental constant
$e^{-iH_0T} = \int e^{iA_c}\,[d\vec{r}(t)]$	$G(T) \equiv \int e^{iA_d}\, J \prod_n d^3r_n\, dt_n$
$e^{-H_0T} = \int e^{-\mathcal{A}_c}\,[d\vec{r}(t)]$	$\mathcal{G}(T) \equiv \int e^{-\mathcal{A}_d}\, J \prod_n d^3r_n\, dt_n$
where $\begin{matrix} A_c \\ \mathcal{A}_c \end{matrix} = \int (\frac{1}{2}\dot{\vec{r}}^2 \mp V)\, dt$	where $\begin{matrix} A_d \\ \mathcal{A}_d \end{matrix} = \sum_n \frac{(\vec{r}_n - \vec{r}_{n-1})^2}{2\epsilon_n} \mp \frac{1}{2}\epsilon_n [V(r_n) + V(r_{n-1})]$
$[d\vec{r}(t)] \equiv \lim_{\substack{T=N\epsilon \\ \epsilon \to 0}} J \prod_n d^3r_n$	
$J \propto (1/\epsilon)^{3N/2}$	$J \propto \prod_n (1/\epsilon_n)^{-3/2}$

In the left column of Table 4 we give the usual continuum Feynman path integration formulation. For the Green's function e^{-iH_0T}, each path carries an amplitude e^{iA_c} where A_c is the continuum action. Because t is just a parameter, the integration is only over the continuous path $[d\vec{r}(t)]$, which is defined to be

$$\lim_{\substack{T=N\epsilon \\ \epsilon \to 0}} J \prod_n d^3r_n$$

where

$$J = \left(\frac{1}{2i\pi\epsilon}\right)^{3N/2} . \tag{4.1}$$

The limit $\epsilon \to 0$ is necessary because in Feynman's formulation the time interval T is divided into N equal intervals of ϵ each. Thus energy is not conserved if $\epsilon \neq 0$. For large T, the matrix element e^{-iH_0T} is oscillatory. It is more convenient to introduce the analytic continuation $T \to -iT$; in that case the Green's function becomes

$$e^{-H_0T} = \int e^{-\mathcal{A}_c} [d\vec{r}(t)] \tag{4.2}$$

where $\mathcal{A}_c$ is the energy integrated over the path $\vec{r}(t)$:

$$\mathcal{A}_c = \int (\tfrac{1}{2}\dot{\vec{r}}^{\,2} + V)\, dt \quad . \tag{4.3}$$

The new Green's function (4.2) is identical to that used in statistical physics and is well-behaved at large T.

In the discrete version, the continuum Green's function e^{-iH_0T} is replaced by $G(T)$ and e^{-H_0T} by $\mathcal{G}(T)$, both of which are given in Table 4. Because t_n are now dynamical variables just like $\vec{r}_n$, the path integrations are over

$$\prod_n d^3r_n\, dt_n$$

where n varies from 1 to N with

$$\frac{N}{T} = \rho \equiv \frac{1}{\ell} = \text{fundamental constant.} \tag{4.4}$$

For clarity, we first let each t_n vary independently from 0 to T. It is convenient to denote their spacings by a set of positive time intervals $\epsilon_1, \epsilon_2, \cdots, \epsilon_N$ so that through a permutation the t_n become chronologically ordered

$$t_1, t_2, t_3, \cdots \rightarrow \epsilon_1, \epsilon_1 + \epsilon_2, \epsilon_1 + \epsilon_2 + \epsilon_3, \cdots . \tag{4.5}$$

In the $(\vec{r}, t)$ plane, the consecutive points that define the discrete path become

$$(\vec{r}_1, \epsilon_1), (\vec{r}_2, \epsilon_1 + \epsilon_2), (\vec{r}_3, \epsilon_1 + \epsilon_2 + \epsilon_3), \cdots . \tag{4.6}$$

Each discrete path carries an amplitude e^{iA_d} or $e^{-\mathcal{A}_d}$ given in Table 4; the path of stationary phase corresponds to the one described by classical discrete mechanics. In the continuum case it is well known that the amplitudes e^{iA_c} and $e^{-\mathcal{A}_c}$ are related to the transformation matrix elements

$$\prod_t < \vec{r}(t + dt) \mid \vec{r}(t) > , \tag{4.7}$$

which describes the fact that the operator $\vec{r}_{op}(t)$ at time t does not commute with $\vec{r}_{op}(t + dt)$. Here the discrete amplitudes e^{iA_d} and $e^{-\mathcal{A}_d}$ can be viewed as representing the corresponding product of matrix elements

$$\prod_n < \vec{r}_{n+1}, t_{n+1} \mid \vec{r}_n, t_n > \tag{4.8}$$

where $t_1, t_2, \cdots$ are assumed to be arranged in a chronological order through the permutation (4.5).

The discreteness in the new mechanics refers to the discrete number of measurements. Each measurement determines the position of the particle $\vec{r}_n$ and the time t_n. Both $\vec{r}_n$ and t_n can take on any of the continuous eigenvalues of the operators $(\vec{r}_n)_{op}$ and $(t_n)_{op}$, which however do not commute with $(\vec{r}_{n+1})_{op}$ and $(t_{n+1})_{op}$, as indicated by (4.8).

In discrete mechanics there is no Hamiltonian or Lagrangian, but only action. What is the physical meaning of these Green's functions? As we shall establish in the next section, for $T \gg \ell$, if we neglect terms $O(e^{-T/\ell})$ or $O(e^{-iT/\ell})$ but keep $O(\ell/T)$, $O(\ell^2/T^2)$, etc., then*

$$G(T) \sim e^{-iHT}$$

and

$$\mathcal{G}(T) \sim e^{-HT} \tag{4.9}$$

where

$$H = H^\dagger = H_0 + \frac{1}{8}\ell^2\left[\nabla^2 - V, [\nabla^2, V]\right] + O(\ell^3) \tag{4.10}$$

and

$$H_0 = -\tfrac{1}{2}\nabla^2 + V \tag{4.11}$$

is the Hamiltonian in the continuum theory.

According to (4.9), $G(T)$ and $\mathcal{G}(T)$ are the Green's functions of a "Schroedinger" equation

$$-\frac{1}{i}\frac{\partial\psi}{\partial T} = H\psi \tag{4.12}$$

or its analytic continuation

$$-\frac{\partial\psi}{\partial T} = H\psi \tag{4.13}$$

where the Hermitian $H = H^\dagger$ plays the role of an effective Hamiltonian. Consequently, the usual physical interpretation of the continuum quantum mechanics can be carried over to the discrete version. For example, the eigenvalues of H are the energy levels of the system and the limit $T \to \infty$ of e^{-iHT} is the S-matrix, which is unitary.

It is not difficult to carry out high-order corrections in ℓ. The details will be given in the next section. For a free particle $V = 0$, the effective Hamiltonian H is identical to H_0; this is then the quantum mechanical version of the classical result: the motion of a free particle is unaltered when the

* The precise mathematical meaning of $\sim$ is given by (4.39) and (4.59) below.

continuum mechanics is replaced by the discrete mechanics.

In the usual continuum quantum mechanics, at any time t we have the following commutation relation between the position operator $x_{op}(t)$ and its conjugate momentum operator $p_{op}(t)$:

$$[p_{op}(t),\ x_{op}(t)] = -i\hbar\ , \tag{4.14}$$

which gives the familiar uncertainty relation

$$\Delta p(t) \cdot \Delta x(t) \geqslant \frac{\hbar}{2}\ . \tag{4.15}$$

Since t is a continuous parameter, as t varies we have an infinite number of (4.15). On the other hand, although the commutator between t and its derivative operator $i\hbar\frac{\partial}{\partial t}$ is also $-i\hbar$, because there is only one t we have just one uncertainty relation of the type

$$\Delta E \cdot \Delta t \geqslant \frac{\hbar}{2}\ . \tag{4.16}$$

This asymmetry is removed in discrete mechanics. There are N relations each of

$$\Delta p_n \cdot \Delta x_n \geqslant \frac{\hbar}{2}$$

and

$$\Delta E_n \cdot \Delta t_n \geqslant \frac{\hbar}{2} \tag{4.17}$$

where p_n and E_n refer to the eigenvalues of $-i\hbar\,\partial/\partial x_n$ and $i\hbar\,\partial/\partial t_n$. The eigenvalue of the effective Hamiltonian H, given by (4.10), represents the coarse grain average of E_n.

4.2. Effective Hamiltonian

We begin with the discrete action $\mathcal{A}_d$ given in Table 4:

$$\mathcal{A}_d = \sum_{n=1}^{N+1} \left\{ \frac{(\vec{r}_n - \vec{r}_{n-1})^2}{2\epsilon_n} + \frac{\epsilon_n}{2}\left[V(\vec{r}_n) + V(\vec{r}_{n-1})\right] \right\} \tag{4.18}$$

where $\vec{r}_{N+1} = \vec{r}_f$, ϵ_n are real > 0, and satisfy

$$\sum_{n=1}^{N+1} \epsilon_n = T . \qquad (4\ 19)$$

In the $(\vec{r}, t)$ space, the N discrete points that define the path are given by (4.6). As before, let the original t-coordinates of these N points be t_1, t_2, $\cdots$, t_n, each of which varies independently between 0 and T. The two sequences ϵ_1, $\epsilon_1 + \epsilon_2$, $\cdots$ and t_1, t_2, $\cdots$ are related by the permutation (4.5).

For a given N, the Green's function $\mathcal{G}_N(T)$ is given by *

$$\mathcal{G}_N(T) \equiv \int e^{-\mathcal{A}_d} \, J \prod_{n=1}^{N} d^3 r_n \, dt_n \qquad (4.20)$$

where

$$J = \prod_{n=1}^{N+1} \left(\frac{1}{2\pi \epsilon_n} \right)^{3/2} . \qquad (4.21)$$

Instead of assuming a definite N, we shall keep only T fixed, but sum over N with a probability distribution proportional to

$$\frac{1}{N!} \left(\frac{1}{\ell} \right)^N e^{-T/\ell} . \qquad (4.22)$$

This technique is familiar in statistical mechanics, corresponding to the transition of a canonical ensemble to a grand canonical ensemble.

In place of the matrix element $\langle \vec{r}_f \mid e^{-TH_0} \mid \vec{r}_0 \rangle$ in the continuum theory, the Green's function in the discrete mechanics is now given by the following sum over all discrete paths leading from $\vec{r}_0$ at $t = 0$ to $\vec{r}_{N+1} = \vec{r}_f$ at $t = T$:

$$\langle \vec{r}_f \mid \mathcal{G}(T, \ell) \mid \vec{r}_0 \rangle \equiv \sum_{N=0}^{\infty} \frac{1}{N!} \left(\frac{1}{\ell} \right)^N e^{-T/\ell} \mathcal{G}_N(T)$$

$$= \sum_{N=0}^{\infty} \frac{1}{N!} \left(\frac{1}{\ell} \right)^N e^{-T/\ell} \int J \prod_{n=1}^{N} d\tau_n \, d^3 r_n \, e^{-\mathcal{A}_d} \qquad (4.23)$$

* We add a subscript N to the Green's function defined in Table 4.

where ℓ^{-1} plays the same role as the fugacity in a grand partition function in statistical mechanics.

Recalling that

$$< \vec{r}' \mid e^{\frac{1}{2}\epsilon\nabla^2} \mid \vec{r} > \; = \; (\frac{1}{2\pi\epsilon})^{3/2} \; e^{-\frac{1}{2\epsilon}(\vec{r}' - \vec{r})^2} , \tag{4.24}$$

we see that the matrix elements of

$$U(\epsilon) \;\equiv\; e^{-\frac{1}{2}\epsilon V} \, e^{\frac{1}{2}\epsilon\nabla^2} \, e^{-\frac{1}{2}\epsilon V} \tag{4.25}$$

are

$$< \vec{r}' \mid U(\epsilon) \mid \vec{r} > \; = \; (\frac{1}{2\pi\epsilon})^{3/2} \; e^{-\frac{1}{2\epsilon}(\vec{r}' - \vec{r})^2 - \frac{1}{2}\epsilon[V(\vec{r}') + V(\vec{r})]} . \tag{4.26}$$

Hence, (4.23) becomes

$$< \vec{r}_f \mid \mathcal{G}(T, \ell) \mid \vec{r}_0 > \; = \; \sum_{N=0}^{\infty} \frac{1}{N!} (\frac{1}{\ell})^N \, e^{-T/\ell} \int \prod_{n=1}^{N} dt_n \cdot < \vec{r}_f \mid U(\epsilon_{N+1}) \, U(\epsilon_N) \cdots U(\epsilon_1) \mid \vec{r}_0 > . \tag{4.27}$$

Because each t_n varies independently from 0 to T, we have

$$\int \prod_{n=1}^{N} dt_n \;=\; T^N .$$

The time-spacings ϵ_n are real and positive, related to the t_n by (4.5). Thus $\int \prod_{n=1}^{N} d\epsilon_n = T^N/N!$. Hence,

$$\int \prod_n dt_n \cdots \;=\; N! \int \prod_n d\epsilon_n \cdots$$

and (4.27) can be written as

$$\mathcal{G}(T, \ell) \;=\; \sum_{N=0}^{\infty} (\frac{1}{\ell})^N \, e^{-T/\ell} \int \prod_{n=1}^{N} d\epsilon_n \, U(\epsilon_{N+1}) \, U(\epsilon_N) \cdots U(\epsilon_1) . \tag{4.28}$$

We note that for a free particle $V = 0$, the continuum Hamiltonian, given by (4.11) is $H_0 = -\frac{1}{2}\nabla^2$; hence, $U(\epsilon) = e^{\frac{1}{2}\epsilon\nabla^2}$ and therefore $\mathcal{G}(T, \ell) = e^{-TH_0}$, the same as the continuum case.

For a general V , we multiply the right hand side of (4.28) by

$$1 = \int d\epsilon_{N+1}\, \delta\Big(\sum_1^{N+1} \epsilon_n - T\Big) = \int_{-i\infty}^{i\infty} \frac{dz}{2\pi i} \int_0^\infty d\epsilon_{N+1}\, e^{Tz - \sum_1^{N+1} \epsilon_n z} \tag{4.29}$$

which takes care of the constraint (4.19), and write $\exp(-T/\ell)$ as $\exp(-\sum_1^{N+1} \epsilon_n/\ell)$. Hence,

$$\mathcal{G}(T, \ell) = \int_{-i\infty}^{i\infty} \frac{dz}{2\pi i}\, e^{Tz} \sum_{N=0}^{\infty} \Big(\frac{M(z)}{\ell}\Big)^N M(z) \tag{4.30}$$

where

$$M(z) = \int_0^\infty d\epsilon\, e^{-\epsilon(\frac{1}{\ell}+z)}\, U(\epsilon) \quad . \tag{4.31}$$

Defining

$$\overline{H}(z) \equiv M^{-1}(z) - z - \frac{1}{\ell} \quad , \tag{4.32}$$

we have

$$\mathcal{G}(T, \ell) = \int_{-i\infty}^{i\infty} \frac{dz}{2\pi i} \frac{e^{Tz}}{z + \overline{H}(z)} \quad . \tag{4.33}$$

Anticipating that the average ϵ_n is $O(\ell)$ which is small, we expand the deviation of $U(\epsilon_n)$ from $e^{-\epsilon_n H_0}$ in powers of ϵ_n. Using (4.11) and (4.25), and writing

$$K \equiv -\tfrac{1}{2}\nabla^2 \quad , \qquad H_0 = K + V$$

and

$$U(\epsilon) = e^{-\frac{1}{2}\epsilon V}\, e^{-\epsilon K}\, e^{-\frac{1}{2}\epsilon V} \quad , \tag{4.34}$$

we find that the expansion

$$U(\epsilon) - e^{-\epsilon H_0} = \sum_1^\infty \frac{\epsilon^n}{n!} h_n \tag{4.35}$$

is given by

$$\begin{aligned} h_1 &= h_2 = 0 \quad , \\ h_3 &= -\tfrac{1}{2}\Big[K + \frac{V}{2}, [K, V]\Big] \quad , \\ h_4 &= \tfrac{1}{2}\Big[K^2 + (K+V)^2, [K,V]\Big] + \tfrac{1}{2}\Big[K, [K, V^2]\Big] - [K,V]^2 \end{aligned} \tag{4.36}$$

etc. Combining (4.31), (4.32) and (4.35), we obtain the power series of $\overline{H}(z)$

$$\overline{H}(z) = H_0 + \ell^2 \overline{H}_2(z) + \ell^3 \overline{H}_3(z) + \ell^4 \overline{H}_4(z) + \cdots \tag{4.37}$$

where

$$\begin{aligned} \overline{H}_2(z) &= -h_3 , \\ \overline{H}_3(z) &= 2z h_3 - h_4 - h_3 H_0 - H_0 h_3 , \\ \overline{H}_4(z) &= -3z^2 h_3 + 3z(h_4 + h_3 H_0 + H_0 h_3) \\ &\qquad - (h_5 + H_0 h_4 + h_4 H_0 + H_0 h_3 H_0) . \end{aligned} \tag{4.38}$$

Substituting (4.37) into (4.33), we shall establish the following asymptotic expression for $\mathcal{G}(T, \ell)$ when T is large:

<u>Theorem.</u> $$\mathcal{G}(T, \ell) = \xi^\dagger e^{-TH} \xi + O(e^{-T/\ell}) , \tag{4.39}$$

where in powers of ℓ

$$\xi = 1 - \ell^3 h_3 - 3\ell^4 (H_0 h_3 + h_3 H_0 + \tfrac{1}{2} h_4) + O(\ell^5) = \xi^\dagger \tag{4.40}$$

and

$$\begin{aligned} H = {} & H_0 - \ell^2 h_3 - \ell^3 (h_4 + 2h_3 H_0 + 2H_0 h_3) \\ & - \ell^4 (h_5 + \tfrac{5}{2} h_4 H_0 + \tfrac{5}{2} H_0 h_4 + 3h_3 H_0^2 + 3H_0^2 h_3 + 4H_0 h_3 H_0) \\ & + O(\ell^5) \end{aligned} \tag{4.41}$$

with $H_0 = -\frac{1}{2}\nabla^2 + V$ as before, and $h_3, h_4, \cdots$ given by (4.35) and (4.36).

<u>Proof.</u> Let us assume that the potential $V(\vec{r})$ is bounded from below. Without any loss of generality, we may set* $V(\vec{r}) \geqslant 0$.

<u>Lemma.</u> In the complex z-plane, if $\mathrm{Re}\, z > 0$ the determinant $| z + \overline{H}(z) | \neq 0$.

---

*A shift $V(\vec{r}) \to V(\vec{r}) + v$ where v is a constant leads, in accordance with (4.25)-(4.27), to $U(\epsilon) \to U(\epsilon)\, e^{-\epsilon v}$ and $\mathcal{G}(T, \ell) \to \mathcal{G}(T, \ell)\, e^{-Tv}$.

Take any normalized state vector ψ. We have $0 < \psi^\dagger e^{-\epsilon K} \psi \leqslant 1$ where $K = -\frac{1}{2}\nabla^2$, as before. Thus, the expectation value of $U(\epsilon) = e^{-\frac{1}{2}\epsilon V} e^{-\epsilon K} e^{-\frac{1}{2}\epsilon V}$ has the same bounds: $0 < \psi^\dagger U(\epsilon)\psi \leqslant 1$. From (4.31) we see that

$$\psi^\dagger M(z)\psi = \int_0^\infty d\epsilon\, e^{-\epsilon(\frac{1}{\ell}+z)} \psi^\dagger U(\epsilon)\psi \quad ;$$

its magnitude satisifes

$$\begin{aligned} |\psi^\dagger M(z)\psi| &\leqslant \int_0^\infty d\epsilon\, \left| e^{-\epsilon(\frac{1}{\ell}+z)} \psi^\dagger U(\epsilon)\psi \right| \\ &= \int_0^\infty d\epsilon\, e^{-\epsilon(\frac{1}{\ell}+\mathrm{Re}\, z)} |\psi^\dagger U(\epsilon)\psi| \\ &\leqslant \frac{\ell}{1+\ell\,\mathrm{Re}\, z} \quad . \end{aligned} \tag{4.42}$$

Hence when $\mathrm{Re}\, z > 0$, we find

$$|\psi^\dagger M(z)\psi| < \ell \quad ,$$

which implies that the magnitude of every eigenvalue of $M(z)$ is less than ℓ, and therefore that of $M^{-1}(z)$ is greater than ℓ^{-1}; consequently the determinant $|M^{-1}(z) - \ell^{-1}| \neq 0$. From (4.32) we have $z + \overline{H}(z) = M^{-1}(z) - \ell^{-1}$ and the lemma is proved.

We now turn to the proof of the theorem.

In the complex z-plane, the roots of $|z + \overline{H}(z)| = 0$ all lie on the half plane $\mathrm{Re}\, z \leqslant 0$. For small ℓ, those roots that are near $|z| = O(\ell^{-1})$ or larger give an $\leqslant O(e^{-T/\ell})$ contribution to $\mathcal{J}(T, \ell)$; they can be ignored when $T \to \infty$. To find the roots that are $O(1)$, we can expand $\overline{H}(z)$ in powers of ℓ.

(i) Neglecting $O(\ell^3)$, (4.37) becomes

$$\overline{H}(z) = H_0 - \ell^2 h_3 \tag{4.43}$$

which is independent of z. Hence, in accordance with (4.33),

$$\mathcal{G}(T, \ell) = \int_{-i\infty}^{i\infty} \frac{dz}{2\pi i} \frac{e^{Tz}}{z + H_0 - \ell^2 h_3} = e^{-TH}$$

where

$$H = H_0 - \ell^2 h_3 + O(\ell^3)$$

in agreement with (4.41).

(ii) Neglecting $O(\ell^4)$, $\overline{H}(z)$ is a linear function of z :

$$\overline{H}(z) = \alpha + \beta z \tag{4.44}$$

where

$$\alpha = H_0 - \ell^2 h_3 - \ell^3 (h_4 + h_3 H_0 + H_0 h_3) + O(\ell^4) \tag{4.45}$$

and

$$\beta = 2\ell^3 h_3 + O(\ell^4) \quad . \tag{4.46}$$

Thus,

$$z + \overline{H}(z) = (1+\beta)^{\frac{1}{2}} (\bar{\alpha} + z) (1+\beta)^{\frac{1}{2}} + O(\ell^4) \tag{4.47}$$

where

$$\bar{\alpha} = (1+\beta)^{-\frac{1}{2}} \alpha (1+\beta)^{-\frac{1}{2}} \quad . \tag{4.48}$$

Substituting (4.47) into (4.33), we obtain

$$\mathcal{G}(T, \ell) = \xi^{\dagger} e^{-TH} \xi$$

where

$$\xi = (1+\beta)^{-\frac{1}{2}} = 1 - \ell^3 h_3 + O(\ell^4)$$

and

$$H = \bar{\alpha} + O(\ell^4) = H_0 - \ell^2 h_3 - \ell^3 (h_4 + 2h_3 H_0 + 2H_0 h_3) + O(\ell^4)$$

in agreement with (4.40) and (4.41).

Similar considerations can be extended to higher order corrections in ℓ .

<u>Comments.</u> Consider a time interval $T \gg \ell$. Neglect $O(e^{-T/\ell})$ but keep $O(\ell/T)$, $O(\ell^2/T^2)$, $O(\ell^3/T^3)$, $\cdots$. According to (4.39)

$$\mathcal{G}(T, \ell) = \xi^{\dagger} e^{-TH} \xi \quad .$$

Analytically continue $T \to iT$ to the "real" time, we have $\mathcal{G}(T, \ell) \to G(T, \ell)$

where $$G(T, \ell) = \xi^{\dagger} e^{-iTH} \xi \quad . \tag{4.49}$$

Regard H as the (effective) Hamiltonian. The corresponding Schroedinger equation with T as the time is

$$-\frac{1}{i}\frac{\partial \psi}{\partial t} = H\psi \quad . \tag{4.50}$$

Given $\psi = \psi(0)$ at $T = 0$, we find

$$\psi(T) = e^{-iTH}\psi(0) \quad . \tag{4.51}$$

The transition amplitude from $\psi(0) = \psi_a$ to

$$\psi(T) = \psi_b$$

is

$$M_{ba}(T) = \psi_b^{\dagger} e^{-iHt} \psi_a \quad . \tag{4.52}$$

Define

$$\psi(T) = \xi\,\phi(T) \, , \quad \text{or} \quad \phi(T) = \xi^{-1}\psi(T) \tag{4.53}$$

then

$$-\frac{1}{i}\frac{\partial \phi}{\partial T} = \mathcal{H}\phi \tag{4.54}$$

where

$$\mathcal{H} = \xi^{-1} H \xi \quad . \tag{4.55}$$

The same transition amplitude $M_{ba}(T)$ can also be written as

$$M_{ba}(T) = \phi_b^{\dagger}\, G(T, \ell)\, \phi_a \tag{4.56}$$

where, in accordance with (4.53),

$$\psi_b = \xi\,\phi_b \qquad \text{and} \qquad \psi_a = \xi\,\phi_a \tag{4.57}$$

and $G(T, \ell)$ is given by (4.49).

The ϕ's are state-vectors in a Hilbert space whose metric is

$$g = \xi^\dagger \xi$$

which is positive. From (4.53) we see that

$$\frac{\partial}{\partial T} (\phi^\dagger g \phi) = 0$$

for $\phi = \phi(T)$ satisfying (4.54). Under the transformation

$$\phi \to \psi = \xi \phi \qquad (4.58)$$

we have the metric

$$g = \xi^\dagger \xi \to g = 1$$

and correspondingly

$$G(T, \ell) \to e^{-iTH} \qquad (4.59)$$

which is the Green's function of the Schroedinger equation $H\psi = i\, \partial\psi/\partial T$.

4.3. Applications

When we make the transition from continuum to discrete, it is of interest to study the lowest order correction to the energy levels in the usual continuum theory. Let $|>$ be an eigenstate of the continuum Hamiltonian $H_0 = -\frac{1}{2m} \nabla^2 + V$:

$$H_0 |> = E_0 |> . \qquad (4.60)$$

From (4.36) and (4.41), we have

$$H = H_0 + \tfrac{1}{2} \ell^2 \left[K + \frac{V}{2}, \, [K, V]\right] + O(\ell^3)$$

where $K = -\frac{1}{2m} \nabla^2$ is the usual kinetic energy operator. Because $K + \frac{V}{2} = H_0 - \frac{V}{2}$ and, on account of (4.60),

$$< | \left[H_0, [K, V]\right] | > = 0 . \qquad (4.61)$$

To $O(\ell^2)$ the corresponding energy in the discrete quantum mechanics for a non-relativistic particle of mass m moving in a potential V is given by ($\hbar = 1$)

$$E = <|H|> + O(\ell^3)$$
$$= E_0 - \frac{1}{4m}\ell^2 <|(\nabla V)^2|> + O(\ell^3) \quad . \qquad (4.62)$$

Examples. (i) In the case of a one-dimensional harmonic oscillator $V = \frac{1}{2}m\omega^2 x^2$, $\nabla V = m\omega^2 x$ we have $E_0 = (n + \frac{1}{2})\,\omega$ and

$$E = (1 - \tfrac{1}{4}\ell^2\omega^2)\,(n + \tfrac{1}{2})\,\omega + O(\ell^3) \qquad (4.63)$$

where $n = 0, 1, 2, \cdots$.

(ii) Let V be the Coulomb potential experienced by an electron of mass m_e at a distance r from a nucleus of charge Z distributed uniformly within a sphere of radius R :

$$V = \begin{cases} -Z\alpha/r & \text{for } r \geqslant R \\ -\frac{1}{2} Z\alpha\left[3 - (\frac{r}{R})^2\right]/R & \text{for } r \leqslant R \end{cases} \qquad (4.64)$$

where $\alpha \cong 1/137$ is the fine structure constant. The additional energy shift ΔE in the discrete mechanics may be estimated by using the non-relativistic wave function $\psi(\vec{r}) = <\vec{r}|>$. We find

$$\Delta E = -\frac{\ell^2}{4m_e} <|(\nabla V)^2|> \cong -\frac{6\pi}{5m_e R}\,|Z\alpha\,\ell\,\psi(0)|^2 . \qquad (4.65)$$

For the $2s$ orbit in hydrogen

$$|\psi(0)|^2 = (8\pi a^3)^{-1} \qquad (4.66)$$

where $a = (m_e\alpha)^{-1}$ is the Bohr radius. Setting the agreement between the experimental observation and the theoretical QED calculation of this level to an accuracy ϵ Rydberg, we obtain

$$\ell m_e < \left[\frac{10}{3}\,(137)^3\,\epsilon\, m_e R\right]^{\frac{1}{2}} \qquad (4.67)$$

which, for $R \sim .8$ fermi and[7] $\epsilon \sim 10^{-11}$, gives

$$\ell < 1.6 \times 10^{-14} \text{ cm} . \tag{4.68}$$

Better limits ($\ell < 10^{-16}$ cm) can be set by using results from the hyperfine splitting,[8] e^+e^- collisions[9] and the forward dispersion relations[10] in πp.

4.4. Jacobian and Supersymmetry

Let us consider a free particle of unit mass. From (4.18) and (4.20) we see that

$$<\vec{r}_f | \mathcal{J}_N | \vec{r}_0> = \int \prod_{1}^{N} d^3r_n \, dt_n \, J \, e^{-<\vec{r}_f | \mathcal{A}_d | \vec{r}_0>} \tag{4.69}$$

where

$$J = \prod_{1}^{N+1} \left(\frac{1}{2\pi\epsilon_n}\right)^{3/2} \tag{4.70}$$

and

$$<\vec{r}_f | \mathcal{A}_d | \vec{r}_0> = \sum_{1}^{N+1} \frac{1}{2\epsilon_n} (\vec{r}_n - \vec{r}_{n-1})^2 \tag{4.71}$$

with $\vec{r}_{N+1} = \vec{r}_f$ and $\sum_{1}^{N+1} \epsilon_n = T$.

The Jacobian J is also related to $\mathcal{A}_d$ by

$$J = \left(\frac{1}{2\pi T}\right)^{3/2} \Big/ \int \prod_{1}^{N} d^3r_n \; e^{-<0 | \mathcal{A}_d | 0>} \tag{4.72}$$

in which $<0 | \mathcal{A}_d | 0>$ is the same matrix element (4.71) but with $\vec{r}_0 = \vec{r}_f = 0$. Thus, apart from an overall multiplicative factor $(2\pi T)^{-3/2}$, the dependence of J on ϵ_n is entirely contained in

$$\left(\int \prod_{1}^{N} d^3r_n \; e^{-<0 | \mathcal{A}_d | 0>}\right)^{-1} . \tag{4.73}$$

The appearance of J is reminiscent of a similar circumstance in statistical mechanics. It was known in classical statistical mechanics that if the ensemble average is over only the coordinate q_n-space, then there is an additional Jacobian in the integral

$$\int J \prod dq_n \cdots \quad . \tag{4.74}$$

On the other hand, if the integration is over the space of both the coordinate q_n and its conjugate momentum p_n then the corresponding Jacobian is 1 ; (4.74) becomes simply

$$\int \prod dq_n dp_n \cdots \quad . \tag{4.75}$$

The true significance was not fully understood until quantum mechanics. Here, the appearance of a multiplicative factor in the integration suggests that we should consider the integration to be performed in a bigger space in which the multiplicative factor is a constant. Only by viewing from the bigger space can we justify the label "Jacobian" for J in the sub-space.

Let us stack the components (x_n, y_n, z_n) of all $\vec{r}_n$ together forming a $\mathcal{N} = 3N$ component column vector:

$$\phi = \begin{pmatrix} x_1 \\ y_1 \\ z_1 \\ \vdots \\ z_N \end{pmatrix} \quad . \tag{4.76}$$

The matrix element $< 0 \mid \mathcal{A}_d \mid 0 >$ can be written as

$$< 0 \mid \mathcal{A}_d \mid 0 > \; = \; \tfrac{1}{2} \tilde{\phi} M \phi \tag{4.77}$$

where $\tilde{\phi}$ is the transpose of ϕ and M is an $\mathcal{N} \times \mathcal{N}$ real symmetric matrix. We introduce $2\mathcal{N}$ elements of a Grassmann algebra: $z_1, \cdots, z_{\mathcal{N}}$ and $\bar{z}_1, \cdots, \bar{z}_{\mathcal{N}}$ which satisfy

$$\{ z_i , z_j \} \; = \; \{ z_i , \bar{z}_j \} \; = \; \{ \bar{z}_i , \bar{z}_j \} \; = \; 0 \tag{4.78}$$

for $i, j = 1, \cdots, \mathcal{N}$. Define

$$< 0 \mid \mathcal{A} \mid 0 > \; \equiv \; \tfrac{1}{2} \tilde{\phi} M \phi + \bar{\psi} M^{\frac{1}{2}} \psi \; = \; < 0 \mid \mathcal{A}_d \mid 0 > + \, \bar{\psi} M^{\frac{1}{2}} \psi \tag{4.79}$$

where

$$\psi = \begin{pmatrix} z_1 \\ \vdots \\ z_{\mathcal{N}} \end{pmatrix} \quad \text{and} \quad \bar{\psi} = (\bar{z}_1 \cdots \bar{z}_{\mathcal{N}}) \ . \tag{4.80}$$

The function $\mathcal{A}$ belongs to a general class of supersymmetrical functions considered by Parisi and Sourlas.[11] It is invariant under the supersymmetric transformation

$$\delta\phi_i = \bar{\psi}_i \varepsilon + \bar{\varepsilon}\psi_i \ ,$$

$$\delta\psi = -\varepsilon M^{\frac{1}{2}}\phi \tag{4.81}$$

and

$$\delta\bar{\psi} = -\tilde{\phi} M^{\frac{1}{2}}\bar{\varepsilon}$$

where ε and $\bar{\varepsilon}$ are anticommuting parameters.

One can readily verify that

$$\int \prod_1^{\mathcal{N}} d\phi_i \, d\bar{z}_i \, dz_i \, e^{-\langle 0|\mathcal{A}|0\rangle} = (2\pi)^{\frac{1}{2}\mathcal{N}} \ . \tag{4.82}$$

Thus, apart from the factor $(2\pi)^{-\frac{1}{2}\mathcal{N}}(2\pi T)^{-3/2}$, (4.69) is identical to

$$\int \prod_n dt_n \, d^3r_n \prod_i d\bar{z}_i \, dz_i \, e^{-\langle \vec{r}_f|\mathcal{A}|\vec{r}_i\rangle} \ , \tag{4.83}$$

where, similar to (4.79)

$$\langle \vec{r}_f|\mathcal{A}|\vec{r}_0\rangle = \langle \vec{r}_f|\mathcal{A}_d|\vec{r}_0\rangle + \bar{\psi} M^{\frac{1}{2}}\psi \ ; \tag{4.84}$$

i.e.,

$$\langle \vec{r}_f|\mathcal{G}_N|\vec{r}_0\rangle = (2\pi)^{-\frac{1}{2}\mathcal{N}}(2\pi T)^{-3/2} \int \prod_n dt_n \, d^3r_n \prod_i d\bar{z}_i \, dz_i \, e^{-\langle \vec{r}_f|\mathcal{A}|\vec{r}_0\rangle} . \tag{4.85}$$

Similar considerations can also be extended to relativistic field theories. The simplifications derived strongly suggest that supersymmetry should play an important and perhaps more basic role in discrete mechanics.

V. SPIN 0 FIELD

5.1. General Discussion

As an example of the relativistic quantum field theory, we first discuss the case of a massless scalar field ϕ interacting with an arbitrary external current j. The comparison between the new discrete theory and the usual continuum formalism is given in Table 5.

TABLE 5

Continuum theory	Discrete theory
$e^{-TH_0} = \int e^{-A_c}\, J \prod_x d\phi(x)$	$\mathcal{G}(T) = \int e^{-A_d}\, J \prod_{i=1}^{N} d^4x_i\, d\phi_i$
$\phi(x)$ = dynamical variable	ϕ_i and x_i are both dynamical variables $(i = 1, 2, \cdots, N)$
x = 4-dimensional euclidean coordinate (parameter)	$\dfrac{N}{\text{volume}} = \rho = \left(\dfrac{1}{\ell}\right)^4$ = fundamental constant
$A_c = \int \left[\frac{1}{2}\left(\dfrac{\partial\phi}{\partial x_\mu}\right)^2 - j\phi\right] d^4x$	$A_d = \sum_{\ell_{ij}} \frac{1}{2}\lambda_{ij}(\phi_i - \phi_j)^2 - \sum_i j_i\phi_i$
Equation of motion: $-\dfrac{\partial^2\phi}{\partial x_\mu^{\,2}} = j(x)$	Equation of motion: $\sum_{\substack{j \text{ linked} \\ \text{to } i}} \lambda_{ij}(\phi_i - \phi_j) = j_i$
Laplace equation	Kirchoff law

For simplicity, we give here only the euclidean version. (See section VIII for the Minkowski version.) In the continuum theory, the field $\phi(x)$ is the dynamical variable; the space-time coordinate x is just a parameter. Hence, the path integration extends only to $d\phi(x)$. In contrast, the corresponding Green's function in the discrete theory consists of integrations over d^4x_i as well as $d\phi_i$. Because of the d^4x_i integration, it is possible to have both the 4-dimensional rotational symmetry and the translational invariance.

In the discrete formulation, given a 4-dimensional volume

$$\Omega = L^3 T ;$$

we postulate that there can be at most N measurements; each determines the space-time position x_i of the observation and the value of the field ϕ_i at x_i . The ratio

$$\rho = \frac{N}{\Omega} \equiv \left(\frac{1}{\ell}\right)^4 \tag{5.1}$$

is a fundamental constant of the theory. For a given set $\{x_i , \phi_i\}$ where $i = 1, 2, \cdots , N$, the action A_d is identical to that of a random lattice. In order to construct A_d , one must solve the following two problems:

(i) Given N points in a volume Ω , it is desirable to couple only nearby points, simulating the local character of the corresponding density of the continuum action. How can we do that?

(ii) Assume that an algorithm is given, so that only neighboring pairs of sites, say i and j , are coupled. Each pair gives a link ℓ_{ij} and contributes to the action a term $(\phi_i - \phi_j)^2$ multiplied by a weight-factor λ_{ij} . The sum over all links

$$\frac{1}{2} \sum_{\ell_{ij}} \lambda_{ij} (\phi_i - \phi_j)^2 \tag{5.2}$$

replaces the integral

$$\frac{1}{2} \int \left(\frac{\partial \phi}{\partial x_\mu}\right)^2 d^4x \tag{5.3}$$

in the continuum theory. Because different links ℓ_{ij} have different lengths and orientations, it is reasonable that they should carry different weights λ_{ij}. What would be the best choice for these weight functions?

We observe that if ϕ_i is identified as the "electric potential" and

$$\lambda_{ij}^{-1} = \text{electric resistance between } i \text{ and } j , \tag{5.4}$$

then the equation of motion

$$\sum_{\substack{j \text{ linked} \\ \text{to } i}} \lambda_{ij}(\phi_i - \phi_j) = j_i \tag{5.5}$$

is identical to the Kirchoff law of an electric circuit, with j_i as the external current entering at point i. In this analog problem, we would like to determine these resistances λ_{ij}^{-1} so that the electric potential ϕ_i gives the "best" discrete approximation to the corresponding continuum solution $\phi(x)$ of the Laplace equation

$$-\frac{\partial^2 \phi}{\partial x_\mu^2} = j(x) . \tag{5.6}$$

The answers to these two problems are given below.

5.2. Simplitial Decomposition in the Euclidean Space

Consider a D-dimensional euclidean space. Given an arbitrary distribution of N points (called lattice sites, or simply sites) in a finite volume Ω, our first task is to decompose Ω into non-overlapping simplices whose vertices are those sites. To avoid complications due to the boundary, we may assume Ω to be a D-dimensional rectangular volume with the standard periodic boundary condition. For $D = 2$, the simplices are triangles; for $D = 3$, they are tetrahedra. In general, a D-simplex consists of $D + 1$ vertices. Between any pair of vertices, we draw a straight line which forms a link. Hence; there are altogether

$$C_2^{D+1} = \tfrac{1}{2}(D+1)\ D$$

different links per simplex. Regarding the links as sides of triangles, we can form

$$C_3^{D+1} = \frac{1}{3!}(D+1)\ D(D-1)$$

different triangles per simplex. Likewise, since the triangles can be viewed as surfaces of tetrahedra, there are

$$C_4^{D+1} = \frac{1}{4!}(D+1)\ D(D-1)(D-2)$$

different tetrahedra per simplex; etc.

For our applications; we wish to draw links only between nearby sites. This is related to the problem of partitioning Ω into non-overlapping simplices. Let us introduce the following concept of clusters of neighboring sites: consider an arbitrary group of $D+1$ random lattice sites. They lie on the surface of a hypersphere in D dimensions, which will be referred to as their circumscribed sphere. If the inside of that sphere is free of all lattice sites, then this group of $D+1$ sites is called a <u>cluster</u> of neighbors. We form a D-simplex for every cluster, using its sites as vertices.

We note that for any $D+1$ infinitesimal volumes $d^D r_1$, $d^D r_2$, $\cdots$, $d^D r_{D+1}$ in Ω , the probability that each volume element should contain a lattice site is

$$\rho^{D+1} \prod_{i=1}^{D+1} d^D r_i \quad ;$$

the probability that there is no lattice point inside their circumscribed sphere is given by the Poisson formula

$$\exp(-v\rho) \quad , \tag{5.7}$$

where ρ is the site-density, v is the volume of the circumscribed sphere, related to its radius R by

$$v = \begin{cases} (2\pi)^{D/2} \; R^D/D!! & \text{if } D \text{ is even,} \\ 2(2\pi)^{(D-1)/2} \; R^D/D!! & \text{if } D \text{ is odd.} \end{cases}$$

Hence, the probability that $d^D r_1$, $d^D r_2$, $\cdots$, $d^D r_{D+1}$ should form a cluster and therefore be the vertices of a D-simplex is

$$\rho^{D+1} \prod_{i=1}^{D+1} d^D r_i \; \exp(-v\rho) \tag{5.8}$$

The factor $\exp(-v\rho)$ insures that the vertices of the simplex cannot be too far apart.

Let us examine, out of the N given sites, all C^N_{D+1} combinations of possible groupings of $D+1$ sites. Whenever a group fulfills the above condition of a cluster, a D-simplex is formed. The simplices thus constructed define our basic random lattice, which has the following properties:

Theorem. (i) There is no overlap between the volumes of any two simplices, and (ii) the volume-sum of all simplices is Ω. [The proof of the theorem is given in the next section.]

In Figure 3, we give an example of a two-dimensional random lattice.

From (5.8), we can compute the various kinematic properties of the random lattice. For $n > m$, let

$$N_{n/m} \equiv \text{average number of } n\text{-simplices per } m\text{-simplex;}$$

i.e., each m-simplex is on the average shared by $N_{n/m}$ n-simplices. For example, each site (0-simplex) is shared by an average of $N_{1/0}$ links (1-simplices) and each link by $N_{2/1}$ triangles (2-simplices), etc. These averages are given in Table 6.

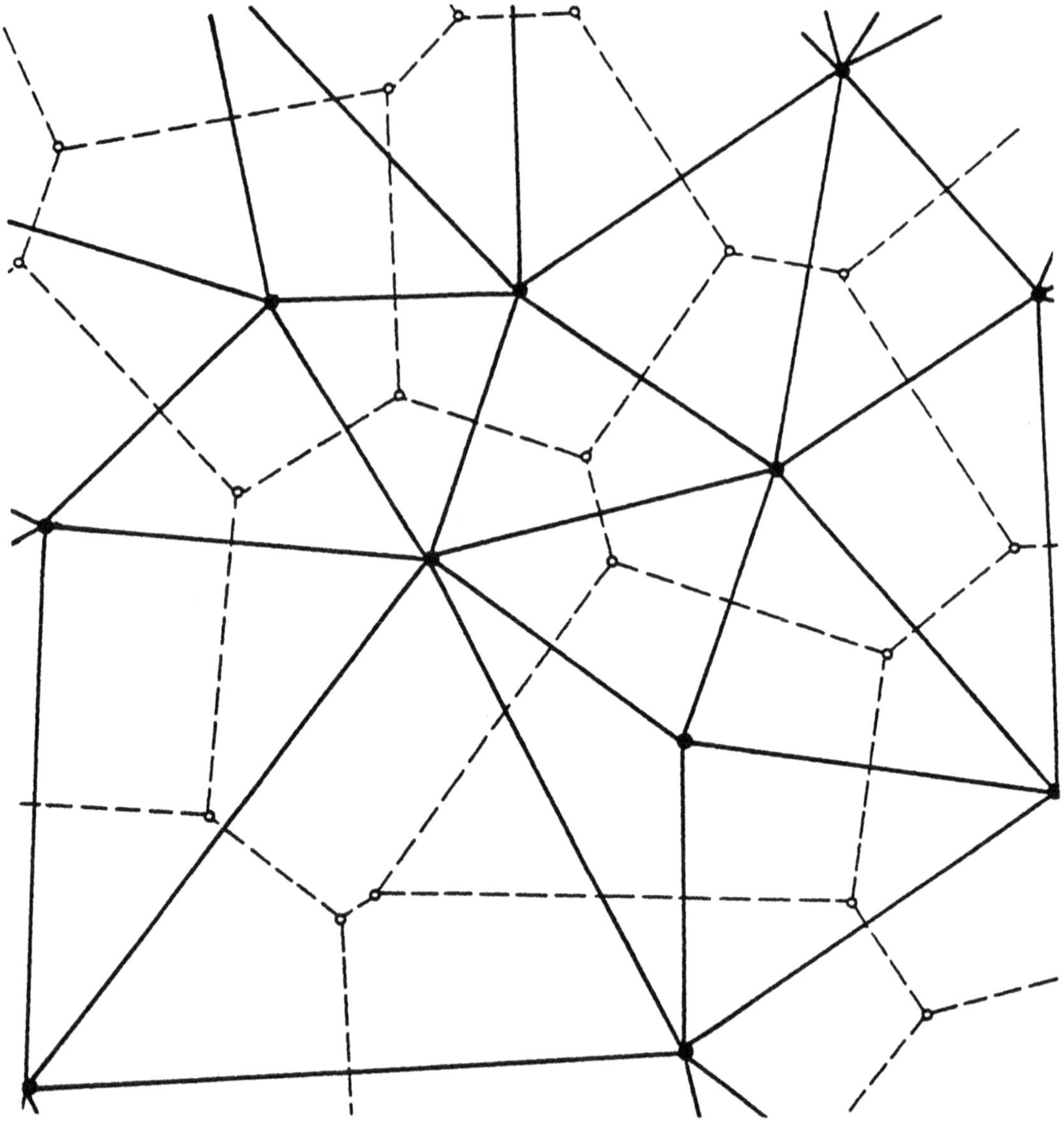

Figure 3. An example of a two-dimensional random lattice (solid lines) and its dual (dashed lines). Dots are the lattice sites, and open circles are the corners of the dual cells. [See section 5.5 for the definition of the dual lattice.]

TABLE 6

		D = 2	D = 3	D = 4
$N_{1/0}$	=	6	$2\left(1 + \frac{24}{35}\pi^2\right)$	$\frac{340}{9}$
$N_{2/0}$	=	6	$\frac{144}{35}\pi^2$	$\frac{590}{3}$
$N_{3/0}$	=		$\frac{96}{35}\pi^2$	$\frac{2860}{9}$
$N_{4/0}$	=			$\frac{1430}{9}$
$N_{2/1}$	=	2	$\frac{144\pi^2}{35 + 24\pi^2}$	$\frac{177}{17}$
$N_{3/1}$	=		$\frac{144\pi^2}{35 + 24\pi^2}$	$\frac{429}{17}$
$N_{4/1}$	=			$\frac{286}{17}$
$N_{3/2}$	=		2	$\frac{286}{59}$
$N_{4/2}$	=			$\frac{286}{59}$
$N_{4/3}$	=			2

We note that

$$N_{D/(D-1)} = 2 ,$$

which is the consequence of the fact that in D-dimensions, each (D-1) - simplex is shared by two D-simplices. In addition, from the same table we have

$$N_{D/(D-2)} = N_{(D-1)/(D-2)} = \begin{cases} 6 & \text{when } D = 2 \\ 5.2276 & \text{when } D = 3 \\ 4.8475 & \text{when } D = 4 \end{cases}$$

which becomes 4 when $D \to \infty$.

When $D = 4$,

$$N_{1/0} = \frac{340}{9} \cong 37.8$$

and

$$N_{2/0} = \frac{590}{3} \cong 197 , \tag{5.9}$$

whereas for a regular "cubic" lattice, $N_{1/0} = 8$ and $N_{2/0} = 24$.

5.3. Proof of the Basic Theorem on the Linking Algorithm

To prove the basic simplitial decomposition theorem stated in the previous section, start with N random lattice sites in a D-dimensional euclidean volume Ω with periodic boundary conditions. Let sites $\{a_1, a_2, \cdots, a_{D+1}\}$ and $\{b_1, b_2, \cdots, b_{D+1}\}$ form two clusters with S_a and S_b as their respective circumscribed hyperspheres. By definition, S_a passes through $a_1, a_2, \cdots, a_{D+1}$, and S_b through $b_1, b_2, \cdots, b_{D+1}$; furthermore, the insides of both S_a and S_b are free of any lattice sites. Hence, S_a cannot be completely inside S_b , nor vice versa. Consequently, S_a and S_b must either be totally separated from each other, which makes (i) an obvious conclusion, or intersect each other.* In the latter case, the intersection I

* The possibility that two different clusters share the same circumsphere (i.e., S_a and S_b are identical) is of zero measure, and is therefore ignored.

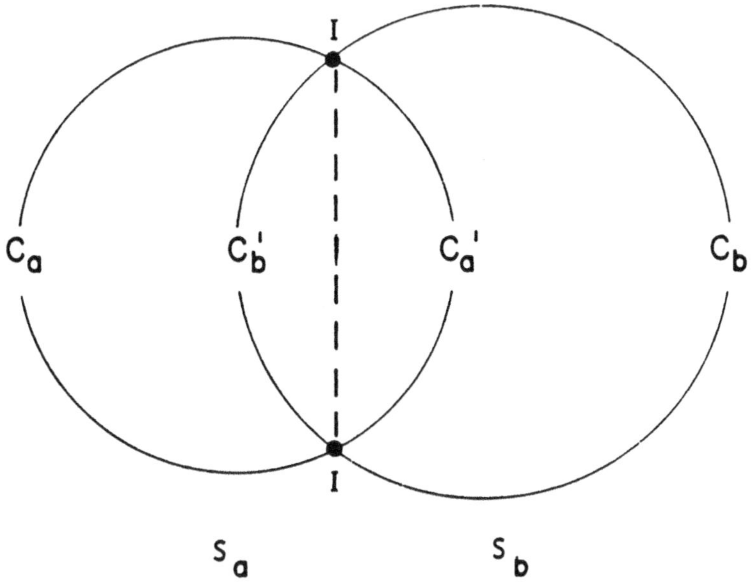

Figure 4. The intersection I of two hyperspheres S_a and S_b cuts each sphere S_i into two caps C_i and C_i' where $i = a$ or b.

of these surfaces S_a and S_b defines a (D−2)-dimensional sphere. As shown in Figure 4, I cuts S_a into two separate caps C_a and C_a' with C_a' lying entirely inside S_b. For definiteness, we exclude the border I from the caps C_a and C_a' so that the sets C_a, C_a' and I are disjoint. The sites in $\{a_i\}$ can be on C_a and I, but not on C_a'; otherwise, some of the a_i would be inside the other sphere S_b, in violation of the assumption. Likewise, I separates S_b into two caps C_b and C_b' with C_b' inside S_a. [Again I is not included in C_b and C_b'.] Hence, none of the b_i can be on C_b'. From $\{a_i\}$ and $\{b_i\}$ we now construct two D-simplices, S_a and S_b. These two simplices must lie on the opposite sides of the (D−1)-dimensional subspace that contains I. Proposition (i) now follows.

We note that S_a and S_b may share some common boundary. If the sets

$\{a_i\}$ and $\{b_i\}$ have a single site in common, then there is a common vertex between S_a and S_b ; if there are two sites in common, then they share a common line, etc. The maximum contact between these two simplices is for $\{a_i\}$ and $\{b_i\}$ to have D sites in common; in that case, S_a and S_b share a common (D-1)-simplex, say s , whose vertices are all on I .

We now turn to the proof of (ii). Every D-simplex is bounded by (D-1)-simplices, each of which has two sides. We shall show that every such (D-1)-simplex is shared by two D-simplices, one on each side. [In two dimensions, this means that every link is shared by two triangles; in three dimensions, every triangle is shared by two tetrahedra, etc.] Assuming that this is not true, then there must be a D-simplex, say τ_a , which has on its surface a (D-1)-simplex, say s, that is not shared by any other D-simplices. Let S_a be the circumscribed sphere of τ_a, and I that of s . We denote by S_b any hypersphere that intersects S_a at I but, as shown in Figure 1, we require the centers of S_a and S_b to lie on opposite sides of s . Consider now the family of all such S_b . By assumption, none of the S_b can contain any lattice points except those on I (i.e., on s). This is clearly false, since S_b can be arbitrarily large and Ω is finite with a periodic boundary condition. [A convenient way to visualize this situation is to consider an array of identical volumes Ω , each containing exactly the same site distribution. These volumes are placed side byside to form an infinite periodic rectangular lattice whose unit cell is Ω . It is then easy to see that the radius of S_b can be arbitrarily large.]

From any D-simplex, we can go through its boundary and reach other D-simplices. By repeating this process, we must eventually cover a volume Ω' which is within Ω and has no boundary. Consequently, $\Omega' = \Omega$, and the proof of the theorem is complete.

A simple example of D = 2 and N = 3 is given in Figure 5.

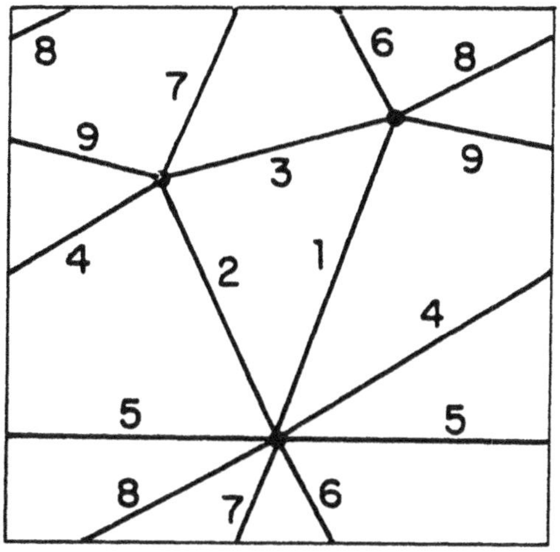

Figure 5. Ω is a two-dimensional square with a periodic boundary condition. There are three sites, nine links and six triangles. The links are marked 1 through 9.

5.4. Generalization of the Theorem

Consider again the problem of N lattice sites randomly distributed in a D-dimensional "rectangular" euclidean volume Ω with periodic boundary conditions. Let $x_1, x_2, \cdots, x_D$ be the cartesian coordinates, and $x_\mu(a)$ the coordinates of the ath site with $\mu = 1, 2, \cdots, D$.

Recall the definition of a "cluster" given in section 5.2. For any group of $D+1$ sites, say $a_1, a_2, \cdots, a_{D+1}$, let S be the $D-1$ dimensional spherical surface (called circumscribed sphere) that passes through these $(D+1)$-sites. To construct S, we may follow a somewhat formal procedure which will however make our subsequent generalization easier. Take an arbitrary point c (which does not have to be a site). With c as the center, consider the sequence of all spherical surfaces $S_R(c)$ of different radii R. Analytically, $S_R(c)$ is represented by

$$\sum_{\mu=1}^{D} [x_\mu - x_\mu(c)]^2 = R^2 \tag{5.10}$$

where $x_\mu(c)$ is the μth component of the position vector of c. By varying R and c, and requiring the $S_R(c)$ to pass through all these $D+1$ sites $a_1, a_2, \cdots, a_{D+1}$, we can find the circumscribed sphere S. According to the definition given in section 5.2, if the interior of S is empty of lattice sites, then the set $\{a_1, a_2, \cdots, a_{D+1}\}$ is called a cluster (of neighbors).

This definition of clusters will now be generalized. Instead of the spherical surfaces $S_R(c)$, we consider the sequence of <u>convex</u> surfaces $\mathcal{S}_R(c)$ of a <u>given shape</u> (with respect to the fixed cartesian coordinate-axes), with c as the center and R as its linear scale. To be specific, we may introduce the standard spherical coordinates $r, \theta_1, \theta_2, \cdots, \theta_{D-2}, \phi$ by

$$\begin{aligned} x_1 - x_1(c) &= r\cos\theta_1 , \\ x_2 - x_2(c) &= r\sin\theta_1 \cos\theta_2 , \\ &\cdots \\ x_{D-1} - x_{D-1}(c) &= r\sin\theta_1 \sin\theta_2 \cdots \sin\theta_{D-2}\cos\phi \\ \text{and} \quad x_D - x_D(c) &= r\sin\theta_1 \sin\theta_2 \cdots \sin\theta_{D-2}\sin\phi \end{aligned} \tag{5.11}$$

where $\theta_1, \theta_2, \cdots, \theta_{D-2}$ vary independently from 0 to π and ϕ from $-\pi$ to π. Instead of (5.10) [which is $r = R$], $\mathcal{S}_R(c)$ is now represented by

$$r = f(\theta_1, \theta_2, \cdots, \theta_{D-2}, \phi)\ R \tag{5.12}$$

where f is any given convex function. The following are two examples in $D = 2$:

(i) $\mathcal{S}_R(c)$ are squares whose sides are parallel to the x_1 and x_2 axes. Equation (5.12) becomes

$$| x_+ - x_+(c) | + | x_- - x_-(c) | = R \tag{5.13}$$

where

$$x_{\pm} \equiv \frac{1}{\sqrt{2}} (x_1 \pm x_2)$$

and (5.14)

$$x_{\pm}(c) = \frac{1}{\sqrt{2}} (x_1(c) \pm x_2(c)) \quad .$$

(ii) $\mathcal{S}_R(c)$ are ellipses given by

$$r^2 = (\cos^2\phi + a^2 \sin^2\phi)^{-1} R^2 \tag{5.15}$$

where a is a <u>fixed</u> constant,

$$x_1 - x_1(c) = r \cos\phi \qquad \text{and} \qquad x_2 - x_2(c) = r \sin\phi \quad .$$

As R changes, (5.15) gives a sequence of ellipses of varying sizes but the same shape.

Given a convex shape function $f(\theta_1, \theta_2, \cdots, \theta_{D-2}, \phi)$, for an arbitrary set of $D+1$ sites, say $a_1, a_2, \cdots, a_{D+1}$, if by varying c and R we can find a surface $\mathcal{S}_R(c)$ that passes through these $D+1$ sites, then that particular $\mathcal{S}_R(c)$ is called the circumscribed surface $\mathcal{S}_a$ of the set $\{a_i\}$. These $D+1$ sites form a cluster* (of neighbors) if, in addition, the interior of $\mathcal{S}_a$ is free of lattice sites. For our problem of N random sites in a volume Ω, we consider all possible groups of $D+1$ sites. Whenever a group is a cluster, we form a simplex using the sites as its vertices. The simplices thus formed satisfy the following theorem.

---

* It is possible that a set of $D+1$ sites $\{a_i\}$ may not have a circumscribed surface, in which case these sites do not form a cluster. For example, in two dimensions if $\mathcal{S}_R(c)$ are the squares given by (5.13) and $\{a_i\}$ consists of three sites all lying on, say, the straight line $x_1 = x_2$, then it is not possible to have a square $\mathcal{S}_R(c)$ passing through these three sites.

Theorem. (i) There is no overlap between the volumes of any two simplices, and (ii) the volume-sum of all simplices is Ω .

Proof. In section 5.3, the entire proof rests only on the convexity of the spherical surfaces. Hence, identical arguments also apply to the present case.

Comments. For any given site-distribution, the simplitial decomposition is not unique, depending on the particular convex function f in (5.12). When f is not a constant (i.e., not spherically symmetric), rotational symmetry may be restored by integrating over the orientation of f . Nevertheless the criterion of using spheres as the circumscribed surface is clearly the simplest choice; it also enjoys the maximum symmetry in a euclidean space. Hence, in Chapters V, VI and VII, throughout our discussions of relativistic theories in the euclidean space we shall always use spheres as the circumscribed surfaces. Different criteria will be adopted only when we discuss gravity in Chapter VIII.

5.5. Weights and the Laplacian

We now return to the discrete action A_d given in Table 5:

$$A_d = \frac{1}{2} \sum_{\ell_{ij}} \lambda_{ij} (\phi_i - \phi_j)^2 - \sum_i j_i \phi_i \quad . \tag{5.16}$$

As discussed in section 5.1, the field equation is obtained by setting $\partial A_d / \partial \phi_i = 0$, which leads to the "Kirchoff Law"

$$\sum_{\substack{j \text{ linked} \\ \text{to } i}} \lambda_{ij} (\phi_i - \phi_j) = j_i \quad . \tag{5.17}$$

Its continuum analog is the Laplace equation

$$- \frac{\partial^2 \phi}{\partial x_\mu^2} = j \quad . \tag{5.18}$$

In (5.17), why can't λ_{ij} be simply a constant, independent of the link ℓ_{ij} ?

Consider, e.g., a one-dimensional random distribution of N sites at positions x_n (with $x_1 < x_2 < \cdots < x_{n-1} < x_n < \cdots$). Assign a field variable ϕ_n to each x_n. Most of us will agree that the approximation

$$dx \frac{d^2\phi}{dx^2} \sim \frac{\phi_{n+1} - \phi_n}{x_{n+1} - x_n} - \frac{\phi_n - \phi_{n-1}}{x_n - x_{n-1}} \tag{5.19}$$

is better than

$$dx \frac{d^2\phi}{dx^2} \sim \lambda(\phi_{n+1} - 2\phi_n + \phi_{n-1}) \tag{5.20}$$

where λ is a constant. This is because when $\phi(x)$ = linear function of x, or correspondingly

$$\phi_n = \text{linear function of } x_n ,$$

both $\frac{d^2\phi}{dx^2}$ and the right hand side of (5.19) are zero, but not the discrete approximation (5.20). This obvious criterion can be readily generalized.

In a multi-dimensional space when $\phi(\vec{r})$ is a linear function of the position vector $\vec{r}$, the Laplace equation is

$$\frac{\partial^2\phi}{\partial x_\mu^2} = 0$$

where x_μ is the μ-component of $\vec{r}$. We require the weights λ_{ij} in the discrete case to satisfy

$$\sum_{\substack{j \text{ linked} \\ \text{to } i}} \lambda_{ij}(\phi_i - \phi_j) = 0 \tag{5.21}$$

at all i for an arbitrary distribution of the site-position vectors $\vec{r}_1, \vec{r}_2, \cdots, \vec{r}_n$ when ϕ_i is a linear function of $\vec{r}_i$; i.e.,

$$\phi_i = a + \vec{b} \cdot \vec{r}_i \tag{5.22}$$

where a and $\vec{b}$ are constants.

To determine λ_{ij}, it is useful to introduce the dual lattice. Let i be an arbitrary site of a D-dimensional random lattice ($i = 1, 2, \cdots, N$). Take any point p in Ω. We say p belongs to i if i is the nearest site to p. The dual to site i is the volume ω_i consisting of all p belonging to i. Each ω_i is a convex polyhedron (called a cell) in D dimensions. The surface of a cell consists of a number of $(D-1)$-dimensional faces, formed by points that belong to two or more sites. The intersections of these $(D-1)$-dimensional faces are $(D-2)$-dimensional faces, which consist of points each belonging to three or more sites, etc. The corners (i.e., vertices) of these cells are single points, each belonging to, and only to, $D+1$ sites. Let c be one of these corners. From our construction we see that c is equidistant from all the $D+1$ sites to which it belongs. Using c as the center, draw a hypersphere through these $D+1$ sites. Since no site is nearer to c than the aforementioned $D+1$ sites, there cannot be any lattice site inside the hypersphere. Hence, these $D+1$ sites satisfy the definition of a cluster, and form a D-simplex, whose dual is the corner c.

The division of Ω into cells gives the dual of our random lattice. Each link in the random lattice is perpendicular to (but need not intersect) a $(D-1)$-dimensional face in its dual; each triangle is perpendicular to a $(D-2)$-dimensional face, etc. This gives a one-to-one correspondence between an n-simplex in the random lattice and a $(D-n)$-dimensional face in its dual. While each D-simplex has a fixed structure (e.g., fixed number of vertices and faces, etc.), this is not so for the cells. For this reason, as we shall discuss, it is often more convenient to formulate the action in terms of simplices, instead of cells. The dual cells are called Voronoi polyhedra in the literature.[12]

In Figure 3 we give a two-dimensional example of a random lattice and its dual.

We set the weights λ_{ij} between any pair of sites i and j to be

$$\lambda_{ij} = \begin{cases} 0 & \text{if } i \text{ and } j \text{ not linked} \\ s_{ij}/\ell_{ij} & \text{if linked} \end{cases} \qquad (5.23)$$

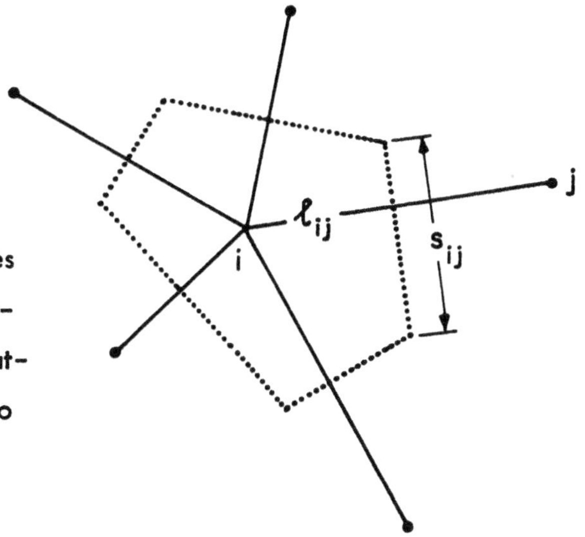

<u>Figure 6.</u> The dotted line indicates the boundary of the Voronoi cell, dual to the lattice site i. The dual to the link ℓ_{ij} is s_{ij}.

where ℓ_{ij} is the length of the link and s_{ij} is the "volume" of the corresponding $(D-1)$-dimensional surface in the dual lattice. When $D = 2$, s_{ij} is the border-length of the dual cell. [See Figure 6.]

<u>Theorem.</u> The λ_{ij} given by (5.23) satisfies

$$\sum_{\substack{j \text{ linked} \\ \text{to } i}} \ell_{ij}^{\mu} \lambda_{ij} = 0 \tag{5.24}$$

at all sites i and

$$L^{\mu\nu} \equiv \sum_{i,j} \ell_{ij}^{\mu} \ell_{ij}^{\nu} \lambda_{ij} = 2\Omega\delta^{\mu\nu} \tag{5.25}$$

where ℓ_{ij}^{μ} is the μ-component of $\vec{\ell}_{ij} = \vec{r}_i - \vec{r}_j$ and $\delta^{\mu\nu} = 0$ if $\mu \neq \nu$ and 1 otherwise.

<u>Proof.</u> When $D = 2$, s_{ij} is simply the border length in the dual lattice, and $\vec{\ell}_{ij} \lambda_{ij}$ is a vector of magnitude s_{ij} in the direction of $\vec{\ell}_{ij}$. A 90° rotation turns it into $\vec{s}_{ij}$. From Figure 6 we see clearly that the sum $\sum \vec{s}_{ij}$ over all j (keeping i fixed) is a null vector, and therefore (5.24) follows.

In any dimension D, two or larger, we may take any constant vector $\vec{V}$. From Gauss' theorem it follows that for any dual cell ω_i

$$0 = \int_{\omega_i} \vec{\nabla} \cdot \vec{V} d^D r = \int \vec{V} \cdot d\vec{S} . \tag{5.26}$$

Because the $(D-1)$-dimensional polyhedra s_{ij} (with i fixed but j varying) form the boundary of ω_i, and because $\hat{\ell}_{ij} \equiv \vec{\ell}_{ij}/\ell_{ij}$ is the normal vector of s_{ij}, (5.26) becomes

$$0 = \vec{V} \cdot \sum_j \hat{\ell}_{ij} s_{ij}$$

which implies the first part of the theorem.

To prove (5.25), we again apply Gauss' theorem, but to the following integral:

$$\omega_i \delta^{\mu\nu} = \int_{\omega_i} \frac{\partial}{\partial x^\nu} (\vec{r} - \vec{r}_i)^\mu \, d^D r = \int (\vec{r} - \vec{r}_i)^\mu dS^\nu$$

$$= \sum_j r_{ic}{}^\mu \, \hat{\ell}_{ij}{}^\nu \, s_{ij} \tag{5.27}$$

where r^ν is the ν-component of $\vec{r}$ and $r_{ic}{}^\mu$ is the μ-component of $\vec{r}_{ic} \equiv \vec{r}_c - \vec{r}_i$. The vector $\vec{r}_c$ is given by

$$\vec{r}_c \equiv \frac{1}{s_{ij}} \int_{s_{ij}} \vec{r} \, d^{D-1} r , \tag{5.28}$$

and is the position vector of the center of mass c of s_{ij}. [See Figure 7.]

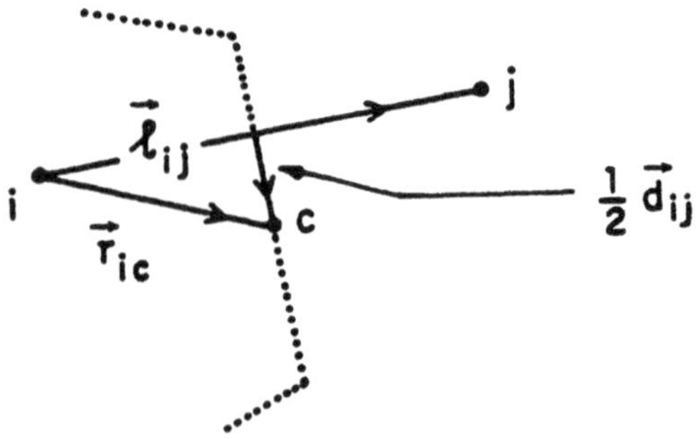

Figure 7. c is the center of the polyhedron s_{ij}, dual to the link ℓ_{ij}.

The link-vector ℓ_{ij} is related to $\vec{r}_{ic}$ and $\vec{r}_{jc}$ by

$$\vec{\ell}_{ij} = \vec{r}_{ic} - \vec{r}_{jc} \quad . \tag{5.29}$$

Define

$$\vec{d}_{ij} \equiv \vec{r}_{ic} + \vec{r}_{jc} \quad . \tag{5.30}$$

Equation (5.27) can then be written as

$$\omega_i \, \delta^{\mu\nu} = \tfrac{1}{2} \sum_j (\vec{\ell}_{ij} + \vec{d}_{ij})^{\mu} \, \hat{\ell}_{ij}^{\nu} \, s_{ij} \quad .$$

Summing over all dual cells ω_i and noting that $\vec{d}_{ij}$ and s_{ij} are even under the exchange of i and j but $\vec{\ell}_{ij}$ and $\hat{\ell}_{ij}$ are odd, we derive

$$\Omega \delta^{\mu\nu} = \sum_i \omega_i \, \delta^{\mu\nu} = \tfrac{1}{2} \sum_{i,j} \ell_{ij}^{\mu} \, \hat{\ell}_{ij}^{\nu} \, s_{ij} \tag{5.31}$$

which establishes (5.25) and completes the proof of the theorem.

From the identities (5.24) and (5.25), we see that when, according to (5.22), ϕ_i is a linear function of $\vec{r}_i$

$$\sum_j \lambda_{ij}(\phi_i - \phi_j) = \vec{b} \cdot \sum_j \lambda_{ij} \, \vec{\ell}_{ij} = 0$$

at all i, , and the sum over all sites i and j

$$\tfrac{1}{2} \sum_{\ell_{ij}} \lambda_{ij}(\phi_i - \phi_j)^2 = \tfrac{1}{4} \sum_{i,j} \lambda_{ij}(\phi_i - \phi_j)^2 = \tfrac{1}{2} \vec{b}^{\,2} \Omega \quad .$$

The former satisfies condition (5.21)-(5.22) for $\sum_j \lambda_{ij}(\phi_i - \phi_j)$ to be a good discrete version of the Laplace operator, and the latter makes the lattice action (5.2) identical to the continuum value (5.3) for an arbitrary lattice before the limit $\ell_{ij} \to 0$, provided that $\phi(\vec{r})$ is a linear function of $\vec{r}$.

In addition, we can show that if the external current is non-zero only at site 0

$$j_i = \delta_{i0} \quad ,$$

then at large distance r_i from 0 the solution ϕ_i becomes the same as that in a continuum; e.g., for $D = 4$ the solution of the Kirchoff equation satisfies

$$\phi_i \rightarrow \frac{1}{4\pi^2 r_i^2} \tag{5.32}$$

when $r_i \rightarrow \infty$. In the corresponding continuum case we have

$$\phi = (4\pi^2 r^2)^{-1} \tag{5.33}$$

which is the solution of

$$-\frac{\partial^2 \phi}{\partial x_\mu^2} = \delta^4(\vec{r}) \ . \tag{5.34}$$

This then insures that the long wave-length limit of the quantum propagator in the discrete theory is the same as the continuum propagator k^{-2}.

The explicit solution of $\phi_i \equiv \phi(r_i)$ for a random lattice of 4^4 sites in $D = 4$ for a site-density $\rho = 1$ is plotted in Figure 8. The average of $\phi(r_i)$ over the position-distribution of lattice sites gives the propagator in the discrete theory. As we can see from Figure 8, even without the average, the result for such a small sample is already relatively smooth, changing from

$$\frac{1}{4\pi^2 r^2} - .00195 \tag{5.35}$$

at large r to a constant $\cong .0165$ at $r = 0$. The constant $- .00195$ in (5.35) is due to the finite size $N = 4^4$ in the sample; it should approach zero and $\phi \rightarrow (4\pi^2 r^2)^{-1}$ as $N \rightarrow \infty$.

Another pleasant feature is that when $\vec{r}_i$ changes, the classical lattice solution ϕ_i and the corresponding action are continuous in $\vec{r}_1, \vec{r}_2, \cdots, \vec{r}_N$ [proved in Ref. 1]. The applicability of continuity concept and the existence of exact equalities, such as (5.24) and (5.25), make it possible to use analytic method in the discrete theory.

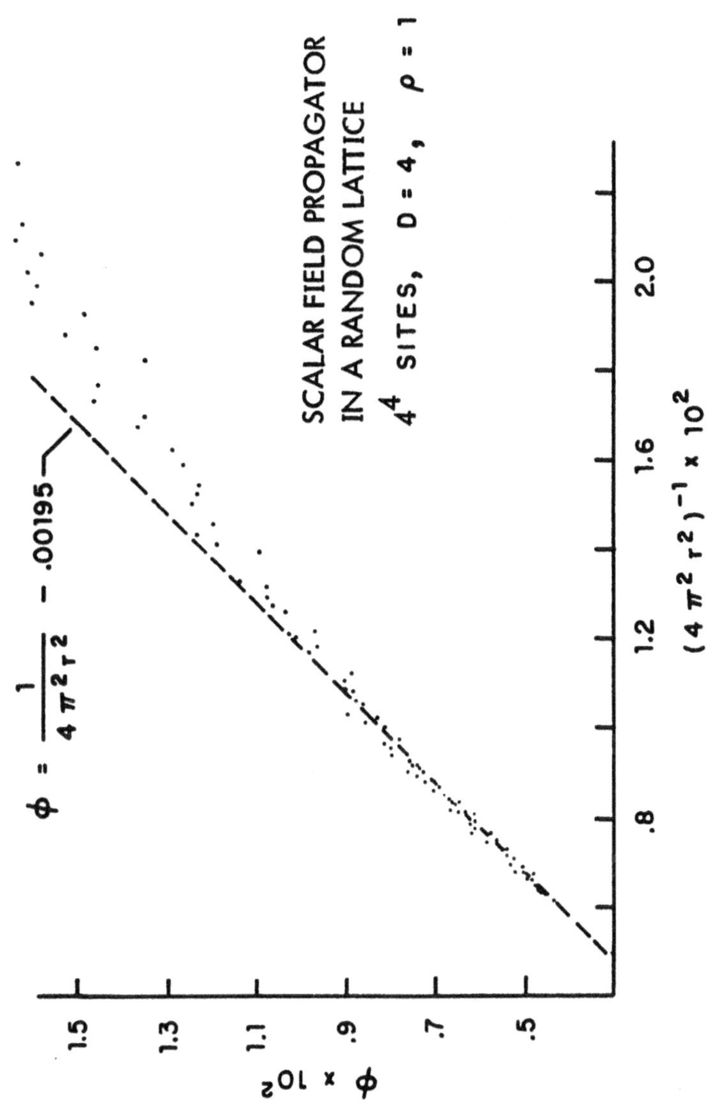

Figure 8. Scalar field propagator in a random lattice.

VI. SPIN - $\frac{1}{2}$ FIELD

The continuum action in a D-dimensional euclidean volume Ω is a hermitian bilinear form given by *

$$A_c(\Omega) = \int_\Omega (\psi^\dagger \gamma^0 \gamma^\mu \frac{\partial}{\partial x^\mu} \psi + m \psi^\dagger \gamma^0 \psi) \, d^D r \tag{6.1}$$

where the dagger denotes hermitian conjugation and $\gamma^0, \gamma^1, \cdots, \gamma^D$ are anti-commuting hermitian matrices whose squares are the unit matrix. The corresponding lattice action is

$$A_d(\Omega) = \frac{1}{2} \sum_{i,j} \psi_i^\dagger \gamma^0 \gamma^\mu \ell_{ij}^\mu \lambda_{ij} \psi_j + m \sum_i \omega_i \psi_i^\dagger \gamma^0 \psi_i \tag{6.2}$$

where λ_{ij} is given by (5.23) and ω_i is the volume of the D-dimensional polyhedron dual to i. Because of (5.24), it follows that

$$\psi_j = \text{a constant spinor} \tag{6.3}$$

satisfies the massless field equation

$$\sum_j \gamma^\mu \ell_{ij}^\mu \lambda_{ij} \psi_j = 0 \tag{6.4}$$

at all sites i.

Next, we consider the function

$$\psi_j = \chi e^{i\vec{p}\cdot\vec{r}_j} \tag{6.5}$$

where χ is a constant spinor; the corresponding lattice action is

$$A_d(\Omega) = \frac{1}{2}(\chi^\dagger \gamma^0 \gamma^\mu \chi) \sum_{i,j} \ell_{ij}^\mu \lambda_{ij} e^{i\vec{p}\cdot\vec{\ell}_{ij}} + m \chi^\dagger \gamma^0 \chi \Omega .$$

For small $\vec{p}$, $e^{i\vec{p}\cdot\vec{\ell}_{ij}} = 1 + i\vec{p}\cdot\vec{\ell}_{ij} - \frac{1}{2}(\vec{p}\cdot\vec{\ell}_{ij})^2 + \cdots$. By using (5.25) and

---

* We exclude other hermitian choices such as $A_c(\Omega) = \int_\Omega (i\psi^\dagger \gamma^\mu \frac{\partial}{\partial x^\mu} \psi + m\psi^\dagger \psi) \, d^D r$ by requiring the field equation to be $(\gamma^\mu \frac{\partial}{\partial x^\mu} + m)\psi = 0$.

neglecting $O(\vec{p}^{\,3})$, we see that

$$A_d(\Omega) = \left[i(\chi^\dagger \gamma^0 \gamma^\mu \chi)\, p^\mu + m\, \chi^\dagger \gamma^0 \chi \right] \Omega$$

which has the same form as in the continuum case.

The lattice action (6.2) may be resolved in terms of its eigenvectors $\psi_i(n)$ and eigenvalues $E(n)$:

$$\tfrac{1}{2} \sum_j \gamma^\mu \ell_{ij}^{\;\mu} \lambda_{ij}\, \psi_j(n) + m\,\omega_i\, \psi_i(n) = E(n)\, \omega_i\, \gamma^0 \psi_i(n) . \qquad (6.6)$$

We write

$$A_d(\Omega) = \sum_{n,i} E(n)\omega_i\, \psi_i(n)^\dagger \psi_i(n) .$$

It is easy to verify that (6.5) is not a solution of the eigenvector equation (6.6). Hence, the low-lying spectrum $E(n)$ of the lattice action is more complicated than the continuum case.

For the zero-mass case, it can be shown that in any dimension D, besides the constant spinor solution (6.3) there is another zero-energy solution which has a rather complicated dependence on $\vec{r}_i$. The full implications of this additional solution and its relation to chiral anomaly are yet to be investigated.

VII. GAUGE THEORY

7.1. General Discussion

As in the conventional treatment[13] of lattice gauge theory, we introduce gauge variables by assigning a group element $U(i,j)$ to a link connecting the points (i.e., sites) i and j in the lattice. For an $SU(N)$ gauge theory, we can set $U(i,j)$ to be an arbitrary $N \times N$ unitary matrix with determinant 1. It is convenient to define both $U(i,j)$ and $U(j,i)$ with $U(j,i) = U(i,j)^{-1}$. Next, with each elementary triangle of vertices i,j and k we associate a group element $U_{ijk} = U(i,j)\, U(j,k)\, U(k,i)$. Recall that in our construction of the random lattice we identify a "cluster" of $D+1$ vertices with the property that the $D-1$ dimensional sphere, on whose surface the cluster of $D+1$ points lies, has no lattice points in its interior. Each pair of points in such a cluster is joined by a link. An elementary triangle Δ_{ijk} is then formed of any three points i, j and k in the same cluster and the links between them.

The discrete action for the lattice is defined using these group variables U_{ijk}:

$$A_d = \frac{1}{g^2} \sum_{\Delta_{ijk}} \kappa_{ijk}\, f(U_{ijk}) \ , \tag{7.1}$$

where g is the coupling constant, and the sum extends over all triangles Δ_{ijk}. The coefficients κ_{ijk} are given by, similar to (5.23),

$$\kappa_{ijk} = \tau_{ijk}/\Delta_{ijk} \ . \tag{7.2}$$

where Δ_{ijk} is the area of the plaquette and τ_{ijk} the $(D-2)$-dimensional volume of its dual. Just as in (5.24) and (5.25) we can prove (Ref. 1) that

$$\sum_k \kappa_{ijk}\, \Delta_{ijk}^{\mu\nu} = 0 \tag{7.3}$$

for all sites i and j that are linked, and

$$T^{\mu\nu\rho\sigma} \equiv \frac{1}{4} \sum_{\Delta_{ijk}} \kappa_{ijk}\, \Delta_{ijk}^{\mu\nu}\, \Delta_{ijk}^{\rho\sigma} = \frac{1}{8} (\delta^{\mu\rho} \delta^{\nu\sigma} - \delta^{\mu\sigma} \delta^{\nu\rho})\, \Omega \ . \tag{7.4}$$

The function $f(U)$ obeys

$$f(uUu^{-1}) \quad = \quad f(U) \tag{7.5}$$

for all group elements u and U ; the simplest choice is

$$f(U) \quad = \quad \tfrac{1}{2}\,\mathrm{tr}\,(I - U) + \text{c.c.} \quad = \quad \mathrm{Re}\,\left[\mathrm{tr}\,(I - U)\right] \quad , \tag{7.6}$$

with I = unit matrix and, for the SU(N) theory, U = an arbitrary $N \times N$ unitary matrix of unit determinant, in accordance with the conventional Wilson formalism.

Next, consider a large loop of area $\mathcal{a}$ bounded by links ℓ_{12}, ℓ_{23}, ..., as shown in Figure 9. [The linear dimension of the loop is assumed to be $O(\mathcal{a}^{\frac{1}{2}})$ in any direction.] Define

$$\exp\,(-T\mathcal{a}) \quad \equiv \quad Z^{-1} \int \left[d^D r_i\right] \left[dU(i,j)\right] \, e^{-A_d/g^2} \, W_L \tag{7.7}$$

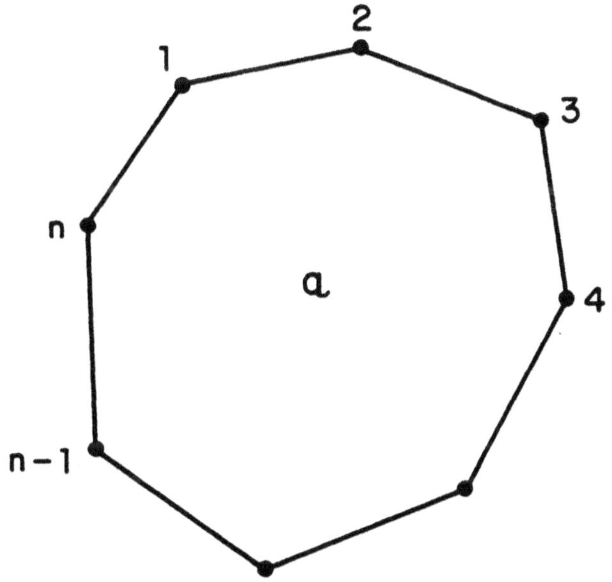

Figure 9. Wilson loop on a random lattice.

where
$$Z \equiv \int [d^D r_i] \, [dU(i,j)] \, e^{-A_d/g^2} , \tag{7.8}$$

T is the string tension, A_d is the action given by (7.1) and

$$W_L = \text{trace } U(1, 2) \; U(2, 3) \cdots , \tag{7.9}$$

with the product extending over the boundary of $\mathcal{a}$.

7.2. Strong Coupling Limit: String Tension, String Thickness and Glueball Spectrum

In the strong coupling limit $(g \to \infty)$, one can show that the theory becomes identical to a relativistic string model. For $D = 4$, we find the string tension T is given by

$$T = 2 \left(\frac{\pi}{\sqrt{3}}\right)^{\frac{1}{2}} (\rho \ln g^2)^{\frac{1}{2}} \tag{7.10}$$

with ρ = site-density. Furthermore, the thickness d of the string is

$$d^2 = \frac{1}{2\pi T} \ln \mathcal{a} . \tag{7.11}$$

Both (7.10) and (7.11) are derived in Ref. 1. The string thickness (7.11) in a random lattice is identical to that given by a relativistic string model.[14] Thus, unlike a rigid "cubic" lattice the strong coupling limit of the random lattice gauge theory is already quite physical, since it carries the phenomenological virtues of a relativistic string model. From (7.10) we see that quarks are confined in the strong coupling limit.

As will be shown, the result of the numerical calculation indicates that there is no phase transition in a random lattice non-Abelian gauge theory. Consequently, when we change the coupling from strong to weak, quarks should remain confined.

The string tension concerns calculations for a given two-dimensional figure (Wilson loop). At the next level of complexity, we consider the glueball

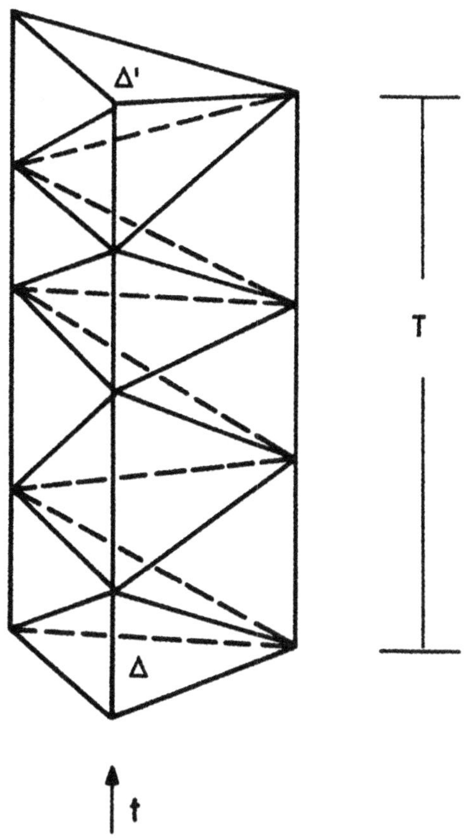

Figure 10. Glueball propagator in the strong coupling limit.

propagator between two plaquettes, separated by a large time interval T. Let Δ and Δ' be these two plaquettes. Then in the strong coupling limit we find that these two plaquettes are connected by tetrahedra of the form given in Figure 10.

When $T \to \infty$, the corresponding propagator has a leading behavior which can be expressed as a superposition of different angular momentum J states, each having an asymptotic behavior

$$\sim e^{-M_J T} \tag{7.12}$$

where M_J is the lowest glueball mass of angular momentum J. By subtracting away the $e^{-M_J T}$ term, the remaining part has an asymptotic behavior $\sim e^{-M_{J*}T}$, where M_{J*} is the second lowest glueball mass of angular momentum J, etc.

For low angular momentum states, we find that in the strong coupling limit[15] the glueball rotational levels are identical to that of a rigid body of moment of inertia I_1, I_2 and I_3 where

$$I_1 = .631\, \rho^{1/4} (\ln g^2)^{5/4}$$

and

$$I_2 = I_3 = .341\, \rho^{1/4} (\ln g^2)^{5/4} .$$

Let M_0 be the ground state mass with $J = 0$ and parity $+$. The low-lying excitation energy $M_J - M_0$ is given by

J	Parity	$M_J - M_0$
1	-	$\frac{1}{2}(I_1^{-1} + I_2^{-1})$
1*	-	I_2^{-1}
2	+	$2I_1^{-1} + I_2^{-1}$
2*	+	$\frac{1}{2}I_1^{-1} + \frac{5}{2}I_2^{-1}$
2**	+	$3I_2^{-1}$

Qualitatively, the three $J = 2$ levels are consistent with the recent experimental results of S. Lindenbaum.[16]

For high angular momentum we have in the strong coupling limit[15]

$$M_J \propto \sqrt{J} \qquad (7.13)$$

so that the system exhibits the typical Regge behavior of a relativistic string in rotation. The proportional constant is exactly that corresponding to a double string (since quark number = 0).

7.3. Numerical Results[4]

Numerical programs for a random lattice gauge theory were set up by Friedberg and Ren at Columbia; the computations were carried out by Ren. In Figures 11a–11b we give the average plaquette energy u and specific heat C vs. $\beta = 1/g^2$ for the $U(1)$ theory.

The corresponding plots for an $SU(2)$ theory are given in Figures 12a–12b. We see that the specific heat has a peak in the $U(1)$ theory, but not in the $SU(2)$ theory. For $U(1)$, the peak becomes steeper when the number of lattice sites increases, suggesting that there is a phase transition. On the other hand, the specific heat curve for $SU(2)$ has no peak, indicating that the passage from strong to weak coupling is a smooth one. Consequently, while both theories are confined in the strong coupling limit, the weak coupling limit is consistent with deconfinement in the $U(1)$ theory (QED), but not in a non-Abelian gauge theory. Extension to $SU(3)$ and also calculations on Wilson string tension are in progress.

In contrast, we give in Figure 13 the recent numerical calculation by N. H. Christ and A. Terrano[17] for the $SU(3)$ gauge theory on a regular lattice. As we can see, there is a sharp peak in the specific heat, suggesting that the transition from strong to weak in a regular lattice is by no means smooth, unlike that in a random lattice.

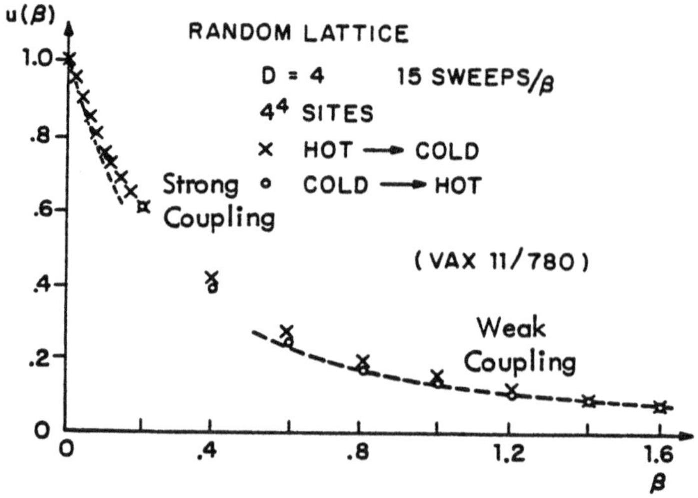

(a)

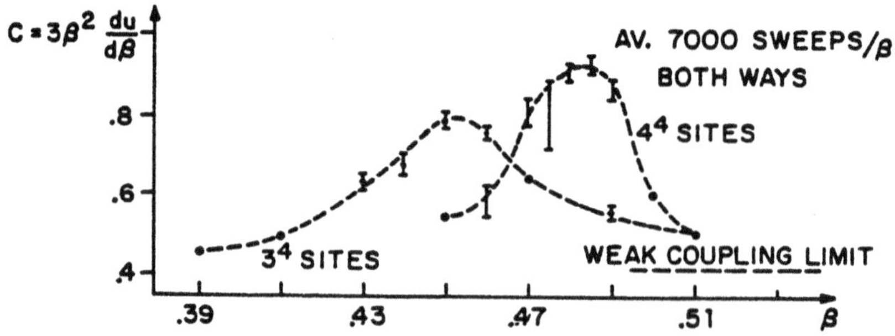

(b)

Figure 11. Numerical results for a four-dimensional U(1) random lattice field theory:

a) Average plaquette energy vs. $\beta = 1/g^2$,

b) Specific heat vs. β . (H. C. Ren)

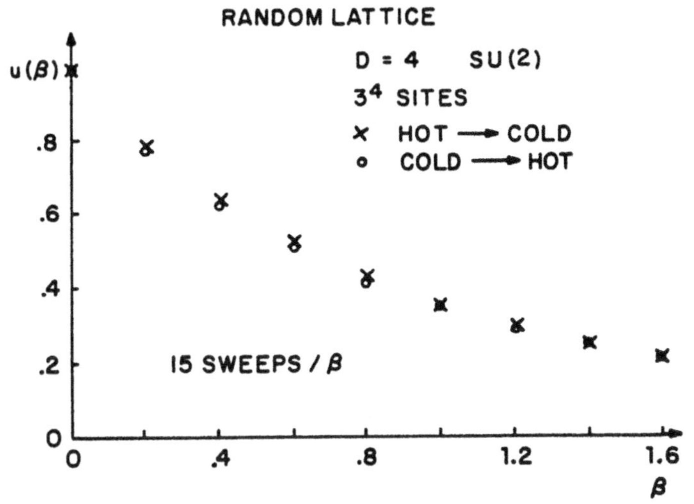

(a)

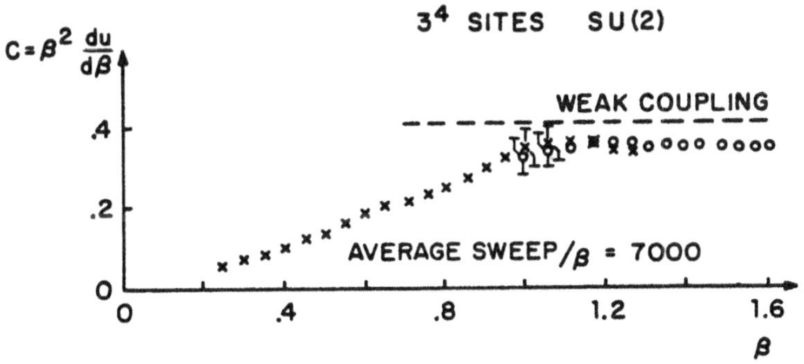

(b)

Figure 12. Numerical results for a four-dimensional SU(2) random lattice field theory:

a) Average plaquette energy vs. β ,

b) Specific heat vs. β . The error bars indicated are applicable to all points. (H. C. Ren)

60.

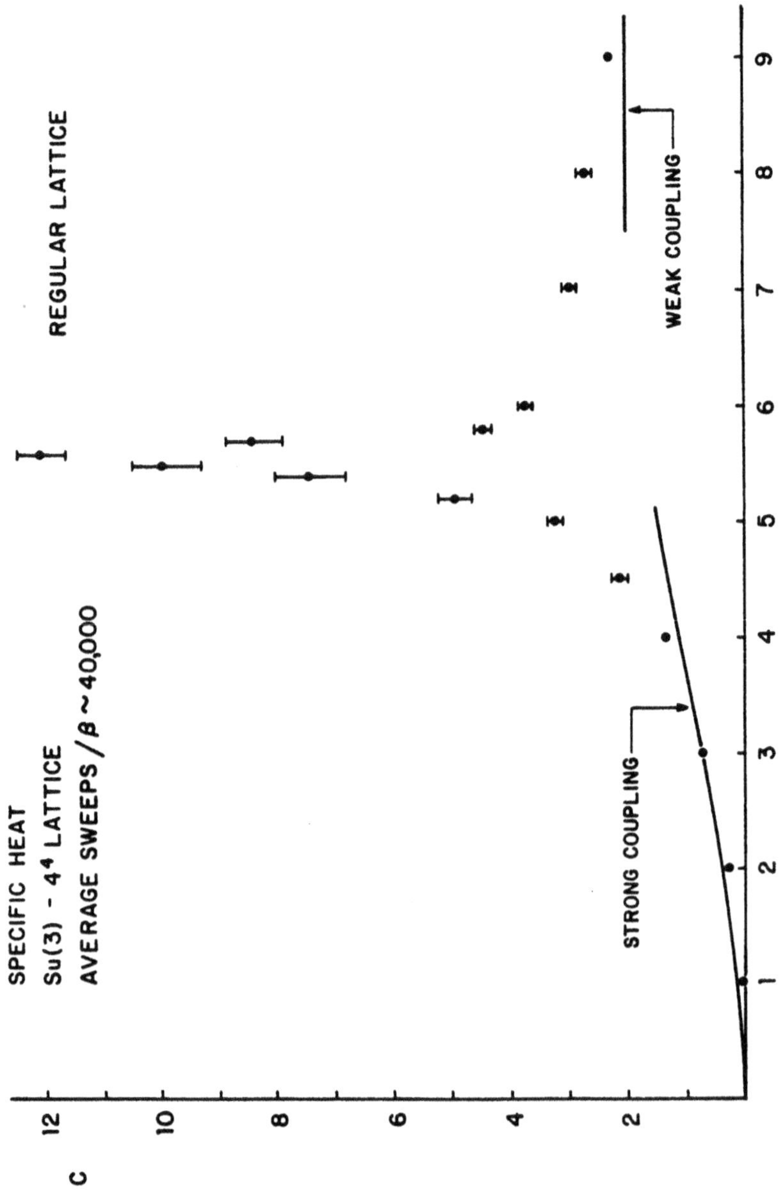

<u>Figure 13.</u> Specific heat vs. β for a four-dimensional SU(3) theory on a "cubic" lattice. (N. H. Christ and A. Terrano)

VIII. QUANTUM GRAVITY

8.1. General Discussion

The technical difficulties of quantizing gravity are well known. Among them the two most prominent are

(i) nonrenormalizability

and (ii) energy not positive definite.

In connection with (ii), the usual Einstein equation in the euclidean space*

$$G_{\mu\nu} = R_{\mu\nu} - \tfrac{1}{2} g_{\mu\nu} R = 0 \tag{8.1}$$

gives solutions that are saddle points of the continuum action

$$A_c = \int \sqrt{|g|}\, R\, d^4x \ ; \tag{8.2}$$

this makes quantization in the euclidean space extremely inconvenient.

Our strategy is to formulate the discrete theory of gravity directly in the minkowski space. In the next section, we shall show how a random lattice can

---

* Throughout this chapter, our conventions for the continuum theory are:

$$ds^2 = g_{\mu\nu}\, dx^\mu\, dx^\nu$$

where the metric for a flat euclidean space is

$$(g_{\mu\nu}) = (\delta_{\mu\nu}) = \begin{pmatrix} 1 & & & \\ & 1 & & \\ & & 1 & \\ & & & 1 \end{pmatrix} \tag{8.3}$$

and for a flat minkowski space is

$$(g_{\mu\nu}) = (\eta_{\mu\nu}) = \begin{pmatrix} -1 & & & \\ & 1 & & \\ & & 1 & \\ & & & 1 \end{pmatrix} . \tag{8.4}$$

The terms "euclidean space" and "minkowski space" are for arbitrary $g_{\mu\nu}$, but with $|g| \equiv \det(g_{\mu\nu}) > 0$ and < 0 respectively.

be set up in the minkowski space. By keeping the density of lattice sites $\rho = (1/\ell)^4$ finite, we avoid the ultraviolet divergence. Because we work directly in the minkowski space, the path integrals become finite. The latter follows from the elementary observation that an integral such as

$$\int_{-\infty}^{\infty}\int_{-\infty}^{\infty} e^{i(ax^2+by^2)}\, dx\, dy$$

exists for any real constants a and b independently of their signs, while the corresponding integral

$$\int_{-\infty}^{\infty}\int_{-\infty}^{\infty} e^{-(ax^2+by^2)}\, dx\, dy$$

exists only for a and b both positive. [A saddle point of the exponent at the origin $x = y = 0$ would have opposite signs for a and b.]

8.2. Random Lattice in a Minkowski Space[18]

Let x^μ be the Cartesian coordinates in a D-dimensional flat minkowski space where the superscript $\mu = 0, 1, 2, \cdots, D-1$ denotes the space-time component with x^0 = time t. The metric $\eta_{\mu\nu}$ is given by (8.4). The invariant volume element is

$$d^Dx = dx^0 dx^1 \cdots dx^{D-1}, \tag{8.5}$$

the same as in a flat euclidean space. Consider a random distribution of N lattice sites in a volume Ω. Clearly, N, Ω and the site-density

$$\rho \equiv \frac{N}{\Omega} \tag{8.6}$$

are all Lorentz invariant.

The immediate problem with setting up a relativistic field theory on a random lattice in a minkowski space is the concept of "neighbors". In order to simulate, e.g., the d'Alambertian operator (or other derivatives) in a continuum theory, the corresponding discrete operator should involve only a few nearby sites (similar to the relation between (5.21) and the Laplace equation in the euclidean space). To be specific, let us consider the example of a free massless scalar field: there should be an expression like (5.21)

$$\sum_{\substack{j \text{ linked} \\ \text{to } i}} \lambda_{ij}(\phi_i - \phi_j) = 0 \tag{8.7}$$

which would approach the d'Alambertian equation

$$-\frac{\partial^2\phi}{(\partial x^0)^2} + \frac{\partial^2\phi}{(\partial x^1)^2} + \cdots + \frac{\partial^2\phi}{(\partial x^{D-1})^2} = 0 \tag{8.8}$$

when the site-density $\rho \to \infty$. In order to construct (8.7) we must first have an appropriate linking algorithm. In the minkowski space, the $(\text{distance})^2$ between any two points p and q is given by

$$\ell_{pq}^{\,2} \equiv -\left[x^0(p) - x^0(q)\right]^2 + \left[x^1(p) - x^1(q)\right]^2 + \cdots + \left[x^{D-1}(p) - x^{D-1}(q)\right]^2 \tag{8.9}$$

where $x^\mu(p)$ and $x^\mu(q)$ are the μth coordinates of p and q. A small $|\ell_{pq}|$ clearly does not imply that $x^\mu(p)$ and $x^\mu(q)$ are near to each other. Yet, in the d'Alambertian (8.8), we only deal with the differences between ϕ at points that are infinitesimally close to each other. Therefore, the linking algorithm in a minkowski space cannot be based simply on the invariant distance ℓ_{pq}, in contrast to the situation in a euclidean space.

In a given random distribution of N lattice sites, we can envisage each site carrying its light cone. Between any two points p and q (which may or may not be the lattice sites), let us define

$$n_{pq} \equiv \text{number of light cones crossed by the straight line } \overline{pq}\,. \tag{8.10}$$

Under a Lorentz transformation, a light cone remains a light cone, and a straight line remains a straight line. If $\overline{pq}$ crosses the light cone of, say, the ith site in one Lorentz frame, this crossing must also be observed in any other Lorentz frame.

Consequently, n_{pq} is invariant. As we shall see, this number n_{pq} can be used to establish the concept of "neighbors" in a minkowski space.

Take the example of two points p and p' lying on each other's light cone. The invariant distance $\ell_{pp'}$ is always zero. However, $n_{pp'}$ varies depending on the relative positions of p and p'. A large $n_{pp'}$ indicates p and p' are far apart. The use of a number (not a distance) to distinguish the nearness or apartness of two points should be familiar to any New Yorker. In New York, for example, from Lincoln Center one has to travel about 55 blocks to Columbia University, but only about 10 blocks to Rockefeller University. Even though the east-west blocks are longer than the north-south ones, since 55 is a much larger number than 10 one may conclude that Lincoln Center is nearer to Rockefeller than to Columbia. In the following, we shall take the example of a two-dimensional minkowski space to illustrate how the number n_{pq} can be used for linking the random lattice sites.

When $D = 2$, the coordinates x^0 and x^1 will be written simply as t and x. Define

$$x_+ \equiv \frac{1}{\sqrt{2}}(x+t)$$

and (8.11)

$$x_- \equiv \frac{1}{\sqrt{2}}(x-t) .$$

The light cone of any point consists of two lines, one parallel to the x_+ axis and the other to the x_- axis. Under a Lorentz transformation

$$\begin{pmatrix} t \\ x \end{pmatrix} \rightarrow \begin{pmatrix} \cosh\theta & \sinh\theta \\ \sinh\theta & \cosh\theta \end{pmatrix} \begin{pmatrix} t \\ x \end{pmatrix} , \tag{8.12}$$

both x_+ and x_- undergo a scale transformation

$$x_+ \rightarrow e^{\theta} x_+ \qquad \text{and} \qquad x_- \rightarrow e^{-\theta} x_- , \tag{8.13}$$

keeping their product x_+x_- invariant.

There is an alternative way of deriving n_{pq}, which is defined by (8.10).

Instead of "moving" from p to q along the straight line $\overline{pq}$, one starts from p, travels along its light cone (say the one that is parallel to the x_+ axis) towards q, crossing n_+ light cones (of other sites) until one reaches the light cone of q; then one makes a turn and travels along the light cone of q, crossing another n_- light cones until one reaches q. It is easy to show that (see Figure 14)

$$n_{pq} = n_+ + n_- . \tag{8.14}$$

Under a Lorentz transformation, the $x_\pm$-coordinates of p, q and all the lattice sites undergo the same scale transformation (8.13); therefore, n_+ and n_- are separately Lorentz invariant.

For convenience, let us distribute the N lattice sites in a box of volume $\Omega = L^2$, which is bounded by the light cones of two points, labeled B and X. These two points are symmetrically placed with respect to the origin O, so that all three points B, O and X lie on one straight line, with O in the middle. The boundary of the box consists of four lines given by

$$\begin{aligned} x_+ &= \frac{L}{2} e^{a} , \qquad & x_- &= \frac{L}{2} e^{-a} , \\ x_+ &= -\frac{L}{2} e^{a} \quad \text{and} & x_- &= -\frac{L}{2} e^{-a} , \end{aligned} \tag{8.15}$$

as shown in Figure 14. The box is therefore "rectangular" in shape, with sides $L_+ \equiv Le^{a}$ and $L_- \equiv Le^{-a}$; its volume is $L_+ L_- = L^2$. The parameter a defines the shape of the box. Under the Lorentz transformation (8.12), the volume of the box is invariant, but the shape changes* with

$$a \rightarrow a + \theta . \tag{8.16}$$

* In a discrete field theory on a flat minkowski-space random lattice, besides integrating over the positions of all lattice sites i we also integrate over the parameter a which defines the box shape. From Figure 14, we see that B is confined to the hyperbola $\ell_{BO}^2 = -\frac{1}{2}L^2 =$ constant, where ℓ_{BO} is the minkowski distance between B and the origin O; the shape of the box is in turn

(continued on page 67)

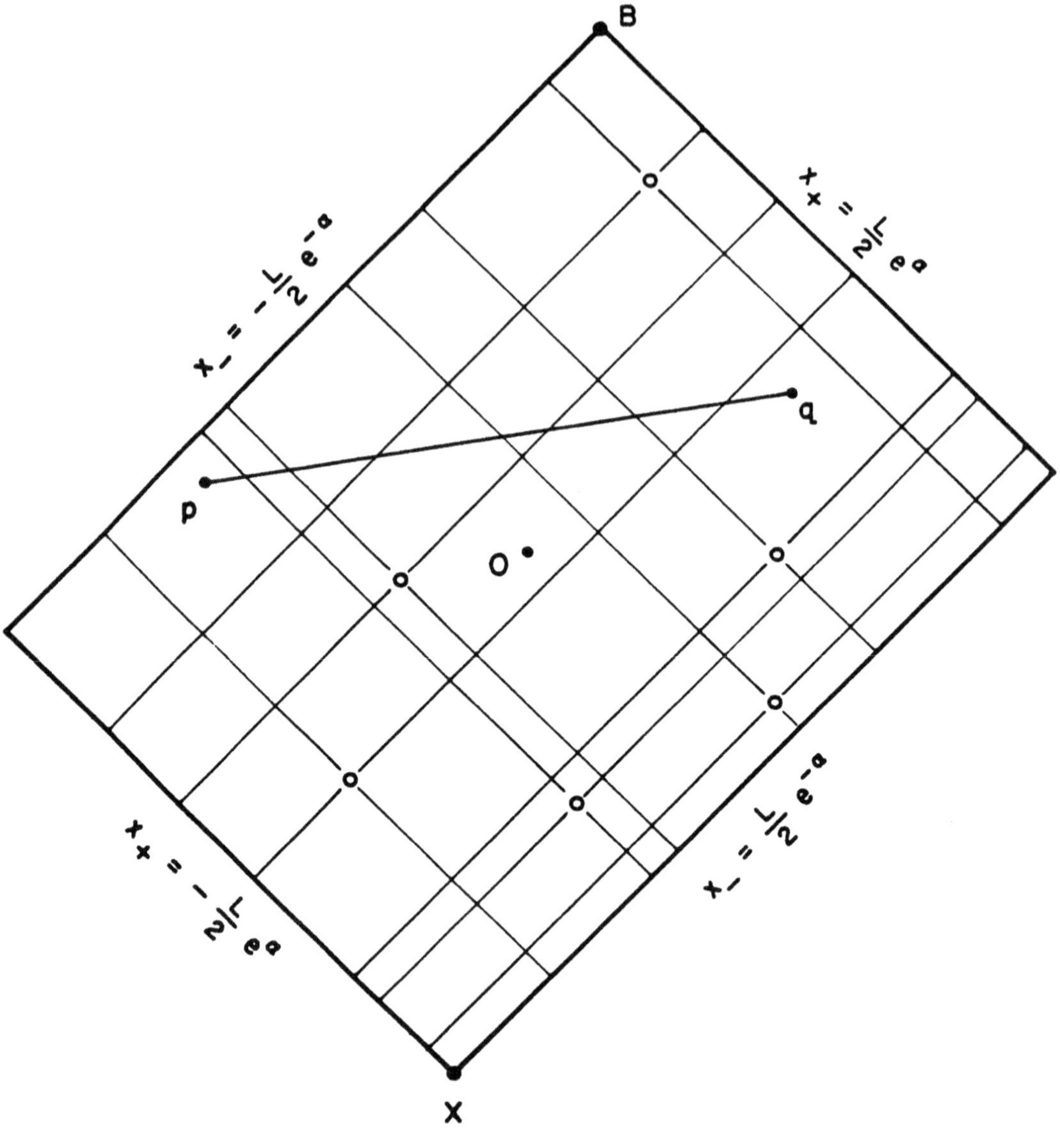

Figure 14. In this example of $N = 6$ sites (represented by circles) in a $D = 2$ box of volume L^2 (bounded by the light cones of B and X), we have $n_{pq} = 7$, $n_+ = 4$ and $n_- = 3$, confirming (8.14). The origin O is the mid-point of the straight line $\overline{BX}$; therefore, the minkowski lengths squared between BO and OX are $\ell^2_{BO} = \ell^2_{OX} = -L^2/2$.

By definition the number n_{pq} is an integer which is somewhat awkward to use. As we shall see, in the limit of $N \to \infty$ and $L \to \infty$, but at a fixed site-density

$$\rho = \frac{N}{L^2} = \text{constant}, \tag{8.17}$$

a continuous parameter ν_{pq} can be derived which is more suitable for our purpose. We define

$$\nu_{pq} \equiv n_{pq}/N^{\frac{1}{2}}$$

and (8.18)

$$\nu_{\pm} \equiv n_{\pm}/N^{\frac{1}{2}} \quad ;$$

(8.14) becomes

$$\nu_{pq} = \nu_{+} + \nu_{-} \quad . \tag{8.19}$$

Let

$$\Delta_{+} \equiv x_{+}(p) - x_{+}(q)$$

and (8.20)

$$\Delta_{-} \equiv x_{-}(p) - x_{-}(q)$$

where $x_{\pm}(p)$ and $x_{\pm}(q)$ are the $x_{\pm}$-coordinates of p and q. Since the total number of light cones crossed by the entire length of the box side $L_{+} = Le^{a}$ is N, the average number of n_{+} is

$$< n_{+} > = \frac{|\Delta_{+}|}{L_{+}} N = \frac{|\Delta_{+}|}{L} N e^{-a} \quad ,$$

with $< >$ denoting the average. Correspondingly,

---

* (continued)

uniquely determined by the position of B on that hyperbola. We may regard B as an extra-lattice site. The integration over the box shape becomes, then, simply an integration over the position of the site B, subject to the constraint that ℓ_{BO}^{2} is fixed.

$$\langle \nu_+ \rangle = \frac{\langle n_+ \rangle}{N^{\frac{1}{2}}} = \frac{|\Delta_+|}{L} N^{\frac{1}{2}} e^{-\alpha} = \rho^{\frac{1}{2}} |\Delta_+| e^{-\alpha} . \quad (8.21)$$

Likewise, we have

$$\langle \nu_- \rangle = \rho^{\frac{1}{2}} |\Delta_-| e^{\alpha} . \quad (8.22)$$

For a random distribution of lattice sites, the fluctuation in $n_\pm$ is given by

$$\langle n_\pm^2 \rangle - \langle n_\pm \rangle^2 = \langle n_\pm \rangle .$$

In the limit of $N \to \infty$ at a constant ρ, we have

$$\langle \nu_\pm^2 \rangle - \langle \nu_\pm \rangle^2 = \frac{\langle n_\pm \rangle}{N} = \frac{\langle \nu_\pm \rangle}{N^{\frac{1}{2}}} \to 0 . \quad (8.23)$$

Thus, we can neglect the fluctuation in $\nu_\pm$, and regard $\nu_\pm$ simply as continuous parameters, equal to $\rho^{\frac{1}{2}} |\Delta_\pm| e^{\mp\alpha}$, which can vary from 0 to ∞. From (8.18)-(8.21) it follows that

$$\nu_{pq} = \rho^{\frac{1}{2}} \left[|x_+(p) - x_+(q)| \; e^{-\alpha} + |x_-(p) - x_-(q)| \; e^{\alpha} \right] . \quad (8.24)$$

Under a Lorentz transformation, $x_+ \to x_+ e^{\theta}$, $x_- \to x_- e^{-\theta}$, $\alpha \to \alpha + \theta$; therefore ν_{pq} is invariant, as expected.

We shall now follow the steps given in section 5.4 for the generalization of the linking algorithm. Take an arbitrary point c as the center. Consider the sequence of curves $\mathcal{S}_R(c)$ with varying R, each of which consists of all points p that satisfy

$$\nu_{pc} = \rho^{\frac{1}{2}} R . \quad (8.25)$$

From (8.24) we see that $\mathcal{S}_R(c)$ are parallelopipes, which fulfill the convexity requirement of (5.12). As in section 5.4, take any group of $D + 1 = 3$ sites. If by varying c and R, we can find an $\mathcal{S}_R(c)$ passing through these three sites and if, in addition, the interior of that $\mathcal{S}_R(c)$ is empty, then these three sites form a cluster of neighbors; they should be linked to each other and become the vertices of a triangle. The theorem proved in section 5.4 insures that by going through all groups of three sites, the triangles thus formed do not intersect each other, and together they fill the entire volume $\Omega = L^2$. Under a Lorentz transformation, sites move and the box-shape changes. However, if two sites are linked in one frame, then they are also linked in any other frame.

For a "square" box (i.e., $a = 0$), (8.24) becomes

$$v_{pq} = \rho^{\frac{1}{2}} \left[\, |x_+(p) - x_+(q)| + |x_-(p) - x_-(q)| \, \right] \; . \qquad (8.26)$$

The curves $\mathcal{S}_R(c)$ are squares whose sides are parallel to the x and t axes. By using (8.25)-(8.26) for the construction of the circumscribed curves $\mathcal{S}_R(c)$, we can decompose a random lattice in the minkowski space (inside a "square" box) into triangles. An explicit example is given in Figure 15a. In Figure 15b, we give the lattice for the same site distribution in the euclidean space, by using circles as the circumscribed curves. It is of interest to compare these two decompositions.

<u>Comments.</u>

1. The method discussed in this section can be readily generalized to $D > 2$. The details of such a minkowski space random lattice field theory will be given elsewhere.[18]

2. At large distances the field propagator in the euclidean space is proportional to r^{2-D} which, for $D > 2$, approaches 0 as $r \to \infty$. Hence for a large euclidean volume, the effect of the boundary can be neglected.

3. The situation is different in the minkowski space. From (8.24)-(8.25) we see that the corresponding convex function f used in (5.12) for simplitial decompositions depends explicitly on the shape parameter a of the box. Lorentz invariance is restored only after the average of the shape parameter.

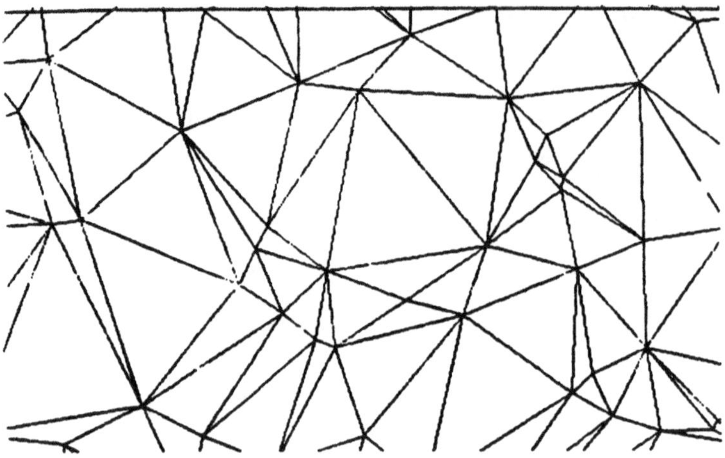

(a)

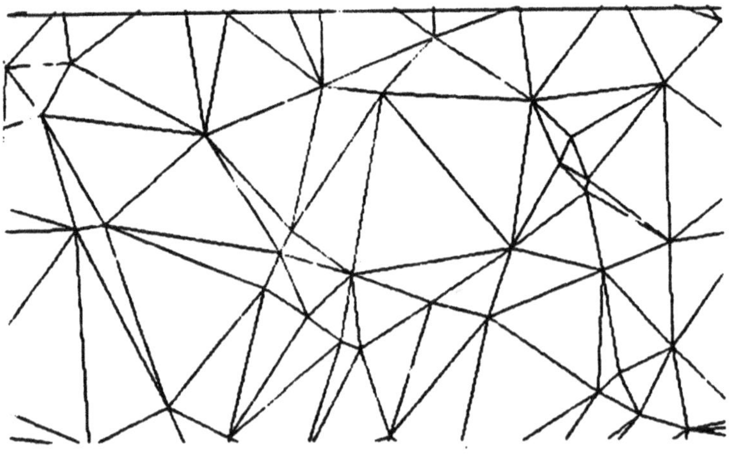

(b)

Figure 15. (a) A two-dimensional minkowski space random lattice (inside a square box).

(b) Same lattice-site distribution, but in a euclidean space.

8.3. Discrete Action

First take a random lattice in a flat D-dimensional minkowski space. Label each site by $i = 1, 2, \cdots$, and each D-simplex by $\overset{\circ}{\tau}(a)$ with $a = 1, 2, \cdots$. Let $\overset{\circ}{x}{}^{\mu}(i)$ be the μth-coordinate of the ith site. For every linked pair of sites i and j, the $(\text{link-length})^2$ is given by the usual expression

$$\overset{\circ}{\ell}{}_{ij}^{2} = \eta_{\mu\nu} \left[\overset{\circ}{x}(i) - \overset{\circ}{x}(j)\right]^{\mu} \left[\overset{\circ}{x}(i) - \overset{\circ}{x}(j)\right]^{\nu} \tag{8.27}$$

where $\eta_{\mu\nu}$ is defined by (8.4).

Consider now an arbitrary variation *

$$\overset{\circ}{\ell}{}_{ij}^{2} \rightarrow \ell_{ij}^{2} \quad . \tag{8.28}$$

Correspondingly, each D-simplex changes from

$$\overset{\circ}{\tau}(a) \rightarrow \tau(a) \tag{8.29}$$

with the same vertices, but different link-lengths. Following Regge,[19, 20] we assume that the interior of each D-simplex remains flat. Hence, inside each simplex $\tau(a)$, we may set up an arbitrary Lorentz frame $\Sigma(a)$. A vector ϕ^{μ} viewed in $\Sigma(a)$ can be written as a column matrix

$$\phi(a) \equiv \begin{pmatrix} \phi^{0}(a) \\ \phi^{1}(a) \\ \vdots \\ \phi^{D-1}(a) \end{pmatrix} \quad . \tag{8.30}$$

Under a Lorentz transformation of $\Sigma(a)$,

---

* For each plaquette Δ_{ijk} in the lattice, the new lengths of its three sides ℓ_{ij}, ℓ_{jk} and ℓ_{ki} are assumed to satisfy the triangular inequality in a flat minkowski space. Likewise, for each tetrahedron, the length-squared of its six sides ℓ_{ij}^{2} satisfies the tetrahedral inequalities, etc.

$$\phi(a) \rightarrow u(a)\ \phi(a) \tag{8.31}$$

where, by definition, the real $D \times D$ matrix $u(a)$ satisfies

$$\tilde{u}(a)\ \eta\ u(a) = \eta \tag{8.32}$$

with $\sim$ denoting the transpose and

$$\eta = \begin{pmatrix} -1 & 0 & 0 & 0 \\ 0 & 1 & 0 & 0 \\ 0 & 0 & 1 & 0 \\ 0 & 0 & 0 & 1 \end{pmatrix} , \tag{8.33}$$

as before.

Each D-simplex, say $\tau(a)$, has $D+1$ vertices and $\frac{1}{2}(D+1)D$ links; viewed in $\Sigma(a)$, apart from the absolute position of its center-of-mass, all the properties of $\tau(a)$ such as shape, volume, orientation, etc. are completely determined by the D independent link-vectors: $\ell_{12}{}^{\mu}$, $\ell_{13}{}^{\mu}$, $\cdots$, $\ell_{1\,D+1}{}^{\mu}$ where $1, 2, \cdots, D+1$ refer to the vertices of $\tau(a)$. Under a Lorentz transformation of $\Sigma(a)$, these vectors transform according to (8.31). From these vectors, we can form a total of $\frac{1}{2}(D+1)D$ independent invariants which may be represented by all the $(\text{link-length})^2$ of $\tau(a)$.

So far, each $\tau(a)$ is separate from all other D-simplices. Suppose that in the original flat minkowski space, two D-simplices, say $\mathring{\tau}(a)$ and $\mathring{\tau}(b)$, share a common $(D-1)$-simplex $\mathring{s}$ whose vertices are, say, sites $1, 2, \cdots, D$. Under the link-length variation (8.28), these two D-simplices transform from

$$\mathring{\tau}(a) \rightarrow \tau(a) \quad \text{and} \quad \mathring{\tau}(b) \rightarrow \tau(b) \ ; \tag{8.34}$$

viewed in $\Sigma(a)$ we have

$$\mathring{s} \rightarrow s(a) \ , \tag{8.35}$$

with each of its $\frac{1}{2}D(D-1)$ link-vectors changing from

$$\overset{\circ}{\ell}{}_{ij}{}^{\mu} \rightarrow \ell_{ij}{}^{\mu}(a) \tag{8.36}$$

where i and j = 1, 2, ⋯ , D. Likewise, viewed in $\Sigma(b)$ we have

$$\overset{\circ}{s} \rightarrow s(b) \tag{8.37}$$

and

$$\overset{\circ}{\ell}{}_{ij}{}^{\mu} \rightarrow \ell_{ij}{}^{\mu}(b) \quad . \tag{8.38}$$

Since in accordance with (8.28)

$$\ell_{ij}(a)^2 = \ell_{ij}(b)^2 = \ell_{ij}{}^2 \tag{8.39}$$

for all i and j from 1, 2, ⋯ , D, there exists a Lorentz transformation $U(b,a)$ which can transform the set $\{\ell_{ij}{}^{\mu}(a)\}$ to $\{\ell_{ij}{}^{\mu}(b)\}$, and therefore $s(a)$ to $s(b)$; i.e.,

$$\ell_{ij}{}^{\mu}(b) = U(b,a)_{\nu}{}^{\mu} \ \ell_{ij}{}^{\nu}(a) \tag{8.40}$$

for all i and j = 1, 2, ⋯ , D. Under Lorentz transformations of both reference frames $\Sigma(a)$ and $\Sigma(b)$, we have on account of (8.31)

$$U(b,a) \rightarrow u(b)\ U(b,a)\ u(a)^{-1} \tag{8.41}$$

where $u(a)$, $u(b)$ and $U(b,a)$, being Lorentz transformations, all satisfy (8.32).

Next we discuss the parallel transport of any vector ϕ. We start with ϕ inside a D-simplex $\tau(a)$. Viewed in $\Sigma(a)$, we write $\phi = \phi(a)$, in accordance with (8.30). Because the interior of $\Sigma(a)$ is flat, the parallel transport of ϕ within $\tau(a)$ is trivial. When we transport ϕ from $\tau(a)$ to one of its adjacent D-simplices, say $\tau(b)$, ϕ becomes $\phi(b)$ [when viewed in $\Sigma(b)$], where $\phi(b)$ is <u>defined</u> to be

$$\phi^{\mu}(b) = U(b,a)_{\nu}{}^{\mu} \ \phi^{\nu}(a) \quad . \tag{8.42}$$

Clearly, $U(b,a)$ does not depend on which point in the interior of $\tau(a)$ we start from, nor on which point in the interior of $\tau(B)$ we end.

Continuing this way, we can parallel transport any vector ϕ along a general path P. It is convenient to think first of the path P in the original flat minkowski lattice. We assume P to be a continuous path starting from the interior of any D-simplex, say $\mathring{\tau}(a_1)$, passing successively through a sequence of D-simplices $\mathring{\tau}(a_2)$, $\mathring{\tau}(a_3)$, $\cdots$, and finally arriving at the interior of $\mathring{\tau}(a_n)$. In going from one D-simplex, say $\mathring{\tau}(a_i)$, to the next one $\mathring{\tau}(a_{i+1})$, the path P always passes through their common $(D-1)$-simplex (i.e., P never intersects any $(D-2)$-simplex, or $(D-3)$-simplex, etc.). After the link-length variation (8.28), P becomes a path from $\tau(a_1)$ through $\tau(a_2)$, $\tau(a_3)$, $\cdots$ to $\tau(a_n)$, the parallel transport of any vector ϕ along P is given by

$$\phi^\mu(a_n) = U(P)_\nu{}^\mu \, \phi^\nu(a_1) \tag{8.43}$$

where

$$U(P) = U(a_n, a_{n-1}) \cdots U(a_2, a_1) \ . \tag{8.44}$$

When $a_n = a_1$, then P is a closed path C, in which case (8.44) becomes

$$U(C) = U(a_1, a_{n-1}) \, U(a_{n-1}, a_{n-2}) \cdots U(a_2, a_1) \ . \tag{8.45}$$

Two different paths P (and therefore also two different C) that pass through the same sequence of $\tau(a_1)$, $\tau(a_2)$, $\cdots$ have the same $U(P)$ (or $U(C)$). Under arbitrary Lorentz transformations of $\Sigma(a_1)$, $\Sigma(a_2)$, $\cdots$ we have from (8.41)

$$U(P) \to u(a_n) \, U(P) \, u(a_1)^{-1}$$

and

$$U(C) \to u(a_1) \, U(C) \, u(a_1)^{-1} \ . \tag{8.46}$$

The departure of $U(C)$ from the unit matrix implies that the space is curved. Note that C is a closed path passing from one D-simplex to the next D-simplex by going through their common $D-1$ simplex. Since the interiors of all D-simplices are flat and since $U(C)$ is independent of the precise passage of

C through these $(D-1)$-simplices, the curvature must rest solely in the $(D-2)$-simplices that C encircles. In any D-dimensional space, a $(D-2)$-simplex, say γ, is shared by several D-simplices, say

$$\tau(a_1),\ \tau(a_2),\ \cdots,\ \tau(a_m) \tag{8.47}$$

in a sequential order; γ is also shared by an equal number of $(D-1)$-simplices which are the common surfaces of these D-simplices, (8.47). For each γ, we introduce a closed path $C(\gamma)$ which encircles γ by following the order (8.47) and ending in $\tau(a_1)$. [As before, it is easiest to visualize the closed path $C(\gamma)$ as first drawn in the original flat minkowski space, before the variation (8.28).] Define

$$U_\gamma \equiv U(C(\gamma)) = U(a_1, a_m)\ U(a_m, a_{m-1}) \cdots U(a_2, a_1) \tag{8.48}$$

which depends only on γ and the reference system $\Sigma(a_1)$. When $D = 2$, γ is a site. When $D = 3$, γ is a link represented by a vector ℓ^μ; when $D = 4$, γ is a plaquette $\Delta^{\mu\nu}$, etc. In general, γ can be represented by a tensor of rank $D-2$:

$$\gamma^{\mu_1 \mu_2 \cdots \mu_{D-2}}, \tag{8.49}$$

whose components are referred to the reference frame $\Sigma(a_1)$.

Given γ and $\tau(a_1)$ in (8.47), there remains the question of the sense of $C(\gamma)$, defined by the sequence (8.47) versus its inverse order. Without any loss of generality, we can set $C(\gamma)$ to be a planar figure; the two-dimensional area that it encloses, when viewed in $\Sigma(a_1)$ and set in the order (8.47), can be represented by a tensor of second rank $C(\gamma)^{\alpha\beta}$. We require that the sense of the curve $C(\gamma)$ and the tensor γ satisfy the usual right-hand relation; i.e.,

$$\epsilon_{\mu_1 \cdots \mu_{D-2}\alpha\beta}\ C(\gamma)^{\alpha\beta}\ \gamma^{\mu_1\mu_2\cdots\mu_{D-2}} \tag{8.50}$$

is positive for all γ , where

$$\epsilon_{\mu_1 \cdots \mu_D} = \begin{cases} 1 & \text{if } \mu_1 \cdots \mu_D \text{ is an even permutation of } 1,2,\cdots, D \\ -1 & \text{if } \mu_1 \cdots \mu_D \text{ is an odd permutation of } 1,2,\cdots, D \\ 0 & \text{otherwise .} \end{cases} \tag{8.51}$$

Let $J^{\rho\sigma} = - J^{\sigma\rho}$ be the D-dimensional real matrix representation of the generator of the Lorentz group

$$\eta^{\sigma\nu} x^\rho \frac{\partial}{\partial x^\nu} - \eta^{\rho\nu} x^\sigma \frac{\partial}{\partial x^\nu} \tag{8.52}$$

where x^ρ is the coordinate of points inside $\tau(a_1)$, with respect to the frame $\Sigma(a_1)$. When $D = 4$,

$$J^{01} = - J^{10} = \begin{pmatrix} 0 & 1 & 0 & 0 \\ 1 & 0 & 0 & 0 \\ 0 & 0 & 0 & 0 \\ 0 & 0 & 0 & 0 \end{pmatrix} ,$$

$$J^{12} = - J^{21} = \begin{pmatrix} 0 & 0 & 0 & 0 \\ 0 & 0 & 1 & 0 \\ 0 & -1 & 0 & 0 \\ 0 & 0 & 0 & 0 \end{pmatrix} , \tag{8.53}$$

etc.

The action of the discrete gravity is defined to be, apart from a proportionality constant,

$$\sum_\gamma \text{trace} \left(\epsilon_{\rho\sigma\mu_1 \cdots \mu_{D-2}} J^{\rho\sigma} \gamma^{\mu_1 \cdots \mu_{D-2}} U_\gamma \right) . \tag{8.54}$$

Because of the trace, the above expression is independent of the frame $\Sigma(a_1)$. Also, if the link-variation (8.28) is merely due to a coordinate change from $\mathring{x}(i) \rightarrow \mathring{x}(i)'$ with $\mathring{\ell}_{ij}{}^\mu \rightarrow \ell_{ij}{}^\mu \equiv [\mathring{x}(i)' - \mathring{x}(j)']^\mu$, then $U_\gamma = 1$ and (8.54) remains zero. The invariances of the action under arbitrary independent Lorentz transformations of all $\Sigma(a)$ and under arbitrary independent translations (when U_γ is near 1) of all sites are the discrete equivalent of Einstein's arbitrary coordinate transformations in the continuum theory.

We note that when $D = 4$ the discrete QCD action in a flat minkowski space (i.e., without gravity) is, similar to (7.6),

$$A_{QCD} = \frac{1}{g^2} \sum_{\Delta_{ijk}} \lambda_{ijk} \; \mathrm{Re} \left[\mathrm{trace} \, (1 - u_{ij} \, u_{jk} \, u_{ki}) \right] \tag{8.55}$$

with u_{ij} the SU(3) matrix associated with link u_{ij} and

$$\lambda_{ijk} = \frac{\text{area of the dual of } \Delta_{ijk}}{\text{area of } \Delta_{ijk}} \quad ; \tag{8.56}$$

the corresponding generating function is

$$Z_{QCD} = \int e^{i A_{QCD}} \left[d^4 x_i \right] \left[du_{ij} \right] \quad . \tag{8.57}$$

When we include gravity, the action A and the generating function Z become

$$A = A_{gravity} + A_{QCD} \tag{8.58}$$

and

$$Z = \int e^{iA} \left[d\ell_{ij}^{\,2} \right] \left[du_{ij} \right] \tag{8.59}$$

where

$$A_{gravity} = \frac{1}{G} \sum_{\Delta_{ijk}} \mathrm{trace} \, (\epsilon_{\rho\sigma\mu\nu} \, J^{\rho\sigma} \, \Delta_{ijk}^{\mu\nu} \, U_{\Delta_{ijk}}) \, , \tag{8.60}$$

$J^{\rho\sigma}$ and $U_{\Delta_{ijk}}$ are both 4×4 real matrices defined by (8.53) and (8.48); A_{QCD} remains given by (8.55), but with the area Δ_{ijk} and its dual-area calculated according to the new length-squares, given by (8.28). The integration $\left[d^4 x_i \right]$ in (8.57) is now replaced by $\left[d\ell_{ij}^{\,2} \right]$ in (8.59). [When some of the $\ell_{ij}^{\,2}$ become too large, the lattice has to be re-linked.] In both (8.57) and (8.60) we keep the total number of lattice sites N and the sum Ω of all flat simplitial volumes fixed; their ratio $N/\Omega = \ell^{-4}$ determines the fundamental length ℓ in the discrete theory.

Comments

1. The coupling g^2 of QCD is dimensionless, but the gravitational coupling G is of dimension $(\text{length})^2$. We have

$$G \sim (\text{Planck length})^2 \; . \tag{8.61}$$

2. In the weak coupling limit, $g^2 \to 0$ and $G \to 0$ the discrete action approaches the usual continuum theory.

3. Only the simplest forms of the discrete actions for QCD and gravity are given in (8.55) and (8.60). If one wishes, one may consider more complicated invariant actions; these would correspond to different discrete theories, although with the same continuum limit, if it exists. [For example, inside the trace we may add higher order polynomials in $u_{\Delta_{ijk}}$ for QCD and $U_{\Delta_{ijk}}$ for gravity.] Further details will be given elsewhere.[21]

IX. CONCLUDING REMARKS

In this series of lectures, I have tried to show how by viewing time as a dynamical variable, a fundamental length ℓ can be introduced which removes all ultra-violet divergences and thereby makes quantization of gravity possible. Because all symmetries, unitary and space-time transformations, are respected, even for renormalizable theories such as QCD the discrete mechanics gives new insight into their solutions. At the minimum this represents an advance in theoretical techniques; at the maximum it may change our basic concept of space-time.

Of course, we do not know how nature really works, and there is no evidence that the discrete mechanics is more appropriate than the usual continuum mechanics. However, once a theoretical possibility of such scope is open, it seems difficult not to face the intellectual challenge. Even at the present stage of enormous imperfection, I find aspects of this new mechanics appealing. I hope, during the past four days, you have shared with me some of my fun in developing this program, and some of my excitement in anticipating its future.

References

1. N. H. Christ, R. Friedberg and T. D. Lee, Nucl. Phys. B202, 89, B210, [FS6], 310, 337 (1982).
2. T. D. Lee, Phys. Lett. 122B, 217 (1983); CU-TP-266 (to be published in the proceedings of the Shelter Island II Conference).
3. R. Friedberg and T. D. Lee, Nucl. Phys. B225 [FS9], 1 (1983).
4. H. C. Ren, CU-TP-272 (Nucl. Phys.); R. Friedberg and H. C. Ren, CU-TP-268 (Nucl. Phys.).
5. C. Itzykson, in Proceedings of the Trieste Workshop, eds. R. Lengo et al. (Singapore, World Scientific, 1983); E. J. Gardner, C. Itzykson and B. Derrida (J. Phys. A); C. Itzykson, "Fields on a Random Lattice," Saclay DPh. G/SPT/83/148.
6. J. C. Maxwell, in A Treatise on Electricity and Magnetism (New York, Dover Publications, 1954).
7. S. R. Lundeen and F. M. Pipkin, Phys. Rev. Lett. 46, 232 (1981).
8. G. Feinberg (private communication); L. Bracci, G. Fiorentini, G. Mezzorani and P. Quarati, Phys. Lett. 133B, 231 (1983).
9. B. Adeva et al., Phys. Rev. Lett. 48, 967 (1982); Proceedings of the 1981 International Symposium on Lepton and Photon Interactions at High Energies, ed. W. Pfeil (Bonn, Physikalisches Institut, 1981).
10. L. A. Fajardo et al., Phys. Rev. D24, 46 (1981); S. J. Lindenbaum, in Particle and Interaction Physics at High Energies (Oxford, Oxford University Press, 1973).
11. G. Parisi and N. Sourlas, Nucl. Phys. B206, 321 (1982).
12. J. M. Ziman, Models of Disorder (New York, Cambridge University Press, 1979); J. L. Meijering, Philips Research Report 8, 270 (1953).
13. K. Wilson, Phys. Rev. D10, 2455 (1974).
14. M. Luscher, Nucl. Phys. B180 [FS2], 1 (1981).
15. N. H. Christ, R. Friedberg and T. D. Lee (to be published).
16. S. Lindenbaum, this proceeding.
17. N. H. Christ and A. E. Terrano, CU-TP-259 (Nuclear Methods and Instrumentation).
18. R. Friedberg, T. D. Lee and H. C. Ren (to be published).

19. T. Regge, Nuovo Cimento 19, 558 (1961). Our treatment is based extensively on Regge's idea.

20. M. Rocek and R. M. Williams, Phys. Lett. 104B, 31 (1981), in Quantum Structure of Space and Time, eds. M. J. Duff and C. J. Isham (Cambridge, Cambridge University Press, 1981).

21. G. Feinberg, R. Friedberg, T. D. Lee and H. C. Ren (to be published).

Nuclear Physics B242 (1984) 145–166

DERIVATION OF REGGE'S ACTION FROM EINSTEIN'S THEORY OF GENERAL RELATIVITY*

R. FRIEDBERG and T.D. LEE

Columbia University, New York, NY 10027, USA

Received 1 March 1984

Regge's action for a discrete lattice is derived from Einstein's continuum formula $\int \sqrt{|g|}\, R\, \mathrm{d}^D x$ in general relativity.

1. Introduction

In the well-known paper of Regge [1], the usual D-dimensional continuous space-time manifold in general relativity is viewed as the limit of a suitable sequence of discrete lattices composed of an increasing number of smaller and smaller simplices. In this approach, the interior of each D-simplex is assumed to be flat; the link lengths l between any pair of its vertices serve as independent variables. Regge proposed that the action for such a discrete lattice in D dimensions is

$$A_{\mathrm{L}} = \sum_s \bar{s}\varepsilon_s\,, \tag{1.1}$$

where s goes over all $(D-2)$-simplices in the lattice, $\bar{s}$ is its volume and ε_s is the "deficit angle" around s. (See ref. [1], or sect. 2, for the definitions of these quantities.) While this expression is quite suggestive, its exact relation to the familiar continuum action

$$A_{\mathrm{C}} = \int \sqrt{|g|}\, R\, \mathrm{d}^D x \tag{1.2}$$

in the theory of general relativity is not clear, where R is the scalar curvature and $|g|$ is the determinant of the metric tensor, assumed to be positive. The lack of a precise formulation relating these two actions has been one of the obstacles to developing the discrete theory of general relativity.

In this paper, we adopt a view that amounts to the converse of Regge's approach. We begin with the usual continuum theory, but regard the lattice as a limit of a sequence of continuum surfaces. As we shall see, Regge's action A_{L} can in fact be derived from Einstein's action A_{C}. This then provides a useful bridge between the discrete and the continuum theories of general relativity.

* This research was supported in part by the US Department of Energy.

For clarity, we discuss first the case that the dimension $D=2$, even though the connection between the total sum (1.1) and the integrated value of (1.2) is known through the Gauss–Bonnet theorem. However, we shall be concerned with the summand in (1.1) and the integrand in (1.2). The relationship between these local quantities lies outside the Gauss–Bonnet theorem. After a brief review of Regge calculus in sect. 2, it will be shown in sect. 3 that the limit of $\sqrt{|g|}\, R\, d^2x$ is indeed the deficit angle ε at a lattice site.

These results will then be generalized in sect. 4 to any dimension D and for arbitrary simplicial lattices. The interplay between the discrete and the continuum theories will be further discussed in sects. 5 and 6. As we shall see, Regge's equation for link lengths can be derived from Einstein's field equation $G_{\mu\nu}=0$.

Although our analysis is carried out in euclidean space, it can be extended to Minkowski space in a straightforward way, except for some minor changes of sign.

2. Regge calculus (review)

We begin with any lattice, regular or random, composed of triangles (2-simplices) in a flat two-dimensional euclidean space R_2. Let l_{pi} be the link length between sites p and i which are vertices of the same triangle in the lattice. Regard l_{pi} as independent dynamical variables which can now take on arbitrary values. In general, with these new link lengths l_{pi}, the lattice can no longer fit in R_2. We then call the discrete space defined by the lattice curved. (The original flat-space lattice provides a convenient map showing how the lattice sites are connected; otherwise, it can be discarded.)

To each triangle $\Delta = pij$ made of lattice sites p, i and j, we require the new link lengths to satisfy the triangular inequalities

$$l_{pi} + l_{ij} > l_{jp}\,, \tag{2.1}$$

etc., so that each Δ by itself still forms a triangle in a flat two-dimensional euclidean space. We assume that the interior of a triangle is always flat. Let $\theta_p(\Delta)$ be the angle at the vertex p in the triangle Δ. The deficit angle ε_p at the site p in the lattice is given by

$$\varepsilon_p \equiv 2\pi - \sum \theta_p(\Delta)\,, \tag{2.2}$$

where the sum extends over all Δ that have the same p as one of their vertices.

The parallel transport of a vector within each triangle Δ is straightforward, since the interior of Δ is flat. To parallel transport the vector from one triangle $\Delta = pij$ to a neighboring one $\Delta' = pik$, we pretend the interior of the entire quadrilateral $pijk$ is flat. If we now parallel transport a vector A^μ along, say, a small circle c around p, then A^μ experiences a rotation of an angle ε_p, indicating that the discrete space may not be flat. All these are, of course, well-known and form the basis of the Regge calculus.

In the usual continuum description, the parallel transport of a vector A^μ, after going around a small circle c, is $A^\mu + \delta A^\mu$ where

$$\delta A^\mu = -R^\mu_{\nu\alpha\beta} A^\nu \, d\sigma^{\alpha\beta}, \tag{2.3}$$

with $R^\mu_{\nu\alpha\beta}$ the Riemann tensor and $d\sigma^{\alpha\beta}$ an infinitesimal surface element bounded by c. Because one deals with finite rotations in the Regge calculus, except for infinitesimal ε_p, it is difficult to give an exact relation between the Riemann tensor and ε_p. The same problem extends also to the Ricci tensor $R_{\alpha\beta}$ and the scalar curvature R. These difficulties will be resolved in the next section.

3. Discrete theory as a limit ($D = 2$)

With the arbitrarily assigned link lengths l_{pi}, the lattice can no longer remain in R_2. As we shall see, it is always possible to embed the lattice in a flat space R_N of a higher dimension N. Since the proof does not depend on (4.8), its choice is optional. In such a space, the total degrees of freedom of a lattice of $\mathcal{N}_0$ sites are $N\mathcal{N}_0$, while the number of constraints imposed by the link lengths is $\mathcal{N}_1$, the total number of links in the lattice. Consequently, the condition on the dimensionality N is

$$N\mathcal{N}_0 - \mathcal{N}_1 \geqslant N + \tfrac{1}{2}N(N-1), \tag{3.1}$$

where the right-hand side refers to the overall translational and rotational degrees of freedom of the lattice in R_N. For a very large lattice, we can neglect the right-hand side, and (3.1) becomes

$$N \geqslant \frac{\mathcal{N}_1}{\mathcal{N}_0}, \tag{3.2}$$

which is 3 for a two-dimensional lattice. For simplicity, we shall first assume $N = 3$. The two-dimensional lattice is then embedded in R_3, with the link length l_{pi} simply given by the distance between p and i.

In R_3, consider the neighborhood of any site, say p, which is linked to sites

$$i = 1, 2, \ldots, n. \tag{3.3}$$

Label the plaquettes (triangles) that are defined by these sites as $\Delta(i)$, where

$$\Delta(1) = pn1, \quad \Delta(2) = p12, \ldots, \qquad \Delta(n) = p\, n-1\, n, \tag{3.4}$$

with the right-hand sides showing the vertices of the triangles. The sites $1, 2, \ldots, n$ are arranged consecutively so that $\Delta(1)$ is sandwiched between $\Delta(n)$ and $\Delta(2)$, $\Delta(2)$ is between $\Delta(1)$ and $\Delta(3)$, etc. We may set up a cartesian coordinate system (x, y, z) with p as the origin. The coordinates of any point on $\Delta(i)$ satisfy

$$n_x(i)x + n_y(i)y + n_z(i)z = 0, \tag{3.5}$$

where $n_x(i)$, $n_y(i)$ and $n_z(i)$ are the components of the normal vector to $\Delta(i)$.

The two-dimensional surface defined by these plaquettes can be described in terms of the coordinates x and y. The geodesic length ds is

$$ds^2 = g_{xx}\,dx^2 + g_{yy}\,dy^2 + 2g_{xy}\,dx\,dy, \tag{3.6}$$

where

$$g_{\mu\nu} = \delta_{\mu\nu} + \frac{\partial z}{\partial x_\mu}\frac{\partial z}{\partial x_\nu}, \tag{3.7}$$

with the subscripts

$$\mu, \nu = x \quad \text{or} \quad y, \tag{3.8}$$

and likewise,

$$x_\mu, x_\nu = x \quad \text{or} \quad y.$$

It is convenient to denote

$$\zeta_x = \frac{\partial z}{\partial x}, \quad \zeta_y = \frac{\partial z}{\partial y}, \tag{3.9}$$

and ζ as the column vector

$$\zeta = \begin{pmatrix} \zeta_x \\ \zeta_y \end{pmatrix}. \tag{3.10}$$

The 2×2 matrix $g_{\mu\nu}$ can be written as

$$g = 1 + \zeta\tilde{\zeta}, \tag{3.11}$$

where $\tilde{\zeta}$ denotes the transpose of ζ. Sometimes it is also useful to represent ζ by a vector $\boldsymbol{\zeta}$ in the (x, y) plane whose components are given by (3.9).

From (3.5) we see that inside $\Delta(i)$, $\boldsymbol{\zeta}$ becomes $\boldsymbol{\zeta}(i)$ whose components are

$$\zeta_x(i) = -\frac{n_x(i)}{n_z(i)}, \quad \zeta_y(i) = -\frac{n_y(i)}{n_z(i)}. \tag{3.12}$$

When we move from one plaquette $\Delta(i)$ to a neighboring one $\Delta(i+1)$, $\boldsymbol{\zeta}(x, y)$ changes discontinuously from $\boldsymbol{\zeta}(i)$ to $\boldsymbol{\zeta}(i+1)$. Because in R_3, the coordinate z changes continuously from plaquette to plaquette, we have the continuity equation for the tangential component of $\boldsymbol{\zeta}$:

$$\boldsymbol{\zeta}(i+1)\cdot\hat{e}_\parallel(i) = \boldsymbol{\zeta}(i)\cdot\hat{e}_\parallel(i), \tag{3.13}$$

where $\hat{e}_\parallel(i)$ is the unit vector parallel to the projection of the link $\overline{pi}$ on the (x, y) plane.

The function $\boldsymbol{\zeta}(x, y)$ is not a continuous one; its derivatives are δ- and δ'-functions. Since we shall deal with nonlinear products of $\boldsymbol{\zeta}$ and its derivatives, it may be useful to envisage $\boldsymbol{\zeta}(x, y)$ as a continuous function but with rapid variations across any link. In this way, we may apply all the usual machinery appropriate for the continuum theory. The discrete lattice surface is viewed as a limit of a series of continuous surfaces; the lattice theory then becomes the limit of the usual continuum theory.

It turns out that direct manipulation of the discontinuous $\boldsymbol{\zeta}(x, y)$ is surprisingly simple. The limit is completely unambiguous and can be carried out without much delicate consideration.

From (3.10), it follows that the inverse of g is

$$g^{-1} = 1 - \frac{1}{|g|}\zeta\tilde{\zeta}, \tag{3.14}$$

$$|g| = 1 + \tilde{\zeta}\zeta. \tag{3.15}$$

The Christoffel symbol is

$$\Gamma_{\mu\alpha\beta} = \zeta_\mu \frac{\partial \zeta_\alpha}{\partial x_\beta}. \tag{3.16}$$

Because of (3.9), we have

$$\frac{\partial \zeta_x}{\partial y} = \frac{\partial \zeta_y}{\partial x} \tag{3.17}$$

and therefore $\Gamma_{\mu\alpha\beta} = \Gamma_{\mu\beta\alpha}$. In addition, the Riemann tensor is

$$R_{xyxy} = |g|^{-1} \frac{\partial(\zeta_x, \zeta_y)}{\partial(x, y)}, \tag{3.18}$$

the Ricci tensor is

$$R_{\mu\nu} = \tfrac{1}{2} g_{\mu\nu} R, \tag{3.19}$$

and the scalar curvature

$$R = 2|g|^{-2} \frac{\partial(\zeta_x, \zeta_y)}{\partial(x, y)}, \tag{3.20}$$

where

$$\frac{\partial(\zeta_x, \zeta_y)}{\partial(x, y)} = \frac{\partial \zeta_x}{\partial x}\frac{\partial \zeta_y}{\partial y} - \frac{\partial \zeta_x}{\partial y}\frac{\partial \zeta_y}{\partial x}. \tag{3.21}$$

Inside each $\Delta(i)$, the vector $\boldsymbol{\zeta}$ and the matrix g are both constants; this follows from (3.11), (3.12) and reaffirms the flatness in the interior of each plaquette. At the boundary of $\Delta(i)$, say link l_{pi}, $\boldsymbol{\zeta}$ varies rapidly from $\boldsymbol{\zeta}(i)$ in $\Delta(i)$ to $\boldsymbol{\zeta}(i+1)$ in $\Delta(i+1)$. The behavior of the curvature tensor near l_{pi} can be most conveniently studied by rotating the axes from x, y to

$$x_\parallel(i) = x \cos\phi_i + y \sin\phi_i,$$
$$x_\perp(i) = -x \sin\phi_i + y \cos\phi_i, \tag{3.22}$$

where ϕ_i is the azimuthal angle of the link l_{pi}. Thus, the unit vector $\hat{e}_\parallel(i)$ introduced in (3.13) can be represented by the column matrix

$$e_{\|}(i)=\begin{pmatrix}\cos\phi_i\\ \sin\phi_i\end{pmatrix};\tag{3.23}$$

its companion unit vector $\hat{e}_\perp(i)$ perpendicular to l_{pi} is

$$e_\perp(i)=\begin{pmatrix}-\sin\phi_i\\ \cos\phi_i\end{pmatrix}.\tag{3.24}$$

Under the rotation (3.22), the components of $\boldsymbol{\zeta}$ change from ζ_x and ζ_y to

$$\zeta_{\|}=\boldsymbol{\zeta}\cdot\hat{e}_{\|}(i)\,,\quad \zeta_\perp=\boldsymbol{\zeta}\cdot\hat{e}_\perp(i)\,.\tag{3.25}$$

From (3.13), we see that $\zeta_{\|}$ does not change from $\Delta(i)$ to $\Delta(i+1)$; only $\zeta_\perp$ varies (rapidly) across the link l_{pi}. Hence, excluding regions near the end-points p and i, we have

$$\frac{\partial\zeta_{\|}}{\partial x_{\|}(i)}=\frac{\partial\zeta_{\|}}{\partial x_\perp(i)}=0\tag{3.26}$$

on the open link l_{pi} as well as inside $\Delta(i)$ and $\Delta(i+1)$. Since

$$\frac{\partial(\zeta_x,\zeta_y)}{\partial(x,y)}=\frac{\partial(\zeta_{\|},\zeta_\perp)}{\partial(x_{\|}(i),x_\perp(i))},$$

(3.26) implies that $\partial(\zeta_x,\zeta_y)/\partial(x,y)$ is zero except near the lattice sites; so also is the Riemann tensor, on account of (3.18). In the limit when $\boldsymbol{\zeta}$ varies suddenly (instead of just rapidly) from plaquette to plaquette, then

$$R_{xyxy}=0\tag{3.27}$$

everywhere except at the lattice sites. We shall call the space defined by this limit the "lattice space", or lattice surface.

To analyze the curvature tensor at the lattice sites, let us consider a small region ω in the (x, y) plane which contains only a single site p, as in fig. 1b.

3.1. THEOREM

$$\int_\omega \sqrt{|g|}\,R\,\mathrm{d}x\,\mathrm{d}y=2\varepsilon_p\,.\tag{3.28}$$

Therefore, by extending the integration over the entire space we have

$$\int\sqrt{|g|}\,R\,\mathrm{d}^2x=2\sum_i\varepsilon_i\,,\tag{3.29}$$

i.e.

$$\sqrt{|g|}\,R=2\sum_i\varepsilon_i\delta(x-x_i)\delta(y-y_i)\,,\tag{3.30}$$

where x_i and y_i are the coordinates of site i.

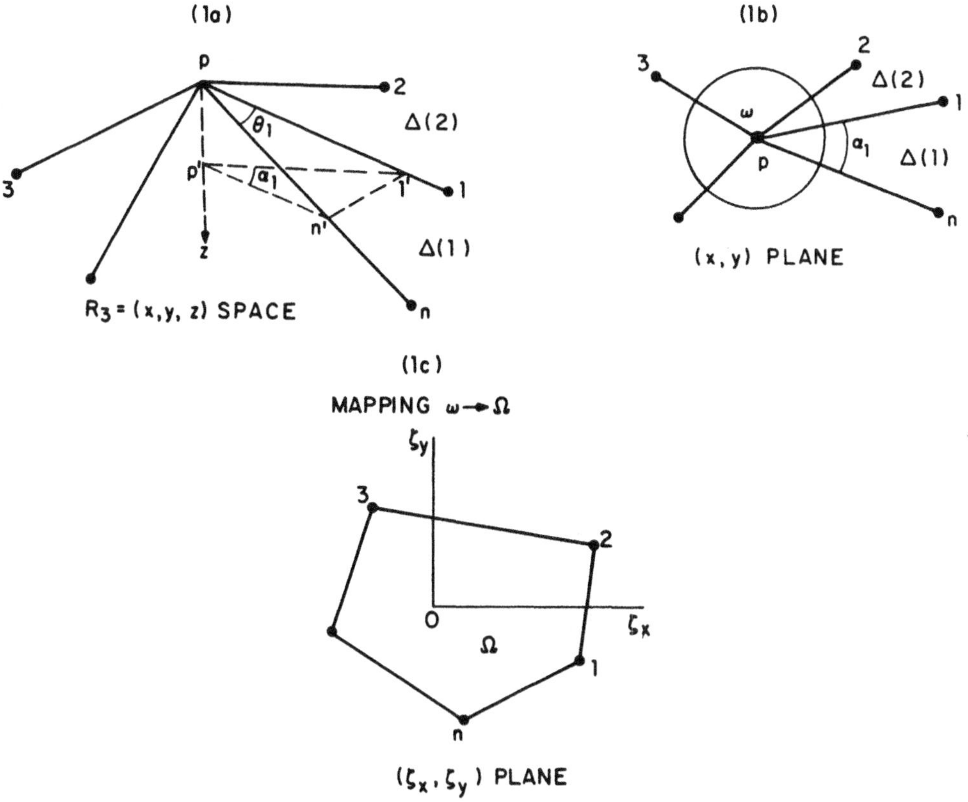

Fig. 1. (a) A two-dimensional lattice surface in R_3. (b) Its projection onto the (x, y) plane. (c) The mapping of ω in the (x, y) plane to Ω in the (ζ_x, ζ_y) plane.

Proof. We begin with the two-dimensional lattice surface embedded in R_3. (See fig. 1a). Take the site p to be the origin, as before. Since the orientation of the z-axis can be arbitrarily chosen, we can always assume the z-axis to be surrounded by plaquettes so that the projection of links $l_{p1}, l_{p2}, \ldots, l_{pn}$ on the (x, y) plane is of the pattern shown in fig. 1b, with p in the middle. As we shall see, the singularity of R_{xyxy} at p can be conveniently analyzed by treating the "dependent" variables ζ_x and ζ_y as independent variables.

Map the region ω in the (x, y) plane to Ω in the (ζ_x, ζ_y) plane. As can be seen from fig. 1c, Ω is a polygon with n corners. The part of ω belonging to $\Delta(i)$ is mapped into the corner i whose coordinates are $\zeta_x(i)$ and $\zeta_y(i)$, and the boundary between $\Delta(i)$ and $\Delta(i+1)$ into the border of Ω connecting corners i and $i+1$.

It is useful to think of the sequence of continuous two-dimensional surfaces in R_3 whose limit is the lattice surface. Before the limit, the direction of the normal to the surface varies rapidly around the z-axis near the origin p; at some point very close to p (which can be p itself) the normal would be parallel to the z-axis, which implies $\boldsymbol{\zeta} = 0$ at that point, because of (3.12). Thus, Ω must contain the origin 0 in the (ζ_x, ζ_y) space, i.e. the point $\boldsymbol{\zeta}=0$. In the limit of the lattice surface, the single site p in the (x, y) plane is mapped into the entire interior of the polygon Ω in the (ζ_x, ζ_y) plane. All curvature tensors are regular in the (ζ_x, ζ_y) plane.

By using (3.15) and (3.20), we see that

$$\int_\omega \sqrt{|g|}\, R \,\mathrm{d}x\, \mathrm{d}y = \int_\Omega \frac{2}{(1+\zeta^2)^{3/2}}\, \mathrm{d}\zeta_x\, \mathrm{d}\zeta_y\,, \tag{3.31}$$

where

$$\zeta^2 = \zeta_x^2 + \zeta_y^2\,. \tag{3.32}$$

This integral is an elementary one. To see how it is related to the deficit angle ε_p, we shall divide Ω into a a sum of triangles

$$\Omega = \sum_1^n \Omega(i)\,, \tag{3.33}$$

where $\Omega(i)$ is the triangle whose vertices are the origin 0 and the corners i and $i+1$. To carry out the integration in $\Omega(i)$, we change the variables from ζ_x, ζ_y to $\zeta_\parallel$ and $\zeta_\perp$, defined by (3.25):

$$\begin{aligned}\int_{\Omega(i)} \frac{\mathrm{d}\zeta_x\, \mathrm{d}\zeta_y}{(1+\zeta^2)^{3/2}} &= \int_0^{\zeta_\parallel(i)} \mathrm{d}\zeta_\parallel \int_{\lambda\zeta_\parallel}^{\mu\zeta_\parallel} \mathrm{d}\zeta_\perp (1+\zeta_\parallel^2+\zeta_\perp^2)^{-3/2} \\ &= \beta_{i+1} - \gamma_i\,,\end{aligned} \tag{3.34}$$

where, in accordance with (3.13) and (3.25),

$$\begin{aligned}&\zeta_\parallel(i) = \boldsymbol{\zeta}(i)\cdot\hat{\boldsymbol{e}}_\parallel(i) = \boldsymbol{\zeta}(i+1)\cdot\hat{\boldsymbol{e}}_\parallel(i)\,,\\ &\lambda = \boldsymbol{\zeta}(i)\cdot\hat{\boldsymbol{e}}_\perp(i)/\zeta_\parallel(i)\,, \quad \mu = \boldsymbol{\zeta}(i+1)\cdot\hat{\boldsymbol{e}}_\perp(i)/\zeta_\parallel(i)\,,\end{aligned} \tag{3.35}$$

$$\begin{aligned}\sin\beta_{i+1} &= \frac{\mu\zeta_\parallel(i)^2\{[1+\boldsymbol{\zeta}(i+1)^2]^{1/2}-1\}}{\boldsymbol{\zeta}(i+1)^2[1+\zeta_\parallel(i)^2]^{1/2}}\,,\\ \sin\gamma_i &= \frac{\lambda\zeta_\parallel(i)^2\{[1+\boldsymbol{\zeta}(i)^2]^{1/2}-1\}}{\boldsymbol{\zeta}(i)^2[1+\zeta_\parallel(i)^2]^{1/2}}\,.\end{aligned} \tag{3.36}$$

For the deficit angle ε_p, we return to the lattice surface in R_3. As in sect. 2, the angle at p in the triangle $\Delta(i) = pi-1i$ is

$$\theta_p(\Delta(i)) \equiv \theta_i\,.$$

Let the projection of θ_i on the (x, y) plane be α_i. At a distance h from the origin p, draw a plane $\parallel (x, y)$, which cuts the links l_{pi} at i' and the z-axis at p', as shown in fig. 1a. By definition, the distance between p and p' is

$$\overline{pp'} = h\,.$$

Because i' is on the link l_{pi} which is shared by both $\Delta(i)$ and $\Delta(i+1)$, the distance between p' and i' is, on account of (3.5), (3.12) and (3.35),

$$\overline{p'i'} = h/\zeta_{\|}(i)\,.$$

From fig. 1b, we see that the triangles $p'n'1'$ and $pn'1'$ share a common side $n'1'$. Thus, θ_1 and α_1 can be related in an elementary way. In the same way we can relate other θ_i and α_i:

$$\cos\theta_i = \frac{\cos\alpha_i + \zeta_{\|}(i)\zeta_{\|}(i-1)}{[1+\zeta_{\|}(i)^2]^{1/2}[1+\zeta_{\|}(i-1)^2]^{1/2}}\,,$$

$$\sin\theta_i = \sin\alpha_i \frac{[1+\boldsymbol{\zeta}(i)^2]^{1/2}}{[1+\zeta_{\|}(i)^2]^{1/2}[1+\zeta_{\|}(i-1)^2]^{1/2}}\,. \tag{3.37}$$

In deriving the latter, we have used the relations

$$\begin{aligned}\hat{\boldsymbol{e}}_{\|}(i) &= \cos\alpha_i \hat{\boldsymbol{e}}_{\|}(i-1) + \sin\alpha_i \hat{\boldsymbol{e}}_{\perp}(i-1)\,,\\ \hat{\boldsymbol{e}}_{\perp}(i) &= -\sin\alpha_i \hat{\boldsymbol{e}}_{\|}(i-1) + \cos\alpha_i \hat{\boldsymbol{e}}_{\perp}(i-1)\,,\end{aligned} \tag{3.38}$$

and therefore

$$\begin{aligned}\boldsymbol{\zeta}(i)^2 &= \zeta_{\|}(i)^2 + [\boldsymbol{\zeta}(i)\cdot\hat{\boldsymbol{e}}_{\perp}(i)]^2\\ &= \zeta_{\|}(i)^2 + \csc^2\alpha_i[\zeta_{\|}(i)\cos\alpha_i - \zeta_{\|}(i-1)]^2\,.\end{aligned} \tag{3.39}$$

By using (3.36) and (3.37), one can verify that (see appendix A)

$$\alpha_i - \theta_i = \beta_i - \gamma_i\,. \tag{3.40}$$

Since $\sum_i \alpha_i = 2\pi$, we derive

$$\varepsilon_p \equiv 2\pi - \sum_i \theta_i = \sum_i (\beta_i - \gamma_i) = \tfrac{1}{2}\int_\omega \sqrt{|g|}\, R\, \mathrm{d}^2x\,, \tag{3.41}$$

which proves the theorem.

To derive the expressions for $R_{\mu\nu\alpha\beta}$, $R_{\mu\nu}$, etc., we define

$$\psi_m(i) \equiv \int_{\Omega(i)} \frac{\mathrm{d}\zeta_x\,\mathrm{d}\zeta_y}{(1+\zeta^2)^m}\,. \tag{3.42}$$

When $m=\frac{3}{2}$, we have according to (3.34) and (3.41)

$$\psi_{3/2}(i) = \beta_{i+1} - \gamma_i\,,$$

and therefore

$$\varepsilon_p = \sum_i \psi_{3/2}(i)\,. \tag{3.43}$$

I.e. in the small region ω containing the origin p

$$\sqrt{|g|}\, R = 2\delta(x)\delta(y)\sum_i \psi_{3/2}(i)\,, \tag{3.44}$$

By using (3.18) and (3.20), we derive the following expressions in ω:

$$R_{xyxy} = \delta(x)\delta(y)\sum_i \psi_1(i)\,,$$

$$R = 2\delta(x)\delta(y)\sum_i \psi_2(i)\,. \tag{3.45}$$

Likewise, $R_{\mu\nu}$ can be obtained by using (3.11) and (3.19).

When the curvature of the lattice space is small, $\psi_m(i)$ can be expanded as a power series in $\zeta(i)^2$:

$$\psi_m(i) = \tfrac{1}{2}\zeta_{\|}(i)^2(\mu-\lambda) - \tfrac{1}{4}m\zeta_{\|}(i)^4[\mu-\lambda+\tfrac{1}{3}(\mu^3-\lambda^3)]$$
$$+\tfrac{1}{12}m(m+1)\zeta_{\|}(i)^6[\mu-\lambda+\tfrac{2}{3}(\mu^3-\lambda^3)+\tfrac{1}{5}(\mu^5-\lambda^5)]+\mathrm{O}(\zeta^8)\,, \tag{3.46}$$

where $\zeta_{\|}(i)$, μ and λ are given in (3.35). Thus, *only* when $\mathrm{O}(\zeta^4)$ is neglected can we approximate

$$R_{xyxy} \cong \varepsilon_p\delta(x)\delta(y)\,,$$
$$R \cong 2\varepsilon_p\delta(x)\delta(y)\,, \tag{3.47}$$

in ω, whereas $\sqrt{|g|}\,R = 2\varepsilon_p\delta(x)\delta(y)$ holds to all orders.

Let $\boldsymbol{n}(i)$ be the normal vector to $\Delta(i)$ in R_3. When $\zeta(i)^2$ is small $\boldsymbol{n}(i)$ is almost parallel to $\hat{z}$, the unit vector along the z-axis. From (3.12), (3.13), and (3.35) we see that

$$\zeta_{\|}(i)^2(\mu-\lambda) = [\boldsymbol{n}(i+1)\times\boldsymbol{n}(i)]\cdot\hat{z}+\mathrm{O}(\zeta^4)\,.$$

Hence, to $\mathrm{O}(\zeta^2)$

$$R_{xyxy} \cong \delta(x)\delta(y)\sum_i \tfrac{1}{2}[\boldsymbol{n}(i+1)\times\boldsymbol{n}(i)]\cdot\hat{z} \tag{3.48}$$

in ω. This result may be understood intuitively. Think of the parallel transport of a vector $\boldsymbol{A}$ on the lattice surface in R_3. The link l_{pi} forms a ridge; on one side the surface is $\Delta(i)$ and on the other $\Delta(i+1)$. When we transport $\boldsymbol{A}$ across the ridge, there is a rotation $\boldsymbol{n}(i+1)\times\boldsymbol{n}(i)$ if $\boldsymbol{A}$ is perpendicular to the ridge, but none if $\boldsymbol{A}$ is parallel. On the average, the rotation is $\tfrac{1}{2}\boldsymbol{n}(i+1)\times\boldsymbol{n}(i)$. Viewed in the (x, y) plane, only the z-component of that rotation is relevant, and the result is (3.48).

3.2. REMARKS

(i) In most of the above considerations, we are concerned only with the neighborhood of a single lattice site, say p. Let

$$S=\{p, i\} \tag{3.49}$$

be the set consisting of p and all the sites $i=1, 2, \ldots, n$ linked to p. Suppose we are only interested in S. As before, assume that S is embedded in R_3, with the link lengths given by

$$l_{pi} = \overline{pi}\,, \quad l_{ii+1} = \overline{ii+1}\,, \tag{3.50}$$

where the right-hand sides denote the distances between sites in R_3.

It is useful to count the degrees of freedom. Altogether there are $3(n+1)$ degrees of freedom for the $n+1$ sites p and i in R_3. Subtract 3 for the overall translational degrees of freedom, 3 for the overall rotation and $2n$ for the constraints (3.50). There are still $n-3$ degrees of freedom left.

When n is >3, these extra degrees of freedom may be represented by the $n-3$ distances, say,

$$\overline{13}, \overline{14}, \ldots, \overline{1\,n-1} \tag{3.51}$$

in R_3 (even though site 1 is not linked to 3, nor to 4, etc.). This particular representation will be useful later on, as we shall see.

(ii) So far the lattice is assumed to be embedded in R_3. Actually, the theorem that we establish remains valid if the underlying flat space is a higher dimensional one, R_N with $N > 3$.

In such a case, the distance between any linked pair of sites, say p and i in R_N, equals the assigned link length l_{pi}. Consider the set $S = \{p, i\}$ where i runs over all the sites linked to p, as in (3.49). In R_N, sites p, 1, 2 and 3 define a tetrahedron

$$v(3) \equiv p123\,. \tag{3.52}$$

Likewise, p, 1, 3 and 4 define another tetrahedron

$$v(4) \equiv p134\,,$$

..., and sites p, 1, $n-1$, n define still another tetrahedron $v(n)$.

We shall show that the lattice in R_N can be mapped into R_3 through a continuous coordinate transformation. Consider the same set of sites S in R_3: arrange the distance between these sites so that (3.50) holds and, in addition, the distances between 1 and the other i-sites are the same as those in R_N. The latter is possible because of (3.51). Thus, the tetrahedron $v(3)$ in R_N is congruent to the corresponding one, $p123$, in R_3, and $v(4)$ in R_N is congruent to $p134$ in R_3, etc. In this way, tetrahedron by tetrahedron, we can map the entire lattice from R_N to R_3. Since $\sqrt{|g|}\, R\, d^2x$ is invariant under such a coordinate transformation, its value is the same whether the underlying flat space is R_3 or R_N. Consequently, (3.28)–(3.30) is valid in general.

4. Generalization to higher dimensions

The discussion for $D > 2$ follows closely that given above. As in sect. 2, it is convenient to start with a simplicial lattice in a flat euclidean space R_D. We then arbitrarily re-assign a new link length l_{pi} to any pair of sites p and i that are linked in the lattice. As a result, in general the lattice can no longer fit in R_D, and the space defined by the lattice becomes curved.

To each D-simplex τ of sites, say $1, 2, \ldots, D+1$, we require the new link lengths l_{12}, $l_{13}, \ldots$ to satisfy all the simplicial inequalities (which include the familiar triangular inequalities (2.1), the tetrahedral inequalities, etc.). Their explicit expressions are given in appendix B. These inequalities insure that each τ by itself, with the new link lengths, can still form a D-simplex in a flat D-dimensional euclidean space.

With these arbitrarily assigned link lengths, it is nevertheless still possible to embed the entire lattice in a flat space R_N, provided that N is larger than the link-site ratio $\mathcal{N}_1/\mathcal{N}_0$, in accordance with (3.1) and (3.2). For a random lattice this ratio [2] is about 7 for $D=3$ and 19 for $D=4$. On the other hand, if we are only interested in a small part of the lattice, then the dimensionality of the underlying flat space can be made much smaller.

Take any $(D-2)$-simplex s in the lattice whose vertices are

$$p_a = p_1, p_2, \ldots, p_{D-1}\,. \tag{4.1}$$

Let

$$i = 1, 2, \ldots, n \tag{4.2}$$

be the sites that are linked to p_a. Define S to be the set of these $D-1+n$ sites:

$$S = \{p_a, i\}\,. \tag{4.3}$$

As we shall see, S can always be embedded in R_{D+1} with the assigned link lengths equal to the distances between sites. Furthermore, excluding the overall translational and rotational degrees of freedom of S in R_{D+1}, there are still $n-3$ degrees of freedom left. This follows because the total degree of freedom of $D-1+n$ sites p_a and i in R_{D+1} is $(D-1+n)(D+1)$. Subtract $(D+1)$ for the overall translational degrees of freedom, $\frac{1}{2}(D+1)D$ for rotation, $\frac{1}{2}(D-1)(D-2)$ for the link-length constraints between p_a and p_b, $(D-1)n$ for those between p_a and i and n for those between i and $i+1$. (Again we arrange $i=1, 2, \ldots, n$ in a consecutive order around s, as in fig. 1b with p replaced by s.) This gives the net degrees of freedom left as $n-3$.

As in (3.51), these extra degrees of freedom may be represented by the distances

$$\overline{13}, \overline{14}, \ldots, \overline{1\,n-1} \tag{4.4}$$

in R_{D+1}. Thus, although the entire lattice is embedded in R_N where N is much larger than $D+1$, we can always map S into R_{D+1} with all link lengths preserved through a continuous coordinate transformation. This can be achieved by setting p_a and i in R_{D+1} with the distances $\overline{p_ap_b}$, $\overline{p_ai}$ and $\overline{i\,i+1}$ equal to the link lengths and with the distances in (4.4) the same as those in R_N. Thus, the $(D+1)$-simplex $v(3)$ whose vertices are p_a, 1, 2 and 3 in R_N is congruent to the corresponding one in

R_{D+1}. Likewise the $(D+1)$-simplex $v(4)=p_a$, 1, 2, 4 in R_N is congruent to the corresponding one in R_{D+1}, etc. The mapping is then obvious.

Once the set S is mapped into R_{D+1}, we may change the representation of the extra $(n-3)$ degrees of freedom to another choice, one that will be adopted in our subsequent analysis. Let

$$x_1, x_2, \ldots, x_D \quad \text{and} \quad z \tag{4.5}$$

be the coordinates in R_{D+1}. Choose, say, site p_{D-1} to be the origin of R_{D+1}; i.e. its coordinates satisfy

$$x_1(p_{D-1})=x_2(p_{D-1})=\cdots=x_D(p_{D-1})=z(p_{D-1})=0\,. \tag{4.6}$$

Next, set the z-coordinates of $S=\{p_a, i\}$ to be

$$z(p_1)=z(p_2)=\cdots=z(p_{D-2})=0\,, \tag{4.7}$$

and, if one wishes,

$$z(1)=z(2)=\cdots=z(n)=h\,. \tag{4.8}$$

That this is possible can again be demonstrated by a direct counting. Condition (4.6) simply eliminates the overall translational degrees of freedom; it does not affect the number of intrinsic degrees of freedom. Condition (4.7) imposes $(D-2)$ constraints and, in addition, reduces the overall rotational degrees of freedom from $\frac{1}{2}(D+1)D$ in R_{D+1} to $\frac{1}{2}D(D-1)$ in R_D. Eq. (4.8) gives an additional $n-1$ constraints (since h is still a free parameter). Putting these together, we see that the net change in degrees of freedom is exactly

$$\tfrac{1}{2}(D+1)D-\tfrac{1}{2}D(D-1)-(D-2)-(n-1)=-(n-3)\,.$$

The invariance of Einstein's action insures that $\sqrt{|g|}\,R\,\mathrm{d}^D x$ is not affected by this change of representation from (4.4) to (4.7), (4.8).

Let ω be a small domain in R_{D+1} which encompasses only the single $(D-2)$-simplex $s=\{p_a\}$ and its neighborhood. As in (3.4), we denote the D-simplices that share s by

$$\begin{aligned}&\tau(1)=sn1\,,\quad \tau(2)=s12,\ldots,\\&\tau(n)=s\,n-1\;n\,.\end{aligned} \tag{4.9}$$

The D-dimensional lattice surface in R_{D+1} can be described by

$$z=z(x_1,\ldots,x_D)\,. \tag{4.10}$$

Within each $\tau(i)$, (4.10) takes on a linear form

$$z=\zeta_1(i)x_1+\zeta_2(i)x_2+\cdots+\zeta_D(i)x_D\,. \tag{4.11}$$

In the following, we adopt the coordinate choice (4.6)–(4.8). The geometry of the D-dimensional lattice surface in ω will be described in terms of $x_1, x_2, \ldots, x_D$.

The $(D-2)$-simplex s defines an R_{D-2}. We may rotate the $x_1, x_2, \ldots, x_D$ axes so that the x_1 and x_2 axes are perpendicular to s; i.e.

$$x_1(p_a) = x_2(p_a) = 0\,. \tag{4.12}$$

Since s is part of each $\tau(i)$, by substituting (4.7) and (4.12) into (4.11) we find

$$\zeta_3(i) = \zeta_4(i) = \cdots = \zeta_D(i) = 0\,. \tag{4.13}$$

Thus, (4.11) becomes

$$z = \zeta_1(i)x_1 + \zeta_2(i)x_2\,. \tag{4.14}$$

The geodesic length ds on the lattice surface is given by

$$ds^2 = g_{\mu\nu}\, dx_\mu\, dx_\nu \tag{4.15}$$

where μ and ν vary from 1 to D. The metric $g_{\mu\nu}$ is $\delta_{\mu\nu} + (\partial z/\partial x_\mu)(\partial z/\partial x_\nu)$. Because of (4.14), se may write the matrix $g_{\mu\nu}$ as

$$g = 1 + \left(\begin{array}{c|c} \zeta\tilde{\zeta} & 0 \\ \hline 0 & 0 \end{array}\right), \tag{4.16}$$

where ζ is a two-component column matrix

$$\zeta = \begin{pmatrix} \dfrac{\partial z}{\partial x_1} \\[2mm] \dfrac{\partial z}{\partial x_2} \end{pmatrix} \tag{4.17}$$

as in (3.10). In each $\tau(i)$, ζ becomes a two-dimensional vector $\boldsymbol{\zeta}(i)$ whose components are $\zeta_1(i)$ and $\zeta_2(i)$.

Let the subscripts

$$a, b = 1 \quad \text{or} \quad 2\,. \tag{4.18}$$

From (4.16), we see that

$$|g| = 1 + \tilde{\zeta}\zeta\,. \tag{4.19}$$

Just as in (3.16)–(3.20), the only non-vanishing elements of $\Gamma_{\mu\lambda\sigma}$, $R_{\mu\nu\lambda\sigma}$ and $R_{\mu\nu}$ are

$$\Gamma_{abc} = \zeta_a \frac{\partial \zeta_b}{\partial x_c} = \zeta_a \frac{\partial \zeta_c}{\partial x_b}\,, \tag{4.20}$$

$$R_{1212} = -R_{1221} = -R_{2112} = R_{2121} = |g|^{-1} \frac{\partial(\zeta_1, \zeta_2)}{\partial(x_1, x_2)}\,, \tag{4.21}$$

$$R_{ab} = \tfrac{1}{2} g_{ab} R\,, \tag{4.22}$$

where the scalar curvature is

$$R = 2|g|^{-2} \frac{\partial(\zeta_1, \zeta_2)}{\partial(x_1, x_2)}\,. \tag{4.23}$$

The Einstein tensor

$$G_{\mu\nu} = R_{\mu\nu} - \tfrac{1}{2} g_{\mu\nu} R \tag{4.24}$$

is then given by the matrix

$$(G_{\mu\nu}) = -\tfrac{1}{2} R \left(\begin{array}{cc|c} 0 & 0 & \\ 0 & 0 & 0 \\ \hline & 0 & 1 \end{array} \right), \tag{4.25}$$

where 1 is the $(D-2)\times(D-2)$ unit matrix. All these expressions are valid in the small region ω, and are expressed in terms of the special coordinates that satisfy (4.6), (4.7) and (4.12). We assume that the link lengths satisfy all the necessary simplicial inequalities given in the appendix.

By following exactly the same argument used in sect. 3, we derive

$$R_{\mu\nu\lambda\sigma} = 0 \tag{4.26}$$

everywhere, except at points on the $(D-2)$-simplices in the lattice space. Furthermore, using the aforementioned special coordinates, we find in ω and on the $(D-2)$ simplex s

$$R_{1212} = \delta(x_1)\delta(x_2) \sum_i \psi_1(i) , \tag{4.27}$$

$$R = 2\delta(x_1)\delta(x_2) \sum_i \psi_2(i) , \tag{4.28}$$

$$\sqrt{|g|}\, R = 2\delta(x_1)\delta(x_2) \sum_i \psi_{3/2}(i) = 2\delta(x_1)\delta(x_2)\varepsilon_s , \tag{4.29}$$

where $\psi_m(i)$ is defined by (3.42) and ε_s is the deficit angle around s. Therefore,

$$\int_\omega \sqrt{|g|}\, R \, \mathrm{d}^D x = 2 \bar{s} \varepsilon_s , \tag{4.30}$$

where $\bar{s}$ is the $(D-2)$-dimensional volume of s. By extending the integration to all space, we derive Regge's formula (1.1) from Einstein's action (1.2):

$$\int \sqrt{|g|}\, R \, \mathrm{d}^D x = 2 \sum_s \bar{s} \varepsilon_s . \tag{4.31}$$

5. Field equation

In this section, we shall derive the field equation in the discrete theory by using Einstein's equation of general relativity

$$G_{\mu\nu} = 0 . \tag{5.1}$$

Under an infinitesimal change

$$g^{\mu\nu} \to g^{\mu\nu} + \delta g^{\mu\nu} , \tag{5.2}$$

the continuum action $A_C=\int\sqrt{|g|}\,R\,d^Dx$ undergoes a variation

$$\delta A_C=\int\sqrt{|g|}\,G_{\mu\nu}\delta g^{\mu\nu}\,d^Dx\,. \tag{5.3}$$

The field equation (5.1) follows from $\delta A_C=0$. For applications to the lattice space, we must restrict $\delta g^{\mu\nu}$ to one that corresponds to a variation in the link length

$$l\to l+\delta l\,, \tag{5.4}$$

so that the space will remain a lattice space.

5.1. $D=3$

As before, consider a small region ω that contains a particular link l (which is the $(D-2)$-simplex s in this case). In the special coordinate system that satisfies (4.6), (4.7) and (4.12), the only nonvanishing component of $G_{\mu\nu}$ is $G_{33}=-\frac{1}{2}R$, on account of (4.25). Assume $\delta g^{\mu\nu}=0$ outside ω; (5.3) becomes

$$\delta A_C=-\tfrac{1}{2}\int_\omega\sqrt{|g|}\,R\,\delta g^{33}\,d^3x\,. \tag{5.5}$$

According to (4.16), $g_{33}=1$. Because the link l is parallel to the x_3-axis, the variation δl can be generated by having a constant $\delta g_{33}=2\delta l/l$ on the link, i.e.

$$\delta g^{33}=-2\delta l/l\,. \tag{5.6}$$

By using (4.29), we see that (5.5) becomes

$$\delta A_C=2\varepsilon_l\delta l\,, \tag{5.7}$$

which leads to the equation in the discrete theory:

$$\varepsilon_l=0 \tag{5.8}$$

for every link l.

5.2. $D=4$

Take any link l in the lattice. Let $\{\Delta(l)\}$ be the set of all plaquettes $\Delta(l)$ that have l as one of their sides. For each $\Delta(l)$ and in its small surrounding region $\omega=\omega_{\Delta(l)}$, the only nonvanishing components of $G_{\mu\nu}$ are $G_{33}=G_{44}=-\frac{1}{2}R$, according to (4.25). Assume $\delta g^{\mu\nu}\neq 0$ only at points on $\Delta(l)$; (5.5) becomes

$$\delta A_C=\sum_{\Delta(l)}-\tfrac{1}{2}\int_{\omega_{\Delta(l)}}\sqrt{|g|}\,R(\delta g^{33}+\delta g^{44})\,d^4x\,. \tag{5.9}$$

From (4.12) and (4.16), we see that in each $\omega_{\Delta(l)}$, $g_{33}=g_{44}=1$, $g_{34}=0$ and $\Delta(l)$ lies on the (x_3, x_4) plane. Assume the matrix $(\delta g_{\mu\nu})$ to be of the following form on $\Delta(l)$:

$$(\delta g_{\mu\nu}) = \left(\begin{array}{c|c} 0 & 0 \\ \hline 0 & \delta g \end{array}\right), \tag{5.10}$$

where δg is a 2×2 constant real symmetric infinitesimal matrix. Let the three sides of $\Delta(l)$ be a, b and l. We must choose δg to correspond only to a change in l, without altering the lengths of the other two sides a and b.

Place $\Delta(l)$ such that its vertex opposite l is the origin (hence, $x_3 = x_4 = 0$) and orient the x_4-axis to bisect the angle ϕ at that vertex, as shown in fig. 2. The two sides a and b can be represented by the column vectors

$$\begin{aligned} \chi_a &\equiv a\begin{pmatrix} \sin\frac{1}{2}\phi \\ \cos\frac{1}{2}\phi \end{pmatrix}, \\ \chi_b &\equiv b\begin{pmatrix} -\sin\frac{1}{2}\phi \\ \cos\frac{1}{2}\phi \end{pmatrix}, \end{aligned} \tag{5.11}$$

in which the upper elements refer to the x_3-components and the lower ones to the x_4-components. Set

$$\delta g = \eta\begin{pmatrix} \cos^2\frac{1}{2}\phi & 0 \\ 0 & -\sin^2\frac{1}{2}\phi \end{pmatrix}. \tag{5.12}$$

We see that

$$\tilde{\chi}_a \delta g \chi_a = \tilde{\chi}_b \delta g \chi_b = 0\,, \tag{5.13}$$

which implies that under $g_{\mu\nu} \to g_{\mu\nu} + \delta g_{\mu\nu}$ the lengths a and b are unchanged. On the other hand, since

$$\tilde{\chi}_a \delta g \chi_b = \tilde{\chi}_b \delta g \chi_a = -\tfrac{1}{2}\eta ab \sin^2\phi \tag{5.14}$$

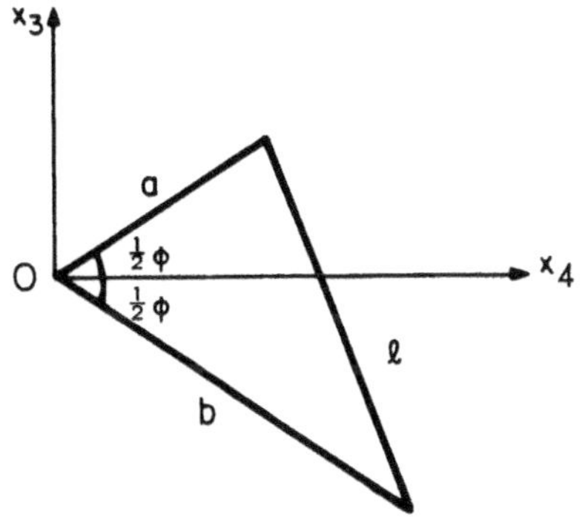

Fig. 2. The triangle $\Delta(l)$ with sides a, b and l.

we have

$$\delta l^2 = \eta ab \sin^2 \phi \, . \tag{5.15}$$

The area of the triangle is

$$\bar{\Delta} = \tfrac{1}{2} ab \sin \phi \, .$$

Using (5.10), (5.12), (5.15), (4.29) and in terms of the deficit angle $\varepsilon_{\Delta(l)}$, we obtain

$$\begin{aligned} \tfrac{1}{2} \int_{\omega_{\Delta(l)}} & \sqrt{|g|}\, R(\delta g^{33} + \delta g^{44})\, \mathrm{d}^4 x \\ &= -\varepsilon_{\Delta(l)}(\cos^2 \tfrac{1}{2}\phi - \sin^2 \tfrac{1}{2}\phi)\eta \tfrac{1}{2} ab \sin \phi \\ &= -l \varepsilon_{\Delta(l)} \cot \phi \delta l \, . \end{aligned} \tag{5.16}$$

Therefore, because of (5.9), the equation in the discrete theory is

$$\frac{\partial A_{\mathrm{C}}}{\partial l} = l \sum_{\Delta(l)} \varepsilon_{\Delta(l)} \cot \phi = 0 \, . \tag{5.17}$$

Since $\partial \bar{\Delta} / \partial l = \frac{1}{2} l \cot \phi$, (5.17) can also be written as

$$\sum_{\Delta = \Delta(l)} \frac{\partial \bar{\Delta}}{\partial l} \varepsilon_{\Delta} = 0 \, . \tag{5.18}$$

The above argument can be readily generalized to any dimension D. The resulting equation in the discrete theory is

$$\frac{\partial A_{\mathrm{C}}}{\partial l} = 2 \sum \frac{\partial \bar{s}}{\partial l} \varepsilon_s = 0 \, , \tag{5.19}$$

where the sum extends over those $(D-2)$-simplices $s = s(l)$ that share the link l. As in (1.1), ε_s is the deficit angle around s and $\bar{s}$ is the $(D-2)$-dimensional volume of s. Because $A_{\mathrm{C}} = 2 \sum \bar{s} \varepsilon_s$, (5.19) implies

$$\sum_s \bar{s} \frac{\partial \varepsilon_s}{\partial l} = 0 \, , \tag{5.20}$$

a relation that was already derived by Regge [1] in his original paper through a pure geometrical consideration. Eqs. (5.8), (5.17) and (5.19) are, of course, exactly the same as Regge's.

Since our derivation is based entirely on the continuum analysis, it reveals an aspect of Regge's equation (5.19) and Regge's identity (5.20) that has hitherto been hidden. Thereby, a firmer connection between the discrete and the continuum theories of general relativity is provided.

6. Invariance

Since for the lattice space

$$\int \sqrt{|g|}\, R \, \mathrm{d}^D x = 2 \sum \bar{s} \varepsilon_s \, , \tag{6.1}$$

the action is clearly invariant under a coordinate transformation

$$x_\mu \to x'_\mu \,. \tag{6.2}$$

Thus, the entire apparatus of coordinate invariance in the usual continuum theory automatically applies to the discrete theory as well.

The discrete theory enjoys yet another class of invariance that lies beyond the usual continuum theory. In the discrete theory, we may regard the coordinates of the lattice sites in R_N (or their equivalent, the link lengths, through (3.1) and (3.2)) as dynamical variables [3]. There could then be symmetries associated with these variables. These dynamical symmetries are outside the kinematic invariances induced by the coordinate transformations (6.2). For example, under a very general condition if these lattices are embedded in a smooth background D-dimensional surface $\mathcal{S}$ of small curvatures, the same action (6.1) is also invariant under an arbitrary motion of the lattice sites, provided that $\mathcal{S}$ and the site density are kept fixed. The details of these dynamical symmetries will be given in a separate paper.

That it is possible to derive the discrete theory from the continuum theory and that the continuum theory can, in turn, be obtained as a limit of the discrete theory reveal a close relationship between these two seemingly opposite descriptions of nature, the depth of which is yet to be uncovered.

We wish to thank G. Feinberg and H.C. Ren for discussions.

Appendix A

PROOF OF EQ. (3.40)

From (3.36), it follows that

$$\cos\beta_i = \frac{\zeta_\parallel(i-1)^2\{[1+\boldsymbol{\zeta}(i)^2]^{1/2}-1\}+\boldsymbol{\zeta}(i)^2}{\boldsymbol{\zeta}(i)^2[1+\zeta_\parallel(i-1)^2]^{1/2}}\,, \tag{A.1}$$

$$\cos\gamma_i = \frac{\zeta_\parallel(i)^2\{[1+\boldsymbol{\zeta}(i)^2]^{1/2}-1\}+\boldsymbol{\zeta}(i)^2}{\boldsymbol{\zeta}(i)^2[1+\zeta_\parallel(i)^2]^{1/2}}\,, \tag{A.2}$$

and therefore

$$\cos(\beta_i-\gamma_i) = N/[\boldsymbol{\zeta}(i)^2]^2[1+\zeta_\parallel(i)^2]^{1/2}[1+\zeta_\parallel(i-1)^2]^{1/2}\,, \tag{A.3}$$

where the numerator is

$$\begin{aligned} N = {} & [\zeta_\parallel(i-1)^2\{[1+\boldsymbol{\zeta}(i)^2]^{1/2}-1\}+\boldsymbol{\zeta}(i)^2][\zeta_\parallel(i)^2\{[1+\boldsymbol{\zeta}(i)^2]^{1/2}-1\}+\boldsymbol{\zeta}(i)^2] \\ & +\zeta_\parallel(i)\zeta_\parallel(i-1)[\boldsymbol{\zeta}(i)\cdot\hat{\boldsymbol{e}}_\perp(i)][\boldsymbol{\zeta}(i)\cdot\hat{\boldsymbol{e}}_\perp(i-1)]\{[1+\boldsymbol{\zeta}(i)^2]^{1/2}-1\}^2\,. \end{aligned} \tag{A.4}$$

By using (3.38), we find

$$[\boldsymbol{\zeta}(i)\cdot\hat{\boldsymbol{e}}_\perp(i)][\boldsymbol{\zeta}(i)\cdot\hat{\boldsymbol{e}}_\perp(i-1)] = \boldsymbol{\zeta}(i)^2\cos\alpha_i - \zeta_\parallel(i)\zeta_\parallel(i-1)\,,$$

which reduces N to

$$N = \{[1+\boldsymbol{\zeta}(i)^2]^{1/2}-1\}^2[\zeta_\parallel(i)\zeta_\parallel(i-1)\boldsymbol{\zeta}(i)^2\cos\alpha_i]$$
$$+\{[1+\boldsymbol{\zeta}(i)^2]^{1/2}-1\}[\zeta_\parallel(i)^2+\zeta_\parallel(i-1)^2]\boldsymbol{\zeta}(i)^2+[\boldsymbol{\zeta}(i)^2]^2$$
$$=\{[1+\boldsymbol{\zeta}(i)^2]^{1/2}-1\}[\zeta_\parallel(i)^2+\zeta_\parallel(i-1)^2-2\zeta_\parallel(i)\zeta_\parallel(i-1)\cos\alpha_i]\boldsymbol{\zeta}(i)^2$$
$$+[1+\zeta_\parallel(i)\zeta_\parallel(i-1)\cos\alpha_i][\boldsymbol{\zeta}(i)^2]^2\,.$$

Because of (3.39), N can be written as

$$N=[\boldsymbol{\zeta}(i)^2]^2\{\cos^2\alpha_i+\sin^2\alpha_i[1+\boldsymbol{\zeta}(i)^2]^{1/2}+\zeta_\parallel(i)\zeta_\parallel(i-1)\cos\alpha_i\}\,. \tag{A.5}$$

Combining (A.3)–(A.5) with (3.37), we derive

$$\cos(\beta_i-\gamma_i)=\cos(\alpha_i-\theta_i)\,. \tag{A.6}$$

Hence

$$\beta_i-\gamma_i=\pm(\alpha_i-\theta_i)\,. \tag{A.7}$$

The sign can be established by examining the case when $\alpha_i-\theta_i$ is small. The result is (3.40).

Appendix B

SIMPLICIAL INEQUALITIES

Given the $\frac{1}{2}(D+1)D$ link lengths l_{ij} between sites

$$i, j = 1, 2, \ldots, D+1,$$

what are the conditions on l_{ij}^2 so that these sites can be vertices of a D-simplex in a flat euclidean space $\mathrm{R_D}$?

(i) When $D=1$, the condition is the trivial one

$$l_{ij}^2>0\,. \tag{B.1}$$

(ii) When $D=2$, the conditions are (B.1) and the triangular inequality

$$\Delta_{ijk}^2\equiv\left(\frac{1}{2!}\right)^2\begin{vmatrix} l_{ij}^2 & c_{i,jk}\\ c_{i,kj} & l_{ik}^2\end{vmatrix}>0\,, \tag{B.2}$$

where

$$c_{i,jk}\equiv\tfrac{1}{2}(l_{ij}^2+l_{ik}^2-l_{jk}^2)=c_{i,kj}\,. \tag{B.3}$$

One can readily verify that the left side of the inequality is identical to the well-known area formula for the triangle ijk:

$$\Delta_{ijk}^2=\sigma(\sigma-l_{ij})(\sigma-l_{jk})(\sigma-l_{ki})\,,$$

where $\sigma = \frac{1}{2}(l_{ij} + l_{jk} + l_{ki})$. Let x_{ij} and y_{ij} be the coordinates of the link vector $\mathbf{l}_{ij}$. The less familiar expression in (B.2) is obtained by equating Δ_{ijk} with the determinant of the matrix

$$M = \frac{1}{2}\begin{pmatrix} x_{ij} & x_{ik} \\ y_{ij} & y_{ik} \end{pmatrix}, \tag{B.4}$$

and therefore

$$\Delta_{ijk}^2 = |\tilde{M}M|. \tag{B.5}$$

The single inequality (B.2) implies

$$l_{ij} + l_{jk} > l_{ki}, \quad l_{jk} + l_{ki} > l_{ij}$$

$$l_{ki} + l_{ij} > l_{jk}. \tag{B.6}$$

(iii) When $D = 3$, there are four sites and six link lengths. The conditions on l_{ij}^2 are (B.1) for every pair i, j, (B.2) for every i, j, k and

$$\tau_{1234}^2 \equiv \left(\frac{1}{3!}\right)^2 \begin{vmatrix} l_{12}^2 & c_{1,23} & c_{1,24} \\ c_{1,32} & l_{13}^2 & c_{1,34} \\ c_{1,42} & c_{1,43} & l_{14}^2 \end{vmatrix} > 0 \tag{B.7}$$

for the set 1, 2, 3 and 4. The left side of the above inequality is essentially Tartaglia's volume formula for a tetrahedron*. This expression can be obtained in the same way as the derivation of (B.2) from (B.5).

That there is an additional inequality (B.7) can be seen by considering the tetrahedron 1234 in R_3. We may vary one of its link lengths, say l_{34}, while keeping the other five fixed. This is equivalent to varying the dihedral angle ϕ between triangles 123 and 124. When $\phi = \pi$, the diagonal $\overline{34}$ of the quadrilateral 1324 imposes a limit on l_{34}.

Condition (B.7) is clearly necessary. From (B.7), we observed that τ_{1234}^2 is a quadratic function of l_{34}^2:

$$\tau_{1234}^2 = \alpha l_{34}^4 + \beta l_{34}^2 + \gamma, \tag{B.8}$$

where

$$\alpha = -\left(\frac{1}{3!}\right)^2 \frac{1}{4} l_{12}^2 < 0. \tag{B.9}$$

The allowed region of l_{34}^2 lies within the two roots of (B.8) = 0. Thus, when both $l_{ij}^2 > 0$ and $\Delta_{ijk}^2 > 0$ are satisfied, $\tau_{1234}^2 > 0$ gives a singly connected region in the l_{ij}^2 space, whose boundary is determined by $\tau_{1234}^2 = 0$. (B.1), (B.2) and (B.7) are therefore necessary and sufficient.

The generalization to higher dimensions is straightforward.

* Nicolo Fontana Tartaglia (1500-1557). We are indebted to Professor J. A. Wheeler for this information.

References

[1] T. Regge, Nuovo Cim. **19** (1961) 558
[2] N.H. Christ, R. Friedberg and T.D. Lee, Nucl. Phys. B202 (1982) 89
[3] T.D. Lee, Discrete mechanics, to be published in Proc. Int. School of Subnuclear Phys. Erice, August 1983

CU - TP - 281

LATTICE GRAVITY NEAR THE CONTINUUM LIMIT

G. Feinberg, R. Friedberg, T. D. Lee and H. C. Ren

Columbia University, New York, N. Y. 10027

This research was supported in part by the U. S. Department of Energy.

ABSTRACT

We prove that the lattice gravity always approaches the usual continuum limit when the link-length $\ell \to 0$, provided that certain general boundary conditions are satisfied. This result holds for any lattice, regular or irregular. Furthermore, for a given lattice, the deviation from its continuum limit can be expressed as a power series in ℓ^2. General formulas for such a perturbative calculation are given, together with a number of illustrative examples, including the graviton propagator. The lattice gravity satisfies all the invariance properties of Einstein's theory of general relativity. In addition, it is symmetric under a new class of transformations that are absent in the usual continuum theory. The possibility that the lattice theory (with a nonzero ℓ) may be more fundamental is discussed.

1. INTRODUCTION

In a recent paper [1] we studied the lattice limit of the usual continuum gravity. It was shown that any given simplicial lattice can be regarded as the limit of a sequence of continuous surfaces with sharper and sharper ridges. In this way, Regge's lattice action [2]

$$A_L = 2 \sum_s \bar{s}\, \epsilon_s \tag{1.1}$$

can be derived from Einstein's continuum action with zero cosmological constant

$$A_C = \int \sqrt{|g|}\; R\, d^D x \tag{1.2}$$

where, as usual, the sum in (1.1) goes over all $(D-2)$-simplices s in the lattice, $\bar{s}$ is its volume, ϵ_s is the deficit angle around s and the integration in (1.2) extends over the continuum surface, with R as the scalar curvature and $|g|$ the determinant of the metric tensor, assumed to be positive.* This derivation is valid for any lattice, regular or irregular, and the distances between lattice sites do not have to be small.

In this paper, we examine the inverse problem, lattice gravity and Regge calculus near their continuum limits. We first raise the question: with what precision can the usual continuous space-time manifold in general relativity be regarded as the

---

* This assumption is not essential; it is adopted here for convenience. All discussions given in Ref. 1 and this paper can be readily extended to minkowski-like space, $|g| < 0$, except for some minor changes of sign.

limit of a sequence of discrete lattices composed of smaller and smaller simplices, each flat in its interior. In particular, for a given D dimensional smooth surface, can its action (1.2) be accurately determined by measuring only the discrete action (1.1) in such a sequence of discrete lattices? Of course, this is the most basic problem of Regge calculus, which was first introduced as an approximation to the continuum theory of gravity. However, to date it is not clear how good the approximation is in general. For example, the Regge calculus has been tested by C. Y. Wong [3] for the Schwarzschild and Reissner-Nordstrom geometries. He found errors ranging from roughly 10% down to less than 1%, depending on the details of his numerical analysis. More recently, M. Roček and R. M. Williams [4] studied the long wavelength limit of Regge's action in the weak gravity approximation. They showed that for a regular hyper-cubic lattice, the usual continuum results can be obtained in the limit. While this is quite satisfactory, one is uncertain whether the result can be generalized to a different lattice or to the case of strong gravity.

As we shall prove, under very general conditions for an arbitrary smooth D dimensional surface $\mathcal{S}$ and for any lattice structure, the limit of the lattice action (1.1) is always the continuum action (1.2), evaluated on $\mathcal{S}$. This holds in any D and for any magnitude of the difference between the metric tensor g_{ij} of $\mathcal{S}$ and that of a flat surface δ_{ij}. Furthermore, for a given lattice L the deviation of its action from the limit can be expressed as a power series in $(\ell/\rho)^2$ where ℓ and ρ refer to the typical dimension of the link-length in L and the radius of curvature in $\mathcal{S}$. The general formula relating these two actions A_L and A_C is given in Sec. 2. In Secs 3 and 4, we illustrate its application by a number of examples. These in turn lead to the discovery of some simple but non-trivial properties in discrete geometry. In Sec. 5, we turn to the physical problem of wave propagation in lattice gravity. For any lattice, regular or random, in the long wavelength limit, the inverse of the graviton propagator can be shown to be always the continuum limit k^2, where k is the wave number. The correction is of the order $k^4 \ell^2$, which can be computed, and the result is given. The general question of invariance and symmetry properties of lattice gravity is discussed in Sec. 6. In Sec. 7 we discuss generalization to Regge's action for discrete gravity.

The most appealing feature of Einstein's theory is its invariance under a general coordinate transformation. It has been shown in Ref. 1 that in terms of a lattice surface formulation, the lattice gravity also satisfies this symmetry, without requiring the link-lengths to be small. As we shall see in this paper, the lattice theory enjoys still another, totally new, class of symmetries which are absent in the usual continuum formulation. Aesthetically, this adds to the attractiveness of the lattice theory. For lattices with nonzero link-lengths there is clearly no ultra-violet divergence. Thus, we avoid this particular difficulty in carrying out quantization. In addition, in the approximation of small link-length, we show that lattice gravity always approaches the usual continuum theory; therefore its agreement with experiment is automatically insured. All of this strongly suggests that the lattice theory with nonzero link-lengths is perhaps more fundamental. The usual continuum theory may well be only an approximation.

2. GENERAL DISCUSSION

Consider a D dimensional lattice consisting of sites

$$p = 1, 2, \cdot\cdot, \mathcal{n} \tag{2.1}$$

embedded* in a flat euclidean space R_N. The entire lattice is composed of D simplices with the lattice sites as their vertices. The link-length ℓ_{pq} between two connected sites p and q is given by their rectilinear distance in R_N:

$$\ell_{pq}^2 \equiv \overline{pq}^2 = [\vec{r}(p) - \vec{r}(q)]^2 \tag{2.2}$$

where $\vec{r}(p)$ denotes the cartesian N dimensional position vector of p. For definiteness, let us assume that these $\mathcal{n}$ sites all lie on a D dimensional smooth enveloping surface $\mathcal{S}$, also embedded in R_N. We will assume that any of the link-lengths ℓ in the lattice is much smaller than the typical radius of curvature ρ of the enveloping surface $\mathcal{S}$; i.e.,

$$\frac{\ell}{\rho} \ll 1 . \tag{2.3}$$

The lattice structure is determined by the linking pattern and the link-lengths ℓ_{pq}, or equivalently the position vectors $\vec{r}(p)$. In the following, we shall consider a <u>fixed</u> $\mathcal{S}$ and envisage a sequence of lattices with increasing site-densities so that

---

* As discussed in Ref. 1 (hereafter referred to as I), this is always possible if N is sufficiently large. For a random lattice, $N \geqslant 3$ for $D = 2$, 7 for $D = 3$ and 19 for $D = 4$. Of course, we assume the link-lengths satisfy all the necessary simplicial inequalities.

the link-lengths become smaller and smaller. According to I, by filling the interior of all simplices in the lattice by points in the flat space R_N, the lattice forms a continuous (but with many sharp ridges) D dimensional surface L, called the lattice surface. On L, the metric tensor $g_{ij}(L)$ is a constant inside each simplex, but changes discontinuously at its boundary when one moves from one simplex to the next.

Take any lattice site, say p. We can choose it to be the origin of R_N and rotate the cartesian coordinate-axes of R_N such that the first D-components of the position vector $\vec{r}$, labeled $y_1, y_2, \cdots, y_D$ coincide with the local coordinates of $\mathcal{S}$ at p, i.e., these D-coordinates form an R_D which is tangent to $\mathcal{S}$ at p. The remaining

$$n = N - D$$

components $z_1, z_2, \cdots, z_n$ are functions of y_i $(i = 1, 2, \cdots, D)$ for points on $\mathcal{S}$:

$$z_\alpha = z_\alpha(y_1, y_2, \cdots, y_D) \tag{2.4}$$

where $\alpha = 1, \cdots, n$. For small y_i, z_α depends quadratically on y_i, and we write

$$z_\alpha = \tfrac{1}{2} c(ij)_\alpha \, y_i y_j + O(y^3) \quad . \tag{2.5}$$

Thus, in the neighborhood of the origin p, the metric tensor on $\mathcal{S}$ is

$$g_{ij}(\mathcal{S}) = \delta_{ij} + h_{ij}(\mathcal{S}) \;, \tag{2.6}$$

$$h_{ij}(\mathcal{S}) = \frac{\partial z_\alpha}{\partial y_i}\frac{\partial z_\alpha}{\partial y_j} = C_{ij,k\ell}\, y_k y_\ell + O(y^3) \qquad (2.7)$$

and

$$C_{ij,k\ell} = \tfrac{1}{2}\left[c(ik)_\alpha\, c(j\ell)_\alpha + c(i\ell)_\alpha\, c(jk)_\alpha\right] \qquad (2.8)$$

where the repeated indices are summed over with the Latin index varying from 1 to D and the Greek from 1 to n. The magnitude of $c(ij)_\alpha$ is $O(\rho^{-1})$ where, as in (2.3), ρ denotes some typical radius of curvature of $\mathcal{S}$.

Since p is the origin, $\vec{r}(p) = 0$. In (2.2), by denoting the components of $\vec{r}(q)$ by $y_i(q)$ and $z_\alpha(q)$, we have

$$\ell_{pq} = \left[y_i(q)\, y_i(q) + z_\alpha(q)\, z_\alpha(q)\right]^{\frac{1}{2}} \qquad (2.9)$$

which differs from the geodesic length on $\mathcal{S}$

$$\widehat{pq} \equiv \int_p^q ds = \int_p^q \left[g_{ij}(\mathcal{S})\, dy_i\, dy_j\right]^{\frac{1}{2}} \qquad (2.10)$$

by an amount $O(\ell^3/\rho^2)$. Let ℓ°_{pq} be the distance between p and the projection of q in the tangent subspace R_D, defined by the $y_1, \cdots, y_D$ axes; i.e.,

$$\ell^{\circ}_{pq} = \left[y_i(q)\, y_i(q)\right]^{\frac{1}{2}} \quad . \qquad (2.11)$$

By using (2.5)-(2.11), one can readily verify that

$$\ell_{pq} \equiv \overline{pq} = \ell^{\circ}_{pq} + \tfrac{3}{4}\left[\widehat{pq} - \ell^{\circ}_{pq}\right] + O(\ell^4/\rho^3) \quad . \qquad (2.12)$$

Consequently, given $\mathcal{S}$ and its local coordinates $y_1, y_2, \cdots, y_D$ at p, although there is a great deal of arbitrariness in the choice of $z_1, z_2, \cdots, z_n$, both in the number n of the additional dimensions and in the precise functional forms of $z_a(y_i)$, the above relation makes an "invariant" connection between the link-lengths $\overline{pq}$ in the lattice space L and the corresponding geodesic distances $\widehat{pq}$ on its enveloping surface $\mathcal{S}$.

To find the metric tensor

$$g_{ij}(L) = \delta_{ij} + h_{ij}(L) \qquad (2.13)$$

on the lattice surface L, let us consider a point p' on L infinitesimally near the origin p. Denote the coordinates of p' in the tangent subspace R_D by $dy_1, dy_2, \cdots, dy_D$. Following I, we see that the cartesian distance $\overline{pp'}$ in R_N when expressed in terms of dy_i determines $g_{ij}(L)$:

$$\overline{pp'}^{\,2} = g_{ij}(L)\, dy_i\, dy_j \quad . \qquad (2.14)$$

Thus, both $h_{ij}(\mathcal{S})$ and $h_{ij}(L)$ are $O(\rho^{-2})$. The same holds for their difference:

$$\delta g_{ij} \equiv g_{ij}(L) - g_{ij}(\mathcal{S}) = O(\rho^{-2}) \quad . \qquad (2.15)$$

In I, it is shown that the lattice action (1.1) is equal to the integral

$$A_L = \int_L \sqrt{|g|}\; R\, d^D x \qquad (2.16)$$

which is carried out over the lattice surface L. The usual continuum action (1.2) is the same integral

$$A_C = \int_{\mathcal{S}} \sqrt{|g|}\; R\, d^D x \qquad (2.17)$$

but over the enveloping surface $\mathcal{S}$. Because of (2.15) we may expand A_L as a Taylor series in δg^{ij} , assuming the same range in x for $\mathcal{S}$ and L :

$$A_L = A_C + \delta A + \frac{1}{2!}\delta^2 A + \frac{1}{3!}\delta^3 A + \cdots \tag{2.18}$$

where δA is $O(\delta g^{ij})$, $\delta^2 A$ is $O((\delta g^{ij})^2)$, etc. The expansion is standard. We find

$$\delta A = \int_{\mathcal{S}} \sqrt{|g|}\; G_{ij}\,\delta g^{ij}\, d^D x + \int_{\mathcal{S}} \frac{\partial}{\partial x_i}\left[\sqrt{|g|}\;(-g^{ij}\,\delta\Gamma^k_{jk} + g^{jk}\,\delta\Gamma^i_{jk})\right] d^D x \tag{2.19}$$

where Γ^k_{ij} is the usual Christoffel symbol given by

$$\Gamma^k_{ij} = \tfrac{1}{2} g^{k\ell}\left(\frac{\partial g_{j\ell}}{\partial x_i} + \frac{\partial g_{i\ell}}{\partial x_j} - \frac{\partial g_{ij}}{\partial x_\ell}\right) , \tag{2.20}$$

and

$$\begin{aligned}\delta^2 A = \int_{\mathcal{S}} \Big\{ \sqrt{|g|}\; \Big[& 2g^{ij}(\delta\Gamma^k_{\ell i}\,\delta\Gamma^\ell_{kj} - \delta\Gamma^k_{k\ell}\,\delta\Gamma^\ell_{ij}) \\ & + \delta g^{ij}\,\delta g^{k\ell}(-g_{ij} R_{k\ell} + \tfrac{1}{2} g_{ik} g_{j\ell} R + \tfrac{1}{4} g_{ij} g_{k\ell} R)\Big] \\ & + \text{t.d.} \Big\}\, d^D x \end{aligned} \tag{2.21}$$

where t.d. stands for the following total derivative:

$$\begin{aligned}\text{t.d.} = \frac{\partial}{\partial x_i}\Big[\sqrt{|g|}\; (& -2\delta g^{ij}\,\delta\Gamma^k_{jk} + 2\delta g^{jk}\,\delta\Gamma^i_{jk} \\ & + g_{ab}\, g^{ij}\,\delta g^{ab}\,\delta\Gamma^k_{jk} - g_{ab}\, g^{jk}\,\delta g^{ab}\,\delta\Gamma^i_{jk} \\ & - g^{ij}\,\delta^2\Gamma^k_{jk} + g^{jk}\,\delta^2\Gamma^i_{jk})\Big] \end{aligned} \tag{2.22}$$

and $\delta^2 \Gamma_{ij}^k$ is the second order expansion of Γ_{ij}^k in δg^{ij}; i.e., when $g^{ij} \to g^{ij} + \delta g^{ij}$, similar to (2.18), the power series expansion of the new Christoffel symbol is given by

$$\Gamma_{ij}^k \to \Gamma_{ij}^k + \delta \Gamma_{ij}^k + \frac{1}{2!} \delta^2 \Gamma_{ij}^k + \frac{1}{3!} \delta^3 \Gamma_{ij}^k + \cdots .$$

Throughout, we employ the customary notation $R_{ijk\ell}$ for the Riemann tensor, R_{ij} for the Ricci tensor, R for the scalar curvature and

$$G_{ij} = R_{ij} - \tfrac{1}{2} g_{ij} R \quad . \tag{2.23}$$

Because of (2.15), we have

$$\delta \Gamma_{ij}^k = O(\rho^{-2}) \quad . \tag{2.24}$$

It is important to note that R, R_{ij} and G_{ij} are also $O(\rho^{-2})$. Thus, for a finite surface $\mathcal{S}$ and not counting the possible ρ dependence in the overall size of the $d^D x$ integration, the zeroth order term A_C in the expansion (2.18) is* $O(\rho^{-2})$. On the right-hand side of (2.19), the first term is $O(\rho^{-4})$ while the second total derivative term is $O(\rho^{-2})$. In the integrand of (2.21), the first term in the square bracket is $O(\rho^{-4})$ and the second is $O(\rho^{-6})$, etc.

For problems in which the integral of the total derivative term in (2.19) is zero, we see that the successive terms in the expansion (2.18) are of the following order of magnitude:

---

* Quite often, we do not exhibit ℓ explicitly. Hence, for most dimensionless intensive quantities $O(\rho^{-2}) = O(\ell^2/\rho^2)$, $O(\rho^{-3}) = O(\ell^3/\rho^3)$, etc.

$$\frac{\delta A}{A_C} = O(\frac{\ell^2}{\rho^2}) \ , \qquad \frac{\delta^2 A}{A_C} = O(\frac{\ell^2}{\rho^2}) \ ,$$

$$\frac{\delta^3 A}{A_C} = O(\frac{\ell^4}{\rho^4}) \ , \qquad \frac{\delta^4 A}{A_C} = O(\frac{\ell^6}{\rho^6}) \tag{2.25}$$

etc. Hence,

$$\lim_{\ell \to 0} A_L = A_C \ . \tag{2.26}$$

From (2.25), it follows that for a given lattice, both δA and $\delta^2 A$ give $O(\ell^2/\rho^2)$ corrections. Thus, to this order we may combine them and write

$$A_L = A_C + \Delta A \tag{2.27}$$

where

$$\Delta A = \int (G_{ij}\, \delta g^{ij} + \delta\Gamma^k_{ij}\, \delta\Gamma^j_{ik} - \delta\Gamma^k_{kj}\, \delta\Gamma^j_{ii})\, d^D x \ . \tag{2.28}$$

It is important to note that even when the integral of the total derivative term in (2.19) is not zero, the expansion (2.18) is still a valid one, provided that ℓ^2/ρ^2 is small. The only modification is that $\delta A/A_C$ is now $O(1)$ instead of $O(\ell^2/\rho^2)$; however, $\delta^2 A/A_C$ remains $O(\ell^2/\rho^2)$ and the other terms in the expansion also do not change; i.e.,

$$\frac{\delta^m A}{A_C} = O\left[(\frac{\ell^2}{\rho^2})^{m-1} \right] \tag{2.29}$$

which differs from (2.25) only when $m = 1$. In this case, when the link-length $\ell \to 0$, (2.26) is replaced by

$$\lim_{\ell \to 0} A_L = A_C + \int_{\mathcal{S}} \frac{\partial}{\partial x_i} (\delta\Gamma_{ijj} - \delta\Gamma_{jji})\, d^D x \tag{2.30}$$

in which the last term can be readily written as a surface integral. There is a variety of problems in which this surface integral is zero, and therefore (2.26)-(2.28) are valid: e.g., when $\mathcal{S}$ is ∞ but the boundary condition at infinity is either flat or periodic, or when $\mathcal{S}$ is finite but satisfies some appropriate boundary conditions, as will be illustrated by a number of examples in the subsequent sections.

3. EXAMPLE: D = 2

For a closed two-dimensional surface, because of the Gauss-Bonnet theorem, the equality between the lattice and continuum actions, A_L and A_C, is well-known. In this section, we shall prove that the same equality also holds for any open two-dimensional lattice surface L and its enveloping continuum surface $\mathcal{S}$:

$$2 \sum_p \epsilon_p = \int_{\mathcal{S}} \sqrt{|g|}\, R\, d^2x \quad . \tag{3.1}$$

Let $\mathcal{C}$ be the boundary of $\mathcal{S}$. The sum extends over all lattice sites p enclosed by $\mathcal{C}$. At first sight, this may appear somewhat strange, since under a variation of $\mathcal{C}$ (and therefore also $\mathcal{S}$), the right-hand side changes gradually while the left-hand side does so discontinuously. Therefore, an important step in our proof is to determine the correct boundary $\mathcal{C}$ for a given L.

When D = 2, both L and $\mathcal{S}$ can be embedded in R_3. Equation (2.4), which describes $\mathcal{S}$, becomes simply

$$z = z(y_1, y_2) \quad . \tag{3.2}$$

Since $ds^2 = dy_1^{\,2} + dy_2^{\,2} + dz^2$, the metric tensor on $\mathcal{S}$ is

$$g_{ij}(\mathcal{S}) = \delta_{ij} + z_i z_j \tag{3.3}$$

where

$$z_i = \frac{\partial z}{\partial y_i} \quad .$$

[Here, y_i does not have to be small.] One can readily verify that

$$\Gamma_{kij} = z_k \frac{\partial z_i}{\partial y_j} = z_k \frac{\partial z_j}{\partial y_i} = \Gamma_{kji} \tag{3.4}$$

and

$$\sqrt{|g|}\ R = 2(1 + z_1^2 + z_2^2)^{-3/2} \frac{\partial(z_1, z_2)}{\partial(y_1, y_2)} \quad . \tag{3.5}$$

Thus, (1.2) becomes

$$A_C = 2 \int_{\mathcal{S}} (1 + z_1^2 + z_2^2)^{-3/2} dz_1\, dz_2 \quad . \tag{3.6}$$

To derive A_L, we follow the discussion given in Sec. 3 of I. Consider first the neighborhood of any lattice site, say p. Assume that p is linked to sites

$$k = 1, 2, \cdots, m \quad . \tag{3.7}$$

Arrange these numbers in consecutive order, so that p is shared by the plaquettes (triangles)

$$\Delta(1) = pm1 \ , \quad \Delta(2) = p12 \ , \cdots ,$$

and (3.8)

$$\Delta(m) = p\ m{-}1\ m$$

with the right hand sides showing the vertices of these triangles. The coordinates of any point p' in $\Delta(k)$ satisfy a linear relation

$$n_1(k)\, y_1 + n_2(k)\, y_2 + n_3(k)\, z = 0 \tag{3.9}$$

where $n_1(k)$, $n_2(k)$ and $n_3(k)$ are the components of the normal vector to $\Delta(k)$ in R_3. In differential form, (3.9) can be written as

$$dz = \zeta_1(k)\, dy_1 + \zeta_2(k)\, dy_2 \tag{3.10}$$

where

$$\zeta_1(k) = -\frac{n_1(k)}{n_3(k)} \quad \text{and} \quad \zeta_2(k) = -\frac{n_2(k)}{n_3(k)} . \tag{3.11}$$

By using (2.14), we see that the metric tensor on the lattice surface L is

$$g_{ij}(L) = \delta_{ij} + \zeta_i \zeta_j \tag{3.12}$$

where ζ_i is a constant inside each plaquette, and it changes from $\zeta_i(k)$ to $\zeta_i(k+1)$ when we move from one plaquette $\Delta(k)$ to a neighboring one $\Delta(k+1)$. By extending these considerations to the entire lattice, ζ_i is defined for all plaquettes in L. [In this section, the same cartesian coordinates y_1, y_2 and z are used for all points in R_3.] A comparison between (3.11)-(3.12) and (3.2)-(3.3) tells us that ζ_i is the same as z_i, but evaluated on L instead of $\mathcal{S}$. The corresponding lattice action is, in accordance with (2.16),

$$A_L = 2 \int_L (1 + \zeta_1^2 + \zeta_2^2)^{-3/2} \, d\zeta_1 \, d\zeta_2 . \tag{3.13}$$

For a finite lattice surface L, its boundary in the (ζ_1, ζ_2) plane is that of a polygon Ω. [See Figure 1.] Comparing (3.6) with (3.13), we see that if the corresponding enveloping surface $\mathcal{S}$ in the (z_1, z_2) plane is the same polygon, then

$$A_L = A_C . \tag{3.14}$$

Thus, through the mapping $y_1, y_2 \rightarrow \zeta_1, \zeta_2$ we have

$$L \rightarrow \Omega ; \tag{3.15}$$

by equating ζ_1, ζ_2 with z_1, z_2 at the boundary of Ω, we obtain through the inverse transformation $z_1, z_2 \to y_1, y_2$,

$$\Omega \to \mathcal{S} . \tag{3.16}$$

The equality (3.14), together with (1.1) and (2.16), gives (3.1):

$$2 \sum_p \epsilon_p = \int_{\mathcal{S}} \sqrt{|g|} \; R d^2 y$$

where the sum extends over all sites p in L. The precise boundary $\mathcal{C}$ of $\mathcal{S}$ is determined by (3.15)-(3.16).

These equalities (3.1) and (3.14) hold for an arbitrary lattice L and its enveloping surface $\mathcal{S}$, <u>without</u> taking any limit. Of course, they are also valid in the limit $\ell/\rho \to 0$ where, as before, ℓ and ρ denote the typical link-length in L and the radius of curvature in $\mathcal{S}$. Define

$$\delta z_i \equiv \zeta_i - z_i \tag{3.17}$$

where z_i and ζ_i are given by (3.3) and (3.11). Because of (3.4), the variation $\delta\Gamma_{kij}$ in (2.30) can be written as

$$\delta\Gamma_{kij} = \delta z_k \frac{\partial z_i}{\partial y_j} + z_k \frac{\partial \delta z_i}{\partial y_j} + \delta z_k \frac{\partial \delta z_i}{\partial y_j} . \tag{3.18}$$

In the sequence of mappings (3.15) and (3.16), $\zeta_i = z_i$ and therefore

$$\delta z_i = 0 \quad \text{at the boundary of } \Omega . \tag{3.19}$$

Hence, at the same boundary (3.18) becomes

$$\delta\Gamma_{kij} = z_k \frac{\partial \delta z_i}{\partial y_j} . \tag{3.20}$$

In (2.30), the integral of the total derivative term is

$$\int_{\mathcal{S}} \frac{\partial}{\partial y_i} (\delta\Gamma_{ijj} - \delta\Gamma_{jij}) d^2y = \oint_{\mathcal{C}} (z_i \frac{\partial \delta z_j}{\partial y_j} - z_j \frac{\partial \delta z_i}{\partial y_j}) ds_i \tag{3.21}$$

where $\mathcal{C}$ is the boundary of $\mathcal{S}$ obtained by the mapping (3.16). Hence, (3.19) implies that $\delta z_i = 0$ on $\mathcal{C}$. Since $\mathcal{C}$ is a closed curve, another partial integration gives

$$(3.21) = 0 .$$

Substituting it into (2.30), we derive (2.26):

$$\lim_{\ell \to 0} A_L = A_C , \tag{3.22}$$

which is hardly surprising since we have already proved $A_L = A_C$ before the limit. Nevertheless, this exercise does have certain virtues. When D is > 2, A_L and A_C are in general different if $\ell/\rho \neq 0$. The above example illustrates that under certain circumstances even for a finite L and without periodic or flat boundary conditions, the integral of the total derivative term in (2.30) can vanish, and thereby establish the equality between A_L and A_C in the limit $\ell/\rho \to 0$. Further examples will be given in the next section.

4. EXAMPLE: $\mathcal{S} = S_D$

In this section we examine the special enveloping surface

$$\mathcal{S} = S_D \quad , \tag{4.1}$$

a D sphere of radius ρ embedded in R_{D+1}, where D is $\geqslant 2$. As we shall see, these discussions will lead us to a number of remarkably simple geometrical properties concerning spherical simplices and deficit angles.

(a) When $D = 2$,

$$\mathcal{S} = S_2 \quad .$$

Consider a small region $\bar{\omega}$ on the lattice surface L ; inside $\bar{\omega}$, assume that there is only a single lattice site p. By using (3.7)-(3.11), we see that under the transformation $y_1, y_2 \rightarrow \zeta_1, \zeta_2$, the mapping (3.15) becomes (as illustrated in Figure 1)

$$\bar{\omega} \rightarrow \Omega \tag{4.2}$$

where, on account of (3.7), Ω is a polygon with m vertices, whose coordinates in the (ζ_1, ζ_2) plane are given by (3.11). Equating $\zeta_i = z_i$ at the boundary of Ω, the inverse transformation (3.16) gives

$$\Omega \rightarrow \omega \tag{4.3}$$

and that determines the corresponding small region ω on S_2. By construction, we have along the boundary of ω,

$$z_i = \frac{\partial z}{\partial y_i} = \zeta_i \quad . \tag{4.4}$$

For $\ell/\rho \ll 1$, the surface of the 2-sphere in the neighborhood near p is described by

$$z = -\frac{1}{2\rho}(y_1^2 + y_2^2) \quad . \tag{4.5}$$

In the (y_1, y_2) plane, it is easy to see that ω is just the Voronoi cell of p in the dual lattice. On S_2 , the scalar curvature R is $2/\rho^2$. Thus, (3.1) gives

$$\epsilon_p = \frac{\omega}{\rho^2}\left[1 + O(\frac{\ell^2}{\rho^2})\right] \tag{4.6}$$

where ϵ_p is the deficit angle at p and ω is the area of its dual cell in the flat (y_1, y_2) space. As we shall show, this simple result can be generalized to any higher dimension $D > 2$. However, before doing that, let us give an alternative but more elementary proof of (4.6).

Let 0 be the center of the 2-sphere which is embedded in R_3 . Points 1, 2 and 3 are on S_2 , forming a spherical triangle (123) on S_2 and a rectilinear triangle Δ_{123} in R_3 . Define

$$\begin{aligned}
&\measuredangle 1 = \text{spherical angle at 1 in (123)} ,\\
&\angle 1 = \text{(rectilinear) angle at 1 in } \Delta_{123} ,\\
&\overline{12}, \overline{13} \text{ and } \overline{01} = \text{length of the respective straight lines in } R_3 , \\
&\overline{123}, \overline{012} \text{ and } \overline{013} = \text{areas of the rectilinear triangles } \Delta_{123}, \Delta_{012} \text{ and } \Delta_{013}
\end{aligned} \tag{4.7}$$

and

$\overline{0123}$ = volume of the rectilinear tetrahedron τ_{0123}, with vertices 0, 1, 2 and 3.

It can be readily shown that

$$\sin \angle 1 = 2 \frac{\overline{123}}{\overline{12} \cdot \overline{13}}$$

and (4.8)

$$\sin \measuredangle 1 = \frac{3}{2} \frac{\overline{0123} \cdot \overline{01}}{\overline{012} \cdot \overline{013}} .$$

The latter follows because $\measuredangle 1$ is equal to the dihedral angle at the edge $\overline{01}$ in τ_{0123}.

Let $0'$ be the circumcenter of Δ_{123} (i.e., $0'$ is equidistant from 1, 2 and 3). The straight line $\overline{00'}$ is $\perp \Delta_{123}$. Denoting its length by $\overline{00'}$, we have

$$\overline{0123} = \frac{1}{3} \overline{00'} \cdot \overline{123} . \tag{4.9}$$

Let A and B be the midpoints of $\overline{12}$ and $\overline{13}$, and a, b and c denote the lengths of the straight lines $\overline{0'A}$, $\overline{0'B}$ and $\overline{0'1}$, as shown in Figure 2. They are related to the lengths $\overline{0A}$, $\overline{0B}$, $\overline{01}$ and $\overline{00'}$ by

$$\overline{0A}^2 = \overline{00'}^2 + a^2 ,$$

$$\overline{0B}^2 = \overline{00'}^2 + b^2 \tag{4.10}$$

and

$$\overline{01}^2 = \overline{00'}^2 + c^2 .$$

Since $\overline{0123} = \frac{1}{3} \overline{00'} \cdot \overline{123}$, $\overline{012} = \frac{1}{2} \overline{0A} \cdot \overline{12}$ and $\overline{013} = \frac{1}{2} \overline{0B} \cdot \overline{13}$, we have from (4.8)

$$\sin \measuredangle 1 = \frac{\overline{00'} \cdot \overline{01}}{\overline{0A} \cdot \overline{0B}} \sin \angle 1 . \tag{4.11}$$

As before, let ℓ be the typical length of any of the sides of the triangle. When ℓ/ρ is $\ll 1$, the difference $\sphericalangle 1 - \angle 1$ is $O(\ell^2/\rho^2)$; hence we may write

$$\sin \sphericalangle 1 = \sin \angle 1 + (\sphericalangle 1 - \angle 1) \cos \angle 1 + O(\frac{\ell^4}{\rho^4}) . \tag{4.12}$$

Combining this with (4.11), we obtain to $O(\ell^2/\rho^2)$

$$\sphericalangle 1 - \angle 1 = (\frac{\overline{OO'} \cdot \overline{OT}}{\overline{OA} \cdot \overline{OB}} - 1) \tan \angle 1 \tag{4.13}$$

in which, by definition, $\overline{OT} = \rho$. To the same order, by using (4.10) we find the quantity inside the parenthesis in (4.13) to be

$$\frac{1}{2\rho^2} (c^2 - a^2 - b^2) . \tag{4.14}$$

From Figure 2, it follows that

$$\tan \angle 1 = \frac{a'b + b'a}{a'b' - ab} \tag{4.15}$$

where $a' = \overline{1A}$ and $b' = \overline{1B}$. Since $c^2 = a^2 + a'^2 = b^2 + b'^2$, we derive, to $O(\ell^2/\rho^2)$,

$$\sphericalangle 1 - \angle 1 = \frac{1}{2\rho^2} (aa' + bb') \tag{4.16}$$

Let 1 be the lattice site p. We sum over all triangles that share p as a common vertex. The sum over all such spherical angles is 2π; the sum of the difference $\sphericalangle 1 - \angle 1$ is the deficit angle ϵ_p and that of $\frac{1}{2}(aa' + bb')$ is precisely the area ω of the Voronoi cell. Hence, (4.6) follows.

If we sum (4.16) over the three angles of Δ_{123} , then we obtain, to $O(\ell^2/\rho^2)$,

$$\measuredangle 1 + \measuredangle 2 + \measuredangle 3 = \pi + \frac{\overline{123}}{\rho^2} \tag{4.17}$$

which is the first-order expansion of the well-known exact relation in elementary spherical trigonometry:

$$\measuredangle 1 + \measuredangle 2 + \measuredangle 3 - \pi = \rho^{-2} \text{ (area of spherical triangle } 123\text{)}. \tag{4.18}$$

This familiar formula, however, has no generalization in a higher dimension. As we shall see, the approximate expressions (4.6) and (4.16) can both be generalized to any dimension.

(b) $D = 3$ and $\mathcal{S} = S_3$

Let 0 be the center of a 3-sphere of radius ρ , embedded in R_4. Points 1, 2, 3 and 4 are on S_3 ; they define a rectilinear tetrahedron $\tau_{1234} \equiv [1234]$ in R_4 and a spherical tetrahedron (1234) on S_3 . Define

$$\begin{aligned}
\measuredangle 12 &= \text{spherical dihedral angle at the arc } \overset{\frown}{12} \text{ in } (1234)\ , \\
\angle 12 &= \text{dihedral angle at } \overline{12} \text{ in } \tau_{1234}\ , \\
\overline{01234} &= \text{4-dimensional volume of the rectilinear 4-simplex } [01234] \\
&\qquad \text{in } R_4, \text{ whose vertices are } 0, 1, 2, 3 \text{ and } 4
\end{aligned} \tag{4.19}$$

and

$$\overline{1234} = \text{3-dimensional volume of } \tau_{1234}\ ;$$

all other notations such as $\overline{01}$, $\overline{12}$, $\overline{012}$, $\cdots$ are the same as in (4.7). We have, as in

(4.8),

$$\sin \angle 12 = \frac{3}{2}\, \frac{\overline{1234} \cdot \overline{12}}{\overline{123} \cdot \overline{124}}$$

and (4.20)

$$\sin \measuredangle 12 = \frac{4}{3}\, \frac{\overline{01234} \cdot \overline{012}}{\overline{0123} \cdot \overline{0124}} \quad .$$

The latter follows since $\measuredangle$ 12 is the "hyper-dihedral" angle at Δ_{012} between the tetrahedra τ_{0123} and τ_{0124} in the rectilinear 4-simplex $[01234]$ in R_4.

Let 0' be the circumcenter of τ_{1234}, A and B the circumcenters of the triangles Δ_{123} and Δ_{124}, and C the center (midpoint) of $\overline{12}$, as shown in Figure 3. All these figures are rectilinear. Just as in (4.11), by using (4.20) we derive

$$\sin \measuredangle 12 = \frac{\overline{00'} \cdot \overline{0C}}{\overline{0A} \cdot \overline{0B}} \sin \angle 12 \quad . \tag{4.21}$$

So far, (4.20) and (4.21) are exact. We shall now assume $\ell/\rho \ll 1$.
By definition, points 0', A, B and C lie on the same plane. As in Figure 2, we label the lengths of $\overline{0'A}$, $\overline{0'B}$, $\overline{AC}$, $\overline{BC}$ and $\overline{0'C}$ by a, b, a', b' and c. Following the steps from (4.10)-(4.16) and replacing 1 by C, we find, to $O(\ell^2/\rho^2)$,

$$\measuredangle 12 - \angle 12 = \frac{1}{2\rho^2} (aa' + bb') \quad . \tag{4.22}$$

Multiply both sides by $\overline{12}$. To the same order of ℓ^2/ρ^2,

$$\overline{12}(\measuredangle 12 - \angle 12) = \frac{3}{\rho^2} (\overline{0'A12} + \overline{0'B12}) \tag{4.23}$$

where $\overline{0'A12}$ and $\overline{0'B12}$ are the volumes of the tetrahedra $\tau_{0'A12}$ and $\tau_{0'B12}$.

Identify $\overline{12}$ with any link ℓ in the lattice, and sum over all tetrahedra that share the same ℓ. The sum of all such spherical dihedral angles $\measuredangle\, 12$ is 2π. Hence (4.23), after the sum, becomes

$$\bar{\ell}\,\epsilon_\ell = \frac{3}{\rho^2}\,\omega_\ell\,\left[1 + O\left(\frac{\ell^2}{\rho^2}\right)\right] \tag{4.24}$$

where ϵ_ℓ is the deficit angle around the link ℓ, $\bar{\ell}$ is its length and ω_ℓ is the 3-dimensional volume of the polyhedron whose vertices are the end-points 1, 2 of the link plus all the corners $0'$, $\cdots$ of its dual. [See Figure 4.]

Instead of the above, we may sum (4.23) over all six dihedral angles in a single tetrahedron τ_{1234}. Let ℓ run over the six links $\overline{12}$, $\overline{13}$, $\cdots$ of the tetrahedron; $\bar{\ell}$ denotes its length and δ_ℓ the difference between the spherical and rectilinear dihedral angles at ℓ:

$$\delta_\ell \equiv \measuredangle\,\ell - \angle\,\ell\,, \tag{4.25}$$

then from (4.23) we see that for any spherical tetrahedron of volume τ, the sum over its six links gives

$$\sum_\ell \bar{\ell}\,\delta_\ell = \frac{3\tau}{\rho^2}\,\left[1 + O\left(\frac{\ell^2}{\rho^2}\right)\right] \tag{4.26}$$

which is closely related to the Schlafli formula [5, 6].

(c) $\mathcal{S} = S_D$ with D arbitrary

The generalization to an arbitrary higher dimension D is straightforward. Let 0 be the center of a D sphere of radius ρ, which is embedded in R_{D+1}. Points

1, 2, · · , D + 1 are on S_D . They define a rectilinear D simplex $[1, 2, \cdots, D+1]$ in R_{D+1} and a spherical D simplex $(1, 2, \cdots, D+1)$ on S_{D+1} . Introduce s to be either the (D-2) rectilinear simplex $[1, 2, \cdots, D-1]$ or the corresponding D-2 spherical simplex $(1, 2, \cdots, D-1)$. As in (4.7) and (4.19), define

$$\begin{aligned} \measuredangle s &= \text{spherical hyper-dihedral angle at } s \text{ in } (1, 2, \cdots, D+1) , \\ \angle s &= \text{hyper-dihedral angle at } s \text{ in } [1, 2, \cdots, D+1] . \end{aligned} \tag{4.27}$$

Let O' be the circumcenter of the rectilinear D simplex $[1, 2, \cdots, D+1]$, A and B the circumcenters of the rectilinear (D-1) simplices $[1, 2, \cdots, D-1, D]$ and $[1, 2, \cdots, D-1, D+1]$, and C the circumcenter of the rectilinear D-2 simplex $[1, 2, \cdots, D-1]$. As in (4.11) and (4.21) we have

$$\sin \measuredangle s = \frac{\overline{OO'} \cdot \overline{OC}}{\overline{OA} \cdot \overline{OB}} \sin \angle s . \tag{4.28}$$

Again, points O', A, B and C lie on the same plane. Labeling the lengths of $\overline{O'A}$, $\overline{O'B}$, $\overline{AC}$, $\overline{BC}$ and $\overline{O'C}$ by a, b, a', b' and c, and following the same steps as in the previous two examples, we find, to $O(\ell^2/\rho^2)$,

$$\measuredangle s - \angle s = \frac{1}{2\rho^2} (aa' + bb') . \tag{4.29}$$

Multiply both sides by $\bar{s}$, the D-2 dimensional volume of s . To the same order of ℓ^2/ρ^2,

$$\bar{s} (\measuredangle s - \angle s) = \frac{D(D-1)}{2\rho^2} \left[\overline{O'A12\cdots(D-1)} + \overline{O'B12\cdots(D-1)} \right] \tag{4.30}$$

where the terms inside the square bracket denote the volumes of the respective rectilinear D simplices.

Identify s with any D-2 simplex in the lattice and sum over all D simplices that share the same s. Because the sum over all spherical hyper-dihedral angles $\measuredangle s$ is 2π, (4.30) leads to

$$\bar{s}\ \epsilon_s = \frac{D(D-1)\ \omega_s}{2\rho^2}\ \left[1 + O(\frac{\ell^2}{\rho^2})\right] \tag{4.31}$$

where ϵ_s is the deficit angle around s, $\bar{s}$ is its D-2 dimensional volume and ω_s is the D dimensional volume of the hyper-polyhedron whose vertices consist of those of s plus all the corners 0', $\cdots$ of its dual.

Equations (4.6), (4.24) and (4.31) give an interesting geometrical interpretation of the deficit angle when ℓ/ρ is small. Summing over all D-2 simplices in the lattice, we establish, once again, the equality between the lattice and continuum actions, A_L and A_C, in the limit $\ell \to 0$.

We may sum (4.30) over all D-2 simplices s in a given D simplex of volume v ; the result is

$$\sum_s \bar{s}\ \delta_s = \frac{D(D-1)}{2\rho^2}\ v\ \left[1 + O(\frac{\ell^2}{\rho^2})\right]\ , \tag{4.32}$$

which is the generalization of (4.26). In (4.32)

$$\delta_s = \measuredangle s - \angle s\ . \tag{4.33}$$

(d) Bianchi identity

The above expressions of deficit angles also lead to a simple geometrical interpretation of the Bianchi identity on the lattice when the enveloping surface is a D sphere, as we shall discuss.

Since the lattice surface L can be viewed as a special type of the continuum [1], the usual Bianchi identity should also be valid on L; that this is the case is well-known, as has been derived by T. Regge [2]. In the limit when the enveloping surface $\mathcal{S}$ has a very large radius of curvature ρ, the lattice Bianchi identity takes on a particularly simple form. For example, when $D = 3$, let p be any lattice site and $k = 1, 2, \cdots, m$ be the sites linked to p, as in (3.7). When $\ell/\rho \ll 1$, the Bianchi identity states

$$\sum_{k=1}^{m} \hat{\ell}^{o}_{pk} \, \epsilon_{pk} = 0 \tag{4.34}$$

where ϵ_{pk} is the deficit angle around the link ℓ_{pk} and $\hat{\ell}^{o}_{pk}$ is the unit vector along the link-vector in the local cartesian (y_1, y_2, y_3) frame at p. [See (2.4)-(2.5) for the definition of the local frame.]

When the enveloping surface $\mathcal{S}$ is S_3, identify the sites 1 and 2 in (4.19)-(4.21) as p and k; (4.22) becomes

$$\measuredangle\, pk - \angle\, pk = \frac{1}{2\rho^2}\,(aa' + bb')\left[1 + O\left(\frac{\ell^2}{\rho^2}\right)\right] . \tag{4.35}$$

Summing over all tetrahedra that share the same link ℓ_{pk}, we obtain

$$\epsilon_{pk} = \frac{\sigma_{pk}}{\rho^2}\left[1 + O\left(\frac{\ell^2}{\rho^2}\right)\right] \tag{4.36}$$

where σ_{pk} is the area dual to the link ℓ_{pk}. [In (4.24), set $\ell = \ell_{pq}$; σ_{pq} is then related to the volume ω_ℓ by $\sigma_{pk} = 3\omega_\ell/\ell$. In Figure 4, the area of the polygon perpendicular to the link ℓ_{12} is σ_{12}.] Thus, by applying the Gauss theorem to the Voronoi cell associated with p in the dual lattice, we derive (4.34). This relationship between the Bianchi identity and the Gauss theorem can be readily generalized to a higher dimension, as we shall see.

Let $\mathcal{S} = S_D$, where D is arbitrary, and assume $\ell/\rho \ll 1$. Select any D-3 simplex τ in the lattice. Any D - 2 simplex $s(\tau)$ with τ as part of its boundaries shares all the lattice sites of τ, plus one more, say k. Choose any point p in τ to be the origin and set up its local cartesian frame $(y_1, y_2, \cdots, y_D)$. Assume the $y_4, y_5, \cdots, y_D$ axes to be parallel to τ. Viewed in the remaining $(y_1, y_2, y_3) \equiv R_3$ subspace, τ becomes a single point p and $s(\tau)$ a line ℓ_{pk}; (4.29) now assumes the same form as (4.35), which leads to (4.36) and the Bianchi identity via the Gauss theorem.

5. GRAVITON PROPAGATOR

We now turn to the physical problem of wave propagation in lattice gravity. For a periodic wave solution, when the link-length $\ell \to 0$, we have (2.26)

$$\lim_{\ell \to 0} A_L = A_C ,$$

and therefore the spectrum of A_L is identical to A_C. To $O(\ell^2)$, the lattice action is

$$A_L = A_C + \Delta A \tag{5.1}$$

where ΔA is given by (2.28). In a perturbation calculation, the $O(\ell^2)$ correction to the spectrum of A_L can be obtained by evaluating the expectation value of A_L, using the zeroth order eigenfunction; i.e., those that satisfy

$$\frac{\delta A_C}{\delta h_{ij}} = \lambda h_{ij} \tag{5.2}$$

where, as before,

$$h_{ij} = g_{ij} - \delta_{ij} .$$

Throughout this section, we adopt the weak gravity approximation $h_{ij} << 1$, and consider only the case $D = 4$.

(a) Zeroth order solution

In the weak gravity approximation, the continuum action can be written as

$$A_C = \int d^D x \left[-\frac{1}{8} \left(\frac{\partial}{\partial x_i} h_{jk} - \frac{\partial}{\partial x_j} h_{ik} \right)^2 + \frac{1}{4} \left(\frac{\partial}{\partial x_i} h_{jj} - \frac{\partial}{\partial x_j} h_{ji} \right)^2 \right] . \tag{5.3}$$

It is convenient to fix the gauge so that h_{ij} satisfies the harmonic coordinate condition

$$\Lambda_j \equiv \frac{\partial}{\partial x_i}(h_{ij} - \tfrac{1}{2}\delta_{ij} h_{kk}) = 0 \ . \qquad (5.4)$$

In this gauge, we can add $-\frac{1}{2}\Lambda_j^2$ to A_C and therefore

$$A_C = A_C - \tfrac{1}{2}\Lambda_j^2 = \int d^4x \left[-\frac{1}{4}\left(\frac{\partial}{\partial x_i} h_{jk}\right)^2 + \frac{1}{8}\left(\frac{\partial}{\partial x_i} h_{jj}\right)^2 \right] . \qquad (5.5)$$

The eigenfunctions of (5.2) are just the wave solutions

$$h_{ij} = \epsilon_{ij}\, e^{ik_a x_a} \ . \qquad (5.6)$$

All these are standard [7, 8]. Here, we give only a very brief review in order to have the notations on hand for the subsequent first-order calculation in $k^2 \ell^2$.

The physical graviton is given by setting the eigenvalue 0 in (5.2). This gives

$$k^2 = 0 \ , \qquad (5.7)$$

the on-mass-shell condition. In the minkowski space, we can have $k^2 = 0$, but $k_i \neq 0$. Because $h_{ij} = h_{ji}$, for a given 4-momentum there are ten independent solutions, characterized by h_{ij}, with $i \leq j$. If we carry out the canonical quantization for the quadratic action (5.5), then of these ten modes, six are of positive metric and four of negative metric: Among the off-diagonal modes, h_{14}, h_{24} and h_{34} are of negative metric, but h_{12}, h_{23} and h_{13} are of positive metric. Only one of the diagonal modes, h_{ii}, is of negative metric; all three other modes are of positive metric. There are four gauge conditions (5.4) with $j = 1, \cdots, 4$. Each condition eliminates a pair of the positive and negative metric modes, leaving only $10 - 2 \cdot 4 = 2$ positive metric modes left which are the zero-mass spin 2 physical graviton solutions.

The off-mass-shell solutions

$$k^2 \neq 0$$

are relevant for the graviton-propagator $D_{ij,ab}(k)$, which is given by

$$k^{-2} \, , \tag{5.8}$$

the inverse of the eigenvalue, multiplied by the (gauge-dependent) tensor

$$\tfrac{1}{2}(\delta_{ia}\,\delta_{jb} + \delta_{ib}\,\delta_{ja} - \delta_{ij}\,\delta_{ab}) \, . \tag{5.9}$$

(b) First-order correction

To evaluate the $O(k^4 \ell^2)$ correction to the propagator, we need the zeroth order eigensolution for $k \neq 0$. Returning now to the euclidean space for convenience, we may choose the 4-vector $\vec{k}$ to be along the x_4-axis. Substituting (5.6) into the harmonic-coordinate condition (5.4), we find in our gauge

$$\epsilon_{14} = \epsilon_{24} = \epsilon_{34} = 0$$

and (5.10)

$$\epsilon_{44} = \epsilon_{11} + \epsilon_{22} + \epsilon_{33} \, .$$

Through a rotation in the (x_1, x_2, x_3) subspace, we can now diagonalize the matrix (h_{ij}). Hence, the zeroth-order wave solution (5.6) can be written as

$$(h_{ij}) = \begin{pmatrix} \epsilon_1 & & & \\ & \epsilon_2 & & \\ & & \epsilon_3 & \\ & & & \epsilon_4 \end{pmatrix} \sin(kx_4 + \theta_0) \tag{5.11}$$

where θ_0 is a constant and, because of (5.10),

$$\epsilon_4 = \epsilon_1 + \epsilon_2 + \epsilon_3 . \tag{5.12}$$

In order to set the lattice sites onto the continuum surface $\mathcal{S}$ whose metric is

$$g_{ij}(\mathcal{S}) = \delta_{ij} + h_{ij} ,$$

we must first embed $\mathcal{S}$ in an appropriate R_N. This can be achieved by assuming $N = 7$. In accordance with the notations used in (2.4), the cartesian coordinates of R_7 are labeled y_1, y_2, y_3, y_4 and z_1, z_2, z_3. We can parametrize the surface $\mathcal{S}$ by

$$z_1 = \tfrac{1}{2}\kappa x_1^2 - \frac{\epsilon_1}{2\kappa} \sin(kx_4 + \theta_0) ,$$

$$z_2 = \tfrac{1}{2}\kappa x_2^2 - \frac{\epsilon_2}{2\kappa} \sin(kx_4 + \theta_0) \tag{5.13}$$

and

$$z_3 = \tfrac{1}{2}\kappa x_3^3 - \frac{\epsilon_3}{2\kappa} \sin(kx_4 + \theta_0)$$

where κ is given by

$$\kappa^2 = \tfrac{1}{2}k^2(\epsilon_1^2 + \epsilon_2^2 + \epsilon_3^2)^{\frac{1}{2}} . \tag{5.14}$$

The y_i are related to the x_i by

$$
\begin{aligned}
y_1 &= x_1\left[1+\tfrac{1}{2}\epsilon_1 \sin(k x_4+\theta_0)\right] - \tfrac{1}{6}\kappa^2 x_1^3 \ , \\
y_2 &= x_2\left[1+\tfrac{1}{2}\epsilon_2 \sin(k x_4+\theta_0)\right] - \tfrac{1}{6}\kappa^2 x_2^3 \ , \\
y_3 &= x_3\left[1+\tfrac{1}{2}\epsilon_3 \sin(k x_4+\theta_0)\right] - \tfrac{1}{6}\kappa^2 x_3^3
\end{aligned}
\qquad (5.15)
$$

and

$$
\begin{aligned}
y_4 &= x_4 - \frac{\epsilon_4}{2k}\left[\cos(k x_4+\theta_0) - \cos\theta_0\right] \\
&\quad - \frac{1}{8}(\epsilon_1^2+\epsilon_2^2+\epsilon_3^2)^{\frac{1}{2}} \left\{x_4 + \frac{1}{2k}\left[\sin 2(kx_4+\theta_0) - \sin 2\theta_0\right]\right\}
\end{aligned}
$$

One can readily verify that

$$
\begin{aligned}
ds^2 &\equiv \sum_{i=1}^{4} dy_i^2 + \sum_{a=1}^{3} dz_a^2 \\
&= \sum_{i=1}^{4} dx_i^2\left[1+\epsilon_i \sin(kx_4+\theta_0) + O(\epsilon^2)\right] \ ,
\end{aligned}
\qquad (5.16)
$$

which agrees with (5.11) in the weak field approximation.

We consider a "rectangular" box of volume

$$
\Omega = \frac{\lambda}{n} \times \frac{\lambda}{n} \times \frac{\lambda}{n} \times \lambda \ . \qquad (5.17)
$$

The center of the box is placed at the origin $x_1 = x_2 = x_3 = x_4 = 0$; its sides are parallel to the x_i axes with the length along the x_4 direction to be the wavelength

$$
\lambda = 2\pi/k \qquad (5.18)
$$

and those along the x_1, x_2 and x_3 axes to be all λ/n, where n is a positive integer, which will take on a wide range of values in our calculations.

(i) Regular lattice

Divide the box (5.17) into n hyper-cubic cells of volume $(\lambda/n)^4$ each. Place a lattice site at each corner of these cells. Thus, the site-density is

$$\sigma \equiv \ell^{-4} \tag{5.19}$$

where

$$\ell = \lambda/n \quad . \tag{5.20}$$

Each cell is in turn divided into 4! simplices, with the same division pattern repeated in all cells. Take any lattice site, say p. Let $y_i(p)$ and $z_\alpha(p)$ be its coordinates calculated from (5.13) and (5.15). The link-length ℓ_{pq} is given by

$$\ell_{pq}^2 = \sum_{i=1}^{4} [y_i(p) - y_i(q)]^2 + \sum_{\alpha=1}^{3} [z_\alpha(p) - z_\alpha(q)]^2 , \tag{5.21}$$

from which the deficit angles and the lattice action can be computed numerically.

For the wave solution (5.11), the continuum action (5.3), or its alternative form (5.5), integrated over the volume (5.17) is

$$A_C = \frac{1}{4} k^2 (\epsilon_1 \epsilon_2 + \epsilon_2 \epsilon_3 + \epsilon_3 \epsilon_1) \, \Omega \tag{5.22}$$

where $\Omega = \lambda^4/n^3$. By varying n from 50 to 200 and ϵ_i from 0.15 to 0.01, we determine the lattice action (1.1) to be

$$A_L = A_C [1 + \beta k^2 \ell^2 + O(\epsilon^2) + O(k^4 \ell^4)] \tag{5.23}$$

where A_C refers to the continuum action in its quadratic approximation (5.22) and

$$\beta \cong -0.0835 \ . \qquad (5.24)$$

The $O(\epsilon^2)$ correction in (5.23) is found to be of the form $a(\epsilon_1\epsilon_2 + \epsilon_2\epsilon_3 + \epsilon_3\epsilon_1)$ with $a \cong 0.103$. Thus, the inverse of the graviton propagator on such a hyper-cubic lattice when its 4-momentum $\vec{k}$ is parallel to one of the sides of the box is modified from k^2, the continuum expression, to

$$k^2 + \beta k^4 \ell^2 + O(k^6 \ell^4) \ . \qquad (5.25)$$

It is of interest to note that for a zero-mass spin 0 field the corresponding parameter in its propagator on a hyper-cubic lattice is

$$\beta_0 = -1/12 = -0.08333 \cdots \ . \qquad (5.26)$$

The difference between these two β is beyond the resolution of our numerical calculation, which suggests that β might be spin-independent.* [This is all the more remarkable, since for different lattice structures, even the functional forms of the $O(k^4 \ell^2)$ correction are not the same. Cf. (5.27) and (5.32) below.] For $\vec{k}$ that lies in an arbitrary direction, we expect (5.25) to be replaced by

$$\sum_{i=1}^{4} \left[k_i^2 + \beta k_i^4 \ell^2 + O(k_i^6 \ell^4) \right] , \qquad (5.27)$$

---

* We wish to thank Dr. Ruth Williams, who has kindly informed us that by using Eq. (D10) of the second paper in Ref. [4] it is possible to establish analytically $\beta = -1/12$ for the graviton propagator.

as in the spin 0 case.

(ii) Random lattice

We distribute in the same box (5.17) $m^4 n$ random lattice sites. [In the previous case of a regular lattice, $m = 1$.] As in (5.19), the site-density is

$$\sigma = \ell^{-4} ,$$

but now with

$$\ell = \lambda / mn \tag{5.28}$$

instead of (5.20). These sites are linked according to the algorithm given in Ref. 9. To avoid unnecessary complications due to boundaries, we set an appropriate number of random lattice sites on the (3-dimensional) surface of the box, so that no link crosses the surface. In this way, we can envisage the whole space filled by an ensemble of such boxes of random lattices, each of identical size (5.17). They are all placed side by side next to each other. Since lattice sites in different boxes are not linked directly, but are connected only through the common sites on the surfaces of these boxes, the action of the ensemble becomes the sum of the individual ones. Clearly, if the total number $m^4 n$ of lattice sites within each box is large, such an arrangement does give a good representation of the random lattice.

We take $m = 3$ and vary n from 20 to 40, ϵ_i from 0.04 to 0.01. The lattice action (1.1) is found to be, for $\epsilon_1 = \epsilon_2 = \epsilon_3$,

$$A_L = A_C \left[1 + \beta k^2 \ell^2 + O(\epsilon^2) + O(k^4 \ell^4) \right] \tag{5.29}$$

where A_C refers, as before, to the continuum action in its quadratic approximation

(5.22) and, after averaging over several random lattices,

$$\beta = -0.14 \pm .02 \quad . \tag{5.30}$$

The dependence of β on m is being investigated and will be reported elsewhere. The inverse of the graviton propagator is now modified from k^2, the continuum limit, to

$$k^2 + \beta k^4 \ell^2 + O(k^6 \ell^4) \quad . \tag{5.31}$$

So far, the direction $\vec{k}$ is assumed to be parallel to one of the sides of each box (5.17). Because the random lattice maintains rotational symmetry, when the direction of $\vec{k}$ is arbitrary, (5.31) becomes

$$k^2 + \beta (k^2)^2 \ell^2 + O(k^6 \ell^4) \tag{5.32}$$

with $k^2 = \sum_1^4 k_i^2$, in contrast to (5.27).

For the $O(k^4 \ell^2)$ correction calculated here, it suffices to use only the zeroth order eigenfunction. The evaluation of the next $O(k^6 \ell^4)$ correction is more complicated, since then it would be necessary to employ the perturbed eigenfunction.

6. SYMMETRY

The quintessence of Einstein's theory of general relativity lies in its invariance under a general coordinate transformation

$$x_i \rightarrow x_i' \tag{6.1}$$

that leaves ds^2 unchanged. As remarked before, this invariance also holds for the discrete gravity action (1.1); the lattice can be completely arbitrary, regular or irregular, and its links can be of any lengths ℓ , long or short. In addition, as we shall see, the lattice theory enjoys still another totally new class of symmetries which does not exist in the usual continuum theory. Aesthetically, this adds greatly to the appeal of lattice gravity. For physical applications, when ℓ is small, our general formula (2.26) insures that all known tests of general relativity are automatically satisfied. Furthermore, by keeping ℓ nonzero, we see that the lattice action A_L per volume possesses only a finite degree of freedom. The normal difficulty of ultra-violet divergence that one encounters in quantum gravity disappears in the lattice theory. All these strongly suggest that the lattice theory with a nonzero ℓ may be more fundamental. The usual continuum theory is very likely only an approximation.

To amplify the aforementioned symmetry properties, let us consider a finite lattice in which there are $\mathcal{N}_0$ sites (0-simplices) and $\mathcal{N}_1$ links (1-simplices). The lattice action (1.1)

$$A_L = 2 \sum_s \bar{s}\ \epsilon_s \tag{6.2}$$

is a function of the $\mathcal{N}_1$ link-lengths ℓ_{pq}. We write

$$A_L = A_L(\ell_{pq}) \quad . \tag{6.3}$$

The same lattice can be embedded in a flat euclidean space R_N, provided that the various simplicial inequalities are satisfied by all link-lengths, and

$$N\,n_0 - n_1 \geq N + \tfrac{1}{2}N(N-1) \tag{6.4}$$

where the right-hand side refers to the overall translational and rotational degrees of freedom of the whole lattice in R_N. When the lattice is very large, the right-hand side can be neglected, and (6.4) reduces to

$$N\,n_0 \geq n_1 \quad . \tag{6.5}$$

For a random D dimensional simplicial lattice, the minimum N is 3 for $D = 2$, 7 for $D = 3$ and 19 for $D = 4$. In this embedding, the link-length ℓ_{pq} becomes equal to the rectilinear distance (2.2):

$$\ell_{pq}^{\;2} = [\vec{r}(p) - \vec{r}(q)]^2 \tag{6.6}$$

where, as before, $\vec{r}(p)$ is the cartesian coordinate vector of the lattice site p in R_N. Hence, A_L can also be viewed as a function of the n_0 coordinate vectors $\vec{r}(p)$, or their $N\,n_0$ components:

$$A_L = A_L(\vec{r}(p)) \quad . \tag{6.7}$$

As in Sec. 2, we can fill the interior of all simplices in the lattice by points in the flat space R_N. The lattice then forms a continuous, but sharply ridged, surface L. In I, it is proved that the lattice action (6.2) is identical to the integral

$$A_L = \int_L \sqrt{|g|}\ R\ d^D x \qquad (6.8)$$

evaluated on the lattice surface L. Consequently, under an arbitrary coordinate transformation (6.1), the lattice action (6.8) enjoys the same invariance as the usual continuum theory.

We may also characterize the lattice by other means of parametrization. As in Sec. 2, we assume these $\mathcal{N}_0$ lattice sites to lie on a D dimensional smooth enveloping surface $\mathcal{S}$, which is also embedded in the same R_N. Analytically, $\mathcal{S}$ can be represented by $n = N - D$ functions z_α, each of which depends on D variables y_i:

$$z_\alpha = z_\alpha(y) \qquad (6.9)$$

with $\alpha = 1, 2, \cdots, n$ and y standing for

$$y_i = y_1, y_2, \cdots, y_D,$$

the D cartesian coordinates of any of the flat subspace R_D in R_N. The z_α are simply the remaining cartesian coordinates of R_N, as in (2.4). Given the positions of the lattice sites p on the surface $\mathcal{S}$ [or, equivalently, their $y_i(p)$ coordinates], we can use (6.9) to evaluate ℓ_{pq}. Equation (6.6) becomes

$$\ell_{pq}^2 = \sum_{i=1}^{D} [y_i(p) - y_i(q)]^2 + \sum_{\alpha=1}^{n} [z_\alpha(y(p)) - z_\alpha(y(q))]^2 . \qquad (6.10)$$

Substituting it into (6.3), we can regard

$$A_L = A_L(z_\alpha(y), y_i(p)) \qquad (6.11)$$

as a functional of n functions $z_a(y)$ and $D\mathcal{N}_0$ variables $y_i(p)$. Since the original A_L is a function only of $\mathcal{N}_1$, or $N\mathcal{N}_0$, discrete variables, this new representation (6.11) greatly increases the number of independent degrees of freedom. This is similar to a gauge theory, in which the number of mathematical variables introduced far exceeds the actual number of physical (i.e., gauge-invariant) quantities. Consequently, we must have additional symmetry, analogous to gauge invariance. We can now consider new transformations such as

$$z_a(y) \rightarrow z_a'(y) \tag{6.12}$$

and

$$y_i(p) \rightarrow y_i'(p) \ . \tag{6.13}$$

The former changes the enveloping surface from $\mathcal{S}$ to $\mathcal{S}'$ and the latter moves the positions of the lattice sites on $\mathcal{S}$. The action $A_L(z_a(y), y_i(p))$ is clearly invariant if these transformations leave all link-lengths ℓ_{pq} unchanged.

When the link-lengths ℓ are much smaller than ρ, the typical radius of curvature of $\mathcal{S}$, this new symmetry can be stated more succinctly. Suppose all $O(\ell^2/\rho^2)$ corrections are neglected. According to (2.26),

$$\lim_{\ell \to 0} A_L = \int_{\mathcal{S}} \sqrt{|g|}\ R\, d^D y \tag{6.14}$$

which is independent of $y_i(p)$ and therefore invariant under the arbitrary motion (6.13) of the lattice sites on $\mathcal{S}$. In addition (6.14) is also invariant under (6.12), $\mathcal{S} \rightarrow \mathcal{S}'$, provided that it leaves $\sqrt{|g|}\ R\, d^D y$ unchanged.

When $O(\ell^2/\rho^2)$ corrections are included, then on account of (2.27)-(2.28), A_L depends on the lattice structure only through certain averages. The invariance property of A_L under (6.12)-(6.13) remains, if they leave these averages unchanged. The independent variations of the lattice points p , detached from those of their underlying manifold $\mathcal{S}$, represent degrees of freedom outside the usual continuum theory.

7. GENERALIZATION

So far, for the lattice theory we have restricted ourselves to the Regge action (1.1), and for the continuum theory to the Einstein action (1.2). On both sides, there are generalizations.

For the continuum theory, we may add to Einstein's action a large variety of invariants, e.g., R^2 or other nonlinear functions of the curvature tensor. By construction, all such modifications enjoy the same general coordinate invariance under (6.1). Their consistency with experimental tests is also guaranteed, since in weak gravity the linear Einstein action R dominates, and that is where all tests concentrate. However, if we regard the lattice theory (with nonzero ℓ) as more fundamental, then Einstein's theory plays a unique role, as will be discussed.

In this context, we should view the continuum theory as an approximation. For a given lattice surface L, we may envisage a series of smooth continuum surfaces which approach L : in the limit

$$\int_L \sqrt{|g|}\ R\, d^D x = 2 \sum_s \bar{s}\, \epsilon_s \quad .$$

By following the same argument given in I, it is straightforward to show that the integrals of $\sqrt{|g|}\ R^2$, $\sqrt{|g|}\ R_{\mu\nu\alpha\beta} R^{\mu\nu\alpha\beta}$, $\sqrt{|g|}\ R^3$, etc. all diverge in the same limit. Therefore, among all such candidates for the continuum approximation, Einstein' theory stands out uniquely. *

---

* For example, the absence of quadratic divergence in $a R^{\mu\nu\alpha\beta} R_{\mu\nu\alpha\beta} + 2b R^{\mu\nu} R_{\mu\nu} + c R^2$ requires $a + b + c = 0$; the absence of linear divergence requires $a = c = -\frac{1}{2} b$ which when $D = 4$ is related to the constant in the Gauss-Bonnet theorem. We wish to thank S. D. Drell for a discussion.

For the lattice theory, Regge's action can also be modified. We may add to $\bar{s}\,\epsilon_s$ any nonlinear functions of ϵ_s. Some explicit candidates are given in Ref. [10]. Since the deficit angles ϵ_s depend only on the link-lengths ℓ_{pq}, any transformations on L, or on $y_i(p)$ and the enveloping surface $\mathcal{S}$, that leave ℓ_{pq} unchanged would also leave A_L invariant. Hence, all these modified discrete actions can have the same invariance properties as the Regge action (1.1). When ℓ/ρ approaches zero, $\epsilon_s = O(\ell^2/\rho^2) \to 0$, $\bar{s}\,\epsilon_s$ dominates and therefore all these different discrete theories approach the same continuum limit, the Einstein action (1.2); the limiting continuum gravity can be of any strength, weak* or strong. The same reasoning also insures that all these discrete theories are consistent with experiment, provided that ℓ is of microscopic length. Hence, the selection of the correct one must depend on other criteria, yet to be uncovered.

ACKNOWLEDGMENT

We wish to thank T. Regge and J. A. Wheeler for discussions during the fall of 1983 when they visited us at Columbia University.

---

* By weak gravity, one means that $h_{ij} = g_{ij} - \delta_{ij}$ is small everywhere with a global choice of coordinates. The ratio ℓ/ρ is between the link-length and the radius of curvature which is determined locally. This is why $\ell/\rho \ll 1$ can be achieved even for strong gravity.

REFERENCES

[1]. R. Friedberg and T. D. Lee, "Derivation of Regge's Action from Einstein's Theory of General Relativity", Nucl.Phys. B242, 145 (1984).

[2]. T. Regge, Nuovo Cimento 19, 558 (1961).

[3]. C. Y. Wong, J.Math.Phys. 12, 70 (1971).

[4]. M. Rocek and R. M. Williams, Phys.Lett. 104B, 31 (1981); Zeits.f.Phys. C21, 171 (1984).

[5]. L. Schläfli, Quarterly J. Math. 2, 269 (1858).

[6]. H. S. M. Coxeter, Non-Euclidean Geometry (Toronto, University of Toronto Press, 1978).

[7]. A. Einstein, Berliner Berichte 1918, 154.

[8]. G. Wentzel, Quantum Theory of Fields (New York, Interscience Publishers, 1949).

[9]. N. H. Christ, R. Friedberg and T. D. Lee, Nucl.Phys. B202, 89 (1982).

[10]. T. D. Lee, "Discrete Mechanics", to be published in the proceedings of the International School of Subnuclear Physics, Erice, 1983.

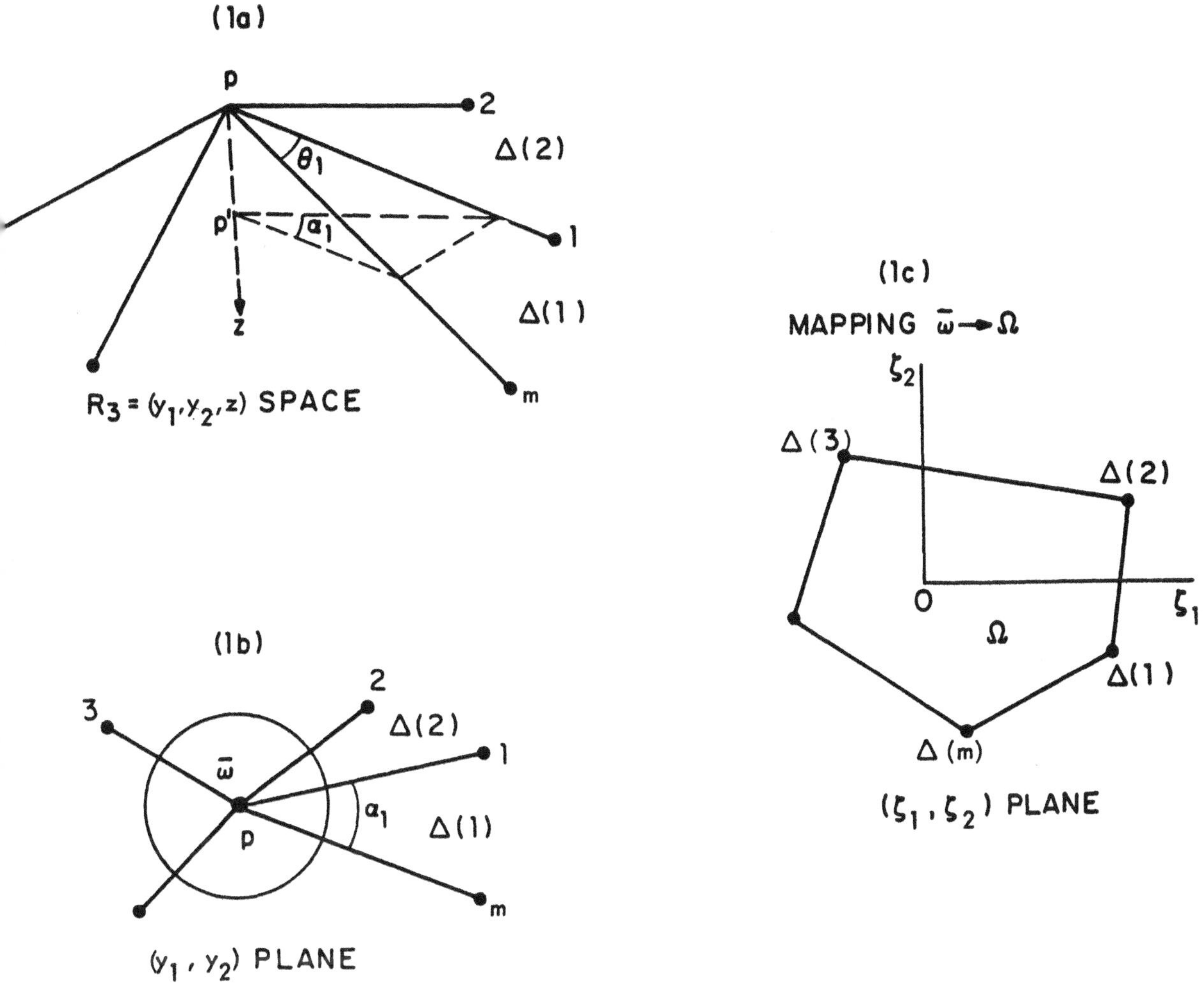

FIGURE 1. Example when the lattice space L is $\bar{\omega}$, a small region near the lattice site p . The deficit angle at p is $\epsilon_p = 2\pi - \sum_{i=1}^{m} \theta_i$.

(1a) $L = \bar{\omega}$ viewed in the flat $R_3 = (y_1, y_2, z)$ space,

(1b) projection of $\bar{\omega}$ onto the (y_1, y_2) plane, and

(1c) the mapping $\bar{\omega} \to \Omega$ from the (y_1, y_2) plane to the (ζ_1, ζ_2) plane. [See (3.15) and (4.2).]

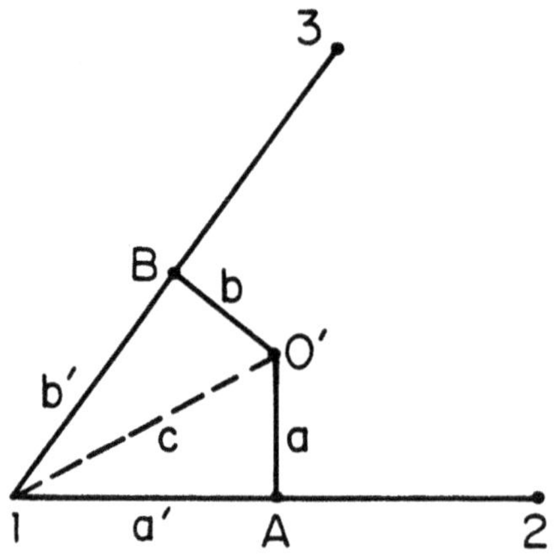

FIGURE 2. Point 0' is the circumcenter of Δ_{123} ; i.e. $\overline{0'1} = \overline{0'2} = \overline{0'3}$.

A and B are midpoints of $\overline{12}$ and $\overline{13}$.

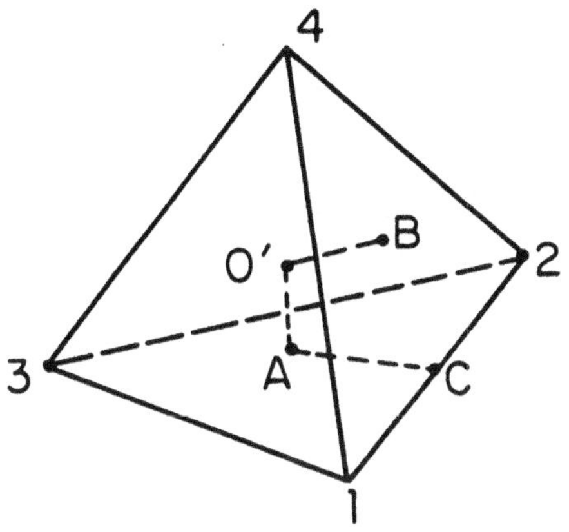

FIGURE 3. Point O' is the circumcenter of the tetrahedron τ_{1234}, points A and B are the circumcenters of Δ_{123} and Δ_{124}, and C is the midpoint of $\overline{12}$.

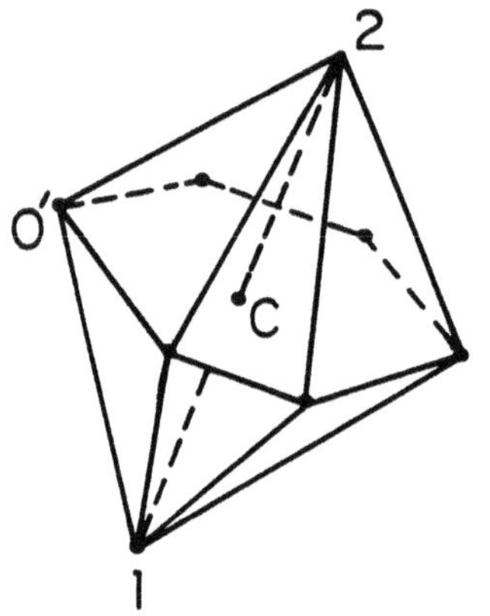

FIGURE 4. Point C is the midpoint of link $\ell = \overline{12}$. 0' is the corner of the polygon which is dual to ℓ_{12}. The polyhedron ω_ℓ in (4.24) refers to the one whose vertices are the end-points 1, 2 of the link ℓ plus the corners 0', $\cdots$ of its dual.

CU - TP - 324

PHYSICS IN TERMS OF DIFFERENCE EQUATIONS

T. D. Lee

Columbia University, New York, N. Y. 10027

A talk given at the Niels Bohr Centenary Symposium, Copenhagen, October 5, 1985.

This research was supported in part by the U.S. Department of Energy.

I. Introduction

The invention of quantum mechanics by Niels Bohr, and others, brought a profound change in our thinking. As a result of quantum mechanics, all dynamics is now described in terms of observables, and each observable is expressed as an operator or a matrix. Its eigenvalues denote the possible outcome of any measurements on the observable, with a probability amplitude given by the appropriate component of the state vector. These ideas were truly revolutionary at the time of their creation. By now they are accepted universally and taught to students everywhere. Mankind will be forever grateful to Niels Bohr, the originator and the developer of quantum mechanics.

Nearly thirty years ago Pauli came to Columbia University to give a seminar on his joint work with Heisenberg. Many people were in the audience, including Niels Bohr. There were a lot of questions, discussions and criticisms, some not so friendly. In the end Pauli said, in an uncharacteristic way: "Perhaps my idea is crazy." Whereupon Bohr stood up, walked a few steps towards Pauli, looked down, pointed a finger at Pauli and said, "The problem is not whether your idea is crazy, but whether it is crazy enough."

It is in this spirit that I wish to analyze further some of the basic concepts of quantum mechanics. Let me first raise two questions:

> In a measurement of, say, the electric field $\vec{E}(\vec{r}, t)$, is it not true that the precise values of $\vec{r}$ and t, like $\vec{E}$, can only be determined after the measurement?
>
> Is the concept of local field theory applicable to very small distances?

1. To take up the first question, consider, say, an experimentalist who proposes to the CERN program committee to measure the electric field in an e^+e^- collision at LEP in 1989. The proposal is approved. The experimentalist and his several hundred collaborators then set up their gigantic instruments. Before the measurement, they have some expectation of the amplitude of the electric field $\vec{E}$ that they are going to measure. They also know the approximate location and the time that the collision will take place. The precise value of $\vec{E}$, the exact location of the reaction and the precise time of the collision can only be determined after his measurement. In other words, in terms of observation, we do treat space-time on the same footing as the electric field. Therefore, in terms of conceptual thinking, shouldn't we also treat space-time on the same footing as the electric field? Wouldn't it be more in the spirit of Niels Bohr to regard space and time both as dynamical variables, the same as the electric field?

In the conventional description of local quantum field theory, we view the fields as the dynamical variables, represented by operators, but space $\vec{r}$ and time t are only parameters. Even in Einstein's general relativity, although the metric is a dynamical variable, the continuous four-dimensional space-time that the metric is embedded in is not. In this talk I shall explore the alternative view, treating space and time as dynamical variables, playing a similar role as the electromagnetic field, the gluon field, etc. As we shall see, this new description has a connection with my second question.

2. Let me present an argument to show that the concept of local field theory is possibly inapplicable to distances of the order of the Planck length ℓ_p. In Figure 1

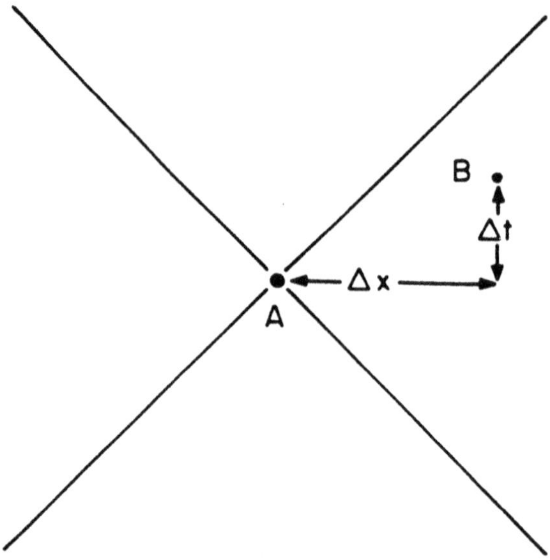

Figure 1.

we consider two points A and B , separated by a spatial distance Δx and a time difference Δt with $(\hbar = c = 1)$

$$\Delta t < \Delta x < \ell_p \sim 10^{-33} \text{ cm}$$

in which the first inequality insures that A and B are outside each other's light cone. Local field theory then states that these two experiments can be done independently of each other, no matter how close the points. Yet, just based on the uncertainty principle, we expect a fluctuation of energy ΔE caused by these two measurements:

$$\Delta E \sim \frac{1}{\Delta t} > \frac{1}{\ell_p} .$$

The gravitational field associated with such a ΔE is very strong at small distances.

Indeed, its Schwarzschild radius (i.e., black hole radius) is

$$R \sim G\,\Delta E > \frac{G}{\ell_p}$$

where $G = \ell_p^{\,2}$ is Newton's gravitational constant. Thus we find

$$R > \Delta x \; !$$

It seems quite unreasonable that these two measurements A and B could be viewed as independent. Consequently the concept of locality very likely breaks down at such distances. If locality is not satisfied at the Planck length, then the correct physical theory must be non-local in character. The fact that the Planck length is small is beside the point. We would like this non-local fundamental theory to retain all the good features of the usual continuum theory: Lorentz invariance, Poincaré invariance, non-abelian gauge symmetries, unitarity and the general coordinate invariance of general relativity. In addition it should not contain divergence difficulties, so that quantization of gravity can be carried out.

There are perhaps two different directions one may follow: one is to add degrees of freedom to the usual local field theory. The other is to subtract degrees of freedom. The former is followed by those working on string theories, and the latter will be the subject of my talk.

II. Time As A Dynamical Variable

In this new theory I shall treat time as a dynamical variable.[1] This will lead to a dynamics which is formulated in terms of difference equations, instead of the usual differential equations. We will first review briefly the classical theory of this new mechanics, called discrete mechanics, and then go over to the quantum theory.

II.1. Classical mechanics

Take the simplest example of a non-dimensional nonrelativistic particle of unit mass moving in a potential $V(x)$. In the usual continuum mechanics the action is

$$A(x(t)) = \int_0^T \left[\tfrac{1}{2}\dot{x}^2 - V(x)\right] dt \quad , \tag{2.1}$$

where $x(t)$ can be any smooth function of the time t. Keeping fixed the initial and final positions, say x_0 and x_f, at $t = 0$ and T, we determine the orbit of the particle by the stationary condition

$$\frac{\delta A}{\delta(x(t))} = 0 \tag{2.2}$$

which leads to Newton's equation

$$\ddot{x} = -dV/dx \quad . \tag{2.3}$$

In the above, x is the dynamical variable and t is merely a parameter. Next, we shall see how this customary approach may be modified in the discrete version.

Let the initial and final positions of the particle be the same

$$x_0 \text{ at } t = 0 \qquad \text{and} \qquad x_f \text{ at } t = T \quad . \tag{2.4}$$

In the discrete mechanics we restrict the usual smooth path $x(t)$ to a "discrete path" $x_D(t)$, which is continuous but piecewise linear, characterized by N vertices (as shown in Figure 2). In 2(a) we have the usual smooth path $x(t)$ of a nonrelativistic particle in classical mechanics. Moving along $x(t)$ from $t = 0$ to $T > 0$, the time t increases monotonically; this property is retained under the constraint restricting $x(t)$ to $x_D(t)$. Thus, as in Figure 2(b), we may label the N vertices of $x_D(t)$ consecutively as $n = 1, 2, \cdots, N$, each of which carries a space-time position x_n and t_n with

$$0 < t_1 < t_2 < t_3 < \cdots < t_N < T . \qquad (2.5)$$

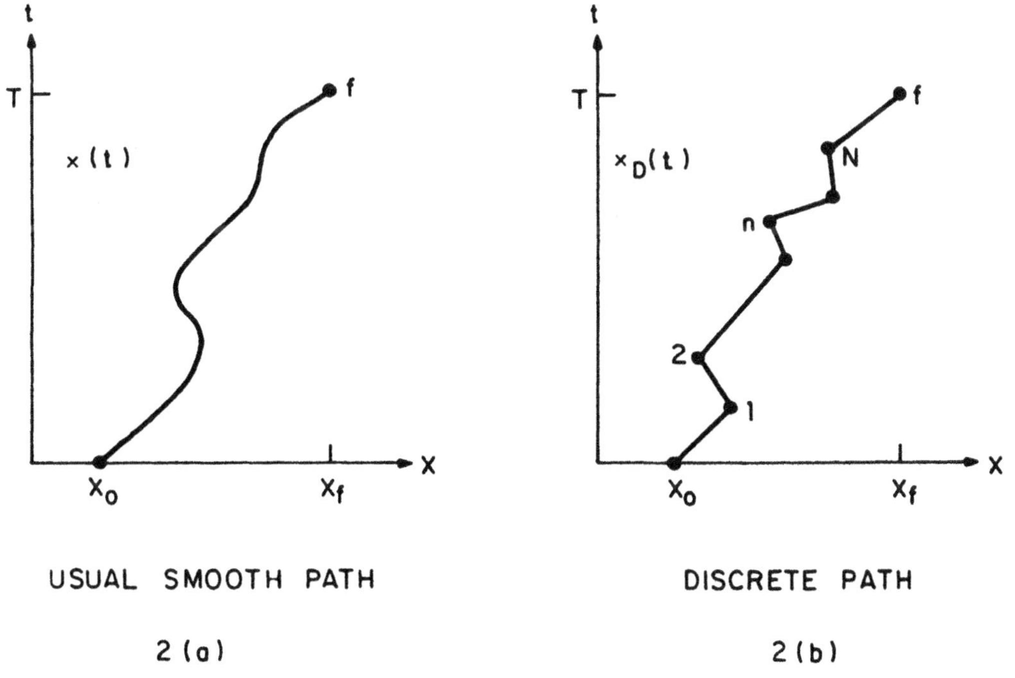

Figure 2.

The nearest neighboring vertices are linked by straight lines, forming the discrete path $x_D(t)$, which also appears as a one-dimensional lattice with n as lattice sites. In 2(b), a variation of the space-time positions of these vertices changes the discrete path $x_D(t)$. However, a mere exchange of any two vertices clearly defines the same $x_D(t)$. This is because only the discrete path with unlabeled vertices has a physical meaning. There is no "individual" identity of any of the vertices. Thus, the time-ordered sequence (2.5) is not an additional restriction, but one that arises naturally when we pass from the usual $x(t)$ to the discrete $x_D(t)$.

In the following, we shall keep the site-density

$$\frac{N}{T} \equiv \frac{1}{\ell} \tag{2.6}$$

fixed, and regard ℓ as a fundamental constant of the theory. The action integral (2.1) evaluated on such a discrete path $x_D(t)$ is

$$A_D = A(x_D(t)) = \sum_n \left[\frac{1}{2}\frac{(x_n - x_{n-1})^2}{t_n - t_{n-1}} - (t_n - t_{n-1})\,\bar{V}(n)\right] \tag{2.7}$$

where

$$\bar{V}(n) = \frac{1}{x_n - x_{n-1}} \int_{x_{n-1}}^{x_n} V(x)\,dx \tag{2.8}$$

is the average of $V(x)$ along the straight line between x_{n-1} and x_n.

Because the path $x_D(t)$ is completely specified by its vertices $n(x_n, t_n)$, a variation in $x_D(t)$ is equivalent to a variation in all the positions of its vertices

$$d\left[x_D(t)\right] = \prod_n \left[dx_n\right]\left[dt_n\right] . \tag{2.9}$$

Correspondingly, the dynamical equation (2.2) becomes the difference equations

$$\frac{\partial A_D}{\partial x_n} = 0 \tag{2.10}$$

and

$$\frac{\partial A_D}{\partial t_n} = 0 \ . \tag{2.11}$$

We see that in this new mechanics the roles of x_n and t_n are quite similar. Both appear as dynamical variables. For each x_n or t_n we have one difference equation, (2.10) or (2.11). The former gives Newton's law on the lattice and the latter gives the conservation of energy

$$E_n \equiv \frac{1}{2}\left(\frac{x_n - x_{n-1}}{t_n - t_{n-1}}\right)^2 + \bar{V}(n) = E_{n+1} \ . \tag{2.11'}$$

In the usual continuum mechanics, conservation of energy is a consequence of Newton's equation. Here, these two equations (2.10) and (2.11) are independent. Altogether there are 2N such equations, matching in number the 2N unknowns x_n and t_n in the problem. Because the action A_D is stationary under a variation in x_n and in t_n for every n, the discrete theory retains the translational invariance of both space and time, and that leaves the conservation laws of energy and momentum intact. *

For a free particle $V(x) = 0$, (2.11) and (2.10) become degenerate; both give

$$v_n = \frac{x_n - x_{n-1}}{t_n - t_{n-1}} = \text{constant.}$$

---

* Here, conservation of momentum means that the change of particle momentum is equal to the "impulse" generated by the potential.

The corresponding trajectory is a straight line, the same as the continuum case.

When $V(X) = gx$ with g a constant, the solution of (2.10) and (2.11) can be readily found. We find in this case the spacing between successive t_n to be independent of n :

$$t_n - t_{n-1} = \epsilon = \text{constant.}$$

Correspondingly, $t_n = t_0 + n\epsilon$ and

$$x_n = x_0 + n v_1 \epsilon - \tfrac{1}{2} n(n-1) g \epsilon^2$$

where v_1 is the initial velocity $(x_1 - x_0) / (t_1 - t_0)$.

When $\ell \to 0$, the site density $\to \infty$ and the discrete path $x_D(t)$ can assume the form of any smooth path $x(t)$; consequently the discrete mechanics approaches the usual continuum mechanics. Introduce

$$\tau \equiv n\ell$$

which varies from 0 to T , as n runs from 0 to N . Regard

$$x_n = x(\tau) \qquad \text{and} \qquad t_n = t(\tau) \ .$$

From (2.10) and (2.11), it can be shown that in the limit $\ell \to 0$, but keeping T fixed (hence, $N \to \infty$),

$$\frac{d^2x}{dt^2} = -\frac{dV}{dx} \tag{2.12}$$

and

$$\left(\frac{dt}{d\tau}\right)^3 \left(\frac{dV}{dx}\right)^2 = \text{constant} \ ; \tag{2.13}$$

the former is Newton's equation, and the latter gives the asymptotic distribution of t_n versus n. The constant in (2.13) is determined by the boundary condition (2.4), so that when τ varies from 0 to T, t also changes from 0 to T. In the usual continuum mechanics, only (2.12) is retained. Therefore, even in this limit, the discrete mechanics contains more information than the usual continuum mechanics. From (2.13), we see that, except for $V(x) = gx$, the spacing $t_n - t_{n-1}$ is not a constant.

It is of interest to examine the distribution $t(\tau)$ near the point $V'(x) \equiv dV/dx = 0$, which occurs at, say, $x = \bar{x}$. Let the particle trajectory in the continuum limit be $x = x(t)$. When $x = \bar{x}$, we have $V'(\bar{x}) = 0$ and, for the solution under consideration, $t = \bar{t}$ so that $\bar{x} = x(\bar{t})$. In the neighborhood x near $\bar{x}$, we may write, with $V''(x) \equiv d^2V/dx^2$ and $\dot{x} \equiv dx/dt$,

$$\begin{aligned} V'(x) &\cong (x - \bar{x})\ V''(\bar{x}) \\ &= (t - \bar{t})\ \dot{x}(\bar{t})\ V''(\bar{x}) \quad . \end{aligned}$$

Substituting this expression into (2.13), we find

$$(t - \bar{t}) \propto (\tau - \bar{\tau})^{3/5} \quad .$$

Hence, as $\tau \to \bar{\tau}$ (correspondingly $n \to \bar{\tau}/\ell$), although $dt/d\tau \to \infty$ one sees that $t \to \bar{t}$ and remains finite. Information such as this is lost if one concentrates only on Newton's equation (2.12).

In the following, we are interested in $\ell \neq 0$, in which case the discrete mechanics is fundamentally different from the continuum theory.

II.2. Nonrelativistic quantum mechanics

When we go over from classical to quantum mechanics, in the usual continuum theory the particle can take on any smooth path $x(t)$; each path carries an amplitude e^{iA} where $A = A(x(t))$ is the same action integral (2.1). In Feynman's path integration formalism, the matrix element of e^{-iHT} in the usual continuum quantum mechanics is given by

$$< x_f \mid e^{-iHT} \mid x_0 > \;=\; \int e^{iA(x(t))}\, d\left[x(t)\right] ,$$

in which all paths $x(t)$ have the same end-points (2.4) and

$$H \;=\; -\tfrac{1}{2}\,\frac{\partial^2}{\partial x^2} + V(x) .$$

Sometimes it is more convenient to consider the analytic continuation of T to $-iT$. The operator e^{-iHT} becomes then e^{-HT}, and its matrix element is given by

$$< x_f \mid e^{-HT} \mid x_0 > \;=\; \int e^{-\mathcal{A}(x(t))}\, d\left[x(t)\right] \tag{2.14}$$

where

$$\mathcal{A}(x(t)) \;=\; \int_0^T \left[\tfrac{1}{2}\dot{x}^2 + V(x)\right] dt . \tag{2.15}$$

In the corresponding discrete theory, we again restrict the particle to move only along the discrete path $x_D(t)$. By using (2.7) and (2.9), we see that (2.12) becomes

$$\int e^{iA_D} \prod_n \left[dx_n\right]\left[dt_n\right] . \tag{2.16}$$

Likewise, (2.14) and (2.15) become

$$< x_f \mid G_N(T) \mid x_0 > \;\equiv\; \int e^{-\mathcal{A}_D} \prod_{n=1}^{N} \left[dx_n\right]\left[dt_n\right] \tag{2.17}$$

where

$$\mathcal{A}_D = \mathcal{A}(x_D(t)) \quad . \tag{2.18}$$

When the vertices $n = 1, 2, \cdots$ are arranged in a time-ordered sequence (2.5), by using (2.15) and (2.18) we see that the discrete action $\mathcal{A}_D$ is given by

$$\mathcal{A}_D = \sum_{n=1}^{N+1} \left[\frac{(x_n - x_{n-1})^2}{2(t_n - t_{n-1})} + (t_n - t_{n-1}) \bar{V}(n) \right] \tag{2.19}$$

with $x_{N+1} = x_f$ and $t_{N+1} = T$, as shown in Figure 2(b).

In the integration over $\prod_n [dt_n]$, whenever t_i appears larger than, say, t_{i+1}, we should re-link the vertices so that the newly linked ones are in a time-ordered sequence. Alternatively, we may re-label them so that (2.5) remains valid; such a relabeling of vertices clearly does not change the discrete path $x_D(t)$. (As explained before, this follows from the usual nonrelativistic continuum mechanics in which the path $x(t)$ is a single-valued function of t.)

In the quantum version of the discrete mechanics it is more convenient to regard the constraint (2.6) as a condition on the average site-density. This can be most easily arranged by considering an ensemble sum over N:

$$\mathcal{G}(T, \ell) \equiv \sum_{N=0}^{\infty} \frac{1}{N!} \left(\frac{1}{\ell}\right)^N G_N(T) \tag{2.20}$$

where $G_N(T)$ refers to the matrix defined by (2.17). One may readily verify that this Green's function satisfies

$$\frac{\partial}{\partial(1/\ell)} \mathcal{G}(T, \ell) = \int_0^T \mathcal{G}(\tau, \ell)\, \mathcal{G}(T - \tau, \ell)\, d\tau , \tag{2.21}$$

from which it follows that for large T the operator $\mathcal{G}(T, \ell)$ becomes

$$\mathcal{G}(T, \ell) \sim e^{-\mathcal{H}T} \tag{2.22}$$

where $\mathcal{H}$ is Hermitian. When $\ell \to 0$, $\mathcal{H}$ reduces to the continuum Hamiltonian H , given by (2.13). The analytic continuation of $\mathcal{G}(T, \ell)$ from T to iT leads at large T to the unitary operator $e^{-i\mathcal{H}T}$, which is the S-matrix of the theory. Therefore, the unitarity of the S-matrix is maintained in the new mechanics.[2]

III. Relativistic Quantum Field Theory

As an example, let $\phi(x)$ be a scalar field in the usual continuum theory with x denoting the space-time coordinates. In the path integration formulation the operator e^{-HT} is given by, similar to (2.14),

$$e^{-HT} = \int e^{-\mathcal{A}} \, [d\phi(x)] \tag{3.1}$$

where H is the Hamiltonian operator, $\mathcal{A}$ the usual continuum action in the euclidean space and T the total "time" interval. (Here, as in (2.14)-(2.15), "time" refers to the euclidean time.) Because in the usual continuum theory the space-time coordinates x are parameters, and only $\phi(x)$ are dynamical variables, the functional integration in (3.1) is over $[d\phi(x)]$, not $[dx]$.

In the discrete version, we impose a constraint on the (average) number N of experiments that can be performed within any given space-time volume Ω, with $N/\Omega \equiv \ell^{-4}$ = fundamental constant. Each measurement determines the field $\phi(i)$ as well as the space-time position $x(i)$ with $i = 1, 2, \cdots, N$. The i will be referred to as lattice sites, as illustrated by Figure 3(a).

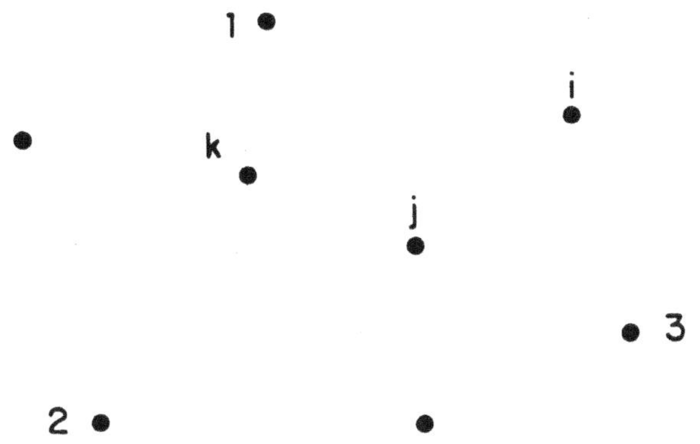

Figure 3(a).

As we shall see, the Green's function (3.1) will be replaced by

$$\int e^{-\mathcal{A}_D} \left[dx(i)\right] \left[d\phi(i)\right] \quad . \tag{3.2}$$

Because $\phi(i)$ and $x(i)$ are all dynamical variables, in the discrete theory we integrate over $\left[d\phi(i)\right]$ as well as $\left[dx(i)\right]$. The latter integration makes it obvious that rotational and translational symmetries can be maintained in the discrete theory.

To simulate the local character of the usual continuum theory, each site in the discrete theory is coupled only to its neighboring sites, as illustrated in Figure 3(b). The whole volume is then divided into triangles if the dimension of $x(i)$ is $d = 2$, tetrahedra if $d = 3$, 4-simplices when $d = 4$, etc. An example of such a simplicial lattice when $d = 2$ appears in Figure 3(b).

x - space
(d - dimension)

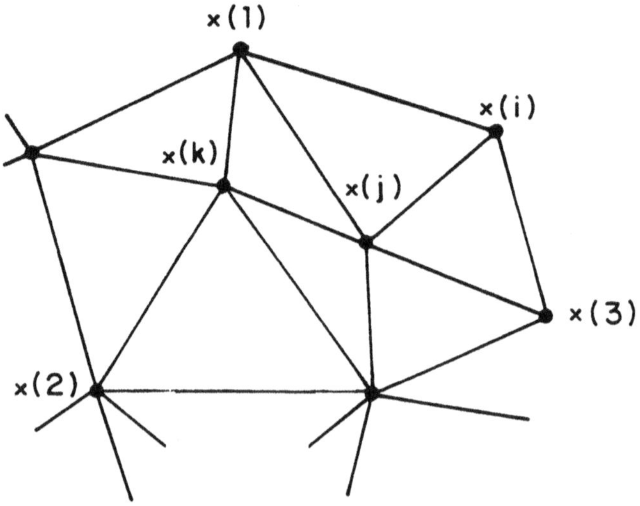

Figure 3(b).

We give the algorithm[3] of linking an arbitrary distribution of sites into a simplicial lattice for any dimension d : select any group of $d+1$ sites. Consider the hypersphere (in the d-dimensional euclidean space) whose surface passes through these $d+1$ sites. If the interior of the sphere is empty of sites, link these sites to form a d-simplex; otherwise, do nothing. Proceed to another group of $d+1$ sites, and repeat the same steps. The d-simplices thus formed never intersect each other, and the sum total of their volumes fills the entire space.

Each site i carries, in addition to its space-time coordinates $x(i)$, also a $\phi(i)$. Viewed in the $x-\phi$ space, the lattice forms a d-dimensional surface represented by $\phi_D(x)$, called the "discrete" function; it is continuous but piece-wise flat within each d-simplex as illustrated in Figure 3(c).

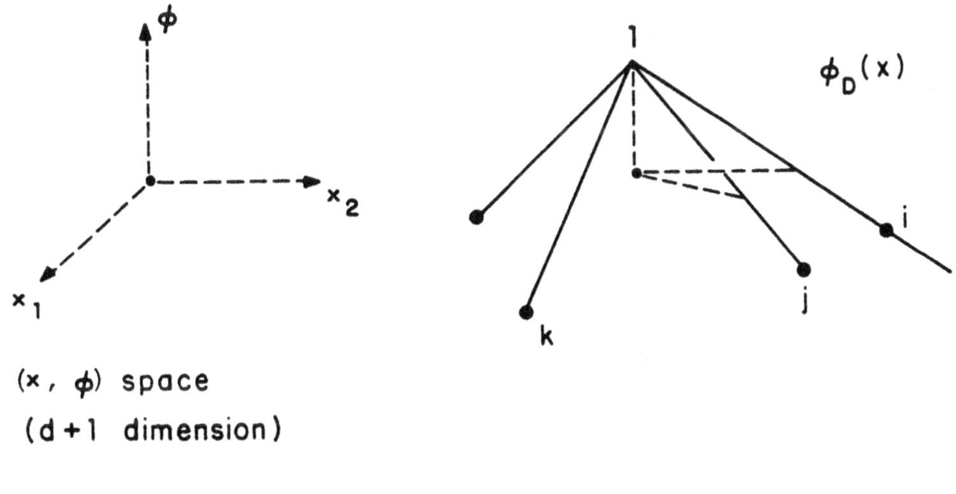

Figure 3(c).

The discrete action $\mathcal{A}_D$ in (3.2) can be readily evaluated by using the usual continuum action $\mathcal{A}(\phi(x))$, but restricting $\phi(x)$ to the discrete function:

$$\mathcal{A}_D \equiv \mathcal{A}(\phi_D(x)) \quad . \tag{3.3}$$

For example, if

$$\mathcal{A}(\phi(x)) = \int \left[\tfrac{1}{2}(\vec{\nabla}\phi)^2 + V(\phi)\right] dx \tag{3.4}$$

where dx is the d-dimensional volume element in the x-space, then setting $\phi(x)$ to be the discrete function $\phi_D(x)$ we find

$$\mathcal{A}_D = \mathcal{A}(\phi_D(x)) = \tfrac{1}{2}\sum_{\ell_{ij}} \lambda_{ij}([\phi(i) - \phi(j)]^2 + \sum_i \omega_i V(\phi(i)) \tag{3.5}$$

where the first sum is over all links ℓ_{ij} and the second over all sites i, ω_i is the volume of the Voronoi cell that is dual to the site i, and[4]

$$\lambda_{ij} = -\frac{1}{d^2} \sum \frac{1}{V(ij)} \vec{\tau}(i) \cdot \vec{\tau}(j) \tag{3.6}$$

in which the sum extends over all d-simplices $V(ij)$ that share the link ℓ_{ij}. In $V(ij)$, each vertex, say k, faces a $(d-1)$-dimensional simplex $\tau(k)$. In (3.6), $V(ij)$ denotes also the volume of the d-simplex and $\vec{\tau}(i)$ is the outward normal vector of $\tau(i)$ times its $(d-1)$ dimensional volume, as illustrated in Figure 4.

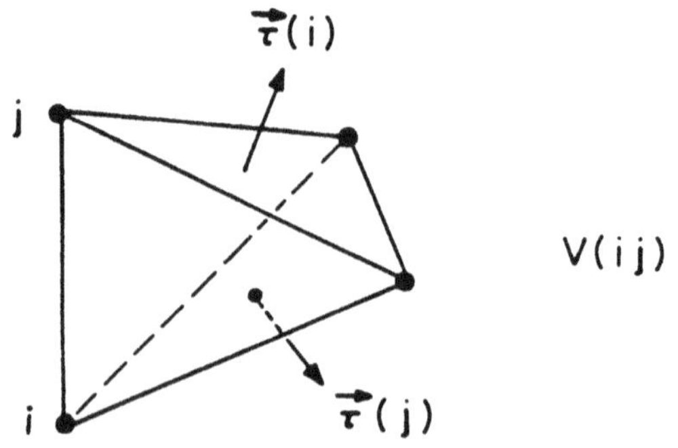

Figure 4.

As in the previous section, mathematically the discrete theory can be regarded as a special case of the usual continuum theory: one in which $\phi(x)$ is restricted to those continuous but piece-wise flat functions $\phi_D(x)$ with a fixed average density of vertices (i.e., lattice sites). Because the site density is an invariant, rotational and translational invariances can both be preserved in the discrete theory.

Since the discrete surface, described by $\phi_D(x)$, is characterized by the positions $\phi(i)$ and $x(i)$ of its vertices, a variation over the functional space $\phi(x)$ in the usual continuum theory becomes

$$\left[d\phi_D(x)\right] = \prod_i \left[d\phi(i)\right] \left[dx(i)\right] \quad . \tag{3.7}$$

Correspondingly (3.1) becomes (3.2). As $x(i)$ changes, the linking algorithm keeps track of how these vertices should be linked, so that the discrete action $\mathcal{A}_D$ is extensive; i.e., $\mathcal{A}_D$ is proportional to the overall space-time volume Ω when Ω is large. Thereby, the unitarity of the S-matrix can be established, as before.

In the usual continuum theory, the equation of motion is given by the partial differential equation

$$\frac{\delta \mathcal{A}(\phi(x))}{\delta \phi(x)} = 0 \quad . \tag{3.8}$$

Here in the discrete version it is replaced by the difference equations

$$\frac{\partial \mathcal{A}_D}{\partial \phi(i)} = 0 \qquad \text{and} \qquad \frac{\partial \mathcal{A}_D}{\partial x(i)} = 0 \quad ; \tag{3.9}$$

the former is the field equation on the lattice and the latter expresses the conservation law of the energy-momentum tensor.

In the integrand of (3.2), the locations of x can be arbitrary. Hence, the discrete action $\mathcal{A}_D$ is identical to that of a random lattice.[5]

IV. Gauge Theory

We review briefly the random lattice results on Abelian (QED) and non-Abelian (QCD) gauge theories.

The lattice gauge theory was introduced by K. Wilson. In the strong-coupling limit (square of coupling constant $g^2 \to \infty$), any lattice gauge theory gives confinement. This holds for both QED and QCD, and for arbitrary space dimension d. The realistic case corresponds, however, to the weak coupling. Thus, a key question is whether the transition from strong to weak coupling is smooth or not. If smooth, then the confinement property of the strong coupling can be carried over to weak coupling; otherwise not. When $\beta = 1/g^2$ changes from 0 (strong coupling) to ∞ (weak coupling), we would like the transition to be smooth for the non-Abelian case, but not smooth for the Abelian, so that the confinement holds for QCD, but not for QED. In a hypercubic lattice, there appears to be a phase transition in β for the U(1) gauge, consistent with the fact that QED is not confined. However, numerical results in SU(2) and SU(3) indicate that the transition from $\beta = 0$ to $\beta = \infty$ is also far from smooth. While there is probably no bona fide phase transition in the non-Abelian case, the change from cubic (when $g^2 = \infty$) to spherical (when $g^2 = 0$) symmetry is sufficiently hazardous that it is difficulty to infer, from the strong-coupling result, that confinement would remain valid in the weak coupling.

On the other hand, for the random lattice, its strong-coupling limit behaves like a relativistic string theory, with full rotational symmetry: the string thickness t is related to the string tension T by

$$t^2 = \frac{1}{2\pi T} \ln a \tag{4.1}$$

where a is the area enclosed by the string. Furthermore, the mass of the glueball m_J for a large angular momentum J varies as

$$m_J \propto \sqrt{J} \quad , \tag{4.2}$$

exhibiting the typical Regge behavior of a rotating relativistic string. Both (4.1) and (4.2) are valid in the strong-coupling limit.

Numerical programs for a random lattice gauge theory were set up by Friedberg and Ren at Columbia; the computations were carried out by Ren.[6] In Figures 5(a)-5(b) we give the average plaquette energy u and specific heat C vs. $\beta = 1/g^2$ for the U(1) theory.

The corresponding plots for an SU(2) theory are given in Figures 6(a)-6(b). We see that the specific heat has a peak in the U(1) theory, but not in the SU(2) theory. For U(1), the peak becomes steeper when the number of lattice sites increases, suggesting that there is a phase transition. On the other hand, the specific heat curve for SU(2) has no peak, indicating that the passage from strong to weak coupling is a smooth one. Consequently, while both theories are confined in the strong coupling limit, the weak coupling limit is consistent with deconfinement in the U(1) theory (QED), but with confinement in a non-Abelian gauge theory (QCD).

In contrast, we give in Figure 7 the numerical calculation by N. H. Christ and A. E. Terrano[7] for the SU(3) gauge theory on a regular lattice. As we can see, there is a sharp peak in the specific heat, suggesting that the transition from strong to weak in a regular lattice is by no means smooth, unlike that in a random lattice.

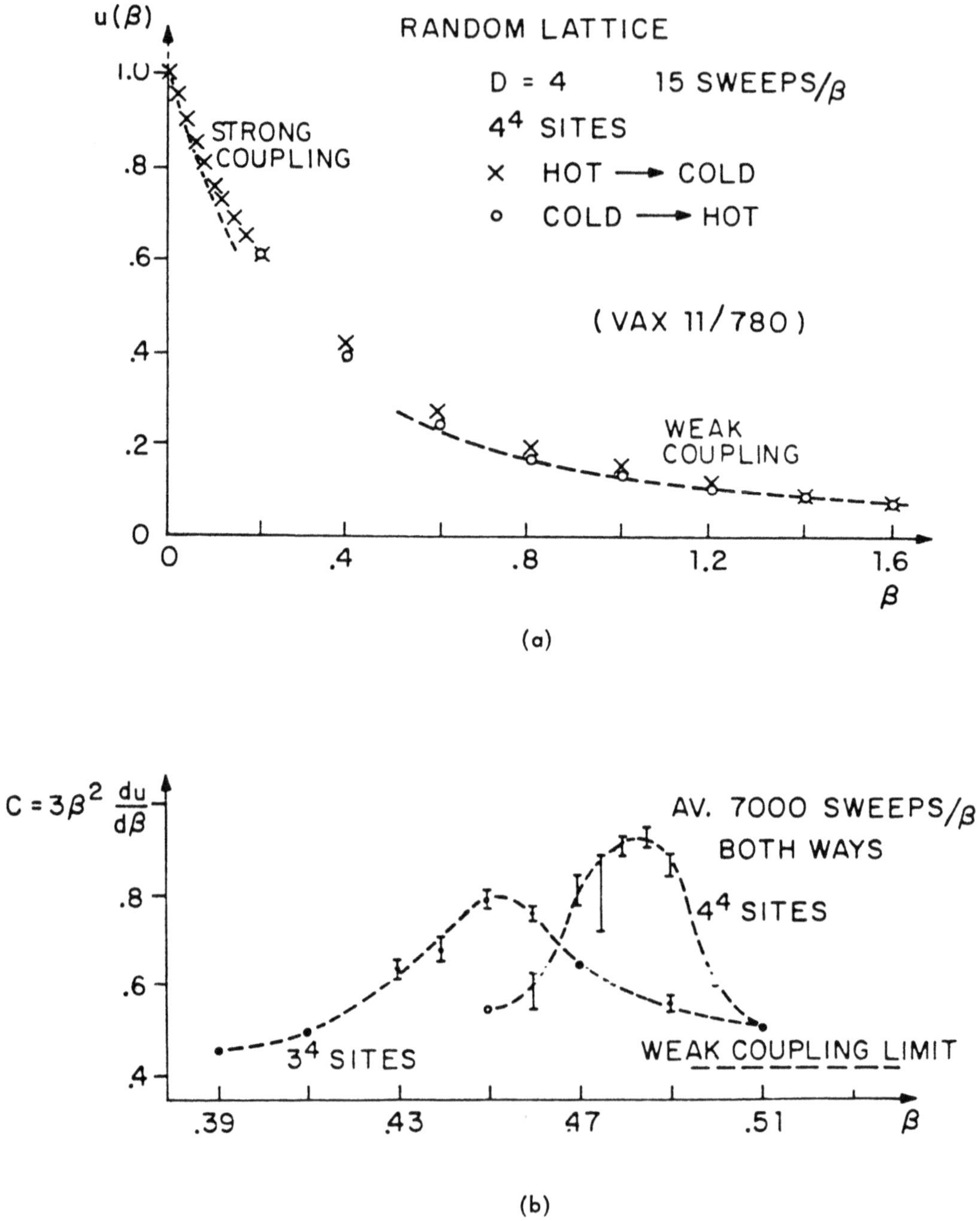
u(β)
RANDOM LATTICE
D = 4 15 SWEEPS/β
4⁴ SITES
× HOT → COLD
o COLD → HOT
STRONG COUPLING
(VAX 11/780)
WEAK COUPLING
1.0
.8
.6
.4
.2
0
0
.4
.8
1.2
1.6
β
(a)
C = 3β² du/dβ
AV. 7000 SWEEPS/β
BOTH WAYS
4⁴ SITES
3⁴ SITES
WEAK COUPLING LIMIT
.8
.6
.4
.39
.43
.47
.51
β
(b)

Figure 5.

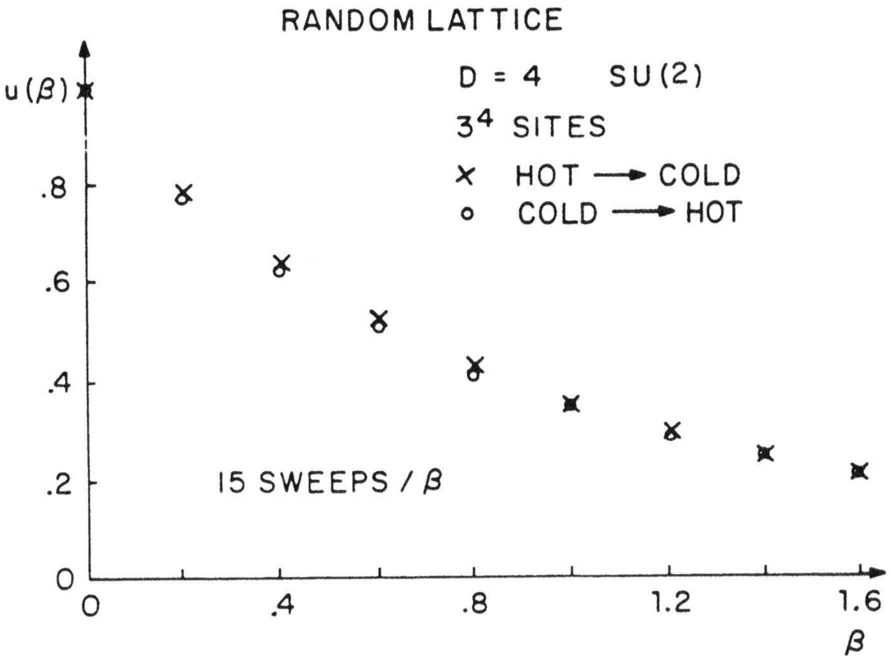
RANDOM LATTICE
D = 4 SU(2)
3⁴ SITES
× HOT → COLD
o COLD → HOT
15 SWEEPS / β
u(β)
.8
.6
.4
.2
0
0
.4
.8
1.2
1.6
β

(a)

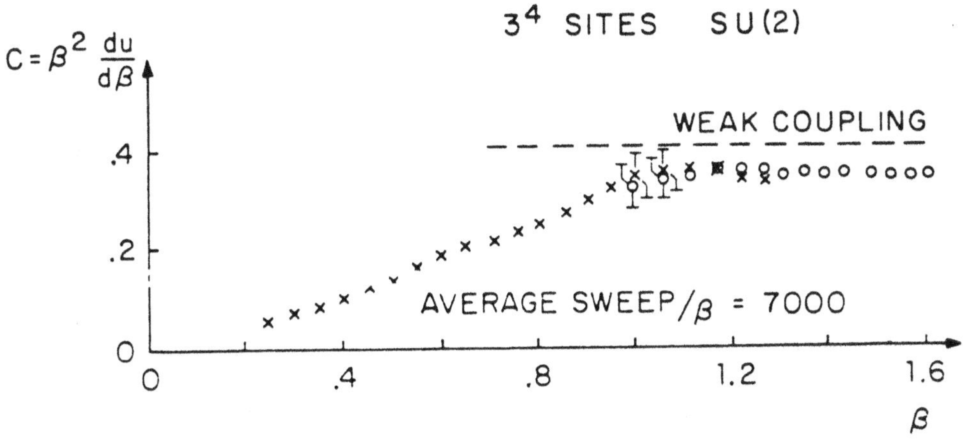
3⁴ SITES SU(2)
C = β² du/dβ
WEAK COUPLING
AVERAGE SWEEP/β = 7000
.4
.2
0
0
.4
.8
1.2
1.6
β

(b)

Figure 6.

23.

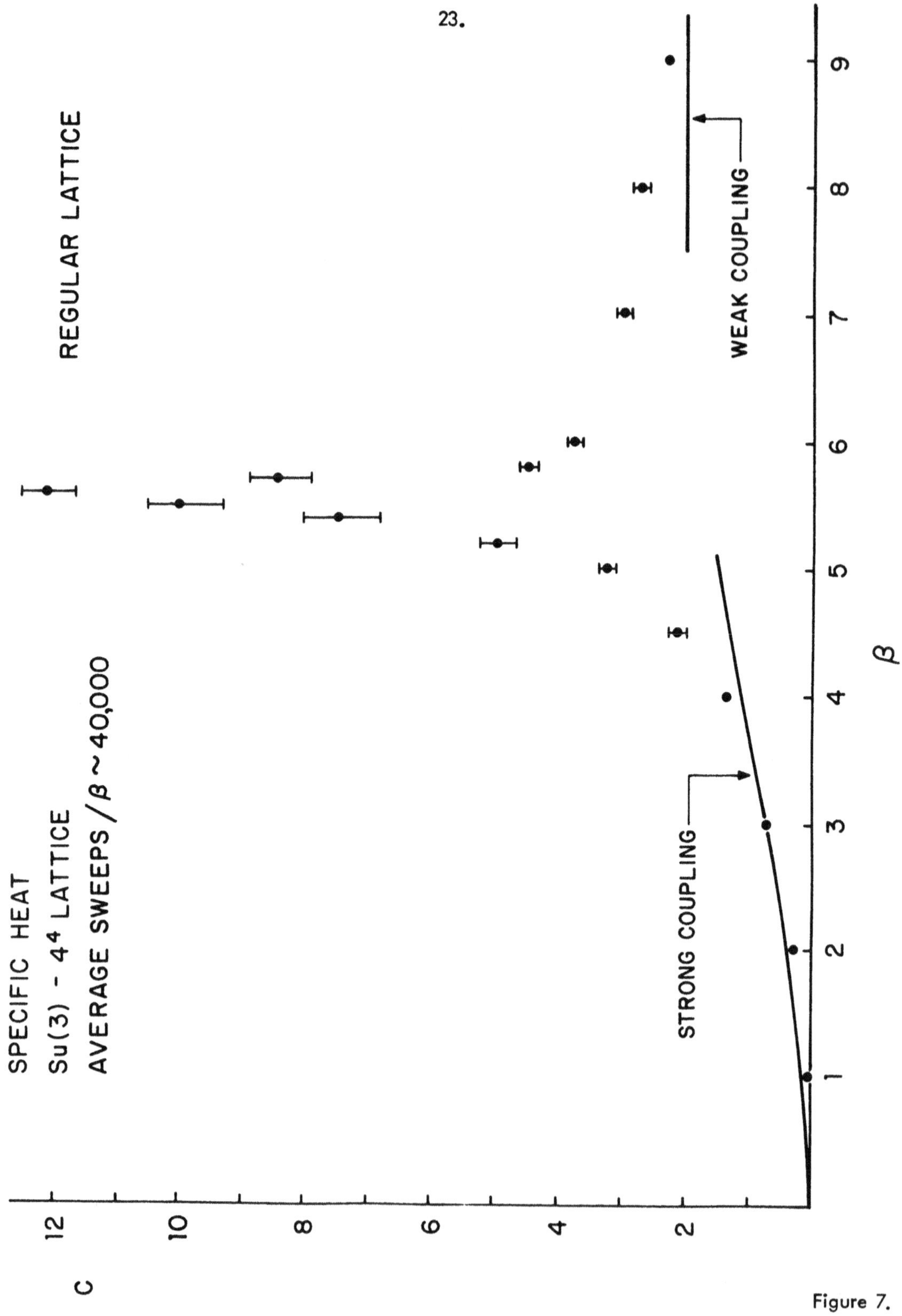

SPECIFIC HEAT
Su(3) - 4^4 LATTICE
AVERAGE SWEEPS / $\beta \sim 40{,}000$
REGULAR LATTICE
C
12
10
8
6
4
2
1
2
3
4
5
6
7
8
9
β
STRONG COUPLING
WEAK COUPLING

Figure 7.

V. Lattice Gravity

The usual Einstein action in general relativity is

$$A(S) = \int_S \sqrt{|g|}\, \mathcal{R}\, dx \tag{5.1}$$

where S is a d-dimensional smooth continuous surface, $|g|$ is the absolute value of the determinant of the matrix of the metric tensor $g_{\mu\nu}$ on S, $\mathcal{R}$ is the scalar curvature and dx is the d-dimensional volume element in the space-time coordinate x.

For lattice gravity, we consider first a (random) lattice $\overset{\circ}{L}$ in a flat d-dimensional euclidean space R_d. Label each site by $i = 1, 2, \ldots$ For every linked pair of sites i and j there is a link-length $\overset{\circ}{\ell}_{ij}$.

Consider now an arbitrary variation

$$\overset{\circ}{\ell}_{ij} \rightarrow \ell_{ij} \quad . \tag{5.2}$$

Correspondingly, each d-simplex, say $\overset{\circ}{\tau}$ in $\overset{\circ}{L}$, becomes a new d-simplex τ with the same vertices, but different link-lengths. These new link-lengths ℓ_{ij} are assumed to satisfy all simplicial inequalities, so that each d-simplex τ, by itself, can still be realized in a flat d-dimensional space R_d. In general the entire new lattice cannot fit into R_d. This then defines[8] a d-dimensional non-flat lattice surface L.

Sometimes, it is convenient to embed L in a flat space R_N. This is possible if

$$N = d + n$$

is sufficiently large; in that case

$$\ell_{ij}^2 = [\vec{r}(i) - \vec{r}(j)]^2 \quad \text{in} \quad R_N \tag{5.3}$$

with $\vec{r}(i)$ the cartesian N-dimensional position vector of the ith site in R_N. Since, as we shall see, we shall deal only with the intrinsic geometric properties of the lattice surface, this embedding is merely a convenience.

Next we wish to evaluate Einstein's action (5.1) when S is restricted to the lattice surface L. At first sight, it might appear difficult because the metric g_{ij} would change discontinuously from simplex to simplex, the Christoffel symbol would then acquire δ-functions, and the scalar curvature δ'-functions. Since Einstein's action is nonlinear in g_{ij}, one might expect the resulting expression to be totally unmanageable. It turns out that this is not so.

It can be shown that the Einstein formula (5.1) evaluated on any d-dimensional lattice space L gives the discrete action[9,10]

$$A(L) \equiv \int_L \sqrt{|g|}\, \mathcal{R}\, dx \tag{5.4}$$

$$= 2 \sum_s s\, \epsilon_s \tag{5.5}$$

where dx is the d-dimensional volume element, s is the volume of the (D-2) simplex, ϵ_s is Regge's deficit angle around s and the sum extends over all s in the lattice. [See ref. 11 for the definition of ϵ_s.] The right-hand side of (5.5) is precisely the formula of Regge calculus.[11]

In Regge's original approach, he considered the discrete action as an approximation to Einstein's continuum action. Here we are reversing the role and regarding the discrete action A(L) as more fundamental. It is therefore satisfying to realize that Regge's action is identical to Einstein's action, but evaluated on L.

The quintessence of Einstein's theory of general relativity lies in its invariance under a general coordinate transformation

$$x \rightarrow x' \tag{5.6}$$

that leaves ds^2 unchanged. Since the action for the lattice space L is the discrete action

$$A_D \equiv A(L) = \int_L \sqrt{|g|}\, \mathcal{R}\, dx = 2 \sum_s s\, \epsilon_s \ , \tag{5.7}$$

the discrete theory clearly remains invariant under the coordinate transformation (5.6). Thus, the entire apparatus of coordinate invariance in the usual continuum theory automatically applies to the lattice theory as well. In addition, as we shall see, the lattice theory enjoys still another totally new class of symmetries which does not exist in the usual continuum theory. Aesthetically, this adds greatly to the appeal of lattice gravity. For physical applications, when the link-length ℓ is small, our general formula (5.7) insures that all known tests of general relativity are automatically satisfied. Furthermore, by keeping ℓ nonzero, we see that the lattice action A_D per volume possesses only a finite degree of freedom. The normal difficulty of ultra-violet divergence that one encounters in quantum gravity disappears in the lattice theory. All these suggest that the lattice theory with a nonzero ℓ may be more fundamental. The usual continuum theory is quite possibly only an approximation.

To amplify the aforementioned symmetry properties, let us consider any lattice L. From (5.7), we see that the discrete action A_D, through its right-hand side, is a function of the link-lengths ℓ_{ij} :

$$A_D = A_D(\ell_{ij}) \ . \tag{5.8}$$

We may also characterize the lattice by other means of parametrization. We assume all the lattice sites i to lie on a d-dimensional smooth enveloping surface S, with $z_\mu(i)$ as the coordinates of the site i on S, where $\mu = 1, 2, \cdots, d$. Embed both S and L in a flat space R_N, which is always possible provided that N is sufficiently large. Because of (5.3), ℓ_{ij} can also be determined by giving S and $z_\mu(i)$. Hence we can also express A_D as a function of the enveloping surface S and the site positions on S:

$$A_D = A_D(S, z_\mu(i)) \quad . \tag{5.9}$$

Thus, we can have new symmetry transformations:

(i) fix $z_\mu(i)$, but vary $S \to S'$,

and (ii) fix S, but vary $z_\mu(i) \to z'_\mu(i)$.

These symmetries are exact if ℓ_{ij} are unchanged; they can be approximate even if ℓ_{ij} does change, provided that, e.g., the link-lengths are sufficiently small and $\sqrt{|g|}\, \mathcal{R}\, dx$ remains the same on the enveloping surface, in which case $A_D \cong A(S)$ of (5.1).

In the usual continuum theory, the physical space-time points and the underlying four-dimensional manifold are the same. Here, they are distinct; the former is related to measurements, while the latter is purely a mathematical artifice (like the choice of gauge in the usual continuum theory of a spin 1, or 2, field).

VI. Concluding Remarks

For more than three centuries we have been influenced by the precept that fundamental laws of physics should be expressed in terms of differential equations. Difference equations are always regarded as approximations. Here, we try to explore the opposite: Difference equations are more fundamental, and differential equations are regarded as approximations.

As we have shown, such a difference equation formulation leads to the discrete mechanics which can also be viewed as the mathematical limit of the usual continuum mechanics, but with a fixed density of lattice sites. Because this is an invariant constraint, the discrete theory shares the same symmetries of the usual continuum theory. In this way, we have succeeded in the creation of theories with finite degrees of freedom, but which retain all the good properties of the usual continuum theory. We suggest that this discrete formulation might be more fundamental.

In this new mechanics, space and time are treated as dynamical variables, on the same footing as electromagnetic fields, gluon fields, etc. In this sense, I hope I am extending further the thinking of Niels Bohr to bring our theoretical concepts closer to actual experimental observation.

References

1. T. D. Lee, Phys.Lett. 122B, 217 (1982); Shelter Island II, ed. R. Jackiw, N. Khuri, S. Weinberg and E. Witten (Cambridge, The MIT Press, 1985), p. 38-64.
2. R. Friedberg and T. D. Lee, Nucl.Phys. B225 [FS9], 1 (1983).
3. N. H. Christ, R. Friedberg and T. D. Lee, Nucl.Phys. B202, 89 (1982).
4. R. Friedberg and T. D. Lee, unpublished.
5. N. H. Christ, R. Friedberg and T. D. Lee, Nucl.Phys. B210 [FS6], 310 (1982).
6. H. C. Ren, "Field Theory On A Random Lattice," Columbia University dissertation, 1984.
7. N. H. Christ and A. E. Terrano, Nuclear Instruments and Methods 222, 534 (1984).
8. For physical application to general relativity, in order to maintain the quasi-local character of the discrete action, we must link only neighboring sites. Thus, when the new link-lengths ℓ_{ij} are too large the sites have to be re-linked. Details will be given elsewhere.
9. R. Sorkin, Phys.Rev. D12, 385 (1975).
10. R. Friedberg and T. D. Lee, Nucl.Phys. B242, 145 (1984).
11. T. Regge, Nuovo Cimento 19, 558 (1961).

VIII

Strong Interaction Models

Commentary

The papers in this section deal mostly with models of the strong interactions. In paper [143], a model of the physical nucleon was introduced involving a bare nucleon coupled to mesons. The model is treated in the Bethe-Salpeter approximation, and it is shown that it can reproduce many aspects of the scaling behavior first described in Feynman's parton model. In paper [144], this model of the nucleon is extended to a more general field theory calculation, including scalar gluons.

Paper [145] uses some of the methods of statistical mechanics to study the multiplicity of mesons produced in a model field theory with trilinear meson-meson interactions.

A number of papers discuss another consequence of the spontaneous symmetry breaking that is analyzed in paper [150]. In paper [152], this is applied to the possibility that under conditions of high particle density, such as inside a nucleus, the expectation value of some scalar field is shifted from its usual vacuum value to a very different one. This would result in a change in the mass of Fermions coupled to the scalar field, such as the nucleons themselves. This change in mass would then show up as a transition to a new form of nuclear matter, with higher density and binding energy than ordinary nuclear matter. This idea is followed up in paper [155], which discusses means of detecting such "abnormal" nuclear states that would be produced in collisions between heavy ions moving at relativistic speeds. Paper [187] discusses the future of relativistic heavy ion physics, including how answers to some physical questions about the structure of the vacuum can be obtained by studying the high density objects that may be produced in such collisions. Papers [156] and [170] discuss in more detail some of the approximation methods that were used to analyze abnormal nuclear states, and related ideas such as pion condensation.

Reprinted from THE PHYSICAL REVIEW, Vol. 135, No. 2B, B467–B477, 27 July 1964
Printed in U. S. A.

Implications of Approximate SU_3 Symmetry and Mass Formulas for the Mesons*

F. GÜRSEY†
Institute for Advanced Study, Princeton, New Jersey

AND

T. D. LEE AND M. NAUENBERG††
Columbia University, New York, New York

(Received 5 March 1964)

The connection between the limit of perfect SU_3 symmetry and the zero mass approximation for the pseudoscalar meson octet is discussed. The well-known quadratic mass formula for the pseudoscalar meson octet and the linear mass formula for the baryons are derived. A simple model of strong interaction is presented which leads to two mass formulas for the nine spin-1 mesons:

$$2M_\phi+M_\rho+M_\omega=4M_{K^*} \quad \text{and} \quad M_\rho\cong M_\omega.$$

The general invariance property of this model which contains, among others, a baryon octet, a pseudoscalar octet, and a vector nonet is examined.

I. GENERAL DISCUSSION

THE question whether the strong interactions are approximately invariant under a SU_3 transformation has been discussed extensively in the recent literature.[1] There exists by now a rather impressive body of experimental evidence[2] supporting such an approximate invariance. Part of this evidence is based on the remarkably accurate "mass formulas" which are obtained[3] by treating, among other assumptions, the symmetry violating interactions as small. On the other hand, since the mass differences among, say, K, π, and η which are members of the same multiplet are not small compared to their actual masses, the violation of SU_3 symmetry is apparently not weak. The additional fact that these mass formulas are linear functions of masses for the baryons, but quadratic functions for the mesons, seems to further veil the foundation of such an approximate symmetry. In this paper, we shall attempt to clarify some of these questions.

The existence of an approximate symmetry under SU_3 for the strong interactions suggests that the total Hamiltonian H contains a part, called primary interaction, H_0, which is invariant under SU_3, and represents the main contribution to the interaction. The remaining part, called secondary interaction, $h=H-H_0$ is small compared to H_0 and has symmetry violating properties. The *typical* energy scale[4] (or level spacing) due to the primary interaction H_0 is $\sim M$ which is assumed to be of the order of a few BeV. The energy shift caused by the secondary interaction h is $\sim m$ which is of the order of a few hundred MeV's. The smallness of the phenomenological dimensionless constant

$$\lambda\equiv(m/M)\ll 1 \tag{1}$$

makes possible the use of h as a perturbation.

We shall further specify the primary interaction H_0 by assuming that it contains eigenstates corresponding to an octet of pseudoscalar *mesons of zero mass*. The perturbation h can be written as

$$h=h_0+h_1, \tag{2}$$

where we assume that h_0 is invariant[5] under SU_3 and h_1 transforms like the isotopic spin=0 component of the octet representation[1] of SU_3. Both H_0 and h transform in the usual way under the inhomogeneous Lorentz transformation and conserve charge, parity, strangeness, etc.

We now apply these considerations to the mass shifts of the pseudoscalar meson octet. Let η, π, K denote the familiar members of this octet and E_η, E_K, E_π their respective energies. By our assumption, these mesons have zero mass in the absence of h (i.e., $\lambda=0$). Therefore, the zeroth-order values of the energies $E_\eta(p)$, $E_K(p)$, $E_\pi(p)$ are given by

$$E_\eta{}^0(p)=E_K{}^0(p)=E_\pi{}^0(p)=p, \tag{3}$$

* Research supported in part by the U. S. Atomic Energy Commission and the National Science Foundation.

† On leave from the Middle East Technical University, Ankara, Turkey.

‡ J. S. Guggenheim Fellow.

[1] M. Gell-Mann, California Institute of Technology Report CTSL-20, 1961 (unpublished). Y. Ne'eman, Nucl. Phys. **26**, 222 (1961). M. Gell-Mann, Phys. Rev. **125**, 1067 (1962). See references in these papers for earlier work on SU_3. Cf. also D. Speiser and A. Tarski, J. Math. Phys. **4**, 588 (1963). [An earlier reference to this work in connection with SU_3 symmetry can be found in footnote 14 of a paper by T. D. Lee and C. N. Yang, Phys. Rev. **122**, 1954 (1961).]

[2] See especially, S. L. Glashow and A. H. Rosenfeld, Phys. Rev. Letters **10**, 192 (1963). V. E. Barnes, P. L. Connolly, D. J. Crennell *et al.*, *ibid.* **12**, 204 (1963).

[3] M. Gell-Mann, Ref. 1, S. Okubo, Progr. Theoret. Phys. (Kyoto) **27**, 949 (1962); **28**, 64 (1962). See also J. J. Sakurai and S. Glashow, Nuovo Cimento **25**, 337 (1962).

[4] It is difficult to give a precise definition of the typical energy scale without a concrete picture of the dynamics. For a more detailed discussion on the meaning of this energy scale M associated with the primary interaction, see the special model given in Sec. II and the related example discussed in Appendix II.

[5] We are reminded of the analogy with the electromagnetic interaction which, while violating the isotopic spin conservation, contains a part that is invariant under the isotopic spin rotation. Similarly, the weak interactions, which do not conserve strangeness, nevertheless have a part, such as the β decay, that does conserve strangeness. In analogy with the relative magnitude of the isoscalar and the isovector terms of the electromagnetic interactions, we assume h_0 and h_1 to be of comparable magnitude.

where p is the magnitude of the three-momentum of the particles. The application of h conserves the three-momentum p but brings an energy shift to each of these particles.

To use perturbation theory we must choose p to be nonzero and fixed. Mathematically, to first order in h, as a result of its assumed transformation property, these energies satisfy the linear relation

$$3E_\eta(p)+E_\pi(p)=4E_K(p). \quad (4)$$

From Lorentz invariance, any first-order correction to $E_i(p)$ must be of the form[6]

$$\delta E_i(p)=E_i(p)-p=m_i^2/(2p), \quad (5)$$

where $i=\eta$, K or π and m_i is the corresponding mass of the particle. Therefore, we obtain

$$3m_\eta^2+m_\pi^2=4m_K^2, \quad (6)$$

which is valid also to the first order in h. Thus, we are led to the mass-squared formula by the requirement of Lorentz invariance and the assumption of zero mass for the pseudoscalar mesons in the unperturbed situation.

In contrast, the linear mass formula for the baryon octet follows if the corresponding baryon mass eigenvalue M_0 of the primary interaction H_0 is large compared to the mass shifts due to the perturbation h. In that case the zeroth-order energy of the baryon octet (N,Λ,Σ,Ξ) is given by

$$E_N{}^0(p)=E_\Lambda{}^0(p)=E_\Sigma{}^0(p)=E_\Xi{}^0(p)=(p^2+M_0{}^2)^{1/2}. \quad (7)$$

To first order in h, the perturbed energies satisfy the following relation:

$$3E_\Lambda(p)+E_\Sigma(p)=2[E_N(p)+E_\Xi(p)]. \quad (8)$$

By using $E_i=(p^2+M_i{}^2)^{1/2}$, the first-order correction in $E_i(p)$ must be of the form

$$\delta E_i=E_i-E_i{}^0=(p^2+M_0{}^2)^{-1/2}M_0\delta M_i, \quad (9)$$

where $M_i=M_0+\delta M_i$ and $i=N$, Λ, Σ, Ξ. Hence, Eq. (8) can be written as

$$3M_\Lambda+M_\Sigma=2(M_N+M_\Xi). \quad (10)$$

[Note that to first order in h, (10) is identical with $3M_\Lambda{}^2+M_\Sigma{}^2=2(M_\Xi{}^2+M_N{}^2)$.]

Similar considerations can be easily extended to other multiplets. The difference between a quadratic and linear mass formula lies solely in the different magnitude of the masses of these multiplets in the absence of the perturbation h.

In this connection it is interesting to note that zero mass pseudoscalar mesons appear also in certain theories with γ_5 invariance.[7] In the absence of h (i.e., $\lambda=0$), because of the zero mass character of pseudoscalar mesons, the axial vector currents in weak decays can be conserved.[8] This conservation law is broken due to the presence of the secondary interaction h. Since the scale of the primary interaction M is also much larger than the nucleon mass m_N, we expect the ratio of the Gamow-Teller coupling constant G_A to the Fermi coupling constant G_V to be of the form[9]

$$G_A/G_V=-1+O(m_N/M), \quad (11)$$

which makes it reasonable to assume the scale M to be of the order[10] of 10 BeV. Furthermore, in the absence of h, the well-known Goldberger-Treiman formula[11] holds.[9] Deviation from the Goldberger-Treiman formula occurs only as a result of the secondary interaction and is therefore relatively small.

II. A MODEL

It seems natural to explore the possibility that the large energy scale M of the primary Hamiltonian H_0 and its invariance character under SU_3 indicate the existence of some massive triplets with very strong interactions which can form compound systems corresponding to the known octets and decuplets.[12]

In this section we give a special model of the primary and the secondary interactions, partly to illustrate some of the discussions given in the previous section. We assume the existence of two massive triplets $(\alpha_0,\alpha_1,\alpha_2)$ and $(\beta_0,\beta_1,\beta_2)$ under the SU_3 transformations. The charges of $(\alpha_0,\alpha_1,\alpha_2)$, and $(\beta_0,\beta_1,\beta_2)$, are given, respectively, by $(q,\ q+1,\ q)$ in units of e. The baryon numbers of the α's and the β's are, respectively, n and $(n+1)$ where q and n are any integers including zero. One of these triplets, say, β_i consists of fermions while the α's represent bosons.[13] Under the isotopic spin rotations, α_0 and β_0 behave like states with $I=0$; (α_1,α_2), and (β_1,β_2) like states with $I=\frac{1}{2}$. Let $\bar{\alpha}_i$ and $\bar{\beta}_i$ be the antiparticles of α_i and β_i. The primary interaction H_0 can be written as

$$H_0=H_{\text{free}}+H_{\text{int}}, \quad (12)$$

[6] For the physical case, in order that Eqs. (4) and (5) be a good approximation we must choose states with $p\gg m_i$. The resulting formula (6) is obviously independent of p. In Appendix I we give a trivial but explicit example illustrating the use of the perturbation method.

[7] See, e.g., Y. Nambu, Phys. Rev. **122**, 345 (1961).

[8] J. C. Taylor, Phys. Rev. **110**, 1216 (1956), J. C. Polkinghorne, Nuovo Cimento **8**, 179 and 781 (1958). Y. Nambu, Ref. 7, Y. Nambu, Phys. Rev. Letters **4**, 380 (1960). F. Gürsey, Nuovo Cimento **16**, 230 (1960); Ann. Phys. (N.Y.) **12**, 91 (1061). M. Gell-Mann and M. M. Lévy, Nuovo Cimento **16**, 705 (1960).

[9] J. Bernstein, S. Fubini, M. Gell-Mann, and W. Thirring, Nuovo Cimento **17**, 757 (1960). F. Gürsey, Ref. 8, Y. Nambu and G. Jona-Lasinio, Phys. Rev. **122**, 345 (1961) and other previous references by Y. Nambu.

[10] Dr. C. S. Wu has kindly informed us that recently C. P. Bhalla (unpublished)has re-examined the *Ft* value of neutron and obtained $(G_A/G_V)=-1.15\pm0.02$. We wish to thank Dr. Wu for this information.

[11] M. L. Goldberger and S. B. Treiman, Phys. Rev. **110**, 1178 (1958).

[12] While this work was in progress, we received a reprint by J. Schwinger in which some similar ideas are discussed. We wish to thank Professor Schwinger for communicating his results to us before publication.

[13] An equally adequate model can be constructed if we exchange the roles of α's and β's.

where

$$H_{\rm free}=\sum_{\mathbf{k},i}\{(M^2+k^2)^{1/2}[a^{\dagger i}(\mathbf{k})a_i(\mathbf{k})+a_i'^{\dagger}(\mathbf{k})a'^i(\mathbf{k})]$$
$$+(M'^2+k^2)^{1/2}[b^{\dagger i}(\mathbf{k})b_i(\mathbf{k})+b_i'^{\dagger}(\mathbf{k})b'^i(\mathbf{k})]\}, \quad (13)$$

M and M' are both of the order of a few BeV, representing the zeroth-order (i.e., $h=0$) masses of α_i and β_i. The $a_i(\mathbf{k})$, $a'^i(\mathbf{k})$, $b_i(\mathbf{k})$, $b'^i(\mathbf{k})$ are respectively the annihilation operators for α_i, $\bar{\alpha}_i$ and β_i, $\bar{\beta}_i$ with momentum $\mathbf{k}$, and $a^{\dagger i}(\mathbf{k})$, $a_i'^{\dagger}(\mathbf{k})$, $b^{\dagger i}(\mathbf{k})$, $b_i'^{\dagger}(\mathbf{k})$ are their Hermitian conjugates:

$$[a_i(\mathbf{k})]^\dagger=a^{\dagger i}(\mathbf{k}),\quad [a'^i(\mathbf{k})]^\dagger=a_i'^\dagger(\mathbf{k}),$$
$$[b_i(\mathbf{k})]^\dagger=b^{\dagger i}(\mathbf{k}) \quad\text{and}\quad [b'^i(\mathbf{k})]^\dagger=b_i'^\dagger(\mathbf{k}), \quad (14)$$

where † denotes Hermitian conjugation and $i=0, 1, 2$. In all these formulas, the spin dependence is suppressed.

The particular form of $H_{\rm int}$ is not too relevant at this point. We only require that there are strong attractive forces between α_i and $\bar{\alpha}_j$, β_i and $\bar{\beta}_j$, and also between $\bar{\alpha}_i$ and β_j. The forces between all other pairs such as (α_i,α_j) or (β_i,β_j) are assumed to be repulsive. As a result of these strong attractive forces the zero mass pseudoscalar meson octet of H_0 can be regarded as a composite system of α_i and $\bar{\alpha}_j$ and is denoted by $(\bar{\alpha}\alpha)_8$ where the subscript 8 indicates the dimension of the group representation. Similarly, we regard all other known strongly interacting particles also as composite systems of α_i and β_j. We denote $(\bar{\beta}\beta)_8$, $(\bar{\beta}\beta)_1$, and $(\bar{\alpha}\beta)_8$ to be the appropriate bound states of $\bar{\beta}_i$, β_j and of $\bar{\alpha}_i$, β_j which transform like the spin-1 meson octet, the spin-1 meson, and the baryon octet,[14] respectively. In contrast to the pseudoscalar meson states, the baryon octet has a large mass M_0 in the absence of the secondary interaction. Furthermore, we assume the force between α_i and $\bar{\alpha}_j$ depends on the representation; as a consequence, the pseudoscalar singlet $(\bar{\alpha}\alpha)_1$, if it exists, may have a different energy from the pseudoscalar octet $(\bar{\alpha}\alpha)_8$.

Let us first consider a particularly simple example for the secondary interaction h where the only violation of SU_3 symmetry occurs through additional mass shifts m_i and m_i' (which may be of the order of a few hundred MeV) for the triplet particles α_i and β_i, respectively. We are led to the following simple expression for h:

$$h=h_\alpha+h_\beta, \quad (15)$$

where

$$h_\alpha=\sum_{\mathbf{k}}\sum_{i=0}^{2}\{(M^2+k^2)^{-1/2}Mm_i \times[a^{\dagger i}(\mathbf{k})a_i(\mathbf{k})+a_i'^\dagger(\mathbf{k})a'^i(\mathbf{k})]\} \quad (16)$$

and h_β is given by

$$\sum_{\mathbf{k}}\sum_{i=0}^{2}\{(M'^2+k^2)^{-1/2}M'm_i' \times[b^{\dagger i}(\mathbf{k})b_i(\mathbf{k})+b_i'^\dagger(\mathbf{k})b'^i(\mathbf{k})]\}, \quad (17)$$

where the isotopic spin conservation is insured by choosing

$$m_1=m_2 \quad\text{and}\quad m_1'=m_2'. \quad (18)$$

Since we have already assumed that the primary SU_3 invariant Hamiltonian H_0 contains, besides the free term (13), also an interaction term that lifts the degeneracy between singlet and octet states for the pseudoscalar mesons and the baryons, it now follows from the transformation property of (15) that the mass formulas such as (6) and (10) hold for the pseudoscalar meson octet $(\alpha\bar{\alpha})_8$ the baryon octet $(\beta\bar{\alpha})_8$, respectively.

To discuss the mass formula for the spin-1 meson states we impose a further condition on the primary interaction. We require that the strong attractive forces between β_i and $\bar{\beta}_j$ that result from H_0 are essentially *independent*[15] of i and j, very much like the Wigner forces for nucleons. Therefore, in the absence of h, the states $(\bar{\beta}\beta)_8$ and $(\bar{\beta}\beta)_1$ are degenerate. For definiteness, we assume the mass of these states to be fairly large.

Most of the degeneracies among these nine states are removed to the first order in h. Let ρ, ω, K^* and ϕ be the physical resonances and $(\bar{\beta}_i\beta_j)$ the bound-state wave function of the system $\bar{\beta}_i$ and β_j. Using the Eqs. (15) and (18) we make the following identifications for the vector meson states:

$$|\phi\rangle=\bar{\beta}_0\beta_0,$$
$$|\omega\rangle=2^{-1/2}(\bar{\beta}_1\beta_1+\bar{\beta}_2\beta_2), \quad (19)$$
$$|\rho^+\rangle=-\bar{\beta}_2\beta_1,$$

and

$$|(K^*)^+\rangle=\bar{\beta}_0\beta_1,$$

where $(\bar{\beta}_i\beta_j)$ has the same transformation property as the state $b_i'^\dagger b^{\dagger j}|0\rangle$. The other five states ρ^-, ρ^0, $(K^*)^0$, etc., can be obtained from ρ^+ and $(K^*)^+$ through the use

[14] The choice of the baryon decuplet is rather free at this stage. In a model it is easy to invent selection rules such that in the absence of the secondary interaction h, the baryon decuplet is completely stable. As an example, we may arbitrarily assume the existence of an SU_3 fermion singlet γ with baryon number 1, but charge -1. The compound state $(\alpha^3\gamma)_{10}$ can represent the baryon decuplet. By adding an appropriate term to Eq. (15) for the secondary interaction h, both the energy shifts and the transitions of the decuplet can be regarded as caused by h. To first order in h, we have the well-known equal level spacing formula for the decuplet. The width of these levels are second order in h. Therefore, the use of perturbation theory could be easily justified. (However, if we compare these results with their experimental values, the fact that the width of N^*, the $I=\frac{3}{2}$ member of the baryon decuplet, is not much smaller than the decuplet energy spacings makes it questionable whether this particular example of the decuplet has much value other than being a simple concrete mathematical example. We wish to thank K. Huang and F. E. Low for pointing out the importance of the large width to us.)

[15] As a model, we may assume the existence of an additional neutral (SU_3) singlet field χ. The primary forces between $\bar{\beta}_i$ and β_j are generated by a Lagrangian of the form $\chi\,\sum_i b^{\dagger i}b_i$ plus a similar term for the antiparticle $\bar{\beta}_i$. Therefore, the primary forces between $\bar{\beta}_i$ and β_j are independent of i and j. On the other hand, because of the difference in spin and statistics the forces between, say, $\bar{\alpha}_i$ and α_j are due to a completely different origin and do depend on the particular singlet or octet representation of $(\bar{\alpha}\alpha)$.

of isotopic spin rotation and charge conjugation. In terms of the two $I=0$ states ω and ϕ, the unitary singlet $(\bar{\beta}\beta)_1$ becomes[16]

$$(\bar{\beta}\beta)_1=(\tfrac{1}{3})^{1/2}(\bar{\beta}_0\beta_0+\bar{\beta}_1\beta_1+\bar{\beta}_2\beta_2)=(\tfrac{1}{3})^{1/2}\phi+(\tfrac{2}{3})^{1/2}\omega\,, \quad (20)$$

while the $I=0$ member of the octet is given by

$$(\tfrac{1}{6})^{1/2}(2\bar{\beta}_0\beta_0-\bar{\beta}_1\beta_1-\bar{\beta}_2\beta_2)=(\tfrac{2}{3})^{1/2}\phi-(\tfrac{1}{3})^{1/2}\omega\,. \quad (21)$$

From (21) and the general transformation property of h [Eq. (2)], it follows (as will be proved under more general conditions in the next section) that the masses of these resonances satisfy[16a]

$$(2M_\phi+M_\omega)+M_\rho=4M_{K^*}\,, \quad (22)$$

which agrees very well with the known experimental values of these masses. [Note that similarly to the baryon mass formula Eq. (10), to first order in h Eq. (22) can also be written as $2M_\phi^2+M_\omega^2+M_\rho^2=4M_{K^*}^2$.] If we use (15) as the explicit form of h, then these masses can be explicitly calculated in terms of m_0' and m_1' of the triplet masses. We find, by using (19),

$$M_\phi=M_1+2m_0'\xi\,,$$
$$M_\omega=M_1+2m_1'\xi\,,$$
$$M_\rho=M_1+(m_1'+m_2')\xi\,,$$

and

$$M_{K^*}=M_1+(m_0'+m_1')\xi\,, \quad (23)$$

where M_1 is the zeroth-order mass of these nine states and ξ is the expectation value

$$\langle\textstyle\sum_{\mathbf{k}}(M'^2+k^2)^{-1/2}M'b^{\dagger 0}(\mathbf{k})b_0(\mathbf{k})\rangle \quad (24)$$

evaluated for the state $(\bar{\beta}_0\beta_0)$ in its center-of-mass system. The precise numerical value of ξ depends on the wave function $(\bar{\beta}_0\beta_0)$. (See Appendix II for an example of such a calculation.) From (23) and (18) we obtain another interesting relation:

$$M_\rho=M_\omega\,, \quad (25)$$

which agrees reasonably well with the experimental values.

In the above, the mass formulas (23) and (25) are derived from the simple form (17) for h_β. Actually these formulas are valid under a general class of secondary interactions. We may consider, in addition to (17), the h_β part of the secondary interaction to contain a general quartic dependence of the $b_i(\mathbf{k})$, $b'^i(\mathbf{k})$ and their Hermitian conjugates. Let h_β be given by

$$h_\beta=(17)+\lambda_1P_0{}^0+\lambda_2(P_1{}^1+P_2{}^2)+\lambda_3Q_0{}^0+\lambda^4(Q_1{}^1+Q_2{}^2)\,, \quad (26)$$

where

$$P_i{}^j=(b^{\dagger l}b_l)b_i'^{\dagger}b'^j+b^{\dagger j}b_i(b_l'^{\dagger}b'^l)\,, \quad (27)$$

and

$$Q_i{}^j=\epsilon_{ikl}\epsilon^{jmn}(b_m'^{\dagger}b'^k)(b^{\dagger l}b_n)\,, \quad (28)$$

in which we suppress the momentum dependence and sum over all repeated indices. The λ_1, λ_2, λ_3, λ_4 are constants and ϵ_{ijk}, ϵ^{ijk} are the third-rank antisymmetric tensors whose only nonvanishing components are ±1 depending on (ijk) being even or odd permutations of (012). It can be readily verified that the mass formula (22) holds if $\lambda_4=0$ and the mass formula (25) holds if, in addition, $\lambda_3=0$. Comparison with the experimental values of these masses indicate $\lambda_4=0$ and $\lambda_3\cong0$.

Physically, the mass formula (22) holds if the secondary forces between $(\bar{\beta}_i\beta_j)$ do not convert the pair $(\bar{\beta}_0\beta_0)$ into any other pairs such as $(\bar{\beta}_1\beta_1)$ and $(\bar{\beta}_2\beta_2)$. The mass formula (25) holds if, in addition, the secondary forces between $(\bar{\beta}_1\beta_1)$, $(\bar{\beta}_2\beta_2)$ and $(\bar{\beta}_2\beta_1)$ are approximately independent of the total isotopic spin $I=0$ or 1 of the compound system.

These massive triplets α_i and β_i, if they exist, can only be strongly produced in pairs by using the octet particles as the beam and the target. Their decays depend on the particular form of the secondary interaction h. If the SU_3 symmetry violating part of this secondary interaction transforms like the isotopic singlet member of an octet, then these two triplets would contain stable members with respect to strong interactions. If, in addition, the usual vector and axial vector currents of the strongly interacting particles that occur in weak interactions also transform like members of an octet, then the basic triplets would contain members that are absolutely stable.

III. FURTHER DISCUSSIONS OF THE MODEL

In this section we wish to investigate in the above model some general conditions under which the primary interaction H_0 should contain nine degenerate $(\bar{\beta}\beta)$ states but only eight degenerate $(\bar{\alpha}\alpha)$ states and eight degenerate $(\bar{\alpha}\beta)$ states. It is useful to introduce the following operators:

$$T_i{}^j=\textstyle\sum_{\mathbf{k}}[a^{\dagger j}(\mathbf{k})a_i(\mathbf{k})-a_i'^{\dagger}(\mathbf{k})a'^j(\mathbf{k})+b^{\dagger j}(\mathbf{k})b_i(\mathbf{k})-b_i'^{\dagger}(\mathbf{k})b'^j(\mathbf{k})]\,, \quad (29)$$

$$U_i{}^j=T_i{}^j-\tfrac{1}{3}\delta_i{}^jT_k{}^k\,, \quad (30)$$

$$S_i{}^j=\eta\textstyle\sum_{\mathbf{k}}[b^{\dagger j}(\mathbf{k})b_i(\mathbf{k})+b_i'^{\dagger}(\mathbf{k})b'^j(\mathbf{k})]\,, \quad (31)$$

and

$$V_i{}^j=S_i{}^j-\tfrac{1}{3}\delta_i{}^jS_k{}^k\,, \quad (32)$$

where

$$\eta=\tfrac{1}{2}[1+(-1)^{N_\alpha}]\,, \quad (33)$$

and

$$N_\alpha=\textstyle\sum_{\mathbf{k}}[a^{\dagger i}(\mathbf{k})a_i(\mathbf{k})-a_i'^{\dagger}(\mathbf{k})a'^i(\mathbf{k})]\,. \quad (34)$$

[16] The ϕ and ω mesons are therefore mixtures of a unitary singlet and the $I=0$ member of an octet. The mixing angle obtained through (20) and (21) agrees numerically with the one calculated on semiempirical ground by J. J. Sakurai [Phys. Rev. **132**, 434 (1963)]. See also M. Gell-Mann (Ref. 1); S. Okubo, Phys. Letters **5**, 165 (1963).

[16a] *Note added in proof.* After this paper was submitted for publication, we received an unpublished report by G. Zweig (An SU_3 model for strong interaction symmetry and its breaking) in which a similar mass formula for the vector mesons is also discussed in the model of a single triplet with fractional charge and baryon number.

All repeated indices are to be summed over. The U_i^j are the generators of the SU$_3$ transformation and satisfy the commutation relation

$$[U_i^j, U_l^k] = \delta_i^k U_l^j - \delta_l^j U_i^k. \quad (35)$$

The invariance property of the primary interaction H_0 is given by the commutation relation

$$[H_0, U_i^j] = 0. \quad (36)$$

For clarity, we will restrict our discussions to those eigenstates of H_0 which consist of only the free single-particle states of either α_i, or β_i (or $\bar{\alpha}_i$, $\bar{\beta}_i$) and the various bound states of these particles. Let the totality of these eigenstates span a subdomain R in the entire Hilbert space. The projection $O(R)$ of any operator O in R is defined by

$$\langle\nu|O(R)|\mu\rangle = \langle\nu|O|\mu\rangle \quad (37)$$

if both states $|\mu\rangle$ and $|\nu\rangle$ are in R; otherwise,

$$\langle\nu|O(R)|\mu\rangle = 0. \quad (38)$$

Theorem 1. If

$$[H_0, V_0^0(R)] = 0, \quad (39)$$

then under the primary interaction, $(\bar{\beta}\beta)_1$ must be degenerate with $(\bar{\beta}\beta)_8$. However, Eq. (39) does not imply any degeneracy between $(\bar{\alpha}\beta)_1$ and $(\bar{\alpha}\beta)_8$, nor between $(\bar{\alpha}\alpha)_1$ and $(\bar{\alpha}\alpha)_8$.

Proof: From (33), it follows that $\eta = 0$ or 1 depending on N_α = odd or even. Let the states $|\phi\rangle$ and $|\omega\rangle$ be defined by Eq. (19). We find

$$V_0^0|\phi\rangle = \tfrac{4}{3}|\phi\rangle,$$
$$V_0^0|\omega\rangle = -\tfrac{2}{3}|\omega\rangle,$$
$$V_0^0|\rho\rangle = -\tfrac{2}{3}|\rho\rangle,$$

and

$$V_0^0|K^*\rangle = \tfrac{1}{3}|K^*\rangle. \quad (40)$$

From the definition Eq. (37), it is clear that identical equations are satisfied by $V_0^0(R)$. We note that the mass splittings given by Eq. (23) are proportional to the values of V_0^0 for these states. Since V_0^0 [therefore, also $V_0^0(R)$] does not commute with either U_0^i or U_i^0, a necessary consequence of the commutation relations (36) and (39) is that $(\bar{\beta}\beta)_1$ must be degenerate with respect to $(\bar{\beta}\beta)_8$ under the primary interaction H_0.

In contrast, $(\bar{\alpha}\beta)_1$, $(\bar{\alpha}\beta)_8$, $(\bar{\alpha}\alpha)_1$, and $(\bar{\alpha}\alpha)_8$ are all eigenstates of V_0^0 and $V_0^0(R)$ with the same eigenvalue 0. Thus, Eq. (39) does not imply any additional degeneracy, and the only degeneracy is that required by the SU$_3$ invariance. Under the primary interaction H_0, there are eight degenerate spin-$\frac{1}{2}$ baryon states $(\bar{\alpha}\beta)_8$, eight degenerate pseudoscalar meson states $(\bar{\alpha}\alpha)_8$ but *nine* degenerate spin-1 meson states $(\bar{\beta}\beta)_1$ and $(\bar{\beta}\beta)_8$.

The introduction of the secondary interaction h splits these degeneracies. According to Eq. (2), under the SU$_3$ transformations generated by U_i^j, the secondary interaction $h = h_0 + h_1$ transforms like a sum of an SU$_3$ invariant term h_0 and another term h_1 which transforms like isotopic spin = 0 member of an octet. For the octets $(\bar{\alpha}\alpha)_8$ and $(\bar{\alpha}\beta)_8$ the mass formulas Eqs. (6) and (10) hold. The following theorem gives a condition under which the mass formula Eq. (22) for the nine spin-1 meson states holds:

Theorem 2. If

$$[h(R), V_0^0(R)] = 0, \quad (41)$$

then to first order in h, the mass formula

$$2M_\phi + M_\omega + M_\rho = 4M_{K^*}$$

holds.

Proof. By using Eqs. (37), (40), and (41), it follows that

$$\langle\omega|h|\phi\rangle = \langle\omega|h(R)|\phi\rangle = 0. \quad (42)$$

Therefore, Eq. (21) remains correct. Theorem 2 is then established by using Eq. (21), together with the general transformation property of h under the SU$_3$ group generated by U_i^j.

So far, our discussions are restricted to the single-particle states and to the bound states of H_0. The extension of some of the above considerations to the multi-particle states leads to several unusual results. Suppose we replace the projection $V_0^0(R)$ in Eq. (39) by the entire operator V_0^0, and assume that

$$[H_0, V_0^0] = 0 \quad (43)$$

is valid. Now, H_0 commutes with U_i^j which, however, does not commute with V_0^0. As proved in Appendix III, Eqs. (36) and (43) necessitate the invariance of the primary interaction H_0 under a group $G = \mathrm{SU}_3 \times \mathrm{SU}_3 \times \mathrm{SU}_3$ which has, among other irreducible representations, a $(\bar{\beta}\beta)$ nonet, a $(\bar{\alpha}\alpha)$ octet and a $(\bar{\alpha}\beta)$ octet. The group G also contains the particular SU$_3$ group whose generators are the nonlocal operators V_i^j given by Eq. (32). It is also shown in Appendix III that several unphysical consequences, such as violation of the asymptotic condition and the violation of crossing symmetry occur as a result. Therefore, Eq. (43) cannot be strictly correct. Nevertheless, it is interesting to explore other physical consequences by assuming the approximate validity of (43). Let v be the eigenvalue of V_0^0. From (43), we have the approximate selection rule

$$\Delta v = 0. \quad (44)$$

The $(\bar{\alpha}\alpha)_8$ states are eigenstates of V_0^0 with eigenvalue $v=0$. Thus, the decay of the $(\bar{\beta}\beta)$ nonet is forbidden under the primary interaction H_0. The decays of these particles such as $\rho \to 2\pi$, etc., occur through the secondary interaction h. Next, let us consider the compound state $(\bar{p}+p)$ which has $v = -\frac{2}{3}$. The conservation law (44) and Eq. (40) yield the following types of approximate selection rules:

$$\bar{p}+p \nleftrightarrow l\pi + mK + n\eta, \quad (45)$$

$$\bar{p}+p \nleftrightarrow K^* + l\pi + mK + n\eta, \quad (46)$$

and

$$\bar{p}+p \nrightarrow \phi+l\omega+mK+n\eta\,, \tag{47}$$

where $l\pi$, mK, $n\eta$ denote arbitrary numbers of π, K (or $\bar{K}$), and η. On the other hand, reactions

$$\bar{p}+p \rightarrow \rho+l\pi+mK+n\eta \tag{48}$$

and

$$\bar{p}+p \rightarrow \omega+l\pi+mK+n\eta \tag{49}$$

are allowed, provided other quantum numbers such as strangeness and isotopic spin are conserved. Similarly, we can consider the compound state $(\pi+p)$ and derive the approximate selection rule

$$\pi+p \nrightarrow (\bar{\beta}\beta)_9+(\bar{\alpha}\beta)_8+l\pi+mK+n\eta\,, \tag{50}$$

where $(\bar{\beta}\beta)_9$ and $(\bar{\alpha}\beta)_8$ represent, respectively, any member of the vector nonet and the baryon octet. The realistic value of such extension of the model is expected to be limited. As mentioned earlier, there are severe inherent difficulties of the commutation rule (43) which must throw doubt on the approximate validity of Eq. (44) for these collision processes. Furthermore, additional violation can occur through the secondary interaction h which can give sizeable contributions to the rates for these reactions. It is, therefore, rather surprising to find some of the above results, Eqs. (47)–(49), concerning the annihilation process $(\bar{p}+p)$ seem to be approximately correct.[17]

IV. CONCLUDING REMARKS

The above discussions on strong interactions several topics have been discussed, which are related but of rather different speculative nature. If the approximate SU_3 symmetry has a fundamental basis, there should exist two entirely different classes of strong interactions: the SU_3 symmetric primary interaction and the nonsymmetric secondary interaction, the former being much stronger than the latter. The fact that the masses of the known pseudoscalar mesons are of the same order of magnitude as their energy differences makes it attractive to regard all these meson masses as generated by the same secondary interaction. The zero mass requirement for the pseudoscalar meson octet gives a clear phenomenological definition of the primary interaction.

The connection between SU_3 symmetry and the approximate zero mass nature of pseudoscalar mesons leads to an understanding of the mass-squared formula for the pseudoscalar meson octet; it also ties in with work done in the last few years by many authors who consider the zero mass approximation independently of SU_3 symmetry (e.g., conserved axial vector current,[8] Goldberger-Treiman formula,[9,11] emission of soft pions,[18] etc.).

As already emphasized in Sec. II, the invariance character of the primary interaction under the SU_3 transformation and the existence of a large energy scale $\sim$ a few BeV's seems to lead rather naturally to a model containing some massive triplets with very strong interactions. In the model, so far as the transformation properties of the known meson and baryon states are concerned, they can be regarded as the same as the composite systems of these triplets. While many different models can be constructed,[19] the fact that there are nine vector states but only eight pseudoscalar states makes it attractive to consider a model which has two different triplets α_i and β_i. The derivation of the mass formula Eq. (22) for the nine vector states gives added support to the general character of this particular model. If the approximate degeneracy of these nine states is regarded as nonaccidental, then the primary interaction should be invariant under a group much larger than a simple SU_3.

ACKNOWLEDGMENTS

We wish to thank R. Serber and G. C. Wick for discussions. One of us (F.G.) would like to acknowledge a grant from the National Science Foundation made available through the Institute for Advanced Study and to express his thanks to Professor J. R. Oppenheimer for his kind hospitality at the Institute. He is also grateful to the U. S. Educational Exchange Commission for the award of a Fulbright travel grant.

APPENDIX I: A TRIVIAL EXAMPLE

In order to illustrate the use of perturbation formula for particles which, in the absence of perturbation, have zero masses, we consider a totally trivial example of an octet of free spin-0 particles. Let ϕ_μ $(\mu=1, 2, \cdots 8)$ represent the Hermitian field operators of these particles. The Lagrangian L_0 for the primary interaction is

$$L_0=-\frac{1}{2}\int \sum_{\lambda=1}^{4}\sum_{\mu=1}^{8}\left(\frac{\partial\phi_\mu}{\partial x_\lambda}\right)d^3r \tag{I1}$$

and the perturbation Lagrangian is

$$-h=-\frac{1}{2}\int \sum_{\mu} m_\mu{}^2\phi_\mu{}^2 d^3r\,, \tag{I2}$$

where h is the Hamiltonian for the secondary interaction and x_λ are the usual space-time coordinates. It is convenient to expand the ϕ_μ in terms of the annihilation operator $a_\mu(k)$ and their Hermitian conjugates $a_\mu{}^\dagger(k)$:

$$\phi_\mu(\mathbf{r})=\sum_{\mathbf{k}}(2\Omega k)^{-1/2}[a_\mu(\mathbf{k})\exp(i\mathbf{k}\cdot\mathbf{r}) + a_\mu{}^\dagger(\mathbf{k})\exp(-i\mathbf{k}\cdot\mathbf{r})]\,, \tag{I3}$$

[17] We wish to thank J. Steinberger for a discussion on these aspects.

[18] Y. Nambu and D. Lurié, Phys. Rev. **125**, 1429 (1962).

[19] M. Gell-Mann, Phys. Letters (to be published). In Appendix IV, we give a list of other simple models. For a formal treatment with two triplets, see also J. J. Sakurai, in *Varenna Lecture Notes, Proceedings of the International School of Physics, Course 26* (Academic Press Inc., New York, 1962).

where Ω is the volume of the system, $k=|\mathbf{k}|$ and $\mathbf{k}$ is the three-momentum conjugate to the space coordinate $\mathbf{r}$. The Hamiltonians H_0 and h for the primary and the secondary interactions become

$$H_0=\sum_{\mathbf{k}}\sum_{\mu} \tfrac{1}{2}k[a_\mu^\dagger(\mathbf{k})a_\mu(\mathbf{k})+a_\mu(\mathbf{k})a_\mu^\dagger(\mathbf{k})] \quad \text{(I4)}$$

and

$$h=\sum_{\mathbf{k}}\sum_{\mu} (4k)^{-1}m^2[a_\mu^\dagger(\mathbf{k})a_\mu(\mathbf{k})+a_\mu(\mathbf{k})a_\mu^\dagger(\mathbf{k}) +a_\mu(\mathbf{k})a_\mu(-\mathbf{k})+a_\mu^\dagger(\mathbf{k})a_\mu^\dagger(-\mathbf{k})]. \quad \text{(I5)}$$

Let $|\text{vac}\rangle_0$ and $|\mu,\mathbf{p}\rangle_0$ be the ground state and the one (μth) particle state with three-momentum $\mathbf{p}$ for the primary interaction. The zeroth-order energy of the μth particle $E_\mu^0(\mathbf{p})$ is given by

$$E_\mu^0(\mathbf{p})=\langle\mu,\mathbf{p}|H_0|\mu,\mathbf{p}\rangle_0-\langle\text{vac}|H_0|\text{vac}\rangle_0. \quad \text{(I6)}$$

By using (I4) we have

$$E_\mu^0(\mathbf{p})=p=|\mathbf{p}|. \quad \text{(I7)}$$

The first-order energy shift produced by h is easily seen to be

$$\delta E_\mu(\mathbf{p})=m_\mu^2/(2p). \quad \text{(I8)}$$

Therefore, a mass formula similar to Eq. (6) results if we identify the eight states $|\mu,\mathbf{p}\rangle$ to be the real 8×8 representation of the SU_3 group and if we choose the appropriate values m_μ such that h transforms according to Eq. (2) under SU_3. The exact energy $E_\mu(\mathbf{p})$ is given by

$$E_\mu(\mathbf{p})=\langle\mu,\mathbf{p}|H_0+h|\mu,\mathbf{p}\rangle-\langle\text{vac}|H_0+h|\text{vac}\rangle, \quad \text{(I9)}$$

where $|\text{vac}\rangle$ and $|\mu,\mathbf{p}\rangle$ are, respectively, the ground state and the one-particle state of the total Hamiltonian (H_0+h). The familiar formula

$$E_\mu(\mathbf{p})=p+\tfrac{1}{2}(m_\mu^2/p)-\tfrac{1}{8}(m_\mu^4/p^3)+\cdots =(p^2+m_\mu^2)^{1/2} \quad \text{(I10)}$$

can be obtained either by applying the usual perturbation series[20] to (I9) or by using the canonical transformation,

$$a_\mu(\mathbf{k})=[\cosh\theta_\mu(k)]b_\mu(\mathbf{k})+[\sinh\theta_\mu(k)]b_\mu^\dagger(-\mathbf{k}), \quad \text{(I11)}$$

where

$$\tanh[2\theta_\mu(k)]=-(2k^2+m_\mu^2)^{-1}m_\mu^2, \quad \text{(I12)}$$

which changes the total Hamiltonian to its diagonal form

$$(H_0+h)=\tfrac{1}{2}\sum_{\mathbf{k}}\sum_{\mu} (k^2+m_\mu^2)^{1/2} \times[b_\mu^\dagger(\mathbf{k})b_\mu(\mathbf{k})+b_\mu(\mathbf{k})b_\mu^\dagger(\mathbf{k})]. \quad \text{(I13)}$$

Identical results can be derived for particles of arbitrary spin by regarding the mass term as the perturbation h. If h is a linear function in the masses, then the lowest order energy shift δE must depend quadratically on h.

APPENDIX II: A NONTRIVIAL BUT UNREALISTIC EXAMPLE

In this Appendix, we shall take advantage of the known solutions of the Bethe-Salpeter equation obtained by Wick[21] and Cutkosky[22] and consider another less trivial but still unrealistic example which, however, will serve to illustrate further the origin of the "mass squared" formula for zero mass bound states. We consider a basic massive triplet (under SU_3) of spin-0 particles (A_0,A_1,A_2) with charge $(q, q+1, q)$ where q is any number including zero. Under the isotopic spin rotation A_0 behaves like $I=0$ and (A_1,A_2) like $I=\frac{1}{2}$. There exist very strong attractive forces between A_i and the antiparticles $\bar{A}_j$ which we assume to be of the same form as that generated through a zero mass zero spin field. Furthermore we suppose that the Bethe-Salpeter equation is the correct equation for the compound system A_i and $\bar{A}_j$.

In the absence of the secondary interaction h, the zeroth-order masses M_i^0 of the triplet A_i are all equal:

$$M_1^0=M_2^0=M_3^0=M. \quad \text{(II1)}$$

Let ψ_i^j be the wave function describing the compound system of A_i and $\bar{A}_j$. The Bethe-Salpeter equation for ψ_i^j is given in this case by

$$(p_i^2+M^2)(p_j^2+M^2)\psi_i^j(p)=\left(\frac{f}{\pi^2}\right)\int\frac{d^4k}{(p-k)^2}\psi_i^j(k), \quad \text{(II2)}$$

where i and j can independently be 0, 1 or 2, p_i^2, and p_j^2 are the (4-momentum)^2 of A_i and $\bar{A}_j$, respectively, p is the relative momentum of the system. As shown by Wick,[21] this equation has zero mass bound state if the coupling constant f is given by

$$f=2M^2. \quad \text{(II3)}$$

There are altogether 9 such zero-mass zero-spin states corresponding to the different values of i and j. In the following, we assume f is indeed given by (II3).

Similar to Eq. (15), we assume the secondary interaction consists of *only* the mass shifts of the basic triplet:

$$m_i=M_i-M_i^0\geqq 0, \quad \text{(II4)}$$

where the inequality is used to exclude bound states with imaginary mass which appear in this unrealistic model. In order to ensure isotopic spin rotation invariance we must have

$$m_1=m_2. \quad \text{(II5)}$$

Using the explicit solutions given in Refs. 21 and 22 one can compute the corresponding mass shifts of these

[20] It is of interest to note that because of the totally trivial structure of this example, the zero-mass approximation for the mesons does not lead to any spurious infrared divergence. For a more complicated case the apparent infrared difficulty can be resolved by using the physical mass of the meson as an infrared cutoff. The only change in the validity of the perturbation series is that, instead of the condition $\lambda\ll 1$, we have $[\lambda\ln(M/m_\mu)]\ll 1$ where m_μ is the mass of the mesons and λ is given by Eq. (1).

[21] G. C. Wick, Phys. Rev. **96**, 1124 (1954).

[22] R. E. Cutkosky, Phys. Rev. **96**, 1135 (1954).

states ψ_i^j. We call these nine (zero spin) eigenstates $I_{1/2}$, I_1, I_0, and I_0', where the subscripts refer to the isotopic spin. To first order in m_i, the corresponding changes of masses of these bound states are given by

$$[m(I_1)]^2=5M[2m_1],$$
$$[m(I_{1/2})]^2=5M[m_0+m_1],$$
$$[m(I_0)]^2=5M[2m_1],$$

and

$$[m(I_0')]^2=5M[2m_0]. \tag{II6}$$

The above results are obtained from the relation

$$f=\tfrac{1}{2}(M_1+M_2)^2-(2/5)m^2, \tag{II7}$$

valid near $m^2=0$ and to first order in M_1-M_2. Here m is the mass of the bound state and M_1, M_2 are the masses of the bound particles. For $M_1=M_2$, this relation (II7) is readily derived from Eq. (54) of Wick (Ref. 21) by a perturbation expansion in m^2, and for $M_1\neq M_2$ by using an appropriate transformation. (See Sec. IV of Cutkosky's paper, Ref. 22.) It is instructive to observe that under the primary interaction ($M_i^0=M$), the octet and the singlet part of ψ_i^j are degenerate. Similar to Eqs. (19)–(21), we find the two isotopic spin=0 states are given by $(I_0)=(1/\sqrt{2})(\psi_1^1+\psi_2^2)$ and $(I_0')=\psi_0^0$, both of which are mixtures of the octet and the singlet representations under SU_3. Because of the zero mass nature of these states (in the absence of the secondary interaction) these masses satisfy the quadratic mass formulas

$$2[m(I_0')]^2+[m(I_0)]^2+[m(I_1)]^2=4[m(I_{1/2})]^2 \tag{II8}$$

and

$$[m(I_0)]^2=[m(I_1)]^2. \tag{II9}$$

APPENDIX III

In this Appendix it will be shown that if R, the restricted Hilbert space defined in Sec. III, is extended to the entire Hilbert space, then the degeneracy of the state $(\bar{\beta}\beta)_1$ with states $(\beta\bar{\beta})_8$, originally expressed by Eq. (39), and now replaced by Eq. (43), implies invariance of the primary Hamiltonian H_0 under a group $G=SU_3\times SU_3\times SU_3$.

The explicit expression of the operator V_0^0 appears in Eq. (43) is

$$V_0^0=\tfrac{1}{3}\eta\sum_{\mathbf{k}}[2b^{\dagger 0}(\mathbf{k})b_0(\mathbf{k})+2b_0'^{\dagger}(\mathbf{k})b'^0(\mathbf{k})-b^{\dagger 1}(\mathbf{k})b_1(\mathbf{k})$$
$$-b_1'^{\dagger}(\mathbf{k})b'^1(\mathbf{k})-b^{\dagger 2}(\mathbf{k})b_2(\mathbf{k})-b_2'^{\dagger}(\mathbf{k})b'^2(\mathbf{k})], \tag{III1}$$

where η is the projection operator given by (33). The fact expressed by (43) that the primary interaction H_0 commutes with V_0^0 and U_i^j, but V_0^0 does not commute with all the U_i^j implies that H_0 is invariant under a group G bigger than SU_3; otherwise $(\bar{\beta}\beta)_1$ would not be degenerate with $(\bar{\beta}\beta)_8$ under H_0.

To investigate the structure of G we first establish the commutation relations:

$$[U_i^j,V_k^l]=\delta_i^l V_k^j-\delta_k^j V_i^l \tag{III2}$$

which merely express the fact that V_k^l possesses tensor character under the group with generators U_i^j. Thus if U_i^j and V_0^0 are part of the generators for G, the latter must also contain all other V_i^j through the repeated operations such as $[U_i^j,V_0^0]$, $[U_l^k,[U_i^j,V_0^0]]$, etc.

For convenience we introduce the operators

$$B_i^j=\eta\sum_{\mathbf{k}}[b^{\dagger j}(\mathbf{k})b_i(\mathbf{k})-\tfrac{1}{3}\delta_i^j b^{\dagger l}(\mathbf{k})b_l(\mathbf{k})] \tag{III3}$$

and

$$B_i'^j=-\eta\sum_{\mathbf{k}}[b_i'^{\dagger}(\mathbf{k})b'^j(\mathbf{k})-\tfrac{1}{3}\delta_i^j b_l'^{\dagger}(\mathbf{k})b'^l(\mathbf{k})] \tag{III4}$$

in terms of which we can write

$$V_i^j=B_i^j-B_i'^j. \tag{III5}$$

Then, we find

$$[V_i^j,V_k^l]=\delta_i^l(B_k^j+B_k'^j)-\delta_k^j(B_i^l+B_i'^l) \tag{III6}$$

so that B_i^j and $B_i'^j$, are also contained in the set of generators for the group G. The analysis of the group structure is further facilitated by the introduction of the operators

$$A_i^j=U_i^j-(B_i^j+B_i'^j). \tag{III7}$$

The A_i^j, B_i^j, and $B_i'^j$ satisfy the commutation relations

$$[A_i^j,B_l^k]=[A_i^j,B_l'^k]=[B_i^j,B_l'^k]=0, \tag{III8}$$
$$[A_i^j,A_l^k]=\delta_i^k A_l^j-\delta_l^j A_i^k, \tag{III9}$$
$$[B_i^j,B_l^k]=\delta_i^k B_l^j-\delta_l^j B_i^k, \tag{III10}$$
$$[B_i'^j,B_l'^k]=\delta_i^k B_l'^j-\delta_l^j B_i'^k \tag{III11}$$

and

$$A_i^i=B_i^i=B_i'^i=0. \tag{III12}$$

Thus the A_i^j, B_i^j, $B_i'^j$ are generators of three independent SU_3 groups. Their direct product $SU_3\times SU_3\times SU_3$ defines the group G. It is important to note that if H_0 commutes with U_i^j and V_0^0, then it must also be invariant under the entire group G and we have

$$[H_0,A_i^j]=[H_0,B_i^j]=[H_0,B_i'^j]=0. \tag{III13}$$

In addition, H_0 transforms like the generator for time translations of the Lorentz group, is invariant under the rotation group, and satisfies the commutation relations

$$[H_0,C]=[H_0,P]=[H_0,N]=[H_0,Q]=0, \tag{III14}$$

where C, P, N, Q are, respectively, the operators for charge conjugation, parity, baryon number, and charge. The operators P, N, and Q commute with the entire group G. On the other hand, the charge-conjugation operator obeys the following relations:

$$CN+NC=0, \tag{III15}$$
$$CQ+QC=0, \tag{III16}$$
$$CB_i^jC^{\dagger}=-B_j'^i, \tag{III17}$$

and

$$CA_i^jC^{\dagger}=-A_j^i. \tag{III18}$$

The eigenstates of H_0 can be labeled by three pairs of numbers, each pair characterizing a definite representation of each independent SU_3 group. For representations with small dimensionality d, d is sufficient to define the irreducible representation apart from the conjugate representation. Hence, in this case, we may represent the eigenstates of H_0 by (x,y,z) where x, y, z are respectively the dimensionality of the irreducible representation under each of the SU_3 groups generated by A_i^j, B_i^j, and $B_i'^j$. Thus, the spin -1 multiplet is represented by (1,3,3), containing $1\times3\times3=9$ states with $N=0$. The pseudoscalar meson octet is represented by (8,1,1,) with $N=0$ and the baryon octet is (8,1,1) with $N=1$. It shoul l be pointed out that only the sufficient condition that G contains the correct varieties of multiplets is established.

The group G has been constructed artificially to accommodate the presently known variety of multiplets. It has many unusual features which we now proceed to discuss.

(1) The separate invariance of H_0 under B_i^j and $B_i'^j$ is a necessary result if the primary force between $\bar{\beta}_i$ and β_j is independent of i and j. For example, in the case of two spin-$\frac{1}{2}$ particles, if the force between these two particles is spin-independent, then the triplet state is degenerate with the singlet. The force is invariant under separate SU_2 transformations for these two spins. Therefore, the relevant group is $(SU_2\times SU_2)$ which has a representation containing 4 states.

(2) The projection operator η has been included in the definition of B_i^j and $B_i'^j$ in order to ensure the compatibility of having nine degenerate vector mesons with the fact that apparently only eight spin-$\frac{1}{2}$ baryons states [identified with $(\bar{\alpha}\beta)_8$] exist. Otherwise η should be replaced by 1 in Eqs. (III3) and (III4).

(3) As far as the properties of the vector meson states $|\Phi_i^j\rangle$ are concerned ,we can take $\eta=1$ and consider only the group $G'=SU_3\times SU_3$, generated by B_i^j and $B_i'^j$. We shall presently show that crossing symmetry is violated in the coupling of $|\Phi_i^j\rangle$ with β_i and $\bar{\beta}_j$. Consider, for example, the particular element of the group G' defined by

$$T(\lambda)=\exp[\tfrac{3}{2}i\lambda V_0^0]. \qquad \text{(III19)}$$

The transformation laws for the states $|\phi\rangle$, $|\beta_{0k}\rangle$, and $|\bar{\beta}_0\rangle$ under $T(\lambda)$ are

$$T|\phi\rangle=\exp(2i\lambda)|\phi\rangle, \qquad \text{(III20)}$$

$$T|\beta_0\rangle=\exp(i\lambda)|\beta_0\rangle, \qquad \text{(III21)}$$

and

$$T|\bar{\beta}_0\rangle=\exp(i\lambda)|\bar{\beta}_0\rangle. \qquad \text{(III22)}$$

It follows that the matrix element $\langle\phi|\beta_0,\bar{\beta}_0\rangle$ is invariant under T while the matrix element $\langle\beta_0|\beta_0,\phi\rangle$ is not. Thus, the process

$$\phi\to\beta_0+\bar{\beta}_0, \qquad \text{(III23)}$$

which is allowed by G', does not lead to the process

$$\beta_0\to\phi+\beta_0, \qquad \text{(III24)}$$

which is forbidden by G', in contradiction with crossing symmetry. Time reversal invariance is however valid since both $\langle\phi|\beta_0,\bar{\beta}_0\rangle$ and $\langle\beta_0,\bar{\beta}_0|\phi\rangle$ are invariant under G'.

It may also be noted that because creation and annihilation operators for the particle ϕ transform differently under (III19), a local Hermitian field operator that one might try to introduce to represent the bound state ϕ, would not have a definite transformation property under G'. Thus the group G' (or G) cannot be applied to local field operators for the nonet.

(4) An extension of G to the many-particle continuum states may lead to paradoxical results. For example, the state of one baryon octet $(\bar{\alpha}\beta)_8$ and one antibaryon octet $(\bar{\beta}\alpha)_8$ at infinity should be represented by (8,1,1) $\times$(8,1,1), with a multiplicity of 64, apart from other spin-momentum dependence. Yet, if we consider the two baryon octets as a single system, since $\eta=1$, the system is also represented by (x,y,z) where $x=1$ or 8 and $y=z=3$. The resulting multiplicity is no longer given by the product $(8,1,1)\times(8,1,1)$. This example illustrates the incompatibility of the group G with the asymptotic condition.

The paradox may be resolved by requiring the group G to be valid only for states which extend over a limited region in space where interaction takes place. Then, if we consider, for example, the collision of a baryon and an antibaryon with the production of a pseudoscalar meson and a vector meson we can write the sequence

$$(\bar{\alpha}\beta)_8+(\bar{\beta}\alpha)_8\to(x,y,z)_{\eta=1}\to(\bar{\alpha}\alpha)_8+(\bar{\beta}\beta)_9, \qquad \text{(III25)}$$

where the intermediate stage refers to the interaction region and $(\bar{\beta}\beta)_9$ denotes the vector meson nonet. In (III25) we must decompose both the initial and the final states into compound states $(x,y,z)_{\eta=1}$ which are representations of the group G. Clearly, only the compound states with $x=8$, $y=3$, and $z=3$ are involved in the process. Some of the particular results for $\bar{p}+p$ reactions have already been discussed in Sec. III.

(5) Restricting ourselves to the β particles and their bound state $|\Phi_i^j\rangle$ representing the nine vector mesons we may interpret the validity of the group $G'=SU_3\times SU_3$ as the mathematical expression of the stability of these bound states. Indeed, if we assume that β and $\bar{\beta}$ are bound by a neutral field χ transforming like a unitary singlet, then the annihilation and recombination process such as

$$\bar{\beta}_0+\beta_0\to\chi\to\bar{\beta}_0+\beta_0 \qquad \text{(III26)}$$

would contribute to the singlet state $(\bar{\beta}\beta)_1$ and not to the state $(\bar{\beta}\beta)_8$, thus lifting the degeneracy between the nine members of the vector meson multiplet. The Bethe-Salpeter type approximations for the two-body amplitude assume implicity that the contributions from the process (III26) are negligible, thus regarding the compound system $(\bar{\beta}_0\beta_0)$ as stable (independent of its mass value). Such an approximation leads immediately to the degeneracy of the nine mesons. The validity of the group G' (and G in general) thus emerges as an approximate dynamical group describing the independence of the $(\beta\bar{\beta})$ binding forces from the unitary spin together with the stability of the nonet.

APPENDIX IV

In this Appendix we list the simplest models involving one fundamental triplet. If no singlets are allowed, one is led to Gell-Mann's "quark" scheme[19] with noninteger charge q and baryon number n (unless the definitions of charge and baryon number are modified for each multiplet). The requirement of integer q and n leads one to consider schemes with at least one singlet α and one triplet β. The triplet components β_0, β_1 and β_2 are assumed to have, respectively, the charges $(q,\ q+1,\ q)$ and q is called the charge of the triplet. The case in which α and β have $q=0$, α being a baryon and β a $n=0$ fermion triplet has already been studied by Gell-Mann.[19] This is model I′ in our table. A variation on this model, in which the baryon number, the charge and the strangeness are all introduced once is the model I with β being a $n=0$ boson. Other simple schemes in which singlets, octets, and decuplets may arise from the direct product $3\times3\times3$ of the fundamental triplet with itself are provided by models II and III. Finally, in model IV, which involves one triplet and two singlets we give an example in which two kinds of octets and decuplets with the usual charge structure may arise, one set from $3\times\bar{3}$ and $(3\times\bar{3})(3\times\bar{3})$ and another set from $(3\times3\times3)$. We now have the possibility of a selection rule that prevents a decuplet belonging to the second set from decaying into two octets from the first set through SU_3 preserving interactions. In such a scheme we can have stable multiplets.

The selection rule is connected with the additional gauge transformation

$$\alpha\to\alpha,\quad \beta\to e^{iu}\beta \tag{IV1}$$

that becomes possible when a fundamental singlet exists as well as a triplet β. As long as u-invariant representations are considered, as in I and I′, this gauge group gives nothing new. However, when baryon multiplets having different "u" charge exist, as in model IV, a new selection rule arises.

Simple assumptions concerning the nature of the binding forces are shown in the table to illustrate how some of the representations that do not seem to occur for mesons and baryons can be eliminated on physical grounds. The other entries in the table are self-explanatory.

In connection with these simple models it should also be remarked that the most general Gell-Mann–Okubo mass splitting within a unitary multiplet is not obtained

TABLE I. A list of other models.

	Models	N	Spin	Q	Forces (R=repulsive, A=attractive)	Mesons ($N=0$)	Baryons ($N=1$)	Remarks
	α^0	1	$\frac{1}{2}$	0	$\beta\beta(R)$	$(\alpha^0\bar{\alpha}^0)_1$, $(\beta\bar{\beta})_1$	$\alpha^0(\beta\bar{\beta})_8$, $\alpha^0(\beta\bar{\beta})_1$	Representations are invariant
I	(singlet)				$\alpha^0\bar{\alpha}^0(A)$, $\beta\bar{\beta}(A)$	$(\beta\bar{\beta})_8$	$\alpha^0[(\beta\bar{\beta})(\beta\bar{\beta})]_{1,8,10,27}$	under the "u" gauge. If $\beta\beta(A)$
	β	0	0	0	$\alpha^0\bar{\beta}(A)$, $\alpha^0\beta(A)$			is allowed new mesons states
	(triplet)							arise: $(\beta\beta)_{\bar{3}}$, $(\beta\beta)_6$, $(\beta\beta\beta)_{1,8,10}$
	α^0	1	$\frac{1}{2}$	0	same	same	same	Discussed by Gell-Mann (Ref. 19)
I′	(singlet)							
	β	0	$\frac{1}{2}$	0				
	(triplet)							
	α^+	2	0	1	$\beta\beta(R)$	$(\alpha^+\bar{\alpha}^+)_1$, $(\beta\bar{\beta})_1$	$\bar{\alpha}^+(\beta\beta\beta)_1$	Possible additional states:
II	(singlet)				$\alpha^+\bar{\alpha}^+(A)$, $\bar{\alpha}^+\beta(A)$	$(\beta\bar{\beta})_8$	$\bar{\alpha}^+(\beta\beta\beta)_8$	$(\alpha^+\bar{\beta})_3$: $N=1$, $Q=(1,0,1)$
	β	1	$\frac{1}{2}$	0	$\beta\bar{\beta}(A)$		$\bar{\alpha}^+(\beta\beta\beta)_{10}$	$\bar{\alpha}^+(\beta\beta)_6$, $\bar{\alpha}^+(\beta\beta)_{\bar{3}}$: $N=0$,
	(triplet)							$Q=(1,\ 0,\ -1)$ containing a meson with $Q=-1$, $Y=-2$
	α^-	1	$\frac{1}{2}$	-1	same	$(\alpha^-\bar{\alpha}^-)_1$, $(\beta\bar{\beta})_1$	$\alpha^-(\beta\beta\beta)_1$	Possible additional states:
III	(singlet)					$(\beta\bar{\beta})_8$	$\alpha^-(\beta\beta\beta)_8$	$(\alpha^-\beta)_3$: $N=1$, $Q=(-1,\ 0,\ -1)$
	β	0	0	0			$\alpha^-(\beta\beta\beta)_{10}$	$\alpha^-(\beta\beta)_6$, $\alpha^-(\beta\beta)_{\bar{3}}$: $N=1$
	(triplet)							
	α^0, α^-	1	$\frac{1}{2}$	0, -1	same	$(\alpha^0\bar{\alpha}^0)_1$, $(\alpha^-\bar{\alpha}^-)_1$	$\alpha^0(\beta\bar{\beta})_1$, $\alpha^0(\beta\bar{\beta})_8$	Example of a forbidden decay:
IV	(singlets)					$(\beta\bar{\beta})_1$, $(\beta\bar{\beta})_8$	$\alpha^-(\beta\beta\beta)_1$, $\alpha^-(\beta\beta\beta)_{8,10}$	$\alpha^-(\beta\beta\beta)_{10}\not\leftrightarrow\alpha^0(\beta\bar{\beta})_8+(\beta\bar{\beta})_8$
	β	0	0	0				
	(triplet)							

by simply giving the $I=0$ member β_0 of the triplet a different mass than the $I=\frac{1}{2}$ members β_1 and β_2. An additional symmetry-breaking interaction Lagrangian is also needed. Furthermore, in any model which has only one triplet it is difficult to understand why there should be nine approximately degenerate spin-1 meson multiplets while there exist only eight approximately degenerate pseudoscalar meson states. For this reason, it appears that the special model discussed in Sec. II is a more realistic one.

Reprinted from:

PHYSICAL REVIEW D VOLUME 5, NUMBER 7 1 APRIL 1972

Scaling Properties and the Bound-State Nature of the Physical Nucleon*

S. D. Drell

Stanford Linear Accelerator Center, Stanford University, Stanford, California 94305

and

T. D. Lee

Columbia University, New York, New York 10027

(Received 20 December 1971)

The scaling property in deep-inelastic electron scattering is established by regarding the physical nucleon as a bound state of a bare nucleon and a bare meson (or a few bare mesons). This bound-state formulation provides a fully relativistic generalization of the "parton" model that is no longer restricted to infinite-momentum frames. It also connects the scaling property in inelastic processes with the rapid decrease of the electromagnetic form factors in elastic scattering. Rigorous statements are derived for specific bound-state solutions of the Bethe-Salpeter equation with the ladder approximation. An Adler sum rule is derived and crossing properties are discussed. A general phenomenological approach is developed which is relativistically covariant and gauge-invariant, and which allows one to correlate directly the observed structure functions and form factors with the appropriate bound-state wave function. If all constituents in the bound state are assumed to be of masses $\lesssim 1$ GeV, the model gives a qualitative understanding as to why the scaling property is experimentally observed at relatively moderate energies.

I. INTRODUCTION

While the validity of the scaling hypothesis[1] has been well established by recent extensive experimental investigations on inelastic electron-proton scattering,[2] its theoretical basis remains in an unsatisfactory state. The original parton idea of Feynman puts a special emphasis on the infinite-momentum frame of reference. It is suggested that in the infinite-momentum frame, the electromagnetic property of the assumed pointlike constituents of the physical nucleon can be treated as that of an assembly of independent free particles. The "infinite-momentum frame," by itself, is clearly not a Lorentz-invariant concept. Furthermore, one can easily show[3] that, in general, the direction of the infinite momentum cannot be arbitrary. It must be limited to a certain restrictive set of directions, depending on the virtual-photon momentum; otherwise, the mass of each of the pointlike constituents has to be lighter than that of the physical proton, and that would be too unphysical. Naturally, this leads to questions of whether such an *ad hoc* rule can be derived from a relativistically invariant theory.

In the literature, there have been several attempts to try to derive the scaling property from the usual relativistic local field theory. So far, the only success has been limited to either the trivial case of free particles (free except for their electromagnetic interaction), or the unphysical case of a superrenormalizable ϕ^3-type theory[4] in which all particles must be of zero spin. For the physically interesting case of spin-$\frac{1}{2}$ charged particles with some nonelectromagnetic interaction, straightforward calculation in lowest-order perturbative expansions leads to a logarithmic deviation from scaling behavior.[5] In order to derive scaling properties for such field theories with renormalizable but not superrenormalizable hadronic interactions, it has been necessary to introduce additional *ad hoc* rules, such as either a transverse momentum cutoff[6] or the so-called "formal manipulation" of current operators,[7] etc. However, at present the theoretical foundation of such rules appears to be quite uncertain. In particular, the transverse momentum cutoff in the field-theoretical derivation of scaling leads to a formalism and a scattering amplitude that are current-conserving only in the infinite-momentum frame and in the scaling region. Therefore, it is difficult to see how one may derive such an *ad hoc* cutoff procedure in a *bona fide* relativistic field theory.

The purpose of this paper is to point out that if one regards the physical nucleon as a bound state, then there exists a large class of relativistic field theories in which, at least for the deep-inelastic electron-nucleon scattering, the scaling property as well as the approach to scaling can be derived by using the conventional field-theoretical rules for bound states, provided radiative corrections are neglected.[8] In constructing the explicit bound-state wave function for the physical nucleon, there is, of course, a certain degree of arbitrariness with respect to both the nature of its constituents and that of the binding forces. In view of the experimental fact that the scaling limit seems to occur at a remarkably low energy range, it seems reasonable that the masses of the constituents and the relevant binding energy should all lie approximately in the 1-GeV range. Therefore, at least in terms of quantum numbers, these constituents should more closely resemble known particles than any unknown particles, such as quarks.[9] As a first example, we assume in Sec. II that the physical nucleon is simply a two-body bound state composed of a "bare" nucleon of spin $\frac{1}{2}$ and a "bare" meson of spin 0. The various quantum numbers (charge, spin, isospin, etc.) of the "bare" nucleon are assumed to be the same as those of the physical nucleon; for simplicity, we may assume its mass to be the same as, or at least comparable to, the physical nucleon mass. Similarly, for reasons of simplicity, the "bare" meson is assumed to be an SU_3 singlet, and its mass as well as other quantum numbers to be the same as those of the physical X^0 meson. There is, however, an important difference between a "bare" particle and the corresponding "physical" one. The electromagnetic form factors of a "bare" particle are always assumed to be *independent* of the 4-momentum transfer. In the language of the parton model, these "bare" particles are the "pointlike" constituents of the physical nucleon. Whether or not such pointlike "bare" particles that are introduced as constituents of a physical nucleon will ever be observed is an open question – as is also the case for "quarks" – at this time. (See, however, the discussion given in Sec. V below.)

To *illustrate* the relation between the scaling limit and the conventional field-theoretical rules for the bound state, we adopt the Bethe-Salpeter equation with the ladder approximation. The covariant potential responsible for the binding is assumed to be of the general form

$$V(q) = \int \frac{\sigma(\kappa^2)\, d\kappa^2}{q^2 + \kappa^2}, \qquad (1.1)$$

where q^2 denotes the square of the 4-momentum transfer. The integral $\int \sigma(\kappa^2)\, d\kappa^2$ is assumed to be convergent; thus,

$$V(q) \sim O(q^{-2}) \text{ as } q^2 \to \infty. \qquad (1.2)$$

As will be shown, this implies that as the relative momentum k between the two "bare" particles in the bound state approaches infinity, the wave function $\phi(k)$ of the Bethe-Salpeter equation has a similar asymptotic behavior[10]; i.e.,

$$\phi(k) \sim O(k^{-2}) \text{ as } k^2 \to \infty. \qquad (1.3)$$

By using this asymptotic property, one can then establish the existence of the scaling limit for the bound-state solution. The underlying picture for deep-inelastic electron scattering that emerges provides a fully relativistic generalization of the "parton" model that is no longer restricted to special infinite-momentum frames. It will be demonstrated that one may view the dynamic process as simply the physical proton dissociating into its bare constituents, with the electrically charged one propagating (invariantly with a Feynman propagator) until the instant when it is scattered onto its mass shell by the incident virtual photon. The final-state rescatterings between the bare constituents will be shown to vanish *in the scaling limit* so that the constituents may be regarded as being scattered independently of one another, just as in the usual impulse approximation. Furthermore, for the physical proton in which the charged constituent is a bare proton of spin $\frac{1}{2}$ so that $m_p W_1 = (1/2x)\nu W_2$, we find that as $x \to 1$ both the W_1 and νW_2 functions approach zero as $(1-x)^3$, where x is the customary scaling variable. This is rather encouraging since it is in good agreement with the present experimental result, unlike most spin-$\frac{1}{2}$ parton models[11] with an *ad hoc* transverse momentum cutoff which leads naturally to a linear $(1-x)$ dependence as $x \to 1$.

The same bound-state wave function can also be used to evaluate the electromagnetic form factors F_1 and F_2 of a physical nucleon. Both these form factors depend on the square of the wave function ϕ. It can be readily shown that, on account of (1.3), as the square of the 4-momentum transfer $q^2 \to \infty$, apart from factors of $\ln q^2$, both F_1 and F_2 are $O(q^{-4})$, in agreement with the theoretical conclusion previously reached by others,[10] and with the threshold rule[12] relating the power of $(1-x)$ in the W_1 function and the power of the q^2 dependence in the elastic form factors. This result has its parallel in the treatment of the nonrelativistic bound-state problem with the Schrödinger equation. Equation (1.3) is the Bethe-Salpeter analog of the condition that the wave function, and hence the charge density, is finite at the origin, from which it also follows that the form factor F_1 is $O(q^{-4})$ as $q \to \infty$.

In Sec. II, we also discuss the forward Compton scattering amplitude and prove an Adler sum rule, which provides a convenient normalization condition for the bound-state solution of the Bethe-Salpeter equation. In addition, the crossing property to the annihilation channel $e + \bar{e} \to \bar{p} +$ anything is discussed in the same section.

While the Bethe-Salpeter equation is useful in illustrating the relation between the scaling property and the bound-state system in a relativistic field theory, it has some obvious limitations for practical applications. Apart from the mathematical complexity of solving the Bethe-Salpeter equation, the covariant potential $V(q)$ is not known, the ladder approximation is not to be trusted, and, furthermore, one cannot expect a simple two-particle bound-state description of the physical nucleon to be an adequate one; due to virtual-meson exchanges, there must be some additional multiparticle components present. Thus, for further insight as well as for practical applications, we adopt in Sec. III a phenomenological approach by assuming that, instead, the bound-state wave function ϕ is known. The W_1 and νW_2 functions can then be directly evaluated in terms of a phenomenological set of diagrams which are based on the usual set of Feynman diagrams for the bound state. In particular, the implications of the physical picture of a proton as a bound state on the singularity structure and asymptotic behavior of these diagrams are explored. This phenomenological description also allows us to generalize the two-particle bound-state concept to include some multiparticle components, which in turn makes it possible for a more detailed comparison between the present experimental results and some simple model calculations.

In Sec. IV, a simple ansatz for the multiparticle final state is made and the results of this model for deep-inelastic electron scattering processes are given. In Sec. V we first consider some further applications of the model, such as the polarization effect, then discuss some open questions, including "Where are these bare constituents to be found in nature?", "What might be the final meson multiplicity?", and "Is scaling an exact law in the infinite-energy limit?", and finally we compare our approach with some related work of others.

II. BETHE-SALPETER EQUATION

In this section we shall discuss some general properties of the bound-state solution of a Bethe-Salpeter equation in the ladder approximation.[10,13] For definiteness, we assume the bound-state solution to represent the physical proton p, and that it is composed of a spin-$\frac{1}{2}$ particle P (the bare proton) of mass M and a spin-0 particle X^0 (the bare meson) of mass μ. It is convenient to use p_ν, P_ν, and X_ν to denote, respectively, the 4-momenta of the particles p, P, and X^0. Therefore, one has

$$p_\nu = P_\nu + X_\nu . \tag{2.1}$$

A. Bound-State Wave Function

The equation for the bound-state "vertex" function $\phi_p(k)$ can be written as[14]

$$\phi_p(k) = i\lambda \int V(k-k') K_p^{-1}(k')\phi_p(k')\, d^4k' , \tag{2.2}$$

where

$$K_p(k) \equiv (-i\gamma \cdot P - M + i\epsilon)(X^2 + \mu^2 - i\epsilon), \tag{2.3}$$

$\epsilon = 0+$, V is the covariant potential given by (1.1), λ is the coupling constant, k is the relative momentum given by the usual expression

$$k_\nu = (M+\mu)^{-1}(\mu P_\nu - MX_\nu), \tag{2.4}$$

the solution $\phi_p(k)$ is a 4-component Dirac spinor function, and the subscript p indicates that the total 4-momentum p_λ acts as a constant parameter for the integral equation. It is useful to work directly with the wave function $\psi_p(k)$, defined by

$$\psi_p(k) \equiv K_p^{-1}(k)\phi_p(k), \tag{2.5}$$

which differs from $\phi_p(k)$ only in the inclusion of the free propagator $K_p^{-1}(k)$. In terms of $\psi_p(k)$, Eq. (2.2) becomes

$$\phi_p(k) = K_p(k)\psi_p(k) = i\lambda \int V(k-k')\psi_p(k')\, d^4k' . \tag{2.6}$$

Both functions $\phi_p(k)$, $\psi_p(k)$ and the free propagator $K_p(k)$ depend implicitly on the total 4-momentum p_λ. For a given V, by setting

$$p^2 = -m_p^2, \tag{2.7}$$

Eq. (2.2), or (2.6), can be regarded as an eigenvalue equation for the coupling constant λ.

The asymptotic behavior of $\phi_p(k)$ at large k can be readily obtained from (2.6) by taking V outside the integral, provided that

$$\int \psi_p(k')\, d^4k' \text{ is finite}. \tag{2.8}$$

That this provision (2.8) is indeed a correct one will be established in Appendix A. From (1.2), (2.6), and (2.8), it follows then[15] that

$$\phi_p(k) \sim O(k^{-2}) \text{ as } k^2 \to \pm\infty . \tag{2.9}$$

(In Appendix A, the next-order correction term to this asymptotic behavior will also be given.) It is also a consequence of the provision (2.8) that in coordinate space the bound-state solution is regular at the origin. [See (A19) in Appendix A.] This is the relativistic analog of the familiar result for a nonrelativistic two-particle system. The bound-state solution of the Schrödinger equation for two particles interacting via a static Yukawa potential is also finite at the origin.

B. Conjugate Solution

In a collision process, if the bound-state system is present in the initial state, then one can simply use the solution $\phi_p(k)$. On the other hand, if the bound-state system is in the final state then one cannot simply use its Hermitian conjugate $\phi_p^\dagger(k)$.

This is because in the bound-state solution, each of the constituents, either P or X, can separately be on its mass shell, and therefore the $i\epsilon$ term in (2.3) is of importance. To derive the correct conjugate solution, called $\overline{\phi}_p(k)$, appropriate for the final-state description, we shall make use of the time-reversal operation.

It is convenient to first introduce a 4×4 matrix $\Gamma_p(k)$ which relates the bound-state wave function $\psi_p(k)$ to that of a free Dirac spinor u_p:

$$\psi_p(k)=\Gamma_p(k)u_p\,, \tag{2.10}$$

where u_p satisfies the free Dirac equation

$$(-i\gamma\cdot p-m_p)u_p=0\,. \tag{2.11}$$

The matrix $\Gamma_p(k)$ satisfies an equation identical to that of $\psi_p(k)$; i.e.,

$$K_p(k)\Gamma_p(k)=i\lambda\int V(k-k')\Gamma_p(k')\,d^4k'\,. \tag{2.12}$$

For definiteness, let us adopt the usual Pauli representation[14] of the Dirac matrices. Under the time-reversal operation, $\Gamma_p(k)$ becomes $[\Gamma_p(k)]_T$, which is related to the complex conjugate $\Gamma_p(k)^*$ by

$$\sigma_2[\Gamma_p(k)]_T\sigma_2=\Gamma_p(k)^*\,. \tag{2.13}$$

From (2.12) and assuming that V is real, one sees that the Hermitian conjugate of $[\Gamma_p(k)]_T$ satisfies

$$[\Gamma_p(k)]_T^\dagger K_p(k)=i\lambda\int[\Gamma_p(k')]_T^\dagger V(k-k')\,d^4k'\,. \tag{2.14}$$

The conjugate solutions $\overline{\phi}_p(k)$ and $\overline{\psi}_p(k)$ are defined to be

$$\overline{\psi}_p(k)\equiv u_p^\dagger\gamma_4[\Gamma_p(k)]_T^\dagger$$

and (2.15)

$$\overline{\phi}_p(k)\equiv u_p^\dagger\gamma_4[\Gamma_p(k)]_T^\dagger K_p(k)\,.$$

Equation (2.14) implies that these conjugate wave functions satisfy

$$\overline{\phi}_p(k)=\overline{\psi}_p(k)K_p(k)=i\lambda\int\overline{\psi}_p(k')V(k-k')\,d^4k'\,. \tag{2.16}$$

In Fig. 1, we give the standard graphical representation of the bound-state vertex function $\phi_p(k)$ and its conjugate solution $\overline{\phi}_p(k)$. It is important to note that these conjugate functions $\overline{\phi}_p(k)$ and $\overline{\psi}_p(k)$ are *not* the time-reversed solutions of $\phi_p(k)$ and $\psi_p(k)$; rather, their relation with $\phi_p(k)$ and $\psi_p(k)$ is the same as that between any outgoing wave and its corresponding incoming wave in a scattering process.

C. Normalization Condition and Some Simple Identities

In terms of the usual Feynman diagram given in Fig. 2, the matrix element of the electromagnetic current operator J_ν between an initial physical proton state of 4-momentum p_ν and a final physical proton state of 4-momentum

$$p'_\nu=p_\nu+q_\nu$$

can be readily derived. Since the electromagnetic vertex of a "bare" proton is simply $ie\gamma_\nu$, one finds for the physical proton

$$\langle p'|J_\nu|p\rangle=-e(2\pi)^{-4}\int d^4k\,(X^2+\mu^2)\overline{\psi}_{p'}(k')\gamma_\nu\psi_p(k)\,, \tag{2.17}$$

where X_ν is related to p_ν and k_ν through (2.1) and (2.4), and k'_ν denotes the relative momentum in the final state which is given by

$$k'_\nu=k_\nu+(M+\mu)^{-1}\mu q_\nu\,. \tag{2.18}$$

Since the physical proton has unit charge e, one must have

$$\langle p'|J_\nu|p\rangle\to e(p_\nu/m_p)\ \text{as}\ q_\nu\to 0\,. \tag{2.19}$$

Therefore, the wave function $\psi_p(k)$ satisfies the normalization condition

$$(2\pi)^{-4}\int d^4k\,(X^2+\mu^2)\overline{\psi}_p(k)\gamma_\nu\psi_p(k)=-(p_\nu/m_p)\,. \tag{2.20}$$

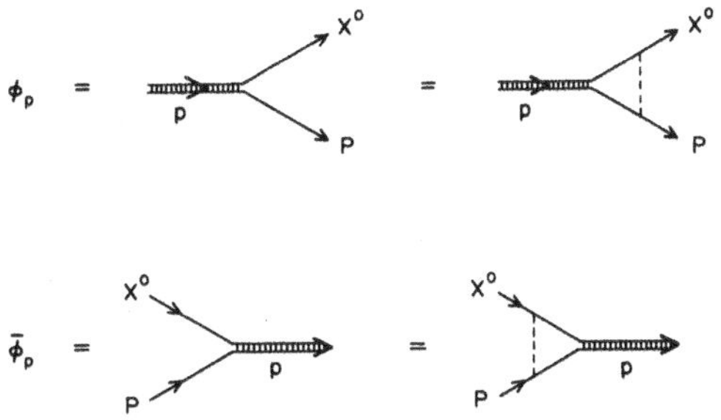

FIG. 1. Graphical representation of the Bethe-Salpeter equation for the bound-state vertex function $\phi_p(k)$ and its conjugate $\overline{\phi}_p(k)$. [See Eqs. (2.6) and (2.16).] The dashed line denotes the covariant potential V.

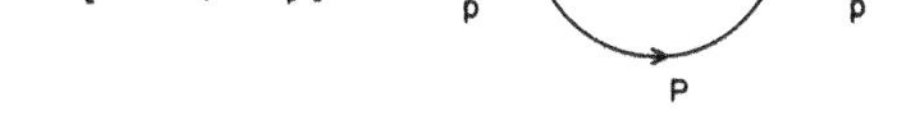

FIG. 2. Diagrams for electromagnetic form factors and the normalization condition (2.22).

In addition, $\psi_p(k)$ or $\phi_p(k)$ satisfies some simple identities which are listed below:

(i)

$$\int \bar{\psi}_p(k)\left[\frac{\partial K_p(k)}{\partial k_\nu}\right]_p \psi_p(k)\, d^4k = 0$$

or (2.21)

$$\int \bar{\phi}_p(k)\left[\frac{\partial K_p^{-1}(k)}{\partial k_\nu}\right]_p \phi_p(k)\, d^4k = 0,$$

where the subscript p outside the square bracket indicates that the total 4-momentum p_ν is kept fixed in the differentiation.

(ii) The normalization condition (2.20) can also be written as

$$i(2\pi)^{-4}\int \bar{\psi}_p(k)K_p(k)\psi_p(k)\, d^4k = [d\ln\lambda/dm_p]^{-1}. \tag{2.22}$$

The proofs of these identities are elementary; for completeness, they are given in Appendix B.

So far as the bound-state equation (2.6) is concerned, the charge of the constituents is irrelevant. Thus, the solution $\psi_p(k)$ or $\phi_p(k)$ can be applied equally well to the case in which the physical proton p is composed of, say, a bare neutron N and a bare π^+ (replacing P and X^0, respectively). Instead of the spin-$\frac{1}{2}$ constituent, it is now the spin-0 constituent that has a nonzero electromagnetic vertex. The first identity (2.21) ensures that the limiting behavior (2.19) remains correct, as it should be. Or, one may also regard the bound-state solution $\psi_p(k)$ or $\phi_p(k)$ as describing a physical neutron n composed of a bare proton P and a bare π^-. The same identity (2.21) ensures that for the physical neutron state

$$\langle n'|J_\nu|n\rangle \to 0 \quad \text{as } q_\nu = (n'-n)_\nu \to 0. \tag{2.23}$$

Of course, if the physical neutron were composed only of an X^0 and a bare neutron N, then $\langle n'|J_\nu|n\rangle$ would be zero identically. The second identity (2.22) makes it possible to express the normalization condition (2.20) in terms of a single-loop diagram without the electromagnetic vertex,[16] as illustrated in Fig. 2.

As we shall see in Sec. II G, the normalization condition (2.20) can be written in still another equivalent form which is related to Adler's sum rule.[17]

D. Electromagnetic Form Factor

To obtain the asymptotic behavior of the electromagnetic form factor for a physical proton state, it is most convenient to express $\langle p'|J_\nu|p\rangle$ in terms of $\phi_p(k)$. Equation (2.17) can be written as

$$\langle p'|J_\nu|p\rangle = -e(2\pi)^{-4}\int d^4X\,(X^2+\mu^2)^{-1}\bar{\phi}_{p'}(k') \times(-i\gamma\cdot P'-M)^{-1}\gamma_\nu(-i\gamma\cdot P-M)^{-1}\phi_p(k), \tag{2.24}$$

where p_ν, k_ν, P_ν, and X_ν are related by (2.1) and (2.4), $P'_\nu = P_\nu + q_\nu$, and therefore

$$p'_\nu = P'_\nu + X_\nu$$

and (2.25)

$$k'_\nu = (M+\mu)^{-1}(\mu P'_\nu - MX_\nu).$$

It can be readily verified by using (2.6) and (2.16) that the current conservation law holds; i.e.,

$$q_\nu\langle p'|J_\nu|p\rangle = 0. \tag{2.26}$$

Thus, one may write

$$\langle p'|J_\nu|p\rangle = ieu^\dagger_{p'}\gamma_4[\gamma_\nu F_1(q^2) + (2m_p)^{-1}\kappa q_\mu\sigma_{\mu\nu}F_2(q^2)]u_p, \tag{2.27}$$

where u_p and $u_{p'}$ are, respectively, the free Dirac spinors with 4-momenta p_ν and p'_ν, $\sigma_{\mu\nu} = (2i)^{-1}\times(\gamma_\mu\gamma_\nu - \gamma_\nu\gamma_\mu)$, κ is the anomalous magnetic moment, and $F_1(q^2)$, $F_2(q^2)$ are the usual charge and Pauli form factors.

We observe that in the integral (2.24) as $q^2\to\infty$, if, say, in the laboratory frame the integration variable X_ν is finite, then $P_\nu = p_\nu - X_\nu$ and $k_\nu = (M+\mu)^{-1}\times\mu p_\nu - X_\nu$ remain finite, but $(-i\gamma\cdot P'-M)^{-1}\phi_{p'}(k')$ is $O(q^{-3})$, on account of (2.9); therefore, the integration over finite regions of X_ν contributes $O(q^{-4})$ to both $F_1(q^2)$ and $F_2(q^2)$, apart from possible factors of $\ln q^2$. If the integration variable X_ν itself is $O(q)$, then the entire integrand in (2.24) is $O(q^{-8})$ but d^4X is $O(q^4)$; therefore, the integration over the $X_\nu = O(q)$ region gives also an $O(q^{-4})$ contribution to $F_1(q^2)$, though an $O(q^{-6})$ contribution to $F_2(q^2)$.

To ascertain the $\ln q^2$ factor, we may assume for large k^2 that $\phi_p(k)$ is proportional to $(k^2+\lambda^2)^{-1}$. A direct calculation then leads to[10]

$$F_1(q^2) = O((\ln q^2)^2/q^4)$$

and (2.28)

$$F_2(q^2) = O(1/q^4).$$

This is in qualitative accord with present data according to which for $q^2 \lesssim 25$ GeV^2

$$G_M(q^2) \equiv F_1 + \kappa F_2 \propto (q^2+m^2)^{-2},$$

where $m^2 \cong 0.71$ GeV^2. The electric form factor $G_E(q^2)$ is known[18] only for $q^2 < 4$ GeV^2, in which region it is proportional to $G_M(q^2)$:

$$(1+\kappa)G_E(q^2) \approx G_M(q^2). \tag{2.29}$$

Since, by definition, $G_E(q^2) \equiv F_1 - (q^2/4M^2)\kappa F_2$, the model we are using predicts a deviation from the relation (2.29). Equation (2.28) may be wrong in detail, based as it is on so crude a model of the proton built of two bare constituents. However, as we shall see in the following, the physical model of a bound-state proton has led to the important general connection of the observed scaling in deep-inelastic scattering with the rapid decrease of the electromagnetic form factors. We have shown here that a rapidly decreasing electromagnetic form factor $F(q^2) \sim O(q^{-4})$ is implied for a bound-state solution of the Bethe-Salpeter equation that is regular at the origin. There is a parallel to this result in the nonrelativistic Schrödinger theory for which the form factor of a bound state that is regular at the origin is also known to decrease at least as fast as $|\vec{q}|^{-4}$ where $\vec{q}$ is the 3-momentum transfer.[19]

E. Structure Functions

Next, we discuss the inelastic ep scattering

$$e+p \rightarrow e+P+X^0 . \tag{2.30}$$

In the simple model of a two-body bound state where the covariant potential V is regarded phenomenologically as a given nonlocal function, this is the only reaction for the deep-inelastic ep scattering process. There are two Feynman diagrams, labeled I_ν and I'_ν in Fig. 3. One finds that (in the usual Feynman gauge)

$$I_\nu = e u_P^\dagger \gamma_4 \gamma_\nu [-i\gamma\cdot(P-q)-M]^{-1}\phi_p(k_{in}) \tag{2.31}$$

and

$$I'_\nu = e(2\pi)^{-4}\int u_P^\dagger \gamma_4 \langle k_f|T_{p+q}|k'\rangle [-i\gamma\cdot P'-M]^{-1}\gamma_\nu[-i\gamma\cdot(p-X')-M]^{-1}(X'^2+\mu^2)^{-1}\phi_p(k_{in})\,d^4X' , \tag{2.32}$$

where q_ν is the virtual photon momentum, k_{in} and k_f denote, respectively, the relative momenta between the P and X^0 in the initial bound state and in the final continuum state, u_P denotes a free Dirac spinor of 4-momentum P_ν, and the matrix T is the two-body scattering matrix defined by

$$\langle k|T_{p+q}|k'\rangle = i\lambda V(k'-k)+(i\lambda)^2(2\pi)^{-4}\int d^4k''\, V(k'-k'')K_{p+q}{}^{-1}(k'')V(k''-k)+\cdots , \tag{2.33}$$

in which the subscript $p+q$ denotes the total 4-momentum of the system, k' and k are, respectively, the initial and final relative momenta for the two-body scattering process.

The diagram I_ν, by itself, is not gauge-invariant. As can be readily checked from the three preceding equations, the sum $I_\nu + I'_\nu$, of course, does satisfy

$$q_\nu(I_\nu + I'_\nu) = 0 .$$

It is useful to decompose the amplitude I_ν into a sum of a longitudinal part $(I_\parallel)_\nu$ and a transverse part $(I_\perp)_\nu$, and similarly $I'_\nu = (I'_\parallel)_\nu + (I'_\perp)_\nu$ such that $q_\nu(I_\perp)_\nu = q_\nu(I'_\perp)_\nu = 0$. While the two longitudinal amplitudes $(I_\parallel)_\nu$ and $(I'_\parallel)_\nu$ are related through the gauge transformation, the two transverse ones are not. As we shall see, in the scaling limit only $(I_\perp)_\nu$ is of importance.

We note that the diagram I_ν is identical in form to a corresponding one in the usual perturbation series in a standard pseudoscalar-coupling meson theory, except that in its evaluation the usual free spinor is now being replaced by the bound-state

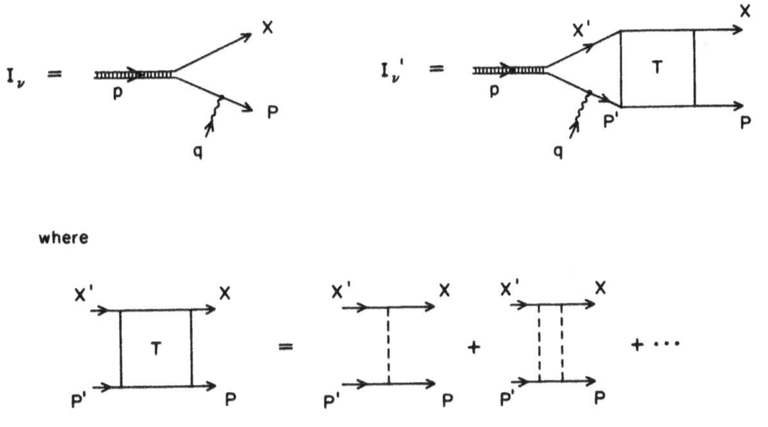

FIG. 3. Diagrams for deep-inelastic ep scattering for the bound-state solution of a simple Bethe-Salpeter equation [See Eqs. (2.31) and (2.32).] The dashed line denotes the covariant potential V.

solution $\phi_p(k_{in})$. Since the usual perturbation diagram fails to scale only by a $\ln\nu$ factor, the presence of $\phi_p(k_{in})$ is sufficient to satisfy the scaling property for the present case because of the asymptotic behavior (2.9). By using (2.9) and (2.31), one can readily verify that in the scaling limit the diagram I_ν gives a finite and nonzero contribution to the transverse structure function W_1, but a zero contribution to the longitudinal structure function W_L which is related to the standard W_1 and νW_2 functions[1,2] by

$$\nu W_2 = 2xm_p(W_1 + W_L)\,. \tag{2.34}$$

To derive the contribution of the diagram I_ν', it is most convenient to use, say, the laboratory frame. The essential point is the observation that the behavior of I_ν' is determined by the configuration in which the virtual momentum P' of the bare proton is $O(q_\nu)$. Since each of the loop integrations in the diagrams for $\langle k|T_{p+q}|k'\rangle$ is a convergent one, and since the additional loop integration over X' in the diagram for I_ν' is a superconvergent one, the integral (2.32) remains convergent if the bare-proton propagator with the large momentum P' is taken outside the integral. (Actually, it remains convergent if, in addition, one takes out also the boson propagator.) It is then easy to verify that in the scaling limit, compared to $(\mathrm{I}_\perp)_\nu$ the transverse part $(\mathrm{I}_\perp')_\nu$ is smaller by a factor $O(q^{-2})$, and consequently the diagram I_ν' does not contribute to W_1. Furthermore, its longitudinal part $(\mathrm{I}_\parallel')_\nu$, like $(\mathrm{I}_\parallel)_\nu$, makes zero contribution to W_L. The entire diagram I_ν' can therefore be neglected in the scaling limit. For completeness, a detailed derivation of this result is presented in Appendix C.

With the vanishing of the rescattering diagram I_ν' in Fig. 3 in the scaling region, we have now derived in the context of this model a fully relativistic generalization of the "parton" model. We find that the amplitude I_ν gives an invariant expression in (2.31) of the bare constituent, or parton, propagating invariantly and scattering independently onto its mass shell in the scaling region. (This replaces the usual parton-theory amplitude of elastic scattering of an almost free parton in the impulse approximation.)

To derive the general functional form of the structure functions in the scaling limit, one needs only the value of the bound-state vertex function $\phi_p(k)$ when one of its constituents, X^0, is on its mass shell. Let us define

$$\lim_{X^2+\mu^2=0} \phi_p(k) \equiv g(K)\gamma_5 u_p\,, \tag{2.35}$$

where, for clarity, the 4-momenta of the two constituents X^0 and P in the bound state are labeled, respectively, by X_ν and K_ν. Thus, in diagram I_ν given in Fig. 3, one has

$$K_\nu = p_\nu - X_\nu = P_\nu - q_\nu\,. \tag{2.36}$$

On account of (2.9), the function g satisfies

$$g \sim O(K^{-2}) \text{ as } K^2 \to \pm\infty\,. \tag{2.37}$$

Furthermore, as will be shown in Appendix A, in this limit $K^2 \to \pm\infty$, g is a c-number function independent of γ matrices. By using (2.31) and (2.35), one finds that in the scaling limit

$$W_L \to 0\,, \tag{2.38}$$

$$W_1 \propto \int_{u_{min}}^{+\infty} \frac{du}{(u+M^2)^2}|g(u)|^2 \times\{(u+M^2)(1-x) + x[(M-m_p)^2-\mu^2]\}\,, \tag{2.39}$$

with

$$u_{min} = \mu^2 x/(1-x) - m_p^2 x$$

and $x = -(2p\cdot q)^{-1}q^2$ is the usual scaling variable. The variable $u = K^2 = (P-q)^2$, and u_{min} is the minimum invariant momentum transfer squared carried by the bare proton in diagram I_ν of Fig. 3. In terms of u, (2.37) becomes

$$g(u) \sim O(u^{-1}) \text{ as } u \to \pm\infty\,. \tag{2.40}$$

We note that, before the scaling limit, the upper limit u_{max} of the integral in (2.39) would be $-2p\cdot q$; because of the above convergence property of the bound-state vertex function $g(u)$, one may set u_{max} to be ∞ in the scaling limit. Now, the lower limit u_{min} grows as $(1-x)^{-1}$ as $x \to 1$; therefore, in the scaling limit, (2.39) shows explicitly that the wave function is being probed for large values of u, or at "small distances." Furthermore, by using (2.40) we find that

$$W_1 \sim O((1-x)^3) \text{ as } x \to 1\,. \tag{2.41}$$

This is in agreement with the threshold relation[12] since $F_1(q^2) \sim O(q^{-4})$ in this model.

In order to show explicitly how the threshold relation emerges in this model, in contrast to the standard parton approach, we rewrite (2.39) in terms of appropriate variables in an infinite-momentum frame. Denoting by $\vec{p} \to \infty$ the momentum of the physical proton along the 3 axis, and by $(1-x)\vec{p}$ and $-\vec{k}_\perp$ the longitudinal and transverse components of momentum of the X^0 in diagram I_ν of Fig. 3, we find

$$u = u_{min} + (1-x)^{-1}k_\perp^2\,.$$

The integration over the momentum transfer in (2.39) can then be written in terms of an integral over $k_\perp^2$ extending from zero to infinity. The damping of the high-momentum contributions evidently scales with $1-x$ since $g(u) \sim u^{-1} \sim (1-x)/k_\perp^2$. It is this scaling of the transverse-momentum distribution with the longitudinal momentum as $x \to 1$, *as required by relativistic covariance*, that provides the cubic powers of $1-x$ in agreement with the threshold relation (2.41), in contrast to the linear power in the original parton models.[11]

The important physical point is that detailed study of W_1 near $x = 1$ measures the bound-state wave function of the nucleon. For example, if $g(u)$ is $\sim O(u^{-n})$ as $u \to \infty$, then, according to (2.39), W_1 would be $\sim O((1-x)^{1+2n})$ as $x \to 1$. This behavior is also in accord with the threshold relation; i.e., F_1 will be $O(q^{-2-2n})$ as $q^2 \to \infty$. To see this, repeat the arguments leading to (2.9), observing that we must replace the potential in (1.2) by $V(q) \sim O(q^{-2n})$ in order to obtain the asymptotic behavior $\phi_p(k) \sim O(k^{-2n})$ from (2.6). By repeating the identical arguments given above (2.28), the stated result can be proved.

In the present simple model where the physical proton consists of only a two-body bound state, one sees that, according to (2.39),

$$W_1 \to \text{constant as } x \to 0, \tag{2.42}$$

and therefore

$$\nu W_2 = 2xm_p W_1 \to 0 \text{ as } x \to 0. \tag{2.43}$$

In the hypothetical case that the "bare" proton is, like the X^0, also of spin 0 and, in addition, the covariant potential $V(q)$ is simply q^{-2}, then the exact bound-state solution is known[20,21] and the corresponding $g(u)$ is given simply by

$$g(u) = \text{const}(u + M^2)^{-1}. \tag{2.44}$$

If in the present case of a spin-$\frac{1}{2}$ bare proton, one makes the *ad hoc* assumption that $g(u)$ remains given by (2.44), then one finds

$$W_1 = \text{const}\frac{(1-x)^3\{(1-x)(M^2 - m_p^2x) + \frac{1}{3}x[2(M-m_p)^2 + \mu^2]\}}{[\mu^2x + (M^2 - m_p^2x)(1-x)]^3}, \tag{2.45}$$

which, of course, satisfies both (2.41) and (2.42).

F. Forward Compton Scattering

We now want to verify directly that the structure functions computed in this model are in fact the absorptive parts of the forward Compton scattering amplitude in the scaling region. That is precisely how W_1 and W_2 are defined in general; however, in calculating these structure functions in Sec. II E we have used only the inelastic reaction (2.30) while ignoring all other final states consisting of $(P + X^0 + \text{gluons})$, where the gluons refer to the quanta of whatever fields are responsible for generating the covariant Yukawa potential between the two constituents P and X^0. Such on-mass-shell states contribute to the absorptive part of the forward Compton amplitude although they are not included in our calculation of the preceding section. Nevertheless, it will be shown that in the scaling limit these additional states are not important in the framework of our Bethe-Salpeter model with the ladder approximation. (For further discussions, see the remark given at the end of this section and also comments D and E in Sec. V.)

In the present simple two-body bound-state model the spin-averaged forward amplitude $F_{\lambda\nu}(p,q)$ for the Compton scattering

$$\gamma + p \to \gamma + p$$

of a virtual (or real) photon of 4-momentum q_ν on a physical proton of 4-momentum p_ν can be written as a sum of four terms

$$F_{\lambda\nu} = A_{\lambda\nu}(p,q) + A_{\nu\lambda}(p,-q) + B_{\lambda\nu}(p,q) + B_{\nu\lambda}(p,-q), \tag{2.46}$$

where $A_{\lambda\nu}(p,q)$ and $B_{\lambda\nu}(p,q)$ represent the corresponding diagrams in Fig. 4, and $A_{\nu\lambda}(p,-q)$ and $B_{\nu\lambda}(p,-q)$ denote, respectively, the same diagrams but with the two external photons exchanged. The explicit expressions of $A_{\lambda\nu}$ and $B_{\lambda\nu}$ are given by

$$A_{\lambda\nu}(p,q) = i\int (2\pi)^{-4} d^4X (X^2+\mu^2)^{-1}\bar{\phi}_p(k_f)[-i\gamma\cdot(p-X) - M]^{-1}\gamma_\nu[-i\gamma\cdot P - M]^{-1}\gamma_\lambda[-i\gamma\cdot(p-X) - M]^{-1}\phi_p(k_{\text{in}}), \tag{2.47}$$

$$B_{\lambda\nu}(p,q) = i\int (2\pi)^{-8} d^4X d^4X' (X^2+\mu^2)^{-1}(X'^2+\mu^2)^{-1}\bar{\phi}_p(k_f)[-i\gamma\cdot(p-X) - M]^{-1} \times\gamma_\nu[-i\gamma\cdot P - M]^{-1}\langle k|T_{p+q}|k'\rangle[-i\gamma\cdot P' - M]^{-1}\gamma_\lambda[-i\gamma\cdot(p-X') - M]^{-1}\phi_p(k_{\text{in}}), \tag{2.48}$$

where X_ν and X'_ν are the 4-momenta of X^0 as labeled in Fig. 4,

$$P = p + q - X, \quad P' = p + q - X', \tag{2.49}$$

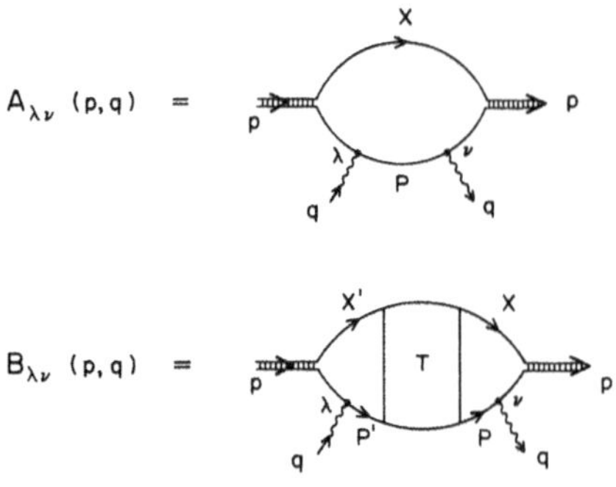

FIG. 4. Diagrams for forward Compton scattering. [See Eqs. (2.47) and (2.48).]

k_{in} and k_f denote the appropriate relative momenta in the bound state, given respectively by

$$k_{in} = (M+\mu)^{-1}\mu p - X'$$

and

$$k_f = (M+\mu)^{-1}\mu p - X\,,$$

k and k' are the relative momenta in the continuum states, related to X, P and X', P' by

$$k = (M+\mu)^{-1}(\mu P - MX)$$

and

$$k' = (M+\mu)^{-1}(\mu P' - MX')\,,$$

and $\langle k | T_{p+q} | k' \rangle$ is given by (2.33). By using the Bethe-Salpeter equation (2.2), one can readily verify that $F_{\lambda\nu}$ satisfies the requirement of current conservation

$$q_\lambda F_{\lambda\nu} = 0\,. \tag{2.50}$$

Thus, $F_{\lambda\nu}$ can be written as

$$F_{\lambda\nu} = \left(\delta_{\lambda\nu} - \frac{q_\lambda q_\nu}{q^2}\right)U_1 + \frac{1}{m_p^2}\left(p_\lambda - \frac{p\cdot q}{q^2}q_\lambda\right)\left(p_\nu - \frac{p\cdot q}{q^2}q_\nu\right)U_2\,, \tag{2.51}$$

where U_1 and U_2 are scalar functions of q^2 and $\nu \equiv -m_p^{-1}(p\cdot q)$. In this section, we are only interested in the case of a spacelike photon $q^2 > 0$ with $\nu > 0$.

As we have already noted, in the present model the absorptive parts (i.e., the imaginary parts) of U_1 and U_2 are, in general, different from the corresponding structure functions πW_1 and πW_2 calculated by using only the inelastic reaction (2.30). However, as we shall prove in Appendix D, in the limit $q^2 \to \infty$, $\nu \to \infty$, but keeping $x = (2m_p\nu)^{-1}q^2$ fixed,

$$\mathrm{Im}U_1 \to \pi W_1$$

and (2.52)

$$\mathrm{Im}(\nu U_2) \to \pi(\nu W_2)\,,$$

where W_1 is the *same* function given by (2.39) and $\nu W_2 = 2xm_p W_1$. Of course, had one included all diagrams, then $\mathrm{Im}U_1 = \pi W_1$ and $\mathrm{Im}U_2 = \pi W_2$ should hold at all values of q^2 and ν. The above expression (2.52) shows that under the ladder approximation these equalities are at least maintained in the scaling limit.

As will be shown in Appendix D, the diagram $B_{\lambda\nu}$ in Fig. 4 can be neglected in the scaling limit. We emphasize that in the Bethe-Salpeter model with the ladder approximation, the gluons are exchanged *only* between *different* bare constituents. This is why in diagram $B_{\lambda\nu}$ these on-mass-shell intermediate $(P + X^0 + \text{gluons})$ states are of no importance in the scaling limit. In a *bona fide* relativistic local field theory there must be other forward Compton diagrams which contain gluon radiative corrections, or renormalizations; in these diagrams there are emissions and absorptions of gluons by the *same* bare constituent. The imaginary part of these diagrams would correspond to gluon radiation in deep-inelastic *ep* scattering; such processes lie outside the present framework of a simple two-body bound-state model, and are therefore *not* included in our calculations on structure functions. As will be discussed in comment E of Sec. V, at least in a perturbation series, these additional gluon-radiation diagrams would lead to violations of the scaling property.

We note that, like the diagram I_ν' in the previous section, the diagram $B_{\lambda\nu}$ contains an s-channel pole at $(p+q)^2 + m_p^2 = 0$. Both diagrams are needed for maintaining gauge invariance, and both can be neglected in the scaling limit.

G. Adler Sum Rule

In this section, we shall show that the normalization condition (2.20) implies that in the scaling limit the νW_2 function should satisfy

$$\int_0^1 x^{-1}dx\,(\nu W_2) = 1\,, \tag{2.53}$$

where $\nu W_2 = 2xm_p W_1$ and W_1 is given by (2.39). Equation (2.53) is (essentially) the Adler sum rule,[17] but applied to the present simple two-body bound-state model of the physical nucleon. By using (2.39), one sees that the above integral is a convergent one. Equation (2.53) then determines the proportionality constant in (2.39). Our interest in deriving (2.53), or, equivalently in our model, in proving the Adler sum rule, stems from the fact that on the basis of (2.20) alone we cannot *in prac-*

tice normalize our structure functions. This is because we have not solved the Bethe-Salpeter equation explicitly for the wave function; we have simply derived those general properties and the asymptotic behavior of $\psi_p(k)$ as required to establish scaling properties, asymptotic behaviors of the elastic form factors, and the threshold theorems. In normalizing the current in (2.20) we have to know the exact wave function $\psi_p(k)$ for all momenta k_μ which is integrated over its entire range, while in evaluating νW_2 we need only $g(u)$, which is, according to (2.35), determined by $\psi_p(k)$ when one of the constituents, X^0, is on its mass shell. The sum rule (2.53) provides the needed normalization of the structure functions for our purposes.

It is useful to define a new scattering amplitude

$$F'_{\lambda\nu}(p,q) \equiv A_{\lambda\nu}(p,q) - A_{\nu\lambda}(p,-q) + B_{\lambda\nu}(p,q) - B_{\nu\lambda}(p,-q), \qquad (2.54)$$

where $A_{\lambda\nu}$ and $B_{\lambda\nu}$ are given, respectively, by (2.47) and (2.48). If one wishes, one may relate $F'_{\lambda\nu}$ to the vector part of the $\bar{\nu}p$ scattering amplitude. However, for the purpose of deriving (2.53), the precise nature of the physical process that is represented by this amplitude $F'_{\lambda\nu}$ is immaterial. By definition, $F'_{\lambda\nu}$ satisfies

$$F'_{\lambda\nu}(p,q) = -F'_{\nu\lambda}(p,-q). \qquad (2.55)$$

By using the Bethe-Salpeter equation (2.2) and the normalization condition (2.20), one can readily establish the divergence condition

$$q_\nu F'_{\lambda\nu} = 2m_p^{-1} p_\lambda, \qquad (2.56)$$

and therefore

$$q_\lambda q_\nu F'_{\lambda\nu} = 2m_p^{-1}(p\cdot q) = -2\nu. \qquad (2.57)$$

Equation (2.56) is the crucial result, since it plays the role of current algebra in the usual derivation. Once we have established the divergence condition, we need only verify that $F'_{\lambda\nu}$ satisfies appropriate convergence requirements (unsubtracted dispersion relations) when $\nu\to\infty$. This is explicitly clear in the discussions of Adler and Bjorken.[17] We reproduce the necessary algebra in Appendix E, commenting here only that the asymptotic behavior of the wave function (1.3) ensures the required convergence properties which establish (2.53). In the literature, the Adler sum rule has been derived for the part of the forward Compton amplitude that is odd under crossing, that is, for the difference of amplitudes for neutrinos and antineutrinos incident on a given target nucleon. In our model with a single charged nucleon in the bound state, the coupling of the z-component isospin I_3 (electromagnetic) current to the proton is just $2^{-1/2}$ in strength relative to that of the usual $I_\pm = 2^{-1/2}(I_1 \pm iI_2)$ current in the weak interaction.

H. Crossing to the Annihilation Channel

The physical process of

$$e + \bar{e} \to \bar{p} + \text{anything} \qquad (2.58a)$$

is related to deep-inelastic scattering

$$e + p \to e + \text{anything} \qquad (2.58b)$$

by crossing symmetry. In a perturbation calculation of these processes, this can be proved by applying the usual substitution rule to each Feynman graph order by order. It has been shown[6] that this symmetry persists in the cutoff field-theory model and that as a consequence the structure functions for (2.58a), $(\overline{W}_1, \overline{W}_2)$, are analytic continuations into the physical region for this process of the structure functions for (2.58b), (W_1, W_2). Moreover, the scaling behavior for these functions has been derived for the deep-inelastic annihilation and scattering regions, respectively. In particular

$$\overline{W}_1(x) = W_1(x)$$

and (2.59)

$$\nu\overline{W}_2(x) = \nu W_2(x);$$

i.e., $\overline{W}_1$ is the continuation of $W_1(x)$ from the physical region for (2.58b) with $x<1$ to the physical region for (2.58a) with $x>1$, where now $1/x$ is the fraction of the total $e\bar{e}$ collision energy deposited on the observed hadron in the collision center-of-mass system.

The practical value of (2.59) is its prediction of the magnitude of the cross section for (2.58a) *near* $x=1$ in terms of scattering measurements. More generally, if we can prove (2.59) for our physical bound-state model, then we have a prediction of the threshold behavior for the inclusive annihilation process in accord with (2.41).

The crossing properties of the scattering amplitude depend on those of the vertex function $g(u)$ appearing in it. Generally, we destroy crossing symmetry when working with the Bethe-Salpeter equation in the ladder approximation. However, for study of the threshold behavior we require only the solution for $g(u)$ for large values of u according to (2.40). For the inelastic scattering near threshold as $1-x\to 0+$, $u\to+\infty$, whereas for the crossed or annihilation process the virtual intermediate constituent is timelike or masslike in Fig. 5 with $u\to-\infty$. In both cases, (2.40) states that $g(u) \sim O(u^{-1})$ as $u\to\pm\infty$. Therefore, (2.59) is valid near $x\sim 1$ for our bound-state model. However, we have not studied the general crossing properties for arbitrary x.

III. PHENOMENOLOGICAL APPROACH

In this section we reconstruct the results of the

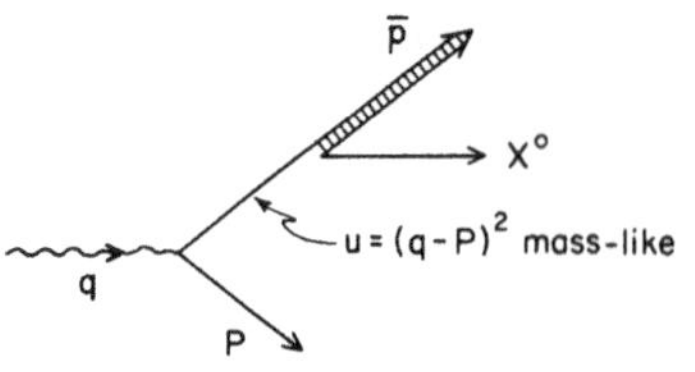

FIG. 5. A diagram for a virtual photon $\rightarrow \bar{p}$ + anything.

Bethe-Salpeter model of the preceding section by developing a phenomenological approach for studying a bound-state model of the proton's structure. Our starting point is the diagram I_ν in Fig. 3 and its corresponding amplitude, by (2.31) and (2.35):

$$I_\nu = eu_{\bar{p}}^\dagger \gamma_4\gamma_\nu[-i\gamma\cdot(P-q)-M]^{-1}\gamma_5 u_p g(u), \quad (3.1)$$

where $g(u)$ is the vertex function at the (pPX^0) vertex and is a function of the mass of the virtual P with both the p and X^0 on their mass shells. In this phenomenological approach, we shall for simplicity assume $g(u)$ to be strictly a c-number function, whereas in our previous discussions on the Bethe-Salpeter equation $g(u)$ is a c-number function only in the limit $u\to\infty$.

A. Current Conservation

Equation (3.1) is not gauge-invariant as noted earlier, and its completion requires that final-state rescattering effects be added. Thus, in the Bethe-Salpeter approach we accomplish this by adding the diagrams I_ν' in Fig. 3. These take into account the fact that the exchange of quanta between the constituents P and X^0 that gave rise to the initial bound state of the physical proton is still occurring after the virtual photon q is absorbed. In the well-known way, all possible insertions of the photon must be made on the charged line in Fig. 3 in order to ensure current conservation.

In constructing the terms which must be added to (3.1) in order to give us an over-all current-conserving amplitude for inelastic scattering, we are guided by two principles: The first is simplicity, and the second is our insistence not to introduce any "unwanted, unphysical" singularities in the amplitude. We may directly write in accord with these principles

$$\text{II}_\nu = +ieu_{\bar{p}}^\dagger\gamma_4\gamma_5 u_p \frac{2p_\nu+q_\nu}{(p+q)^2+m_p^2}g(u), \quad (3.2)$$

so that $q_\nu(I_\nu+\text{II}_\nu)=0$. This form has an s-channel pole at the bound-state proton mass, as it should. In the simple phenomenological approach, it seems sensible to choose only functions with simple poles. This is to be contrasted with our previous Bethe-Salpeter approach in which II_ν is replaced by I_ν' of (2.32) which has in addition to the same bound-state s-channel pole also cuts due to the continuum states. Of course, Eq. (3.2) is not unique and one can readily construct a more general term that, for simplicity, still retains the spin structure of (3.2) by assuming II_ν to be given by

$$ieu_{\bar{p}}^\dagger\gamma_4\gamma_5u_p(\lambda_1 p_\nu+\lambda_2 q_\nu+\lambda_3 P_\nu), \quad (3.3)$$

where λ_i are scalar functions of the three independent scalars

$$s\equiv(p+q)^2,\quad u\equiv(P-q)^2,\quad t\equiv(p-P)^2,$$

with (3.4)

$$s+t+u=-(m_p^2+M^2+\mu^2-q^2).$$

Current conservation requires that the functions λ_i satisfy the restriction

$$\lambda_1 q\cdot p+\lambda_2 q^2+\lambda_3 q\cdot P=g(u). \quad (3.5)$$

Such a relation has no unique solution; we can add to any particular solution of (3.5) arbitrary amounts, $\delta\lambda_1$, such that

$$\delta\lambda_1 q\cdot p+\delta\lambda_2 q^2+\delta\lambda_3 q\cdot P=0. \quad (3.6)$$

This is equivalent to the observation that the restriction of current conservation fixes the Amperian-current interactions, but not the nonminimal terms proportional to electromagnetic field strengths and their derivatives. These are automatically gauge-invariant. In our model we rule out all such nonminimal terms. For example, their presence in the interaction Lagrangian of a relativistic field-theory calculation generally leads to nonrenormalizable theories. If included in the electromagnetic interaction of the constituents (or partons), they violate scaling.

Returning to (3.5), we look for that solution which introduces no additional u-channel singularities since these are already contained in (3.1), i.e., the graph I_ν in Fig. 3. We also rule out t-channel singularities, or by (3.4) poles in $(X-q)^2$, since the X^0 is taken to be neutral with no direct electromagnetic interaction. We find therefore that (3.5) gives

$$\lambda_1=2\lambda_2=\frac{2g(u)}{(p+q)^2+m_p^2},\quad \lambda_3=0 \quad (3.7)$$

and (3.3) reduces to (3.2). We have now the inelastic scattering amplitude according to our criteria of simplicity, of minimal electromagnetic couplings, and of no unwanted singularities in the u or t channels.

B. Bound-State Versus Elementary-Particle Model of the Proton

The difference between a physical model of the proton as a bound state of two constituents, and an elementary-particle picture of the proton lies in

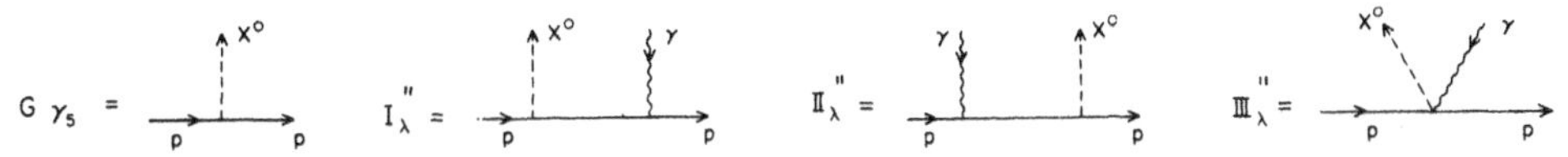

FIG. 6. Feynman diagrams for $p \to p + X^0$ and "γ" $+ p \to p + X^0$, assuming that the physical proton is an elementary particle instead of a bound-state composite. [See Eqs. (3.10), (3.11), and (3.14).]

the behavior of the vertex function $g(u)$. As we have already shown for a particular bound-state model in the ladder approximation to the Bethe-Salpeter equation, as illustrated in Fig. 3, the vertex function behaves as in (2.40); i.e.,

$$g(u) \sim O(1/u), \quad \text{for } u \to \pm\infty \tag{3.8}$$

where u is the virtual mass of either of the constituents forming the bound state. On the other hand, if the nucleon is elementary and there is an elementary vertex interaction, as illustrated in Fig. 6, in place of $g(u)\gamma_5$ one has the vertex function $G\gamma_5$ where G is a constant in the lowest-order perturbation calculation. Furthermore, even if one includes final-state interactions with a potential given by (1.1), the effective vertex function G for the three-point function $p \to p + X^0$, after summing over all ladder diagrams in the final state of pX^0, remains dominated by this nonvanishing lowest-order diagram at least at large u; therefore, in the ladder approximation,

$$G \to \text{constant for large } u\,. \tag{3.9}$$

This contrast between bound-state and elementary-particle descriptions of the nucleon has been analyzed and discussed in some detail by Ball and Zachariasen.[10]

From the view of the elementary-particle model, one may, of course, invoke the possibility that the final-state strong interaction can be quite complicated, and that for some unknown reason the three-point vertex for $p \to p + X^0$ is given phenomenologically by $G\gamma_5$ where G does not satisfy (3.9); instead, it can be of a general form

$$G = G(p^2, P^2, X^2), \tag{3.10}$$

where p and P denote the initial and final momenta of the same elementary particle p, and X that of the boson as before. Because of crossing symmetry,

$$G(p^2, P^2, X^2) = G(P^2, p^2, X^2)\,.$$

As we shall see, even then there is an important difference between the elementary-particle approach and the bound-state approach.

As shown in Fig. 6, there are at least two standard diagrams I_λ'' and II_λ'' for the process

$$\gamma + p \to p + X^0$$

in the elementary-particle approach. In these two diagrams, the electromagnetic vertex is assumed to be given by the usual minimal interaction $e\gamma_\lambda$. If G is a constant, or if G satisfies (3.9), then the result does not scale as noted before. With G given by (3.10), the sum of the first two amplitudes in Fig. 6 is just

$$\begin{aligned} I_\lambda'' + II_\lambda'' = ieu_{p'}^\dagger\gamma_4\Big[&\gamma_5 \frac{1}{-i\gamma\cdot(p+q)-m}\gamma_\lambda G(s, -m^2, -\mu^2) \\ &+ \gamma_\lambda \frac{1}{-i\gamma\cdot(p'-q)-m}\gamma_5 G(-m^2, u, -\mu^2)\Big]u_p\,, \end{aligned} \tag{3.11}$$

where m is the nucleon mass, q is the photon momentum, and p and p' denote, respectively, the initial and final proton momenta. If G is not a constant, then (3.11) has the defect of failing to conserve electromagnetic current, viz.,

$$\begin{aligned} q_\lambda(I_\lambda'' + II_\lambda'') = -eu_{p'}^\dagger\gamma_4\gamma_5 u_p[&G(s, -m^2, -\mu^2) \\ &- G(-m^2, u, -\mu^2)]\,. \end{aligned} \tag{3.12}$$

One is led to remedy this defect in the usual manner by adding to the amplitude (3.11) a four-point contact interaction as the diagram III_λ'' in Fig. 6 which is suitably tailored so that the additional contribution, III_λ'', restores current conservation to the over-all amplitude:

$$q_\lambda(I_\lambda'' + II_\lambda'' + III_\lambda'') = 0\,. \tag{3.13}$$

As in our prior discussion of (3.3) and (3.5), there is no unique solution to this condition, but one usually searches for a "guess" that is both simple and free of unwanted singularities. In this elementary nucleon example, the latter condition means in particular that III_λ'' should contain neither s- nor u-channel poles since the direct and exchanged nucleon poles are already included in the two terms of (3.11). In contrast, for a bound-state physical model an s-channel pole was introduced in the gauge term in (3.2), since it represented the entire s-channel contribution. As we shall see, in the bound-state approach it is the presence of this growing denominator factor at high energies, $s \to \infty$, that leads to the limiting behavior that the gauge term vanishes in the scaling region. Without the large denominator, the gauge terms generally will grow and eventually destroy scaling behavior, as would be the case in the elementary-particle approach. For example, we may choose

$$\mathrm{III}_\lambda'' = eu_{p'}^\dagger\gamma_4\gamma_5 u_p\left[\frac{G(s,-m^2,-\mu^2)-G(-m^2,-m^2,-\mu^2)}{s+m^2}(2p+q)_\lambda + \frac{G(-m^2,u,-\mu^2)-G(-m^2,-m^2,-\mu^2)}{u+m^2}(2p'-q)_\lambda\right], \tag{3.14}$$

which satisfies (3.13), has no unwanted poles, and is also the result of the standard minimal electromagnetic coupling; i.e., as can be readily verified, if one represents the three-point vertex $G\gamma_5$ by an effective Lagrangian, the formal replacement of $\partial/\partial x_\lambda$ by $\partial/\partial x_\lambda - ieA_\lambda$ for the charged field in the Lagrangian would lead to a four-point function given by (3.14). By choosing a sufficiently convergent function G, the contribution to the structure functions from the square of the first two amplitudes, $\mathrm{I}_\lambda''+\mathrm{II}_\lambda''$ in (3.13), can be easily made to scale in the scaling limit. However, the third amplitude III_λ'' survives in the scaling limit due to the constant vertex $G(-m^2,-m^2,-\mu^2)$ that was added to both terms to remove the s- and u-channel poles and leads to a logarithmic violation of scaling for the transverse function W_1 and a linear one for the longitudinal function W_L, and therefore also for νW_2:

$$\nu W_2 \sim O(\nu)$$

and

$$W_1 \to \mathrm{const}\int_{u_{\min}}^{2m\nu}\frac{u\,du}{(u+m^2)^2}G^2(-m^2,-m^2,-\mu^2) + \text{finite terms} \propto \ln(\nu/m)+\cdots, \tag{3.15}$$

where $u_{\min}=\mu^2x/(1-x)-m^2x$. We can trace the origin of the difficulties leading to (3.15) to the appearance of an elementary s-channel pole term in the diagram II_λ'' of Fig. 6. No such diagram appears in the bound-state model because there is never a first moment for the electromagnetic current to interact with an elementary charged constituent "before" it finds itself bound into the physical proton. In the bound-state model, the s-channel pole term is introduced only through the final-state interactions between the emerging constituents after they have been struck by the photon. In place of the amplitude $\mathrm{II}_\lambda''+\mathrm{III}_\lambda''$, one has the amplitude II_ν given by (3.2) in the simple phenomenological approach [or I_ν', given by (2.32) in the Bethe-Salpeter approach].

For a bound-state proton with the vertex function (3.8) it is easy to show that the contribution to the structure functions W_1 and νW_2 coming from the added "gauge" term, or II_ν in (3.2), vanishes as $1/q^2$ in the scaling limit. This is because in the scaling region of large ν and q^2 the energy denominator in this amplitude is large, $\sim\nu$ or s, and cannot be canceled by large numerator factors in integrating over the final two-particle phase space when the region of large u values is damped as in (3.8). Both the square of the gauge term II_ν and its interference with I_ν can be computed for studying the approach to scaling, but the entire contribution in the scaling region comes only from the amplitude I_ν in (3.1). This conforms to the simple "parton" picture in that the relevant mechanism is the "dissociation" of a physical nucleon into its constituents as it interacts with the electromagnetic field. Whereas the parton model is limited to the infinite-momentum frame, we have now a covariant and current-conserving description in terms of Feynman propagators. As expected, our final result in the phenomenological approach scales and coincides with (2.39), which was obtained previously in the Bethe-Salpeter approach.

We recall from Sec. II that the same bound-state model of a proton that leads to a scaling behavior for the inelastic structure functions also predicts a rapid decrease of the electromagnetic form factors:

$$F_1(q^2) \sim O((\ln q^2)^2/q^4)$$

and (3.16)

$$F_2(q^2) \sim O(1/q^4)$$

in qualitative accord with present data. Since the physical model constructed here with a phenomenological approach has the same vertex-function behavior for large momenta, i.e., $g(u)\sim O(1/u)$ for large u according to (3.8), we again find as in Sec. II that $W_1\sim O((1-x)^3)$ as $x\to 1$. As noted before, this is in agreement with the threshold relation by (3.16). Moreover, the crossing relation to the annihilation channel described in Sec. II remains valid. Finally, we note that the gauge-invariant amplitude $(\mathrm{I}_\nu+\mathrm{II}_\nu)$ can also lead to a sum rule (2.53) of the Adler type for normalizing the structure functions in the phenomenological approach.

While the simple phenomenological approach described in this section is guided by the Bethe-Salpeter equation, it has a greater heuristic value than our previous Bethe-Salpeter approach. It allows us to dissociate the rigid connection between the vertex function $g(u)$ and the solution of a particular class of equations based on the ladder approximation. For any given $g(u)$, the amplitude $(I_\nu + II_\nu)$ is relativistically invariant and gauge-invariant, thus allowing us to calculate directly the structure functions in the scaling limit as well as the approach to scaling.

IV. SIMPLE MODEL CALCULATIONS

The description that a physical nucleon is only a two-body bound state is clearly an over-simplified one. Through virtual-meson exchanges there also must be multimeson channels connected to the physical nucleon state. Furthermore, as we shall see, the existence of such multimeson channels can alter the limiting behavior of the structure functions as $x \to 0$; in particular, νW_2 may approach a constant, instead of zero as given by (2.43) for a two-body bound state.

A. First Example

In order to illustrate such possibilities, we take as a first example a simple ansatz that the physical nucleon p or n consists of (i) a two-body state composed of a bare nucleon P or N and an SU_3 singlet X^0, and (ii) a multibody state composed of, at least, a bare SU_3 octet baryon and two SU_3 octet mesons, say $P\pi^0\pi^0$, or $\Sigma^+K^-\pi^+$, etc.; formally, one may write

$$p = (PX^0) + (P2\pi) + (\Sigma K\pi) + \cdots , \tag{4.1}$$

etc. [The possibility that there should also be, in addition, a two-body $(P\pi)$ channel will be discussed later.] While the actual bound-state calculation for a multiparticle system is complicated, a phenomenological description, based on the bound-state idea, can be easily given. Following the discussion given in Sec. III, we may represent the reaction

$$\gamma + p \to P + X^0 \tag{4.2}$$

by diagrams I_λ and II_λ in Fig. 7, the reaction

$$\gamma + p \to P + \pi^0 + \pi^0 \tag{4.3}$$

by diagrams III_λ and IV_λ in the same figure, etc. The amplitudes I_λ and II_λ are given by (3.1) and (3.2), respectively; i.e.,

$$I_\lambda = e u_p^\dagger \gamma_4 \gamma_\lambda [-i\gamma\cdot(P-q) - M]^{-1} \gamma_5 u_p g \tag{4.4}$$

and

$$II_\lambda = ie u_p^\dagger \gamma_4 \gamma_5 u_p [(p+q)^2 + m_p^{\,2}]^{-1} (2p+q)_\lambda g. \tag{4.5}$$

In complete analogy, the diagrams III_λ and IV_λ may be represented by

$$III_\lambda = ie u_p^\dagger \gamma_4 \gamma_\lambda [-i\gamma\cdot(P-q) - M]^{-1} u_p g' \tag{4.6}$$

and

$$IV_\lambda = -e u_p^\dagger \gamma_4 u_p [(p+q)^2 + m_p^{\,2}]^{-1} (2p+q)_\lambda g', \tag{4.7}$$

where as before g denotes the two-body bound-state vertex function when X^0 is on the mass shell, and similarly g' denotes the three-body bound-state vertex function when both mesons are on their mass shells. Thus, g depends only on one variable, say $u = (P-q)^2$, but g' depends on three variables, say $(p-\pi_a)^2$, $(p-\pi_b)^2$, and $u = (P-q)^2$, where π_a and π_b denote the final momenta of the two pions. It is easy to see that for arbitrary g and g' one has

$$q_\lambda (I_\lambda + II_\lambda) = 0$$

and (4.8)

$$q_\lambda (III_\lambda + IV_\lambda) = 0.$$

In a phenomenological approach, g and g' may be assumed to be any reasonable vertex functions. We note that according to (3.8)

$$g(u) \sim O(u^{-1}) \text{ as } u \to \infty. \tag{4.9}$$

For definiteness, $g(u)$ may be assumed to be given by (2.44); i.e.,

$$g(u) \propto (u + M^2)^{-1}, \tag{4.10}$$

which we recall is the rigorous solution of the Bethe-Salpeter equation if all particles are of zero spin and if $V(q^2)$ is q^{-2}. To assign a specific form to g', there is at present very little theoretical guidance. Purely for reasons of computational simplicity, we choose as an ansatz for g'

$$g' = g'(u), \tag{4.11}$$

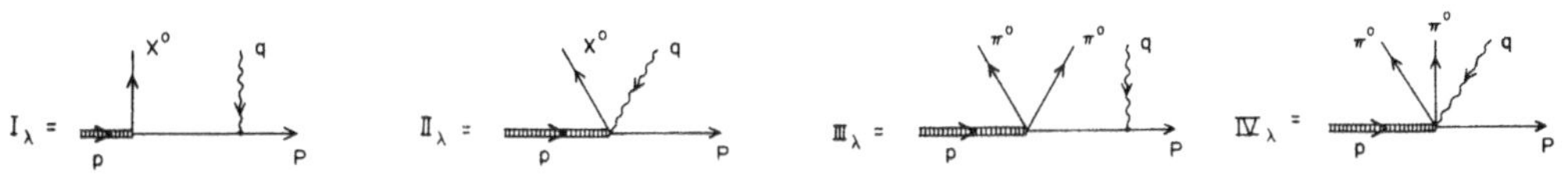

FIG. 7. Diagrams for deep-inelastic *ep* scattering in the simple phenomenological model calculation. [See Eqs. (4.4)–(4.7).]

which depends *only* on u. Physically, this corresponds to the assumption that the 2π system behaves like a composite of a variable mass κ, and that as an extremely crude approximation g' is taken to be independent of the mass κ and the other internal variable of the 2π system. It is then straightforward to evaluate the structure function by using (4.6) and (4.7). We find that in the scaling limit, diagrams III_λ and IV_λ lead to a W_1 function given by

$$W_1 \propto \int_{4\mu^2}^{\infty} \sigma(\kappa)\, d\kappa^2 \int_{u_{\min}}^{\infty} \frac{du}{(u+M^2)^2} |g'(u)|^2 \times \{(u+M^2)(1-x) + x[(M-m_p)^2 - \kappa^2]\}, \tag{4.12}$$

where μ is the pion mass,

$$u_{\min} = \kappa^2 x/(1-x) - m_p{}^2 x, \tag{4.13}$$

and $\sigma(\kappa)$ denotes the density of states for the 2π system which is given by

$$\sigma(\kappa) = \kappa^{-1}(\kappa^2 - 4\mu^2)^{1/2}. \tag{4.14}$$

As expected, the integrand inside the first integral in (4.12) is identical to that in (2.39) except for the replacement of g by g' and μ by κ.

From (4.12), one can readily verify that if

$$g'(u) \to O(u^{-n}), \quad \text{as } u \to \infty \tag{4.15}$$

when $n \geq 1$, then

$$W_1 \to O((1-x)^{1+2n}) \quad \text{as } x \to 1 \tag{4.16}$$

and

$$W_1 \to O(x^{-1}) \quad \text{as } x \to 0. \tag{4.17}$$

This latter behavior is closely connected with the property that the density of states satisfies

$$\sigma(\kappa) \to \text{constant as } \kappa \to \infty. \tag{4.18}$$

From a phenomenological point of view, the precise origin of the approximation (4.11) is not of immediate concern. We may regard (4.12) as representing the combined contributions of all multibody states, not just the three-body state, provided that (4.11) and (4.18) serve as reasonable approximations.

To assign a specific value for n, we may be guided by the aforementioned crude physical picture of the $(P2\pi)$ system. Let us neglect the π-π interaction. The two pions, say π_a and π_b, are then bound only through the two-body πP potential V. The most elementary Feynman diagram for the three-body potential U should consist of two bonds: One links π_a to P and the other π_b to P. Each bond gives a factor V, and therefore U is proportional to the square of V. Since as $q^2 \to \infty$, V is $\sim O(q^{-2})$, one may expect U to be $\sim O(q^{-4})$ where q denotes the 4-momentum transfer between P and the (2π) system, assuming that q is of the same order as the 4-momentum transfer between P and each individual pion. This crude picture suggests that as a pure trial form one may assume $n=2$ in (4.15); i.e.,

$$g' \sim O(u^{-2}) \quad \text{as } u \to \infty.$$

For definiteness of calculation, we make the further *ad hoc* assumptions that g is given by (4.10) with $M = m_p$, i.e., apart from normalization factors,

$$g = (u + m_p{}^2)^{-1},$$

and similarly g' is given by

$$g' = (u + m_p{}^2)^{-2}. \tag{4.19}$$

The result is then, for the physical proton,

$$(W_1)_p = \beta_p[F_{\mathrm{II}}(x) + \alpha_p F_{\mathrm{I}}(x)], \tag{4.20}$$

where α_p and β_p are constants, the contribution due to the two-body channel is proportional to

$$F_{\mathrm{I}}(x) = (1-x)^3 \frac{m_p{}^4[m_p{}^2(1-x)^2 + \frac{1}{3} m_\chi{}^2 x]}{[m_p{}^2(1-x)^2 + m_\chi{}^2 x]^3}, \tag{4.21}$$

and the contribution due to the multibody channel is proportional to

$$F_{\mathrm{II}}(x) = x^{-1}(1-x)^5 h(x), \tag{4.22}$$

in which

$$h(x) = \frac{1}{64(1-x)^6}\Big(135 - 33\xi^2 - 23\xi^4 - 15\xi^6 - \frac{3}{2\xi}(1-\xi^2)(45 + 19\xi^2 + 11\xi^4 + 5\xi^6) \times \ln\frac{1+\xi}{1-\xi}\Big) \tag{4.23}$$

and

$$\xi = [(1-x)^2 + 4x(\mu^2/m_p{}^2)]^{-1/2}(1-x). \tag{4.24}$$

The function $h(x)$ satisfies

$$h(0) = 1 \quad \text{and} \quad h(1) = \frac{12}{105}\left(\frac{m_p}{2\mu}\right)^6. \tag{4.25}$$

Similarly to (4.1), one may regard the physical neutron as being composed of

$$n = (NX^0) + (N2\pi) + (\Sigma K\pi) + \cdots, \tag{4.26}$$

which can be obtained from (4.1) through the usual charge-symmetry operation. In this simple model of the physical neutron, the constituents in the two-body channel are all neutral; thus, only the multibody channel can contribute to structure functions in the deep-inelastic en scattering. One finds

$$(W_1)_n = \beta_n F_{\mathrm{II}}(x). \tag{4.27}$$

The parameter β_n is related to the probability of finding a charged baryon in the multiparticle channel of the physical neutron state, and the parameter β_p is related to the corresponding probability in the physical proton state. The ratio (β_p/β_n) depends on the detailed SU_3 structure of the multiparticle channel. For example, in the three-body channel, if the two mesons are in a pure octet state then (β_p/β_n) should lie within a range $0.58 < (\beta_p/\beta_n) < 1.8$.

In the multiparticle channel, since the bare mesons can carry charge, there must also be diagrams in which the photon is absorbed by the bare meson. These diagrams lead to a nonvanishing longitudinal structure function W_L. The magnitude of W_L depends on the bound-state amplitude g'' in which only the relevant bare charged meson is off the mass shell while all other constituents, mesons and baryons, are on their mass shells. The ratio

$$\eta = \left\langle \frac{g''}{g'} \right\rangle_{av} \tag{4.28}$$

is a free parameter in the simple phenomenological approach, though it should be computable if one knew the correct potential and, in addition, could solve the corresponding Bethe-Salpeter equation. Since the present experimental value of W_L, at least for ep scattering, is quite small and uncertain, we shall assume, for simplicity,

$$|\eta| \ll 1. \tag{4.29}$$

For a comparison with the present experimental value, we set in F_{I}, Eq. (4.21), the value for m_X to be simply the observed X^0 mass $\cong 958$ MeV, and in F_{II}, Eq. (4.22), the value for (μ^2/m_p^2) to be

$$\frac{\mu^2}{m_p^2} = \frac{3m_\pi^2 + 4m_K^2 + m_\eta^2}{3m_\Sigma^2 + 2m_N^2 + 2m_\Xi^2 + m_\Lambda^2} \cong 0.12, \tag{4.30}$$

which represents in an approximate way its SU_3 average over different three-body states $P\pi\pi$, $\Sigma K\pi$, etc. In Fig. 8, the theoretical curve for $(W_1)_p$ is plotted by choosing

$$\beta_p = 0.22 \quad \text{and} \quad \alpha_p = 1.4. \tag{4.31}$$

The corresponding ratio for $(W_1)_n/(W_1)_p$ is plotted in Fig. 9 by assuming as a simple choice

$$\beta_p = \beta_n, \tag{4.32}$$

which means that in the multiparticle channel, the probability of finding a charged bare baryon in the physical nucleon state is almost independent of the total charge of the physical state p or n.

We recall that, as shown in Sec. II, in the simple model where the physical proton consists of only the two-body channel PX^0, the (νW_2) function satisfies an Adler sum rule (2.53); furthermore, since in that simple model $\nu W_2 = 2x m_p W_1$, Eq. (2.53) may also be written as

$$2m_p \int_0^1 dx\, (W_1)_p = 1. \tag{4.33}$$

In the present case, because of multiparticle channels, the integral $\int_0^1 dx\, (W_1)_p$ diverges, and therefore Eq. (4.33) cannot possibly hold. (Apart from other reasons, the assumption of no subtraction constant for the dispersion integral is clearly not valid here, unlike in the simple case of a two-body bound state discussed in Sec. II.) However, if one chooses $\beta_p = \beta_n$, then the difference $(W_1)_p - (W_1)_n$ depends only on the PX^0 channel; in this case one may define

$$\zeta \equiv 2m_p \int_0^1 dx\, [(W_1)_p - (W_1)_n]. \tag{4.34}$$

Heuristically, ζ may be regarded as the probability for finding the PX^0 channel in the physical proton state. Thus, ζ should lie between 0 and 1. For the choice of parameters given by (4.31) and

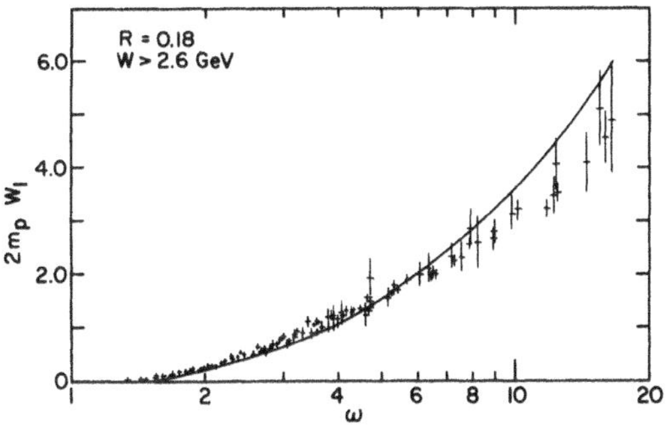

FIG. 8. $2m_p W_1$ versus $\omega = x^{-1}$. The experimental data are taken from Bloom *et al.*, Ref. 2, and the theoretical curve is plotted by using Eq. (4.20) and choosing $\alpha_p = 1.4$ and $\beta_p = 0.22$.

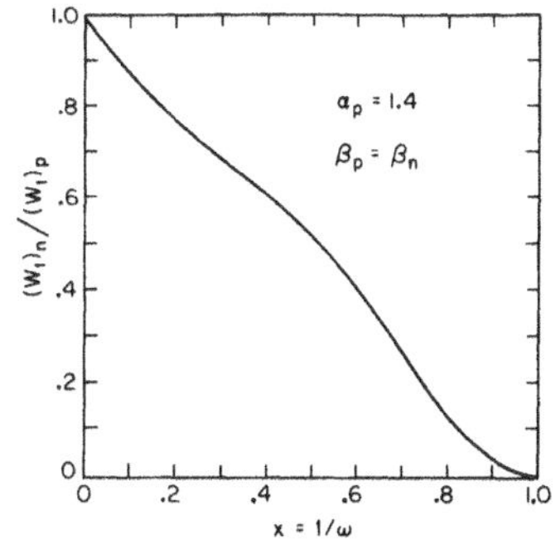

FIG. 9. The ratio $(W_1)_n/(W_1)_p$ versus x. The theoretical curve is obtained by using Eqs. (4.20) and (4.27), and choosing $\beta_p = \beta_n$ and $\alpha_p = 1.4$.

(4.32), $\zeta \cong 0.17$. In the simple model where the physical nucleon consists of only the two-body channel PX^0 or NX^0, one has $(W_1)_n = 0$ and $\zeta = 1$. Equation (4.34) then reduces to the sum rule (4.33).

B. Second Example

So far we have neglected the possibility of a two-body channel in which the meson can also be charged, such as $(N\pi^+)$, $(\Sigma^0 K^+)$, etc. The inclusion is straightforward; it leads to, instead of (4.20) and (4.27),

$$(W_1)_p = \beta_p[F_{II}(x) + \alpha_p F_I(x) + \alpha_p' F_I'(x)] \tag{4.35}$$

and

$$(W_1)_n = \beta_n[F_{II}(x) + \alpha_n' F_I'(x)], \tag{4.36}$$

where α_p' and α_n' are new parameters,

$$F_I'(x) = (1-x)^3 \frac{m_p^4[m_p^2(1-x)^2 + \frac{1}{3}\mu^2 x]}{[m_p^2(1-x)^2 + \mu^2 x]^3} \tag{4.37}$$

and μ^2/m_p^2 is given by (4.30). We note that, as $x \to 1$, $(W_1)_n/(W_1)_p$ now approaches a *nonzero* constant. The presence of such new two-body states will also give a new contribution to the W_L function. The magnitude of this W_L function depends on the two-body vertex function g''' in which the charged meson is off the mass shell, but the baryon is on the mass shell. It is easy to see that g''' should have a similar asymptotic behavior as that of $g(u)$ given in (4.9). If one assumes, similarly to (4.10),

$$g''' \propto (u + \mu^2)^{-1}, \tag{4.38}$$

then the resulting W_L function is proportional to

$$F_L(x) = \frac{x(1-x)^2[m_p^2 x^2 + \frac{1}{3}\mu^2(1-x)]}{[m_p^2 x^2 + \mu^2(1-x)]^3}. \tag{4.39}$$

Equation (4.39) shows that the threshold relation which we have derived for spin-$\frac{1}{2}$ charged constituents and the transverse structure function is not, however, valid for spin-0 constituents for the longitudinal structure function. For interactions with both nucleon and pion currents, the elastic form factor decreases[10] as $1/q^4$; but Eq. (4.39) shows only a *quadratic* threshold behavior $(1-x)^2$ for W_L as $x \to 1$. That W_L must be an even function of $(1-x)$ near threshold was already proved in Ref. 6. At present, the experimental data is consistent with $\alpha_p' \cong \alpha_n' \cong 0$.

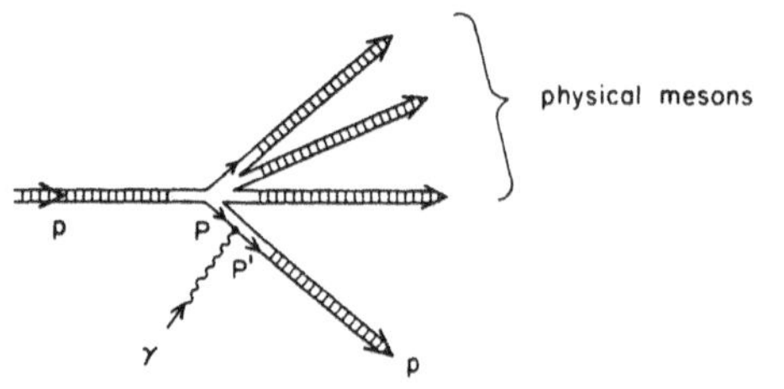

FIG. 10. A schematic diagram for "γ"+ $p \to p$ +mesons.

We showed in our Bethe-Salpeter analysis that structure functions measure the wave function of the bound state. In this section we have now shown that a simple guess of the wave function can provide a qualitative fit to the data.

V. COMMENTS

(A) The physical bound-state model that we have developed in Secs. II and III avoids some of the formal difficulties of the parton model. It is Lorentz-invariant and relativistically covariant at each step of the development, not just at the infinite-momentum scaling limit. It provides a physical connection between the observations of scaling of the structure functions in deep-inelastic scaling and of rapid decrease of the electromagnetic form factors in elastic scattering. It also reveals the relation between the bound-state vertex function of the proton for large momenta, i.e., $g(u) \sim O(u^{-1})$, and the threshold behavior of $W_1 \sim O((1-x)^3)$ near $x=1$ as well as its crossing property to the annihilation channel. In addition, since the bound state is constructed of a few (two or three) constituents of masses $\lesssim 1$ GeV, there is no large parameter > (1 GeV)^2 against which to measure $m_p\nu$ or q^2 in approaching the scaling region. This model gives then a qualitative understanding of the fact that in *ep* inelastic scattering the scaling property is already observed at relatively moderate energies.

In the context of this model we may ask what will be the behavior of inelastic transitions induced by very virtual photons to specific final states (e.g., "γ" $+p \to \Delta$, or $\to p + \pi$, etc.). In order to form such states it is necessary to add *specific* final-state interaction channels such as illustrated schematically in Fig. 10 for "γ" $+p \to p +$ some fixed meson channel. Aside from spin factors and their accompanying numerator polynomials, the amplitude for this graph will have the same asymptotic behavior for large q^2 as in Fig. 2 for the elastic form factor. The large momentum q can ride along the virtual bare proton line, marked P', in both diagrams since the proton's decreasing wave functions introduce more than enough powers of momentum to converge the loop integrals. Therefore, the added propagator and vertices connected with meson wave functions in Fig. 10 have no effect on the power dependence of the asymptotic behavior; aside from spin factors,[22] the structure functions for such "exclusive" reactions will show form factors decreasing with[23] increasing q^2.

(B) The bound-state model constructed in the previous sections can also be applied to calculating additional observables – in particular spin-dependent effects for scattering of polarized electrons from polarized targets[24] and neutrino and antineutrino deep-inelastic cross sections. Near the threshold region, i.e., near $x=1$, the two-particle state of (PX^0) will dominate in our model due to the threshold factor of $(1-x)^3$ relative to the $(1-x)^5$ factor for the three-particle contributions. In the scaling region it contributes a large polarization asymmetry A, defined as the difference divided by the sum of cross sections for parallel and antiparallel spin directions of the incident electron relative to the target nucleons. Let E_1 and E_2 be, respectively, the electron energies in the laboratory before and after scattering, and θ the scattering angle. One finds using $g=(u+m_p^2)^{-1}$

$$A=\frac{d\sigma_{\uparrow\uparrow}-d\sigma_{\uparrow\downarrow}}{d\sigma_{\uparrow\uparrow}+d\sigma_{\uparrow\downarrow}}=\left[\frac{E_1+E_2}{\nu+xm_p\cos^2(\frac{1}{2}\theta)}\right]\left[\frac{3(1-x)^2-x(\mu^2/m_p^2)}{3(1-x)^2+x(\mu^2/m_p^2)}\right]. \tag{5.1}$$

The first factor is essentially kinematical; the second one is limited in magnitude to be less than unity by positivity conditions[25] on the structure functions when $W_L=0$, as it is in this model with a spin-$\frac{1}{2}$ charged constituent. One sees that the asymmetry depends sensitively on the parameter (μ/m_p). For $(\mu/m_p)\ll 1$, the second factor is close to unity for $0\leqslant x<1-O(\mu/m_p)$, dropping rapidly to zero at $x\cong 1-3^{-1/2}(\mu/m_p)$ and then reversing sign. For a single *free* spin-$\frac{1}{2}$ constituent we would expect the asymmetry to be unity; the above behavior reflects the very massive virtual character of the intermediate constituent that is apparent in (2.39) as $x\simeq 1$. For small x values we must also include the three-particle wave function and our model predicts maximal asymmetry if there are no charged zero-spin constituents and $W_L=0$. For neutrino processes we find near $x\sim 1$, in analogy with the proton-neutron ratio shown in Fig. 9, that the ν to $\bar{\nu}$ ratio on proton targets vanishes. This is because the bare proton constituent can lower but not raise its charge by one unit by absorbing a negatively charged intermediate boson W^- as $\bar{\nu}\to e^+ + W^-$, but not a positively charged intermediate boson W^+ as $\nu\to e^- + W^+$. The prediction of spin asymmetries and of $\nu,\bar{\nu}$ differences can be further refined in terms of the ratio, R, of longitudinal to transverse photon interactions which we have for simplicity set to zero in the present analysis.

(C) As we shall discuss in the following, there remain several important questions whose answers lie outside the scope of the present model. In common with all parton (and quark) models, there exists the difficult problem of "Where and what are the constituents that are bound together forming the physical nucleon?" Since these constituents are assigned pointlike electromagnetic vertices, in order to give a meaningful calculation for the structure functions and electromagnetic form factors of the physical nucleon, they differ from observed hadrons (viz., nucleons and their low-lying resonances; pions, etc.). We have no complete or persuasive theory of the nature and reality of these constituents. Nevertheless, it seems reasonable to think that they should "dress" themselves before emerging, as illustrated by the schematic diagram given in Fig. 10. According to this picture, the asymptotic state describing the initial hadronic state at time $T\to-\infty$ is the physical proton. At any finite time T when the virtual photon hits, it interacts only with the bare charged constituents. These bare constituents are not, however, contained in the complete spectrum of asymptotic free-particle states emerging at $T\to+\infty$ after the scattering. To answer the question as to what is going on dynamically in the outgoing region, one needs to construct a more complete theory, and this we are not proposing to do here.

The problem that we are facing is not a new one in physics. It occurs, for example, in nuclear physics. The deuteron is a stable bound state, and has as its bound constituents the neutron and proton; however, when the neutron emerges from a disintegrated deuteron, it decays in ≈18 min. Because of the weak binding of the deuteron (~0.1% packing fraction) and the long decay time of the neutron due to the weakness of the β-decay coupling, there is no difficulty in understanding the physical processes involved. However, when we study the dynamics of the proton's structure we are up against strong binding (≈100% packing fraction) and presumably very short "dressing times" or "decay times" for the constituents so that simple approximate intuitive physical pictures fail us. Nevertheless, as a formal solution, one may adopt the naive approach that the bare constituents are simply ordinary unstable particles. For example, this instability may be caused by postulating the bare constituents to be somewhat heavier, say within a factor of 2, than the corresponding physical particles; in addition, their decays are assumed to be extremely fast so as to escape detection. Such an approach is clearly formally sound, though physically unsatisfactory, or at least incomplete.

As an alternative suggestion (though it may, in fact, be an equivalent one), we may adopt the physically attractive picture that the relation between a bare constituent and its corresponding physical "composite" particle is almost identical to that between a bare particle and a physical "elementary" particle in the usual field theory. The bare particles have pointlike vertices, while the physical particles have q^2-dependent form factors; both the spectrum of all bare particles and the spectrum of all physical particles are complete, and therefore any calculation of the total cross section summing over all final channels of physical particles can also be obtained by summing over all final channels of bare particles. The bare-particle concept is useful in the Heisenberg equation of motion description, since on account of their pointlike vertices the interactions between bare particles are local; however, in terms of observations in the asymptotic region of a collision process, only the physical particles actually emerge. At present, it is still an open question as to how one might develop a complete theory, so that such relations could indeed exist between the bare constituents and their corresponding physical composites.

(D) The question of final-state multiplicity is an important one, though it is clear that no complete answer is possible unless the question raised in the preceding comment is resolved. Nevertheless, it is conceivable that some partial answers could be derived by using, for example, the simple (and artificial) field-theory model in which all bare particles are described by elementary fields and the bare nucleon is regarded simply as an unstable particle. (If one wishes, one may either regard the bare mesons as the physical ones except for renormalizations, or as unstable particles like the bare nucleon.) There are then at least three different mechanisms that might contribute to final-state multiplicity: (i) the multiparticle bound-state channels in the physical nucleon state, such as the $(P\pi\pi)$, $(\Sigma K\pi)$ channels discussed in Sec. IV, (ii) the decay of the bare nucleon after its separation from the physical composite system, and (iii) the radiation of additional quanta (called gluons for convenience, as in Sec. II F) associated with whatever fields are responsible for the binding potential for the bound state. For large values of $s=-(p+q)^2$, the first two mechanisms are expected to lead to a final-state multiplicity independent of s, while the third mechanism, at least for soft quanta radiation and in the eikonal approximation, might give a final-state multiplicity that increases logarithmically with s. One may then conjecture that the total final-state multiplicity N in say ep inelastic scattering is approximately given by, for large s,

$$N=a(x)+b(x)\ln(s/m), \tag{5.2}$$

where $a(x)$ and $b(x)$ are independent of s but depend on the scaling variable x as well as on the nature of the particular physical particle under consideration, such as the pion multiplicity or the kaon multiplicity.

In the simple bound-state model considered in this paper only the above mechanism (i) is explicitly taken into account. It is clear that mechanism (ii) does not alter the (total) deep-inelastic cross section, since one sums over all final hadronic channels. With respect to mechanism (iii) the over-all picture has a familiar analog in atomic physics, in which case the quantum responsible for the binding force is the electromagnetic radiation. One may either simply ignore the final photon radiation and directly evaluate the cross section for say the photodisintegration of the atom, or one may do the same calculation, but properly take into account all final soft photon radiation. Under the Bloch-Nordsieck approximation, these two approaches lead to the same cross section, provided all final channels are summed over.

The above remark is correct if one includes only soft radiation. In the case of atomic physics, the inclusion of hard photons in the final state would lead to departures from the Bloch-Nordsieck approximation; in the present case, as we shall discuss, it may lead to violations of the scaling property.

(E) We now turn to another important question: whether the scaling property is an exact law of nature in the high-energy (scaling) limit, or just an approximate one which holds only in some intermediate energy range. For definiteness, let us consider the forward Compton scattering amplitude. As shown in Sec. II F, under the ladder approximation, the absorptive part due to those on-mass-shell intermediate states consisting of the bare constituents plus a number of gluons can be neglected in the scaling limit. As a generalization of the ladder approximation we may also include crossed graphs; i.e., diagrams in which the gluons being exchanged between *different* bare constituents are absorbed and emitted in arbitrary sequence. If the contribution of these crossed graphs is calculated iteratively using a bound-state wave function based on a ladder model that satisfies (1.3), it is not difficult to see that the same conclusion holds and the radiation of real gluons can be neglected.

If these were the *only* diagrams, then the scaling property could be an exact law. However, there are other diagrams, such as those connected with renormalizations in which the gluons are exchanged between the *same* bare constituent. Unlike the case of either ladder diagrams or soft radiation, these gluon renormalization and radiative-correction diagrams are typically divergent in the ultraviolet region.

If or when these gluon renormalization contributions become important we may expect deviations to occur from the scaling prediction. On the one hand, they may behave in analogy to the case of electromagnetic radiative corrections, where it is known that there should be deviations from the scaling property $\sim O(\alpha \ln(q^2/m^2))$; in the case that the gluon is, say, a spin-0 meson, since the renormalization diagrams have the same logarithmic divergence as the corresponding electromagnetic ones, one may expect a similar deviation from the scaling property $\sim O(\epsilon \ln(q^2/m^2))$. The magnitude of ϵ is proportional to the probability of *hard* gluon (or meson) radiation, which can be rather small. The present experimental data is consistent with $\epsilon \lesssim 10^{-1}$. In any case, just based on the electromagnetic radiative correction, one should have $\epsilon \gtrsim \alpha$. *The scaling property is, therefore, not an exact one in the mathematical high-energy limit, but is only approximately valid in the range including the one presently accessible to experimental study*

$$1 < q^2 \text{ in } (\text{GeV})^2 < \exp(\epsilon^{-1}) \leq \exp(\alpha^{-1}) . \qquad (5.3)$$

Alternatively, the possibility remains that the sum total of the gluon radiative corrections may be, simply, to give rise to a form factor of the constituent itself that decreases with increasing momentum transfer q and tends to zero for large q^2 in analogy to the observed hadronic structure. If such is the case, we may expect to observe modifications from scaling leading to structure functions decreasing at sufficiently high momentum transfers q such that $q^2\langle R^2\rangle$ is no longer negligible where $\langle R^2\rangle$ denotes a mean square radius of the charge constituents (i.e., "bare" particles). Evidently the extension of the kinematic range of experimental study for the structure functions will be of exceedingly great interest and importance.

(F) Although the physical description and motivation of the bound-state model described in this paper are very different, there are related studies of the proton's structure in deep-inelastic scattering that have similar mathematical features. In particular, Landshoff, Polkinghorne, and Short[26] present a parton model in which the virtual photon is assumed to be absorbed on a pointlike parton current and the physical input is given in terms of the properties of the parton-proton scattering amplitude. Their major assumption is that the off-mass-shell parton-proton scattering amplitude decreases sufficiently rapidly with increasing parton mass. In their model this assumption provides the necessary convergence properties that we introduced via the vertex function $g(u) \sim 1/u$ in our bound-state model.

The phenomenological model of West[27] is very much in the same spirit as this. Again it introduces electromagnetic interactions with pointlike partons and makes a convergence or "smoothness" assumption on the parton-proton scattering amplitude. West derives in this way the same relation between elastic and inelastic structure functions that we have given earlier. What is introduced as the dependence of the parton-proton scattering amplitude on the mass of the virtual parton in West's approach corresponds to the bound-state wave function of the physical proton in our approach.

Note added in proof. After submitting this paper for publication we learned that similar ideas relating scaling to the bound-state structure of the physical proton have been developed independently by C. H. Woo. We wish to thank Professor Woo for kindly informing us of his work.

ACKNOWLEDGMENTS

We wish to thank J. D. Bjorken and S. B. Treiman for stimulating discussions. One of us (T.D.L.) wishes to thank E. M. Henley and other members of the Physics Department at the University of Washington for helpful discussions and for their kind hospitality extended to him in the summer of 1971, during which the early part of this work was done.

APPENDIX A

In this appendix, we shall show that the Fourier transform of the ground-state wave function $\psi_p(k)$ is regular at the origin in the coordinate space $x=0$, and therefore

$$\int \psi_p(k)\, d^4k \quad \text{is finite} . \qquad (A1)$$

It is convenient to start with the Bethe-Salpeter equation (2.12). For simplicity, let us consider the special case

$$\mu = M \quad \text{and} \quad m_p = 0 ; \qquad (A2)$$

i.e., the bound state is of zero mass and its two constituents P and X^0 are of the same mass M. (It is straightforward, though somewhat tedious, to extend the arguments given below to the general case $\mu \neq M$ and $m_p \neq 0$.)

Following Wick,[20] we first rotate the real axis in the complex k_0 plane counterclockwise to the imaginary axis, thus changing k_0 real to $k_0 = ik_4$ imaginary. By using (A2) and by setting the total 4-momentum $p_\nu = 0$, one may write Eq. (2.12) in the Euclidean space as

$$\left(\gamma_\nu \frac{\partial}{\partial x_\nu} + M\right)\left(\frac{\partial^2}{\partial x_\mu \partial x_\mu} - M^2\right) \bar{\Gamma}(x) = -\lambda \bar{V}(x)\bar{\Gamma}(x) , \qquad (A3)$$

where $\tilde{V}(x)$ and $\tilde{\Gamma}(x)$ are, respectively, the Fourier transforms of $V(k)$ and $\Gamma_{p=0}(k)$,

$$\tilde{V}(x) = \int V(k) e^{ik\cdot x}[d^4k], \tag{A4}$$

$$\tilde{\Gamma}(x) = \int \Gamma_{p=0}(k) e^{ik\cdot x}[d^4k], \tag{A5}$$

and $[d^4k] = dk_1 dk_2 dk_3 dk_4$ is real. The spinor dependence of $\tilde{\Gamma}(x)$ can be separated by introducing two c-number functions $f_1(x)$ and $f_2(x)$:

$$\tilde{\Gamma}(x) = f_1\gamma_5 + M^{-1}\left(\gamma_\mu \frac{\partial}{\partial x_\mu}\right) f_2\gamma_5, \tag{A6}$$

in which the factor γ_5 is due to the assumed pseudoscalar nature of X^0. If X^0 were a scalar, then γ_5 would be replaced by unity. For the ground state, f_1 and f_2 depend only on

$$r = (x_1^2 + x_2^2 + x_3^2 + x_4^2)^{1/2}. \tag{A7}$$

Equation (A2) reduces then to two coupled ordinary differential equations

$$\left(\frac{d^2}{dr^2} + \frac{3}{r}\frac{d}{dr} - M^2\right)\left[M^2 f_1 + \left(\frac{d}{dr} + \frac{3}{r}\right) f_2'\right] = -\lambda M \tilde{V} f_1$$

and (A8)

$$\left(\frac{d^2}{dr^2} + \frac{3}{r}\frac{d}{dr} - M^2 - \frac{3}{r^2}\right)\left[\frac{df_1}{dr} + f_2'\right] = -\lambda M^{-1}\tilde{V} f_2',$$

where

$$f_2' = \frac{df_2}{dr}. \tag{A9}$$

From (A6), one sees that $\tilde{\Gamma}(x)$ depends only on f_1 and f_2', but not on f_2 explicitly. Since both equations in (A8) are of third order in (d/dr), there should be six independent solutions for $\tilde{\Gamma}(x)$.

According to the standard procedures for the eigenvalue problem of a Sturm-Liouville equation, in order to determine whether the eigenstate $\tilde{\Gamma}(x)$ is regular at $r=0$ or not, one has to show that (A8) has three regular solutions at $r=0$, say Γ_1, Γ_2, Γ_3, and three regular solutions at $r=\infty$, say Γ_a, Γ_b, and Γ_c. The bound state $\tilde{\Gamma}(x)$ can be written as a sum

$$\tilde{\Gamma}(x) = \alpha\Gamma_1 + \beta\Gamma_2 + \Gamma_3. \tag{A10}$$

One then adjusts the three parameters α, β, and the eigenvalue λ so that $\tilde{\Gamma}(x)$ is regular at $r=\infty$; i.e.,

$$\tilde{\Gamma}(x) = a\Gamma_a + b\Gamma_b + c\Gamma_c. \tag{A11}$$

To show that there are three regular solutions at $r=\infty$, let us assume, for simplicity, the covariant potential V to be superpositions of Yukawa-like forms with nonzero masses [i.e., in (1.1), $\sigma(\kappa^2)$ is not proportional to $\delta(\kappa^2)$]. Therefore, as $r\to\infty$, $\tilde{V}\to 0$ exponentially. At large r (A8) becomes simply the free equation. One can readily verify that for $V=0$ the six independent solutions consist of three regular ones

(a) $f_1 = r^{-1}H_1^{(1)}(iMr) \to \text{const}\, r^{-3/2}\exp(-Mr)$,

$f_2' = -\dfrac{df_1}{dr}$,

(b) $f_1 = 0$,

$f_2' = r^{-1}H_2^{(1)}(iMr) \to \text{const}\, r^{-3/2}\exp(-Mr)$,

(c) $f_1 = r^{-1}G_1^{(1)}(iMr) \to \text{const}\, r^{-1/2}\exp(-Mr)$,

$f_2' = -\dfrac{df_1}{dr}$;

and three irregular ones

(d) $f_1 = r^{-1}H_1^{(2)}(iMr) \to \text{const}\, r^{-3/2}\exp(Mr)$,

$f_2' = -\dfrac{df_1}{dr}$,

(e) $f_1 = 0$,

$f_2' = r^{-1}H_2^{(2)}(iMr) \to \text{const}\, r^{-3/2}\exp(Mr)$,

and

(f) $f_1 = r^{-1}G_1^{(2)}(iMr) \to \text{const}\, r^{-1/2}\exp(Mr)$,

$f_2' = -\dfrac{df_1}{dr}$,

where $H_1^{(1)}$, $H_1^{(2)}$, $H_2^{(1)}$, and $H_2^{(2)}$ are the standard Hankel functions,[28] and $G_1^{(i)}$ ($i=1$ or 2), besides satisfying the above asymptotic behavior, is also the solution of the following inhomogeneous Bessel equation:

$$\left[\frac{d^2}{dz^2} + \frac{1}{z}\frac{d}{dz} + \left(1 - \frac{1}{z^2}\right)\right] G_1^{(i)}(z) = H_1^{(i)}(z). \tag{A12}$$

Our next task is to investigate the behavior at small r. On account of (1.2), as $r\to 0$, $\tilde{V}$ is proportional to r^{-2}. Without any loss of generality, we may choose $M=1$ and define the coupling constant λ so that

$$\tilde{V} \to 4r^{-2} \quad \text{as } r\to 0. \tag{A13}$$

As we shall see, as $r\to 0$ there are indeed three regular solutions given by

(i) $f_1 \to c_1 r^2$, $\quad f_2' \to c_2 r^3$,

(ii) $f_1 \to c_1 r^2 \ln r$, $\quad f_2' \to c_2 r$,

and

(iii) $f_1 \to c_1$, $\quad f_2' \to c_2 r \ln r$;

in addition there are three irregular solutions

(iv) $f_1 \to c_1 \ln r$, $\quad f_2' \to c_2 r^{-1}$,

(v) $f_1 \to c_1 r^{-2}$, $\quad f_2' \to c_2 r^{-1}\ln r$,

and

(vi) $f_1 \to c_1 r^{-4}\ln r\,, \quad f_2' \to c_2 r^{-3}\,,$

where c_1 and c_2 denote the appropriate constants.

The derivation of these results follows the standard method. The first regular solution is obtained through the usual indicial equation by expanding f_1 and f_2' as power series in r. One finds

$$f_1 = s(r) = r^2 \sum_0^\infty s_n r^{2n}\,, \qquad f_2' = t(r) = r^3 \sum_0^\infty t_n r^{2n}\,. \tag{A14}$$

To obtain the second regular solution, one assumes f_1 and f_2' to be given, respectively, by $\ln r$ times the first regular solution plus a power series in r. The result is

$$f_1 = (\ln r)s + r^4 \sum_0^\infty a_n r^{2n}$$

and (A15)

$$f_2' = (\ln r)t + r \sum_0^\infty b_n r^{2n}\,,$$

where s and t are given by (A14) and a_n, b_n can be easily determined by substituting (A15) into (A8).

To derive the third regular solution, one assumes

$$f_1 = (\ln r)^2 s + (\ln r)u_1 + v_1$$

and (A16)

$$f_2' = (\ln r)^2 t + (\ln r)u_2 + v_2\,,$$

where u_i and v_i ($i = 1$ and 2) are power series in r. It can be readily verified that these power-series expansions are of the form

$$u_1 = r^4 \sum_0^\infty (u_1)_n r^{2n}\,, \quad v_1 = \sum_0^\infty (v_1)_n r^{2n}\,, \qquad u_2 = r\sum_0^\infty (u_2)_n r^{2n}\,, \quad v_2 = r\sum_0^\infty (v_2)_n r^{2n}\,. \tag{A17}$$

More explicitly, this third regular solution is given by, as $r \to 0$,

$$f_1 \to 1 + \tfrac{1}{8}\lambda^2 r^2 (\ln r)^2 + O(r^2)$$

and (A18)

$$f_2' \to -\tfrac{1}{2}\lambda r \ln r + \tfrac{1}{8}(-3\lambda + 4)r - \tfrac{1}{96}\lambda^2(\lambda+2)r^3(\ln r)^2 + O(r^3 \ln r)\,.$$

According to (A10), the bound-state wave function is a linear function of the above three regular solutions, (A14)–(A16). From the explicit form of these solutions, one sees that as $r \to 0$ the third regular solution dominates over the other two; thus, (A18) holds for the bound-state wave function as well. By using (A6) and (A18) one sees that $\tilde{\Gamma}(0)$ exists, and therefore, on account of (2.10),

$$\psi_p(k) = \Gamma_p(k) u_p\,,$$

the Fourier transform of the bound-state wave function $\psi_p(k)$

$$\tilde{\psi}_p(x) \equiv \int \psi_p(k) e^{ik\cdot x}\, d^4k \tag{A19}$$

is regular at the origin; consequently, (2.8) and (2.9) hold.

We note that in a 4-dimensional Euclidean space the Fourier transform of any function $h(r)$, regular at both $r = 0$ and ∞, satisfies

$$\int e^{ik\cdot x} h(r)\, d^4r = -4\pi^2 k^{-4} \int_0^\infty J_0(kr)[3h' + 5rh'' + r^2 h''']\,dr\,, \tag{A20}$$

where J_0 is the zeroth-order Bessel function. Thus, for example, by setting $h = f_2$, which according to (A18) is

$$c - \tfrac{1}{4}\lambda r^2 \ln r + O(r^2)\,,$$

where c is a constant, and by using (A20), one finds that, through partial integration, as $k^2 \to \infty$

$$\int e^{ik\cdot x} f_2\, d^4r = -16\pi^2\lambda k^{-6} + O(k^{-8})\,. \tag{A21}$$

It is then straightforward to show that as $k^2 \to \infty$ the bound-state solution $\Gamma_p(k)$ satisfies

$$\Gamma_p(k) = \mathrm{const}(-i\gamma\cdot k)\gamma_5 k^{-6} + O(k^{-6}\ln k^2)\,. \tag{A22}$$

The explicit form (A18) enables one to determine not only the leading behavior, but also the form of the next-order correction term $O(k^{-6}\ln k^2)$. By using (A22), one may also directly establish (2.9) and (2.37).

APPENDIX B

To establish the identity (2.21), one may first differentiate both (2.6) and (2.16) with respect to k_ν, keeping p_ν fixed; next multiply them, respectively, by $\bar{\psi}_p(k) = \bar{\phi}_p(k)K_p^{-1}(k)$ on the left and $\psi_p(k) = K_p^{-1}(k)\phi_p(k)$ on the right; and then add these two equations together. This leads to

$$\int \bar{\phi}_p(k)\left[\frac{\partial K_p^{-1}(k)}{\partial k_\nu}\right]_p \phi_p(k)\, d^4k = -i\lambda \int \bar{\psi}_p(k)\psi_p(k')\left[\frac{\partial}{\partial k_\nu'} + \frac{\partial}{\partial k_\nu}\right] V(k-k')\, d^4k\, d^4k' = 0\,, \tag{B1}$$

and, therefore, (2.21) is proved.

To establish (2.22), one may first differentiate (2.6) with respect to p_ν, keeping k_ν fixed; next multiply on the left by $\bar{\psi}_p(k)$; and then integrate over d^4k. After using (2.16), one finds

$$\int \bar{\psi}_p(k)\left[\frac{\partial K_p(k)}{\partial p_\nu}\right]_k \psi_p(k)\,d^4k = i\frac{\partial\lambda}{\partial p_\nu}\int \bar{\psi}_p(k)V(k-k')\psi_p(k')\,d^4k\,d^4k' = \frac{\partial \ln\lambda}{\partial p_\nu}\int \bar{\psi}_p(k)K_p(k)\psi_p(k)\,d^4k\,. \quad \text{(B2)}$$

Since $\lambda = \lambda(m_p^2)$ and

$$\left[\frac{\partial}{\partial p_\nu}K_p(k)\right]_k + \frac{\mu}{M+\mu}\left[\frac{\partial}{\partial k_\nu}K_p(k)\right]_p = \left[\frac{\partial}{\partial P_\nu}K_p(k)\right]_X = -i\gamma_\nu(X^2+\mu^2)\,,$$

(2.22) follows on account of (B2), (2.20), and (2.21).

APPENDIX C

In this appendix, we give the detailed derivation of why the amplitude I'_ν, defined by (2.32), can be neglected in the scaling limit. It is useful to determine in (2.32) the integration domain in which the integrand does not approach zero in the scaling limit. Let us first examine the region in which the denominator remains finite. We observed that because of the vertex function $\phi_p(k_{\text{in}})$ and the propagators $[-i\gamma\cdot(p-X')-M]^{-1}(X'^2+\mu^2)^{-1}$ in the integrand in (2.32) this domain is restricted to

$$X'^2 = O(m^2) \quad \text{and} \quad (p-X')^2 = O(m^2)\,, \quad \text{(C1)}$$

and because of the denominator in the propagator

$$[-i\gamma\cdot P'-M]^{-1} = -(P'^2+M^2)^{-1}[-i\gamma\cdot P'+M] \quad \text{(C2)}$$

there is the additional constraint

$$P'^2 = (p+q-X')^2 = O(m^2)\,, \quad \text{(C3)}$$

where m denotes collectively the relevant masses m_p, M, and μ. In the laboratory frame, the first two constraints (C1) imply that each of the four components of X'_μ should be restricted to

$$X'_\mu = O(m)\,. \quad \text{(C4)}$$

The additional constraint (C3) then implies that in the scaling limit of $q^2\to\infty$, $-(p\cdot q)\to\infty$ but keeping the scaling variable

$$x = q^2/(-2p\cdot q) \text{ finite}\,, \quad \text{(C5)}$$

the components X'_3 and X'_0 are further related by

$$X'_0 - X'_3 = (1-x)m_p + O(m^3/q^2)\,, \quad \text{(C6)}$$

where the z axis is chosen to be parallel to $\vec{q}$. These constraints (C4) and (C6) restrict the integration region of interest to a volume

$$\int d^4X' = O(m^6/q^2)\,; \quad \text{(C7)}$$

in this region, all denominators in the integrand of (2.32) are finite in the scaling limit. Next, we have to determine in this region the magnitude of the numerator in (2.32). On account of (2.33) and (1.2), at large 4-momentum transfer squared $(k'-k)^2$ the scattering matrix $\langle k|T_{p+q}|k'\rangle$, like $V(k'-k)$, decreases as $(k'-k)^{-2}$. This, together with (C4) and (C6), implies that in the laboratory frame the final 4-momenta are restricted to

$$X_\mu = O(m) \quad \text{and} \quad P_\mu = q_\mu + O(m)\,.$$

Since the virtual momentum $P'_\mu = q_\mu + O(m)$ as well, in the integral (2.32), after using (C2), one has the numerator factor

$$u_p^\dagger\gamma_4\langle k|T_{p+q}|k'\rangle(-i\gamma\cdot P'+M) = \langle k|T_{p+q}|k\rangle u_p^\dagger\gamma_4(-i\gamma\cdot P+M) + O(m)\,,$$

which is also $O(m)$. Therefore, one finds that the integrand in (2.32) remains finite in the integration region (C7) and, consequently, the integral over this region is proportional to the volume which approaches zero in the scaling limit. Outside this region (C7), the integrand itself approaches zero at least as $O(q^{-2})$; furthermore, the corresponding integral is dominated by the region in which (C4) remains valid, but, in place of (C3), $P'^2 = O(q^2)$. Since the integration over X' is a superconvergent one, the integral (2.32) remains convergent (in fact remains superconvergent) if the propagator with the large virtual momentum P' is taken outside the integral. It then follows that the entire integral I'_ν goes to zero in the scaling limit and therefore can be neglected.

APPENDIX D

To establish (2.52) it is convenient to start from (2.51) and to use the laboratory frame with the z axis parallel to $\vec{q}$. One has then

$$U_1 = F_{11} = F_{22}$$

and

$$\nu U_2 = \frac{\nu q^2}{\nu^2+q^2}\left(F_{11} + \frac{q^2}{\nu^2}F_{33}\right)\,. \quad \text{(D1)}$$

It is easy to see that in the scaling limit F_{33} remains finite, and therefore $\nu U_2 \to 2xm_pU_1$. Consequently, one needs only to investigate F_{11}, or A_{11} and B_{11}. Because of the rapid convergence property of the bound-state vertex function, the integral (2.47) for A_{11} is determined only by its value over the domain in which

$$X^2 = O(m^2) \quad \text{and} \quad (p-X)^2 = O(m^2)\,. \quad \text{(D2)}$$

These constraints imply that in the laboratory frame all components of X_μ are $O(m)$. Similarly, in the same laboratory frame the integral (2.48) for B_{11} is determined only by its value over the domain in which

$$X_\mu = O(m) \text{ and } X'_\mu = O(m). \quad \text{(D3)}$$

For A_{11} the magnitude of the integrand in (2.47) is proportional to the appropriate matrix element of

$$[-i\gamma\cdot P - M]^{-1} = (P^2+M^2)^{-1}[i\gamma\cdot q + O(m)] \quad \text{(D4)}$$

in which the numerator on the right-hand side is $O(\nu)$. If $P^2 = O(q^2)$, then (D4) is $O(m^{-1})$, and the integrand in (2.47) remains finite in the scaling limit. On the other hand, if $P^2 = O(m^2)$, then (D4) is $O(\nu/m^2)$. However, in the latter case, on account of $P^2 = (p+q-X)^2$, the components of X_μ in the laboratory frame must be further restricted to

$$X_0 - X_3 = (1-x)m_p + O(m^2/\nu); \quad \text{(D5)}$$

consequently, although the integrand in (2.47) is $O(\nu)$, the relevant integration volume $\int d^4X$ is $O(m^5/\nu)$. Thus, one finds that A_{11} remains finite and nonzero in the scaling limit.

To study the magnitude of B_{11}, it is convenient to use (2.33) and formally expand B_{11} as a power series in the coupling constant λ of the binding potential V. We write

$$B_{11} = \sum_{n=1}^{\infty} \lambda^n B_{11}^n \quad \text{(D6)}$$

in which, e.g., the first term λB_{11}^1 is given by the same expression (2.48) except that $\langle k | T_{p+q} | k'\rangle$ is replaced by $i\lambda V(k'-k)$. The magnitude of the resulting integrand for B_{11}^1 is proportional to the appropriate matrix element of the product

$$\begin{aligned}[-i\gamma\cdot P - M]^{-1}[-i\gamma\cdot P' - M]^{-1} &= (P^2+M^2)^{-1}(P'^2+M^2)^{-1} \\ &\quad\times[-i\gamma\cdot P + M][-i\gamma\cdot P' + M]. \end{aligned} \quad \text{(D7)}$$

Because of (2.49) and (D3), the numerator on the right-hand side of (D7) is equal to $-q^2 + O(m) \times[-i\gamma\cdot q] + O(m^2)$ which is $O(q^2)$ in the scaling limit. If $P^2 = O(q^2)$ and $P'^2 = O(q^2)$, then (D7) is $O(q^{-2})$. If $P^2 = O(m^2)$ but $P'^2 = O(q^2)$, then (D7) is $O(m^{-2})$ but, in addition to (D3), the components of X_μ are further constrained by (D5), so that the relevant integration volume $\int d^4X$ is $O(m^5/\nu)$. Similarly, if $P'^2 = O(m^2)$ then the relevant integration volume $\int d^4X'$ is reduced to $O(m^5/\nu)$. Putting all these together, one finds that

$$B_{11}^1 \sim O(\nu^{-1})$$

in the scaling limit. By following the same reasoning, it can be readily established that for $l \geq 1$, apart from possible $\ln(\nu/m)$ factors,

$$B_{11}^{2l} \sim O(\nu^{-l}) \text{ and } B_{11}^{2l-1} \sim O(\nu^{-l}).$$

Thus, in the scaling limit

$$F_{11} \to A_{11}(p,q) + A_{11}(p,-q), \quad \text{(D8)}$$

from which, by taking the imaginary part, (2.52) follows. We note that, like the diagram I'_ν in Fig. 3, the diagram $B_{\lambda\nu}$ in Fig. 4 is needed for maintaining gauge invariance but can be neglected in the scaling limit.

APPENDIX E

To complete the proof for the sum rule (2.53), we follow closely the steps taken by Adler,[17] making sure that all the assumptions used in the usual derivation are indeed valid in the present case. In the laboratory frame, the amplitude $F'_{\lambda\nu}$, defined by (2.54), may be written as

$$\begin{aligned} F'_{ij}(p,q) &= \delta_{ij}a_1(q^2,\nu) + q_iq_ja_2(q^2,\nu), \\ F'_{i4}(p,q) &= q_i b(q^2,\nu), \\ F'_{4i}(p,q) &= q_i b(q^2,-\nu), \end{aligned} \quad \text{(E1)}$$

and

$$F'_{44}(p,q) = c(q^2,\nu),$$

where the subscripts i and j denote the space components. These four functions a_1, a_2, b, and c are related through the divergence condition (2.56). By setting the subscript λ in (2.56) to be, respectively, the time and the appropriate space components one finds that

$$\nu c - i(q^2+\nu^2)b = 2 \quad \text{(E2)}$$

and

$$a_1 + (q^2+\nu^2)a_2 + i\nu b = 0 \quad \text{(E3)}$$

in which, on account of (2.55), $a_1(q^2,\nu)$, $a_2(q^2,\nu)$, and $c(q^2,\nu)$ are odd in ν, but $b(q^2,\nu)$ is even in ν.

It is convenient to define

$$f(q^2,\nu) \equiv \nu^{-1}q^2(a_1 + q^2a_2) \quad \text{(E4)}$$

which is even in ν. By using the explicit expressions (2.47) and (2.48), one can readily prove that a_1, a_2, b, and c are all regular at $\nu = 0$. At a fixed q^2, as $\nu \to 0$ Eq. (E3) implies $\nu^{-1}q^2(a_1+q^2a_2) \to -ibq^2$ which, in turn, on account of (E2), approaches 2. Thus, one finds that at $\nu = 0$

$$f(q^2, 0) = 2. \quad \text{(E5)}$$

Next, we shall show that at a fixed q^2

$$f(q^2,\nu) \sim O(\nu^{-2}) \text{ as } \nu \to \infty. \quad \text{(E6)}$$

This can be established by following almost exactly the same arguments used in Appendix D for A_{11}

and B_{11}, except that instead of the scaling limit, one now takes the limit of $\nu \to \infty$ while keeping q^2 fixed. We note that (D3) remains valid. Since in the laboratory frame $P_\mu = q_\mu + O(m)$, the numerator on the right-hand side of (D4) remains $O(\nu)$, as before. Thus, (D4) is $O(m^{-1})$ if $P^2 = O(m\nu)$, in which case the corresponding integration volume $\int d^4X = O(m^4)$. If $P^2 = O(m^2)$, then (D4) is $O(\nu)$, but the corresponding integration volume is $O(m^5/\nu)$, because of (D5). Consequently, in the laboratory frame $A_{11}(p, q)$ is finite in the limit $\nu \to \infty$. Furthermore, the same limit holds if $\nu \to -\infty$. The difference $A_{11}(p, q) - A_{11}(p, -q)$ is, therefore, $O(\nu^{-1})$. Similar arguments and conclusions can be readily extended to the differences $A_{33}(p, q) - A_{33}(p, -q)$, $B_{11}(p, q) - B_{11}(p, -q)$, and $B_{33}(p, q) - B_{33}(p, -q)$. By using (2.54), (2.57), and (E3), one establishes the asymptotic condition (E6).

Since $f(q^2, \nu)$ is an even function in ν, one may consider the complex $z = \nu^2$ plane, and regard f as a function of z, keeping q^2 fixed. In this complex plane, f has a pole at

$$z = \nu_0^2 \equiv [q^2/(2m_p)]^2 \tag{E7}$$

and a cut along the real axis extending from $\nu_{\min}^2$ to ∞ where

$$\nu_{\min} = (2m_p)^{-1}[q^2 + (M+\mu)^2 - m_p^2]\,. \tag{E8}$$

To study the discontinuity of f across the cut, it is only necessary to investigate the absorptive part of $F'_{\lambda\nu}$ for $\nu > 0$. By using (2.47) and (2.48), one sees that for $\nu > 0$ the absorptive parts of $F'_{\lambda\nu}$ and $F_{\lambda\nu}$ are equal where $F_{\lambda\nu}$ is the forward Compton amplitude defined by (2.46). It follows then that the discontinuity of f across the cut is simply $2i\,\mathrm{Im}(\nu U_2)$, where U_2 is defined by (2.51). The standard Cauchy theorem and the asymptotic condition (E6) lead to, for z away from these singularities,

$$f(q^2, z) = f(q^2, \nu_0^2) + \pi^{-1}\int_{\nu_{\min}^2}^{\infty} d\nu^2(\nu^2 - z)^{-1}\,\mathrm{Im}(\nu U_2)\,. \tag{E9}$$

Upon setting $z = 0$ and changing the variable ν to $x = q^2/(2m_p\nu)$ while keeping q^2 fixed, (E9) becomes, on account of (E5),

$$1 = \tfrac{1}{2}f(q^2, \nu_0^2) + \pi^{-1}\int_0^{x_{\max}} x^{-1}dx\,\mathrm{Im}(\nu U_2)\,, \tag{E10}$$

where $x_{\max} = 1 - q^{-2}[(M+\mu)^2 - m_p^2]$. By repeating exactly the same argument used in Appendix D, one can show that (2.52) can also be written, at a fixed x, as

$$\mathrm{Im}(\nu U_2) = \pi(\nu W_2)[1 + O(m^2 q^{-2})]\,.$$

As $q^2 \to \infty$, one has $f(q^2, \nu_0^2) \to 0$ since it is proportional to the square of the elastic electromagnetic form factor. Moreover, $x_{\max} \to 1$ and at fixed x

$$\mathrm{Im}(\nu U_2) \to \pi(\nu W_2)\,;$$

(E10) becomes then simply the Adler sum rule (2.53).

---

*Research supported in part by the U. S. Atomic Energy Commission.

[1]J. D. Bjorken, Phys. Rev. 179, 1547 (1969); R. P. Feynman (unpublished); Phys. Rev. Letters 23, 1415 (1969); in *High Energy Collisions*, Conference held at the State University of New York, Stony Brook, 1969, edited by C. N. Yang *et al*. (Gordon and Breach, New York, 1969); J. D. Bjorken and E. A. Paschos, Phys. Rev. 185, 1975 (1969).

[2]E. D. Bloom *et al*., MIT-SLAC Report No. SLAC-PUB-796, 1970 (unpublished), presented at the Fifteenth International Conference on High Energy Physics, Kiev, U.S.S.R., 1970; Henry W. Kendall, in *Proceedings of International Symposium on Electron and Photon Interactions, 1971*, edited by N. B. Mistry (Cornell Univ. Press, Ithaca, N.Y., 1972), p. 247. See also various references mentioned in these papers.

[3]For simplicity, let us assume that in the infinite-momentum frame the 3-momentum $\vec{k}$ of the parton is parallel to the 3-momentum $\vec{p}$ of the physical proton; i.e., in the frame $\vec{p} \to \infty$, $\vec{k} = x\vec{p}$. Therefore, one may write for the 4-momentum $k_\lambda = xp_\lambda + O(|\vec{p}|^{-1})$. In the evaluation of the νW_2 function, in order to replace $\delta(q^2 + 2k\cdot q)$ by $\delta(q^2 + 2xp\cdot q)$, one must have either the time component q_0 of the virtual photon to remain finite as $\vec{p} \to \infty$, or the difference $k_\lambda - xp_\lambda$ to be zero identically, not just $O(|\vec{p}|^{-1})$. The former restricts the direction $\vec{p}$ of the infinite-momentum frame, and the latter implies that the parton has no strong interaction and, furthermore, it is of mass xm_p, where m_p denotes the proton mass.

[4]Sidney D. Drell and Tung-Mow Yan, Ann. Phys. (N.Y.) 66, 578 (1971).

[5]Y. S. Tsai (unpublished); R. Jackiw and G. Preparata, Phys. Rev. Letters 22, 975 (1969); S. L. Adler and Wu-Ki Tung, *ibid*. 22, 978 (1969). Cf. also T. K. Gaisser and J. C. Polkinghorne, Nuovo Cimento 1A, 501 (1971).

[6]See, e.g., S. D. Drell, Donald J. Levy, and Tung-Mow Yan, Phys. Rev. 187, 2159 (1969); Phys. Rev. D 1, 1035, (1970); 1, 1617 (1970).

[7]See, e.g., S. B. Treiman and D. Gross, Phys. Rev. D 4, 1059 (1971).

[8]The electromagnetic radiative corrections are known to introduce deviations from the scaling property $\sim O(\alpha \ln q^2)$. The question whether there might be similar deviations due to nonelectromagnetic radiative corrections associated with meson fields is, as yet, an open one. As we shall see in the following sections, at least in the ladder approximation of the Bethe-Salpeter equa-

tion, the scaling property can be shown to hold rigorously for a large class of relativistic field theories. (See comment E in Sec. V for further discussions.)

[9]For parton models based on quarks as constituents, see Bjorken and Paschos, Ref. 1; J. Kuti and V. F. Weisskopf, Phys. Rev. D 4, 3418 (1971). Cf. also P. Landshoff and J. C. Polkinghorne, Nucl. Phys. B28, 240 (1971).

[10]J. S. Ball and F. Zachariasen, Phys. Rev. 170, 1541 (1968); D. Amati, L. Caneschi, and R. Jengo, Nuovo Cimento 58, 783 (1968); D. Amati, R. Jengo, H. R. Rubinstein, G. Veneziano, and M. A. Virasoro, Phys. Letters 27B, 38 (1968); M. Ciafaloni and P. Menotti, Phys. Rev. 173, 1575 (1968); M. Ciafaloni, *ibid.* 176, 1898 (1968).

[11]A linear $(1-x)$ dependence is obtained by Bjorken and Paschos (Ref. 1) for their three-quark model. The same linear dependence is also obtained by Drell, Levy, and Yan (Ref. 6) in their field-theoretical calculation with an *ad hoc* transverse momentum cutoff.

[12]S. D. Drell and Tung-Mow Yan, Phys. Rev. Letters 24, 181 (1970). Cf. G. B. West, *ibid.* 24, 1206 (1970). See also the comment below Eq. (4.39) in Sec. IV.

[13]There exists a substantial amount of literature on the Bethe-Salpeter equation. Among the early ones are E. Salpeter and H. Bethe, Phys. Rev. 84, 1232 (1951); M. Gell-Mann and F. Low, *ibid.* 84, 350 (1951); G. C. Wick, *ibid.* 96, 1124 (1954); R. Cutkosky, *ibid.* 96, 1135 (1954); S. Mandelstam, Proc. Roy. Soc. (London) A233, 248 (1955); A. Klein and C. Zemach, Phys. Rev. 108, 126 (1957); R. Blankenbecler and L. F. Cook, Jr., *ibid.* 119, 1745 (1960); R. E. Cutkosky and M. Leon, *ibid.* 135, B1445 (1964). For further references, see, e.g., those given by C. H. Llewellyn Smith, Nuovo Cimento 60A, 348 (1969).

[14]Throughout the paper, we adopt the convention that $d^4k = d^3k\, dk_0$, the scalar product between any two 4-vectors, say k_ν and q_ν, is $k\cdot q = \vec{k}\cdot\vec{q} - k_0 q_0$, and the Dirac γ matrices are all Hermitian. For the time-reversal operation used in Eq. (2.13) below, these γ matrices are assumed to be explicitly given by the usual Pauli representation: $\gamma_i = \rho_2\sigma_i$, $\gamma_4 = \rho_3$ and therefore $\gamma_5 = -\rho_1$ where ρ_i and σ_i each represents a set of the usual three Pauli matrices.

[15]See also the discussions given by Ball and Zachariasen, Ref. 10, and by Amati *et al.*, Ref. 10.

[16]Cutkosky and Leon, Ref. 13.

[17]S. L. Adler, Phys. Rev. 143, 1144 (1966); see also J. D. Bjorken, *ibid.* 148, 1467 (1966).

[18]See rapporteur's talk by Richard Wilson, in *Proceedings of the Fifteenth International Conference on High Energy Physics, Kiev, U.S.S.R., 1970* (Atomizdat, Moscow, 1971).

[19]For discussions of the nonrelativistic problem, cf. S. D. Drell, A. Finn, and M. H. Goldhaber, Phys. Rev. 157, 1402 (1967).

[20]Wick, Ref. 13.

[21]Cutkosky, Ref. 13.

[22]See in particular the analyses of Amati *et al.*, Ref. 10.

[23]J. D. Walecka and P. A. Zucker, Phys. Rev. 167, 1479 (1968); P. L. Pritchett, J. D. Walecka, and P. A. Zucker, *ibid.* 184, 1825 (1969); P. L. Pritchett and P. A. Zucker, Phys. Rev. D 1, 175 (1970).

[24]J. D. Bjorken, Phys. Rev. D 1, 1376 (1970); J. Kuti and V. F. Weisskopf, to be published.

[25]M. G. Doncel and E. de Rafael, IHES report, 1971 (unpublished); F. Niedermayer, thesis, Eötvös University, 1971 (unpublished); L. Gálfi, P. Gnädig, J. Kuti, F. Niedermayer, and A. Patkós, in *Proceedings of the Fifteenth International Conference of High Energy Physics, Kiev, U.S.S.R., 1970* (Atomizdat, Moscow, 1971); ITP-Budapest Report No. 281, 1971 (to be published).

[26]P. V. Landshoff, J. C. Polkinghorne, and R. D. Short, Nucl. Phys. B28, 225 (1971).

[27]G. B. West, Phys. Rev. Letters 24, 1206 (1970).

[28]See, e.g., E. Jahnke and F. Emde, *Tables of Functions* (Dover, New York, 1945), 4th ed.

Reprinted from:

PHYSICAL REVIEW D VOLUME 6, NUMBER 4 15 AUGUST 1972

Simple Bound-State Model of the Physical Nucleon*

T. D. Lee
Columbia University, New York, New York 10027

(Received 1 May 1972)

A simple, but bona fide, local field-theoretic model is presented which illustrates the recently discussed bound-state concept of the physical nucleon. The scaling properties of W_1 and νW_2 functions in the deep-inelastic electron scattering are shown to be valid in the energy range in which the laboratory virtual-photon energy ν, in units of GeV, is larger than $O(1)$ but less than $O(\exp(\epsilon^{-1}))$ where ϵ is a small, but nonzero, free parameter, assumed to be $\lesssim O(10^{-1})$. In this model, one of the constituents of the physical proton p is a local spin-$\frac{1}{2}$ charged field operator $\psi(x)$, and the usual "pointlike" assumption for the electromagnetic vertex of the constituent emerges simply as the standard minimal electromagnetic interaction in a local field theory.

I. INTRODUCTION

In a recent paper,[1] it was suggested that the scaling property in deep-inelastic electron scatterings can be established by regarding the physical nucleon p as a bound state of a "bare" nucleon and a "bare" meson (or a few bare mesons). In particular, rigorous statements concerning scaling and its connection with elastic form factors can be derived for specific bound-state solutions of the Bethe-Salpeter equation with the ladder approximation. An important question is to inquire whether scaling holds outside the ladder approximation. It was pointed out in the same paper (Sec. V E) that the scaling property remains valid if one includes also crossed diagrams; however, deviations from scaling could arise due to higher-order radiative corrections in which virtual gluons (or mesons) are emitted and absorbed by the same "bare" nucleon. For example, let $\phi(x)$ denote the spin-0 gluon field through which the covariant binding potential is generated, and let f be the renormalized coupling constant between $\phi(x)$ and the spin-$\frac{1}{2}$ bare nucleon field $\psi(x)$. In a perturbation calculation, there should be a fractional deviation of W_1 and νW_2 structure functions from their scaling forms $F_1(\omega)$ and $F_2(\omega)$ of the order of $(4\pi)^{-1}f^2 \times \ln(\nu/m_p)$. From the present SLAC data, one requires the deviation to be small:

$$\epsilon \equiv \frac{f^2}{4\pi} \lesssim O(10^{-1}). \quad (1)$$

As we shall discuss, this small deviation from the scaling property can also be connected, at least qualitatively, to the smallness of the "hard"-meson multiplicity; i.e., the number of final mesons emitted with large transverse momentum. While the true origin of such a weak renormalized (or effective) coupling constant may stem from a rather complicated set of underlying mechanisms, Eq. (1) seems to give a reasonable phenomenological description of the present physical situation.

However, even on a qualitative level, questions can arise as to whether such a weak coupling constraint is compatible with (a) the strong binding force that is necessary in the model, and (b) the observed strong coupling between the physical nucleon and the physical mesons. In addition, the weak coupling suggests that the usual perturbation considerations could be applicable to the "bare" field $\psi(x)$. It would then appear that $\psi(x)$ might lead to an unstable state, one with a very narrow width because of the weak coupling. This would seem to be in conflict with (c) the absence of any very narrow physical nucleon resonance state in the GeV range. The main purpose of this paper is to give a simple model, based on the local field theory, to illustrate the compatibility between these three conditions (a), (b), (c) and the weak-coupling constraint (1).

The model that we shall discuss consists of only three fields: a spin-$\frac{1}{2}$ charged field $\psi(x)$, a pseudoscalar neutral meson field $\pi(x)$, and a scalar neutral gluon field $\phi(x)$. (The model can, of course, be easily extended to include charged meson fields.) The interaction Lagrangian density is assumed to be

$$\mathcal{L}_\kappa + \mathcal{L}_f ,$$

where

$$\mathcal{L}_\kappa = \kappa_0 \pi^2 \phi , \quad (2)$$

$$\mathcal{L}_f = f_0 \psi^\dagger \gamma_4 \psi \phi ; \quad (3)$$

f_0 and κ_0 are unrenormalized coupling constants. Since $\mathcal{L}_\kappa$ is a superrenormalizable interaction and $\mathcal{L}_f$ is a renormalizable one, the usual renormalization process can be readily carried out. Throughout the paper, we denote M, m_π, and m_ϕ to be, respectively, the renormalized masses of ψ, π, and ϕ. The renormalized coupling constant f is assumed to be small, so that Eq. (1) holds. In contrast, we assume the renormalized coupling constant κ to be large, so that there would be a bound state p, the physical proton, formed out of the composite $(\psi\pi)$ through the exchange of gluons ϕ. In order that the physical proton p is the only stable spin-$\frac{1}{2}$ particle, we require its mass m_p to satisfy

$$m_p < M - m_\pi , \quad (4)$$

so that M lies above the threshold $m_p + m_\pi$ of the continuum $p + \pi$.

As an order-of-magnitude estimation of how large κ should be, we may consider a hypothetical case in which the gluon field ϕ is generated by a static ψ located at the origin; i.e., replacing $\psi^\dagger\gamma_4\psi$ by $\delta^3(\vec{r})$. At a distance r, ϕ is given by

$$\phi_{\text{stat}} = (4\pi r)^{-1} f e^{-m_\phi r} .$$

The equation of motion of π is then determined by the Klein-Gordon equation

$$(-\Box + m_\pi^2 - 2\kappa\phi_{\text{stat}})\pi = 0 .$$

In this static-potential case, we shall replace (4) by requiring the binding energy $= m_\pi$, and therefore the field π is time-independent. One finds then, for $m_\phi \lesssim O(m_\pi)$,

$$(4\pi)^{-1} f\kappa \sim O(m_\pi). \quad (5)$$

[If $m_\phi = 0$, then $(4\pi)^{-1} f\kappa = m_\pi$ implies the binding energy $= m_\pi$.] If, in addition, one assumes $(4\pi)^{-1}f^2 \sim O(10^{-1})$, as required by (1), then

$$\kappa/(4\pi m_\pi) \sim O(1). \quad (6)$$

The details of the model will be discussed in the subsequent sections. In this model, the physical proton p is assumed to be the $S_{1/2}$ bound state of the composite $\psi\pi$. The pseudoscalar nature of π implies that ψ and p are of opposite parity. Consequently, one has the selection rule

$$\psi \not\to p .$$

The electromagnetic interaction of the charged field $\psi(x)$ is introduced by the usual minimal principle; such an interaction is, of course, a local one, and is therefore q^2-independent. In contrast, the physical proton p has electromagnetic form factors that decrease rapidly with increasing q^2.

In the following, the strong-interaction Lagrangian $\mathcal{L}_\kappa$ will be analyzed in Secs. II and III. The bound-state description of the physical nucleon will be given in Sec. IV. The scaling property and the final meson distribution in ep scattering will be discussed in Sec. V. As we shall see, in this simple model, the scaling property for deep-inelastic ep scattering holds only in the energy range in which the laboratory photon energy ν, in units of GeV, is larger than $O(1)$ but less than $O(\exp(\epsilon^{-1}))$, where ϵ is the small number defined in Eq. (1). In the scaling region, at a fixed value of the scaling variable $\omega = 2m_p \nu/q^2$, the model leads to a final hadron multiplicity which is ν-independent, and a transverse momentum distribution of the final hadrons that decreases as $k_\perp^{-6} d^2k_\perp$ at large transverse momentum $k_\perp$.

II. STRONG INTERACTION

For clarity, we shall first consider the problem without $\mathcal{L}_f$. The entire Lagrangian density is then given by

$$\mathcal{L}_0 = \mathcal{L}_{\text{free}} + \mathcal{L}_\kappa + \mathcal{L}_{\text{counter}} \tag{7}$$

in which the free Lagrangian density is

$$\mathcal{L}_{\text{free}} = -\frac{1}{2}\left[\left(\frac{\partial \pi}{\partial x_\nu}\right)^2 + m_\pi^2 \pi^2 + \left(\frac{\partial \phi}{\partial x_\nu}\right)^2 + m_\phi^2 \phi^2\right], \tag{8}$$

the interaction $\mathcal{L}_\kappa$ is given by (2) and the counterterms used for renormalization are included in

$$\mathcal{L}_{\text{counter}} = -\tfrac{1}{2}[(m_\pi^0)^2 - m_\pi^2]\pi^2 - \tfrac{1}{2}[(m_\phi^0)^2 - m_\phi^2]\phi^2 + \lambda_0\phi, \tag{9}$$

where m_π^0 and m_ϕ^0 are the appropriate unrenormalized masses, and λ_0 is determined by the condition

$$\langle \text{vac} | \phi | \text{vac} \rangle = 0. \tag{10}$$

In this theory, there are only three divergent diagrams given in Fig. 1; these divergences can be removed respectively by the three corresponding terms in $\mathcal{L}_{\text{counter}}$.

To derive the power-series expansion, we shall regard $\mathcal{L}_\kappa + \mathcal{L}_{\text{counter}}$ as the perturbation. The resulting series is finite to all orders in κ_0. From (7), it follows that the Hamiltonian does not have a lower bound. By using, e.g., the WKB approximation, one sees that the instability of the theory depends on factors proportional to powers of $\exp(-\kappa_0^{-2})$ which approaches zero extremely fast as $\kappa_0 \to 0$. Thus, one expects the perturbation series to be at least an asymptotic series. In the following, we shall assume that this perturbation series describes, in fact, the correct (asymptotic) power-series expansion of a well-defined theory.[2]

The unrenormalized propagators D_π^0 and D_ϕ^0 of π and ϕ are both finite. We may write

$$D_i^0(q) = -i\int (q^2 + \mu^2)^{-1}\rho_i(\mu^2)d\mu^2, \tag{11}$$

where the subscript i can be either π or ϕ. The spectral function $\rho_i(\mu^2)$ is positive definite. By using the usual canonical commutation relations, one finds[3]

$$\int \rho_i(\mu^2)d\mu^2 = 1 \tag{12}$$

and

$$\int \mu^2\rho_i(\mu^2)d\mu^2 = (m_i^0)^2. \tag{13}$$

The physical masses m_π and m_ϕ are determined by the poles of the corresponding propagator; i.e., as $\mu^2 \to m_i^2$,

$$\rho_i(\mu^2) \to Z_i\delta(\mu^2 - m_i^2) \tag{14}$$

where Z_i is the usual wave-function renormalization which satisfies

$$0 < Z_i \leq 1. \tag{15}$$

The integral (13) is infinite. Since the only divergent diagrams for m_π^0 and m_ϕ^0 are given in Fig. 1, the asymptotic behavior of $\rho_i(\mu^2)$ at large μ^2 is determined completely by these divergent graphs. Thus, one establishes that to any arbitrarily high, but finite, order in κ_0, as $\mu^2 \to \infty$,

$$\rho_\pi(\mu^2) \sim (4\pi^2)^{-1}\kappa_0^2\mu^{-4}$$

and

$$\rho_\phi(\mu^2) \sim (8\pi^2)^{-1}\kappa_0^2\mu^{-4}. \tag{16}$$

Let $2\kappa_0\Gamma^0$ denote the sum of all proper vertex diagrams for

$$\pi \to \pi + \phi.$$

One may define the vertex renormalization Z_1 by requiring for both on-mass-shell initial and final π,

$$\text{as } q^2 \to 0, \quad \Gamma^0(q) \to Z_1^{-1}, \tag{17}$$

where q denotes the 4-momentum of ϕ. In contrast, to any arbitrarily high, but finite, order in κ_0,

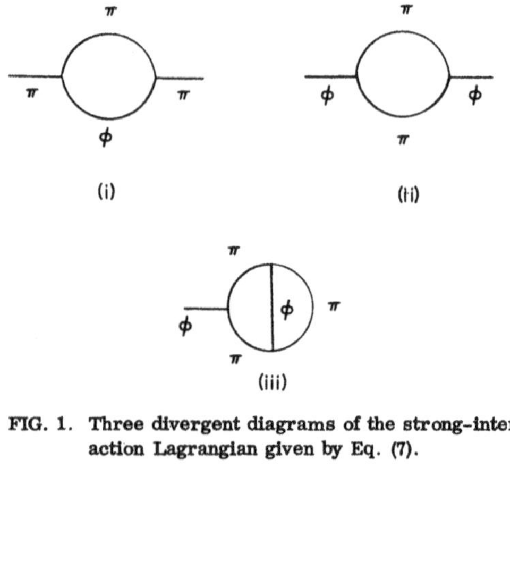

FIG. 1. Three divergent diagrams of the strong-interaction Lagrangian given by Eq. (7).

$$\text{as } q^2 \to \infty, \quad \Gamma^0(q) \to 1. \tag{18}$$

The renormalized coupling constant κ may be defined by

$$\kappa = \kappa_0 Z_\pi Z_\phi^{1/2}/Z_1. \tag{19}$$

As will be shown in Appendix A, there is a Ward-like relation between the vertex diagrams and the propagator diagrams. Such a relation can lead to, among other results, the identity

$$\frac{\partial (m_\pi^0)^2}{\partial m_\pi^2} = Z_1/Z_\pi. \tag{20}$$

If one wishes, one may introduce the renormalized propagators D_π, D_ϕ and the renormalized vertex function Γ in the usual way:

$$D_\pi^0 = Z_\pi D_\pi, \quad D_\phi^0 = Z_\phi D_\phi,$$

and (21)

$$\Gamma^0 = Z_1^{-1}\Gamma.$$

In this theory, both κ_0 and κ are finite. The coupling-constant renormalization and the wave-function renormalizations are introduced only for convenience, since they are all convergent. By definition, the coupling strength between π and ϕ in the low-momentum region is best represented by the renormalized constant κ, but in the high-momentum region, on account of (12) and (18), by the unrenormalized constant κ_0.

As an order-of-magnitude estimation, one likes to know the approximate ratio between κ and κ_0 for the strong-coupling value given by (6). If in the evaluation of Z_ϕ, Z_π, and Z_1 only the single-loop diagrams are included, then at $m_\phi = m_\pi$ and $(4\pi m_\pi)^{-1}\kappa = 1$, the ratio between κ and κ_0 is given by

$$\kappa/\kappa_0 \cong 1.4 \tag{22}$$

and correspondingly

$$Z_\phi^{1/2} \cong 0.9. \tag{23}$$

Further details are given in Appendix B.

III. MESON EMISSIONS

Next, we consider the emission of ϕ by an external source, say

$$a \to b + \phi, \tag{24}$$

where a and b denote the appropriate single-particle, or continuum, states of the external source, which for simplicity is assumed to be unrelated to either π or ϕ. Let $T_\phi^0(q)$ be the emission amplitude of a real, or virtual, ϕ with a 4-momentum q, but for $\kappa_0 = 0$; i.e., in the *absence* of the strong interaction $\mathcal{L}_\kappa$, the cross section for (24) is given by

$$d\sigma^0(a \to b+\phi) = \int |T_\phi^0(q)|^2 d^4q\, \delta(q^2 + m_\phi^2)$$
$$= (2q_0)^{-1} |T_\phi^0(q)|^2 d^3q \quad \text{at } q^2 + m_\phi^2 = 0. \tag{25}$$

Because of the strong-interaction $\mathcal{L}_\kappa$, the observed amplitude for (24) will be different from T_ϕ^0. The correct cross section with all effects of $\mathcal{L}_\kappa$ included should be reduced from (25) by a factor Z_ϕ. One has (assuming that $m_\phi < 2m_\pi$ so that ϕ is stable)

$$d\sigma(a \to b+\phi) = Z_\phi d\sigma^0(a \to b+\phi). \tag{26}$$

In addition, there will be multimeson final states. We may write

$$a \to b + C,$$

where C denotes an arbitrary complex of π and ϕ, such as ϕ, $\pi\pi$, $\pi\pi\phi$, etc. It can be readily established that the differential cross section, after summing over all possible final states $C(\mu)$ with the same invariant mass between μ and $\mu + d\mu$, is given by

$$\sum_C d\sigma(a \to b + C(\mu))$$
$$= d\mu^2 \int |T_\phi^0(q)|^2 \rho_\phi(\mu^2) d^4q\, \delta(q^2+\mu^2)$$
$$= d\mu^2 d^3q (2q_0)^{-1} |T_\phi^0(q)|^2 \rho_\phi(\mu^2), \tag{27}$$

where $q_0 = (\vec{q}^{\,2} + \mu^2)^{1/2}$. The matrix element $T_\phi^0(q)$ depends on, besides the explicit variable q, also the states a and b and their 4-momenta. The differential cross section $d\sigma_{\text{total}}$ summing over all possible final channels C is given by the integral of (27) over $d\mu^2$; i.e.,

$$d\sigma_{\text{total}} = d^3q \int d\mu^2 (2q_0)^{-1} |T_\phi^0(q)|^2 \rho_\phi(\mu^2). \tag{28}$$

In the following, for most cases of practical interest, $T_\phi^0(q)$ is not an increasing function of q^2. Assuming that the magnitude of $T_\phi^0(q)$ for $a \to b + C$ is not larger than that for the real emission $a \to b + \phi$, one finds, on account of (12), (25), and (28),

$$d\sigma_{\text{total}} \leq d\sigma^0(a \to b+\phi); \tag{29}$$

that is, the inclusion of the strong interaction $\mathcal{L}_\kappa$ *diminishes*, rather than enhances, the total cross section. By using (26), one may also express the above inequality in an alternative form

$$d\sigma_{\text{total}} \leq Z_\phi^{-1} d\sigma(a \to b+\phi),$$

where, on account of the superrenormalizability, Z_ϕ^{-1} is finite.

Identical conclusions can be derived by considering the emission $a \to b + \pi$, instead of $a \to b + \phi$.

IV. BOUND STATE

We now turn on the weak-coupling interaction $\mathcal{L}_f$. The total Lagrangian density becomes

$$\mathcal{L} = \mathcal{L}_0 + \mathcal{L}_f + \mathcal{L}'_{\text{counter}}, \tag{30}$$

where $\mathcal{L}_0$ and $\mathcal{L}_f$ are, respectively, given by (7) and (3), and $\mathcal{L}'_{\text{counter}}$ consists of the additional counterterms required for the renormalization of $\mathcal{L}_f$. [To lowest order in $\epsilon = (4\pi)^{-1}f^2$, we may neglect $\mathcal{L}'_{\text{counter}}$.] In a formal perturbation expansion in f, each Feynman diagram can be expressed in terms of the renormalized functions D_ϕ, D_π, and Γ, defined by (21). Among these, one has all the usual ladder diagrams, and also the crossed diagrams for the composite $\psi\pi$. The physical nucleon p is assumed to be the bound state of $\psi\pi$. The Bethe-Salpeter equation for the bound-state wave function $\Psi_p(k)$ in the ladder approximation is now given by

$$K_p(k)\Psi_p(k) = i(2f\kappa)\int V(k-k')\Psi_p(k')d^4k', \tag{31}$$

where p is the total 4-momentum, k is the relative momentum related to the 4-momenta P of $\psi(x)$ and π of $\pi(x)$ by

$$P = (M+m_\pi)^{-1}Mp + k,$$

and

$$\pi = (M+m_\pi)^{-1}m_\pi p - k,$$

$$K_p(k) = -i(-i\gamma\cdot P - M + i\epsilon)D_\pi^{-1}(\pi), \tag{32}$$

$$V(q) = iD_\phi(q)\,\Gamma(q),$$

and f is the renormalized coupling constant, related to the unrenormalized coupling constant f_0 by

$$f = f_0 Z_\phi^{1/2}, \tag{33}$$

with Z_ϕ defined by (14). Since $\epsilon = (4\pi)^{-1}f^2$ is small, $\lesssim O(10^{-1})$, we may neglect $O(\epsilon)$ corrections in (31) and (33). In (31), the potential $V(q)$ does not depend only on the explicitly exhibited variable $q = k - k'$, since the vertex function $\Gamma(q)$ also depends on the on- and off-mass variation of π. [If π is charged, then in the ladder approximation, it is best to use only the function $\Gamma(q)$ in which the initial and final π are both on mass shell, so that $V(q)$ does depend only on the momentum transfer q. Because of the gauge condition, it is more convenient to include the off-mass-shell variation of the vertex function with respect to π in the discussions of other nonladder diagrams.]

From (14), (16), (18) and the definition (21), one sees that at large 4-momenta

$$K_p(k) \to Z_\pi k^2(-i\gamma\cdot k) \text{ and } V(q) \to Z_1 Z_\phi^{-1}q^{-2}.$$

Since all these renormalization constants are finite, by following exactly the same argument used in Ref. 1, one can readily establish that the Bethe-Salpeter vertex function $\Phi_p(k)$, defined by

$$\Phi_p(k) = K_p(k)\Psi_p(k), \tag{34}$$

satisfies

$$\Phi_p(k) \sim O(k^{-2}) \text{ as } k^2 \to \pm\infty. \tag{35}$$

The "effective" coupling constant associated with the covariant binding potential is $f\kappa$. As already noted in the Introduction, the apparent difficulty with the weak-coupling constraint (1) and the strong-binding condition (a) can be resolved by requiring the product

$$(4\pi)^{-1}f\kappa \sim O(m_\pi),$$

as given by (5). This is analogous to the familiar example of a mesonic atom which consists of, say, a muon and a heavy nucleus of charge Z. We may identify the coupling constants Ze and e in the case of mesonic atoms with, respectively, (κ/m_π) and f in this model. The strong binding force implies that $Z\alpha$ is of the order of unity, and the weak-coupling condition (1) on ϵ corresponds to the fine structure $\alpha \ll 1$. It is of interest to observe that, on account of (22) and (23), at the assumed strong-coupling region $(4\pi m_\pi)^{-1}\kappa \cong 1$, the magnitudes of the unrenormalized constants κ_0 and f_0 are, respectively, of the same order as those of the renormalized constants κ and f. Throughout the paper, we assume that for the simple model there

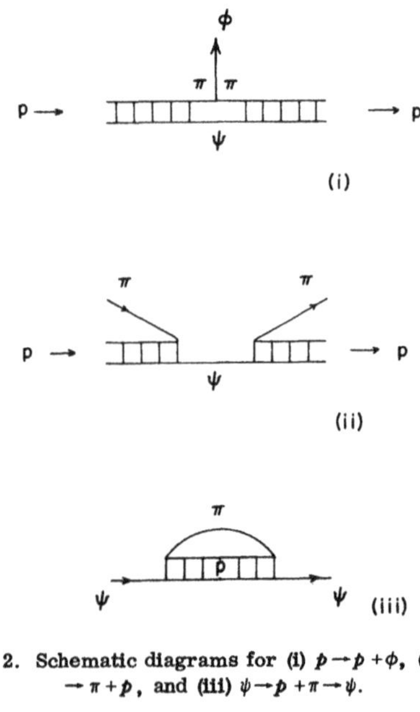

FIG. 2. Schematic diagrams for (i) $p \to p + \phi$, (ii) $\pi + p \to \pi + p$, and (iii) $\psi \to p + \pi \to \psi$.

is indeed such a strong-binding solution of the Bethe-Salpeter equation.

We note that while the coupling constant for $\psi\to\psi+\phi$ is weak, the transition

$$p\to p+\phi$$

is dominated by the absorption or emission of the ϕ meson by the constituent π in the bound state. As explicitly illustrated by diagram (i) in Fig. 2, the coupling for $p\to p+\phi$ is $O(\kappa/m_\pi)$, which is strong. Thus, there is no contradiction between the weak-coupling constraint (1) and the strong physical nucleon-meson coupling condition (b), mentioned in the Introduction.

Similar considerations can be extended to

$$p\to p+\pi\,,$$

provided that an additional small coupling

$$f'\psi^\dagger\gamma_4\gamma_5\psi\pi \qquad (36)$$

is introduced to the model where f' is of the same order as f. With this additional coupling, the bound state p will be a coherent mixture of $\psi\pi$ in the $S_{1/2}$ state (with ϕ as the gluon) and $\psi\phi$ in the $P_{1/2}$ state (partly with π as the gluon). While the transition amplitude for $\psi\to\psi+\pi$ may be small, the corresponding amplitude for $p\to p+\pi$ can be quite large, since the external pion can be absorbed by an appropriate strong $\pi\pi\phi$ vertex.

As already mentioned in the Introduction, in this model the parity of p is different from that of ψ due to the pseudoscalar nature of π. Consequently, there is the selection rule

$$\psi\not\to p\,. \qquad (37)$$

We observe that without the additional coupling (36), the parity of π is not determined by the original Lagrangian (30). However, there exists, then, a multiplicatively conserved quantum number $(-1)^n$ which can be +1 or −1, depending on whether the number n of π is even or odd. The bound state p belongs to n = odd, but the field ψ belongs to n = even. Therefore, the selection rule (37) remains valid.

In the ψ sector, because of (4), there is no stable physical single-particle state; all physical states belong to the continuum $p\pi$, $p\pi\phi$, etc. The existence of $\psi(x)$ should lead to a resonance at the rest energy M in the scattering

$$p+\pi\to p+\pi\,, \qquad (38)$$

where $M>m_p+m_\pi$. Within the ladder approximation, the resonant scattering is determined by diagram (ii) in Fig. 2. The width of the resonance is given by the product of

$$\sum_{\text{spin}}|u_P^\dagger\gamma_4\Phi_p(k)|^2 \qquad (39)$$

times the usual phase space factor, where u_P is a free Dirac spinor with a 4-momentum P, which satisfies $P^2+M^2=0$ and is related to the 4-momenta p, π and the relative momentum k by

$$P=p+\pi$$

and

$$-\pi=(M+m_\pi)^{-1}m_\pi p-k\,.$$

It is clear that although $\epsilon=(4\pi)^{-1}f^2\lesssim O(10^{-1})$ is small, the matrix element (39) is $\sim O(1)$. The width of the resonance is by no means narrow. Thus, the existence of a weakly coupled elementary field $\psi(x)$ does not imply that there must be a narrow-resonance state. This then resolves the apparent conflict between the weak-coupling constraint (1) and the condition (c) mentioned in the Introduction. [If one wishes, one could identify the resonance energy M with the 1535-GeV baryon level, which is of isospin $\frac{1}{2}$, spin $\frac{1}{2}$, and parity odd.]

The propagator $S(P)$ of the field ψ can be written as

$$S(P)=i[-i\gamma\cdot P-M+\Sigma(P)]^{-1}\,. \qquad (40)$$

On the physical sheet, the propagator has no pole, but a cut along the real axis in the complex $-P^2$ plane, which begins at $(m_p+m_\pi)^2$. To the same ladder approximation, $\Sigma(P)$ is given by diagram (iii) of Fig. 2. The imaginary part of $\Sigma(P)$ is given by the same expression (39), multiplied by the appropriate phase-space factor, but keeping $-P^2$ arbitrary. We note that at $p^2+m_p{}^2=0$ and $\pi^2+m_\pi{}^2=0$, as $-P^2\to\infty$, expression (39) becomes $\sim O(P^{-4})$, and therefore

$$\mathrm{Im}\Sigma(P)\sim O(P^{-3})$$

and (41)

$$S(P)\to(-\gamma\cdot P)^{-1}\,;$$

that is, in the usual spectral representation

$$S(P)=i\int(-i\gamma\cdot P-m+i\epsilon)^{-1}\sigma(m^2)dm^2$$

the spectral function satisfies, in the same approximation,

$$\int\sigma(m^2)dm^2=1 \qquad (42)$$

and

$$\sigma(m^2)\sim O(m^{-6})\ \text{ as } m^2\to\infty\,. \qquad (43)$$

In principle, we should have used $S(P)^{-1}$ instead of $-i(-i\gamma\cdot P-M+i\epsilon)$ in $K_p(k)$ for the original Bethe-

Salpeter equation (31). Because of (41), we would reach the same conclusion on the asymptotic behavior of $\Phi_p(k)$ as that given by (35).

So far we have considered only the ladder-diagram approximation for the bound state. We shall now briefly discuss the systematics of other diagrams. It is important to note that the exchange amplitude of a gluon ϕ between the spin-$\frac{1}{2}$ field ψ and itself is $O(\epsilon)$, which is small since $\epsilon=(4\pi)^{-1}f^2$ is assumed to be $\lesssim O(10^{-1})$. All other diagrams in which the gluons are exchanged only between π and ψ and/or between π and π are convergent; their contributions can be evaluated iteratively by using the ladder approximation as the zeroth-order description for the bound state. By simple power counting, one can show that, neglecting all $O(\epsilon)$ diagrams, the above conclusions (35) and (43) on the asymptotic behaviors remain valid with the inclusion of an arbitrary number of such additional diagrams.

V. SCALING PROPERTY

In the simple model in which only ψ is charged, the standard minimal electromagnetic interaction is

$$i\psi^\dagger\gamma_4\gamma_\lambda\psi A_\lambda,$$

where A_λ denotes the electromagnetic field. (The model can, of course, be straightforwardly extended to the case where π is charged.) In the "parton" language,[4] one would say that the charged constituent (or parton) ψ has a "pointlike" electromagnetic structure. Just as in Ref. 1, the evaluation of the electromagnetic form factors of the physical nucleon can be readily carried out. One finds the familiar result[5] that both $F_1(q^2)$ and $F_2(q^2)$ decrease as q^{-4}, apart from $\ln q^2$ factors, at large q^2.

In the inelastic electron scattering

$$e+p\to e+p+\cdots, \tag{44}$$

the final hadron state "$\cdots$" can consist of any number of mesons, π or ϕ. In Fig. 3, we give two typical diagrams (i) and (ii) for (44). It is convenient to define a scalar vertex function g for the bound-state solution:

$$\Phi_p(k)=g(-\pi^2,-P^2)u_p, \tag{45}$$

where $\Phi_p(k)$ is given by (34), u_p is the free Dirac spinor of 4-momentum p, and the function g depends on two variables $-\pi^2$ and $-P^2$, related to p and k by (32). Because of (35), as $\pi^2\to\pm\infty$ at fixed P^2

$$g(-\pi^2,-P^2)\sim O(\pi^{-2}),$$

and as $P^2\to\pm\infty$ at fixed π^2

$$g(-\pi^2,-P^2)\sim O(P^{-2}). \tag{46}$$

By following the same arguments used in Ref. 1, one can readily establish the scaling property for the deep-inelastic scattering, provided that all $O(\epsilon)$ diagrams are neglected.

Diagrams (i) and (ii) in Fig. 3, by themselves, are not gauge-invariant. Gauge invariance requires the presence of all other diagrams that can be generated from (i) and (ii) by moving the external photon vertex onto different internal spin-$\frac{1}{2}$ charged lines. For example, from (i), one can generate diagrams of the form (i′) and (i″) in Fig. 3. Just as in Appendix C of Ref. 1, one can verify directly that the ratio of the amplitude of (i′) to that of (i) approaches zero in the scaling limit; similarly, one can show that the ratio of the amplitude of (i″) to that of (i) also approaches zero. These discussions can be easily generalized to other diagrams. Therefore, in the scaling limit, one need only consider diagrams of the types (i) and (ii) in Fig. 3. For these diagrams it is not difficult to sum over all different final hadron states.

As we shall see, in the scaling region, as $q^2\gg m_p^2$ and $-2p\cdot q=2m_p\nu\gg m_p^2$, these diagrams lead to the following longitudinal and transverse structure functions:

$$W_L\to 0$$

and

$$W_1\propto\int\rho_\pi(\mu^2)d\mu^2\,G(\mu^2,x), \tag{47}$$

where $x=(2m_p\nu)^{-1}q^2$,

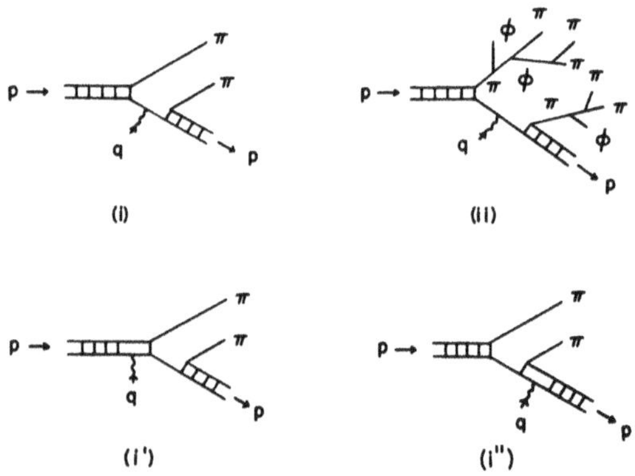

FIG. 3. Examples of "γ" $+p\to p+$mesons. Diagrams (i′) and (i″) are related to diagram (i) by gauge invariance, but can be neglected in the scaling limit.

$$G(\mu^2, x) = \int_{u_{\min}}^{\infty} \frac{du}{(u+M^2)^2} |g(\mu^2, u)|^2 \times\{(u+M^2)(1-x) + x[(M-m_p)^2 - \mu^2]\}, \tag{48}$$

with

$$u_{\min} = \frac{\mu^2 x}{1-x} - m_p{}^2 x \tag{49}$$

and $\rho_\pi(\mu^2)$ given by (11). To derive (47), we first consider only the transition $\gamma + p \to \psi + \pi$. This transition rate has been calculated explicitly in Ref. 1, and the resulting structure functions in the scaling limit are $W_L = 0$ and $W_1 \propto G(\mu^2, x)$, where μ denotes the invariant mass of the external π line. It is important to observe that $G(\mu^2, x)$ does not depend on m, the invariant mass of the external ψ line. To include the subsequent emissions of $\psi \to p +$mesons and $\pi \to$ mesons, we follow the same argument as that used in Sec. III. We note that the interference terms between the mesons emitted from ψ and those from π vanish in the scaling limit. The resulting W_1 is then given by, in the approximation where only diagrams of the types (i) and (ii) in Fig. 3 are included,

$$W_1 \propto \int \sigma(m^2) dm^2 \int \rho_\pi(\mu^2) d\mu^2\, G(\mu^2, x),$$

which reduces to (47), on account of (42). [Equations (47) and (48) are identical in form[6] to (4.21) of Ref. 1.]

In the scaling limit, because $W_L = 0$, the usual νW_2 function is related to W_1 by

$$\nu W_2 = 2xm_p W_1 .$$

By using (16), (46), and (47), one sees that

$$W_1 \sim O((1-x)^3) \text{ as } x \to 1, \tag{50}$$

$$W_1 \sim O(1) \text{ as } x \to 0, \tag{51}$$

and therefore

$$\nu W_2 = 0 \text{ at } x = 0. \tag{52}$$

As has been pointed out in Sec. IV of Ref. 1, in order to derive $\nu W_2 \neq 0$ at $x=0$, one needs in (47) a spectral function $\rho_\pi(\mu^2)$ which should approach a nonzero constant as $\mu^2 \to \infty$. In the present simple model, on account of (16), $\rho_\pi(\mu^2) \to 0$ as $\mu^2 \to \infty$; consequently, in the above approximation, νW_2 is zero at $x=0$.

The Bethe-Salpeter wave function $\Psi_p(k)$ is normalized by the condition

$$\langle p | J_\mu | p\rangle = ep_\mu / m_p \text{ at } q_\mu = 0, \tag{53}$$

where J_μ is the electromagnetic current operator, q_μ is the 4-momentum transfer, and p_μ is the 4-momentum of the physical proton state $|p\rangle$. By following the arguments given in Secs. II F and II G of Ref. 1, but with some minor modifications, one can readily prove that, in the approximation which includes only diagrams of the type (i) and (ii) in Fig. 3, the above normalization condition (53) is *identical* to requiring in the scaling limit the νW_2 function to satisfy an Adler-type sum rule[7]

$$\int_0^1 x^{-1} dx (\nu W_2) = 1 . \tag{54}$$

Another interesting problem is the final-state hadron distribution. From diagram (ii) in Fig. 3, one sees that there are two distinct groups of hadrons, N_π and N_p, where N_π refers to the hadrons emitted from the spin-0 constituent field π and N_p those emitted from the spin-$\frac{1}{2}$ constituent field ψ. In the scaling limit, the group N_π has a finite total longitudinal momentum in the laboratory system, while the group N_p has a total longitudinal momentum $=\vec{q}$. The transverse momentum distribution of any one of these hadrons decreases at least as fast as

$$k_\perp{}^{-5} dk_\perp \tag{55}$$

at large $k_\perp$, where $k_\perp$ denotes the magnitude of the momentum component perpendicular to $\vec{q}$. This can be shown by noting that, in the scaling region, the variable u in (48) is related to the sum $K_\perp \equiv |\sum \vec{k}_\perp|$ of the transverse momenta of all hadrons in the N_π group (or the N_p group, since both groups are of the same $K_\perp$) by

$$u = u_{\min} + (1-x)^{-1} K_\perp{}^2, \tag{56}$$

and therefore, because of (46) and (48), at fixed μ the distribution in $K_\perp$ is $\sim K_\perp{}^{-5} dK_\perp$ at large $K_\perp$; in addition, on account of (16), (47), and (49), the distribution in μ is $\sim \mu^{-7} d\mu$ at large μ. Since the magnitude $k_\perp$ of each individual hadron in the N_π group is $\leq (K_\perp + \frac{1}{2}\mu)$, one obtains (55) as the upper bound for the $k_\perp$ distribution. Since as we shall see, the final hadron multiplicity is finite in this model, consequently (55) also gives the average asymptotic behavior of the $k_\perp$ distribution.

Let $m_{\min}$ be the smallest mass among m_p, m_π, and m_ϕ. The average number of hadrons produced by the spin-$\frac{1}{2}$ constituent field clearly has an upper bound, given by

$$N_p < m_{\min}{}^{-1} \int m \sigma(m^2) dm^2 .$$

Because of (43), one sees that N_p is finite. Similarly, because of (16) and (47), N_π is finite. Therefore, in this model, the average number of

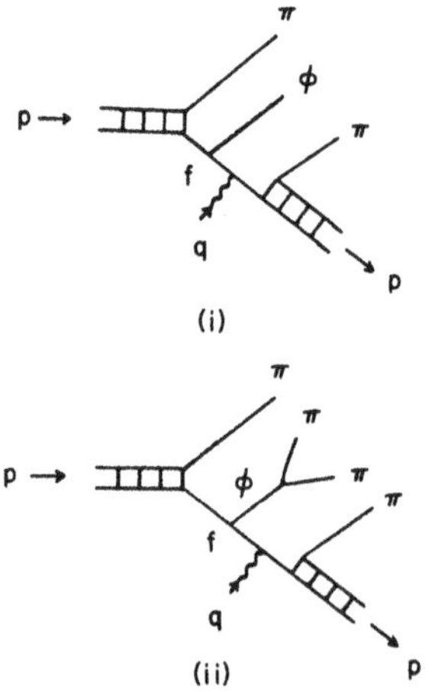

FIG. 4. Examples of $O(\epsilon)$ diagrams for "γ" $+p \rightarrow p$ + mesons. Both diagrams give a fractional violation of the scaling property $\sim O(\epsilon \ln(\nu/m_p))$ where $\epsilon = (4\pi)^{-1}f^2$ is a small number $\lesssim O(10^{-1})$.

hadrons produced remains finite in the scaling limit,[8] provided that all $O(\epsilon)$ diagrams are neglected. Furthermore, from diagrams (i) and (ii) in Fig. 3, it follows that, in the scaling limit, there must be at least three hadrons (including p) in the final state; i.e.,

$$N_\pi + N_p \geq 3. \qquad (57)$$

So far, we have neglected all $O(\epsilon)$ diagrams. Examples of such $O(\epsilon)$ diagrams are given by diagrams (i) and (ii) in Fig. 4. It is straightforward to show that, in the scaling limit, diagram (i) introduces a fractional deviation of the W_1 function from its scaling form (47) of the order of

$$\epsilon \ln(\nu/m_p). \qquad (58)$$

Diagram (ii) can be derived from diagram (i) by including additional meson emissions from its external ϕ line. If one sums over all such meson emissions, by following the discussions given in Sec. III, one can easily show that the resulting fractional deviation remains of the order given by (58). Since, according to (1), the parameter ϵ is assumed to be small, we can neglect these scaling-violation diagrams provided that ν/m_p is not too large. Therefore, in this model *the scaling property holds only in the energy range where ν/m_p is greater than $O(1)$ but less than $O(\exp(\epsilon^{-1}))$.* Another characteristic of these scaling-violation diagrams is the presence of some final hadrons whose transverse momentum distribution is $\sim k_\perp^{-1} dk_\perp$ at large $k_\perp$. Thus, as already remarked in the Introduction, the small deviation from the scaling property is connected to the smallness of the "hard" meson multiplicity, i.e., the smallness in the number of final mesons with large transverse momentum.

VI. REMARKS

In the present model, because of the selection rule (37), the simplest local operator that can be used as the interpolating field for the physical proton state p is the bilinear product $\psi(x)\pi(x)$. Within the context of a specific Lagrangian field theory,[9] we may define a physical single-particle state to be a composite, if there is not a single canonical field operator in the Lagrangian that has the same set of quantum numbers as the physical state. In this sense, we should regard the physical proton p in this model as a composite, not an elementary particle state. The experimental consequences of p being a composite are, as mentioned before, the scaling property in the deep-inelastic scattering and the rapidly decreasing electromagnetic form factors at large q^2.

In contrast, the interpolating field for the $P_{1/2}$ continuum state ($p\pi$) in this model can be simply $\psi(x)$. This has the interesting consequence that, neglecting all $O(\epsilon)$ diagrams, the electromagnetic vertex function for the appropriate ($p\pi$) system

$$(\pi + p) \rightarrow (\pi + p) + \gamma \qquad (59)$$

does *not* approach zero as the square of the virtual photon momentum $q^2 \rightarrow \pm\infty$. The same conclusion applies also to

$$\Gamma \rightarrow \Gamma' + \gamma \qquad (60)$$

where Γ and Γ' can be any two complexes each of which has the same set of quantum numbers as the field $\psi(x)$. (Therefore, neither Γ nor Γ' is p.) At any finite invariant masses of Γ and Γ', the vertex (60) also does not approach zero as $q^2 \rightarrow \pm\infty$, provided that all $O(\epsilon)$ diagrams are neglected.

ACKNOWLEDGMENTS

I wish to thank S. Brodsky, G. Chew, N. Christ, and A. H. Mueller for discussions.

APPENDIX A

In this Appendix, we shall establish a Ward-like identity between the unrenormalized π propagator D_π^0 and the unrenormalized vertex function Γ^0 for the Lagrangian $\mathcal{L}_0$, given by (7). We may write

$$[D_\pi^0(p)]^{-1} = i[p^2 + (m_\pi^0)^2 + \Sigma_\pi]. \qquad (A1)$$

It is convenient to introduce a momentum cutoff Λ; the unrenormalized masses m_π^0, m_ϕ^0, and the counterterm λ_0, introduced in (9), are then finite.

In the following, we shall keep λ_0, m_ϕ^0, and Λ fixed, and regard Σ_π as a function of p^2 and $(m_\pi^0)^2$, i.e.,

$$\Sigma_\pi = \Sigma_\pi((m_\pi^0)^2, p^2). \tag{A2}$$

Let $2\kappa_0\,\Gamma^0(p, q)$ be the sum of all proper vertex diagrams for

$$\pi(p) \to \pi(p-q) + \phi(q)$$

in which the arguments p, $p-q$, and q denote the appropriate momenta of the particles. By expanding Σ_π as an infinite sum over all proper self-energy diagrams, one can readily establish that, diagram by diagram,

$$\Gamma^0(p, q=0) = 1 + \left(\frac{\partial\Sigma_\pi}{\partial(m_\pi^0)^2}\right)_{p^2}. \tag{A3}$$

The renormalized mass m_π is defined by

$$-m_\pi^2 + (m_\pi^0)^2 + \Sigma_\pi((m_\pi^0)^2, p^2 = -m_\pi^2) = 0. \tag{A4}$$

By differentiating (A4) with respect to m_π^2, and by using

$$Z_\pi^{-1} = 1 + \left(\frac{\partial\Sigma_\pi}{\partial p^2}\right)_{(m_\pi^0)^2}$$

and

$$Z_1^{-1} = \Gamma^0(p, q=0) \quad \text{at } p^2 + m_\pi^2 = 0, \tag{A5}$$

we derive Eq. (20),

$$\frac{\partial(m_\pi^0)^2}{\partial m_\pi^2} = \frac{Z_1}{Z_\pi}.$$

As the momentum cutoff $\Lambda \to \infty$, although $(m_\pi^0)^2$ becomes ∞, its derivative with respect to m_π^2 remains finite.

APPENDIX B

In the one-loop approximation, diagrams (i) and (ii) in Fig. 1 are the only proper self-energy diagrams for the π and ϕ propagators. It is convenient to introduce a momentum cutoff Λ, so that the free π and ϕ propagators are given by

$$-i(p^2 + m_a^2)^{-1}(p^2 + \Lambda^2)^{-1}\Lambda^2, \tag{B1}$$

where the subscript a denotes π or ϕ. These diagrams can be easily evaluated. The unrenormalized propagators $D_\pi^0(p)$ and $D_\phi^0(p)$ are, in the one-loop approximation, given by

$$[D_a^0(p)]^{-1} = i(p^2 + m_a^2 + \delta m_a^2 + \Sigma_a), \tag{B2}$$

where $\delta m_a^2 = (m_a^0)^2 - m_a^2$,

$$\Sigma_\pi = (4\pi^2)^{-1}\kappa_0^2[\ln(\Lambda^2/m_\pi m_\phi) + g(-p^2, m_\pi^2, m_\phi^2)], \tag{B3}$$

$$\Sigma_\phi = (8\pi^2)^{-1}\kappa_0^2[\ln(\Lambda^2/m_\pi^2) + g(-p^2, m_\pi^2, m_\pi^2)], \tag{B4}$$

$$g(-p^2, m_\pi^2, m_\phi^2) = (2p^2)^{-1}[(m_\pi^2 - m_\phi^2)\ln(m_\pi^2/m_\phi^2) + Q^{1/2}\ln R], \tag{B5}$$

$$Q = (p^2)^2 + 2p^2(m_\pi^2 + m_\phi^2) + (m_\pi^2 - m_\phi^2)^2, \tag{B6}$$

and

$$R = \frac{p^2 + m_\pi^2 + m_\phi^2 - Q^{1/2}}{p^2 + m_\pi^2 + m_\phi^2 + Q^{1/2}}. \tag{B7}$$

In the complex $z = -p^2$ plane, $g(z, m_\pi^2, m_\phi^2)$ is analytic everywhere, except for a cut along the real axis from $z = (m_\pi + m_\phi)^2$ to ∞. Furthermore,

$$g(z, m_\pi^2, m_\phi^2)^* = g(z^*, m_\pi^2, m_\phi^2).$$

Therefore, g is real for z real $< (m_\pi + m_\phi)^2$. Along the cut, for $\mathrm{Im}\,z = 0+$, g acquires an imaginary part given by

$$\mathrm{Im}\,g = (\pi/z)\,|Q^{1/2}|. \tag{B8}$$

The vertex function Γ^0 at zero momentum transfer can be evaluated from Σ_π through the identity (A3). By using (B3) and (B4), one establishes (16), which, as remarked earlier in Sec. II, holds to *all* orders in κ_0^2.

In the one-loop approximation, one finds that for $m_\phi = m_\pi = \mu$

$$(m_\pi^0)^2 = \mu^2 + (4\pi^2)^{-1}\kappa_0^2\left[\ln(\Lambda^2/\mu^2) - \frac{\pi}{\sqrt{3}}\right], \tag{B9}$$

$$(m_\phi^0)^2 = \mu^2 + (8\pi^2)^{-1}\kappa_0^2\left[\ln(\Lambda^2/\mu^2) - \frac{\pi}{\sqrt{3}}\right], \tag{B10}$$

$$Z_\pi^{-1} = 1 + (4\pi^2\mu^2)^{-1}\kappa_0^2\left[-1 + \frac{2\pi}{3\sqrt{3}}\right], \tag{B11}$$

$$Z_\phi^{-1} = 1 + (8\pi^2\mu^2)^{-1}\kappa_0^2\left[-1 + \frac{2\pi}{3\sqrt{3}}\right], \tag{B12}$$

and

$$Z_1^{-1} = 1 + (12\pi\mu^2\sqrt{3})^{-1}\kappa_0^2. \tag{B13}$$

If one assumes

$$(4\pi\mu)^{-1}\kappa = 1, \tag{B14}$$

then the value for $(\kappa/\kappa_0) = Z_1^{-1} Z_\phi^{1/2} Z_\pi$ is given by (22); i.e.,

$$\kappa/\kappa_0 \cong 1.4. \tag{B15}$$

Correspondingly, Z_ϕ is given by (23); i.e.,

$$Z_\phi^{1/2} \cong 0.9. \tag{B16}$$

To test the sensitivity of these order-of-magnitude estimations, we may set, instead of (B14), $(4\pi\mu)^{-1}\kappa_0 = 1$; this changes the values for κ/κ_0 and $Z_\phi^{1/2}$ slightly, from (B15) and (B16) to $(\kappa/\kappa_0) \cong 1.56$ and $Z_\phi^{1/2} \cong 0.84$.

---

*This research was supported in part by the U. S. Atomic Energy Commission.

[1]S. D. Drell and T. D. Lee, Phys. Rev. D 5, 1738 (1972). The fields $P(x)$ and $X(x)$ used in this reference correspond, respectively, to $\psi(x)$ and $\pi(x)$ in the present paper. See also the various other references mentioned therein.

[2]Formally, one may consider the Lagrangian of this well-defined theory to be of the form $\mathcal{L}_0 - \eta\pi^4$, where $\mathcal{L}_0$ is given by (7) and η is a function of κ_0 satisfying $\lim_{\kappa_0 \to 0} \kappa_0^n \eta \to 0$ for all negative powers n. As an example, one may choose, say, $\eta = \exp(-b/\kappa_0^2)$ where b is a positive constant. The corresponding Hamiltonian now clearly has a lower bound. By choosing the appropriate renormalization counterterms, one can easily arrange in this theory to have an (asymptotic) expansion in powers of κ_0 which is determined by $\mathcal{L}_0$ only.

[3]K. Johnson, Nucl. Phys. 25, 435 (1961).

[4]R. P. Feynman, Phys. Rev. Letters 23, 1415 (1969); in *High Energy Collisions*, Third International Conference held at the State University of New York, Stony Brook, 1969, edited by C. N. Yang *et al.* (Gordon and Breach, New York, 1969); J. D. Bjorken and E. A. Paschos, Phys. Rev. 185, 1975 (1969).

[5]J. S. Ball and F. Zachariasen, Phys. Rev. 170, 1541 (1968); D. Amati, L. Caneschi, and R. Jengo, Nuovo Cimento 58, 783 (1968); D. Amati, R. Jengo, H. R. Rubinstein, G. Veneziano, and M. A. Virasoro, Phys. Letters 27B, 38 (1968); M. Ciafaloni and P. Menotti, Phys. Rev. 173, 1575 (1968); M. Ciafaloni, *ibid.* 176, 1898 (1968).

[6]Compare also P. V. Landshoff, J. C. Polkinghorne, and R. D. Short, Nucl. Phys. B28, 225 (1971), and S. Brodsky, F. E. Close, and J. F. Gunion, Phys. Rev. D 5, 1384 (1972), for a similar expression especially for the spin-0 particle.

[7]S. L. Adler, Phys. Rev. 143, 1144 (1966); see also J. D. Bjorken, *ibid.* 148, 1467 (1966).

[8]Dr. A. Mueller has kindly pointed out to me that, in the scaling region, the final meson multiplicity remains finite in this model provided the scaling variable $x \neq 0$.

[9]Without a specific Lagrangian, it would be difficult to make a sharp distinction between a composite and an elementary-particle state. See the discussions given by W. Zimmermann, Nuovo Cimento 10, 597 (1958).

Reprinted from:

PHYSICAL REVIEW D VOLUME 6, NUMBER 12 15 DECEMBER 1972

Multiplicity Distribution in the Multiperipheral Model and a One-Dimensional Gas*

T. D. Lee

Columbia University, New York, New York 10027

(Received 17 August 1972)

The multiplicity distribution at high energy in the multiperipheral model for the ϕ^3 theory is shown to be identical to the grand canonical ensemble distribution of a particular one-dimensional gas with only repulsive forces, which can be decomposed into two-body, three-body, and other multibody forces. The specific form of these forces and the corresponding virial expansion of the gas system are discussed. An alternative systematic expansion method is developed which is different from the virial series but appears to be of a greater practical value for this particular class of physical problems.

I. INTRODUCTION

In this paper, we shall discuss the exact gas-analog problem in statistical mechanics that corresponds to the multiplicity distribution at high energy in the multiperipheral model of Amati, Fubini, and Stanghellini[1] for the ϕ^3 theory (hereafter referred to either as the ϕ^3-multiperipheral model or simply as the multiperipheral model). The interaction Lagrangian is assumed to be

$$(3!)^{-1} m g \phi^3, \tag{1}$$

where ϕ is a scalar field, m denotes its mass, and g is the dimensionless coupling constant. In the ϕ^3-multiperipheral model, the two-body elastic scattering is given simply by the sum of all t-channel ladder diagrams; the corresponding absorptive parts then give the multiplicity distribution. Such a sum of ladder diagrams is of interest since, as is well known, it represents on the one hand the sum of all "leading" diagrams in a perturbation expansion of the ϕ^3 theory at high energy, and on the other hand, it gives the simplest prototype of field-theoretic models that exhibit Regge behavior for elastic scattering,[2] and a $\ln s$ dependence for multiplicity.[1] There exists already quite a sizable literature[3-6] which discusses the similarity between the meson distribution in a multiperipheral-type model and the ensemble distribution of a gas system in statistical mechanics. However, as yet, the precise formulation and the explicit interaction of the gas-analog system have not been given. The purpose of this note is to provide this needed information in order to complete the connection between these two different types of physical problem.

In Sec. II, the ϕ^3-multiperipheral model is briefly reviewed. The equivalence between the multiperipheral model and its one-dimensional gas analog is discussed in Sec. III (and proved in Appendix A). As we shall see, in the gas analog the potential energy U_N between N atoms of the gas is ≥ 0 everywhere; it can be decomposed into a sum of a two-body potential V_2 between only the nearest neighbors, a three-body potential V_3 between only the nearest and the next-nearest neighbors, etc.,

$$U_N = V_2 + V_3 + V_4 + \cdots .$$

The explicit forms of $V_2, V_3, \ldots,$ are given, and the corresponding virial series discussed.

In Sec. IV an alternative systematic expansion method is developed which is different from the virial series, but appears to be of a greater practical value, especially if the gas pressure p is not too small. In the first approximation of this new expansion method, we set the potential U_N to consist of only the two-body potentials,

$$U_N \cong V_2 = \sum v(x_i),$$

where x_i is the distance between the ith and the $(i+1)^{\text{th}}$ atoms. The corresponding equation of state is shown to be given by the simple formula

$$z^{-1} = \int_0^\infty dx \exp[-v(x) - px], \tag{2}$$

where z is the fugacity. By using the functional dependence of p on z, one can readily determine the multiplicity distribution. The subsequent approximations of including also the three-body potential V_3, and then the four-body potential V_4, etc., are discussed in the same section (and also in Appendix C). The method developed is of a rather general character, not restricted to the specific ϕ^3-multiperipheral model. It is suggested that expressions such as (2) may be used to analyze phenomenologically the meson multiplicity problem in realistic cases of high-energy collisions. From the observed multiplicity distributions (in the so-called "central region"), one can determine an effective "potential," which may be partly attractive and partly repulsive; by using the same potential, one can then calculate the m-body correlation functions and compare the results with measurements on various inclusive reactions.

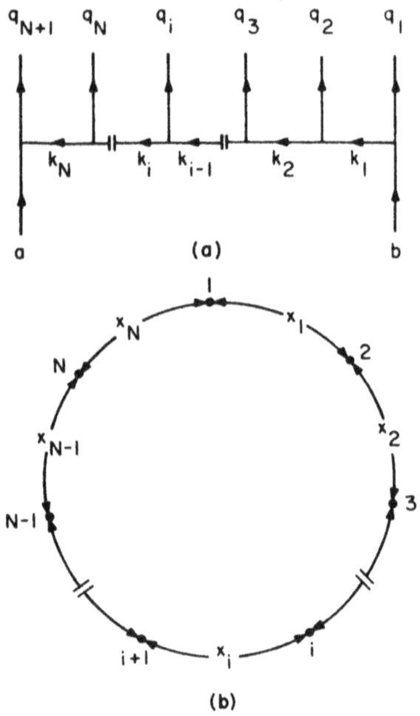

FIG. 1. (a) Feynman diagram for $(N+1)$-meson emission in the ϕ^3-multiperipheral model. (b) Corresponding diagram for N atoms on a ring in the one-dimensional gas model.

II. MULTIPERIPHERAL MODEL

For definiteness, let us consider the Lagrangian (1) and discuss the physical process of producing $(N+1)$ mesons,

$$\phi(a) + \phi(b) \to (N+1)\phi,$$

where N may vary from unity to infinity, and a and b denote respectively the two initial 4-momenta. In the ϕ^3-multiperipheral model, the rate of this reaction is determined by only one Feynman diagram, which is given by Fig. 1(a); in this diagram $q_1, q_2, \ldots, q_{N+1}$ denote respectively the 4-momenta of the final mesons, and $k_1, k_2, \ldots, k_N$ the 4-momenta of the N virtual-meson lines. The Feynman amplitude of Fig. 1(a) is

$$M_{N+1} = \prod_1^N (k_i^2 + m^2)^{-1}, \tag{3}$$

and the corresponding cross section is

$$\sigma_{N+1} = 4\pi^4 (a_0 b_0 v)^{-1} (mg)^{2N+2} \int |M_{N+1}|^2 \prod_1^{N+1} \left[(16\pi^3)^{-1} \frac{d^3 q_j}{(q_j)_0} \right] \delta^4\left(\sum_1^{N+1} q_j - a - b \right), \tag{4}$$

where the subscript 0 denotes the energy component; v is the relative velocity between the two initial mesons; and s is the square of their center-of-mass energy. The sum of σ_{N+1} gives the total cross section

$$\sigma_{\rm tot} = \sum_{N=1}^{\infty} \sigma_{N+1} . \qquad (5)$$

Throughout the paper, we are interested in the meson multiplicity problem only in the limit $\ln s \to \infty$. In this limit, as is well known[2] (and as will also be proved by using the gas analog discussed in the next section), the total cross section exhibits a Regge-pole behavior:

$$\sigma_{\rm tot} \sim s^{\alpha-1} . \qquad (6)$$

It is convenient to use the laboratory frame $\Sigma_{\rm lab}$, in which $\phi(a)$ is at rest; i.e.,

$$a_0 = m$$

and

$$b_0 = (2m)^{-1} s - m .$$

Let us introduce a set of N positive variables $x_1, x_2, \ldots, x_N$, defined by

$$(k_1)_0 = b_0 e^{-x_1} ,$$

and for $1 < i \le N$

$$(k_i)_0 = (k_{i-1})_0 e^{-x_i} , \qquad (7)$$

where the subscript 0 denotes the energy component of the relevant 4-momenta in $\Sigma_{\rm lab}$. The energy of the final mesons in $\Sigma_{\rm lab}$ is then given by

$$(q_1)_0 = b_0 (1 - e^{-x_1}) ,$$

and for $1 < i \le N$

$$(q_i)_0 = b_0 (1 - e^{-x_i}) \exp\left(-\sum_1^{i-1} x_j\right) . \qquad (8)$$

In the integration (4), one may trivially eliminate the $d^3 q_{N+1}$ integration by using the three-dimensional δ function; the remaining δ function of energy conservation becomes simply

$$\delta\left((q_{N+1})_0 - m - b_0 \exp\left(-\sum_1^N x_i\right)\right) . \qquad (9)$$

Since $k_N = q_{N+1} - a$, one finds

$$(q_{N+1})_0 = (2m)^{-1} (k_N{}^2 + m^2 - q_{N+1}{}^2) . \qquad (10)$$

Because of the square of the Feynman propagator $(k_i{}^2 + m^2)^{-2}$, in the integral (4) the value of any $k_i{}^2$, including $k_N{}^2$, is on the average $O(m^2)$. Therefore, the on-mass-shell condition $q_{N+1}{}^2 + m^2 = 0$, together with (10), implies that the value of $(q_{N+1})_0$ is on the average $O(m)$.

To carry out the gas analog, it is more convenient to impose the on-mass-shell condition

$$q_i{}^2 + m^2 = 0$$

only for $1 \le i \le N$. For the $(N+1)$th meson, we shall replace its on-mass-shell condition by

$$(q_{N+1})_0 = (1+\lambda) m , \qquad (11)$$

where λ is a positive constant, independent of s. In the integral (4), on account of (10), the important integration region now becomes one in which $q_{N+1}{}^2 \sim O(m^2)$ [instead of $(q_{N+1})_0 \sim O(m)$ and $q_{N+1}{}^2 = -m^2$]. The δ function (9) then becomes

$$b_0^{-1} \exp\left(\sum_1^N x_i\right) \delta\left(\sum_1^N x_i - L\right) , \qquad (12)$$

where

$$L = \ln\left(\frac{b_0}{\lambda m}\right) \to \ln\left(\frac{s}{m^2}\right) \qquad (13)$$

as $\ln s \to \infty$. The limiting value of L is therefore independent of λ. (If one wishes, one may choose λ so that $\langle q^2 + m^2 \rangle_{\rm av} = 0$, where $\langle\ \rangle_{\rm av}$ denotes some suitably defined average.)

It is clear that the above modification does not alter the multiplicity distribution in the $\ln s \to \infty$ limit.

III. A ONE-DIMENSIONAL GAS

Next, we consider a one-dimensional classical Boltzmann gas of N identical but distinguishable atoms on a ring of length L. Let U_N be the potential energy. For our purpose, we need only consider isotherms; therefore, the natural unit for energy is[7]

$$kT = 1 ,$$

where k is the Boltzmann constant and T the absolute temperature. To evaluate the partition function Q_N, it is only necessary to consider an *ordered* set,[8] say atoms $1, 2, \ldots, N$ distributed in a strictly sequential order with x_1 as the absolute distance between atoms 1 and 2, x_2 that between atoms 2 and 3, etc., as illustrated by Fig. 1(b). One has

$$Q_N = \int \delta\left(\sum_1^N x_i - L\right) \exp(-U_N) \prod_1^N dx_i , \qquad (14)$$

where $x_i \ge 0$. The grand partition function $\mathfrak{Q}$ is given by

$$\mathfrak{Q} = \sum_N z^N Q_N , \qquad (15)$$

where z denotes the fugacity. The thermodynamical pressure of the system is related to $\mathfrak{Q}$ by

$$p = \lim_{L\to\infty} L^{-1} \ln \mathcal{Q}.$$

By using the dependence of the pressure p on the fugacity z, one can compute directly the ensemble distribution of N as $L \to \infty$; e.g., the density ρ is given by

$$\frac{dp}{d\ln z} = \rho = \lim_{L\to\infty} L^{-1}\langle N\rangle;$$

the number fluctuation $\langle N^2\rangle - \langle N\rangle^2$ is given by

$$\frac{d^2p}{d(\ln z)^2} = \lim_{L\to\infty} L^{-1}[\langle N^2\rangle - \langle N\rangle^2],$$

etc. Since the relative probability of finding N particles at a given length L is $z^N Q_N$, as L becomes large the relative probability approaches asymptotically the product of z^N multiplied by

$$(2\pi i)^{-1} \oint z^{-(N+1)} e^{pL} dz, \tag{16}$$

where the contour can be any counterclockwise small closed curve around the origin in the complex z plane. [For a finite L, (16) holds if p is replaced by $p_L \equiv L^{-1}\ln \mathcal{Q}$, without taking the limit $L\to\infty$.]

As we shall prove in Appendix A, the multiplicity distribution in the ϕ^3-multiperipheral model is identical to the above number distribution in the grand canonical ensemble, provided that the potential U_N is of a specific form [determined by Eq. (A14) in Appendix A]. We may decompose

$$U_N = V_2 + V_3 + V_4 + \cdots, \tag{17}$$

where V_2 is a sum of only two-body nearest-neighbor potentials,

$$V_2 = \sum_{i=1}^{N} v_2(x_i); \tag{18}$$

V_3 is a sum of three-body potentials between only the nearest and the next-nearest neighbors,

$$V_3 = \sum_{i=1}^{N} v_3(x_i, x_{i+1}); \tag{19}$$

and V_4 is a sum of four-body potentials between only the nearest and the next two nearest neighbors, etc. The explicit form for the two-body potential is given by

$$\exp[-v_2(x)] = \frac{1-e^{-x}}{1-e^{-x}+e^{-2x}}\left[1 + \frac{e^{-x}-e^{-2x}}{1-e^{-x}+e^{-2x}}\ln(e^{-x}-e^{-2x})\right], \tag{20}$$

where $x = x_i$ denotes the absolute value of the distance between any two nearest-neighbor atoms, say i and $i+1$. The three-body potential between any three neighboring atoms, say i, $i+1$, and $i+2$, is given by

$$\exp[-v_3(x,x')] = \left(1+\epsilon\ln\frac{\epsilon}{1+\epsilon}\right)^{-1}\left(1+\epsilon'\ln\frac{\epsilon'}{1+\epsilon'}\right)^{-1} \times\left[1+\lambda\ln\frac{(1+\epsilon)\lambda}{1+(1+\epsilon)\lambda} + \epsilon(1-\lambda)\ln\frac{\epsilon}{1+\epsilon} + \frac{\epsilon\lambda^2}{1+\lambda}\ln\frac{\epsilon\lambda}{1+(1+\epsilon)\lambda}\right], \tag{21}$$

where

$$\epsilon = \frac{e^{-x}-e^{-2x}}{1-e^{-x}+e^{-2x}},$$

$$\epsilon' = \frac{e^{-x'}-e^{-2x'}}{1-e^{-x'}+e^{-2x'}},$$

$$\lambda = \left(\frac{\epsilon'}{\epsilon}\right)e^{-x},$$

$x = x_i$ is the absolute value of the distance between atoms i and $i+1$, and $x' = x_{i+1}$ is that between atoms $i+1$ and $i+2$. The general expression of v_m is somewhat complicated, and will be discussed in Appendix A.

From (20) and (21), one can establish that both v_2 and v_3 are repulsive; i.e., $v_2(x) \geq 0$ at arbitrary $x \geq 0$, and $v_3(x, x') \geq 0$ at arbitrary $x \geq 0$ and $x' \geq 0$. In Appendix A, it is shown that the total potential U_N is also repulsive; we find

$$U_N \geq 0 \tag{22}$$

for arbitrary $x_i \geq 0$, and $U_N \to \infty$ as any single $x_j \to 0$. From (20) and (21), one also sees that v_2 and v_3 represent short-range forces; i.e., $v_2(x) \to 0$ exponentially as $x\to\infty$, and $v_3(x,x') \to 0$ exponentially as either $x\to\infty$ or $x'\to\infty$. In Appendix A, as will be shown by Eq. (A22), similar short-range properties hold for other v_m's as well, provided that m is finite, independent of N. For $m \sim O(N)$, the corresponding m-body force is clearly long-range. As will also be shown in Appendix A [Eq. (A28)], for configurations near the average one,

$$x_1 = x_2 = \cdots = x_N = L/N,$$

the total potential U_N has an upper bound given by

$$U_N = V_2 + V_3 + \cdots + V_N < N \times \text{constant}, \tag{23}$$

where the constant denotes a finite function of the average interatomic distance L/N. Because this upper bound is linear in N, the presence of long-range interactions, such as V_N, $V_{N-1}, \ldots$, does not jeopardize the thermodynamical limit. We note that due to the one-dimensional character of the gas and the pure repulsive nature of the forces, there should not be any phase transition for this particular system.

The detailed correspondence between this one-dimensional gas and the multiperipheral model is given in Table I. For the gas system, the average number of particles is clearly proportional to the length L at large L. By using Table I, or Eq. (13), one derives the familiar result[1] that the average meson multiplicity in the multiperipheral model increases linearly with $\ln s$ at large s. For the gas system, $\ln \mathfrak{L}$ is also proportional to L at large L. By using Table I, or Eq. (A13) in Appendix A, one obtains Eq. (6) which gives the Regge behavior for the multiperipheral model.

The explicit form of $V_2, V_3, \ldots$ enables one to directly evaluate the virial series

$$p=\sum_1^\infty b_l z^l \tag{24}$$

by using the well-known cluster-expansion technique, developed by Mayer and Mayer.[9] For example,

$$b_1=1,$$

$$b_2=\int_0^\infty f(x)\,dx, \tag{25}$$

where

$$f(x)=e^{-v_2(x)}-1;$$

$$b_3=b_2^2-\int_0^\infty xf(x)\,dx+\int_0^\infty dx\int_0^\infty dx'\,F(x,x'), \tag{26}$$

where

$$F(x,x')=[1+f(x)][1+f(x')][e^{-v_3(x,x')}-1],$$

and so on. For the specific $v_2(x)$ given by (20), one finds

$$b_2=-\frac{8}{3\sqrt{3}}\sum_1^\infty m^{-2}\sin(\tfrac{1}{3}m\pi)=-1.5626, \tag{27}$$

in which the sum equals the Clausen's integral[10] $f(\theta)$ at $\theta=\frac{1}{3}\pi$. Similarly, b_3 can be evaluated, and the approximate numerical value is[11]

$$b_3=1.9573, \tag{28}$$

in which the ratio between the contribution of the three-body potential v_3 and that of the two-body potential v_2 is $\sim -\frac{1}{4}$.

From Table I, one may derive the dependence of the Regge-pole power α on g^2 for the ϕ^3-multi-

TABLE I. Correspondence between the one-dimensional gas model and the multiperipheral model in the high-energy limit [where s = (center-of-mass energy)^2 and α = Regge-pole power for forward elastic scattering].

Gas system	Multiperipheral model (ϕ^3 theory)
Fugacity z	$=(g/4\pi)^2$
Pressure p	$=\alpha+1$
Grand partition function $\mathfrak{L}$	$\sim s^2\sigma_{\text{tot}}$
Length L	$\sim \ln(s/m^2)$
$L\dfrac{dp}{d\ln z}$	$\sim$ average multiplicity $\langle N\rangle$
$L\dfrac{d^2p}{d(\ln z)^2}$	$\sim \langle N^2\rangle-\langle N\rangle^2$

peripheral model from the virial expansion:

$$\alpha=-1+\left(\frac{g}{4\pi}\right)^2+b_2\left(\frac{g}{4\pi}\right)^4+b_3\left(\frac{g}{4\pi}\right)^6+\cdots. \tag{29}$$

The coefficients of $(g/4\pi)^2$ and $(g/4\pi)^4$ have been calculated in the literature,[2,12] and of course they agree with the above virial-expansion results.

IV. AN ALTERNATIVE EXPANSION METHOD

For most applications, since at high energy $\sigma_{\text{tot}}\sim s^{\alpha-1}\sim$ constant, one is interested in multiplicity problems in which $\alpha\cong 1$; this corresponds to a gas system with its pressure $p\cong 2$. At such high pressure, the virial series $p=\sum b_l z^l$ does not offer the most practical method for evaluating the function $p=p(z)$. As we shall see, for the class of problems in which the potential U_N is a sum $U_N = V_2+V_3+\cdots$, where V_2 consists of only two-body interactions between the nearest neighbors, and V_3 only three-body interactions between the nearest and the next-nearest neighbors, ..., there exists an alternative new systematic expansion method which is different from the virial series, but which appears to be more useful for practical applications. This new systematic expansion method is applicable to any one-dimensional gas with such a potential, not restricted to the specific multiperipheral model discussed in the previous two sections.

We observe that for large values of L and N, the function $\delta(\sum x_i - L)$ in (14) may be replaced by[13]

$$\exp[-\beta(\textstyle\sum x_i - L)]. \tag{30}$$

The partition function Q_N can then be written as

$$Q_N=e^{\beta L}h^N, \tag{31}$$

where h is a function of β and N given by

$$h^N = \int \exp(-U_N - \beta \sum x_i) \prod dx_i \,, \tag{32}$$

in which each x_i is integrated independently from zero to infinity. Correspondingly, the grand partition function becomes

$$\mathcal{Q} = \sum_N (zh)^N e^{\beta L} \,, \tag{33}$$

where β must be regarded as a *function* of N and L, determined by

$$\left(\frac{\partial \ln h}{\partial \beta}\right)_N = -N^{-1} L \,. \tag{34}$$

In the grand canonical ensemble, the relative probability distribution $(zh)^N e^{\beta L}$ has a maximum at $N = \overline{N}(L)$, which can be obtained by setting the derivative of the relative probability with respect to N to be zero. One finds that at $N = \overline{N}(L)$

$$h = z^{-1} \,. \tag{35}$$

As $L \to \infty$, on account of (33) and (35), the value of β, evaluated at $N = \overline{N}(L)$, approaches the thermodynamical pressure p. Thus, by taking the logarithm of (35), we derive the basic equation

$$-\ln z = \lim_{N\to\infty} N^{-1} \ln \int \exp\left(-U_N - p \sum_1^N x_i\right) \prod dx_i \,, \tag{36}$$

in which, as in (32), all x_i are integrated independently from zero to infinity.

The new expansion method consists of first neglecting all interactions, then including only V_2, then only $V_2 + V_3$, etc., as follows.

(1) In the zeroth approximation, we set

$$U_N = 0 \,.$$

The system satisfies the perfect gas law

$$p = z = \rho \,.$$

From (16), it follows that the number distribution is given by the familiar Poisson formula. For the ϕ^3-multiperipheral model, since U_N is positive, this zeroth approximation is also an upper bound; i.e., with the inclusion of $U_N \geq 0$,

$$p \leq z \,, \tag{37}$$

or,[14] by using Table I,

$$\alpha \leq -1 + \left(\frac{g}{4\pi}\right)^2 , \tag{38}$$

where the equality holds only in the weak coupling limit.

(2) The first approximation is to set

$$U_N = V_2 = \sum_1^N v_2(x_i) \,. \tag{39}$$

By using (36), we find for the arbitrary two-body potential $v_2(x)$

$$z^{-1} = \int_0^\infty dx \exp[-v_2(x) - px] \,. \tag{40}$$

If one wishes, one may also expand p as a power series of z:

$$p = \sum b_l z^l \,.$$

From the closed expression (40), it follows directly that $b_1 = 1$, b_2 is given by (25), and b_3 is given by (26) with $F = 0$, etc. At large z, only the value of $\exp[-v_2(x)]$ near $x = 0$ is of importance. We may expand

$$\exp[-v_2(x)] = a_0 + a_1 x + \cdots \,.$$

Equation (40) implies that

$$z^{-1} = a_0 p^{-1} + a_1 p^{-2} + \cdots \,.$$

If the potential is infinitely repulsive at $x = 0$, as is the case in the ϕ^3-multiperipheral model, then $a_0 = 0$. As $z \to \infty$, $p \sim (a_1 z)^{1/2}$. For the ϕ^3-multiperipheral model, $v_2(x)$ is given by (20), which gives $a_1 = 1$, and therefore

$$p \sim z^{1/2} \text{ as } z \to \infty \,.$$

As will be shown in Appendix A [Eq. (A34)], for the ϕ^3-multiperipheral model the inclusion of all other $V_3, V_4, \ldots$ forces always increases the value of the repulsive potential, i.e.,

$$U_N \geq V_2 \,.$$

Therefore the pressure p_1 determined by the first approximation (where the subscript 1 is added for clarity) also forms an upper bound for the rigorous pressure p, which is calculated with the entire U_N without any approximation; we derive then at any given $z \geq 0$ the inequality

$$p \leq p_1(z) \,, \tag{41}$$

where, according to (20) and (40), $p_1(z)$ is given by

$$z^{-1} = \int_0^\infty \frac{1 - e^{-x}}{1 - e^{-x} + e^{-2x}} \times \left[1 + \frac{e^{-x} - e^{-2x}}{1 - e^{-x} + e^{-2x}} \ln(e^{-x} - e^{-2x})\right] e^{-p_1 x} dx \,. \tag{42}$$

The inequality (41) is, of course, a better inequality than (37). As $z \to \infty$, (42) gives $p_1 \sim z^{1/2}$, and therefore (41) implies

$$p \leq z^{1/2} \text{ as } z \to \infty \,, \tag{43}$$

or, by using Table I,

$$\alpha \leq (4\pi)^{-1} g \tag{44}$$

in the strong-coupling limit, in agreement with the bound derived by Tiktopoulos and Treiman.[14]

(3) In the second approximation, we equate

$$U_N = V_2 + V_3 = \sum_{i=1}^{N} [v_2(x_i) + v_3(x_i, x_{i+1})] . \quad (45)$$

It is convenient to consider a Hilbert space of base vectors $\psi_1(x), \psi_2(x), \psi_3(x), \ldots$, which satisfy the usual orthonormal relation

$$\int_0^\infty \psi_i(x)\psi_j(x)\, dx = \delta_{ij} . \quad (46)$$

Among these, $\psi_1(x)$ is chosen to be

$$\psi_1(x) = (z_1)^{1/2} \exp\{-\tfrac{1}{2}[v_2(x) + px]\} , \quad (47)$$

where z_1 is the normalization constant, defined by

$$z_1^{-1} = \int_0^\infty dx \exp[-v_2(x) - px] ; \quad (48)$$

the other base vectors $\psi_2(x), \psi_3(x), \ldots$ can be arbitrary real functions that satisfy (46) and, together with $\psi_1(x)$, the completeness theorem. Let us define two real matrices ϵ and H_0 in this Hilbert space:

$$\epsilon_{ij} \equiv \int_0^\infty dx \int_0^\infty dy\, \psi_i(x) \langle x|\epsilon| y\rangle \psi_j(y), \quad (49)$$

$$\langle x|\epsilon| y\rangle = \psi_1(x)[e^{-v_3(x,y)} - 1]\psi_1(y), \quad (50)$$

$$(H_0)_{11} \equiv 1, \quad (51)$$

and

$$(H_0)_{ij} \equiv 0 \text{ for all other } i \text{ and } j . \quad (52)$$

Thus, the integral in (36) becomes

$$\int \exp\left(-U_N - p\sum_1^N x_i\right) \prod dx_i = z_1^{-N} \operatorname{Tr}(H_0 + \epsilon)^N . \quad (53)$$

The logarithm of (53), at fixed N, can be readily evaluated as a power series in ϵ. By using (36) and taking the limit $N \to \infty$, one can verify directly that

$$-\ln z = -\ln z_1 + \epsilon_{11} + (\epsilon^2)_{11} - \tfrac{3}{2}(\epsilon_{11})^2 + (\epsilon^3)_{11} - 4(\epsilon^2)_{11}\epsilon_{11} + \tfrac{10}{3}(\epsilon_{11})^3 + \cdots . \quad (54)$$

Since z_1 and

$$\epsilon_{11} = \int \psi_1(x)\langle x|\epsilon| y\rangle \psi_1(y)\, dx\, dy,$$

$$(\epsilon^2)_{11} = \int \psi_1(x)\langle x|\epsilon|y\rangle\langle y|\epsilon|z\rangle \psi_1(z)\, dx\, dy\, dz,$$

etc., are functions of p, Eq. (54) determines $z = z(p)$.

Equation (54) can also be derived by a simpler method, without any direct calculations. We observe that the general form of the series (54) is *independent* of whether the matrix ϵ is symmetric or not. Thus, we may consider the special case of a symmetric three-body potential $v_3(x, y) = v_3(y, x)$, and therefore $\epsilon_{ij} = \epsilon_{ji}$. Let $\lambda = \lambda(p)$ be the largest eigenvalue of $H_0 + \epsilon$. By using (36) and (53), we find the closed expression

$$\lambda(p) = z_1/z . \quad (55)$$

The series expansion can then be obtained by noting that H_0 has only one eigenvalue equal to unity, while all its other eigenvalues are zero. Thus, as $\epsilon \to 0$, $\lambda \to 1$ and (55) reduces to (40). For $\epsilon \neq 0$, the power-series expansion of λ is given by the familiar perturbation formula

$$\lambda = 1 + \epsilon_{11} + \sum_{i\neq 1} \lambda^{-1}\epsilon_{1i}\epsilon_{i1} + \sum_{i\neq 1;\, j\neq 1} \lambda^{-2}\epsilon_{1i}\epsilon_{ij}\epsilon_{j1} + \cdots ,$$

which, together with (55), leads to (54).

The higher-order approximations including $V_4, V_5, \ldots$ can be carried out in a similar manner. The details are given in Appendix C.

V. REMARKS

As noted earlier, the method developed in this section is of a rather general character, not restricted to the specific multiperipheral model discussed in the previous two sections. For practical applications, it seems reasonable to try first the approximation of only two-body nearest-neighbor forces. Equation (40) can be used phenomenologically to determine an effective two-body potential $v_2(x)$ from the observed meson multiplicity distributions in high-energy collisions, provided that $\ln(s/m^2)$ is sufficiently large and that the average multiplicity and its fluctuation are indeed linear in $\ln(s/m^2)$. Within this approximation, one may apply the same effective "potential" to evaluate the m-body correlation functions, which can then be compared with various inclusive reactions.

ACKNOWLEDGMENTS

I wish to thank A. Mueller for giving me some lectures on topics related to the multiperipheral model. A major part of this work was done when I was visiting the Scuola Normale Superiore, Pisa, Italy in the early summer of 1972; I wish to thank G. Bernardini and L. A. Radicati for their kind hospitality.

APPENDIX A

To derive the explicit potential of the gas system, we start with Eq. (4) for the Feynman diagram in Fig. 1(a). In the laboratory frame, let $\hat{k}_{i-1}$ be the unit vector parallel to the 3-momentum of the virtual meson line k_{i-1}. The three-momentum $\vec{q}_i$ of the ith final meson can be written as

$$\vec{q}_i = \vec{\delta}_i + [(q_i)_0 - \tfrac{1}{2}(q_i)_0^{-1}(m^2 + \vec{\delta}_i^{\,2})]\hat{k}_{i-1}, \quad \text{(A1)}$$

where $\vec{\delta}_i \perp \hat{k}_{i-1}$, and the energy $(q_i)_0$ is assumed to be $\gg m^2$ and $\vec{\delta}_j^{\,2}$. Since, as noted before, k_i^2 is of the order of m^2, $(k_i)_0$ may be assumed to be $\gg |k_i^2|^{1/2}$. By using $k_i = k_{i-1} - q_i$ and (7), one has

$$k_i^2 + m^2 = (1-e^{-x_i})^{-1} \times [\vec{\delta}_i^{\,2} + m^2 + k_{i-1}^2 e^{-x_i}(1-e^{-x_i})], \quad \text{(A2)}$$

where $k_i^2 \equiv \vec{k}_i^2 - (k_i)_0^2$. Through induction, (A2) can be written as

$$k_i^2 + m^2 = (1-e^{-x_i})^{-1}\left(\sum_{a=1}^{i} A_{ia}\vec{\delta}_a^{\,2} + m^2 B_i\right), \quad \text{(A3)}$$

where

$$A_{ii} = 1,$$

$$A_{ia} = (1-e^{-x_a})^{-1}(1-e^{-x_i})\exp\left(-\sum_{j=a+1}^{i} x_j\right) \quad \text{(A4)}$$

for $a<i$, and

$$B_i = \sum_{a=1}^{i} A_{ia}(1 - e^{-x_a} + e^{-2x_a}).$$

The usual parametric form of Feynman propagator gives

$$\prod_i (k_i^2 + m^2)^{-2} = \int e^{-E} \prod_i (1-e^{-x_i})^2 l_i\, dl_i \quad \text{(A5)}$$

where each l_i varies independently from zero to infinity, and

$$E = \sum_{a=1}^{N} \vec{\delta}_a^{\,2}\left(\sum_{j=a}^{N} l_j A_{ja}\right) + m^2 \sum_{j=1}^{N} l_j B_j .$$

The integration of (A5) over $\prod d^2\delta_i$ is an elementary one. One finds

$$\int \prod_{i=1}^{N} (k_i^2+m^2)^{-2} d^2\delta_i = \pi^N \prod_{i=1}^{N} (1-e^{-x_i})^2 \int \prod_{a=1}^{N}\left(\sum_{j=a}^{N} l_j A_{ja}\right)^{-1} l_a\, dl_a \exp\left(-m^2 \sum_{i=1}^{N} l_i B_i\right). \quad \text{(A6)}$$

It is convenient to change the variables from $l_1, l_2, \ldots, l_N$ to $\xi_1, \xi_2, \ldots, \xi_N$, defined by

$$(1 - e^{-x_a} + e^{-2x_a})^{-1}\xi_a \equiv m^2 \sum_{j=a}^{N} l_j A_{ja} . \quad \text{(A7)}$$

Therefore,

$$m^2 \sum_i l_i B_i = \sum_i \xi_i . \quad \text{(A8)}$$

From (A4), one can readily establish $A_{ab}A_{bc} = A_{ac}$ for arbitrary a, b, and c that satisfy $a \geq b \geq c$. The inverse transformation from ξ_i to l_i is

$$m^2 l_N = (1 - e^{-x_N} + e^{-2x_N})^{-1}\xi_N$$

and (A9)

$$m^2 l_a = (1 - e^{-x_a} + e^{-2x_a})^{-1}(\xi_a - \lambda_{a+1}\xi_{a+1}),$$

for $a = 1, 2, \ldots, N-1$, where

$$\lambda_{a+1} = \left(\frac{e^{-x_{a+1}} - e^{-2x_{a+1}}}{1-e^{-x_a}}\right)\left(\frac{1-e^{-x_a}+e^{-2x_a}}{1-e^{-x_{a+1}}+e^{-2x_{a+1}}}\right). \quad \text{(A10)}$$

By using (4), (12), (A6), and

$$\prod_{j=1}^{N} (q_j)_0^{-1} d^3 q_j = \prod_i (1-e^{-x_i})^{-1} e^{-x_i}\, dx_i\, d^2\delta_i ,$$

one finds that, apart from a multiplicative constant *independent* of N and s, the cross section σ_{N+1} in the ϕ^3-multiperipheral model is given by

$$\sigma_{N+1} \propto s^{-2} z^N Q_N = s^{-2} z^N \int e^{-\Delta}\, \delta\left(\sum_1^N x_i - L\right) \prod_{i=1}^{N} \frac{(1-e^{-x_i})dx_i}{1-e^{-x_i}+e^{-2x_i}} , \quad \text{(A11)}$$

where $z = (4\pi)^{-2} g^2$ and Δ is a function of $x_1, x_2, \ldots, x_N$, defined by the integral

$$e^{-\Delta}=\int(\xi_1-\lambda_2\xi_2)(\xi_2-\lambda_3\xi_3)\cdots(\xi_{N-1}-\lambda_N\xi_N)\xi_N\prod_{a=1}^{N}\xi_a^{-1}e^{-\xi_a}d\xi_a\,, \tag{A12}$$

in which the integration domain extends over $\xi_1\geq\lambda_2\xi_2$, $\xi_2\geq\lambda_3\xi_3,\ldots,$ $\xi_i\geq\lambda_{i+1}\xi_{i+1},\ldots,$ and $\xi_N\geq 0$. On account of (A11), as $\ln s\to\infty$, the total cross section σ_{tot} satisfies

$$\ln\sigma_{\text{tot}}+2\ln s\to\ln\mathfrak{Q}, \tag{A13}$$

where

$$\mathfrak{Q}=\sum_N z^N Q_N\,.$$

The function Q_N denotes the partition function of the gas analog, and $\mathfrak{Q}$ is the corresponding grand partition function. Upon comparing (A11) with (14), we find the potential U_N for the gas system to be

$$e^{-U_N}=e^{-\Delta}\prod_i\frac{1-e^{-x_i}}{1-e^{-x_i}+e^{-2x_i}}\,. \tag{A14}$$

To derive the two-body potential $v_2(x_i)$, we need only consider, at a fixed x_i, the limit of all other $x_j\to\infty$. (To avoid problems with the boundary, we choose $i\neq 1$.) In this limit, all three-body forces, four-body forces, etc., are, by definition, zero; therefore,

$$U_N\to V_2\to v_2(x_i)\,. \tag{A15}$$

On the other hand, in the same limit, according to (A10),

$$\lambda_j\to 0\quad\text{for all } j\neq i,$$

and

$$\lambda_i\to\epsilon_i\equiv\frac{e^{-x_i}-e^{-2x_i}}{1-e^{-x_i}+e^{-2x_i}}\,. \tag{A16}$$

Thus,

$$e^{-\Delta}\to\int_0^\infty d\xi_{i-1}\int_0^{\epsilon_i^{-1}\xi_{i-1}}d\xi_i\,\xi_{i-1}^{-1}(\xi_{i-1}-\epsilon_i\xi_i)\exp(-\xi_{i-1}-\xi_i)=1+\epsilon_i\ln\frac{\epsilon_i}{1+\epsilon_i}\,, \tag{A17}$$

which, together with (A14) and (A15), lead to the explicit two-body force given by (20).

To derive the three-body potential, we keep x_i and x_{i+1} fixed (where $i\neq 1$), and then consider the limit of all other $x_j\to\infty$. In this limit, by definition,

$$U_N\to v_2(x_i)+v_2(x_{i+1})+v_3(x_i\,,x_{i+1})\,. \tag{A18}$$

According to (A10), in the same limit λ_{i+1} remains fixed; $\lambda_i\to\epsilon_i$ [which is given by (A16)]; and all other $\lambda_j\to 0$; therefore, the integral (A12) becomes

$$e^{-\Delta}\to\int(\xi_{i-1}\xi_i)^{-1}(\xi_{i-1}-\epsilon_i\xi_i)(\xi_i-\lambda_{i+1}\xi_{i+1})\exp(-\xi_{i-1}-\xi_i-\xi_{i+1})\,d\xi_{i-1}\,d\xi_i\,d\xi_{i+1}\,, \tag{A19}$$

where the integration domain is

$$\xi_{i-1}\geq\epsilon_i\xi_i,\quad \xi_i\geq\lambda_{i+1}\xi_{i+1},\quad\text{and}\quad\xi_{i+1}\geq 0\,.$$

By using (A14), (A18), and (A19), one obtains the explicit form (21) for $v_3(x_i\,,x_{i+1})$. In a similar manner, we can derive explicitly the four-body interaction, the five-body interaction, etc.

In general, to derive the m-body potential $v_m(x_i\,,x_{i+1},\ldots,x_{i+m-2})$, we keep $x_i\,,x_{i+1},\ldots,x_{i+m-2}$ fixed (where $i\neq 1$), and consider the limit of all other $x_j\to\infty$. In this limit, $\lambda_i\to\epsilon_i$ [which is given by (A16)]; $\lambda_{i+1},\ldots,\lambda_{i+m-2}$ remain unchanged; and all other $\lambda_j\to 0$. The potential U_N becomes

$$U_N\to U_m(x_i\,,x_{i+1},\ldots,x_{i+m-2})\equiv\sum_{k=i}^{i+m-2}v_2(x_k)+\sum_{k=i}^{i+m-3}v_3(x_k,x_{k+1})+\cdots+v_m(x_i\,,x_{i+1},\ldots,x_{i+m-2})\,, \tag{A20}$$

and

$$e^{-\Delta}\to\int(\xi_{i-1}-\epsilon_i\xi_i)(\xi_i-\lambda_{i+1}\xi_{i+1})\cdots(\xi_{i+m-3}-\lambda_{i+m-2}\xi_{i+m-2})\xi_{i+m-2}\prod_{k=i-1}^{i+m-2}\xi_k^{-1}e^{-\xi_k}d\xi_k\,, \tag{A21}$$

where the integration extends over the domain

$$\xi_{i-1} \geq \epsilon_i \xi_i, \quad \xi_i \geq \lambda_{i+1}\xi_{i+1}, \ldots,$$

$$\xi_{i+m-3} \geq \lambda_{i+m-2}\xi_{i+m-2}, \quad \text{and} \quad \xi_{i+m-2} \geq 0.$$

The explicit form of v_m can then be derived by using (A14), (A20), and (A21). We note that if $x_i \to \infty$, then $\epsilon_i \to 0$ exponentially; therefore

$$U_m(x_i, x_{i+1}, \ldots, x_{i+m-2}) \to U_{m-1}(x_{i+1}, \ldots, x_{i+m-2}).$$

Similarly, if $x_{i+s} \to \infty$ $(s \geq 1)$, then $\lambda_{i+s} \to 0$ exponentially, and therefore

$$U_m(x_i, x_{i+1}, \ldots, x_{i+m-2}) \to U_{s+1}(x_i, \ldots, x_{i+s-1}) U_{m-s-1}(x_{i+s+1}, \ldots, x_{i+m-2}).$$

Together, these relations imply the short-range nature of v_m:

$$v_m(x_i, x_{i+1}, \ldots, x_{i+m-2}) \to 0 \text{ exponentially} \tag{A22}$$

as any single $x_k \to \infty$, where $i \geq k \geq i+m-2$, provided that m is finite, independent of N.

Inequalities

(1) We note that while the range of v_m is short for any finite m, for m of the order of N (therefore, also of the order of L) the corresponding m-body force has to be long-range. In order to establish a well-defined thermodynamical limit, we shall show that, at a constant density N/L and for regions near the average configuration

$$x_1 = x_2 = \cdots = x_N = L/N, \tag{A23}$$

the total potential energy U_N has an upper bound which increases linearly with N. For the configuration (A23), one has, on account of (A10),

$$\lambda_2 = \lambda_3 = \cdots = \lambda_N \equiv \lambda = \exp[-N^{-1}L] < 1; \tag{A24}$$

furthermore, because of the inequality

$$\xi^{-1} > e^{-\xi}, \tag{A25}$$

$$e^{-\Delta} > \int (\xi_1 - \lambda\xi_2)(\xi_2 - \lambda\xi_3) \cdots (\xi_{N-1} - \lambda\xi_N)\xi_N \prod e^{-2\xi_i} d\xi_i. \tag{A26}$$

The right-hand side of (A26) can be readily evaluated. By taking the logarithm of (A26), one obtains

$$\frac{1}{N}\Delta < 2\ln 2 - 2\ln(1-\lambda) + \frac{2}{N}\sum_{m=1}^{N} \ln(1-\lambda^m). \tag{A27}$$

Since $0 < \lambda < 1$, $\sum_{m=1}^{N} \ln(1-\lambda^m)$ is larger than $(1-\lambda)^{-1}\ln(1-\lambda)$ but less than zero. One finds, as $N \to \infty$,

$$\frac{1}{N}\sum_{1}^{N} \ln(1-\lambda^m) \to 0.$$

Because of (A14), (A27) can be written as (for $N \gg 1$)

$$\frac{1}{N} U_N < 2\ln 2 - 3\ln(1-\lambda) + \ln(1-\lambda+\lambda^2). \tag{A28}$$

The inequality (A28) can be easily extended to any configuration $x_1, x_2, \ldots, x_N$ in which x_i's can be unequal, but the maximum value λ_{max} of the corresponding $\lambda_2, \lambda_3, \ldots, \lambda_N$ is less than unity. From (A12), one can verify that, keeping $\lambda_{j \neq i}$ fixed,

$$\frac{\partial\Delta}{\partial\lambda_i} > 0. \tag{A29}$$

Consequently, Δ satisfies the same inequality (A27), provided that λ is replaced by λ_{max}. We note that the inequality (A25) can be easily improved, since

$$\xi^{-1} > e^{-\xi}\left(1 + \frac{1}{2!}\xi + \frac{1}{3!}\xi^2 + \cdots\right);$$

therefore, the upper bound (A28) can also be improved.

(2) Since $\xi_a^{-1}(\xi_a - \lambda_{a+1}\xi_{a+1}) \leq 1$, (A12) implies the inequality

$$e^{-\Delta} \leq 1, \tag{A30}$$

and therefore

$$e^{-U_N} \leq \sum_i \frac{1-e^{-x_i}}{1-e^{-x_i}+e^{-2x_i}}. \tag{A31}$$

Because the right-hand side of (A31) is ≤ 1, we establish the repulsive nature of U_N; i.e., for arbitrary $x_i \geq 0$,

$$U_N \geq 0. \tag{A32}$$

Furthermore, as any single x_i approaches zero, the right-hand side of (A31) approaches zero, and therefore $U_N \to \infty$.

(3) The inequalities (A30) and (A32) can be readily improved. We shall show that

$$e^{-\Delta} \leq \prod_i \left[1 + \frac{e^{-x_i} - e^{-2x_i}}{1 - e^{-x_i} + e^{-2x_i}} \ln(e^{-x_i} - e^{-2x_i})\right] \tag{A33}$$

and therefore

$$U_N \geq V_2, \tag{A34}$$

where V_2 is the two-body interaction, given by (18) and (20).

Proof. From (A10), it follows that

$$\lambda_a \geq \epsilon_a \equiv \frac{e^{-x_a} - e^{-2x_a}}{1 - e^{-x_a} + e^{-2x_a}}, \tag{A35}$$

which, together with (A29), implies

$$e^{-\Delta} \leq F, \tag{A36}$$

where

$$F \equiv \int_\Omega (\xi_1 - \epsilon_2\xi_2)(\xi_2 - \epsilon_3\xi_3) \cdots (\xi_{N-1} - \epsilon_N\xi_N)\xi_N \prod_{a=1}^{N} \xi_a^{-1} e^{-\xi_a} d\xi_a \tag{A37}$$

and the integration volume Ω extends over

$$\xi_2 \leq \epsilon_2^{-1}\xi_1, \quad \xi_3 \leq \epsilon_3^{-1}\xi_2, \ldots,$$
$$\xi_{N-1} \leq \epsilon_{N-1}^{-1}\xi_{N-2}, \quad \text{and} \quad 0 \leq \xi_N \leq \epsilon_N^{-1}\xi_{N-1}. \tag{A38}$$

It is useful to define

$$H(\epsilon_2, \epsilon_3, \ldots, \epsilon_N) \equiv \left[F - \prod_{i=2}^{N} f(\epsilon_i)\right] \prod_{j=2}^{N} \epsilon_j^{-1}, \tag{A39}$$

where

$$f(\epsilon) \equiv 1 + \epsilon \ln\frac{\epsilon}{1+\epsilon}. \tag{A40}$$

By a straightforward calculation, one can verify that

$$\frac{\partial}{\partial\epsilon^{-1}}[\epsilon^{-1}f(\epsilon)] = (1+\epsilon)^{-1}$$

and

$$\frac{\partial^{N-1}}{\prod \partial\epsilon_i^{-1}}\left(F\prod \epsilon_j^{-1}\right) = (1+\epsilon_2)^{-1}(1+\epsilon_3+\epsilon_3\epsilon_2)^{-1}(1+\epsilon_4+\epsilon_4\epsilon_3+\epsilon_4\epsilon_3\epsilon_2)^{-1} \cdots (1+\epsilon_N+\epsilon_N\epsilon_{N-1}+\cdots+\epsilon_N\epsilon_{N-1}\cdots\epsilon_2)^{-1},$$

where the subscripts i and j vary independently from 2 to N. Consequently,

$$D_{N-1}H \equiv \frac{\partial^{N-1}}{(\partial\epsilon_2^{-1})(\partial\epsilon_3^{-1})\cdots(\partial\epsilon_N^{-1})} H \leq 0 \tag{A41}$$

in the entire physical region $\epsilon_i \geq 0$. As $\epsilon_2^{-1} \to 0$, $f(\epsilon_2)$ and the integration volume (A38) approach zero; therefore,

$$D_{N-2}H \equiv \frac{\partial^{N-2}}{\partial(\epsilon_3^{-1})\partial(\epsilon_4^{-1})\cdots\partial(\epsilon_N^{-1})} H = 0 \quad \text{at } \epsilon_2^{-1} = 0,$$

which, together with (A41), implies

$$D_{N-2}H \leq 0$$

in the entire physical region. Similarly, one shows that

$$D_{N-3}H \equiv \frac{\partial^{N-3}}{\partial(\epsilon_4^{-1})\cdots\partial(\epsilon_N^{-1})} H = 0 \quad \text{at } \epsilon_3^{-1} = 0,$$

and therefore in the entire physical region

$$D_{N-3}H \leq 0.$$

By induction, one finds, for arbitrary $\epsilon_i \geq 0$,

$$H \leq 0, \tag{A42}$$

which, together with (A36) and (A39), leads to

$$e^{-\Delta} \leq F \leq \prod_{i=2}^{N} f(\epsilon_i) . \tag{A43}$$

Thus, we complete the proof for (A33) and (A34). [The fact that the factor $f(\epsilon_1)$ is absent from the above product is relevant only to the boundary condition at $i=1$; it has no effect on any of the thermodynamical properties that we are interested in.]

As noted in Sec. IV, the inequality (A34) implies (41):

$$p \leq p_1(z) , \tag{A44}$$

where $p_1(z)$ is given by (42).

(4) It may be of interest to compare the above inequality with one derived by Tiktopoulos and Treiman[14] for the multiperipheral model. We note that, on account of (20), the two-body potential V_2 satisfies

$$e^{-V_2} \leq \prod_i (1-e^{-x_i}) , \tag{A45}$$

which implies, via (A34),

$$e^{-U_N} \leq \prod_i (1-e^{-x_i}) . \tag{A46}$$

Now, let us consider a hypothetical case[15] in which the corresponding function e^{-U_N} is this upper bound $\prod(1-e^{-x_i})$. By using (40), one finds that the gas pressure p' of the hypothetical case is related to the fugacity z by

$$z^{-1} = \int_0^\infty (1-e^{-x}) e^{-p' x} dx ,$$

or

$$p'^2 + p' - z = 0 .$$

Thus, we obtain

$$p \leq p'(z) = \tfrac{1}{2}[(1+4z)^{1/2} - 1] . \tag{A47}$$

By using Table I one can also write (A47) in the form given by Tiktopoulos and Treiman,[14]

$$\alpha \leq -\tfrac{3}{2} + [\tfrac{1}{4} + (g/4\pi)^2]^{1/2} . \tag{A48}$$

Because of (A45), one has

$$p_1(z) \leq p'(z) ; \tag{A49}$$

therefore, the inequality (A44) is a better one than (A47). As $z \to \infty$, both inequalities reduce to (43): $p \leq z^{1/2}$. As $z \to 0$,

$$p_1(z) = z - 1.5626 z^2 + O(z^3) ,$$

while

$$p'(z) = z - z^2 + O(z^3) .$$

As expected, the second virial b_2 is correctly given by $p_1(z)$, but not by $p'(z)$.

APPENDIX B

In Sec. IV, the partition function Q_N is evaluated after the replacement

$$\delta\left(\sum_1^N x_i - L\right) \to \exp\left[-\beta\left(\sum_1^N x_i - L\right)\right] . \tag{B1}$$

It is well known that such a replacement leads to the correct thermodynamical limit; i.e., it gives the correct $O(N)$ term in $\ln Q_N$, as $N \to \infty$ at a fixed finite density N/L. In this appendix, we shall show how this method may be extended to derive the correction term, which will turn out to be $O(\ln N)$. Since, on account of (A11), $s^{-2} Q_N$ is proportional to the cross section σ_{N+1} in the multiperipheral model, this correction term is related to the question of $\ln s$ dependence of the cross section.

For simplicity, we shall consider only two-body forces. The total potential U_N is assumed to be

$$U_N = \sum_1^N v(x_i) . \tag{B2}$$

The partition function $Q_N(L)$ is given by (14):

$$Q_N(L) = \int e^{-U_N} \delta\left(\sum x_i - L\right) \prod dx_i , \tag{B3}$$

where, as before, each x_i is ≥ 0. From (B3), one obtains the recursion formula

$$Q_{N+1}(L) = \int_0^L e^{-v(x)} Q_N(L-x)\, dx . \tag{B4}$$

Theorem. For $N \gg 1$, but $(N/L) \sim O(1)$,

$$\ln Q_N(L) = N \ln h(\beta) + L\beta - \tfrac{1}{2} \ln N + O(1) , \tag{B5}$$

where

$$h(\beta) = \int_0^\infty e^{-v(x) - \beta x} dx \tag{B6}$$

and β is a function of L/N, determined by

$$-\frac{d \ln h(\beta)}{d\beta} = \frac{L}{N} . \tag{B7}$$

We note that the replacement (B1) leads directly to the first two terms on the right-hand side of (B5), both of which are $O(N)$. The theorem gives also the correction term $-\frac{1}{2}\ln N$. As we shall see by following the proof, one can systematically calculate the remaining $O(1)$ term as well.

Proof. Let us define α and γ to be the solutions of β at

$$-\left[\frac{d \ln h(\beta)}{d\beta}\right]_{\beta=\alpha} = \frac{L}{N+1} \tag{B8}$$

and

$$-\left[\frac{d \ln h(\beta)}{d\beta}\right]_{\beta=\gamma} = \frac{L-x}{N} \equiv \eta , \tag{B9}$$

and assume $Q_N(L)$ to be of the form

$$Q_N(L)=[h(\beta)]^N e^{\beta L} A_N\,, \tag{B10}$$

where A_N is to be determined. It is convenient to introduce the Fourier transform of $e^{-v(x)}$ in the physical region $x \geq 0$:

$$e^{-v(x)}=\int_{-\infty}^{\infty} C_\omega e^{i\omega x}\,d\omega\,. \tag{B11}$$

(The behavior of the Fourier integral in the unphysical region $x<0$ is immaterial to our discussion.) Equation (B4) can then be written as

$$[h(\alpha)]^{N+1} e^{\alpha L} A_{N+1}=NA_N\int_{-\infty}^{\infty} C_\omega e^{i\omega L}\,d\omega \int_0^{N^{-1}L}[H_\omega(\eta)]^N d\eta\,, \tag{B12}$$

where

$$H_\omega(\eta)=h(\gamma)\,e^{(\gamma-i\omega)\eta} \tag{B13}$$

and $\gamma=\gamma(\eta)$ is defined by (B9).

Let β be defined by (B7). It is straightforward to expand α around β:

$$\alpha=\beta+\left[\frac{d^2\ln h(\beta)}{d\beta^2}\right]^{-1}\frac{L}{N(N+1)}+\cdots.$$

The logarithm of the left-hand side of (B12) is then given by

$$(N+1)\ln h(\beta)+\beta L+\ln A_{N+1}-\frac{1}{2N}\left[\frac{d^2\ln h(\beta)}{d\beta^2}\right]^{-1}\left(\frac{L}{N}\right)^2+O\left(\frac{1}{N^2}\right). \tag{B14}$$

To evaluate the right-hand side, we use the relation

$$\int H_\omega{}^N d\eta=\frac{H_\omega{}^{N+1}}{N+1}\frac{d\eta}{dH_\omega}-\frac{H_\omega{}^{N+2}}{(N+1)(N+2)}\frac{d}{dH_\omega}\left(\frac{d\eta}{dH_\omega}\right)+\cdots. \tag{B15}$$

Since $e^{-v(x)}$ is regular at the origin, as $\beta\to\infty$, both $h(\beta)$ and $d\ln h(\beta)/d\beta$ approach zero. By using (B9), one sees that as $\eta\to 0$, $\gamma\to\infty$ and $H_\omega\to 0$. Let us consider the definite integral (B15) from $\eta=0$ to $\eta=N^{-1}L$. The right-hand side of (B15) is zero at the lower limit $\eta=0$; by substituting its value at the upper limit $\eta=N^{-1}L$ to the right-hand side of (B12), we find, e.g., the first term $(N+1)^{-1}H_\omega{}^{N+1}(d_\eta/dH_\omega)$ in (B15) leads to

$$\frac{NA_N}{N+1}[h(\beta)]^N e^{\beta L}\int_{-\infty}^{\infty}\frac{C_\omega d\omega}{\beta-i\omega}=\frac{NA_N}{N+1}[h(\beta)]^{N+1}e^{\beta L}\,. \tag{B16}$$

By carrying out the same operation for the remaining term, one can verify that the logarithm of the right-hand side of (B12) is

$$(N+1)\ln h(\beta)+\beta L+\ln A_N-\frac{1}{2Nh(\beta)}\left[\frac{d^2\ln h(\beta)}{d\beta^2}\right]^{-1}\frac{d^2h(\beta)}{d\beta^2}+O\left(\frac{1}{N^2}\right). \tag{B17}$$

By using (B14) = (B17), we obtain, for $N\gg 1$ and neglecting the $O(N^{-2})$ terms,

$$\frac{d\ln A_N}{dN}=-\frac{1}{2N}$$

or

$$A_N=N^{-1/2}\times\text{constant}\,, \tag{B18}$$

which completes the proof.

Remarks. In the sum for the grand partition function $\mathcal{Q}=\sum z^N Q_N$, the maximal value of $z^N Q_N$ occurs at $N=\overline{N}$, determined by

$$\frac{\partial \ln Q_N}{\partial N}+\ln z=0 \quad\text{at } N=\overline{N}. \tag{B19}$$

Therefore, as $L\to\infty$ at a constant $\overline{N}/L$, (B19) reduces to (40):

$$z^{-1}=\int_0^{\infty} e^{-v(x)-px}\,dx\,,$$

where

$$p=\beta \text{ evaluated at } L^{-1}N=L^{-1}\overline{N}\,. \tag{B20}$$

In terms of the maximal value, one may write

$$z^N Q_N \cong z^{\overline{N}} Q_{\overline{N}}\exp[-(N-\overline{N})^2/\Lambda^2]\,, \tag{B21}$$

where Λ^2 denotes the fluctuation. Thus,

$$\mathcal{Q}\cong\int z^N Q_N\,dN=z^{\overline{N}}Q_{\overline{N}}\Lambda\pi^{1/2}$$

or, since Λ^2 is proportional to $\overline{N}$,

$$\ln\mathcal{Q}=\overline{N}\ln z+\ln Q_{\overline{N}}+\tfrac{1}{2}\ln\overline{N}+O(1)\,. \tag{B22}$$

By using the theorem and (B20), one finds

$$\ln \mathcal{Q} = pL + O(1). \tag{B23}$$

The $-\frac{1}{2}\ln N$ term in (B5) is canceled by the $+\frac{1}{2}\ln N$ term in (B22). From Table I, one sees that the total cross section σ_{tot} is proportional to $s^{\alpha-1}$. [Without the $-\frac{1}{2}\ln N$ term in (B5), one would obtain an incorrect multiplicative factor ln$s$ for σ_{tot}.]

APPENDIX C

To illustrate how the method developed in Sec. IV can be generalized to include four-body, five-body, ... forces, let us assume the total potential energy U_N to be

$$U_N = V_2 + V_3 + V_4, \tag{C1}$$

where $V_2 + V_3$ is given by (45), and

$$V_4 = \sum_i v_4(x_i, x_{i+1}, x_{i+2}). \tag{C2}$$

We define the matrix $\langle x_i, x_{i+1} | H | x_{i+2}, x_{i+3}\rangle$ by

$$\langle x_i, x_{i+1} | H | x_{i+2}, x_{i+3}\rangle \equiv \exp(-\phi_0 - \phi_1), \tag{C3}$$

where

$$\phi_0 = \tfrac{1}{3}(x_i + 2x_{i+1} + 2x_{i+2} + x_{i+3})p + \tfrac{1}{3}[v_2(x_i) + 2v_2(x_{i+1}) + 2v_2(x_{i+2}) + v_2(x_{i+3})] \tag{C4}$$

and

$$\phi_1 = \tfrac{1}{2}[v_3(x_i, x_{i+1}) + 2v_3(x_{i+1}, x_{i+2}) + v_3(x_{i+2}, x_{i+3})] + v_4(x_i, x_{i+1}, x_{i+2}) + v_4(x_{i+1}, x_{i+2}, x_{i+3}). \tag{C5}$$

The integral in (36) then becomes

$$\int \exp\left(-U_N - p\sum_1^N x_i\right) \prod dx_i = \mathrm{Tr} H^{N/2}, \tag{C6}$$

where any power of the matrix H is defined according to the usual rule:

$$\langle x, y | H^m | s, t\rangle \equiv \int_0^\infty du \int_0^\infty dv \langle x, y | H^{m-1} | u, v\rangle \times \langle u, v | H | s, t\rangle.$$

To evaluate the trace of $H^{N/2}$, we may separate

$$H = H_0 + \epsilon,$$

where the matrix element of H_0 is equal to $\exp(-\phi_0)$ and that of ϵ is equal to $[\exp(-\phi_1) - 1] \times \exp(-\phi_0)$. With only minor changes, the discussion following Eq. (53) in Sec. IV can be directly applied to the present case. Similarly, one can extend the method to five-body, six-body, ... forces.

---

*Research supported in part by the U. S. Atomic Energy Commission.

[1]D. Amati, S. Fubini, and A. Stanghellini, Nuovo Cimento 26, 896 (1962); G. F. Chew and A. P. Pignotti, Phys. Rev. 176, 2112 (1968).

[2]B. W. Lee and R. F. Sawyer, Phys. Rev. 127, 2266 (1962); J. C. Polkinghorne, J. Math. Phys. 4, 503 (1963).

[3]R. P. Feynman (unpublished); K. Wilson, Cornell Report No. CLNS-131, 1970 (unpublished).

[4]A. H. Mueller, Phys. Rev. D 4, 150 (1971); M. Bander, Phys. Rev. D 6, 164 (1972); J. D. Bjorken, SLAC Report No. SLAC-PUB-974 (unpublished).

[5]D. K. Campbell and S.-J. Chang, Phys. Rev. D 4, 1151 (1971); S.-J. Chang, T.-M. Yan, and Y.-P. Yao, *ibid.* 4, 3012 (1971); S.-S. Shei and T.-M. Yan, *ibid.* 6, 1744 (1972); D. K. Campbell, *ibid.* 6, 2658 (1972).

[6]R. C. Arnold, Phys. Rev. D 5, 1724 (1972); R. C. Arnold and J. Steinhoff, ANL Report No. ANL/HEP 7219 (unpublished). For further related work, see other references quoted in these papers.

[7]Usually, a thermodynamical system has three extensive variables: the number of particles N, the length (or volume) L and the energy E; corresponding to these three extensive variables there are three intensive variables: the fugacity z, the pressure p, and the temperature T. In the present case there are only two extensive variables: N and $L \sim \ln(s/m^2)$; correspondingly, there are also only two intensive variables: z and p. Hence, kT can be chosen to be unity.

[8]The permutation due to different ordering of these particles is canceled by the usual $[(N-1)!]^{-1}$ factor due to Boltzmann statistics. [If the length L has ends, then the Boltzmann factor is $(N!)^{-1}$.]

[9]See, e.g., J. E. Mayer and M. G. Mayer, *Statistical Mechanics* (Wiley, New York, 1940).

[10]See Table 27.8 in *Handbook of Mathematical Functions*, Appl. Math. Ser., Vol. 55, edited by M. Abramowitz and I. A. Stegun (National Bureau of Standards, Washington, D. C., 1964).

[11]I wish to thank M. Y. Chen for kindly checking the numerical integration for b_3.

[12]See T. L. Trueman and T. Yao, Phys. Rev. 132, 2741 (1963); and Chang, Yan, and Yao, Ref. 5.

[13]That such a replacement gives the correct thermodynamical limit, as L and N approach infinity, is well known. In Appendix B, we shall show how this method can be extended to obtain also the correction term, when L and N are large but not infinite.

[14]The upper bounds (38) and (A48) have been derived by G. Tiktopoulos and S. B. Treiman [Phys. Rev. 135, B711 (1964)]. For a still better upper bound, see (41) and (42). [See also G. Tiktopoulos and S. B. Treiman, Phys. Rev. 137, B1597 (1965) for discussions on lower bounds for models involving zero-mass particles.]

[15]If there were only one space dimension, instead of three, in the ϕ^3-multiperipheral model, then the corresponding two-body potential would be $1 - e^{-x}$. This expression follows directly from the result of Campbell and Chang, (Ref. 5); it can also be derived by using arguments similar to those given in Appendix A.

Abnormal nuclear states and vacuum excitation*†

T. D. Lee

Physics Department, Columbia University, New York, New York 10027

We examine the theoretical possibility that at high densities there may exist a new type of nuclear state in which the nucleon mass is either zero or nearly zero. The related phenomenon of vacuum excitation is also discussed.

CONTENTS

I. INTRODUCTION

In this talk, I would like to discuss some of my recent theoretical speculations, made in collaboration with G. C. Wick. As you shall see, these speculations suggest the possible existence of some rather interesting physical objects, hitherto unobserved.[1] An effective way to search for these new objects is through the use of high-energy heavy ions, which is the subject matter of this meeting.

To begin with, we assume the existence of a strongly interacting neutral spin 0 even parity meson field $\phi(x)$. Such a field may simply be a phenomenological description of a composite 0+ state of other particles, say $\pi\pi$, or $K\bar{K}$.[2] Through the transformation $\phi(x) \rightarrow \phi(x) +$ constant, we can always choose for the normal vacuum state

$$\langle \text{vac} \mid \phi(x) \mid \text{vac} \rangle = 0 \quad \text{everywhere.} \tag{1}$$

The state $\mid\,\rangle$ that we are interested in is an excited state; it has an abnormal expectation value of $\phi(x)$ in a relatively large volume Ω:

$$\langle \mid \phi(x) \mid \rangle \begin{cases} = \text{constant} \neq 0 & \text{inside } \Omega \\ = 0 & \text{outside } \Omega \\ \text{has rapid variation near the surface of } \Omega. \end{cases} \tag{2}$$

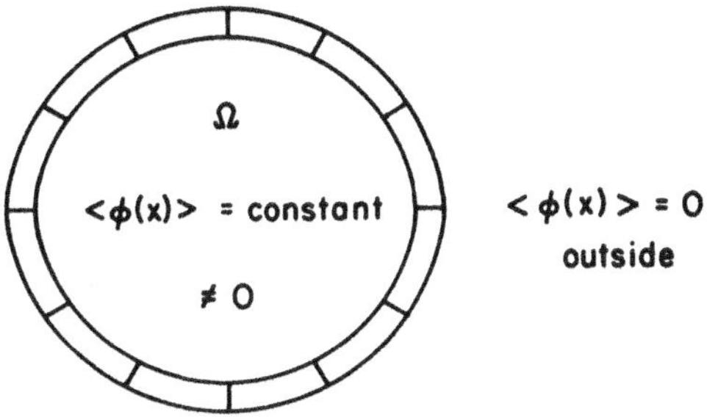

FIG. 1. A pictorial representation of an excited state in which $\langle \phi(x) \rangle$ differs from its vacuum value inside a macroscopic volume Ω.

Pictorially, we may visualize the expectation value $\langle \phi(x) \rangle$ in such an excited state as represented by Fig. 1. The linear dimension of Ω is assumed to be much larger than the usual microscopic length in particle physics (say, $\Omega^{1/3}$ is $\sim$, or $>$, 10^{-13} cm). Thus, much of the description of the field becomes almost classical.

In the following, we shall discuss two different circumstances:

(i) Ω is filled with nuclear matter, and

(ii) Ω does not contain any nuclear matter.

As we shall see, case (i) may lead to abnormal nuclear states and case (ii) to pure vacuum excitation states.

II. ABNORMAL NUCLEAR STATES

We first consider case (i). For definiteness, the Lagrangian density is assumed to be

$$\mathcal{L} = -\tfrac{1}{2}(\partial\phi/\partial x_\mu)^2 - U(\phi) - \psi^\dagger\gamma_4[\gamma_\mu(\partial/\partial x_\mu) + (m_N + g\phi)]\psi, \tag{3}$$

where

$$U(\phi) = \tfrac{1}{2}m_\phi^2\phi^2 + \cdots, \tag{4}$$

m_ϕ is the 0+ meson mass, m_N is the nucleon mass, and ψ is the nucleon field. In (4) the precise form of "$\cdots$" depends on the theory; for a renormalizable theory, it contains cubic and quartic powers of ϕ.

Let the volume Ω of the state $\mid\,\rangle$, given by Eq. (2), be simply that of a super-heavy nucleus. Outside the nucleus, $\langle \phi(x) \rangle = 0$ and the nucleon mass is m_N. But inside the

---

* Based on an invited talk given at the "Annual Bevatron Users Meeting", Lawrence Berkeley Laboratory, January 19, 1974.

† This research was supported in part by the U.S. Atomic Energy Commission.

[1] For some earlier speculations on related subjects, cf. E. Feenberg and H. Primakoff, 1946, Phys. Rev. **70**, 980; A. R. Bodmer, 1971, Phys. Rev. D **4**, 1601; A. B. Migdal, 1972, Zh. Eksp. Teor. Fiz. **63**, 1993; Y. Ne'eman, 1972, in *Physics of Dense Matter* I. A. U. Symposium, Boulder, Colorado.

[2] From a theoretical point of view, one knows that at least in the low and intermediate energy region, the chiral $SU_2 \times SU_2$ symmetry is a reasonably good approximation, as supported by the Adler–Weisberger relation, the Goldberger–Treiman relation, and the various soft-pion relations. The chiral symmetry leads naturally to either $\pi\pi$ correlations or a 0+ field, such as in the σ-model.

nucleus, the "effective" nuclear mass is $m_{\rm eff}$, determined by

$$m_{\rm eff}^2 = [m_N + g\langle\phi\rangle]^2, \tag{5}$$

which may be quite different from m_N. In particular, if we assume

$$\langle\phi(x)\rangle \cong -m_N/g \quad \text{inside } \Omega, \tag{6}$$

then the "effective" nucleon mass inside the nucleus would be $\cong 0$. The energy difference Δ between such an "abnormal" state and the normal state may be estimated approximately

$$\Delta \sim -Nm_N + U(-m_N/g)\Omega + \text{surface energy} \tag{7}$$

where N is the total number of nucleons.[3] For a sufficiently heavy nucleus, the surface energy may be neglected. Since the negative term in Eq. (7) is proportional to N, while the positive term is proportional to the volume Ω, the energy difference Δ in this simple system becomes negative if the nucleon density $n \equiv N/\Omega$ is sufficiently high. To have a rough idea of the order of magnitude, we may take $U(\phi) \sim \frac{1}{2}m_\phi^2\phi^2$; the abnormal state becomes the lower energy state if the nucleon density n is greater (or much greater) than a critical value n_c where

$$n_c \sim m_\phi^2 m_N/2g^2. \tag{8}$$

If we assume $m_\phi \sim m_N$ and its coupling g to be of the same order as the $\pi - N$ coupling $(4\pi)^{-1}g^2 \sim 15$, then at the critical density, the internucleon distance is

$$n_c^{-1/3} \sim 1.5 \times 10^{-13} \text{ cm} \tag{9}$$

which is of the same order as the distance between nucleons in the existing nuclei.[4]

Of course, the above estimation, Eq. (8) is quite crude, since it neglects the nuclear forces, the relativistic motion of nucleons, etc. While a complete analysis is difficult, some simple model calculations can be readily made.

III. σ-MODEL

For definiteness, let us assume the well-known σ-model for the meson field.[5] In this model, besides the usual isovector 0− pion field $\boldsymbol{\pi}(x)$ there is also an isoscalar 0+ field $\sigma(x)$. In terms of the customary notation of $\sigma(x)$ used in the literature, the aforementioned field $\phi(x)$ is given by

$$\phi = \sigma - (m_N/g), \tag{10}$$

and its mass m_ϕ = σ-meson mass m_σ. In the normal vacuum, we adopt the convention $\langle \text{vac} \mid \phi(x) \mid \text{vac}\rangle = 0$ as before; consequently,

$$\sigma_0 \equiv \langle \text{vac} \mid \sigma(x) \mid \text{vac}\rangle = (m_N/g). \tag{11}$$

Since $\boldsymbol{\pi}(x)$ is a pseudoscalar field, we have

$$\langle \text{vac} \mid \boldsymbol{\pi}(x) \mid \text{vac}\rangle = 0. \tag{12}$$

In the σ-model, the potential energy density U_σ is given by

$$U_\sigma = \tfrac{1}{4}\lambda^2(\sigma^2 + \boldsymbol{\pi}^2 - \mu^2/\lambda^2)^2 - (m_\pi^2\sigma_0)\sigma, \tag{13}$$

where σ_0 is given by Eq. (11), and the constants λ and μ are related to the σ-mass m_σ and the π-mass m_π by

$$2\mu^2 = m_\sigma^2 - 3m_\pi^2 \quad \text{and} \quad 2\lambda^2\sigma_0^2 = m_\sigma^2 - m_\pi^2. \tag{14}$$

If one neglects m_π, the σ-model is symmetric under the chiral $SU_2 \times SU_2$ transformation. So far as the meson fields are concerned, this chiral transformation is the same as the four-dimensional orthogonal transformation between $\sigma(x)$ and $\boldsymbol{\pi}(x)$. Consequently, the σ-nucleon coupling is equal to the π-nucleon coupling. One has

$$g^2/4\pi \cong 15. \tag{15}$$

Thus, in the σ-model there is *only one unknown parameter* m_σ.

Let us apply the σ-model to the problem of abnormal states in a large nucleus, considered in the previous section. The effective nuclear mass is now given by

$$m_{\rm eff} = g(\langle\sigma\rangle^2 + \langle\boldsymbol{\pi}\rangle^2)^{1/2}, \tag{16}$$

where $\langle\sigma\rangle$ and $\langle\boldsymbol{\pi}\rangle$ are the expectation values of σ and $\boldsymbol{\pi}$ inside the nucleus, and both are assumed to be constants.[6] For simplicity, let us assume the nucleons to be described by a degenerate Fermi distribution with a top Fermi momentum k_F. In the simple case of an equal number of protons and neutrons, k_F is related to the nucleon density n by

$$k_F = (3\pi^2 n/2)^{1/3}. \tag{17}$$

The kinetic energy density of nucleons is given by

$$U_N = (2/\pi^2)\int_0^{k_F} k^2(k^2 + m_{\rm eff}^2)^{1/2}\, dk. \tag{18}$$

In addition, there is the usual short-range nuclear interaction. As a first model calculation, we shall assume that because of the short-range interaction (especially if the repulsive force is particularly strong) the nuclear matter resembles an incompressible fluid. Therefore, we may keep the nucleon density *n fixed*; the energy density of the system is then assumed to be given by

$$\mathcal{E} \equiv U_\sigma + U_N - nm_N \tag{19}$$

plus an additive constant that may depend on the fixed parameter n.[7]

---

[3] For simplicity, we neglect the relativistic correction here; a detailed calculation is given in the next section.

[4] Here, as well as in the following, the critical density n_c depends only on the ratio m_ϕ^2/g^2.

[5] For references on the σ-model, see B. W. Lee, 1972, *Chiral Dynamics*.

[6] As we shall see, in order to minimize energy, $\langle\boldsymbol{\pi}\rangle = 0$, and therefore $m_{\rm eff}^2 = g^2\langle\sigma\rangle^2$ which reduces to Eq. (5) because of Eq. (10).

[7] Later, in a more realistic model-calculation, the short-range repulsive interaction will be considered explicitly. See the discussion given in this section after Eq. (23) and in the Appendix.

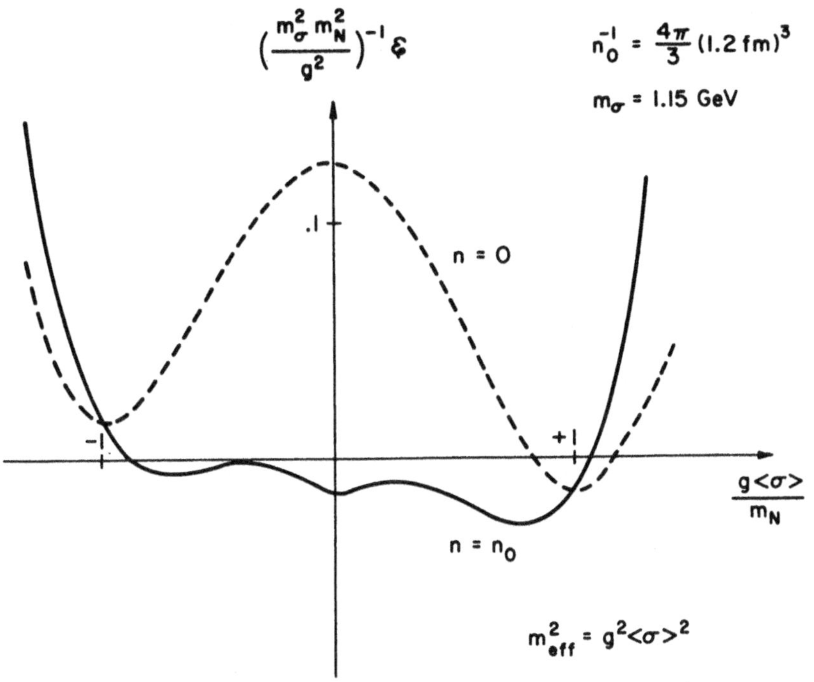

FIG. 2. Energy density ϵ in the σ model vs $g\langle\sigma\rangle/m_N$.

We shall assume the volume Ω of the nucleus to be sufficiently large so that the surface energy can be neglected. The solution that we are interested in is one in which $\langle\sigma\rangle$ and $\langle\boldsymbol{\pi}\rangle$ are constants inside the nucleus. Because of $m_\pi \neq 0$, from Eq. (13) it follows that in the minimum energy state the expectation value $\langle\boldsymbol{\pi}\rangle = 0$; therefore

$$m_{eff}^2 = g^2\langle\sigma\rangle^2. \tag{20}$$

At a fixed nucleon density n, we may plot the energy density $\mathcal{E}$ against $g\langle\sigma\rangle/m_N$ (see Fig. 2).

In this figure m_σ is assumed to be 1.15 GeV.[8] The dashed line is for zero nucleon density $n = 0$, and the solid line is for n equal to the density in the existing heavy nuclei. The slight asymmetry between $\langle\sigma\rangle$ positive and $\langle\sigma\rangle$ negative is due to $m_\pi \neq 0$. One sees that at zero nucleon density, the minimum of $\mathcal{E}$ is at $m_{eff} = m_N$. As n increases, the minimum slowly decreases from $m_{eff} = m_N$ to $m_{eff} \sim .8m_N$ at $n = [(4\pi/3)(1.2\ \mathrm{fm})^3]^{-1}$. The point $m_{eff}/m_N \cong 0$ is a local maximum of $\mathcal{E}$ at $n = 0$, but it becomes a local minimum of $\mathcal{E}$ when n increases to near the density of the existing heavy nuclei.

In the above figure, the unit of energy density is $(m_\sigma m_N/g)^2$. At the density $n = [(4\pi/3)(1.2\ \mathrm{fm})^3]^{-1}$, this unit energy density corresponds to $\sim$5.2 GeV/nucleon; the energy difference between the normal state ($m_{eff} \cong 0.8\ m_N$) and the abnormal state ($m_{eff} \cong 0$) is about 80 MeV/nucleon. There is a local maximum of $\mathcal{E}$ at about $m_{eff} \cong .25m_N$. The energy difference between this local maximum and the abnormal state is about 15 MeV/nucleon which is quite a sizable potential barrier. Thus, the abnormal state $m_{eff} \cong 0$ is a metastable state, if n is constrained at the fixed value $[(4\pi/3)(1.2\ \mathrm{fm}^3)]^{-1}$. At the same fixed value of nuclear density, in order to reach the abnormal state from the normal state, one has to pass through a potential barrier of maximum height $\sim 15 + 80 = 95$ MeV/nucleon.

If the nucleon density can be further increased to exceed a critical value n_c,[9] then the abnormal state becomes the absolute minimum state. Consequently for the minimum energy state, the value m_{eff}/m_N makes a discontinuous jump at $n = n_c$ (see Fig. 3). In the abnormal state ($n > n_c$), the effective nucleon mass m_{eff} is almost zero, but not exactly zero; this is due to $m_\pi \neq 0$. In the zero pion-mass limit, $m_{eff} = 0$ in the abnormal state.

The conditions for the production of such abnormal nuclear states are

1. $N^{1/3} \gg 1$, so that the surface energy can be neglected, and

[8] As we shall see, in the hard-sphere model calculation given in the Appendix, this value for m_σ gives about the correct binding for the normal nuclear state. See T. D. Lee and M. Margulies, Phys. Rev. D (in press) for results in which different values of m_σ and g are assumed.

[9] Here n_c is $\sim 1.16 \times [(4\pi/3)(1.2\ \mathrm{fm})^3]^{-1}$ for $m_\sigma = 1.15$ GeV, and $(4\pi)^{-1}g^2 = 15$

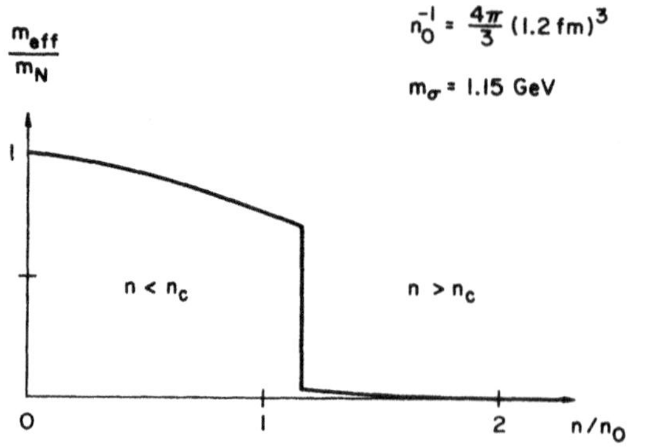

FIG. 3. A discontinuous transition in the effective mass m_{eff} between the normal state ($n < n_c$) and the abnormal state ($n > n_c$).

2. $n > n_c$. The precise value of n_c depends on m_σ and g; neither is really known with any degree of certainty. For different values of m_σ and g, the critical density n_c varies as m_σ^2/g^2, or more explicitly

$$n_c \cong 12(m_\sigma/m_N)^2(4\pi/g^2)n_0, \tag{21}$$

where

$$n_0^{-1} = (4\pi/3)(1.2\ \mathrm{fm})^3. \tag{22}$$

Thus, one expects that by using *high energy collisions between heavy ions* (say Pb on Pb at $\sim\frac{1}{2}$ GeV per nucleon in the center-of-mass system), one may double the nucleon density, and thereby produce such abnormal states, provided that m_σ is not too much higher than 1 GeV.

So far we have discussed only the minimum energy state at a fixed nucleon density n. Next (still assuming $N^{1/3} \gg 1$), we would like to examine the optimum nucleon density n, and inquire whether the minimum energy state is a normal state $n < n_c$, or an abnormal state $n > n_c$?[10] In order to answer this question, one must leave the "incompressible fluid" model. The real physics problem is tied closely to the strength and the range of the short-range repulsive force between nucleons.[11]

For definiteness, let us assume that the attraction between nucleons is provided by the long-range interaction through $\langle\sigma\rangle$, and the short-range repulsion between two nucleons (proton or neutron) is of the simple form

$$u(r) = \begin{cases} u_0 & \text{for } r < a, \\ 0 & \text{for } r > a, \end{cases} \tag{23}$$

where r is the relative distance between the two nucleons. We shall consider in the following, two extreme cases (i) $\frac{1}{2}$ GeV $\gg u_0 \gg$ 20 MeV and (ii) $u_0 = \infty$. The correct physical situation probably lies somewhere in between.

[10] The condition $N^{1/3} \gg 1$ excludes practically all existing nuclei.

[11] For the abnormal state, since one deals with a large-scale energy change $\sim O$ (GeV) per nucleon, the usual short-range attractive force does not play any important role.

(i) Soft-core interaction

We assume Eq. (23) with $\frac{1}{2}$ GeV $\gg u_0 \gg$ 20 MeV for the short-range interaction. Thus, so far as the normal nuclear states are concerned, the repulsive interaction resembles the hard-sphere interaction with a = diameter of the hard sphere. However, in the abnormal state, since one deals with energy change $\sim O$ (1 GeV) per nucleon, as a zeroth approximation this repulsive interaction can be neglected. As shown in the Appendix, in order for the normal nuclear state to have a nucleon density $n = [(4\pi/3)(1.2\ \mathrm{fm})^3]^{-1}$ and a (volume) binding energy $\sim$16 MeV per nucleon,[12] one finds

$$a \cong 0.56\ \mathrm{fm} \quad \text{and} \quad m_\sigma \cong 1.15\ \mathrm{GeV}.^{13} \tag{24}$$

For the abnormal state, one has $m_{eff} \cong 0$ and therefore the total energy E is given by

$$E/N = \tfrac{3}{4}k_F + m_\sigma^2 m_N^2/8ng^2 + \text{surface energy} \tag{25}$$

where, as before, N is the total number of nucleons and n is the nucleon density, related to the top Fermi momentum k_F by Eq. (17). In Eq. (25), the first term $\frac{3}{4}k_F$ is the average kinetic energy per nucleon, and the second term $m_\sigma^2 m_N^2/8ng^2$ is simply the value of U_σ/N at $\sigma = \pi = 0$ where U_σ is given by Eq. (13). For simplicity, we have set $m_\pi = 0$ in Eq. (25) and also neglect the soft-core repulsion on account of $u_0 \ll \frac{1}{2}$ GeV. The minimum of Eq. (25) is at $(\partial/\partial n)(E/N) = 0$, which implies that, after neglecting the surface energy,

$$\tfrac{1}{4}k_F - m_\sigma^2 m_N^2/8ng^2 = 0; \tag{26}$$

this gives for $g^2/4\pi = 15$

$$(k_F/m_N) \cong 0.45(m_\sigma/m_N)^{1/2}. \tag{27}$$

From Eqs. (25) and (26) one sees that

$$E/N = k_F, \tag{28}$$

and therefore the binding energy per nucleon is $(m_N - k_F)$. If one assumes $(4\pi)^{-1}g^2 \cong 15$ and $m_\sigma \cong 1.15$ GeV, as given by Eq. (24), then one finds for the abnormal state

binding energy $\cong$ 475 MeV/nucleon

and

$$n^{-1} \cong (4\pi/3)(0.65\ \mathrm{fm})^3. \tag{29}$$

Since the short-range repulsion is completely neglected in this calculation, the above large binding energy can only be regarded as an *upper bound* of the actual value.

(ii) Hard-core interaction

Next, we consider the other extreme case that the short-range interaction is given by Eq. (23) with $u_0 = +\infty$. The description of the normal nuclear state is the same as that

[12] See, e.g., A. Bohr and B. R. Mottelson, 1969, *Nuclear Structure*, Vol. 1, p. 142.

[13] One notes that if the hard sphere has a radius $\sim(\omega\text{-mass})^{-1}$, then its diameter a is $\sim$.5 fm.

in the previous case of a soft-core repulsive interaction, and therefore Eq. (24) holds. As shown in the Appendix, the binding energy and the nucleon density of the abnormal state can be calculated approximately. For a hard-sphere interaction with a diameter $\cong 0.56$ fm, $g^2/4\pi \cong 15$ and $m_\sigma \cong 1.15$ GeV, one finds for the abnormal state

binding energy $\cong$ 130 MeV/nucleon,

and

$$n^{-1} \cong (4\pi/3)(0.91 \text{ fm})^3. \quad (30)$$

In either case, if $N^{1/3}$ is sufficiently large so that one may neglect the surface energy, then the abnormal state is stable; it is the lowest energy state with a substantially larger binding energy per nucleon than that in the normal state.

Remarks

1. The above calculations are approximations serving only to illustrate the general features of the abnormal states. Within such approximation, one may ask: suppose that m_σ is >1.15 GeV; would the abnormal state remain the lowest energy state? If one assumes a soft-core repulsion, then from Eqs. (27) and (28), one sees that for $(4\pi)^{-1}g^2 = 15$, the abnormal state is stable provided that m_σ is less than $\sim 4.9 m_N$, and $N^{1/3} \gg 1$. In the case of hard-core repulsion, for $(4\pi)^{-1}g^2 = 15$, the abnormal state is stable only if $m_\sigma \lesssim 1.5 m_N$; the corresponding range for metastable abnormal states is, of course, much wider. Since only the ratio m_σ^2/g^2 enters into the calculation, the above range in m_σ also gives the corresponding latitude in g^2.

2. So far, we have assumed the short-range repulsive interaction to be an isoscalar. There is, in addition, an isovector part due to, e.g., ρ exchange. This isovector part makes the short-range repulsion between nn and pp larger than that between np. This difference in repulsion plus the role of statistical weight lead one to expect that in the abnormal state the average number Z of protons is $\sim \frac{1}{2}N$, so as to minimize the total energy.[14]

3. Keeping $Z \sim \frac{1}{2}N$, one finds the Coulomb energy to be $\sim(1/10)m_N$ when $N \sim 10^4$. Thus, for $N \gg 10^4$, the abnormal state is most likely unstable.[15] When N is of astronomical size, the abnormal state may again be stable because of gravitation.

4. The general feature of the "abnormal nuclear states" is not sensitive to the details of the σ-model; it depends only on the existence of a strongly interacting 0+ resonance, whose long wavelength aspect may be represented by a field. Since we are only interested in the long wavelength limit, the microscopic details of the 0+ resonance should not be important. This is analogous to the phenomena of Bose-Einstein condensation and superfluidity. He^4 is a composite composed only of Fermions p, n and e^-, yet it exhibits superfluidity and undergoes Bose-Einstein condensation. Through the B.C.S. pair-correlation, the electrons in metals exhibit similar phase transitions that give rise to superconductivity. Recently, it has been observed that even He^3 exhibits superfluidity. Likewise, the abnormal state that we are interested in is a similar *condensed phase* of the long wavelength limit of a 0+ field.

5. As remarked earlier, in order to produce such abnormal states it is best to collide heavy ions on heavy ions at high energy, so as to maximize both $N^{1/3}$ and n. If such abnormal states do exist, one may observe in the final state a *stable, or metastable, nucleus of very large baryon number*, say $N \gtrsim 400$. Both the binding energy and the radius of these abnormal states are quite different from the usual extrapolations derived from the normal states. The nucleons in the abnormal states behave like zero-mass particles, and that should produce rather distinct physical characteristics in the dynamics of these abnormal states (e.g., in its interaction with the electromagnetic field).

6. It may be of interest to examine a strongly interacting 0+ field theory different from the σ-model. As an extreme example, one may assume the potential $U(\phi)$ in Eq. (3) to consist of only $\frac{1}{2}m_\phi^2\phi^2$ without any nonlinear interaction. In such a case, as the nuclear density n increases, the transition from the normal state $m_{\text{eff}} \sim m_N$ to the abnormal state $m_{\text{eff}} \sim 0$ becomes continuous, though rapid. Once away from the transition region, even in this extreme example, the overall characteristics of the nuclear state remain similar to those in the σ-model; e.g., as seen from Eq. (8), when n is $\gg m_\phi^2 m_N/2g^2$, the effective nucleon mass does become near zero and the nuclear state becomes abnormal, just as in the σ-model.

IV. PURE VACUUM EXCITATION

We now turn to our next topic; a pure vacuum excitation state. For definiteness, let us consider again a strongly interacting 0+ field ϕ, with a Lagrangian density given by Eq. (3). Like the σ model, the potential $U(\phi)$ is of the form (see Fig. 4). However, unlike in the σ-model, we assume the potential U to have a *second local minimum* at $\phi = \phi_{\text{vex}} \neq 0$, as shown in Fig. 4.[16]

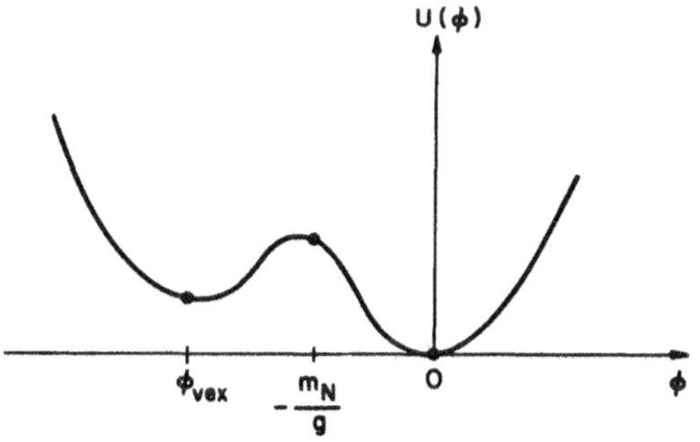

FIG. 4. An example of a potential function $U(\phi)$ with two minima.

[14] See also the discussion given at the end of the Appendix.

[15] When Coulomb energy becomes important, the abnormal state can create e^+e^- pairs; the e^+ will be sent to infinity, but the e^- will be kept within the abnormal nucleus. As Z increases, the number of e^- also increases. The interplay between the added Fermi energy of e^- and the Coulomb energy may eventually bring the abnormal state to the point of instability.

[16] In the σ-model, because of the π field, U_σ has only one local minimum at $\sigma \simeq (m_N/g)$. The point $\sigma \simeq -(m_N/g)$ is a saddle point, *not* a local minimum.

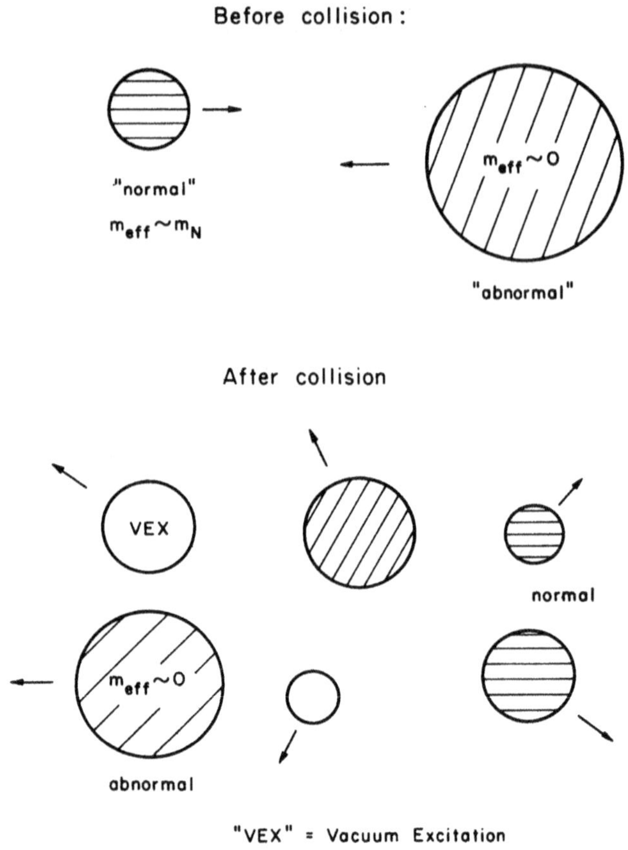

FIG. 5. Possible production of a vacuum excitation state from collisions between normal and abnormal nuclear states.

From our previous discussion, one sees that the condition for producing the "abnormal" nuclear state in the present case is identical to that in the σ model. Now, suppose that we have created such an "abnormal" nuclear state in which $m_{eff} \sim 0$, and therefore $\langle\phi\rangle \sim -m_N/g$. Let us consider the collision between an "abnormal" nucleus ($\langle\phi\rangle \sim -m_N/g$) and a normal nucleus ($\langle\phi\rangle \sim 0$) (see Fig. 5). After the collision, there will be various fragments, some normal and some abnormal. However, among the fragments, occasionally there can appear a new physical object called the "vacuum excitation" state. The vacuum excitation state occupies a volume Ω, and it carries a 4-momentum P_μ. Inside Ω, one has

$$\langle\phi\rangle = \phi_{vex} \tag{31}$$

except near the surface region. Outside Ω, one has the normal vacuum $\langle\phi\rangle = 0$. The rest mass M_{vex} of the vacuum excitation state is given by

$$M_{vex}^2 = -P_\mu^2 = [\Omega U(\phi_{vex}) + \text{surface energy}]^2. \tag{32}$$

The baryon number of the vacuum excitation is zero. Thus, if one wishes, one may view the vacuum excitation state as a gigantic meson "blob." The linear dimension $\Omega^{1/3}$ is assumed to be much greater than any of the usual microscopic lengths in particle physics.

Next, we examine the question of the width of the vacuum excitation state. It is important to differentiate two cases:

1. Degenerate case

We assume in this case $U(\phi)$ at $\phi = \phi_{vex}$ to be degenerate with that at $\phi = 0$; i.e.,

$$U(\phi_{vex}) = 0 \tag{33}$$

in which we have adopted the usual convention that for the normal vacuum $\phi = 0$ one sets $U(0) = 0$. Thus, the rest mass M_{vex} is entirely due to the surface energy. The vacuum excitation state can decay through meson emission by contracting its surface. From relativity, one knows that the lifetime τ is of the order

$$\tau \sim O(\Omega^{1/3}). \tag{34}$$

Thus, the ratio of the width to the mass is

$$(\text{width/mass})_{vex} \sim O(1/m_\phi{}^3\Omega) \ll 1 \tag{35}$$

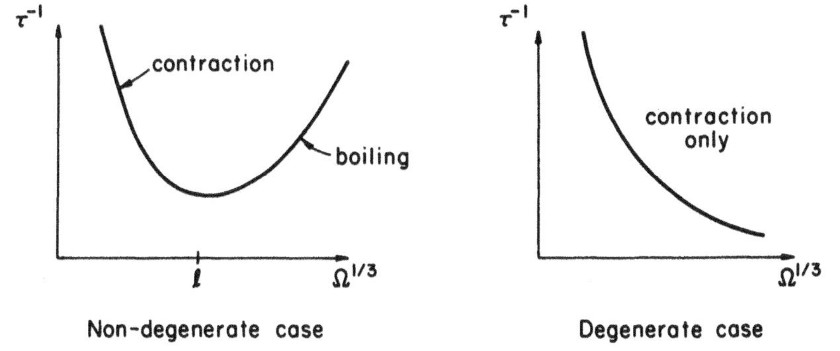

FIG. 6. Lifetime τ of a vacuum excitation state vs its linear dimension $\Omega^{1/3}$.

in which, as well as in the following, we use m_ϕ^{-1} as a typical microscopic length in the problem. From Eq. (35), one sees that the larger the volume Ω is, the sharper is the vacuum excitation state defined. There is no limit to the volume Ω; it can even be of astronomical size.

2. Non-degenerate case

In this case, we have instead of (33)

$$U(\phi_{\text{vac}}) > 0. \tag{36}$$

It turns out that for $\Omega^{1/3} \gg m_\phi^{-1}$, but $\ln(\Omega^{1/3}m_\phi)$ not too large, the vacuum excitation state decays via the same "contraction" mechanism as in the degenerate case, in which through surface contraction, mesons are emitted near the surface. However, when $\ln(\Omega^{1/3}m_\phi) \gg 1$, there is another decay mechanism, called "boiling" in which mesons are produced in the interior of the volume, on account of Eq. (36). In this case there is a limit to the volume Ω. We may plot schematically the width τ^{-1} versus the linear size $\Omega^{1/3}$ for these two cases (see Fig. 6).

In the non-degenerate case, the minimum width occurs at $\Omega^{1/3} = l$ where $\ln(lm) \gg 1$. In some typical examples, l is found to be $\sim O$ (mm). Thus, in both cases, the size of the vacuum excitation can reach macroscopic dimensions.[17]

V. REMARKS

1. In the literature, there have been extensive theoretical discussions of the "spontaneous symmetry breaking" mechanism and other related topics, such as Goldstone bosons, Higgs phenomenon, etc. In all these discussions, the interaction is assumed to be symmetric under a certain group of transformations (or nearly symmetric, as in the case of the σ model). The observed asymmetry is due to the specific expectation value of a certain spin 0 field $\phi(x)$ in the physical vacuum state. In such a theory, it is necessary that there must exist other states which are degenerate (or nearly degenerate) with the physical vacuum state. A natural question to ask is what happens to the other degenerate (or nearly degenerate) states in which $\langle\phi(x)\rangle$ is different from its normal vacuum value. It has been often argued that since we are dealing with an infinite system, only one vacuum state can be realized in our world, all the other states are unphysical, even if they are degenerate with the physical vacuum state.

This situation is rather similar to the example of Heisenberg's infinite ferromagnet. The existence of a ferromagnet clearly defines a specific direction in space, but it does not imply a violation of rotational symmetry of the physical law. If the ferromagnet is of infinite dimension, then it will be physically impossible to rotate all the spins of an infinite ferromagnet.

However, as is well known, there can be "domain structure." In the case of the ferromagnet, such a "domain" may be created by applying a local magnetic field over a relatively large region in space. Here, we may make the analog:

$$\langle\phi(x)\rangle \leftrightarrow \text{spin}$$

$$\psi^\dagger\gamma_4\psi \leftrightarrow \text{magnetic field.}$$

Since nuclear matter interacts linearly with $\phi(x)$, by having a sufficient amount of nuclear density over a large volume in space, we may hope to create a similar "domain structure" with respect to the vacuum state.

2. The question whether we live in a "medium" or in a "vacuum" dates back to the beginning of physics. From relativity, we know that the "vacuum" must be Lorentz-invariant, but that does not mean the "vacuum" is simple. From Dirac's hole theory, one has learned that the vacuum, though Lorentz-invariant, can be rather complicated. However, so long as all of its properties cannot be changed, so long as, e.g., the value of vacuum polarization cannot be modified, then it is purely a question of semantics whether the vacuum should be called a medium or not.

What we try to suggest is that if we do indeed live in a medium, then there should be ways through which we may change the properties of that medium.

Hitherto, in high-energy physics we have concentrated on experiments in which we distribute a higher and higher amount of energy into a region with smaller and smaller

[17] For further details and the question of quantum fluctuations, see T. D. Lee and G. C. Wick, 1974, Phys. Rev. D 9, 2291.

dimensions. In order to study the question of "vacuum," we must turn to a different direction; we should investigate some "bulk" phenomena by distributing high energy over a relatively large volume. *The fact that this direction has never been explored should*, by itself, *serve as an incentive for doing such experiments*. As we have discussed, there are possibilities that abnormal states may be created, in which the nucleon mass may be very different from its normal value. It is conceivable that inside the volume of the abnormal state, some of the symmetry properties may become changed, or even that the usual roles of strong and weak interactions may become altered. If indeed the properties of the "vacuum" can be transformed, we may eventually be led to some even more striking consequences than those that have been discussed in this lecture.

APPENDIX: HARD-SPHERE GAS MODEL

In this appendix, we simply approximate the short-range nuclear force by the hard-sphere interaction of diameter a. The attraction between nucleons is assumed to be provided by the long-range interaction through the expectation value of the σ field. The result should give us at least a qualitative understanding of the abnormal state. For self-consistency of the model, the parameters a and m_σ are to be determined by fitting the known properties of normal nuclear states.

A. Normal nuclear states[18]

In the normal state, we may consider the nucleons to be nonrelativistic; furthermore, the nonlinear aspect of the σ field may be neglected, since σ does not deviate too much from its vacuum expectation value σ_0. The energy per nucleon is given by

$$E/N = m_{\text{eff}} + (1/2m_{\text{eff}})\tfrac{3}{5}P_F^2 + \tfrac{1}{2}m_\sigma^2(\sigma - \sigma_0)^2(\Omega/N) \tag{A1}$$

where, for simplicity, we have neglected the surface energy,

$$\sigma_0 = m_N/g, \qquad m_{\text{eff}} = g\sigma, \tag{A2}$$

Ω is the nuclear volume

$$\Omega = (4\pi/3)r^3N, \tag{A3}$$

and P_F is the top Fermi momentum for the hard sphere gas, related to the nuclear density $n = N/\Omega$ by

$$P_F = [(3\pi^2/2)n]^{1/3}[r/(r - 0.8a)]. \tag{A4}$$

The above Fermi momentum differs from that of a free gas[19] by a factor $r/(r - 0.8a)$, showing that the effective nuclear radius available to the hard spheres is smaller than r by a factor $1 - 0.8(a/r)$. The coefficient 0.8 is chosen, so that for a dilute system of hard spheres, the first-order energy correction in $(n^{1/3}a)$ agrees with the *exact* calculation; the second order energy correction turns out to be too large, about 1.5 times the exact value, which implies that this "Van der Waals type" approximation perhaps overestimates the repulsive energy.[20]

By setting

$$(\partial/\partial\sigma)(E/N) = 0 \quad \text{and} \quad (\partial/\partial r)(E/N) = 0,$$

we derive

$$m_N(1 - m_{\text{eff}}^{-1}T) = (m_N - m_{\text{eff}})^{-1}2m_Nu_\sigma \tag{A5}$$

and

$$T = \tfrac{5}{4}u_\sigma[1 - 0.8(a/r)], \tag{A6}$$

where T is the average kinetic energy per nucleon, and u_σ is the average σ-field energy per nucleon;

$$T = 3P_F^2/10m_{\text{eff}} \tag{A7}$$

and

$$u_\sigma = \tfrac{1}{2}m_\sigma^2(\sigma - \sigma_0)^2\Omega/N. \tag{A8}$$

The binding energy (b.e.) per nucleon is given by

$$\text{b.e.} = m_N - N^{-1}E. \tag{A9}$$

If we set b.e. $\cong$ 16 MeV, $r \cong 1.2$ fm, and $(4\pi)^{-1}g^2 = 15$, then we obtain

$$a \sim .56 \text{ fm}, \qquad m_{\text{eff}} \sim .85m_N \tag{A10}$$

and

$$m_\sigma \sim 1.15 \text{ GeV}. \tag{A11}$$

The corresponding value of the average kinetic energy T is $\sim$60 MeV per nucleon and the average field energy $u_\sigma \sim 65$ MeV per nucleon, which together with $m_{\text{eff}} - m_N \sim -141$ MeV lead to the 16 MeV (bulk) binding energy per nucleon.[21]

B. Abnormal nuclear states

In the abnormal state, one has $\sigma = 0$ and $m_{\text{eff}} = 0$. For the hard-sphere interaction, upon neglecting the surface energy and setting $m_\pi = 0$, one has, instead of Eq. (A1),

$$E/N = \tfrac{3}{4}P_F + m_N^2m_\sigma^2/8g^2n, \tag{A12}$$

where P_F is given by (A4) and $n = N/\Omega = [(4\pi/3)r^3]^{-1}$.[22] By setting

$$(\partial/\partial r)(E/N) = 0,$$

[18] I wish to thank R. Serber for discussions. For further details, see T. D. Lee and M. Margulies, Phys. Rev. D (in press). Cf., also S. A. Chin and J. D. Walecka, 1974, Phys. Lett. 52B, 24 for a related but independent calculation of the normal nuclear states. I wish to thank Dirk Walecka for sending me a copy of his paper before publication.

[19] Compare Eq. (17).

[20] See, for example, p. 256 of Bohr and Mottelson (1969).

[21] Compare pp. 142 and 245 of Bohr and Mottelson (1969).

[22] If one has the soft-core repulsion, then P_F is replaced by p_F, and (A11) becomes (25).

one finds for the abnormal state

$$\text{b.e.} \sim 130 \text{ MeV/nucleon} \quad \text{(A13)}$$

and

$$r \sim .91 \text{ fm}. \quad \text{(A14)}$$

The corresponding value of $\frac{3}{4}P_F$ is ~490 MeV and the field energy per nucleon is ~320 MeV; together, they lead to a binding energy of about 130 MeV per nucleon, provided that the surface energy can be neglected.

If one is dealing with a pure neutron system ($Z = 0$), then instead of Eq. (A4), P_F is given by

$$P_F = (3\pi^2 n)^{1/3}[r/(r - 0.8a)]. \quad \text{(A15)}$$

Because of the isovector part of the repulsive force (e.g., due to ρ exchange), one expects the hard-core diameter between neutrons to be greater than that given by Eq. (A10). As an illustration, we may assume $a > .6$ fm for a pure neutron system; then by using Eq. (A12), we find the minimum of (E/N) for the corresponding abnormal state to be greater than $m_N + 25$ MeV; consequently, the abnormal state for a pure neutron system has a *higher* energy than the (unbound) normal state.

REFERENCES

Bodmer, A. R., 1971, Phys. Rev. D **4**, 1601.
Bohr, A. and B. R. Mottelson, 1969, *Nuclear Structure* (Benjamin, New York), Vol. 1, p. 142.
Chin, S. A., and J. D. Walecka, Phys. Lett. 1974, Phys. Lett. **52B**, 24.
Feenberg, E., and H. Primakoff, 1946, Phys. Rev. **70**, 980.
Lee, B. W., 1972 *Chiral Dynamics* (Gordon and Breach, New York).
Lee, T. D., and G. C. Wick, 1974, Phys. Rev. D **9**, 2291.
Lee, T. D. and M. Margulies, 1974, Columbia University preprint CO-2271-33 (Phys. Rev. D, in press).
Migdal, A. B., 1972, Zh. Eksp. Teor. Fiz. **63**, 1993.
Ne'eman, Y., 1972, in *Physics of Dense Matter*, International Astronomical Union Symposium No. 53, August, 1972, Boulder, Colorado, edited by C. J. Hansen (Reidel, Boston).

A Possible New Form of Matter at High Density*

T.D. Lee
Columbia University, New York, New York 10027

I. INTRODUCTION

In this talk, I would like to discuss some of my recent theoretical speculations, made in collaboration with Gian-Carlo Wick. Over the past year, we have tried to investigate the structure of the vacuum. It is through this investigation that the possibilities of vacuum excitation states and abnormal nuclear states have been suggested. Before coming to the main topic, whether or not there may be the possibility of a new form of matter at high density, perhaps I should first digress on questions related to the vacuum.

In physics, one defines the vacuum as the lowest energy state of the system. By definition, it has zero 4-momentum. In most quantum field-theoretic treatments, quite often the vacuum state is used only to enable us to perform the mathematical construct of a Hilbert space. From the vacuum state, we build the one-particle state, then the two-particle state, . . . ; hopefully, the resulting Hilbert space will eventually resemble our universe. From this approach, different vacuum state means different Hilbert space, and therefore different universe.

Nevertheless, one may ask: What is this vacuum state? Does it have complicated structure? If so, can a part of this structure be changed? Ever since the formulation of relativity, after the downfall of the classical aether concept, one learns that the vacuum is Lorentz invariant. At least, one knows that just running around and changing the reference system won't alter the vacuum. However, Lorentz invariance alone does not insure that the vacuum is necessarily simple. For example, the vacuum can be as complicated as the product or sum of any scalar field or other scalar object at the zero 4-momentum limit:

$$\text{vacuum} \sim \phi^n \quad \text{or} \quad (\bar{\psi}\psi)^m \quad \text{at} \quad k_\mu = 0\,. \tag{1}$$

From Dirac's hole theory, one knows that the vacuum, though Lorentz-invariant, can be rather complicated. That this complicated structure of the vacuum may in part be changeable is suggested by the large variety of broken symmetries, found especially over the past two decades.

If we consider symmetry quantum numbers such as the isospin **I**, the strangeness S, the parity P, . . . , we find

$$\frac{d}{dt}\begin{Bmatrix} \mathbf{I} \\ S \\ P \\ C \\ CP \\ \vdots \end{Bmatrix}_{\text{matter}} \neq 0\,. \tag{2}$$

*This research was supported in part by the U.S. Atomic Energy Commission.

Aesthetically, this may appear disturbing. Why should nature abandon perfect symmetry? Physically, this also seems mysterious. What happens to these missing quantum numbers? Where do they go to? Can it be that matter alone does not form a closed system? If we also include the vacuum, then perhaps symmetry may be restored:

$$\frac{d}{dt}\begin{Bmatrix} \mathrm{I} \\ S \\ P \\ C \\ CP \\ \vdots \end{Bmatrix}_{\text{matter}+\text{vacuum}} = 0 . \tag{3}$$

This is, of course, the basic idea underlying the general heading "Spontaneous Symmetry Breaking." On the other hand, unless we have other links connecting matter with the vacuum, how can we be sure that this idea is right, and not merely a tautology?

A way out of this dilemma is to realize that the restriction $k_\mu=0$ for the vacuum state is only a mathematical idealization. After all, the universe has a finite radius, and k_μ is never strictly zero. So far as the microscopic system of particle physics is concerned, there is little difference between $k_\mu=0$ and k_μ nearly 0; the latter corresponds to a state that varies only very slowly over a large space-time extension, say of dimension

$$L \gg (\text{microscopic length}) . \tag{4}$$

The typical microscopic length may be ~ 1 fm for a nuclear system, and smaller for a subnuclear system. Thus, it seems worth while to search for solutions in a quantum field theory which vary only over a very large space-time volume. As we shall see, excited states of the vacuum may be viewed as examples of such solutions characterized by macroscopic scales.

With these remarks in mind, we now turn to the question of how to generate the vacuum excitation states.

II. VACUUM EXCITATION

According to (1), the vacuum is $\sim\phi^n$ or $(\bar{\psi}\psi)^m$ at the zero momentum limit. Its excitation can be viewed simply as a change in the expectation value of the corresponding operator ϕ^n or $(\bar{\psi}\psi)^m$ over a large volume. The simplest phenomenological description is to adopt the language of a scalar local field ϕ. The vacuum state is the lowest energy state in such a local field theory. Through the transformation $\phi\rightarrow\phi+$ constant, one may always assume as a convention

$$\langle \text{vac} \,|\, \phi \,|\, \text{vac} \rangle = 0 . \tag{5}$$

A vacuum excitation is represented by an excited state in which the expectation value

$$\bar{\phi} \equiv \langle \phi(x) \rangle \cong \text{constant} \neq 0 \qquad \text{inside } \Omega$$
$$= 0 \qquad \text{outside } \Omega \tag{6}$$

where the linear dimension $\Omega^{1/3}$ is $\gg$ microscopic distance.

How can we produce such changes in $\langle \phi \rangle$? If produced, how can we detect them? The problem is analogous to the formation of "domain structures" in a ferromagnet. We may draw the analog (see Figure 1):

$$\langle \phi(x) \rangle \leftrightarrow \text{spin}\,,$$
$$J = \text{matter source} \leftrightarrow \text{magnetic field}\,. \tag{7}$$

In the case of a very large ferromagnet, because its spin interacts linearly with the magnetic field, a domain structure can be created by applying an external magnetic field over a large volume. Furthermore, after domains are created, one may remove the external field; depending on the long-range forces, the surface energy, and other factors, such domain structure may persist even after the external magnetic field is removed. Similarly, by applying over a large volume any matter source J which has a linear interaction with $\phi(x)$, one may hope to create a domain structure in $\langle \phi(x) \rangle$. Depending on the dynamical theory, such domains may also remain as physical realities, even after the matter source J is removed. It is important to note that high temperature tends to destroy domain structures. Therefore, throughout this talk, I will concentrate only on zero temperature or low temperature phenomena.

As an illustration, we may consider a local scalar field theory. The Lagrangian density is

$$\mathcal{L}_\phi = -\frac{1}{2}\left(\frac{\partial \phi}{\partial x_\mu}\right) - U(\phi)\,. \tag{8}$$

If the theory is renormalizable, then U is a fourth-order polynomial in ϕ:

$$U(\phi) = \frac{a\phi^2}{2} + \frac{b\phi^3}{3!} + \frac{c\phi^4}{4!}\,. \tag{9}$$

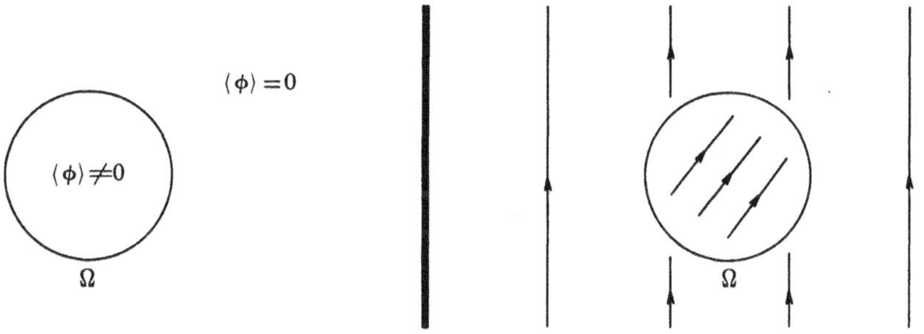

Figure 1.

Since we are interested in the long wavelength limit of the field, the scalar field $\phi(x)$ is used only as a phenomenological description. It can be any kind of $0+$ resonance between particles. The details of its microscopic structure do not concern us, nor is the renormalizability an important factor. All we are interested in is that its zero 4-momentum limit exists; once such a limit exists, then the expectation value $\langle \phi \rangle$ becomes automatically and inextricably connected to the description of the vacuum and the question of vacuum excitations.

In (9), we require the constant $c>0$ so that U has a lower bound, and $a \geqslant \frac{1}{3} b^2/c > 0$ so that the absolute minimum of U is at $\phi = 0$, in accordance with our convention (5). Since we are interested only in a slowly varying field over a large volume, we may neglect both the surface energy and the energy due to $(\partial\phi/\partial x_\mu)^2$. Therefore, the classical description of $\phi(x)$ should be a reasonably good approximation. The energy density of the field is then simply $U(\phi)$.

A. Constant Source

Let us now introduce an external source $J(x)$. As a first example, we consider the simplest case that J is a constant inside a large volume Ω, but zero outside. The energy of the system becomes

$$[U(\phi) + J\phi]\Omega\,.$$

The graphs in Figure 2 illustrate how in the lowest energy state, inside Ω

$$\bar{\phi} = \langle \phi(x) \rangle \tag{10}$$

can be changed under the influence of a constant J.

In case (ii), as J increases, there is a critical value at which $\bar{\phi}$ makes a sudden jump. Once a domain of an abnormal value of $\bar{\phi}$ is created, if J is subsequently removed, in case (i), $\bar{\phi}$ will return to 0; but in case (ii), depending on the magnitude of the abnormal value produced, even if J is removed $\bar{\phi}$ may not return immediately to 0.

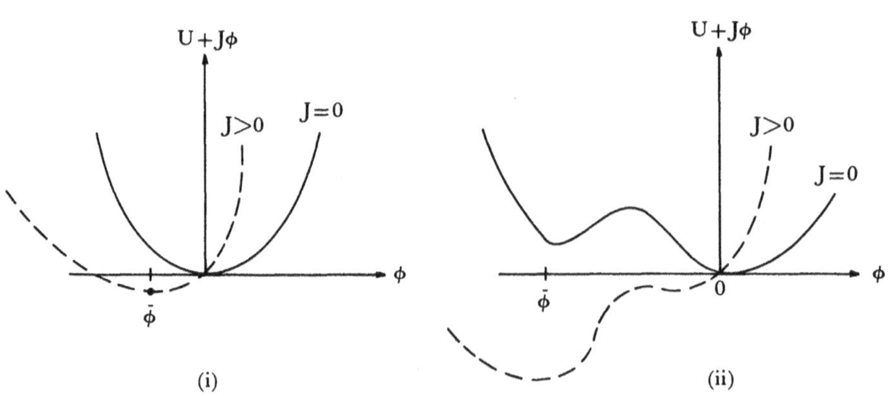

Figure 2.

B. Matter Source

Let us consider the more realistic case that the source is not a constant, but consists of, say, some physical fermions, represented by a Dirac field ψ. (Of course, if one wishes, one can also use bosons instead of fermions.) For our later discussions on abnormal nuclear states, ψ is the nucleon field; for other purposes such as the "bag model," ψ can be the quark field. The Lagrangian density becomes

$$\mathcal{L}=\mathcal{L}_\phi-\psi^\dagger\gamma_4[\gamma_\mu(\partial/\partial x_\mu)+(m+g\phi)]\psi \qquad (11)$$

where $\mathcal{L}_\phi$ is given by (8), m is the mass of the free fermion, and g is the coupling constant. An important feature is that if $\bar{\phi}$ is a constant $\neq 0$ over a large volume Ω, then inside Ω the effective mass of the fermion becomes

$$m_{\text{eff}}=m+g\bar{\phi}\,. \qquad (12)$$

Thus, by measuring m_{eff}, one can detect whether there is any change in $\bar{\phi}$. In the above Lagrangian, so far as the fermion field is concerned, there is an equivalence between

$$m\rightarrow m+\delta m \quad \text{and} \quad \phi\rightarrow\phi-(\delta m/g)\,. \qquad (13)$$

Up to a point, this transformation resembles both the gauge invariance in electrodynamics

$$\psi\rightarrow e^{i\theta}\psi\,, \qquad A_\mu\rightarrow A_\mu-\frac{1}{e}\frac{\partial\theta}{\partial x_\mu}\,, \qquad (14)$$

and the equivalence between gravitation and acceleration in general relativity. However, in both electromagnetism and gravitation, because of the zero mass of photon and graviton we have exact symmetries. Here, because there is no spin 0 particle of zero mass, transformation (13) is not an exact symmetry in the physical world. (It would, of course, be disastrous if it were, since the inertia mass would then become a non-observable.) As we shall see, this transformation can be used to produce changes in m and ϕ. Thereby, it may lead to a new form of matter at high density.

III. ABNORMAL NUCLEAR STATES

Let the volume Ω be filled with nucleons of density n,

$$n\equiv\Omega^{-1}\int\langle\psi^\dagger\psi\rangle\, d^3r\,. \qquad (15)$$

As we shall see, if the nucleon density n is sufficiently high, the system may exist in an "abnormal nuclear state," in which the effective nucleon mass becomes zero, or nearly zero (instead of the free nucleon mass m_N). The simplest way to see why such an abnormal state may develop is to examine the quasi-classical solution.

Let us assume that the nucleons form a Fermi gas of a uniform density and ϕ is a classical field. In such a quasi-classical treatment, the lowest energy state is one in which ϕ is a constant and the Fermi gas is completely degenerate. The corresponding energy density $\xi(n)$ is given by

$$\xi(n) = U(\phi) + \frac{2}{\pi^2} \int_0^{k_F} (k^2 + m_{\rm eff}^2)^{1/2} k^2 dk \tag{16}$$

where $m_{\rm eff}$ is the effective mass of the nucleon, related to $m_N \cong 940$ MeV by

$$m_{\rm eff} = m_N + g\phi\,, \tag{17}$$

and k_F is the Fermi momentum given by

$$k_F = (3\pi^2 n/2)^{1/3}\,. \tag{18}$$

For simplicity, we assume the nuclear matter to be composed of half protons and half neutrons. We shall also assume Ω to be sufficiently large that the surface energy can be neglected.

Throughout our discussion, we define the *normal nuclear state* to be one in which $m_{\rm eff} \cong m_N$ and the *abnormal nuclear state* to be one in which $m_{\rm eff} \cong 0$. As n increases, the Fermi-sea contribution to the energy becomes increasingly more important. Thus, independent of the detailed form of $U(\phi)$, in the high-density limit one finds

$$\lim_{n\to\infty} \phi = -(m_N/g) \quad \text{and} \quad \lim_{n\to\infty} m_{\rm eff} = 0\,; \tag{19}$$

i.e., the state becomes abnormal. On the other hand, in the low-density limit, because of (5), one must have

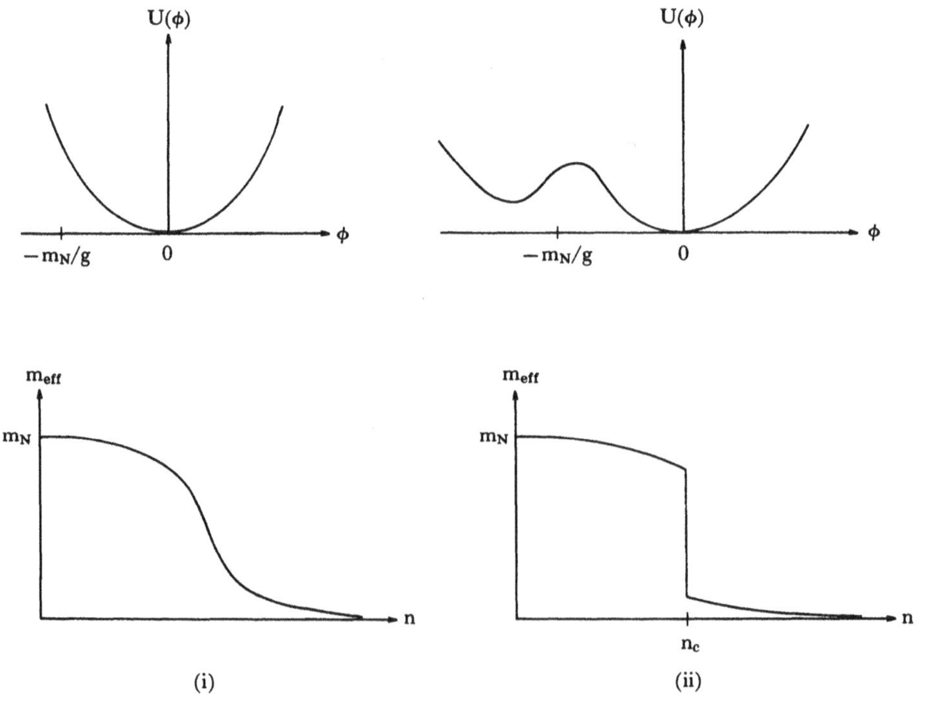

Figure 3.

$$\lim_{n\to 0} \phi = 0 \quad \text{and} \quad \lim_{n\to 0} m_{\text{eff}} = m_N ; \tag{20}$$

i.e., the state is normal. As illustrated in Figure 3, depending on the parameters in the theory, the transition from the low-density "normal" solution ($m_{\text{eff}} \cong m_N$) to the high-density "abnormal" solution ($m_{\text{eff}} \cong 0$) may or may not be a continuous one. The mechanism of the transition is similar to that in the case of a constant J discussed before. There is however one important difference: because the nucleon energy depends only on $(m_N + g\phi)^2$, its minimum occurs at $\phi = -m_N/g$. Hence, $\text{Lim}\ \phi = -m_N/g$ as $n \to \infty$, while in the case of a constant J one has $\text{Lim}\ \phi = -\infty$ as $J \to \infty$.

IV. σ-MODEL

A particularly interesting example of a discontinuous transition is the well-known σ-model. For a fairly wide range of the parameters, the critical density n_c for the transition is found to be approximately given by

$$n_c \cong 11.6 \left(\frac{m_\sigma}{m_N}\right)^2 \left(\frac{g^2}{4\pi}\right)^{-1} n_0 \tag{21}$$

where

$$n_0^{-1} = \tfrac{4}{3}\pi (1.2\,\text{fm})^3 ,$$

and m_σ is the σ-meson mass. In Figure 4, we map out the region in $g^2/4\pi$ and m_σ for $n_c \leqslant 2n_0$.

At present, there are no reliable data on either m_σ or g^2. If we identify σ to be the broad 2π resonance, then

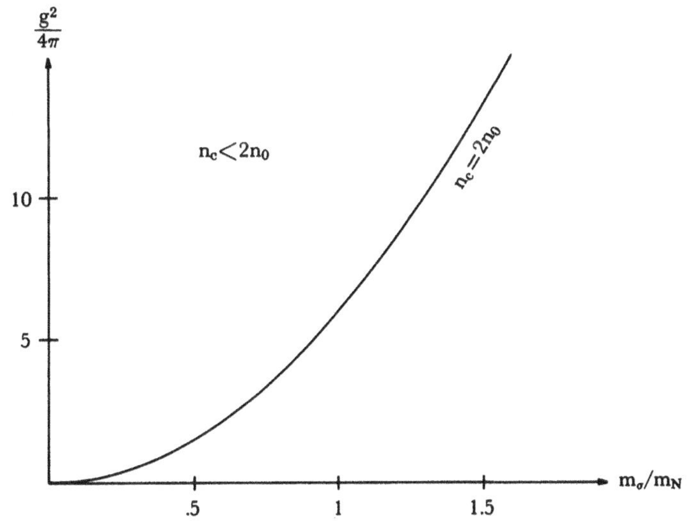

Figure 4.

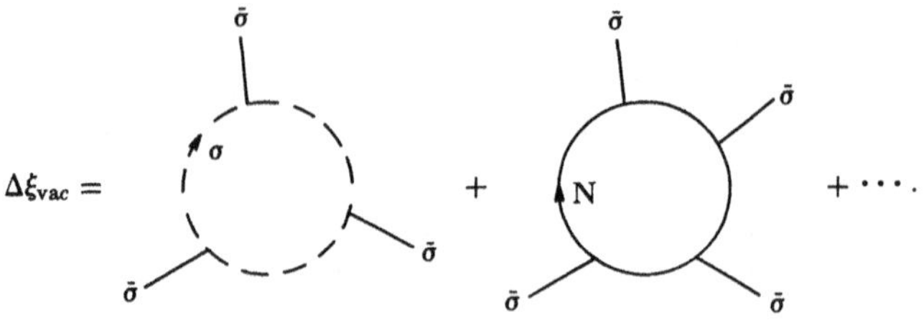

Figure 5.

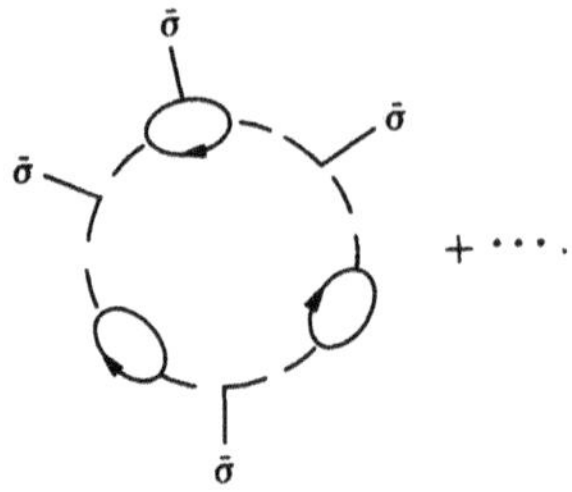

Figure 6.

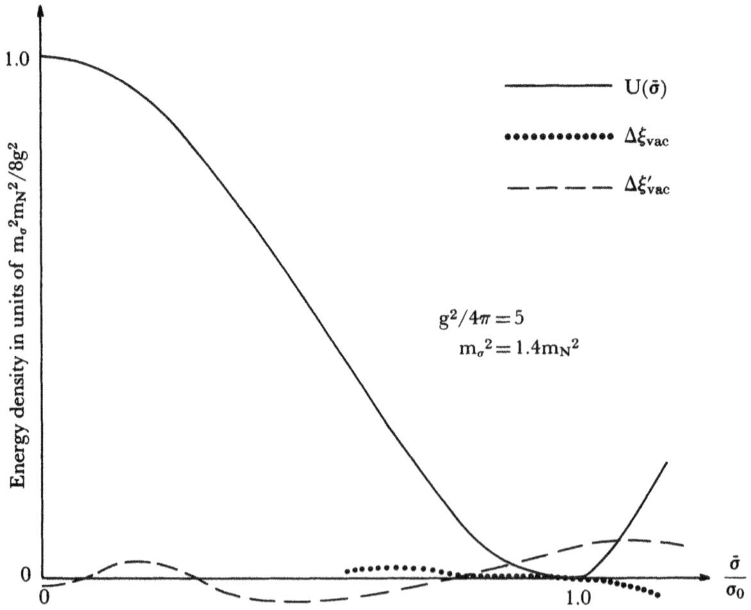

Figure 7.

$$m_\sigma \sim 700 \text{ MeV}. \tag{22}$$

Extensive calculations have been made in the literature to fit the central attractive part of the nuclear forces from such a 2π exchange. Depending on the assumption of repulsive forces (due to vector meson exchanges), different authors obtained different coupling constants for the $0+$ channel:

$$g^2/4\pi \text{ varies from 2.5 to 15.} \tag{23}$$

Even with such a wide range of estimated values, it seems possible that the physical reality may be located at the left of the curve $n_c = 2n_0$. Thus, by doubling the present nucleon density one may produce the abnormal nuclear state.

So far all our discussions are carried out only in the so-called tree approximation. One may wonder about the higher-order loop diagrams. This problem has been systematically investigated recently. All the one-loop and two-loop diagrams have been explicitly calculated; in addition, some of the three- and more-than-three-loop diagrams have also been evaluated by one of my students, Mr. M. Margulies. Figure 5 gives the result of all one-loop diagrams for $m_\sigma^2 = 1.4\ m_N^2$ and $g^2/4\pi = 5$. For the σ-model

$$U(\sigma) = \frac{m_\sigma^2 m_N^2}{8g^2}\left[1 - \left(\frac{\sigma}{\sigma_0}\right)^2\right]^2 \tag{24}$$

where $\sigma_0 = m_N/g$, and $\bar\sigma = \sigma_0$ in the vacuum state.

$\Delta\xi_{\text{vac}}$ is the sum of all one-loop diagrams (Figure 5); and $\Delta\xi'_{\text{vac}}$ includes also the diagram shown in Figure 6. At $\bar\sigma^2 < \frac{1}{3}\sigma_0^2$, $\Delta\xi_{\text{vac}}$ becomes complex; one must use $\Delta\xi'_{\text{vac}}$ instead (see Figure 7). Because the meson-loop diagrams and the nucleon-loop diagrams are always of opposite sign, large cancellations may occur, especially when $m_\sigma^2 \cong 1.4\ m_N^2$. This explains why even for $g^2/4\pi = 5$, $\Delta\xi_{\text{vac}}$ and $\Delta\xi'_{\text{vac}}$ (calculated for $n = 6n_0$) are both rather small.

V. PRODUCTION AND DETECTION

In order to produce the abnormal nuclear state, we must consider reactions in which (i) a large number of nucleons are involved, so that the surface energy can be neglected, and (ii) the nucleon density can be increased by a significant factor ~ 2. One is, therefore, led to considerations of high energy collisions between heavy ions, say

$$\text{U} + \text{U} \rightarrow Ab + \ldots$$

where Ab denotes the abnormal nuclear state. If Ab is produced, what are its characteristics? How can it be detected?

Because of the high density involved, the equation of state of the abnormal state depends sensitively on the short-range repulsive force. For definiteness, we may assume it to be a simple hard-sphere repulsion. Let d be the diameter of the hard spheres. In Figure 8, we plot the (volume) binding energy per nucleon,

$$\text{b.e.} \equiv m_N - (\xi/n)\,, \quad \text{versus} \quad d \equiv \text{hard-sphere diameter.}$$

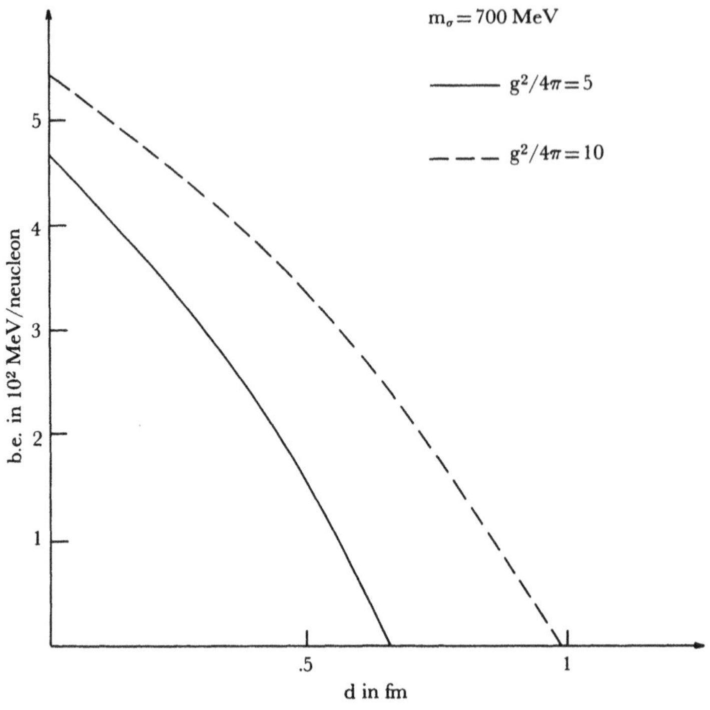

Figure 8.

As an example, we may assume $m_\sigma = 700$ MeV, $g^2/4\pi = 5$, and $d = 2/m_\omega \cong 0.5$ fm where $m_\omega = \omega^0$ meson mass. In this case, per nucleon the kinetic energy is ~490 MeV and the σ-field energy is ~300 MeV; therefore, one has a binding energy ~(940−490−300)~150 MeV/nucleon for the abnormal state. The nuclear radius is ~(0.9 fm)$A^{1/3}$. If one neglects both the surface energy and the Coulomb energy, the abnormal nuclear state is *stable*! (For a pure neutron system, the kinetic energy is much higher because of the exclusion principle; in addition, because of ρ^0-exchange, one expects the repulsive force between neutrons to be greater than that between neutron and proton. Consequently, most likely, the abnormal state of a pure neutron system is unstable.)

A clear signal of reaction (24) is the detection of a stable (or metastable) nucleus of very large baryon number A, say, ~400. For a charge Z~200, the Coulomb energy is not important; therefore, we expect $Z \sim \frac{1}{2}A$. Once an abnormal state is produced, through successive neutron absorption one may hope to increase its A and Z gradually. When Z increases, Coulomb energy becomes important; furthermore, the abnormal state can create e^+e^- pairs. The e^+ will be sent to infinity, but a fair fraction of the e^- will be kept within the abnormal nucleus. As Z increases, the number of e^- also increases. The interplay between the added Fermi energy of e^- and the Coulomb energy may eventually bring the abnormal state to the point of instability when A reaches $\lesssim 10^4$.

There remains the problem of the production mechanism of the abnormal state. Only a brief discussion will be given, since we are not able to evaluate the various complexities involved. Let us consider reaction (24) at $\sim \frac{1}{2}$ to 1 GeV/nucleon in the center-of-mass system.

(i) ***Penetration problem:*** Although the mean free path of a nucleon inside the nuclear matter is relatively short, because of the high energy, in most of the nucleon-nucleon collisions one expects the two final nucleons to move more or less along their initial general direction, with about 1 or 2 soft mesons created per nucleon-nucleon collision. Thus, provided the impact parameter is small, these two U nuclei should penetrate each other, and thereby increase the nuclear density from its normal value n_0 to $\sim 2n_0$. In addition, there is a large number of soft mesons produced which carry away a large part of the excess energy.

(ii) ***Response time:*** The collision time of (24) is about

$$\tau_{\text{coll}} \sim 10^{-12}\ \text{cm}/c$$

where c is the velocity of light (provided that the energy is not very high so that Lorentz contraction can be neglected). The order of magnitude of the response time of the meson field is about

$$\tau_{\text{resp}} \sim \left(\frac{h}{m_\phi c}\right) \Big/ c \sim 3 \times 10^{-14}\ \text{cm}/c\,.$$

This follows from the simple Klein-Gordon equation for a scalar field ϕ:

$$-\frac{\partial^2}{\partial x_\mu{}^2}\phi + m_\phi{}^2 \phi = -g n_S + \ldots$$

where ... denotes the nonlinear part of the meson-meson force and $n_S = \langle \psi^\dagger \gamma_4 \psi \rangle$.

Since

$$\tau_{\text{resp}} \ll \tau_{\text{coll}}\,,$$

as n_S changes, the value of the meson field responds almost adiabatically. When n_S reaches $\sim m_\phi{}^2 m_N/g$, the meson field becomes $\phi \sim -m_N/g$, which forms the abnormal state.

(iii) ***Potential barrier:*** After the two U nuclei penetrate each other, and ϕ develops the abnormal value, there is still the question: would the nucleons keep on moving away from each other, so that the whole system becomes separated? We note that as the meson field assumes the abnormal value $\phi \sim -(m_N/g)$, the effective nucleon mass becomes zero. The energy of each nucleon is $|\mathbf{p}|$ inside the nucleus, but $(\mathbf{p}^2 + m_N{}^2)^{1/2}$ outside. Thus, there is a potential barrier B holding the nucleons together in the abnormal state. So long as $|\mathbf{p}|$ is $\lesssim m_N$, most of the nucleons will collide with the barrier and remain inside. Through such collisions, additional mesons will be emitted, and the nucleons may gradually readjust their wave functions to those of the abnormal state.

While an accurate calculation of the production probability is difficult, from the above discussion one sees that, in order to produce Ab, the collision energy of the heavy ions should not be small; otherwise penetration becomes difficult. The collision energy should also be not too high; otherwise the collision time τ_{coll} may become too small because of Lorentz contraction, and the barrier B becomes $\sim \frac{1}{2} |\mathbf{p}|^{-1} m_p^2$, which may also be too low to hold the nucleons together. The best value may perhaps be $\gtrsim 100$ to several hundred MeV/nucleon in the center-of-mass system.

VI. REMARKS

The question whether we live in a "medium" or in a "vacuum" dates back to the beginning of physics. From relativity, we know that the "vacuum" must be Lorentz-invariant. As remarked before, Lorentz invariance by itself does not mean that the "vacuum" is simple. From Dirac's hole theory, one has learned that the vacuum, though Lorentz-invariant, can be rather complicated. However, so long as all of its properties cannot be changed, so long as, e.g., the value of vacuum polarization cannot be modified, then it is purely a question of semantics whether the vacuum should be called a medium or not.

What we try to suggest is that if we do indeed live in a medium, then there should be ways through which we may change the properties of that medium.

It may be worth while to emphasize once again how limited have been our experiences in either nuclear physics or particle physics. So far almost all our nuclear physics experiments have been restricted to nuclei at a constant density. We have never really ventured out to study nuclear physics at any densities other than the normal one. Likewise, in particle physics, we have a similar tradition in specializing along a fairly narrow direction. Take, for example, high energy physics. Hitherto, we have concentrated only on experiments in which we distribute a higher and higher amount of energy into a region with smaller and smaller dimensions. In order to study the question of "vacuum," or the possibility of the abnormal states, we must turn to a different direction; we should investigate some "bulk" phenomena by distributing high energy or high nucleon density over a relatively large volume. *The fact that such directions have never been explored should*, by itself, *serve as an incentive for doing such experiments*. As we have discussed, there are possibilities that abnormal states may be created, in which the nucleon mass may be very different from its normal value. It is conceivable that inside the volume of the abnormal state some of the symmetry properties may become changed, or even that the usual roles of strong and weak interactions may become altered. If indeed the properties of the "vacuum" can be transformed, we may eventually be led to some even more striking consequences than those that have been discussed in this lecture.

Reprinted from
Brookhaven National Laboratory Report 50445, 1–12 (1974)

Interaction of a dense fermion medium with a scalar-meson field*

T. D. Lee and M. Margulies
Columbia University, New York, New York 10027
(Received 2 December 1974)

The interaction between a dense fermion medium and a scalar-meson field is studied. It is shown that in the quasiclassical approximation, independently of the details of the theory, at low fermion density the lowest energy state is normal (i.e., effective fermion mass $m_{\text{eff}} \cong$ free mass), but at high density the state is abnormal (i.e., $m_{\text{eff}} \cong 0$). The nature of the transition is analyzed at zero and low temperatures. Our main concern is to examine the problem of quantum fluctuation. Both the one- and two-loop diagrams are calculated. By developing a variational formalism involving only two-line irreducible diagrams, we derive a suitable high-density expansion for the energy. The quasiclassical solution emerges as the lowest-order term in this expansion. Therefore, when the expansion is valid, the overall description of the transition given by the quasiclassical solution remains correct with the inclusion of quantum corrections.

I. INTRODUCTION

Recently, it has been suggested[1,2] that if there exists a strongly interacting 0^+ meson, or resonance, then when the nucleon density becomes sufficiently high over an extended volume, there is the possibility of abnormal nuclear states, hitherto unobserved. A basic feature of the abnormal nuclear state is that in such a state the nucleon mass becomes zero (or nearly zero). Under suitable conditions, the abnormal nuclear states may be stable, or metastable. Furthermore, depending on the particular meson theory, as the nucleon density increases, the transition from the normal nuclear state to the abnormal nuclear state can be a sudden one, exhibiting the standard characteristics of a phase transition. So far, because of computational difficulties, the theoretical discussion of the abnormal nuclear state is restricted to a quasiclassical treatment only, similar to that used in a Thomas-Fermi model for the atomic system. Questions naturally arise as to other effects not included in such a quasiclassical solution, especially those related to quantum fluctuations because of multiloop diagrams. The main purpose of this paper is to give a systematic analysis of these loop diagrams.[3]

In order to understand the essential features of the problem involved, we shall consider only a simple renormalizable field theory in which there are two fields: a boson field ϕ of spin 0 and a fermion field ψ of spin $\frac{1}{2}$. (For applications to the nuclear system, ψ represents the nucleon field and ϕ some strongly interacting 0^+ meson, or resonance.) The Lagrangian density is

$$\mathcal{L} = -\frac{1}{2}\left(\frac{\partial\phi}{\partial x_\mu}\right)^2 - U(\phi) - \psi^\dagger\gamma_4\left[\gamma_\mu\frac{\partial}{\partial x_\mu} + (m+g\phi)\right]\psi + \text{counterterms}, \qquad (1.1)$$

where ϕ is Hermitian,

$$U(\phi) = \tfrac{1}{2}a\phi^2 + (3!)^{-1}b\phi^3 + (4!)^{-1}c\phi^4, \qquad (1.2)$$

the parameters a, b, c, m, and g are the appropriately defined renormalized constants, and the counterterms are for renormalization purposes. Through the transformation $\phi \to \phi + \text{constant}$, one may always assume for the vacuum state

$$\langle \text{vac}|\phi(x)|\text{vac}\rangle = 0, \qquad (1.3)$$

where the vacuum state is defined to be the lowest eigenstate of the system with fermion number 0. The total fermion number $N \equiv \int \psi^\dagger\psi d^3r$ is, of course, conserved. Note that $U(\phi)$ does not contain a linear term in ϕ; consequently, there is a linear term in the counterterms. We assume the constant c to be positive so that $U(\phi)$ has a lower bound and the constants a and b to satisfy

$$b^2 \leq 3ac$$

so that the absolute minimum of $U(\phi)$ is at $\phi = 0$. For convenience, we shall quantize the system in a box of a finite volume Ω with the periodic boundary condition, and then take the limit $\Omega \to \infty$ in the end.

Let $|n\rangle$ be the *lowest energy eigenstate* of the total Hamiltonian H that satisfies the constraint

$$\Omega^{-1}\int d^3r\langle n|\psi^\dagger\psi|n\rangle \equiv n, \qquad (1.4)$$

where n denotes the average fermion density. A useful concept is the energy-density function $\mathcal{E}(n)$, related to the total Hamiltonian H by

$$\mathcal{E}(n) \equiv \lim_{\Omega\to\infty}\Omega^{-1}\langle n|H|n\rangle. \qquad (1.5)$$

For practical applications to a finite but large nucleon system, $\mathcal{E}(n)$ denotes its volume energy. The simplest way to see why, for n sufficiently high, the system may exist in the abnormal state is to examine the quasiclassical solution.

Let us assume the fermions to form a degenerate gas and ϕ to be described by a classical field. In such a quasiclassical treatment, we first examine the case that the fermion density n is kept uniform in Ω. In the lowest energy state, the classical field ϕ must, then, be a constant. The corresponding energy density $\mathcal{E}(n)$ is given by

$$\mathcal{E} = U(\phi) + (4\pi^3)^{-1}\kappa \int_F d^3p(\vec{p}^2 + m_{\text{eff}}^2)^{1/2}, \tag{1.6}$$

where m_{eff} is the effective mass of the fermion, related to the free particle mass m by

$$m_{\text{eff}} = |m + g\phi|, \tag{1.7}$$

the subscript F next to the integral sign indicates that the integral extends only over the Fermi sea $|\vec{p}| \leq k_F$, and k_F is related to n by

$$k_F = (3\pi^2 n/\kappa)^{1/3}, \tag{1.8}$$

with κ denoting the fermion degeneracy due to its internal symmetry. For example,

$$\kappa = \begin{cases} 1 & \text{for a neutron medium}, \\ 2 & \text{for a nuclear medium}, \end{cases} \tag{1.9}$$

where, for simplicity, we assume equal numbers of protons and neutrons for the nuclear medium.

Throughout our discussion, we define the *normal state* to be one in which $m_{\text{eff}} \cong m$, and the *abnormal state* to be one in which $m_{\text{eff}} \cong 0$. As n increases, the Fermi-sea contribution to the energy becomes increasingly more important. Thus, for any given set of parameters a, b, c, g, and m, in the high-density limit one finds by using (1.6),

$$\lim_{n\to\infty} \phi = -(m/g) \text{ and } \lim_{n\to\infty} m_{\text{eff}} = 0\,; \tag{1.10}$$

i.e., the state becomes abnormal. On the other hand, in the low-density limit, because of (1.3), one must have

$$\lim_{n\to 0} \phi = 0 \text{ and } \lim_{n\to 0} m_{\text{eff}} = m_N\,;$$

i.e., the state is normal. As illustrated in Fig. 1, depending on the parameters in the theory, the transition from the low-density "normal" solution ($m_{\text{eff}} \cong m$) to the high-density "abnormal" solution ($m_{\text{eff}} \cong 0$) may or may not be a continuous one. The details will be given in Sec. II. It will be shown there that one of the necessary conditions for a discontinuous transition is to have a sizable ϕ^3 coupling

$$b^2 > 2ac\,. \tag{1.11}$$

Of particular interest is the case of the σ model,[4] in which $b^2 = 3ac$. It has been shown in Refs. 1 and 2 that when the fermion density n increases, the quasiclassical solution of the σ model does produce a discontinuous transition from the normal state to the abnormal state. Further details concerning the nature of the phase transition will be given in Sec. II and in Appendix A.

Our main concern is to investigate the quantum correction to the quasiclassical solution. As will be shown, in a fully quantum-mechanical treatment, the role of the constant classical field in the above quasiclassical solution is replaced by the average value

$$\overline{\phi} = \Omega^{-1} \int d^3r \langle n|\phi(x)|n\rangle\,. \tag{1.12}$$

In the vacuum state ($n = 0$), one has $\overline{\phi} = 0$; when $n \neq 0$, in the normal state, one has $\overline{\phi} \cong 0$, but in the abnormal state $\overline{\phi}$ is shifted to $\cong -(m/g)$. This shift in $\overline{\phi}$ is smallest in the strong-coupling limit, but largest in the weak-coupling limit. Thus, it seems reasonable that the quantum correction may also be smaller in the strong-coupling limit. Unfortunately, at present there exists no reliable technique that is capable of handling a relativistic quantum field theory with strong coupling (except in the totally fictitious two space-time dimensions, and even there only for a very few special examples[5]). As $g \to \infty$, it is not even clear whether the vacuum state does exist or not. The lack of knowledge concerning the vacuum state makes it difficult to discuss the quantum correction to the abnormal state in the strong-coupling limit.

On the other hand, from the quasiclassical solution, the abnormal state is expected to exist even for a weakly coupled meson field, provided the fermion density is sufficiently high. Thus, it may be worthwhile to develop systematically the *high-density* but weak-coupling expansion of a fermion medium interacting with a 0^+ meson field. Some insight into the physical nature of the abnormal nuclear state and the related phase transition phenomenon may then be obtained. The actual task of establishing a systematic high-density, though weak-coupling, expansion turns out to be a nontrivial one. This is not surprising, since in the abnormal state the shift in $\overline{\phi}$ is $\cong -(m/g)$, which approaches infinity as g becomes infinitesimal. The usual perturbation series is applicable only if the coupling is weak and the fermion density remains relatively low. A suitable rearrangement of the usual perturbation series is therefore necessary in order to derive a high-density expansion.

We recall that in this problem both the energy density $\mathcal{E}(n)$ and the average field $\overline{\phi}$ depend on six parameters: a, b, c, m, and g, and the Fermi momentum k_F. Among these, only c and g^2 are di-

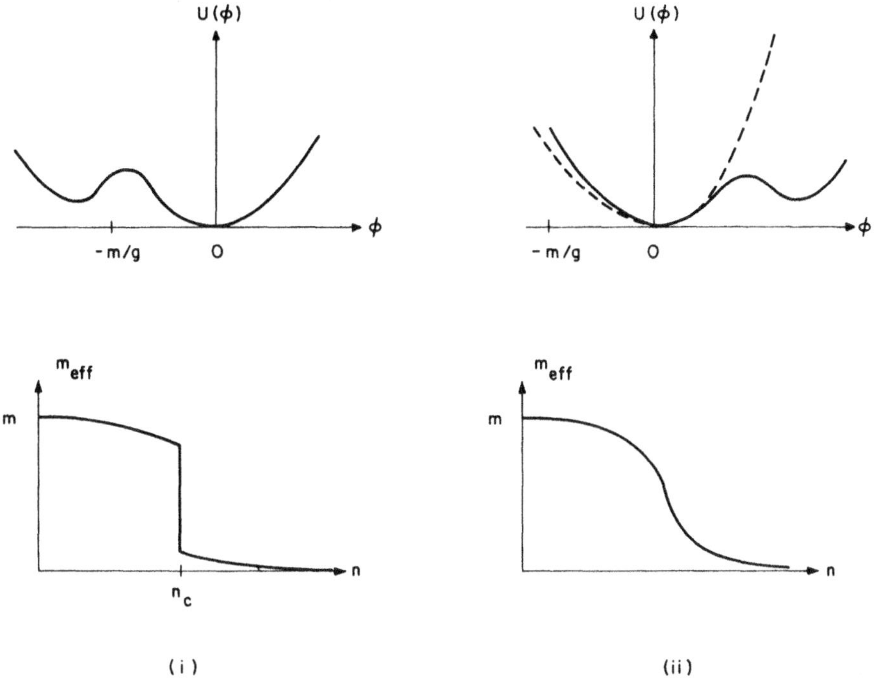

FIG. 1. Examples of (i) a discontinuous transition and (ii) a continuous transition between the normal state ($m_{eff} \cong$ free fermion mass m) and the abnormal state ($m_{eff} \cong 0$). The necessary conditions for a discontinous transition [i.e., case (i)] are $b^2 > 2ac$ and b of the same sign as g; both conditions are satisfied in the σ model. In case (ii), the function $U(\phi)$ can have either two minima (solid curve), or only one minimum (dashed curve).

mensionless. Out of these six parameters, one may construct five dimensionless parameters:

$$g^2, \ \frac{g^2 k_F^2}{a}, \ \frac{b^2}{ac}, \ \frac{a}{m^2}, \text{ and } \frac{c}{g^2}. \qquad (1.13)$$

In the usual perturbation series, one expands in powers of g^2, $(gk_F)^2$, b, and c. [If k_F is $\ll m$, then $(gk_F)^2$ should be replaced by $(gk_F)^2(k_F/m)$.] For the abnormal state, such a perturbation expansion is obviously inadequate. This is clear especially if we recall the condition on these parameters in the interesting case of a discontinuous transition. From (1.11), one sees that

$$\frac{b^2}{ac} \sim O(1).$$

Furthermore, the Fermion density must be quite high, since, on the one hand, $\bar{\phi} \cong -(m/g)$ in the abnormal state, and, on the other hand, $\bar{\phi} \sim -(gk_F^2 m/a)$ from the field equation; it follows therefore

$$\frac{g^2 k_F^2}{a} \sim O(1).$$

The appropriate high-density weak-coupling expansion that we are seeking is one in which, out of the five dimensionless parameters in (1.13), *only g^2 acts as the small expansion parameter.* The other four parameters in (1.13) are all regarded as $O(1)$; i.e.,

$$c \sim O(g^2) \ll 1, \quad b^2 \sim O(ac),$$

and (1.14)

$$a \sim O(m^2) \sim O(g^2 k_F^2).$$

We may formally write in ascending powers of g^2

$$\mathcal{E}(n) = \mathcal{E}_0 + \mathcal{E}_1 + \mathcal{E}_2 + \cdots, \qquad (1.15)$$

where

$$\begin{aligned} &\mathcal{E}_0 \sim O(k_F^4), \\ &\mathcal{E}_1 \sim O(g^2 k_F^4), \\ &\mathcal{E}_2 \sim O(g^4 k_F^4) \text{ or } O(g^4 k_F^4 \ln g^2), \\ &\text{etc.} \end{aligned} \qquad (1.16)$$

To each order in g^2, or $g^2 \ln g^2$, the dependence of $\mathcal{E}(n)$ on the last four parameters in (1.13) is supposed to be evaluated *exactly*. Because $k_F \sim O(m/g) \gg O(m)$, this weak-coupling expansion is also a high-density expansion.

To derive such an expansion, we are helped by methods developed in many-body problems and statistical mechanics.[6] We will start, in Sec. III, from the usual perturbation series, and then sum over all one-line reducible graphs to obtain the well-known expansion in terms of the prototype diagrams.[7] The series of prototype diagrams is then further rearranged through a second summation over all two-line reducible graphs. The result is a variational formalism of $\mathcal{E}(n)$ in terms of the average field $\overline{\phi}$ and the two full propagators: $D'(k)$ of the meson field and $S'(p)$ of the fermion field. The details are given in Sec. IV. This variational formalism can then be used to give the desired high-density weak-coupling expansion.

As will be shown in Sec. IVC, if we neglect $O(g^4k_F^4)$ and $O(g^4k_F^4\ln g^2)$ in the loop diagrams, the energy-density function becomes

$$\mathcal{E} = U(\overline{\phi}) + (4\pi^3)^{-1}\kappa\int_F d^3p\,[\vec{p}^{\,2} + (m+g\overline{\phi})^2]^{1/2} + (64\pi^4)^{-1}\kappa g^2k_F^4\,, \tag{1.17}$$

where U is given by (1.2), and $\overline{\phi}$ is determined by the minimum of $\mathcal{E}$. We note that in the abnormal state $\overline{\phi}\sim -(m/g)$; therefore, $U(\overline{\phi})\sim O(am^2/g^2)$, which is $\sim O(g^2k_F^4)$ according to (1.14). In the above expression, the first two terms are exactly the same as the quasiclassical result provided that $\overline{\phi}$ replaces the classical field. The third term $(64\pi^4)^{-1}\kappa g^2k_F^4$ is a new term, representing the quantum correction; more specifically, it is due to the change in the fermion self-energy on account of the Fermi sea. [See (4.37) for the modification if there are pions in addition to the 0^+ meson.] Because it is independent of $\overline{\phi}$, (1.17) leads to the same dependence of $\overline{\phi}$ on the fermion density as that in the quasiclassical solution. Thus, the overall description of the transition given by the quasiclassical solution remains correct in a quantum-mechanical treatment, at least when the expansion (1.15) is a valid one.

II. QUASICLASSICAL SOLUTION

In this section, we shall briefly review some of the properties of the quasiclassical solution.

A. General discussion

In the quasiclassical solution, the fermions are assumed to form a completely degenerate gas. We examine first the case that the fermion density n is uniform inside the volume Ω. (The case of a nonuniform density will be discussed later in Sec. IIB.) The meson field $\phi(x)$ is assumed to be classical. Since the fermion density is kept the same everywhere, it is easy to see that in the lowest energy state the meson field must also be a constant. The energy density $\mathcal{E}(n)$ is given by (1.6), which may be written as

$$\mathcal{E}(n) = U_\phi + U_F\,, \tag{2.1}$$

where

$$U_\phi \equiv U(\phi) = \tfrac{1}{2}a\phi^2 + (3!)^{-1}b\phi^3 + (4!)^{-1}c\phi^4\,, \tag{2.2}$$

$$U_F = (4\pi^3)^{-1}\kappa\int_F d^3p\,(\vec{p}^{\,2} + m_{\rm eff}{}^2)^{1/2}\,; \tag{2.3}$$

$m_{\rm eff}$ and κ are given respectively by (1.7) and (1.9).

As already stated in the Introduction, from the above expression of $\mathcal{E}(n)$ it follows that, independent of the details of U_ϕ, at low density the system is normal (i.e., $\phi\cong 0$ and $m_{\rm eff}\cong m$), while at sufficiently high density the system must become abnormal (i.e., $\phi\cong -m/g$ and $m_{\rm eff}\cong 0$). The nature of the transition from the low-density "normal" solution to the high-density "abnormal" solution can be best analyzed by examining the derivative of $\mathcal{E}(n)$. At any given n, the physical value of ϕ is determined by minimizing $\mathcal{E}(n)$. Thus, it must be given by one of the solutions of

$$\frac{dU_\phi}{d\phi} = -\left(\frac{\partial U_F}{\partial\phi}\right)_n\,. \tag{2.4}$$

Depending on the parameters a, b, c, m, g, and n, Eq. (2.4) may have one, or three, or five solutions. We note that the left-hand side, $dU_\phi/d\phi$, is a cubic function of ϕ, independent of n, while the right-hand side, $-(\partial U_F/\partial\phi)_n$, is both an odd function of $m+g\phi$ and a monotonically decreasing function of ϕ, varying between the limiting value gn at $\phi=-\infty$ and $-gn$ at $\phi=+\infty$. As $n\to 0$, the minimum of $\mathcal{E}(n)$ is given by the solution $\phi\to 0$, and as $n\to\infty$ it is given by the solution $\phi\to -(m/g)$. When n varies, the transition between these two solutions may or may not be discontinuous. The necessary and sufficient condition for a discontinuous transition is to have a density n at which there should be two or more solutions of (2.4) in the open interval between $\phi=0$ and $\phi=-(m/g)$. The proof is quite straightforward, as will be illustrated by the following examples:

(1) If $b^2<2ac$, the function U_ϕ has no point of inflection. Thus $dU_\phi/d\phi$ is a monotonically increasing function of φ, varying from $-\infty$ at $\phi=-\infty$ to $+\infty$ at $\phi=+\infty$. Equation (2.4) has only one solution, which gives the minimum of $\mathcal{E}(n)$; therefore, the physical value of ϕ varies continuously from $\phi=0$ at $n=0$ to $\phi=-(m/g)$ at $n=\infty$.

(2) If b is of the opposite sign from g, the function $dU_\phi/d\phi$ is a monotonically increasing function at least in the region $g\phi<0$; it varies from $-\infty$ at $g\phi=-\infty$ to 0 at $\phi=0$. Thus, in the interval between $\phi=0$ and $\phi=-(m/g)$, there is again only

one solution of (2.4). Just like in the previous case, as n increases the physical value of ϕ varies continuously from 0 to $-(m/g)$. [Both cases (1) and (2) are represented by (ii) in Fig. 1.]

(3) Next, we examine the more interesting case when $b^2 > 2ac$ and b is of the same sign as g. Without any loss of generality, we may choose g positive, and therefore b also positive. There are two points of inflection of U_ϕ, called A and B, where

$$\phi_A = c^{-1}[-b + (b^2 - 2ac)^{1/2}]$$

and (2.5)

$$\phi_B = c^{-1}[-b - (b^2 - 2ac)^{1/2}].$$

If b^2 is also $> \frac{8}{3} ac$, then the curve $dU_\phi/d\phi$ has three zeros, at $\phi = 0$, ϕ_+, and φ_-, where

$$\phi_\pm = 3(2c)^{-1}[-b \pm (b^2 - \tfrac{8}{3}ac)^{1/2}]. \qquad (2.6)$$

Because of (1.3), U_ϕ has its absolute minimum at $\phi = 0$, a local maximum at ϕ_+, and a local minimum at ϕ_-. As shall be shown, provided that (m/g) is sufficiently large, there is always a discontinuous jump in ϕ as n increases. For clarity, we discuss separately the following possibilities:

(i) If $b^2 > \frac{8}{3} ac$ and $\phi = -(m/g)$ lies to the left of $\phi = \phi_+$, then the transition from the low-density solution, $\phi \to 0$, to the high-density solution, $\phi \to -(m/g)$, must be a discontinuous one.

In this case, at sufficiently low density, Eq. (2.4) has three solutions, labeled α, β, and γ, which may be arranged in a descending order according to their ϕ values; i.e.,

$$0 > \phi_\alpha > \phi_\beta > \phi_\gamma . \qquad (2.7)$$

In the interval between the origin and $\phi = -(m/g)$, depending on the position of $\phi = \phi_-$, one may have all these three solutions, or only two, α and β, as shown in case (i) of Fig. 2. In either case, as $n \to 0$, the solution α approaches the origin; therefore, the minimum of $\mathcal{E}(n)$ is given by α. But as n increases, α approaches β. At a certain density, α and β coalesce, and above that density α disappears. Consequently, there must be a discontinuous transition.

(ii) If $b^2 > \frac{8}{3} ac$, but $\phi = -(m/g)$ lies to the right of the point of inflection $\phi = \phi_A$, than as n increases, the physical value of ϕ varies continuously from 0 to $-(m/g)$.

In this case, between the origin and $\phi = -(m/g)$, Eq. (2.4) has only one solution, called α, as shown in case (ii) of Fig. 2. It is not difficult to establish graphically that the shaded area $\alpha\beta A$ is always larger than the shaded area $\beta\gamma B$. Consequently, the minimum of $\mathcal{E}(n)$ is always given by α (never γ), and therefore, there is no discontinuity as n changes.

(iii) Similar considerations extend to the case when $b^2 > \frac{8}{3} ac$, but $\phi = -(m/g)$ lies in between ϕ_A and ϕ_+. As n varies, the transition from the solution $\phi = 0$ to $\phi = -(m/g)$ is a discontinuous one provided that at some density Eq. (2.4) has three solutions α, β, and γ between the origin and $\phi = -(m/g)$, as shown in case (iii) of Fig. 2. The discontinuity occurs when the two shaded areas $\alpha\beta A$ and $\beta\gamma C$ become equal. At low density, the absolute minimum of $\mathcal{E}(n)$ is given by α, but at high density it is given by γ.

It can be shown that in this case a sufficient condition for a discontinuous transition is $(m/g) > \frac{1}{3}(b^2 - 2ac)^{-1}(b^3/c)$.

(iv) If $2ac < b^2 < \frac{8}{3} ac$, the transition from $\phi = 0$ to $\phi = -(m/g)$ may still be discontinuous, provided that $-(m/g)$ is sufficiently negative, as shown in case (iv) of Fig. 2. The discontinuity occurs when the shaded areas $\alpha\beta A$ and $\beta\gamma B$ become equal.

B. σ model

Of particular interest is the σ model.[4] For simplicity, we shall discuss the model only in the zero pion-mass limit. The σ model consists of a spin-$\frac{1}{2}$ nucleon field ψ, a spin-0 even-parity field σ, and the usual pseudoscalar field $\vec{\pi}$. The Lagrangian density is given by

$$\mathcal{L} = -\psi^\dagger \gamma_4 \gamma_\mu \frac{\partial \psi}{\partial x_\mu} - g\psi^\dagger \gamma_4 (\sigma + i\vec{\pi}\cdot\vec{\tau}\gamma_5)\psi - \frac{1}{2}\left[\left(\frac{\partial \sigma}{\partial x_\mu}\right)^2 + \left(\frac{\partial \vec{\pi}}{\partial x_\mu}\right)^2\right] - \frac{1}{4}\lambda^2(\sigma^2 + \vec{\pi}^2 - \sigma_0{}^2)^2 + \text{counterterms}. \qquad (2.8)$$

In the tree approximation, the free nucleon mass m_N and the free σ-meson mass m_σ are related to the parameters σ_0 and λ by

$$m_N = g\sigma_0 \text{ and } m_\sigma = \sqrt{2}\,\lambda\sigma_0 . \qquad (2.9)$$

In the quasiclassical treatment, we may regard σ and $\vec{\pi}$ both as classical fields. As before, we first examine the case of a uniform nucleon density n. In the lowest energy state, both σ and $\vec{\pi}$ are then also uniform. The energy-density function is given by

$$\mathcal{E}(n) = U_\sigma + U_N , \qquad (2.10)$$

where

$$U_\sigma = \tfrac{1}{4}\lambda^2(\sigma^2 + \vec{\pi}^2 - \sigma_0{}^2)^2 , \qquad (2.11)$$

$$U_N = (4\pi^3)^{-1}\kappa \int_F d^3p(\vec{p}^2 + m_{\text{eff}}{}^2)^{1/2} , \qquad (2.12)$$

$$m_{\text{eff}}{}^2 = g^2(\sigma^2 + \vec{\pi}^2), \qquad (2.13)$$

and κ is given by (1.9). Equations (2.10)–(2.12) reduce, respectively, to the previous expressions

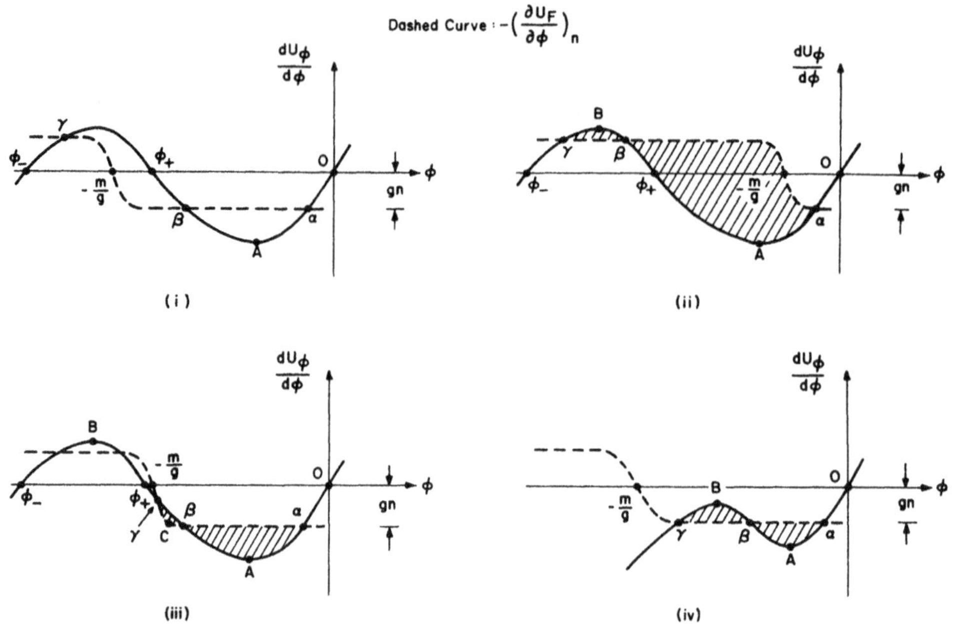

FIG. 2. Schematic drawings of $dU_\phi/d\phi$ and $-(\partial U_F/\partial\phi)_n$ versus ϕ for the cases (i)–(iv) discussed in Sec. II A.

(2.1)–(2.3) if we set

$$\phi = (\sigma^2 + \vec{\pi}^2)^{1/2} - \sigma_0, \quad b^2 = 3ac,$$
$$c = 6\lambda^2, \text{ and } a = 2\lambda^2\sigma_0^2 = m_\sigma^2. \tag{2.14}$$

[See (3.30) for the definition of a when radiative corrections are included.] By following the same argument given in Sec. II A, one finds that as the nucleon density n increases from zero, there is a discontinuous transition[1,3] from the normal state ($m_{\text{eff}} \cong m_N$) to the abnormal state ($m_{\text{eff}} \cong 0$), as illustrated by case (i) in Fig. 1. Let n_c be the critical density. From (2.10)–(2.12), and by using a simple scale transformation, one sees that n_c in units of m_N^3 depends only on a single combination of parameters:

$$m_N^3 n_c^{-1} = f(\kappa g^2 m_N^2/m_\sigma^2). \tag{2.15}$$

It is not difficult to calculate the explicit form of f. In Fig. 3, we map out the region in $g^2/4\pi$ and m_σ for $n_c \leq 2n_0$, where n_0 is the nuclear density in existing nuclei, given by

$$n_0^{-1} = \tfrac{4}{3}\pi(1.2 \text{ fm})^3. \tag{2.16}$$

For simplicity, we set the degeneracy factor $\kappa = 2$ for the nuclear matter. Although at present there is no reliable data on either m_σ or g^2, the range given in Fig. 3 is not incompatible with most of the available discussions of 0^+ resonances in the literature.[4,8]

Since it is perhaps possible that through heavy-ion collisions the nucleon density may increase from n_0 to about $2n_0$, this suggests that experimental investigations of abnormal nuclear states may become feasible.

Next, we investigate the lowest energy state of the system by allowing the nucleon density n to be

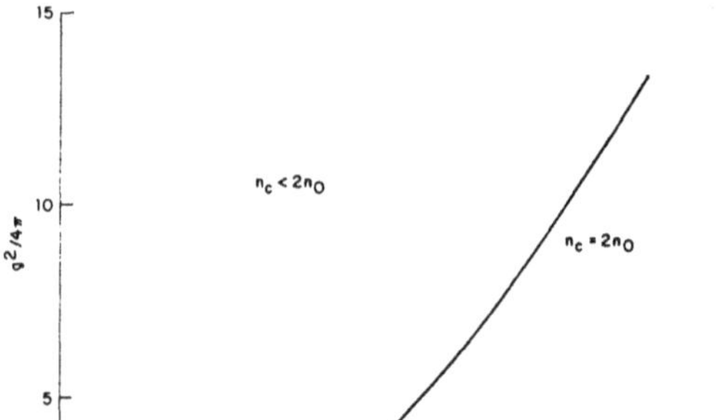

FIG. 3. The solid curve denotes $n_c = 2n_0$, where $n_0 = [\frac{1}{3}4\pi(1.2 \text{ fm})^3]^{-1}$ and n_c is the critical density in the σ model. Over a fairly wide range of m_σ and g, $n_c \cong 11.6\,(m_\sigma/m_N)^2(4\pi/g^2)n_0$.

nonuniform, of course keeping the total baryon number fixed. As the volume of the system $\Omega \to \infty$, the only way a nonuniform density may perhaps further lower the energy is to develop a two-phase region. This problem is closely connected to the question of the binding energy of the abnormal state. By setting $\sigma = 0$, one finds that for a uniform density, the total energy of the abnormal state is

$$E \equiv \Omega \mathcal{E} = \tfrac{3}{4} N k_F + \tfrac{1}{8}\Omega m_N^2 (m_\sigma^2/g^2) , \tag{2.17}$$

where $N = \Omega n$. At $(\partial E/\partial \Omega)_N = 0$, one obtains for the minimum of E

$$N^{-1}E_{\min} = k_F = (\tfrac{3}{4})^{1/4} (m_N m_\sigma \pi/g)^{1/2} . \tag{2.18}$$

The binding energy per nucleon in the abnormal state is $m_N - k_F$. Thus, the system binds if

$$\frac{g^2}{4\pi} > \frac{3\pi}{16}\left(\frac{m_\sigma}{m_N}\right)^2 . \tag{2.19}$$

Furthermore, one can show that in this case $\sigma = 0$ always corresponds to the absolute minimum of E. Let n_{Ab} be the corresponding density of the abnormal state at zero pressure. Because of (2.18), n_{Ab} is given by

$$\begin{aligned} n_{\mathrm{Ab}} &= \tfrac{1}{3}(2k_F{}^3/\pi^2) \\ &= (2\pi\sqrt{3}\,)^{-1/2} (m_N m_\sigma/g)^{3/2} . \end{aligned} \tag{2.20}$$

Assuming that (2.19) is satisfied, for an average nucleon density $\Omega^{-1}N < n_{\mathrm{Ab}}$, and Ω sufficiently large, one finds (now allowing the density to be nonuniform) the *lowest* energy state to consist of a *two-phase region*: a part of the volume, $\Omega - (N/n_{\mathrm{Ab}})$, is empty and the rest in the abnormal state. The corresponding thermodynamic p-v diagram is plotted in Fig. 4 [diagram (i)], where p is the pressure and v is the specific volume,

$$v = N^{-1}\Omega . \tag{2.21}$$

In this case, at zero temperature there is a first-order phase transition at zero pressure. For $(4\pi)^{-1}g^2 < (3\pi/16)(m_\sigma/m_N)^2$, the abnormal state does not bind; even then, there can still be a phase transition which, at zero temperature occurs only under a finite pressure. Both possibilities are illustrated by diagrams (i) and (ii) in Fig. 4.

Similar considerations can be readily extended to other zero-spin meson fields, different from the σ model. It is important to note that high temperature[9] tends to destroy coherence. Throughout our discussions, we will concentrate only on zero, or low, temperature phenomena. A brief description of the thermodynamics functions at low but non-zero temperature is given in Appendix A. [See (A15)-(A18) for the expression of the two-phase region in the σ model.]

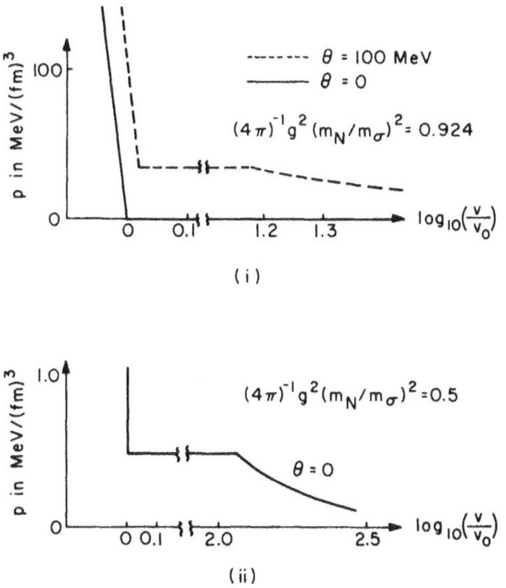

FIG. 4. p-v diagram in the σ model. θ = Boltzmann constant times the absolute temperature, p = pressure, v = specific volume, and v_0 is given by (A19). Case (i) is for $(4\pi)^{-1}g^2 > (16)^{-1}3\pi(m_\sigma/m_N)^2$: at $\theta = 0$ the transition occurs at $p = 0$. The graph is plotted for a choice of parameters such that the binding energy u = 100 MeV per fermion in the abnormal state (i.e., liquid phase). Case (ii) is for $(4\pi)^{-1}g^2 < (16)^{-1}3\pi(m_\sigma/m_N)^2$: at $\theta = 0$ the transition occurs at $p > 0$.

So far we have not considered any of the usual short-range nuclear forces. To make a thorough investigation of this important problem is quite beyond the scope of our present paper. In Appendix B, a simplified "hard sphere" model is given which we hope may illustrate in a qualitative way some of the problems involved.

III. LOOP DIAGRAMS

The Lagrangian density (1.1) may be written as

$$\mathcal{L} = \mathcal{L}_0 + \mathcal{L}_1 + \text{counterterms} , \tag{3.1}$$

where

$$\mathcal{L}_0 = -\frac{1}{2}\left(\frac{\partial\phi}{\partial x_\mu}\right)^2 - \tfrac{1}{2}a\phi^2 - \psi^\dagger\gamma_4\left(\gamma_\mu \frac{\partial}{\partial x_\mu} + m\right)\psi , \tag{3.2}$$

$$\mathcal{L}_1 = -(3!)^{-1}b\phi^3 - (4!)^{-1}c\phi^4 - g\phi\psi^\dagger\gamma_4\psi , \tag{3.3}$$

and

$$\text{counterterms} = -(\delta J)\phi - \tfrac{1}{2}(\delta a)\phi^2 - (3!)^{-1}(\delta b)\phi^3 - (4!)^{-1}(\delta c)\phi^4 - (\delta m)\psi^\dagger\gamma_4\psi - (\delta g)\phi\psi^\dagger\gamma_4\psi$$
$$- \tfrac{1}{2}(\delta Z_\phi)\left(\frac{\partial\phi}{\partial x_\mu}\right)^2 - (\delta Z_\psi)\psi^\dagger\gamma_4\gamma_\mu\frac{\partial\psi}{\partial x_\mu}\ . \qquad (3.4)$$

The fields ϕ and ψ are the renormalized fields and the parameters a, b, c, g, and m are the appropriately defined renormalized constants. In (3.4), δJ is determined by (1.3); the determinations of other counterterms depend on the precise definitions of the renormalized constants, which will be discussed later.

A. Perturbation series

To analyze the systematics of the loop diagrams, we shall begin with the usual perturbation series by regarding ($\mathfrak{L}_1$ + counterterms) as the perturbation. The zeroth-order meson propagator $D(q)$ is, as usual,

$$D(q) = -i\,(q^2 + a - i\epsilon)^{-1}\ , \qquad (3.5)$$

where $q^2 = \vec{q}^{\,2} - q_0^{\,2}$ and $\epsilon = 0+$. However, for a state with a nonzero fermion density, the zeroth-order nucleon propagator $S(p)$ is given by[10]

$$S(p) = i[-i\gamma_\mu p_\mu - m + i\epsilon_n(p)]^{-1}\ , \qquad (3.6)$$

where $\epsilon_n(p)$ is a noncovariant function of the four-momentum $p_\mu = (\vec{p}, ip_0)$ and the Fermi momentum k_F;

$$\epsilon_n(p) = 0-, \quad \text{if } |\vec{p}| \le k_F \text{ and } p_0 > 0$$
$$\text{and} \qquad (3.7)$$
$$\epsilon_n(p) = 0+ \quad \text{otherwise.}$$

In the zeroth-order solution of the ground state $|n\rangle$ for a fermion density $n \ne 0$, all fermion levels with energy $\omega_p \equiv (\vec{p}^{\,2} + m^2)^{1/2} > 0$ and $|\vec{p}| \le k_F$ are occupied. Thus, in the complex p_0 plane, for $|\vec{p}| > k_F$ the poles of $S(p)$ are located as usual at $p_0 = \pm(\omega_p - i\epsilon)$, but for $|\vec{p}| \le k_F$ they are displaced to $p_0 = (\pm\omega_p) + i\epsilon$.

At zero fermion density $n = 0$, the ground state is $|\text{vac}\rangle$ and therefore, because of (1.3), $\bar{\phi} = 0$, where $\bar{\phi}$ is defined by (1.12). If $n \ne 0$, then $\bar{\phi}$ is $\ne 0$. In Fig. 5, we give the usual perturbation series expansion for $\bar{\phi}$ and $\mathcal{E}(n)$. In these diagrams, a solid line gives a factor $S(p)$, a dashed line a factor $D(k)$, and the various vertices give factors according to the explicit expressions of $\mathfrak{L}_1$ and the counterterms. From (3.3), one sees that there is a ϕ^3 vertex carrying a factor $-ib$, a ϕ^4 vertex carrying a factor $-ic$, and another three-point $\phi\psi^\dagger\gamma_4\psi$ vertex carrying a factor $-ig$. In addition, there are vertices due to the counterterms given by (3.4). It is convenient to display explicitly the fermion-number conservation by assigning each solid (fermion) line an arrow, indicating the flow of the fermion number, but none to the dashed (meson) line. In addition, each diagram carries a factor s^{-1}, where s is its symmetry number, and each fermion loop gives a factor (-1).

To derive the *symmetry number*, let us consider a diagram with l_F internal fermion lines and l_M internal meson lines. The total number of internal lines is

$$l = l_F + l_M\ . \qquad (3.8)$$

We first assign to each end of every internal line a different integer: 1, or 2, ..., or $2l$. The resulting diagram is called a "numbered diagram." We then consider the $(2l)!$ permutations of these $2l$ integers. The total number of permutations that leave the "numbered diagram" unchanged is the symmetry number. [Because the fermion lines carry arrows but the meson lines do not, there is an alternative way to derive the symmetry number. We may assign only one number to every fermion line, but as before, two numbers to every meson line, one for each end. There are altogether $l_F + 2l_M$ numbers assigned. Next we consider the product of the $l_F!$ permutations be-

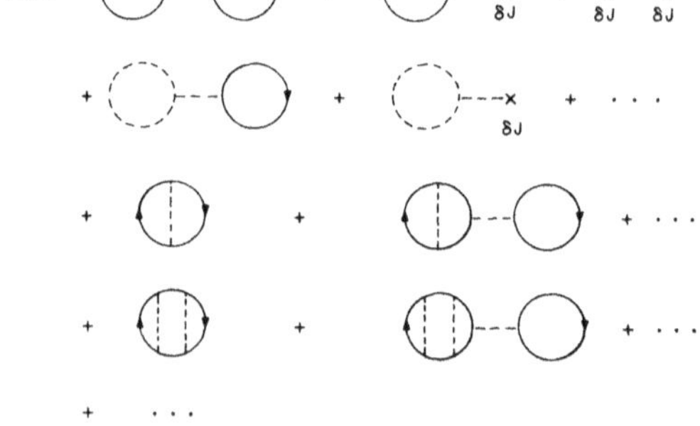

FIG. 5. Graphical representation of the usual perturbation-series expansion of $\bar{\phi}$ and $\mathcal{E}(n)$.

tween the numbers associated with the fermion lines times the $(2l_M)!$ permutations between the numbers associated with the ends of the meson lines. Among the $l_F! \times (2l_M)!$ permutations, the total number of permutations that leave the "numbered diagram" invariant is the same symmetry number s defined before.] See Fig. 6 for the symmetry number of some sample diagrams.

It is useful to introduce the concept of "L-line reducibility." (Actually, we are only interested in the special cases of $L=1$ and 2.) We call a diagram "L-line reducible" if it can be separated into two or more disconnected parts by cutting L different internal lines open; otherwise, it is called "L-line irreducible." From this definition, it follows that if a diagram is two-line irreducible, then it is also one-line irreducible; the converse is, of course, not true. For example, in Fig. 5, the first two rows of diagrams for $\mathcal{E}(n)$ are all one-line reducible, therefore also two-line reducible; on the third row, only the first one is two-line irreducible (and therefore also one-line irreducible); on the fourth row, the first one is two-line reducible, though one-line irreducible.

The partial sum over all one-line reducible diagrams can be readily carried out,[6] which leads to the familiar expression in terms of the *prototype diagrams*.[7] Each prototype diagram is one-line irreducible, and vice versa.

B. Prototype diagrams

A simple way to derive the series expansion of $\mathcal{E}(n)$ in terms of the prototype diagrams is to introduce

$$\chi \equiv \phi - \bar{\phi}\,, \tag{3.9}$$

so that in the ground state $|n\rangle$

$$\bar{\chi} \equiv \Omega^{-1}\int d^3r\,\langle n|\chi(x)|n\rangle = 0\,. \tag{3.10}$$

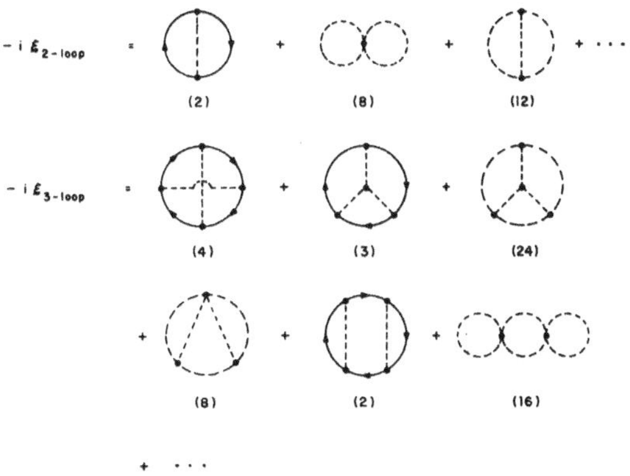

FIG. 6. $\mathcal{E}_{l\text{-loop}}$ in terms of prototype diagrams for $l=2$ and 3. The symmetry number of each diagram is listed under the diagram. In the case of $l=2$, there are only three prototype diagrams (except for diagrams involving counterterms, which are represented by "...").

The Lagrangian density (3.1) can then be written as

$$\begin{aligned}\mathcal{L} = &-\frac{1}{2}\left(\frac{\partial\chi}{\partial x_\mu}\right)^2 - U(\bar{\phi}) - \tfrac{1}{2}\bar{a}\chi^2 \\ &- (3!)^{-1}\bar{b}\chi^3 - (4!)^{-1}\bar{c}\chi^4 \\ &- \psi^\dagger\gamma_4\left(\gamma_\mu\frac{\partial}{\partial x_\mu} + \bar{m} + g\chi\right)\psi + \text{counterterms}\,,\end{aligned} \tag{3.11}$$

where

$$\begin{aligned}&\bar{a} = d^2U/d\bar{\phi}^2 = a + b\bar{\phi} + \tfrac{1}{2}c\bar{\phi}^2\,,\\ &\bar{b} = d^3U/d\bar{\phi}^3 = b + c\bar{\phi}\,,\\ &\bar{c} = c, \text{ and } \bar{m} = m + g\bar{\phi}\,.\end{aligned} \tag{3.12}$$

The counterterms are now given by

$$\begin{aligned}\text{counterterms} = &-\delta U(\bar{\phi}) - (\delta\bar{J})\chi - \tfrac{1}{2}(\delta\bar{a})\chi^2 - (3!)^{-1}(\delta\bar{b})\chi^3 - (4!)^{-1}\delta\bar{c}\chi^4 - (\delta\bar{m})\psi^\dagger\gamma_4\psi - (\delta g)\chi\psi^\dagger\gamma_4\psi \\ &- \tfrac{1}{2}\delta Z_\phi\left(\frac{\partial\chi}{\partial x_\mu}\right)^2 - (\delta Z_\psi)\psi^\dagger\gamma_4\gamma_\mu\frac{\partial\psi}{\partial x_\mu}\,,\end{aligned} \tag{3.13}$$

where

$$\begin{aligned}&\delta U = (\delta J)\bar{\phi} + \tfrac{1}{2}(\delta a)\bar{\phi}^2 + (3!)^{-1}(\delta b)\bar{\phi}^3 \\ &\qquad + (4!)^{-1}(\delta c)\bar{\phi}^4\,,\\ &\delta\bar{J} = d(U+\delta U)/d\bar{\phi}, \quad \delta\bar{a} = d^2(\delta U)/d\bar{\phi}^2\,,\\ &\delta\bar{b} = d^3(\delta U)/d\bar{\phi}^3, \quad \delta\bar{c} = \delta c\,,\end{aligned} \tag{3.14}$$

and

$$\delta\bar{m} = \delta m + (\delta g)\bar{\phi}\,.$$

Since the above Lagrangian density depends explicitly on $\bar{\phi}$, so would the energy levels. The physical value of $\bar{\phi}$ is determined by the condition (3.10).

By using (3.11), one can readily expand $\mathcal{E}(n)$ in a new perturbation series by regarding $\bar{b}$, $\bar{c}$, and g as the perturbations. The zeroth-order meson propagator and fermion propagator are now respectively given by, instead of (3.5) and (3.6),

$$\bar{D}(q) = -i(q^2 + \bar{a} - i\epsilon)^{-1} \quad (3.15)$$

and

$$\bar{S}(p) = i[-i\gamma_\mu p_\mu - \bar{m} + i\epsilon_n(p)]^{-1}, \quad (3.16)$$

where $\epsilon = 0+$ and $\epsilon_n(p)$ is given by (3.7).

To derive the zeroth-order energy, we adopt the usual convention that the vacuum energy is zero. Since the vacuum state is defined to be the ground state in the case of zero fermion density (i.e., $n=0$), we have

$$\mathcal{E}(0) = 0 . \quad (3.17)$$

By using (3.11), one finds that to *zeroth order* in $\bar{b}$, $\bar{c}$, and g the ground-state energy density $\mathcal{E}(n)$ for $n \neq 0$ is given by

$$\mathcal{E}_{\text{q.cl.}} + \Delta\mathcal{E}_{\text{vac}}, \quad (3.18)$$

where $\mathcal{E}_{\text{q.cl.}}$ is the previously obtained quasiclassical expression

$$\mathcal{E}_{\text{q.cl.}} = U(\bar{\phi}) + (8\pi^3)^{-1}\int_F (2\kappa)d^3p\,(\vec{p}^2 + \bar{m}^2)^{1/2} \quad (3.19)$$

and $\Delta\mathcal{E}_{\text{vac}}$ is the change in the vacuum energy due to $\bar{\phi} \neq 0$,

$$\Delta\mathcal{E}_{\text{vac}} = -(8\pi^3)^{-1}\int(2\kappa)d^3k[(\vec{k}^2 + \bar{m}^2)^{1/2} - (\vec{k}^2 + m^2)^{1/2}] + (8\pi^3)^{-1}\int \tfrac{1}{2}d^3k[(\vec{k}^2 + \bar{a})^{1/2} - (\vec{k}^2 + a)^{1/2}] + \text{subtraction}, \quad (3.20)$$

in which the first term is the energy change of the "fermion negative sea" and the second term is the change in the "zero-point meson energy;" both are divergent, but finite results can emerge if we subtract out an appropriate fourth-order polynomial in $\bar{\phi}$ due to renormalization. [See (3.22) below.]

The higher-order corrections can be derived in terms of diagrams by using (3.11), (3.15), and (3.16). We may write for the energy density

$$\mathcal{E}(n,\bar{\phi}) = \mathcal{E}_{\text{q.cl.}} + \Delta\mathcal{E}_{\text{vac}} + \sum_{l=2}^{\infty}\mathcal{E}_{l\text{-loop}}, \quad (3.21)$$

in which $\mathcal{E}$ depends explicitly on $\bar{\phi}$. Because $\bar{\chi} = 0$, it can be readily seen that all loop diagrams are *one line irreducible*. In Fig. 6 examples of two-loop and three-loop diagrams, together with their symmetry numbers, are listed. The counterterms $\delta\bar{m}$, δZ_ψ, $\delta\bar{a}$, ... can always be reexpressed as sums of loop diagrams in terms of the renormalized constants $\bar{a}$, $\bar{b}$, $\bar{c}$, $\bar{m}$, $\bar{g}$, and $\bar{\phi}$. In (3.21), for diagrams involving counterterms, we find it convenient to define "l-loop" to also include these loops in the counterterms. Because of this definition, there is no one-loop diagram in the sum over $\mathcal{E}_{l\text{-loop}}$. (An otherwise "one-loop" diagram with counterterms $\delta\bar{m}$, $\delta\bar{a}$, ... as vertices is here considered as a two-, or more, loop diagram.)

It may be instructive to compare the above series (3.21), which is in terms of one-line irreducible diagrams (i.e., prototype diagrams), with the original perturbation-series expansion discussed in Sec. III A. By using Fig. 5, one can express $\mathcal{E}(n)$ in terms of $\bar{\phi}$; one finds then, as usual, $U(\bar{\phi})$ comes from the sum of the tree diagrams, while $\Delta\mathcal{E}_{\text{vac}}$ comes from the sum of all one-loop diagrams. In (3.21), however, both terms come from the zeroth-order energy calculation of the Lagrangian (3.11). The explicit form of the vacuum energy change is[1,7]

$$\Delta\mathcal{E}_{\text{vac}} = -(16\pi^2)^{-1}\kappa\bar{m}^4\ln(\bar{m}^2/m^2) + (64\pi^2)^{-1}\bar{a}^2\ln(\bar{a}/a) + \text{a finite fourth-order polynomial in } \bar{\phi}, \quad (3.22)$$

where $\bar{m} = m + g\bar{\phi}$ and $\bar{a} = a + b\bar{\phi} + \tfrac{1}{2}c\bar{\phi}^2$. There is a certain arbitrariness in the choice of the fourth-order polynomial, depending on the precise definitions of the renormalized constants a, b, and c. Different polynomials correspond simply to different choices of a, b, and c. Among all possible conventions, the following is a particularly simple one.

Let us first consider the σ model. Its Lagrangian (2.8) is invariant under a discrete O_4 rotation:

$$\sigma \to -\sigma, \quad \vec{\pi} \to -\vec{\pi}, \quad \text{and} \quad \psi \to \gamma_5\psi . \quad (3.23)$$

A similar invariance exists even if there is no pion, provided that the constants a, b, and c for the 0^+ field $\phi \equiv \sigma - \sigma_0$ satisfy (2.14), where $\sigma_0 = (m/g)$. The corresponding symmetry becomes, instead of (3.23), simply a reflection:

$$\sigma \to -\sigma \text{ and } \psi \to \gamma_5\psi . \quad (3.24)$$

In either case, $\Delta\mathcal{E}_{\text{vac}}$ should be, like

$$U(\bar{\sigma}) = \tfrac{1}{4}\lambda^2(\bar{\sigma}^2 - \sigma_0^2)^2, \quad (3.25)$$

an even function of $\bar{\sigma}$, which is the expectation value of the σ field. Since $\bar{m} = g\bar{\sigma}$,

$$\bar{a} = \lambda^2(3\bar{\sigma}^2 - \sigma_0^2) \quad (3.26)$$

and $\bar{\phi} = \bar{\sigma} - \sigma_0$, one sees that in order to have

$\Delta\mathcal{E}_{vac}$ as an even function of $\bar{\sigma}$, the fourth-order polynomial in (3.22) must be even in $\bar{\sigma}$. On account of (3.26), this is mathematically equivalent to writing (3.22) in the form

$$\Delta\mathcal{E}_{vac} = -(16\pi^2)^{-1}\kappa\bar{m}^4\ln(\bar{m}^2/m^2) + (64\pi^2)^{-1}\bar{a}^2\ln(\bar{a}/a) + \text{a quadratic function in } \bar{a}\,, \qquad (3.27)$$

in which the quadratic function can be determined by requiring at $\bar{\phi}=0$, $\Delta\mathcal{E}_{vac}=0$

$$\frac{d}{d\bar{\phi}}(\Delta\mathcal{E}_{vac})=0 \text{ and } \frac{d^2}{d\bar{\phi}^2}(\Delta\mathcal{E}_{vac})=0\,. \qquad (3.28)$$

Thus, when the fermion density $n=0$, $\bar{\phi}=0$ is a minimum of $\mathcal{E}$. (In the σ model, $\bar{\phi}=0$ corresponds to $\bar{\sigma}=\sigma_0$.) Just as in Ref. 1, we shall require (3.27) and (3.28) to hold in *general*,[11] not just in the σ model. The result for an arbitrary $U(\phi)$ is

$$\Delta\mathcal{E}_{vac} = -(16\pi^2)^{-1}\kappa\bar{m}^4\ln(\bar{m}^2/m^2) + (64\pi^2)^{-1}\bar{a}^2\ln(\bar{a}/a) + \tfrac{1}{2}\alpha(\bar{a}-a)^2 + \beta(\bar{a}-a)\,, \qquad (3.29)$$

where

$$\alpha = (8\pi^2 b)^{-1}[\kappa g(m/b)^2(7gb-mc) - \tfrac{3}{8}b]$$

and

$$\beta = (8\pi^2 b)^{-1}(\kappa g m^3 - \tfrac{1}{8}ab)\,.$$

Throughout our discussion, the renormalization is carried out at $k_\mu=0$. Thus, the renormalized constants m and a are defined by

$$m = -i\lim_{k_\mu=0}\lim_{n=0}[S'(k)]^{-1}$$

and $\qquad (3.30)$

$$a = -i\lim_{k_\mu=0}\lim_{n=0}[D'(k)]^{-1}\,,$$

where $S'(k)$ and $D'(k)$ denote respectively the *full* renormalized fermion and meson propagators. Because of radiative corrections m and $a^{1/2}$ are *not* the same as the physical free fermion and free meson masses.

For the σ model, (3.29) becomes

$$\Delta\mathcal{E}_{vac} = -(8\pi^2)^{-1}(g\bar{\sigma})^4\ln(\bar{\sigma}^2/\sigma_0^2) + (64\pi^2)^{-1}\bar{a}^2\ln(\bar{a}/a) + (16\pi^2)^{-1}3g^4[1-\tfrac{9}{32}(a^2/m^4)](\bar{\sigma}^2-\sigma_0^2)^2 + (8\pi^2)^{-1}g^2m^2[1-\tfrac{3}{16}(a^2/m^4)](\bar{\sigma}^2-\sigma_0^2)\,, \qquad (3.31)$$

where $\bar{a}$ is given by (3.26) and, for simplicity, we set the degeneracy factor $\kappa=2$. It is of interest to observe that for $a\approx 1.4m^2$, even for a relatively large $(4\pi)^{-1}g^2$, $\Delta\mathcal{E}_{vac}$ remains relatively small over a fairly wide range of $\bar{\sigma}$. (See Fig. 7.)

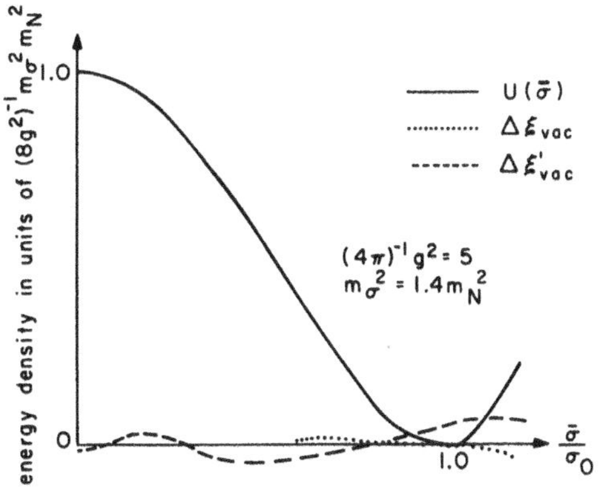

FIG. 7. $U(\bar{\sigma})$, $\Delta\mathcal{E}_{vac}$, and $\Delta\mathcal{E}'_{vac}$ in units of $(8g^2)^{-1}m_\sigma^2 m_N^2$ versus $(\bar{\sigma}/\sigma_0)$. In this unit, $U(\bar{\sigma})=[1-(\bar{\sigma}/\sigma_0)^2]^2$, shown by the solid curve. The dotted curve denotes $\Delta\mathcal{E}_{vac}$ [Eq. (3.31)] and the dashed curve denotes $\Delta\mathcal{E}'_{vac}$ [Eq. (3.35)]; the latter is plotted for $n=6n_0$, where $n_0^{-1}=\frac{1}{3}4\pi(1.2\text{ fm})^3$.

For $\bar{\sigma}^2<\frac{1}{3}\sigma_0^2$, $\bar{a}$ becomes <0 and therefore $\Delta\mathcal{E}_{vac}$ is complex. As will be shown in Sec. IV, one should have used the full meson propagator $D'(k)$ instead of $\bar{D}(k)$, and this would lead to replacing $\bar{a}=-i[\bar{D}(0)]^{-1}$ by the positive-definite

$$a' \equiv -i[D'(0)]^{-1}\,. \qquad (3.32)$$

That

$$a'>0 \qquad (3.33)$$

can be established by using

$$D'(k) = \int d^4x \exp(ik_\mu x_\mu)\langle n|T\chi(x)\chi(0)|n\rangle \qquad (3.34)$$

in the Heisenberg representation, where $\chi=\sigma-\bar{\sigma}$ in the σ model (or $\phi-\bar{\phi}$ in the general case) and T denotes the time-ordered product; (3.33) then follows from the usual positivity requirement of the spectral representation of $D'(k)$. At zero nucleon density, $a'=a$ according to (3.30). To have a rough idea of the variation of vacuum energy change versus $\bar{\sigma}$ for $\bar{\sigma}^2<\frac{1}{3}\sigma_0^2$, we may simply replace $\bar{a}$ by a' in (3.31), but (rather arbitrarily) without including any other changes; $\Delta\mathcal{E}_{vac}$ becomes then

$$\Delta\mathcal{E}'_{vac} \equiv -(8\pi^2)^{-1}(g\bar{\sigma})^4\ln(\bar{\sigma}^2/\sigma_0^2) + (64\pi^2)^{-1}a'^2\ln(a'/a) + (16\pi^2)^{-1}3g^4[1-\tfrac{9}{32}(a^2/m^4)](\bar{\sigma}^2-\sigma_0^2)^2 + (8\pi^2)^{-1}g^2m^2[1-\tfrac{3}{16}(a^2/m^4)](\bar{\sigma}^2-\sigma_0^2)\,, \qquad (3.35)$$

where, as can be readily verified, if one neglects higher-order radiative corrections,

$$a' = \bar{a} + (2\pi^3)^{-1} g^2 \int_F (\vec{p}^2 + g^2\bar{\sigma}^2)^{-3/2} \vec{p}^2 d^3 p \ . \quad (3.36)$$

In Fig. 7, besides $\Delta\mathcal{E}_{vac}$, we also plot $\Delta\mathcal{E}'_{vac}$ versus $\bar{\sigma}$ for $a = 1.4m^2$, $(4\pi)^{-1}g^2 = 5$, and $n = 6n_0$, where $n_0^{-1} = \frac{1}{3}4\pi(1.2\ \text{fm})^3$. These plots illustrate the cancellation between the vacuum energy change of the mesons and that of the fermions when a is near $1.4m^2$.

Returning now to the general problem of a 0^+ meson ϕ with an arbitrary fourth-order polynomial $U(\phi)$, we note that since the Lagrangian density (3.11) depends on $\bar{\phi}$, so does the energy density $\mathcal{E}(n, \bar{\phi})$. As mentioned before, the physical value of $\bar{\phi}$ is determined by the requirement $\bar{\chi} = 0$. The following theorem shows that this is equivalent to minimizing $\mathcal{E}(n, \bar{\phi})$ with respect to $\bar{\phi}$.

Theorem 1. At the physical value of $\bar{\phi}$, one has

$$\frac{\partial}{\partial \bar{\phi}} \mathcal{E}(n, \bar{\phi}) = 0 \quad (3.37)$$

and

$$\frac{\partial^2}{\partial \bar{\phi}^2} \mathcal{E}(n, \bar{\phi}) > 0 \ , \quad (3.38)$$

where $\mathcal{E}(n, \bar{\phi})$ is given by (3.21), and in the differentiation the original renormalized constants, a, b, c, m, and g, together with the original counterterms δa, δb, δc, ..., are all kept fixed.

Proof. Under the variation $\bar{\phi} \to \bar{\phi} + d\bar{\phi}$, because of (3.12), (3.15), and (3.16),

$$\begin{aligned} &\bar{D}(k) \to \bar{D}(k) - i\bar{b}[\bar{D}(k)]^2 d\bar{\phi} \ , \\ &\bar{S}(p) \to \bar{S}(p) - ig[\bar{S}(p)]^2 d\bar{\phi} \ , \\ &\bar{b} \to \bar{b} + c(d\bar{\phi}) \ , \end{aligned} \quad (3.39)$$

and similar variations for the counterterms $\delta\bar{a}$, $\delta\bar{b}$, etc. By definition, the loop diagrams for $\mathcal{E}(n, \bar{\phi})$ are all without external lines. From (3.39), one sees that under the $\bar{\phi}$ variation, to first order in $d\bar{\phi}$, the loop diagrams all turn into diagrams with a single external meson line, which carries an amplitude $d\bar{\phi}$ and a zero four-momentum. Upon comparison with the original power-series expansion of $\bar{\phi}$, given in Fig. 5, one can readily verify that (3.37) and $\bar{\chi} = 0$ are the same condition. Similarly, one finds that to $(d\bar{\phi})^2$, the same variation in $\bar{\phi}$ turns every loop diagram into a meson-propagator diagram at zero four-momentum. Through a straightforward diagram counting, one can show that

$$\frac{\partial^2}{\partial \bar{\phi}^2} \mathcal{E}(n, \bar{\phi}) = -i[D'(0)]^{-1} \ . \quad (3.40)$$

The inequality (3.38) follows on account of (3.32) and (3.33). Theorem 1 is then proved.

The evaluation of the two-loop diagrams is straightforward, though tedious. From Fig. 6 we see that, apart from diagrams involving counterterms, there are only three two-loop prototype diagrams. Among these, the second and the third do not explicitly depend on k_F. These two diagrams have already been calculated; they are given by Eq. (3.17) of Ref. 1. The evaluation of the first two-loop diagram in Fig. 6 turns out to be rather complicated. We may separate the result into a sum of three terms, in descending order of their explicit dependence on k_F: $O(k_F^4) + O(k_F^2) + O(k_F^0)$. The complete expression is quite long. Here, we give only the term proportional to k_F^4. For a single 0^+ field ϕ, with arbitrary constants a, b, and c, we find the two-loop contribution to be

$$[\mathcal{E}_{2\text{-loop}}]_\phi = (64\pi^4)^{-1}\kappa g^2 k_F^4 + O(k_F^2) \ . \quad (3.41)$$

For the σ model, after including the effect of pion fields and setting $\kappa = 2$, we find

$$[\mathcal{E}_{2\text{-loop}}]_\sigma = -(16\pi^4)^{-1} g^2 k_F^4 + O(k_F^2) \ . \quad (3.42)$$

The details are given in Appendix C. As will be seen there, (3.41) and (3.42) are due to the change in the fermion self-energy because of the exclusion principle and the presence of the Fermi sea.

IV. TWO-LINE IRREDUCIBLE DIAGRAMS

For theories with a sizable $b\phi^3$ coupling, e.g., the σ model, in the abnormal state $\bar{\phi} \cong -(m/g)$ the sign of $\bar{a} = d^2U(\bar{\phi})/d\bar{\phi}^2$ may become negative [see case (i) of Fig. 1]. The series in terms of prototype diagrams is then obviously inadequate, since its zeroth-order meson propagator $\bar{D}(k)$ would have a tachyon pole. On the other hand, as discussed in Sec. III, the full meson propagator $D'(k)$ always satisfies the positivity requirement: $a' = -i/D'(0)$ is always positive, even though $\bar{a} = -i/\bar{D}(0)$ may not be. Thus, it seems sensible to expand the energy-density function $\mathcal{E}(n)$ in terms of the full meson propagator $D'(k)$.

A. Variational principle

The technique of expansion in terms of the full propagator has been developed[6] for quite some time for problems in statistical mechanics and many-body systems. The same technique can be readily extended to the present problem.[3] We may start from the Lagrangian density (3.11), and expand $\mathcal{E}(n)$ formally first in terms of the prototype diagrams, as before. These diagrams are all without external lines. We then perform a partial sum over all two-line reducible diagrams. This leads to a formulation of $\mathcal{E}(n)$ as a sum over only *two line irreducible diagrams*. In each diagram, the Feynman rule is modified: Every meson (dashed) line now carries a factor $D'(k)$, the full renormalized meson propagator, and every fermion (solid) line carries a factor $S'(p)$, the full renormalized fermion propagator. By definition, there is no two-point vertex in a two-line irreducible diagram; there are only three- and four-point vertices in these diagrams. These vertices carry the same factors as the corresponding ones in the prototype diagrams; i.e., there are the $\bar{b}\chi^3$, $\bar{c}\chi^4$, and $g\psi^\dagger\gamma_4\psi\chi$ vertices, as well as other three- and four-point vertices due to the counterterms, given by (3.13). For simplicity, in the following we shall consider the sum $\bar{b}+\delta\bar{b}$ as one vertex, and similarly for $\bar{c}+\delta\bar{c}$ and $g+\delta g$.

Definition 1. We first define a functional f_l of D' and S' by

$$-if_l(D',S') \equiv \text{sum of all } l\text{-loop two line irreducible diagrams} . \tag{4.1}$$

There are only three $l=2$ two-line irreducible diagrams, all given in Fig. 6, and four $l=3$ two-line irreducible diagrams, which are given by the first four three-loop diagrams in Fig. 6.

Definition 2. Next, we define another functional

$$F(D',S') \equiv U(\bar{\phi}) + (8\pi^3)^{-1}\int_F (2\kappa)d^3p(\vec{p}^{\,2}+m_0^2)^{1/2} - i(2\pi)^{-4}\int d^4p\, \mathrm{tr}\{\ln[S'(p)/S_0(p)] - [S'(p)/\bar{S}_0(p)]+1\}$$
$$+\tfrac{1}{2}i(2\pi)^{-4}\int d^4k\{\ln[D'(k)/D_0(k)] - [D'(k)/\bar{D}_0(k)]+1\} + \sum_{l=2}^{\infty} f_l(D',S') + \text{subtraction term} , \tag{4.2}$$

where $U(\bar{\phi})$ is given by (1.2), m_0 is the unrenormalized mass

$$m_0 = (1+\delta Z_\psi)^{-1}(m+\delta m) , \tag{4.3}$$

and S_0, D_0, ... are various unrenormalized propagators defined by

$$S_0(p) = i[(1+\delta Z_\psi)(-i\gamma_\mu p_\mu) - m - \delta m + i\epsilon_n]^{-1} , \tag{4.4}$$

$$D_0(k) = -i[(1+\delta Z_\phi)k^2 + a + \delta a - i\epsilon]^{-1} , \tag{4.5}$$

$$\bar{S}_0(p) = i[(1+\delta Z_\psi)(-i\gamma_\mu p_\mu) - \bar{m} - \delta\bar{m} + i\epsilon_n]^{-1} , \tag{4.6}$$

and

$$\bar{D}_0(k) = -i[(1+\delta Z_\phi)k^2 + \bar{a} + \delta\bar{a} - i\epsilon]^{-1} . \tag{4.7}$$

In (4.2), the subtraction term is a fourth-order polynomial in $\bar{\phi}$; it differs from δU [given by (3.14)] by an additive constant:

$$\text{subtraction term} = \delta U + \text{constant} , \tag{4.8}$$

where the constant is to be chosen such that when, later on, F is used for evaluating $\mathcal{E}(n)$, one has, at $n=0$, $\mathcal{E}(0)=0$.

Theorem 2. The correct full renormalized propagators $D'(k)$ and $S'(p)$ are determined by

$$\frac{\delta}{\delta D'(k)} F(D',S') = 0 \tag{4.9}$$

and

$$\frac{\delta}{\delta S'(p)} F(D',S') = 0 , \tag{4.10}$$

in which $\bar{\phi}$ and n are kept fixed. Furthermore, when D' and S' assume their correct forms, the functional $F(D',S')$ is equal to the energy-density function $\mathcal{E}(n)$. [The matrix $S'(p)$ may be written as $A+(-i\gamma_\mu p_\mu)B$, where A and B are c-number functions. In (4.10), it is understood that $\delta S' = \delta A + (-i\gamma_\mu p_\mu)\delta B$.]

Proof. The proof of this theorem follows closely that of a similar theorem[6] in statistical mechanics and many-body problems. From the definition of F, it follows that (4.9) and (4.10) imply

$$[D'(k)]^{-1} - [\bar{D}_0(k)]^{-1} = 2i(2\pi)^4 \frac{\delta}{\delta D'(k)} \sum_{l=2}^{\infty} f_l(D',S') \tag{4.11}$$

and

$$[S'(p)]^{-1} - [\bar{S}_0(p)]^{-1} = -i(2\pi)^4 \frac{\delta}{\delta S'(p)} \sum_{l=2}^{\infty} f_l(D',S') . \tag{4.12}$$

Through iterations, the solutions D' and S' can be expressed in terms of $\bar{D}_0$ and $\bar{S}_0$, or $\bar{D}$ and $\bar{S}$ [introduced in (3.15) and (3.16)]. By comparing with the series expansion of D' and S' in terms of prototype diagrams, one can readily establish that the solutions of (4.11) and (4.12) are indeed the full renormalized propagators D' and S'.

It is useful to define

$$G(D', S') \equiv -i(2\pi)^{-4}\int d^4p\,\mathrm{tr}\{\ln[S'(p)/\bar{S}_0(p)] - [S'(p)/\bar{S}_0(p)] + 1\}$$

$$+\tfrac{1}{2}i(2\pi)^{-4}\int d^4k\{\ln[D'(k)/\bar{D}_0(k)] - [D'(k)/\bar{D}_0(k)] + 1\} + \sum_{l=2}^{\infty} f_l(D', S'). \tag{4.13}$$

The difference between $F(D', S')$ and $G(D', S')$ is

$$F - G = U(\bar{\phi}) + (8\pi^3)^{-1}\int_F (2\kappa)d^3p(\vec{p}^2 + m_0^2)^{1/2} - i(2\pi)^{-4}\int d^4p\,\mathrm{tr}\{\ln[\bar{S}_0(p)/S_0(p)]\} + \tfrac{1}{2}i(2\pi)^{-4}\int d^4k\,\ln[\bar{D}_0(k)/D_0(k)]$$

$$+ \text{subtraction term}. \tag{4.14}$$

By using the identity

$$-i(2\pi)^{-1}\int_{-\infty}^{\infty} dk_0 \ln\frac{k^2 + \beta - i\epsilon}{k^2 + \alpha - i\epsilon} = (\vec{k}^2 + \beta)^{1/2} - (\vec{k}^2 + \alpha)^{1/2}, \tag{4.15}$$

one finds

$$F - G = U(\bar{\phi}) + (8\pi^3)^{-1}\int_F (2\kappa)d^3p(\vec{p}^2 + \bar{m}_0^2)^{1/2} - (8\pi^3)^{-1}\int (2\kappa)d^3p[(\vec{p}^2 + \bar{m}_0^2)^{1/2} - (\vec{p}^2 + m_0^2)^{1/2}]$$

$$+ \tfrac{1}{2}(8\pi^3)^{-1}\int d^3k[(\vec{k}^2 + \bar{a}_0)^{1/2} - (\vec{k}^2 + a_0)^{1/2}] + \text{subtraction term}, \tag{4.16}$$

where m_0 is given by (4.3),

$$a_0 = (1 + \delta Z_\phi)^{-1}(a + \delta a),$$

$$\bar{m}_0 = (1 + \delta Z_\psi)^{-1}(\bar{m} + \delta\bar{m}), \tag{4.17}$$

and

$$\bar{a}_0 = (1 + \delta Z_\phi)^{-1}(\bar{a} + \delta\bar{a}).$$

Because $F - G$ is independent of D' and S', if one wishes, one may express the stationary conditions (4.9) and (4.10) in terms of G: $(\delta G/\delta D') = (\delta G/\delta S') = 0$. This leads to the same equations (4.11) and (4.12). We may substitute the iterative solutions of D' and S' back into G, and then express G as a sum of l-loop diagrams ($l \geq 2$) in terms of $\bar{D}$ and $\bar{S}$. By comparing with the sum $\sum \mathcal{E}_{l\text{-loop}}$ in (3.21), we find that the difference between $\sum \mathcal{E}_{l\text{-loop}}$ and $G(D', S')$ is equal to the sum of all "one-loop" prototype diagrams which have only the counterterms $\delta\bar{m}$, $\delta\bar{a}$, δZ_ψ, and δZ_ϕ as vertices. These diagrams are included in $\sum \mathcal{E}_{l\text{-loop}}$ but not in G. [See the discussion given after (3.21) in Sec. III B.] By using the identity (4.15), one finds that this difference is also equal to that between the right-hand side of (4.16) and $\mathcal{E}_{\text{q.cl.}} + \Delta\mathcal{E}_{\text{vac}}$ [given by (3.19) and (3.20)]. Thus,

$$\mathcal{E}(n, \bar{\phi}) = F(D', S'), \tag{4.18}$$

where $\mathcal{E}(n, \bar{\phi})$ is given by (3.21). Theorem 2 is then proved.

From theorem 1, one sees that F is also stationary with respect to $\bar{\phi}$. The physical value of $\bar{\phi}$ corresponds to a minimum of F.

B. Approximate solutions

The variational formulation allows a convenient basis for making approximations.

(1) As a first example, we may consider the simplest approximation in which all $f_l = 0$, but keeping $\bar{\phi} \neq 0$ in the definition of the functional F. From (4.11) and (4.12), one finds $D'(k) = \bar{D}_0(k)$ and $S'(p) = \bar{S}_0(p)$. It is clear that since in this approximation there is no (two-line irreducible) loop diagram in F, all counterterms should be zero; i.e., $\delta Z_\phi = \delta Z_\psi = \delta a = \delta b = \delta c = \delta m = \delta g = 0$. Therefore

$$D_0(k) = D(k) = -i(k^2 + a - i\epsilon)^{-1},$$

$$D'(k) = \bar{D}_0(k) = \bar{D}(k) = -i(k^2 + \bar{a} - i\epsilon)^{-1},$$

$$S_0(p) = S(p) = i(-i\gamma_\mu p_\mu - m + i\epsilon_n)^{-1}, \tag{4.19}$$

and

$$S'(p) = \bar{S}_0(p) = \bar{S}(p) = i(-i\gamma_\mu p_\mu - \bar{m} + i\epsilon_n)^{-1}.$$

After substituting these expressions into (4.2) and by using (4.15) and (4.18), one obtains

$$\mathcal{E}(n, \bar{\phi}) = \mathcal{E}_{\text{q.cl.}} + \Delta\mathcal{E}_{\text{vac}}, \tag{4.20}$$

where $\mathcal{E}_{\text{q.cl.}}$ and $\Delta\mathcal{E}_{\text{vac}}$ are given by (3.19) and (3.20), respectively. [Note that the identity (3.40) is not satisfied in this approximation.]

(2) As a second example, we consider the approximation in which only the first two-loop diagram in Fig. 6 is included; i.e., we set $f_l = 0$ for $l \geq 3$ and

$$-if_2 = \tfrac{1}{2}(g + \delta g)^2(2\pi)^{-8}$$

$$\times\int d^4p\,d^4q\,D'(p - q)\,\mathrm{tr}[S'(p)S'(q)]. \tag{4.21}$$

Equations (4.11) and (4.12) become

$$[D'(k)]^{-1}-[\bar{D}(k)]^{-1}=-(g+\delta g)^2(2\pi)^{-4}\times\int d^4p\,\mathrm{tr}[S'(p)S'(p-k)]+i[(\delta Z_\phi)k^2+\delta\bar{a}] \qquad (4.22)$$

and

$$[S'(p)]^{-1}-[\bar{S}(p)]^{-1}=(g+\delta g)^2(2\pi)^{-4}\int d^4k\,D'(k)S'(p-k)-i[(\delta Z_\psi)(-i\gamma_\mu p_\mu)-\delta\bar{m}], \qquad (4.23)$$

where $\bar{D}$ and $\bar{S}$ are given by (3.15) and (3.16), respectively. In these equations, $\bar{\phi}$ enters into the propagators $\bar{D}$ and $\bar{S}$ through

$$\bar{m}=m+g\bar{\phi} \text{ and } \bar{a}=a+b\bar{\phi}+\tfrac{1}{2}c\bar{\phi}^2.$$

We regard $\bar{\phi}$ as an independent parameter, and assume $g\bar{\phi}\sim O(m)$. Depending on the value of $\bar{\phi}$, $\bar{a}$ may be negative.

Let us discuss the solutions of these two coupled equations in the weak-coupling but high-density limit; i.e., we assume

$$\frac{g^2}{4\pi}\ll 1 \text{ and } g^2k_F{}^2\sim O(a)\sim O(m^2). \qquad (4.24)$$

In this limit, just by power counting, one sees that the right-hand side of (4.22) is typically $O(g^2k_F{}^2)$, while that of (4.23) is $O(g^2k_F)\ll m$. Thus, the zeroth-order solution of $S'(p)$ is $\bar{S}(p)$, and consequently (4.22) becomes

$$[D'(k)]^{-1}=[\bar{D}(k)]^{-1}-g^2(2\pi)^{-4}\int d^4p\,\mathrm{tr}[\bar{S}(p)\bar{S}(p-k)]+i[(\delta Z_\phi)k^2+\delta a], \qquad (4.25)$$

since to this order $\delta g=\delta b=\delta c=0$, and therefore $\delta\bar{a}=\delta a$. The solution is

$$D'(k)=-i(k^2+\bar{a}+\Pi_{\mathrm{vac}}+\Pi_F)^{-1}, \qquad (4.26)$$

where Π_{vac} is the usual vacuum-polarization correction to the renormalized mass operator of the meson field (in the absence of the fermion density), and Π_F is the correction due to the fermion density.

It is not difficult to evaluate Π_F. We find

$$\Pi_F(\vec{k}^2,k_0)=\int_0^\infty(k_0{}^2-E^2+i\epsilon)^{-1}\times[-\rho_+(\vec{k}^2,E)+\rho_-(\vec{k}^2,E)]dE^2, \qquad (4.27)$$

$$\rho_+(\vec{k}^2,k_0)=g^2\kappa(8\pi^3)^{-1}\times\int_F d^3q\,(\omega_q\omega_p)^{-1}(-2\bar{m}^2+k_0\omega_q-\vec{k}\cdot\vec{q})\times\delta(k_0-\omega_q-\omega_p), \qquad (4.28)$$

and

$$\rho_-(\vec{k}^2,k_0)=g^2\kappa(8\pi^3)^{-1}\times\int_F d^3q\,(\omega_q\omega_p)^{-1}(2\bar{m}^2+k_0\omega_q+\vec{k}\cdot\vec{q})\times\delta(k_0+\omega_q-\omega_p)\theta(|\vec{p}|-k_F), \qquad (4.29)$$

where $\theta(x)$ is a step function, which is 1 if $x>0$ and 0 if $x<0$, $\omega_q=(\vec{q}^2+\bar{m}^2)^{1/2}$, and $\vec{p}=\vec{k}-\vec{q}$. Both ρ_+ and ρ_- satisfy the positivity condition,

$$\rho_\pm(\vec{k}^2,E)\geq 0. \qquad (4.30)$$

Because in (4.28) and (4.29) the integration extends only over $|\vec{q}|\leq k_F$, one finds $\rho_\pm(\vec{k}^2,E)\neq 0$ only within a limited interval in E. The E integration in (4.27) is, therefore, convergent. Physically, $-\rho_+$ is due to the "absence" of the pair-creation process into $N_{\vec{q}}^+$ and $\bar{N}_{\vec{p}}^+$ when $\vec{q}$ lies inside the Fermi sea (N denotes the fermion and $\bar{N}$ the antifermion), whereas ρ_- is due to the absorption process of converting a fermion from $-\vec{q}$ in the Fermi sea to $\vec{p}$ out of the Fermi sea (as insured by the step function θ). At $\vec{k}=0$, one has

$$\rho_-(0,E)=0, \qquad (4.31)$$

and

$$\rho_+(0,E)=\begin{cases}g^2\kappa(8\pi^2E)^{-1}(E^2-4\bar{m}^2)^{3/2}, & \text{if } 2\omega_F>E>0\\ 0, & \text{otherwise}\end{cases} \qquad (4.32)$$

where $\omega_F=(k_F{}^2+\bar{m}^2)^{1/2}$. By using (4.27), one finds at $k_\mu=0$

$$\Pi_F(0,0)=g^2\kappa(4\pi^3)^{-1}\int_F(\vec{q}+\bar{m}^2)^{-3/2}\vec{q}^{\,2}\,d^3q. \qquad (4.33)$$

In the high-density limit, $(\Pi_{\mathrm{vac}}/\Pi_F)$ is typically $\sim O(m^2/k_F{}^2)\ll 1$. Thus,

$$D'(k)\cong -i(k^2+\bar{a}+\Pi_F)^{-1}. \qquad (4.34)$$

If, in addition, we are interested in the region in which both $|\vec{k}|$ and k_0 are $\ll k_F$, then we have

$$D'(k)\cong -i[k^2+\bar{a}+\Pi_F(0,0)]^{-1} \qquad (4.35)$$

which, through (3.32) and for $\kappa=2$, leads to the same expression as (3.36).

C. High-density weak-coupling expansion

We now turn to the case that (1.14) holds; i.e., $a \sim O(m^2) \sim O(g^2 k_F^2)$, $b^2 \sim O(ac)$, and $c \sim O(g^2) \ll 1$. Formally, one may expand in the form (1.15):

$$\mathcal{E}(n) = \mathcal{E}_0 + \mathcal{E}_1 + \mathcal{E}_2 + \cdots ,$$

where $\mathcal{E}_0 \sim O(k_F^4)$, $\mathcal{E}_1 \sim O(g^2 k_F^4)$, and $\mathcal{E}_2 \sim O(g^4 k_F^4)$, etc. In this case, neglecting higher-order corrections in g^2, one finds

$$S'(p) = \bar{S}(p) = i(-i\gamma_\mu p_\mu - \bar{m} + i\epsilon_n)^{-1}, \qquad (4.36)$$

but $D'(k)$ is given by (4.34), not $\bar{D}(k)$. Substituting these expressions into (4.2), one sees that there is only one two-loop diagram $\sim O(g^2 k_F^4)$, and that is the first two-loop diagram in Fig. 6. By using (4.34), (4.36), and the result derived in Appendix C, we find that to $O(g^2 k_F^4)$, the sum over all loop diagrams is given by (3.41), or (3.42). Therefore, in the case of a single ϕ field, to $O(g^2 k_F^4)$ the energy-density function is given by (1.17). In the case of the σ model, because of pion exchanges, we have, instead of (1.17),

$$\mathcal{E} = \tfrac{1}{4}\lambda^2(\bar{\sigma}^2 - \sigma_0^2)^2 + (2\pi^3)^{-1}\int d^3p(\vec{p}^{\,2} + g^2\bar{\sigma}^2)^{1/2} - (16\pi^2)^{-1} g^2 k_F^4. \qquad (4.37)$$

In either case, we derive the same expression for the dependence of $m_{\text{eff}}^2 = (m + g\bar{\phi})^2$, or $(g\bar{\sigma})^2$, on the fermion density as that given by the quasiclassical solution.

The evaluation of the next order, $O(g^4 k_F^4)$, term is straightforward but complicated. The details will be given in a separate publication.

V. REMARKS

From the above discussion, it is clear that, at least when the expansion (1.15) is valid, the general character of the quasiclassical solution remains correct with the inclusion of quantum corrections. For physical applications to the abnormal nuclear states, the parameter $(4\pi)^{-1}g^2$ is, in general, not a small one. Nevertheless, as already remarked in the Introduction, since the shift of $\bar{\phi}$ from its normal value 0 to the abnormal value $-(m/g)$ is smallest in the strong-coupling limit,[12] so also may be the difference between the quantum correction in the abnormal state and that in the vacuum state. On the other hand, the lack of information concerning the vacuum state itself makes it difficult to perform calculations in the strong-coupling limit.

There is a certain analogy between the abnormal state and the Bose-Einstein condensation. Mathematically, the Bose-Einstein transition can be rigorously treated only for ideal bosons and for bosons with either weak interactions or small scattering lengths. For physical applications, He^4 is known to have strong interactions; yet, it does undergo Bose-Einstein condensation and exhibits superfluidity. Through the BCS pair-correlation, the electrons in metals exhibit similar phase transitions that give rise to superconductivity. Recently, it has been observed that even He^3 exhibits superfluidity. Likewise, it seems reasonable to expect that the transition to the abnormal state is a fairly general phenomenon, not restricted to the validity of the particular expansion (1.15).

ACKNOWLEDGMENTS

One of us (T.D.L.) wishes to thank G. C. Wick and A. Kerman for discussions.

APPENDIX A

In this appendix, we examine the temperature effect on the system at low temperature. Let

$$\theta = \text{Boltzmann constant} \times \text{absolute temperature}. \qquad (A1)$$

We assume the temperature to be sufficiently low, $\theta \ll k_F$ and $a^{1/2}$, so that the fermions remain degenerate and there are no free mesons. In the quasiclassical treatment, in which ϕ is assumed to be constant, the Helmholtz free energy F is given by

$$\Omega^{-1}F = nkT\ln z + U(\phi) - \pi^{-2}\kappa\theta\int_0^\infty k^2 dk \ln[1 + ze^{-\omega/\theta}], \qquad (A2)$$

where $U(\phi)$ and κ are given by (1.2) and (1.9), respectively,

$$\omega = [k^2 + (m + g\phi)^2]^{1/2}, \qquad (A3)$$

and z is the fugacity, related to the Gibbs thermodynamic potential μ per fermion by

$$z = e^{\mu/\theta}. \qquad (A4)$$

The average number of fermions with wave number k (and at a given spin and, say, isospin state) is

$$n_k = (z^{-1}e^{\omega/\theta} + 1)^{-1}, \qquad (A5)$$

and the fermion density is

$$n = \pi^{-2}\kappa\int k^2 dk\, n_k. \qquad (A6)$$

The physical value of ϕ is determined by the minimum of $\Omega^{-1}F$, keeping n and θ fixed. At low temperature, by using (A2)–(A6), we find

$$\Omega^{-1}F = U(\phi) + \pi^{-2}\kappa\int_0^{k_F} q^2\omega dq - \tfrac{1}{6}\kappa\theta^2 k_F\omega_F + O(\theta^4), \tag{A7}$$

where

$$k_F \equiv (3\pi^2 n/\kappa)^{1/3} \tag{A8}$$

and

$$\omega_F = [k_F^2 + (m+g\phi)^2]^{1/2}. \tag{A9}$$

For the σ model interacting with a nucleon field, (A7) becomes

$$\Omega^{-1}F = \tfrac{1}{4}\lambda^2(\sigma_0^2 - \sigma^2)^2 + \frac{2}{\pi^2}\int_0^{k_F} q^2\omega dq - \tfrac{1}{3}\theta^2 k_F\omega_F + O(\theta^4), \tag{A10}$$

where

$$\omega = [q^2 + (g\sigma)^2]^{1/2} \text{ and } \omega_F = [k_F^2 + (g\sigma)^2]^{1/2}.$$

Let us consider the special case that at zero temperature the abnormal state (i.e., $\sigma = 0$) is stable; according to (2.19), one has

$$\frac{g^2}{4\pi} > \frac{3\pi}{16}\left(\frac{m_\sigma}{m_N}\right)^2. \tag{A11}$$

The density of the abnormal state at zero temperature is given by (2.20):

$$n_{Ab} = (2\pi\sqrt{3})^{-1/2}(m_N m_\sigma/g)^{3/2}. \tag{A12}$$

From (A10), one sees that for $\theta \ll k_F$, and at density $n > n_{Ab}$, the minimum of F remains given by $\sigma = 0$; i.e., for the abnormal phase

$$\Omega^{-1}F = \frac{m_\sigma^2 m_N^2}{8g^2} + \frac{k_F^4}{2\pi^2} - \tfrac{1}{3}\theta^2 k_F^2 + O(\theta^4). \tag{A13}$$

The corresponding thermodynamic pressure p is [after neglecting $O(\theta^4)$]

$$p = \tfrac{1}{4}(\tfrac{3}{2}\pi^2)^{1/3}(n^{4/3} - n_{Ab}^{4/3}) + \tfrac{1}{9}(\tfrac{3}{2}\pi^2)^{2/3}\theta^2 n^{2/3}. \tag{A14}$$

When $n < n_{Ab}$, as the density decreases, because of evaporation the system no longer stays in a single abnormal phase. In the two-phase region, we may define the "liquid phase" to be the abnormal state ($\sigma = 0$), while the "gas phase" is the normal state ($\sigma = \sigma_0$); in the gas phase the nucleons form a nondegenerate gas. Because of the low density, the gas phase can be approximately described by the perfect gas law. By equating the fugacities in these two phases, one finds that in the two-phase region p is given approximately by

$$p = 4\theta\left(\frac{m_N\theta}{2\pi}\right)^{3/2} e^{-u/\theta}, \tag{A15}$$

where u is the binding energy per nucleon in the abnormal state; i.e., according to (2.18),

$$u = m_N - k_F \cong m_N\left[1 - \left(\frac{3\pi^2 m_\sigma^2}{4g^2 m_N^2}\right)^{1/4}\right]. \tag{A16}$$

Let v_g and v_l be, respectively, the specific volumes in the gas and liquid phases in the two-phase region. From the perfect gas law, v_g is related to the pressure p by

$$v_g = p^{-1}\theta, \tag{A17}$$

and from (A14), v_l is related to the same p by

$$v_l \cong v_0\left[1 + \frac{\pi^2}{2}\left(\frac{\theta}{k_0}\right)^2 - \frac{9\pi^2}{2}\frac{p}{k_0^4}\right], \tag{A18}$$

where v_0 and k_0 are the specific volume and the Fermi momentum in the abnormal state at zero temperature; i.e.,

$$v_0 = n_{Ab}^{-1} = (2\pi\sqrt{3})^{1/2}(m_N m_\sigma/g)^{-3/2} \tag{A19}$$

and

$$k_0 = (3\pi^2 n_{Ab}/2)^{1/3}. \tag{A20}$$

See Fig. 4 [case (i)] for a plot of the two-phase region.

APPENDIX B

We discuss here a simplified "hard sphere" model. Let us consider a nuclear medium of density n, and define r by

$$n^{-1} = \tfrac{4}{3}\pi r^3. \tag{B1}$$

For free nucleons, the top Fermi momentum is related to r by the usual expression (for $\kappa = 2$)

$$k_F = \tfrac{1}{2}(9\pi)^{1/3}/r. \tag{B2}$$

Because of the short-range repulsive force between nucleons, the average de Broglie wavelength of each nucleon would be shorter than that in the free case; consequently, the nuclear momentum should be correspondingly increased. Nevertheless, phenomenologically we may still regard the nucleons as forming a degenerate Fermi sea, but with its top Fermi momentum raised from k_F to[13]

$$K_F \equiv \tfrac{1}{2}(9\pi)^{1/3}/(r - 0.8d) \tag{B3}$$

where d is a phenomenological parameter. The average kinetic energy per nucleon is then given by

$$u_N = \frac{\int_0^{K_F} k^2(k^2 + m_{\text{eff}}^2)^{1/2} dk}{\int_0^{K_F} k^2 dk},$$

or

$$u_N = \tfrac{3}{4}K_F\{(1+x^2)^{1/2}(1+\tfrac{1}{2}x^2) - \tfrac{1}{2}x^4\ln[x^{-1} + (x^{-2}+1)^{1/2}]\}, \tag{B4}$$

where $x = m_{eff}/K_F$ and m_{eff} is given by (1.7). The combination $0.8d$ in (B3) is chosen such that for a system of hard spheres, in the dilute (therefore also nonrelativistic) limit, (B4) gives the exact energy expression to $O(n^{1/3}d)$, provided[13]

$$d = \text{diameter of the hard sphere.} \tag{B5}$$

In addition to u_N, there is also the potential energy $U(\phi)$ of the spin-0 field. In the quasiclassical approximation, the total energy $E \equiv \Omega\mathcal{E}$ of the system is given by

$$A^{-1}E \cong u_N + n^{-1}U(\phi) , \tag{B6}$$

where A is the total nucleon number.

1. Abnormal state

In (B6), except through the average ϕ field, all other effects due to the usual short-range attractive force between nucleons are not included. Such an approximation may not be too unreasonable (though definitely crude) for the abnormal state, since the basic energy scale there is rather large $\sim O(m_N)$. In the abnormal state, we may set $\phi \cong -(m/g)$; therefore $m_{eff} \cong 0$ and

$$A^{-1}E \cong \tfrac{3}{4}K_F + n^{-1}U_0 , \tag{B7}$$

where

$$U_0 \equiv U(\phi) \text{ at } \phi = -(m/g) . \tag{B8}$$

By varying the volume Ω at a fixed A, we find that the minimum E occurs at

$$r(r-0.8d) = \tfrac{1}{4}\left(\frac{3}{2U_0}\right)^{1/2}\left(\frac{3}{\pi}\right)^{1/3} , \tag{B9}$$

and its value is

$$A^{-1}E_{min} \cong K_F[1 + \tfrac{1}{5}(r-0.8d)^{-1}d] . \tag{B10}$$

In the σ model, $U_0 = (8g^2)^{-1}m_\sigma^2 m_N^2$. As an illustration, in Fig. 8 the binding energy per nucleon,

$$u \equiv m_N - A^{-1}E_{min} , \tag{B11}$$

is plotted versus d for $m_\sigma = 700$ MeV and $(4\pi)^{-1}g^2 = 5$ and 10. We may take, as a further example, $d = 2/m_\omega \cong 0.5$ fm, where m_ω is the mass of the ω^0 meson; for $m_\sigma = 700$ MeV and $(4\pi)^{-1}g^2 = 5$, we find $r \cong 0.86$ fm, $n^{-1}U_0 \cong 300$ MeV, $\frac{3}{4}K_F \cong 490$ MeV, and therefore the binding energy $u \cong 150$ MeV per nucleon.

2. Normal state

If the σ model is to be regarded as an approximation to reality, then the normal nuclear state should appear as a minimum, either local or absolute, of the (volume) energy E. Thus, it is of interest to search for the condition that E has two minima: one for the normal state ($\sigma \cong \sigma_0$) and the other for the abnormal state ($\sigma \cong 0$). On the other hand, in the normal state, the usual short-range attractive force between nucleons should be important. Thus, (B6) is not expected to be a reasonable approximation.[14] Nevertheless, purely as a *mathematical* model, one may ask under what condition does E, given by (B6), have two minima? By using (2.11) and (2.14), one sees that in the σ model (B6) may be written as (after requiring $\vec{\pi} = 0$)

$$A^{-1}E \cong u_N + (8g^2n)^{-1}m_\sigma^2 m_N^2(y^2-1)^2 , \tag{B12}$$

where $y = \sigma/\sigma_0$, u_N is given by (B4) with $x = |ym_N/K_F|$. Set $m_N = 940$ MeV; $A^{-1}E$ depends only on two parameters: $(m_\sigma/g)^2$ and d. By varying $A^{-1}E$ with respect to σ and r, we may search for its minima. In Fig. 9, we map out the region in which $A^{-1}E$ has two minima, and both minima have a positive binding energy. The region turns out to be extremely narrow, because of the requirement that there also be a "bound" normal state in this simple model.

APPENDIX C

In Fig. 6, we see that there are only three two-loop diagrams, apart from other related diagrams which involve counterterms. Among these three diagrams, the second and the third diagrams have been evaluated in Ref. 1 [see Eq. (3.17) of Ref. 1]. In this Appendix, we discuss only the first two-loop diagram, hereafter referred to as $\mathcal{G}$, in Fig. 6.

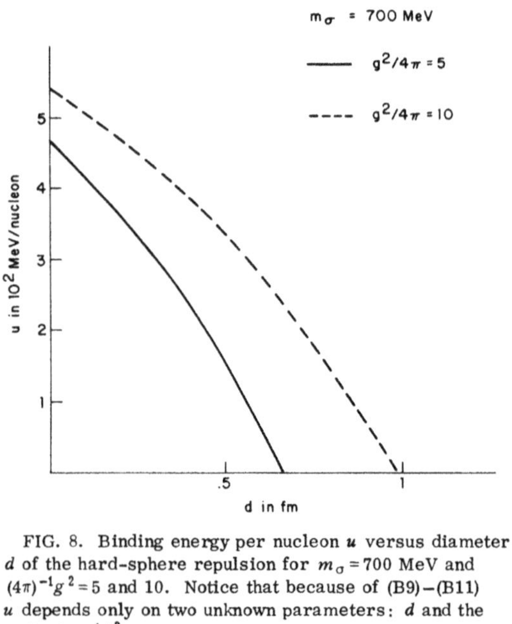

FIG. 8. Binding energy per nucleon u versus diameter d of the hard-sphere repulsion for $m_\sigma = 700$ MeV and $(4\pi)^{-1}g^2 = 5$ and 10. Notice that because of (B9)–(B11) u depends only on two unknown parameters: d and the ratio $(m_\sigma/g)^2$.

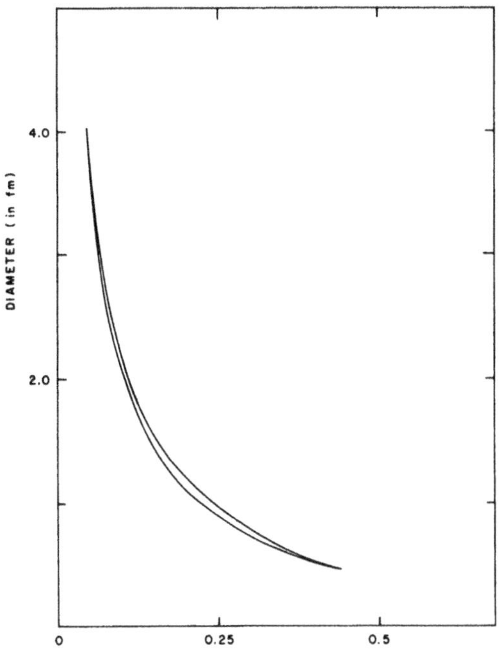

$\frac{m_\sigma}{m_N}\left(\frac{4\pi}{g^2}\right)^{1/2}$

FIG. 9. The region in which $A^{-1}E$, given by (B12) in the simplified "hard sphere" model, has two minima, and both minima have a positive binding energy. One of the minima corresponds to the normal state ($\sigma \cong \sigma_0$) and the other corresponds to the abnormal state ($\sigma = 0$).

1. Prototype diagram (a single 0^+ boson field)

Let us first consider the case that $\mathcal{G}$ is a prototype diagram and there is only a single 0^+ boson field besides the fermions. Its contribution to the energy density is (also denoted by $\mathcal{G}$, for simplicity)

$$\mathcal{G} = \tfrac{1}{2} i g^2 (2\pi)^{-8} \operatorname{tr} \int d^4p\, d^4q\, \overline{D}(p-q)\overline{S}(p)\overline{S}(q), \tag{C1}$$

where $\overline{D}$ is given by (3.15) and $\overline{S}$ by (3.16). The p_0 and q_0 integrations can be carried out by converting their contours from the real axes to those around the cuts in their respective upper half planes. In either the p_0 or q_0 complex plane, immediately above the real axis there are two such cuts: One is from $|\overline{m}|$ to $(k_F^2+\overline{m}^2)^{1/2}$ due to the positive-energy Fermi sea, and the other is from $-\infty$ to $-|\overline{m}|$ due to the negative-energy Dirac sea, where $\overline{m}$ is given by (3.12). Thus, (C1) can be written as

$$\mathcal{G} = \mathcal{G}_{++} + \mathcal{G}_{+-} + \mathcal{G}_{--}, \tag{C2}$$

where $\mathcal{G}_{++}$ (or $\mathcal{G}_{--}$) is due entirely to the positive (or negative) energy cuts in both the p_0 and q_0 planes, while $\mathcal{G}_{+-}$ is due to the positive energy cut in one plane and the negative energy cut in the other. It is easy to see that at high density $\mathcal{G}_{++} \sim O(g^2k_F^4)$, $\mathcal{G}_{+-} \sim O(g^2k_F^2m^2)$, and $\mathcal{G}_{--} \sim O(g^2m^4)$. Thus to $O(g^2k_F^4)$, only $\mathcal{G}_{++}$ is of importance; i.e., $\mathcal{G}$ is dominated by

$$\mathcal{G}_{++} = \tfrac{1}{2} g^2 \kappa (2\pi)^{-6} \times \int_F d^3p\, d^3q [(p-q)^2 + \overline{a}]^{-1} (\vec{p}^2 + \overline{m}^2)^{-1/2} \times (\vec{q}^2 + \overline{m}^2)^{-1/2} [\tfrac{1}{2}(p-q)^2 + 2\overline{m}^2], \tag{C3}$$

in which both $\vec{p}$ and $\vec{q}$ lie inside the Fermi sea. The dominant contribution in (C3) is due to $\vec{p}^2 \sim O(k_F^2)$, $\vec{q}^2 \sim O(k_F^2)$, and $(p-q)^2 \sim O(k_F^2)$. Thus, we have

$$\mathcal{G} \cong \mathcal{G}_{++} = (64\pi^4)^{-1}\kappa g^2 k_F^4 + O(g^2k_F^2m^2 \ln(k_F/m)). \tag{C4}$$

The physical origin of the above $O(g^2k_F^4)$ term can be understood by first calculating the self-energy change ΔE_p for a fermion of momentum $\vec{p}$ in the presence of a Fermi sea (because of the exclusion principle). One can readily verify that for $k_F \gg \overline{m}$ and $\overline{a}^{1/2}$,

$$\Delta E_p \cong (16\pi^2)^{-1} g^2 k_F^2 (\vec{p}^2 + \overline{m}^2)^{-1/2},$$

which leads to an energy-density change of an amount

$$(8\pi^3)^{-1}\, 2\kappa \int_F d^3\vec{p}\,(\tfrac{1}{2}\Delta E_p) \cong (64\pi^4)^{-1}\kappa g^2 k_F^4. \tag{C5}$$

To derive the lower-order expression in k_F^2, the integral (C3) should be evaluated exactly. We find

$$\mathcal{G}_{++} = (4\pi)^{-4}\kappa g^2 \overline{m}^4 [(\sinh\omega - \omega)^2 + (1 - \tfrac{1}{2}\lambda) f], \tag{C6}$$

where

$$f = 32 \int_0^{\omega/2} d\theta \sinh\theta \{(\lambda \cosh\theta) \ln[\lambda^{-1}(\cosh 2\theta + \lambda - 1)] - 2\theta(\lambda - 1)\sinh\theta + 2\lambda(\lambda - 2)(\csc A \sinh\theta)\tan^{-1}[\tanh\theta \tan(\tfrac{1}{2}A)]\}, \tag{C7}$$

$\omega = 2\sinh^{-1}(k_F/\bar{m})$, $\lambda = \frac{1}{2}\bar{a}/\bar{m}^2$, and $\cos A = \lambda - 1$. In the high-density limit, $\omega \gg 1$; thus, $f \sim 2\omega^2 e^{\omega}$, and consequently

$$\mathcal{G}_{++} \sim (4\pi)^{-4}\kappa g^2 \bar{m}^4 \sinh^2\omega$$
$$\sim 4(4\pi)^{-4}\kappa g^2 k_F^4,$$

which agrees with (C4). We have also evaluated the other two terms $\mathcal{G}_{+-}$ and $\mathcal{G}_{--}$ in (C2). The calculation is tedious though straightforward. The final expression is quite long; it will be given in a separate publication.

2. Prototype diagram (σ model)

Next, we consider the σ model in which there is the $\vec{\pi}$ field, besides the 0^+ σ field. Instead of (C4), we find (after setting $\kappa = 2$ and summing over both σ exchange and $\vec{\pi}$ exchange)

$$\mathcal{G} \cong \mathcal{G}_{++} = -(16\pi^4)^{-1} g^2 k_F^4 + O(g^2 k_F^2 m^2 \ln(k_F/m)). \tag{C8}$$

Since in Fig. 6, according to Eq. (3.17) of Ref. 1, except for $\mathcal{G}$ the other two diagrams are both $\sim O(a^2)$, from (C4) and (C8) we derive (3.41) and (3.42).

3. Two-line irreducible diagram

Lastly, we assume the diagram $\mathcal{G}$ to be a two-line irreducible diagram. According to the rules given in Sec. IV, instead of (C1) we have

$$\mathcal{G} = \tfrac{1}{2} i g^2 (2\pi)^{-8} \operatorname{tr} \int d^4p\, d^4q\, D'(p-q) S'(p) S'(q). \tag{C9}$$

By using (4.27), (4.34), and (4.36), we find that (C4) remains valid if there is only a single 0^+ meson field, and that (C8) remains valid if there is in addition the $\vec{\pi}$ field. Thus, (1.17) and (4.37) are both established.

---

*This research was supported in part by the U. S. Atomic Energy Commission.

[1]T. D. Lee and G. C. Wick, Phys. Rev. D 9, 2291 (1974).

[2]See also a talk given by T. D. Lee at the Annual Bevatron Users Meeting, Lawrence Berkeley Laboratory, 1974 (Rev. Mod. Phys., to be published).

[3]A preliminary report of this work was given by T. D. Lee, in *Proceedings of the XVII International Conference on High Energy Physics, 1974*, edited by J. R. Smith (Rutherford Laboratory, Chilton, Didcot, Berkshire, England, 1974), p. III-81.

[4]For references on the σ model, see B. W. Lee, *Chiral Dynamics* (Gordon and Breach, New York, 1972).

[5]W. Thirring, Ann. Phys. (N.Y.) 3, 91 (1958). See also J. M. Luttinger, J. Math. Phys. 4, 1154 (1963), and D. C. Mattis and E. H. Lieb, *ibid.* 6, 304 (1965).

[6]T. D. Lee and C. N. Yang, Phys. Rev. 117, 22 (1960).

[7]S. Coleman and E. Weinberg, Phys. Rev. D 7, 1888 (1973).

[8]See, e.g., the review article by G. Kramer, *Springer Tracts in Modern Physics*, edited by G. Höhler (Springer, New York, 1970), Vol. 55, p. 158; the review article by the Particle Data Group in Rev. Mod. Phys. 45, S1 (1973); and also the article by S. Protopopescu *et al.*, in *Experimental Meson Spectroscopy—1972*, proceedings of the third international conference on experimental meson spectroscopy, Philadelphia, 1972, edited by Kwan-Wu Lai and Arthur H. Rosenfeld (A.I.P., New York, 1972).

[9]See S. Weinberg, Phys. Rev. D 9, 3357 (1974); L. Dolan and R. Jackiw, *ibid.* 9, 3320 (1974).

[10]V. M. Galitskii and A. B. Migdal, Zh. Eksp. Teor. Fiz. 34, 139 (1958) [Sov. Phys.—JETP 7, 96 (1958)].

[11]In the absence of the spin-$\frac{1}{2}$ field, Eqs. (3.27) and (3.28) of this paper reduce to Eq. (3.8) of Ref. 1 through the relation $\bar{a} = a(1+\Delta)$.

[12]As $g \to \infty$, both the shift $\bar{\phi} = -(m/g)$ and the critical density approach zero. In this connection, see also some related work on quark models by W. A. Bardeen, M. S. Chanowitz, S. D. Drell, M. Weinstein, and T.-M. Yan [Phys. Rev. D 11, 1094 (1974)], and M. Creutz [Phys. Rev. D 10, 1749 (1974)]. See also P. Vinciarelli, Nuovo Cimento Lett. 4, 905 (1972).

[13]See, e.g., A. Bohr and B. R. Mottelson, *Nuclear Structure* (Benjamin, New York, 1969), Vol. 1, p. 256.

[14]In the normal state, if $U(\phi)$ is replaced by its quadratic approximation, $U(\phi) \cong \frac{1}{2}a\phi^2$, then (B6) does give a surprisingly good fit to many of the properties of the existing nuclei. For details, see the Appendix of Ref. 2, and also R. Serber (unpublished).

PHYSICAL REVIEW D VOLUME 12, NUMBER 12 15 DECEMBER 1975

Erratum

Erratum: Interaction of a dense fermion medium with a scalar-meson field [Phys. Rev. D 11, 1591 (1975)]

T. D. Lee and M. Margulies

Equation (3.42) should read

$$[\mathcal{E}_{2-\mathrm{loop}}]_\sigma = +(8\pi^4)^{-1} g^2 k_F{}^4 + O(k_F{}^2).$$

Equation (4.37) should read

$$\mathcal{E} = \tfrac{1}{4}\lambda^2(\bar{\sigma}^2 - \sigma_0{}^2)^2 + (2\pi^3)^{-1}\int_F d^3p\,(\vec{p}^2 + g^2\bar{\sigma}^2)^{1/2} + (8\pi^4)^{-1} g^2 k_F{}^4.$$

Equation (C8) should read

$$\mathcal{G} \cong \mathcal{G}_{++} = +(8\pi^4)^{-1} g^2 k_F{}^4 + O(g^2 k_F{}^2 m^2 \ln(k_F/m)).$$

12 4008

Mesons in Nuclei
Eds. M. Rho and D.H. Wilkinson

ABNORMAL NUCLEAR STATES

T.D. LEE AND G.C. WICK

Columbia University, New York, N.Y. 10027

1. Introduction

The main purpose of this article is to discuss and summarize the present state of the question, raised by the present authors early in 1974, about the possible existence of a state of nuclear matter of anomalously large density and binding energy. This suggestion was made in connection with a study of the role of scalar fields in theories of strong interactions endowed with a degenerate vacuum; we found that, in the presence of nuclear matter, a sudden shift in the average value of the scalar field away from its ordinary vacuum value could occur when the nucleon density attains a certain critical value (Lee and Wick 1974). This would be accompanied, of course, by a sudden change in the binding energy and other properties of the system. In terms of particles, rather than fields, the alleged phenomenon could be described as the sudden release by nuclear matter of *scalar* particles (particles of 0^+-spin) in a coherent state. A similar observation with regard to pseudoscalar (0^-) particles, specifically *pions*, had been made somewhat earlier by Migdal (1971, 1972) whose investigation was also originally inspired by questions about the "stability of the vacuum". This phase-transition, "pion-condensation", was also independently postulated slightly later by Sawyer (1972) and Scalapino (1972) in connection with problems about the interior of neutron stars. Since pion-condensation is fully covered in other articles in this book (Migdal ch. 25, Sawyer ch. 26, Baym and Campbell ch. 27, Backman and Weise ch. 28), we shall restrict ourselves to a few comments expressing our views on a possible comparison between the two alleged transitions, the point at issue being, of course, how reliably one can predict the essential parameters, e.g. the critical density (or *even the mere existence*), of either process.

The field we postulated is a Lorentz-scalar and isoscalar field. We shall recall briefly in the next section the reasons why such a field may indeed play an important role in strong interactions. The objections that have been raised against this assumption will also be mentioned and discussed briefly. Probably the main source of misgivings, with regard to any theory in which a scalar field plays a major role, is the checkered history of the analysis of s-wave

References p. 887

resonances in $\pi\pi$ scattering. The S*, the $\epsilon(1200)$ and their predecessors have always been poor competitors with their highly visible pseudoscalar cousins. There is little doubt about the degree of respectability that a solid citizen like the pion confers upon the corresponding field. This is somewhat ironical, since the notion of a strict correspondence between basic fields and elementary particles is by now a relic from the days of second quantisation. And what about quark theory, partons, SU(4) and all that? Still, if we want to be fair, we must grant that the existence of a considerable body of knowledge about pions and their interactions with nucleons gives "pion-condensation" theory a more solid basis to start from.

This, however, turns out to be a mixed blessing; indeed, what we know experimentally about the pion–nucleon interaction is that the s-wave interaction is rather weak and on the whole repulsive; it is in p-waves that one finds a strong attractive interaction, which favors the condensation phenomenon. This fits in nicely with the pseudoscalar nature of the pion, but turns "pion-condensation" into a rather delicate question; the problem, in other words, looks easy at the beginning but gets harder as one goes on*.

In a certain sense, the case of the scalar field has the opposite virtues and defects. The initial assumptions are based on speculation, and the essential parameters are subject to considerable uncertainty; on the other hand, the nature of the interaction is such, that if one assumes suitable values for the basic parameters (values which are, by the way, quite compatible with anything we know), it is easy to develop a rather simple and reasonable approximation. Even so, it goes without saying that under the circumstances it is not possible to arrive at a firm prediction; as any sensible person would know from the very beginning, and as our friends have reminded us repeatedly, nuclear matter is a particularly nasty many-body problem. It involves many complicated effects (tensor force, spin–orbit coupling, excited baryons, etc.) which we did not consider explicitly, and which are really quite difficult to take into account in a reliable manner. In a matter as complicated and thorny as this, opinions may differ. In the summary at the end we shall try to state as objectively as we can the situation as we see it.

2. Role of the scalar field

Our basic assumption is that an important component of the strong *attractive* interactions between nucleons is mediated by a scalar (i.e. spin zero, even

*It is perhaps unfair to quote the many conflicting statements made by the various authors on this subject. See, however, for some highlights: Migdal et al. (1974, 1976) and Brown and Weise (1976).

parity) neutral meson field $\phi(x)$ (what we mean by this field will be discussed below).

Naturally, this cannot be the whole story: presumably there are strong repulsive forces at short distances, due to vector meson exchanges; excited nucleons or hyperons may occur as virtual intermediate states; one-pion exchange will play some role, and so on and so forth. Many of these mechanisms are not expected to contribute significantly to nuclear binding, and even if some do, the only assumption we need is that the particular mechanism postulated above provides an *important fraction* of the binding energy. Specifically, we assume that the field ϕ is coupled in the natural way, i.e. linearly, to the scalar nucleon density $\bar{\psi}\psi$, with a *large* coupling constant g (see eq. (3) below).

This is not an unreasonable assumption. There is no equally simple alternative explanation of the strong attraction between two nucleons in a ^{1}S state; as is well-known, this attraction contributes quite significantly to nuclear binding energy in calculations based on static potentials (Bethe 1971). In addition, the assumption receives considerable support from widely accepted general ideas about strong interactions, e.g. from chiral symmetry*, most directly, of course, in the specific form of the theory known as the σ-model (Schwinger 1957, Polkinghorne 1958, Gell-Mann and Lévy 1960). For purposes of illustration with numerical examples we shall in fact often assume that our field ϕ is, aside from an additive constant, just the σ-field of that model with the value of the coupling constant g completely determined by the model (see below). Having done this we shall then (in the low-order calculations) ignore the terms referring to the π-mesons in the Lagrangian.

We may refer to this as the "reduced σ-model". One should notice that other possible realizations of chiral symmetry could easily yield similar results. In particular, our approach is quite compatible with the assumption that the field ϕ merely represents a convenient way to approximate the effects of the emission and reabsorption by nucleons of one or several kinds of particle pairs (e.g. $\pi\pi$ or $K\bar{K}$ pairs) in a state of relative motion with zero angular momentum. In view of this, the mass parameter for the meson field that will appear in the Lagrangian (see e.g. m_σ in eq. (8) below) should be treated as a phenomenological constant.

Once these assumptions are accepted, either simple intuition, or rather elementary calculations indicate that if the density of nuclear matter is increased above its normal value n_0, a transition will occur at some density $n_C > n_0$ to another phase with very different properties. Somewhat more elaborate calculations, described later, indicate that the existence of this phenomenon may well survive the introduction of complications, such as

*For a summary of chiral symmetric field theories see B.W. Lee (1972).

References p. 887

repulsion at very short distances, finite size effects (surface energy) and, finally, higher order radiative corrections. These somewhat more realistic calculations (sects. 4, 5) suggest the possible existence of superheavy "anomalous" nuclei of large binding energy. It should be pointed out that the aim of our study was from the beginning the investigation of this possibility, and that in the anomalous state, if it exists, the importance of the interaction mechanism we postulate is very much emphasized; as a result, approximations that may seem quite crude in the case of ordinary nuclei are in fact much better in the situation which is the main object of this study.

Some comment is necessary about the relationship of this effort with "ordinary" nuclear theory, and possibly with other questions such as the interior of neutron stars. An ideal theory, of course, should provide a framework within which it is possible to describe accurately the properties of any nuclear system, starting from a single proton or neutron up to the heaviest known nuclei. No such thing, of course, exists today. What we have are elaborate calculations connecting, to be sure, a vast amount of data not to basic laws but to an intermediate theoretical structure of a phenomenological nature. This structure consists of bits of phenomenological information, such as static potentials, which can be fitted to describe quantitatively, binding energies, nuclear sizes and charge distributions, and so on. After a lot of painstaking effort, reasonably good fits have been obtained; the intricacies of the many-body problem and strong interaction physics being what they are, the connection between the fitted parameters and more fundamental constants remains somewhat obscure.

A field theoretic description, such as the one we have attempted, or somewhat similar efforts by other authors (Walecka 1974, 1975, Chin 1976), must of course also be regarded today as largely phenomenological: it ignores completely a deeper level (quarks? gluons?) where the key to the structure of nucleons resides. The starting point, however, is so different from that of much ordinary nuclear theory and the approximations involved in making comparisons are so elaborate and often so doubtful, that one should be extremely careful in drawing sweeping conclusions. Some authors have expressed a firm conviction that the success of calculations of nuclear binding energies and other properties, based on theories which do not involve an interaction with a scalar field, leaves no room for the existence of such an interaction. We find this kind of certainty hard to understand. Let us, in particular, emphasize again that the model we described was not originally intended for the purpose of accurately describing the details of nuclear physics. This much should be obvious*. Work with this as its primary purpose

*See, however, Serber (1976) where a satisfactory description of various features of nuclei is given without introducing any new ad hoc assumptions.

may possibly continue to find that the best phenomenological fits are obtained from an entirely different set of assumptions.

3. A naive model

Let us now make the first step towards a quantitative formulation by writing down the most simplified form of a Lagrangian density embodying the main ideas mentioned above. It contains only two fields, i.e. the scalar meson field ϕ and the isospinor nucleon field ψ. It should hardly be necessary, but it is probably wise, to point out immediately a very unrealistic feature of this model: it does not make any provision for a short-range repulsion between nucleons. It is only to be expected that under the strong attraction due to the scalar field the model will lead to quite unreasonable binding energies. Only in very small nuclei would this tendency be counterbalanced to some extent by surface effects. Nevertheless, the model is introduced as a first step in order to make a certain point.

The kinetic terms for the two fields have a familiar form. We assume that an additive constant in the field ϕ is arranged so that the expectation value of the field in the vacuum is zero

$$\langle 0|\phi(x)|0\rangle = 0. \tag{1}$$

Denoting by p and n, respectively, the proton and neutron components of the ψ field, we have

$$\psi = \begin{pmatrix} p \\ n \end{pmatrix}, \qquad \bar{\psi} = (\bar{p}, \bar{n}), \tag{2}$$

where $\bar{p} = p^\dagger \gamma^4$, $\bar{n} = n^\dagger \gamma^4$ (p or n is, of course, a 4-component Dirac field, and $\dagger$ denotes Hermitean conjugation); we shall write

$$\mathcal{L} = -\tfrac{1}{2}(\partial\phi/\partial x_\mu)^2 - U(\phi) - \bar{\psi}[\gamma^\mu(\partial/\partial x_\mu) + (m_N + g\phi)]\psi, \tag{3}$$

where m_N is the nucleon mass and $U(\phi)$ is a "potential-energy term" for the meson field ϕ which, in addition to a "mass term" $\frac{1}{2}m_\phi^2\phi^2$, may include self-interaction terms $\sim\phi^3$ and ϕ^4*. Qualitatively the behavior of $U(\phi)$ could correspond to any of the curves a, b, c in fig. 1. In any case U is assumed to possess an absolute minimum at $\phi = 0$, which is then the value of the field ϕ in the vacuum when the theory is developed entirely at the classical level. The field equations

$$\Box\phi - U'(\phi) = g\bar{\psi}\psi, \tag{4}$$

*Terms of even higher order in ϕ are not included, so as to preserve "renormalizability"; see e.g. B.W. Lee (1972).

References p. 887

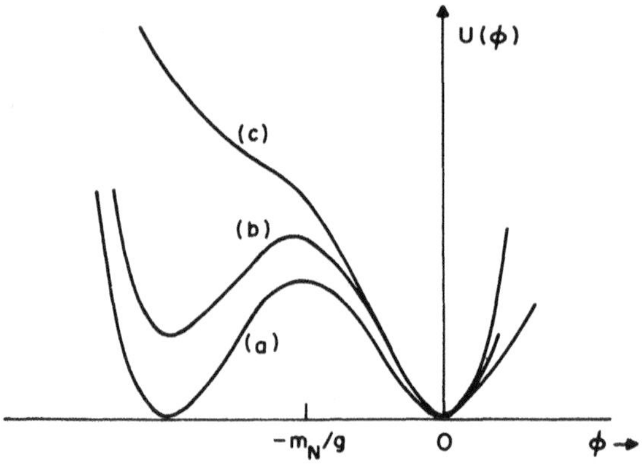

Fig. 1.

$$\gamma^\mu \partial\psi/\partial x_\mu + (m_N + g\phi(x))\psi = 0, \tag{5}$$

which are valid also at the quantum level*, indicate that: (i) the "scalar nucleon density" $\bar{\psi}\psi$ acts as a source term for the scalar field ϕ, (ii) the field $\phi(x)$, multiplied by g, acts as a "variable mass term" in the Dirac equation for the nucleon field ψ. In particular, if we assume that in a certain region $\phi(x)$ can be approximated by a constant, then in that region the Dirac equation (5) corresponds to free particles of mass $|m_N + g\phi|$.

Before we discuss quantum effects let us be more specific about the σ-model. The full σ-model, of course, has additional symmetries and furthermore requires the addition to (3) of terms involving the isovector π-meson field. But if we then ignore these terms we revert to (3) with special values for the coefficients of $U(\phi)$ (see e.g. Lee (1974)). Specifically, in this model, the coupling constant g is the usual "strong" pion–nucleon coupling:

$$g^2/4\pi \approx 15 \tag{6}$$

while the coefficients of the polynomial $U(\phi)$ depend on a single adjustable parameter; this may be chosen to be the coefficient m_σ^2 of the quadratic term; an equally convenient choice is the coefficient λ^2 of the fourth-order term. The two are related via

$$m_\sigma^2 = 2\lambda^2\sigma_0^2 + m_\pi^2, \tag{7}$$

where $\sigma_0 = m_N/g$ is the vacuum value of the σ-field. Then

$$U(\phi) = \tfrac{1}{2}m_\sigma^2\phi^2 + \lambda^2\sigma_0\phi^3 + \tfrac{1}{4}\lambda^2\phi^4. \tag{8}$$

If in eq. (7) one adopts the approximation $m_\pi = 0$, then $U(\phi)$ becomes symmetric about the point $\phi = -\sigma_0$ (curve a in fig. 1). It has then two

*In this case certain ordering assumptions must be made.

degenerate minima. The assumption $m_\pi \neq 0$, of course, removes the degeneracy (curve b), but the actual value of m_π is so small that the second minimum is still very much in evidence. We must warn the reader, however, that this does not play a prominent role in the following. Numerically we notice the approximate relations: $g = 13.7$ (from eq. (6)), $\sigma_0 \simeq 68.5$ MeV. Hence $m_\pi \simeq 2\sigma_0$ and $\lambda^2 \simeq 2((m_\sigma^2/m_\pi^2) - 1)$.

Quantum effects in theories of this kind have been discussed mainly by perturbation methods. The customary divergences can be removed* by the addition, to the Lagrangian (3), of suitable renormalization terms. When these are included, the mass- and interaction-parameters ultimately appearing in the field equations are unrenormalized (unphysical) masses and coupling constants. These peculiar complications of renormalization theory do not play any role in the following calculations†; they are mentioned here only to set the stage for the following remark: we have attributed measured or *physical* (i.e., in standard terminology, renormalized) values to the parameters m_N and g appearing in the initial (unrenormalized) Lagrangian, eq. (3). This is not only correct, provided the values of the renormalization terms are suitably chosen, but also in agreement with well-established procedures‡.

Having said this, we shall forget renormalization and proceed to a discussion of the "naive" eqs. (4) and (5) by a semiclassical approximation method in which the field $\phi(x)$ is treated as a c-number. This approximation has been used by various authors under the name of "average field method". It is, of course, analogous to procedures quite often used (Thomas–Fermi method, Hartree or Hartree–Fock approximations). The Thomas–Fermi approximation seems most suited for a first exploration of the problem. In fact, the physical parameters of the problem are not sufficiently well-known at present to justify more elaborate methods.

We consider then the ground state of a system of many nucleons, containing (for simplicity) an equal number of neutrons and protons (this assumption can be easily modified later) and take the ground state expectation value of both sides of eq. (4). This takes care, in the usual way, of the difficulty that the anticommutative field ψ itself connot be approximated classically. The bilinear scalar density, however, in a certain sense can. Its expectation value over the ground state is, of course, time independent, and so is the field ϕ that it generates. Needless to say, this involves some approximation also on the left-hand side of eq. (4), since the expectation values of ϕ^2 and ϕ^3 in $U'(\phi)$

*The σ-model is renormalizable. See B.W. Lee (1972).

†Needless to say, renormalization terms *are* always included in the discussion of higher-order (closed loops) effects, such as those mentioned later.

‡A quite readable popularized version of the rules of this rather complicated game may be found in Coleman (1975). We note in passing that a suitable renormalization term (linear in ϕ) is also responsible for rescuing the validity of eq. (1) from "tadpole" effects.

References p. 887

must be replaced by the square and cube of the expectation value. The rest of the procedure is equally familiar. Assuming ϕ is known, eq. (5) yields the stationary levels of a nucleon and the corresponding wave functions. Filling the levels up to a top Fermi level yields the scalar density as a sum over occupied levels

$$\langle\bar{\psi}\psi\rangle = \sum_i \bar{\psi}_i\psi_i. \tag{9}$$

We must remind ourselves of the difference between the familiar "nucleon density" operator $\psi^\dagger\psi$ and the scalar density $\bar{\psi}\psi = \psi^\dagger\beta\psi$, which counts particles per unit volume with a velocity-dependent weight*, which tends to zero when the velocity v of the particle approaches c. In the Fermi–Thomas approximation, stationary solutions of the Dirac equation (5) are approximated locally by plane waves with momenta k varying up to a maximum $k_F(x)$ which is determined by the local value of the mass coefficient $M(x) = m_N + g\phi(x)$. The expectation value of the scalar density in the ground state of the Fermi gas is

$$\langle\bar{\psi}\psi\rangle = \frac{2}{\pi^2}\int_0^{k_F} k^2\,dk \cdot M(k^2+M^2)^{-1/2}. \tag{11}$$

The local value of k_F is determined by an energy condition

$$(k_F^2 + M^2(x))^{1/2} = E_F \tag{12}$$

where the constant E_F is the maximum total energy of a nucleon in the Fermi gas. The equation for the field ϕ, which is obtained by substituting the values (11), (12) on the r.h.s. of eq. (4) was obtained in Lee and Wick (1974) (see especially eqs. (5.37) and (5.33)) in a different way (a variational principle) and solved by numerical methods. In order to understand what is to be expected, let us first look at a limiting case.

*Some obvious facts may be worth recalling. Consider first a Dirac Hamiltonian: $H = \boldsymbol{\alpha}\cdot\boldsymbol{p} + M\beta$, with M a constant, and the standard (Dirac) representation of the matrices $\boldsymbol{\alpha}$ and β. It describes, of course, free fermions of mass $|M|$. Then, after the usual symmetrization manoeuvers for charge conjugation, the usual density $\psi^\dagger\psi$ counts particles (holes) in positive (negative) energy states with weight $+1$ (-1). The scalar-density operator, instead, counts particles (holes) of momentum k with a weight $w(k)$ $(-w(k))$ where

$$w(k) = M(k^2+M^2)^{-1/2}. \tag{10}$$

In the Thomas–Fermi method $\phi(x)$ is treated as slowly variable and the local value of the mass-coefficient $M(x) = m_N + g\phi(x)$ is inserted in $w(k)$ thus leading to eq. (11). For positive M, $w(k)$ is just the Lorentz-contraction factor $(1-v^2)^{1/2}$ (v being the velocity in relativistic units). Although the circumstance does not occur in our numerical examples, it is amusing to notice that particles will be weighted with the *negative* value of the square root if ϕ becomes so large and negative that M becomes <0. (Reversing the sign of M does not affect the mass of the particles, but it does interchange the roles of "upper" and "lower" components of the ψ-function.)

This is the limit of infinite size*, that is to say: extended homogeneous "nuclear matter" of variable density n (assuming an external pressure is applied). The upper limit k_F, corresponding to the usual relation

$$k_F = (3\pi^2 n/2)^{1/3} \tag{13}$$

is now x-independent; likewise, the ground-state expectation value $\langle\phi(x)\rangle$ of the meson field generated by the nucleons is a constant $\bar{\phi}$, and the local mass factor $M(x)$ becomes a constant effective mass $M = m_N + g\bar{\phi}$. In this case, neglecting quantum fluctuations as before, $\bar{\phi}$ obeys an equation like (4), with the derivative terms omitted, and the density on the r.h.s. replaced by (11). The alternative form of the equation, considered in Bethe (1971), is now a minimum energy condition

$$\frac{d}{d\bar{\phi}}\mathscr{E} \equiv \frac{d}{d\bar{\phi}}[U(\bar{\phi}) + U_N(\bar{\phi})] = 0. \tag{14}$$

The derivative is taken at constant n (constant k_F) and $U_N(\bar{\phi})$ is the total kinetic energy of the Fermi gas

$$U_N(\bar{\phi}) = \frac{2}{\pi^2}\int_0^{k_F} k^2(k^2 + M^2)^{1/2}\, dk. \tag{15}$$

To see what is involved, we plot as an example in the full curve fig. 2 the energy density $\mathscr{E}$ ($= U(\phi) + U_N(\bar{\phi}) - nm_N$; the constant rest energy is subtracted for convenience) as a function of $\bar{\phi}$ for $n = n_0$ (ordinary nucleon density) and for

*This limit exists, of course, only because we have neglected Coulomb interactions.

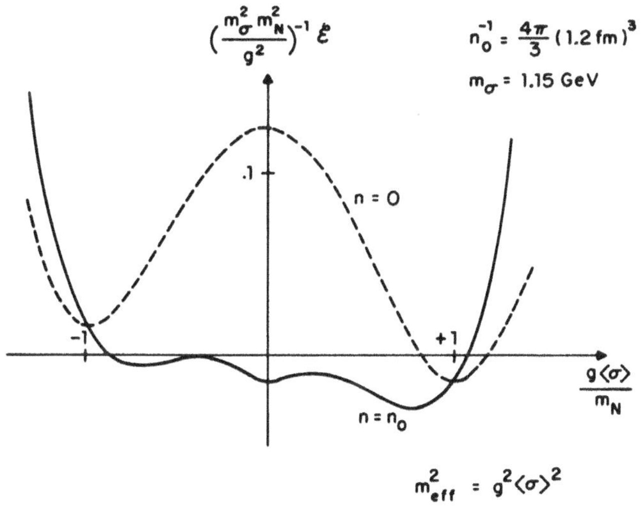

Fig. 2. Energy density $\mathscr{E}$ in the σ model vs. $g\langle\sigma\rangle/m_N$.

References p. 887

the reduced σ-model; actually the abscissa is $1 + (\bar{\phi}/\sigma_0)$. Included in the figure is also the dashed curve corresponding to the vacuum; it exhibits the small asymmetry due to $m_\pi \neq 0$ and requires otherwise no comment. Of course, a particular value of the mass m_σ was chosen, but qualitatively the results would not be very different if m_σ were, say, $m_\sigma = 1$ GeV or 700 MeV. The figure shows that, as was to be expected, the energy minimum which is at $\bar{\phi} = 0$ for the vacuum is shifted in nuclear matter towards negative values of $\bar{\phi}$, so that the effective mass M is reduced ($M = 0.8\, m_N$ at the absolute minimum). If only the nucleon term $U_N(\bar{\phi})$ were present, the best value, of course, would be $M = 0$. At $n = n_0$ we are still far from that, but it is interesting to note that the energy curve already exhibits a second minimum near $M = 0$. If n is increased to exceed a critical value n_C, this minimum will drop below the minimum at the right. Consequently this (no doubt oversimplified) model of nuclear matter undergoes a discontinuous transition at $n = n_C$, see fig. 3. Above $n = n_C$ nuclear matter would be in an "abnormal state" in which the effective nucleon mass M is almost zero (the slight residual mass being due to the chiral asymmetry $m_\pi \neq 0$). Naturally the discontinuous jump in M is accompanied by other discontinuities in total energy, pressure, etc.

The precise value of n_C, of course, depends on both g^2 and m_σ. In this naive model, over a fairly wide range of g^2 and m_σ, we find $n_C \cong 11.6\,(m_\sigma/m_N)^2 \cdot (4\pi/g^2) n_0$, where $m_N \cong 940$ MeV is the nuclear mass and $n_0 \cong 3(4\pi)^{-1}(1.2\text{ fm})^{-3}$ is the normal nuclear density. A plot of the domain $n_C \leqq 2n_0$ in the (g^2, m_σ) plane is given in fig. 4.

It would not be hard to pursue the investigation of this particular model of nuclear matter even further and to show that indeed the energy per nucleon attains a minimum at a density above n_C; that is to say, the equilibrium state of nuclear matter has "abnormal" density and binding energy. Similar results

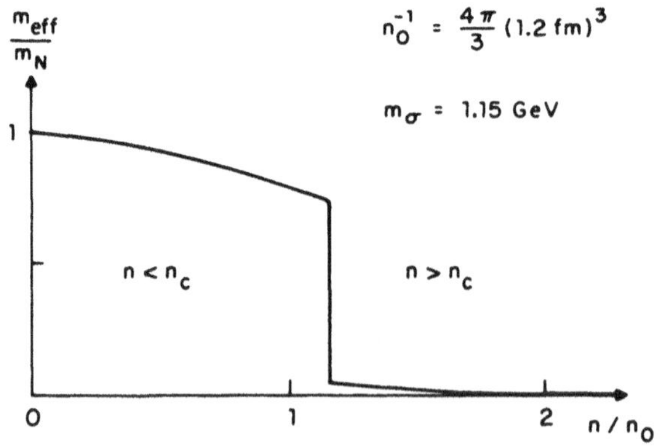

Fig. 3. Discontinuous transition of the field as a function of density. The ordinate is proportional to $m_N + g\bar{\phi}$.

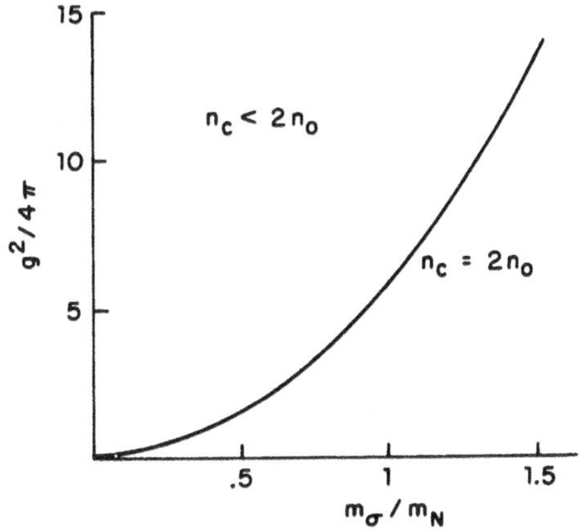

Fig. 4. Illustrates relationship between critical density and the parameters m_σ and g. The solid curve separates the regions where $n_C > 2n_0$ and $n_C < 2n_0$ respectively, $n_0 = [\frac{1}{3}4\pi(1.2\ \text{fm})^3]^{-1}$ being the nucleon density of ordinary nuclei.

were also obtained in the numerical calculations on finite nuclei mentioned earlier. Having fulfilled our first purpose, however, we shall proceed without further ado to introduce some additional features which are designed to bring the model closer to reality.

4. The effect of short-range repulsion

As we said before, the most serious deficiency of the Lagrangian (3) is the total absence in it of a field that would account for the short-range repulsion between nucleons; there is evidence for the existence of this repulsion from scattering experiments, and its presence is often attributed to the interaction of nucleons with vector fields, corresponding to spin-one particles such as, for instance, the ω^0-meson. The fairly large mass ($m_\omega = 783$ MeV) and, in addition, the somewhat more singular character of the interaction generated by such a field would account for the short-range nature of the repulsion.

In principle, the addition to our Lagrangian of terms corresponding to a vector field would be a simple matter and would also be entirely consistent with our approach. Other authors (Walecka 1974 for example) have done precisely this in dealing with the equation state of ordinary nuclear matter and of neutron star matter; having adopted the "average field" approximation for the scalar field (just as we do), these authors then, naturally enough, proceed to apply the same method to the vector field. It is in fact hard to see what else one could do, without unreasonable effort, in a purely field-theoretic approach. The main objection to this method is that it does not take into account positional correlations between neighbouring particles, an error which must

References p. 887

necessarily become much more serious in the case of the repulsive force with its strong r-dependence at short distances. For example, in the case of a many-body system with strong repulsive potential interaction V of range a, the energy change is proportional to a, *not* to the average value of V, which is ∞ for a system of hard spheres.

It is for this reason that one of the present writers (see Lee 1974b) adopted a less fundamental, but probably more accurate, method of estimating the effect of short-range repulsion, based on a suggestion by Bohr and Mottelson (1969). Essentially it consists in revising the dependence of the top Fermi momentum k_F on the density n of the Fermi gas. Qualitatively the nature of this correction is obvious: owing to the short-range repulsion, the space available to each particle is reduced and the Fermi momentum is increased correspondingly. Let us denote the "volume per particle" as usual by

$$n^{-1} = \tfrac{4}{3}\pi r^3. \tag{16}$$

We shall assume the dependence on r of the revised Fermi momentum to be

$$K_F = \tfrac{1}{2}(9\pi)^{1/3}(r - r_0)^{-1}, \tag{17}$$

where r_0 is a constant. Equation (13) corresponds to $r_0 = 0$. Equation (17) may be regarded as an approximation to well-known results on a Fermi-gas of hard spheres; if d is the diameter of the hard spheres, setting $r_0 = 0.8\,d$, eq. (17) reproduces correctly the average kinetic energy of the gas including terms of order d/r. Comparison of higher order terms suggests that, in the case of hard spheres, the approximation over-estimates the effect of repulsion. Since r_0 is here treated as an adjustable constant, some compensation of this will occur automatically. The reference to hard spheres, in order to justify eq. (17), has been criticized by Pandharipande and Smith (1975) as unrealistic. We surmise that there probably is not a semi-empirical formula containing a single adjustable parameter that does a better job than eq. (17). At the present state of the art, introducing more parameters would not do us any good, since we would not know how to determine them.

There is no reasonable way to determine the parameter r_0 from scattering data. Its definition, as given implicitly in eq. (17), indicates that r_0 should be determined by fitting data on ordinary nuclear matter. In so doing we also have m_σ as an additional free parameter; we shall also allow ourselves some additional freedom by adjusting the coupling constant g; this is not unreasonable if we take into account the familiar dependence of renormalized coupling constants on the renormalization point.

It should be clear that the purpose of our next calculation is *not* to show how well the above model can fit ordinary nuclear matter, with its very delicate balance between positive and negative contributions to the binding energy. If that were our main purpose, we would know a priori that our model is still too simple; for instance additional contributions to the attractive force would be required*, the normal state will be for us only a means to verify that the model we use, and the attendant parameters, are not so far removed from reality as to be of no predictive value whatsoever.

Assuming certain values for the constants m_σ, g and r_0, the binding energy per nucleon: $n^{-1}\mathscr{E}$ is obtained from $\mathscr{E} = U(\bar{\phi}) + U_N(\bar{\phi}) - nm_N$ if in (15) we replace k_F by the value (17). We must look for the existence of minima of $n^{-1}\mathscr{E}$ as a function of the two variables $\bar{\phi}$ and r. In addition we require these minima to have positive binding energy. The question has been investigated numerically (see Lee and Margulies 1975). If a minimum is found with σ close to zero ($\bar{\phi}$ close to $-m_N/g$) and with rather a large density, we call this an abnormal state; fig. 5 gives the result of calculations for $m_\sigma = 700$ MeV and for two different values of g. It shows, as expected, that the minimum disappears when r_0 exceeds a value dependent on g. Nevertheless, the abnormal state exists over a fairly large region of values of the parameters; this is because the relevant scale of binding energies is of the order of 100–200 MeV; the abnormal minimum is not easily destroyed.

The normal state, if the model exhibits one, should appear as another

*The additional parameters involved in the new terms are not entirely free, but a certain amount of adjustment, compatible with scattering experiments on free nucleons, is certainly possible. Since our crude model comes pretty close to an adequate description of the normal state, we surmise that with additional terms it could give a quite good fit to the normal state if one *tries hard enough*. We shall freely admit that contrary opinions have been expressed in the literature (Pandharipande and Smith 1975, Nyman and Rho 1976).

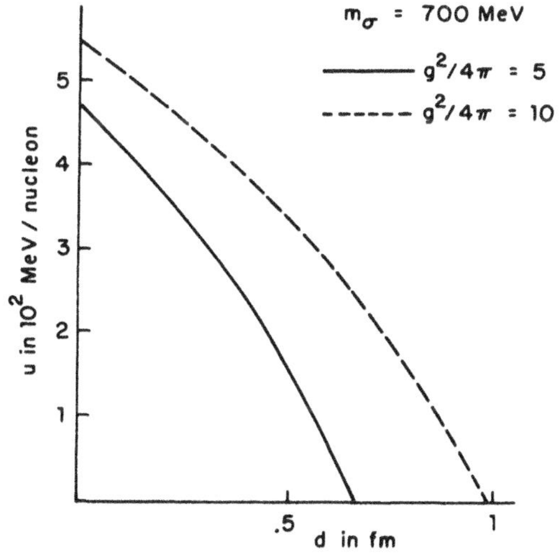

Fig. 5. Binding energy per nucleon u versus the constant r_0 of eq. (17). In abscissa: the "sphere diameter" $d = 1.25\, r_0$.

References p. 887

minimum with $\bar{\phi}$ close to zero. This second minimum exists, but only within a very narrow range in the (m_σ, r_0) plane. We would not interpret this as meaning that this is a good way to determine the parameters. One should probably relax the conditions somewhat. The positive binding energy of the normal state, as pointed out before, is too delicate an effect for these crude calculations; it can be easily destroyed. Let us nevertheless mention that, for instance, $r_0 = 0.5\,\text{fm}$, $m_\sigma = 750\,\text{MeV}$ and $g^2/4\pi = 5$, or $m_\sigma \simeq 1100\,\text{MeV}$ and $g^2/4\pi = 10$ lie in the region where the second minimum exists. These values seem reasonable.

5. Higher order corrections

Our discussion so far has been restricted to a semiclassical treatment in analogy with the Thomas–Fermi approximation in the case of atomic systems. We must now briefly examine what is known about the most thorny aspect of the problem, namely the effect of quantum fluctuations, or, in the language of perturbation theory, the effect of diagrams involving arbitrary closed loops.

Before we go into any details, let us discuss briefly what benefits one can expect a priori from such an endeavor. Needless to say, it is important in calculations that may involve considerable time and effort to keep some perspective on the value of what one is doing. The a priori objections to any attempt to discuss strong coupling theories by methods involving an expansion in powers of the coupling constants have been repeated ad nauseam; often they are stated as a ritual exorcism preliminary to ignoring them. Now there are, of course, well-known and useful procedures (we shall, in fact, refer to them shortly) whereby in the treatment of certain effects we can by-pass the limitations of the perturbation method. What is left, however, in the general problem, after various selective summations and ingenious rearrangements, is still an infinite expansion, and if the basic dimensionless parameters are large we have no right to assume that any finite number of terms will bear a close relationship to the total sum. (Even greater caution is suggested by the following remark: An intriguing feature of the perturbation expansion is that, in calculable examples, it often exhibits an oscillating behavior; for strong coupling, the *oscillations of course increase in size as one goes on.*)

In connection with our model theory of abnormal nuclear matter one can now pose the following question: Is there some limiting case in which (a) the model exhibits a phase transition to an abnormal state, and (b) a consistent treatment by perturbation theory is possible? As Lee and Margulies (1975) have pointed out, this case exists when g is small ($g^2/4\pi \ll 1$) but the nucleon density is large enough to compensate for the smallness of g (it must generate an average field $\bar{\phi}$ of the order of $\bar{\phi} = -m_N/g$). Clearly this case is not "realistic", but at least it lends itself to a consistent mathematical treatment. There is some hope, in this way, to gain some insight into the physical nature of the alleged phase transition of nuclear matter to an abnormal state.

The calculations are too complicated to be reproduced here, but we shall outline the general nature of the results and the procedure employed to obtain them. When our Lagrangian, see eq. (3) and eq. (20) later, is treated by perturbation methods the immediate result is an expansion in powers not only of g, but also in powers of the parameters b and c. There are, however, certain relations between these parameters and , on closer inspection, one can regroup the terms of the expansion essentially in powers of g^2. More specifically the energy density $\mathscr{E}(n)$ for a given nucleon density n (i.e. for a given value k_F of the Fermi momentum) can be expanded

$$\mathscr{E}(n) = \mathscr{E}_0 + \mathscr{E}_1 + \mathscr{E}_2 + \dots, \tag{18}$$

where

$$\begin{aligned} &\mathscr{E}_0 \sim O(k_F^4), \\ &\mathscr{E}_1 \sim O(g^2 k_F^4), \\ &\mathscr{E}_2 \sim O(g^4 k_F^4) \qquad \text{or} \qquad O(g^4 k_F^4 \ln g^2) \end{aligned} \tag{19}$$

etc. Since $k_F \sim O(m/g)$ is very large, this is a weak-coupling but high-density expansion.

Before we apply perturbation theory, however, one must face the (relatively easy) problem of grafting the perturbation theoretic treatment on the "semiclassical" solution corresponding to the abnormal state. Stated another way, since we are aiming at a situation in which $\bar{\phi} \approx -m_N/g$ it is not immediately clear what an expansion in powers of g means. We anticipate that the answer lies in one of the rearrangement.procedures so often used in field theory.

Let us in fact assume that our model is treated, to begin with, by means of the conventional perturbation theoretic expansion. In Feynman language, this means summing over graphs of arbitrary complexity, with two kinds of lines representing free propagators, i.e. continuous lines (say) for nucleons and dashed lines for mesons. The meson lines are not *oriented* (do not carry an arrow), but a nucleon line representing the propagator $G(x, x')$ is oriented from x' to x. All these lines, of course, connect *vertices*, of which in our theory there are three types, corresponding (in the usual way) to the meson nucleon interaction $g\phi\bar{\psi}\psi$ and to the cubic and quartic terms in $U(\phi)$. In addition there will be vertices corresponding to renormalization terms, but we shall not discuss these here. It is more important to recall certain small differences between the problems of ordinary field theory, where the lowest state is the *vacuum*, and the present case. As in the usual many-body problem of a Fermi gas, our starting point is an (unperturbed) ground state Φ_0 of free

References p. 887

nucleons filling all positive energy states up to a maximum Fermi momentum k_F. In addition, since the treatment is relativistic we must treat the "negative sea" as filled, that is, there are no *holes*. The state Φ_0 replaces the "mathematical vacuum" of ordinary field theory, and consequently the nucleon lines in a graph represent a propagator $G(x, x')$ defined as an expectation value relative to Φ_0. Thus the existence of occupied positive energy states gives rise to certain differences in the unperturbed propagator (see Galitskii and Migdal 1958); another important difference, of course, is that the nucleon density and correspondingly the scalar density are not zero in the ground state Φ_0.

In spite of these differences, many things we are used to remain unchanged, or nearly so; for instance if we ignore (or indicate by dots) the $i\epsilon$-terms (which *are* changed), we can write for the propagator in momentum space the usual $(i\gamma p + m + \ldots)^{-1}$. Likewise, the nucleon density, or the scalar density, must be related as usual to the trace of $\gamma^4 G(x, x)$, or respectively $G(x, x)$. That this requires taking a limit $x' \to x$ in $G(x, x')$ is not a serious problem; we know what the answer must be (a sum over positive occupied states) and take the limit accordingly*.

Let us note parenthetically that, in a diagram, $G(x, x)$ corresponds to an oriented line bent into a closed loop (or as our learned friends say: "a snake eating its own tail"). This is the only closed loop that does not represent a quantum effect, and already shows up in the semiclassical approximation. Its non-zero value is, of course, the trigger that generates an infinite sequence of tree-graphs (with loops attached to all terminals except one) representing the expectation value of the field ϕ in the presence of the nucleon Fermi-gas (fig. 6). Instead of trying to sum this series, and following Lee and Margulies (1975), we use the standard device to introduce the expectation value $\bar{\phi}$ as an

*In a higher order of approximation polarization effects related to the negative sea require some care as usual.

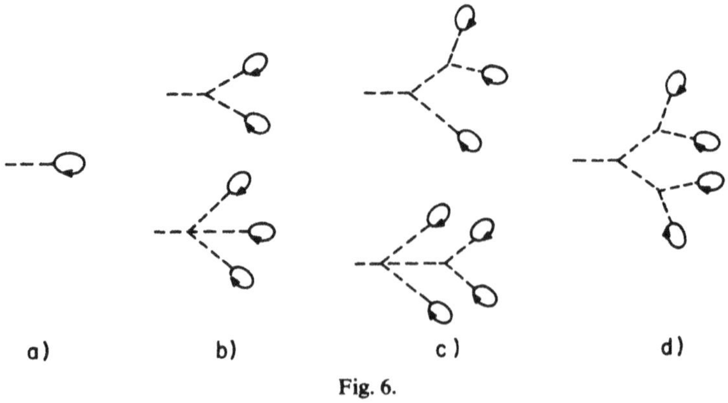

Fig. 6.

unknown quantity and again "shift" the field by taking $\chi = \phi - \bar{\phi}$ as a new variable. Later on, we shall determine $\bar{\phi}$ by the condition $\langle\chi\rangle = 0$ *in the presence of nuclear matter*; this is, of course, the analogue of (1).

We must now replace ϕ by $\chi + \bar{\phi}$ in the Lagrangian and start all over again. We spare the reader the details. Notice only that we will have for example

$$U(\phi) = U(\bar{\phi}) + U'(\bar{\phi})\chi + \tfrac{1}{2}\bar{a}\chi^2 + \frac{1}{3!}\bar{b}\chi^3 + \frac{1}{4!}\bar{c}\chi^4,$$

$$m_N + g\phi = M + g\chi, \qquad (M = m_N + g\bar{\phi}), \tag{20}$$

where $\bar{a} = U''(\bar{\phi})$, etc. We have a new perturbation theory with different parameters, but very similar to the old one. Since M is a constant we shall use it as the effective nucleon mass in the nucleon propagator. The closed loops (scalar density) will now be calculated with this value of the mass. The free meson propagator also gets modified; the square of the original meson mass, e.g. m_σ^2 in eq. (8) is now replaced by $\bar{a}$. Finally we still have 3-meson and 4-meson vertices, but with different coefficients. The perturbation expansion is now an expansion in powers of these new coefficients as well as in powers of g.

It is now easy to see that in this new expansion of $\mathscr{E}(n)$ the term of zero order in g, b, c, is of the form $\mathscr{E}_{qcl} + \Delta\mathscr{E}_{vac}$ where $\mathscr{E}_{qcl} = U(\bar{\phi}) + U_N(\bar{\phi})$ is precisely the expression for the energy density that was obtained in the quasiclassical treatment of sect. 3; see eqs. (14) and (15). (We have therefore achieved our first aim of bridging the gap between that treatment and perturbation theory.) Already at this zero order, however, we have somewhat more subtle effects, represented by the term $\Delta\mathscr{E}_{vac}$, which, as the notation indicates, correspond to shifts in the energy of the vacuum due to the presence of an average field $\bar{\phi}$. Part of $\Delta\mathscr{E}_{vac}$ is due to the change in mass of the meson field ($\bar{a}$ is the new mass^2), and part of it is due to the change in mass of the nucleons.

Terms of higher order in g, b, c are to be calculated by graphs containing two or more closed loops. For details on these calculations the reader is referred to Lee and Margulies (1974) and Margulies (1976). We shall merely reproduce here, in fig. 7, the results of two different estimates of $\Delta\mathscr{E}_{vac}$ (the one denoted $\Delta\mathscr{E}'_{vac}$ includes in a rough way certain higher order effects). The potential $U(\bar{\phi})$ is also shown for comparison. (In the figure the field $\sigma = \sigma_0 + \phi$ is used instead of ϕ). Perhaps the most striking feature of the result is the smallness of these higher order effects (in spite of the rather large value of g), due to the fact that fermion loops and boson loops give correction terms of opposite signs, so that one is tempted to conclude that the overall description of the state of nuclear matter may not be affected in a significant way by them.

References p. 887

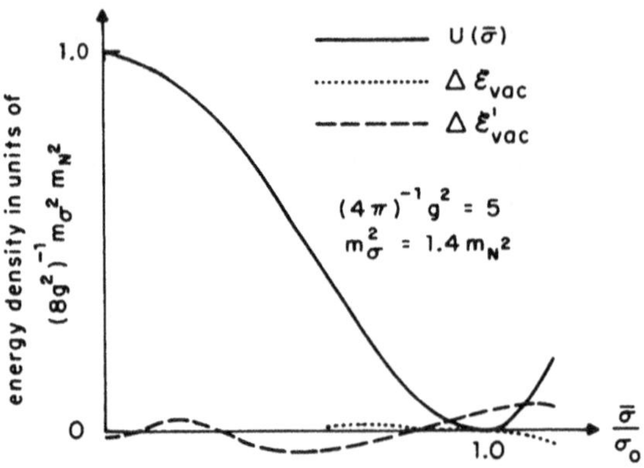

Fig. 7. Various pieces of the energy density vs. $\bar{\sigma}/\sigma_0$. In units of $(8g^2)^{-1}m_\sigma^2 m_N^2$, $U(\bar{\sigma}) = [1-(\bar{\sigma}/\sigma_0)^2]^2$, shown by the solid curve. In the same units the dotted curve denotes $\Delta\mathcal{E}_{vac}$ (see text). The dashed curve denotes $\Delta\mathcal{E}'_{vac}$, a corrected form of $\Delta\mathcal{E}_{vac}$ including higher order corrections.

6. Final remarks

The investigation of higher order effects, described above, indicates that, at least when the expansion method employed is valid, the general character of the quasiclassical solution is not seriously affected by the inclusion of quantum corrections. Needless to say the limits of validity of the expansion are such that the case for the existence of an abnormal state is not strengthened very much by these calculations.

On the other hand one has certainly the right to derive some assurance from another kind of consideration, of a more qualitative kind. There is a rather obvious analogy between the abnormal state and the Bose–Einstein condensation phenomenon. Now it is well-known that a satisfactory treatment of this phenomenon is only possible for rather idealized situations (bosons with weak interactions or small scattering lengths). The actual situation in ^{4}He is quite different from the idealized one. Nevertheless the phenomenon actually occurs and the condensed system exhibits superfluidity. Recently even ^{3}He has been found to exhibit superfluidity. The much wider field of superconductivity is another case in point. A priori one would speculate that the occurrence of such "coherence" phenomena requires a rather critical set of conditions. The lesson one seems to get from experience is that they tend to occur if only given a chance. There is often a limit beyond which a priori calculations cannot go. For the moment at least it would not seem wise to expect theory to tell us with certainty ahead of time what we will find in this unexplored area of experimental research.

References

Bethe, H., 1971, Ann. Rev. Nucl. Sci. **21**, 93.

Bohr, A. and B.R. Mottelson, 1969, Nuclear structure, vol. 1 (Benjamin, New York) p. 256.

Brown, G.E. and W. Weise, 1976, Phys. Rep. **27C**, 1.

Chin, S.A., 1976, Phys. Lett. **62B**, 263.

Coleman, S., 1975, Secret symmetry: an introduction to spontaneous symmetry breakdown and gauge fields, in Laws of hadronic matter, 1973 lectures, International school of subnuclear physics (ed. A. Zichichi) (Academic Press, London).

Galitskii, V.M. and A.B. Migdal, 1958, Zh. Eksp. Teor. Fiz. **34**, 139. (Transl. Sov. Phys. JETP **7**, 96.)

Gell-Mann, M. and M. Lévy, 1960, Nuovo Cimento **16**, 705.

Lee, B.W., 1972, Chiral dynamics (Gordon and Breach, New York).

Lee, T.D., 1974a, Abnormal nuclear states, in Proceedings of the 17th int. conf. on high energy physics (ed. J.R. Smith) (Science Research Council) sect. 3, p. 81.

Lee, T.D., 1974b, Rev. Mod. Phys. **47**, 267.

Lee, T.D., 1975, Abnormal nuclear states and vacuum excitations, in Lepton and hadron structure, vol. 12 (ed. A. Zichichi) (Academic Press, London) p. 696.

Lee, T.D. and M. Margulies, 1975, Phys. Rev. **D11**, 1591. (See also erratum, Phys. Rev. **D12**, 4008.)

Lee, T.D. and G.C. Wick, 1974, Phys. Rev. **D9**, 2291.

Lee, T.D. and C.N. Yang, 1960, Phys. Rev. **117**, 22.

Margulies, M., 1976, Phys. Rev. **D13**, 1621, 1642.

Migdal, A.B., 1971, Zh. Eksp. Teor. Fiz. **61**, 2210.

Migdal, A.B., 1972, JETP **34**, 1184.

Migdal, A.B., O.A. Markin and I.N. Mishustin, 1974, Zh. Eksp. Teor. Fiz. **66**, 443.

Migdal, A.B., O.A. Markin and I.N. Mishustin, 1976, Zh. Eksp. Teor. Fiz. **70**, 1592.

Nyman, E.M. and M. Rho, 1976, Nucl. Phys. **A268**, 408.

Pandharipande, V.R. and R.A. Smith, 1975, Phys. Lett. **59B**, 15.

Polkinghorne, J.C., 1958, Nuovo Cimento **8**, 179, 781.

Sawyer, R.F., 1972, Phys. Rev. Lett. **29**, 382.

Scalapino, D.J., 1972, Phys. Rev. Lett. **29**, 386.

Schwinger, J., 1957, Ann. Phys. **2**, 407.

Serber, R., 1976, Phys. Rev. **C14**, 718.

Walecka, J.D., 1974, Ann. Phys. **83**, 491.

Walecka, J.D., 1975, Phys. Lett. **59B**, 109.

CU - TP - 173

QCD AND THE BAG MODEL OF HADRONS

T. D. Lee

Columbia University, New York, N. Y. 10027

A talk delivered at the Guangzhou (Canton) Conference on Theoretical Particle Physics, January 1980.

This research was supported in part by the U.S. Department of Energy.

Abstract

By regarding the QCD vacuum as a perfect, or nearly perfect, dia-electric medium, we show how the quark confinement mechanism can be understood. The color dielectric constant κ characterizes the long-range order effect of QCD. The hadrons are viewed as soliton solutions with $\kappa = 1$ inside and $\kappa = 0$, or $\cong 0$, outside. The hadron masses and other dynamic properties can then be calculated in terms of the power series expansion of $\alpha = g^2/4\pi$ around these soliton solutions, where g is a coupling inside the hadron.

1. Vacuum As a Color Dia-electric

In this talk I would like to discuss some of my recent efforts to link QCD as a fundamental theory to the various bag models, such as the MIT bag[1] and the SLAC bag[2]. My approach will essentially be phenomenological.

To give QCD a well-defined meaning, we first contain the whole system within a volume of size L^3 to avoid infrared difficulties. In QCD there are quark fields ψ_q^a where q, the flavor index, $= u, d, s, \cdots$, and a, denoting the color, $= 1, 2, 3$. In addition, we have the gauge fields V^i where $i = 1, 2, \cdots, 8$ representing the color index of the gauge-field quanta. Let g_L be the renormalized coupling constant in the long wavelength limit, momentum $k \sim L^{-1}$. It is possible to prove under rather general assumptions and valid to all orders in $(\text{coupling})^2$

$$g_L > g_\ell \qquad \text{if} \quad L > \ell \quad . \tag{1.1}$$

This result is connected with the familiar "asymptotic freedom" property[3] of the theory: When $\ell \to 0$, g_ℓ decreases to 0. As there is no scale in the theory (assuming that all quarks are of zero mass), this implies that when ℓ increases, g_ℓ must also increase, which leads to (1.1). The difficulty lies in the infrared limit. When $L \to \infty$, it seems likely that g_L may $\to \infty$, or at least $\gg 1$. Since the true physical system is one with $L = \infty$, we may always be in the ultra-strong coupling limit. As we shall see, this difficulty can be resolved by regarding the vacuum as a color dia-electric medium.

Let us introduce κ_L which is called the color dielectric constant of the vacuum in a volume L^3. As a convention, we shall adopt a standard renormalized coupling constant g , defined by

$$g = g_\ell \quad \text{when } \ell = \text{some arbitrarily chosen length, say. the proton radius.} \tag{1.2}$$

The color dielectric constant κ_L will then be defined as

$$g_L^2 = \frac{g^2}{\kappa_L} \quad . \tag{1.3}$$

Consequently, in accordance with (1.2)

$$\kappa_\ell = 1 \quad \text{when } \ell = \text{proton radius}. \tag{1.4}$$

Equation (1.1) now implies

$$\kappa_L < \kappa_\ell \quad \text{if } L > \ell \quad . \tag{1.5}$$

To the second order[3] we have,

$$\frac{\kappa_L}{\kappa_\ell} = \frac{1}{1 + \frac{1}{2\pi}\frac{g^2}{4\pi}(11 - \frac{2}{3} n_f) \ln \frac{L}{\ell} + O(g^4)} \tag{1.6}$$

where n_f is the number of quark flavor varieties (assumed to be < 17) . With the convention (1.4) we find that the color dielectric constant of the vacuum in an infinite volume $L = \infty$ is

$$\kappa_\infty < 1 \tag{1.7}$$

which is valid to all orders in g^2 . The ultra-strong coupling difficulty mentioned before corresponds to

$$\kappa_\infty = 0 \quad , \tag{1.8}$$

or

$$\kappa_\infty \ll 1 \quad . \tag{1.9}$$

In the former we call the vacuum in QCD a perfect color dia-electric medium, in the latter a nearly perfect color dia-electric medium.

2. A Hypothetical Problem

To see how this will lead to quark confinement and bag models, let us consider a hypothetical problem in classical electromagnetism. In quantum electrodynamics, our usual convention is to set the dielectric constant of the vacuum state $\kappa_{vac} = 1$. It is then possible to prove that all physical media have their dielectric constants $\kappa > 1$. This can be seen most easily by using the familiar formula

$$\vec{D} = \vec{E} + 4\pi\vec{P} \qquad (2.1)$$

where $\vec{D}$ is the displacement vector, $\vec{E}$ the electric field, and $\vec{P}$ the polarization vector. Since under $\vec{E}$, all atoms have their polarization $\vec{P}$ in the same direction as $\vec{E}$, so as to produce a screening effect, we have $\kappa > 1$.

In this section, we shall consider a hypothetical problem. Let us imagine that in classical electromagnetism, but without the quantum theory of atoms, there could be a medium with its dielectric constant

$$\kappa \equiv \kappa_{med} << 1 , \quad \text{or} \quad \cong 0 , \qquad (2.2)$$

i.e., this hypothetical medium is anti-screening. Now, suppose we place a small charge distribution ϵ in the medium. As we shall show, no matter how small ϵ is, the medium will develop a hole surrounding the charge. Inside the hole, we have $\kappa = 1$, but outside $\kappa = \kappa_{med}$, as shown in Fig. 1. To see this, let us assume that such a hole is formed. Because of the anti-screening nature of the medium, the induced charge on the inner surface of the hole is of the same sign as ϵ.

Consequently, if we want to reduce the size of the hole, we must do work to overcome the repulsion between ϵ and the induced charge. That work is infinite if the hole is to be eliminated in toto. Hence the hole will not disappear. [This

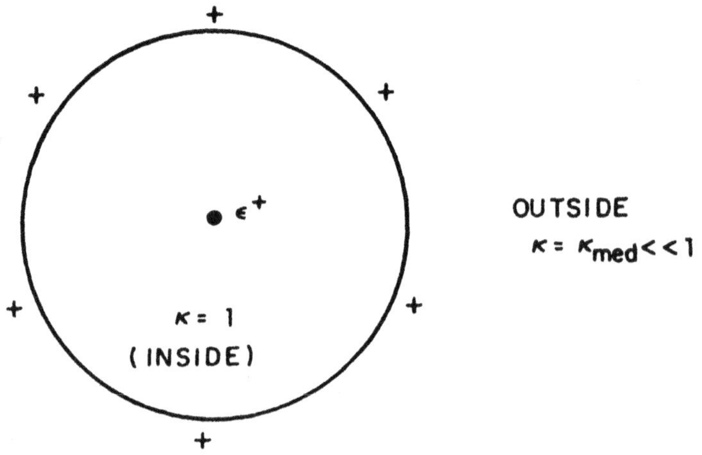

Fig. 1

situation is completely different for a normal medium whose dielectric constant is > 1. Imagine that a similar hole were created. The corresponding induced charge would be of the opposite sign to ϵ. The hole would automatically shrink to 0, resulting in the usual Coulomb field distribution around the charge. The medium will remain homogeneous if ϵ is sufficiently small.] Here, no matter how small ϵ is, when $\kappa_{med} \to 0$ inhomogeneity in the medium in the form of holes must occur.

We may estimate the radius R of such a hole. Let D_{out} and E_{out} be, respectively, the normal components of $\vec{D}$ and $\vec{E}$ outside the hole when $r = R$. Similarly, let D_{in} and E_{in} be the corresponding components inside the hole when $r = R$. For a spherical hole we have

$$D_{in} = E_{in} = D_{out} = \frac{\epsilon}{R^2}$$

and

$$E_{out} = \frac{\epsilon}{\kappa_{med} R^2} ,$$

where ϵ is the total charge of this small charge distribution, which is assumed to have a non-zero extension. The electric energy inside the hole is finite and independent of κ_{med}. The electric energy outside the hole is given by the volume integral of $\vec{D} \cdot \vec{E}$, which is $\propto \kappa_{med}^{-1}$. Therefore, when $\kappa_{med} \to 0$, the total electric energy U_{el} is

$$U_{el} \sim \frac{1}{2} \frac{\epsilon^2}{\kappa_{med} R} . \tag{2.3}$$

In addition, there is the energy U_{hole}, needed to create such a hole. The amount U_{hole} is a function of R. When R is large, U_{hole} should be proportional to the volume plus a term proportional to the surface, etc. We may write

$$U_{hole} = \frac{4\pi}{3} R^3 p + 4\pi R^2 s + \cdots \tag{2.4}$$

where $p, s, \cdots$ are positive constants. The total energy of the system is given by

$$M = U_{el} + U_{hole} , \tag{2.5}$$

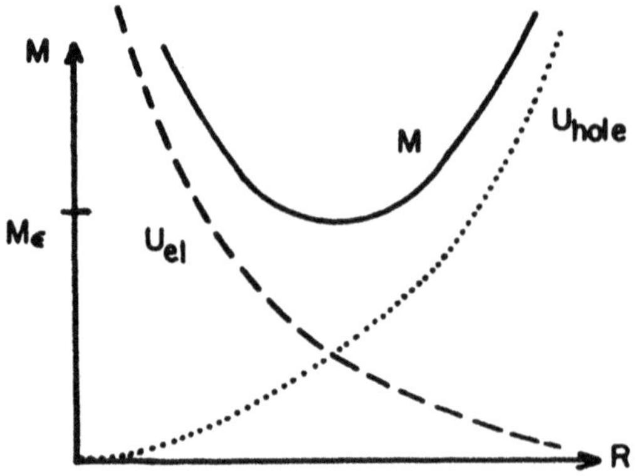

Fig. 2

which is drawn schematically in Fig. 2. Its minimum, as determined by $dM/dR = 0$, gives (in the approximation $U_{hole} \cong \frac{4\pi}{3} R^3 p$)

$$M \equiv M_\epsilon \sim \frac{4}{3} \left(\frac{\epsilon^2}{2\kappa_{med}} \right)^{3/4} (4\pi p)^{1/4}$$

from which we can draw the conclusion that, if the total charge $\epsilon \neq 0$, then $R \neq 0$. Furthermore, the total energy

$$M_\epsilon \to \infty \quad \text{when} \quad \kappa_{med} \to 0 . \tag{2.6}$$

It is not difficult to show that the same conclusion can also be reached without the approximation $U_{hole} \cong \frac{4\pi}{3} R^3 p$.

Next, we replace the single charge distribution by a dipole distribution; i.e. two small charge distributions of total charge ϵ^+ and ϵ^-. It is easy to see that when κ_{med} is sufficiently small, the minimal energy state again requires the formation of a hole around these two charges. As before, inside the hole $\kappa = 1$, and outside $\kappa = \kappa_{med}$. When $\kappa_{med} \to 0$, at the surface of the hole the electric field inside should be parallel to the surface so that $\vec{D}$ is 0 outside, as shown in Fig. 3. Thus, U_{el} remains finite, as does the total energy $M_{\epsilon^+\epsilon^-}$:

$$M_{\epsilon^+\epsilon^-} = \text{finite} \quad \text{when} \quad \kappa_{med} \to 0 . \qquad (2.7)$$

If we try to separate the two charges ϵ^+ and ϵ^- by an infinite distance, then because of (2.6) the work required must also be infinite. This is the analog to "quark confinement" in our hypothetical problem.

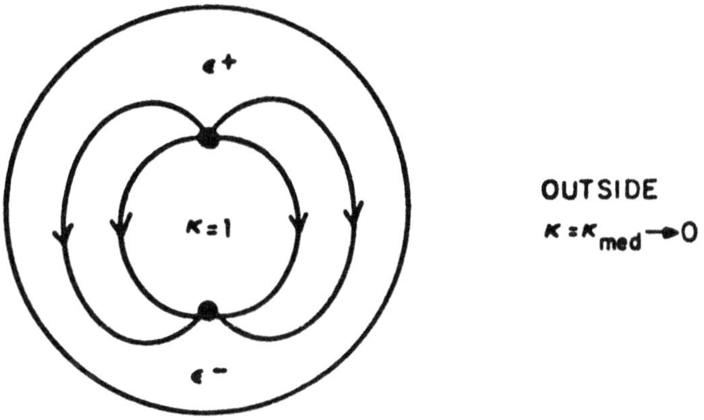

Fig. 3

3. Quark Confinement

We now return to the corresponding problem in QCD, discussed in Section 1. The vacuum is assumed to be a perfect (or nearly perfect) color dia-electric medium. By following the same argument given in the above section, we see that whenever there are quarks or antiquarks there must be inhomogeneity in the space surrounding these particles. We may call these either bags[1,2], or domain structures, or nontopological soliton solutions[4,5], or (if the total color is zero) simply "hadrons". which is 0, or << 1. If the total "color" is non-zero, the mass of such a bag would be infinite when $\kappa_\infty \to 0$, in analogy with (2.6). However, just as in (2.7), the bag mass remains finite when $\kappa_\infty \to 0$, if inside the bag one has a color singlet such as

$$\text{meson:} \quad \bar{q}^a q^a \quad , \quad \text{or} \quad \text{baryon:} \quad \epsilon_{abc}\, q^a q^b q^c$$

where a, b, c are color indices which can vary from 1 to 3. (See Fig. 4.)

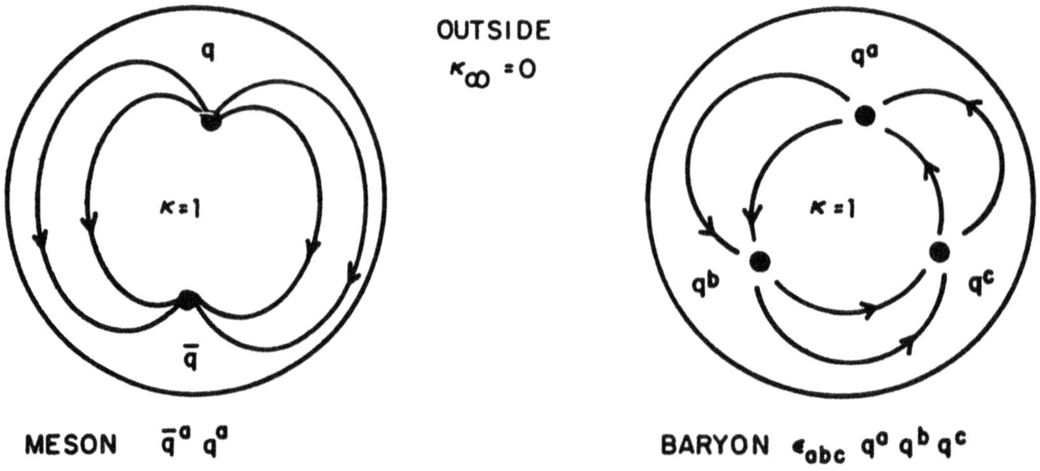

Fig. 4

From this one can derive that the work required to separate the quarks to a large distance r is approximately proportional to r. The quark confinement is then "explained" by the assumption that the vacuum in QCD is a perfect (or nearly perfect) color dia-electric.

Of course, at present it is not known that the mass M_q of a truly free quark is indeed infinite. However, the fact that no free quark has been observed so far in the final state of any high energy collision sets a lower bound $M_q > 5$ GeV for $q = u, d$ and s. It can be shown that this lower bound implies[6]

$$\kappa_\infty < .013 \left(\frac{g^2}{4\pi}\right) \qquad (3.1)$$

where g is defined by (1.2).

4. A New Power Series Expansion

One may wonder in what sense we have by-passed the difficulty of ultra-strong coupling, mentioned previously. To understand this, let us consider other problems in physics in which interactions are also ultra-strong and yet there are no mathematical difficulties. A good example is the interaction potential $V(r)$ between two He atoms. As shown in Fig. 5, for $r < a$, the potential V is $\cong$ infinite;

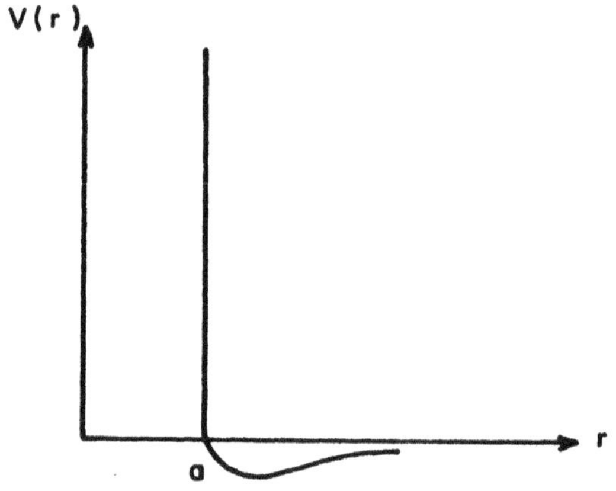

Fig. 5

for $r > a$, V is small. Because the strong potential is of a repulsive nature, the two He atoms will avoid the strong interaction region, and therefore this strong interaction does not present any difficulty. The usual method is to replace the strong repulsive part of V by a hard-sphere potential which leads to a new boundary value problem: The distance r between the two He atoms is restricted to a region $r \geq a$ in which the potential $V(r)$ is relatively weak, and therefore may be regarded as a perturbation.

From Fig. 4, we see that when $\kappa_\infty \to 0$, i.e. $g_\infty^2 = g^2/\kappa_\infty \to \infty$, the quarks are confined to a domain in which $\kappa = 1$. Thus the ultra-strong region, $g_\infty \to \infty$, exerts a repulsion against the quarks and antiquarks. Just as in the above problem of He-He interaction, the particles automatically stay away from the ultra-strong interaction region. Because of Lorentz invariance, the corresponding boundary value problem in the present case requires the relativistic soliton solution, as has been given in the literature.[5,6] When one expands around these soliton solutions, only the coupling g inside the soliton is the relevant parameter, and that resolves the difficulty. [It is only when one attempts to make a simple-minded plane wave expansion around a homogeneous background, that one has to use g_∞ as the expansion parameter. Exactly the same kind of difficulty would appear if one were to force a similar plane-wave expansion in the aforementioned He - He problem.] Because QCD is asymptotically free, the quarks inside the hadron behave approximately like free particles; furthermore, their effective masses can be relatively small.

In the following we shall discuss how to obtain such a power series expansion in terms of the g^2 inside the hadron (not the g_∞^2 outside). For example, the hadron mass M can be expanded

$$M = M_0 + \alpha M_1 + \alpha^2 M_2 + \cdots \tag{4.1}$$

where $\alpha = g^2/4\pi$ and g^2 is defined by (1.2).

Since our theory is relativistic invariant, the velocity of light $c/(\kappa\mu)^{\frac{1}{2}}$ in the vacuum must remain c. Hence

$$\kappa\mu = 1 \tag{4.2}$$

where κ is the color dielectric constant and μ the corresponding "magnetic" susceptibility. When $\kappa = \kappa_\infty \to 0$ one must have $\mu = \mu_\infty \to \infty$.

With (4.2), we can consider κ to be a Lorentz-invariant quantity. It may be useful to introduce a phenomenological spin-0 field σ which depends linearly on κ:

$$\sigma \propto 1 - \kappa \quad . \tag{4.3}$$

Inside the hadron we have $\kappa = 1$ and, therefore, $\sigma = 0$. Outside the hadron, κ is κ_∞ and σ is σ_{vac}. In the case that $\kappa_\infty = 0$, (4.3) becomes simply

$$\sigma = \sigma_{vac}(1 - \kappa) \quad .$$

The phenomenological Lagrangian density is assumed to be

$$\mathcal{L} = -\frac{1}{4}\kappa V^a_{\mu\nu} V^a_{\mu\nu} - \psi^\dagger \gamma_4(\gamma_\mu D_\mu + f\sigma + m)\psi - \frac{1}{2}\left(\frac{\partial\sigma}{\partial x_\mu}\right)^2 - U(\sigma) + \text{counter terms} \tag{4.4}$$

where $\dagger$ denotes the Hermitian conjugation, $x_\mu = (\vec{r}, it)$,

$$V^a_{\mu\nu} = \frac{\partial}{\partial x_\mu} V^a_\nu + \frac{\partial}{\partial x_\nu} V^a_\mu + g\, C^{abc} V^b_\mu V^c_\nu \quad ,$$

$$D_\mu = \frac{\partial}{\partial x_\mu} - \frac{1}{2} i g \lambda^a V^a_\mu \quad , \tag{4.5}$$

the λ^a's are the standard Gell-Mann matrices related to C^{abc} by the commutation relation

$$[\lambda^a, \lambda^b] = 2i\, C^{abc} \lambda_c \quad ,$$

the γ_μ's are the usual Hermitian Dirac matrices and the function $U(\sigma)$ has an absolute minimum at $\sigma = \sigma_{vac}$ and a local minimum at $\sigma = 0$ with

$$U(\sigma_{vac}) = 0$$

and $$U(0) \equiv p > 0 \quad , \qquad (4.6)$$

as shown in Fig. 6.

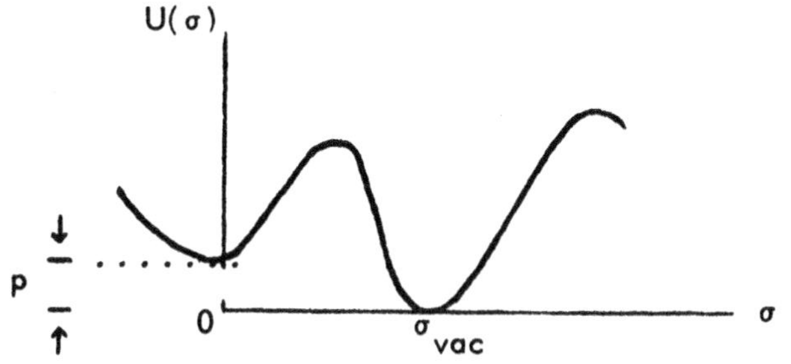

Fig. 6

As we shall see, the detailed form of $U(\sigma)$ is not important. If one wishes, one may assume $U(\sigma)$ to be simply a quartic function of σ. In (4.4) ψ denotes the quark field which besides being a color triplet also has F flavors, m is the mass matrix for quarks inside the hadron, and g and f are both renormalizable coupling constants. Since σ is only a phenomenological field, describing the long-range collective effects of QCD, its short wavelength components do not exist in reality. The counter terms in $\mathcal{L}$ are for renormalization; they consist only of those due to loop diagrams of the vector gauge field V_μ^a and the quark field ψ. In the following, we shall ignore all σ-loops; i.e., σ will be approximated by a classical field.

In our phenomenological Lagrangian density, the σ-field is coupled to the quarks in two ways: one is through the vector field via the quark-V_μ^a coupling g and $-\frac{1}{4}\kappa V_{\mu\nu}^a V_{\mu\nu}^a$ which, because of (4.3), couples V_μ^a with σ, the other is a direct quark-σ coupling f. So far as the quark confinement problem is concerned, there is no need to have the direct quark-σ coupling; it suffices to have the vacuum be a perfect dia-electric $(\kappa_\infty = 0)$. The origin of the f-coupling lies in the dichotomy that only inside the hadron is the coupling between the quark and the vector gauge field really g, which is relatively small; outside the hadron, because of (1.3) it is actually $g/\kappa_\infty^{\frac{1}{2}}$. For $\kappa_\infty \ll 1$, $g/\kappa_\infty^{\frac{1}{2}}$ becomes very large; on the one hand this has the desirable effect of preventing the quarks from moving outside the hadron, but on the other hand it also presents a technical difficulty for a diagram analysis in powers of g^2. In (4.4) the direct quark-σ coupling f with

$$f\,\sigma_{vac} \gg \text{hadron mass} \tag{4.7}$$

is introduced phenomenologically to give a convenient alternative formulation of the same physical effect, but bypassing the above difficulty. As we shall see, because of (4.7) the f-coupling restricts the quarks to staying always inside the hadron, and that enables us to take full advantage of the relative smallness of the quark-V_μ^a coupling g inside the hadron. With the f-coupling we may now expand any physical observable, say the hadron mass M, in a power series of the form (4.1). Phenomenologically, if f were zero, it would be difficult to derive the zeroth-order term M_0, since when $\alpha = 0$ both g and $g/\kappa_\infty^{\frac{1}{2}}$ should be zero; hence, without the f-coupling quarks would be unconfined.

5. Zeroth Order

With the direct quark $-\sigma$ coupling f, we can now in the zeroth-order calculation neglect the exchange of V_μ^a. The description of the hadron reduces to that of a simple soliton model, consisting of the scalar σ field and the quark ψ field. The relevant part of the Lagrangian density (4.4) now consists simply of

$$-\psi^\dagger \gamma_4 (\gamma_\mu \frac{\partial}{\partial x_\mu} + f\sigma + m)\psi - \tfrac{1}{2}(\frac{\partial \sigma}{\partial x_\mu})^2 - U(\sigma) \quad . \quad (5.1)$$

As mentioned before, since σ is only a phenomenological field that has no short wavelength components, we shall neglect all σ-loop diagrams. The remaining σ-diagrams are all tree diagrams, which correspond to the quasi-classical approximation that has been extensively studied in the literature[5]. Here we give only a summary of the results for light quark hadrons, with $m = 0$. In this case, ψ and σ can be reduced to c. no. functions which satisfy

$$(-i\vec{\alpha}\cdot\vec{\nabla} + f\beta\sigma)\psi = \epsilon\psi$$

and

$$-\nabla^2\sigma + U'(\sigma) = -fN\psi^\dagger\beta\psi \qquad (5.2)$$

where $U'(\sigma) = dU/d\sigma$, $\vec{\alpha}$ and β are the usual Dirac matrices, $\epsilon > 0$, ψ satisfies $\int \psi^\dagger\psi\, d^3r = 1$ and N is the total number of quarks and antiquarks,

$$N = 2 \qquad \text{for mesons}$$

and

$$N = 3 \qquad \text{for baryons.}$$

As shown in Ref. 5 for $f\sigma_{vac} \gg$ hadron mass and under very general assumptions, the hadron acquires a well-defined surface $\mathcal{S}$. One has

$$\psi = 0$$

and

$$\sigma = \sigma_{vac} \qquad \text{outside} \qquad ; \tag{5.3}$$

on the inner side of the surface $\mathcal{S}$,

$$-i\beta\vec{\alpha}\cdot\hat{n}\,\psi = \psi \tag{5.4}$$

where $\hat{n}$ is the unit vector normal to $\mathcal{S}$, and inside $\mathcal{S}$ the field σ is $\cong 0$, with its deviation from zero proportional to $\psi^{\dagger}\beta\,\psi$. For the $s_{\frac{1}{2}}$ orbit, $\mathcal{S}$ is a spherical surface of radius R. Inside $\mathcal{S}$, (5.2) can be further reduced, through scaling, to a simple system of two coupled first-order differential equations:

$$\frac{du}{d\rho} = (-1 + u^2 - v^2)\,v$$

and

$$\frac{dv}{d\rho} + \frac{2v}{\rho} = (1 + u^2 - v^2)\,u \tag{5.5}$$

where ρ is related to the radius r by $\rho = \epsilon r$,

$$\psi \propto \begin{pmatrix} u \\ i(\vec{\sigma}\cdot\vec{r}/r)\,v \end{pmatrix} \zeta \,,$$

$$\zeta = \begin{pmatrix} 0 \\ 1 \end{pmatrix} \quad \text{or} \quad \begin{pmatrix} 1 \\ 0 \end{pmatrix} \,,$$

$\vec{\sigma}$ is the usual Pauli spin matrix, and the variables ρ, u and v are all dimensionless. Although (5.5) does not contain any explicit parameters, its solutions form a one-parameter family. As $\rho \to 0$, one has $v \to 0$ and $u \to u(0)$. For every $u(0)$ between 0 and a critical value $u_c = 1.7419$, there is a solution of (5.5). The solution can be obtained by direct integration from $\rho = 0$ to $\rho = \rho_0$. At $\rho = \rho_0$, one has $u(\rho_0) = v(\rho_0)$ and therefore the boundary condition (5.4) is satisfied. The radius of the hadron is given by

$$R = \rho_0/\epsilon \,.$$

A convenient parameter to label these solutions can be either $u(0)$ or the integral

$$n \equiv \int (u^2 + v^2)\, d^3\rho \tag{5.6}$$

over the region $\rho \leq \rho_0$. As the initial value $u(0) \to 0$, one has $n \to 0$, but as $u(0) \to u_c = 1.7419$, $n \to \infty$. Examples of these two limiting solutions are given in Fig. 7.

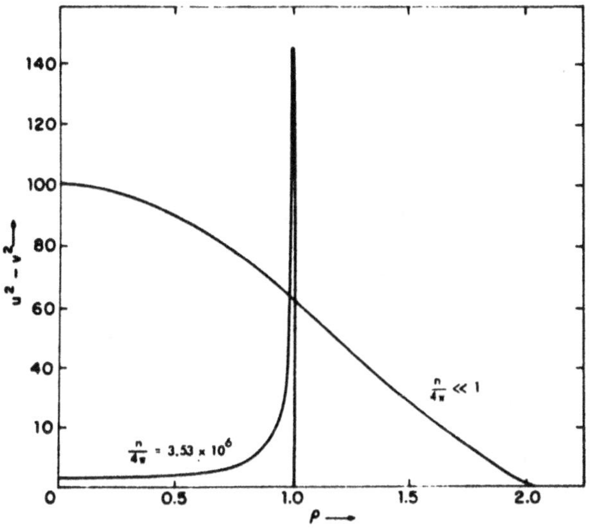

Fig. 7

The physical description of the soliton solution then resembles that of a gas bubble (i.e., the hadron) inside a medium (i.e., the vacuum). The hadron mass is determined by three parameters:

$$p\,, \quad s \quad \text{and} \quad n \tag{5.7}$$

where p is defined by (4.6) which represents the pressure of the medium on the bubble,

s is the surface tension which arises because σ changes from 0 to σ_{vac} across the soliton surface, and n determines the gas pressure inside the bubble which is due to both the kinetic energy of quarks and the excitation energy of σ . In either of the limits $n \to 0$ or $n \to \infty$ the hadron mass has the form, in the notation of (4.1)

$$M_0 = \frac{N \rho_0}{R} + \frac{1}{3} 4\pi R^3 p + 4\pi R^2 s \tag{5.8}$$

where

$$\rho_0 = 2.0428 \quad \text{when} \quad n \to 0$$

and

$$\rho_0 = 1 \quad \text{when} \quad n \to \infty \quad . \tag{5.9}$$

In Table 1 we give the theoretical result for various physical observables when two of the parameters in (5.7) are 0. The first column $n \to 0$ and $s \to 0$ corresponds to the MIT bag[1], the last column $n \to \infty$ and $p \to 0$ gives the SLAC bag[2]. From Table 1 we conclude that for applications to hadrons, the parameter n could be either O(1) or smaller. In any case, it should be away from the $n \to \infty$ limit. Otherwise, g_A/g_V would be $\frac{5}{9}$ and the charge density would be distributed only on the surface of the soliton; both features seem to be quite different from those of the physical nucleon.

The low-lying hadron states are made of light quarks, u, d and s, and their antiparticles. All hadrons are color singlets. In the zeroth approximation there are no spin-dependent interactions, and for convenience we may set the masses of these three light quarks to be 0 inside the hadron. The mesons are $q\bar{q}$ compounds; their lowest energy states have a 36-fold degeneracy consisting of the pseudoscalar nonet, π, K, η and η', and the vector nonet, ρ, K_V, ω, ϕ (each having a 3-fold

Physical observable	Experimental value	Theoretical value in some limiting cases			
		$n \to 0$		$n \to \infty$	
		$s \to 0$	$p \to 0$	$s \to 0$	$p \to 0$
r_p	$3.86/m_N$	$4.25/m_N$	$4.78/m_N$	$2.86/m_N$	$3.21/m_N$
μ_p	$2.79/(2m_N)$	$2.36/(2m_N)$	$2.66/(2m_N)$	$1.90/(2m_N)$	$2.14/(2m_N)$
μ_n	$-0.685\,\mu_p$	$-\frac{2}{3}\mu_p$	$-\frac{2}{3}\mu_p$	$-\frac{2}{3}\mu_p$	$-\frac{2}{3}\mu_p$
g_A/g_V	1.25	1.09	1.09	5/9	5/9

Table 1. Root mean-squared charge radius r_N, magnetic moment μ_N and g_A/g_V of the nucleon N. The parameter n is defined by (5.6). The first column $n \to 0$, $s \to 0$ corresponds to the MIT bag, the last column $n \to \infty$, $p \to 0$ to the SLAC bag.

degeneracy due to spin). The baryons are qqq compounds; their lowest energy states have a 56-fold degeneracy consisting of the spin-$\frac{1}{2}$ octet N, Λ, Σ and Ξ, and the spin-$\frac{3}{2}$ decuplet Δ, $\Sigma^*(1385)$, $\Xi^*(1530)$ and Ω. These states exhibit SU(6) symmetry. The 36-meson states consist of two SU(6) representations, $\underline{35}$ and $\underline{1}$; the 45-baryon states form a single SU(6) representation. In order to split the SU(6) degeneracy, we must include the spin-dependent interaction due to the exchanges of gauge quanta, which are the QCD corrections.

6. Gauge Field Propagator

To carry out the QCD corrections we need the propagator $D_{\mu\nu}(x, x')$ of the gauge field inside a hadron. Let $\mathcal{S}$ be the surface of the hadron; inside $\mathcal{S}$ the color dielectric constant $\kappa = 1$ and outside $\kappa = \kappa_\infty \to 0$. In the Coulomb gauge we can decompose $D_{\mu\nu}$ into transverse and longitudinal components:

$$D_{\mu\nu} = \begin{cases} (D_{ij})_{tr} & \mu = i \neq 4 \\ & \nu = j \neq 4 \\ (D_{44})_{long} & \mu = \nu = 4 \\ 0 & \text{otherwise} \end{cases}$$

where, when expressed in powers of κ_∞

$$(D_{44})_{long} = O(\frac{1}{\kappa_\infty}) + O(1) + O(\kappa_\infty) + \cdots ,$$

$$(D_{ij})_{tr} = O(1) + O(\kappa_\infty) + \cdots .$$

For example, if $\mathcal{S}$ is a sphere of radius R then

$$(D_{44}(x, x'))_{long} = \delta(t-t') \frac{1}{4\pi} \Big[(\frac{1}{\kappa_\infty} - 1) \frac{1}{R} + \frac{1}{|\vec{r} - \vec{r}'|}$$
$$+ \sum_{\ell=1}^{\infty} \frac{\ell + 1}{\ell} (\frac{rr'}{R^2})^{\ell} \frac{1}{R} P_\ell (\cos\theta) \Big] + O(\kappa_\infty) \qquad (6.1)$$

where $x_\mu = (\vec{r}, it)$, $x'_\mu = (\vec{r}', it')$ and $\theta = \angle(\vec{r}, \vec{r}')$. From (6.1) one sees that as $\kappa_\infty \to 0$, $D_{44} \to \infty$. Because of this, the matrix element of the S-matrix between any two non-color-singlet states $|a\rangle$ and $|b\rangle$ diverges when $\kappa_\infty \to 0$. Nevertheless, it can be shown that the matrix element between any two color singlets satisfies

$$\langle b | S | a \rangle = O(1)$$

when $\kappa_\infty \to 0$. Furthermore, between color singlets (6.1) can be replaced by $\delta(t-t')\, G(\vec{r}, \vec{r}\,')$ where

$$G(\vec{r}, \vec{r}\,') = \frac{1}{4\pi}\left[\frac{1}{|\vec{r}-\vec{r}\,'|} + \sum_{\ell=1}^{\infty} \frac{\ell+1}{\ell}\left(\frac{rr'}{R^2}\right)^{\ell} \frac{P_\ell(\cos\theta)}{R}\right] + O(\kappa_\infty) \quad . \tag{6.2}$$

Likewise $(D_{ij})_{tr}$ can be worked out explicitly.

Our gauge field propagator $D_{\mu\nu}$ depends on κ_∞ and the shape of the surface $\mathcal{S}$. Only for color singlet states and in the limits $\kappa_\infty \to 0$ and 4-momentum transfer $\to \infty$, does the propagator $D_{\mu\nu}$ reduce to the usual one used in the literature.

7. First Order

We now turn to the spectroscopy of hadron levels, such as those listed in Table 2.

π (140)	η (549)	η' (958)
ρ (770)	ω (783)	ϕ (1020)
N (940)	Δ (1232)	etc.

Table 2.

As mentioned in Section 5, the mesons have a 36-fold degeneracy and the baryons a 56-fold degeneracy. Part of the lifting of this degeneracy is due to the fact that the s quark "mass" is heavier than those of u and d by ~ 100 MeV. That breaks the flavor-SU(3) symmetry and leads to the familiar Gell-Mann-Okubo mass formula. The exchange of gauge-field quanta brings in a spin-dependent interaction and that lifts the degeneracy of states with different spins, such as those between π and ρ and between N and Δ. For simplicity, we shall concentrate only on these four states.

To evaluate the first-order correction in α, we include only the diagram (1) in Fig. 8 for mesons, and the corresponding one in Fig. 9 for baryons. For low-lying hadron states, the quarks are all in the $s_{\frac{1}{2}}$ state. In the approximation of

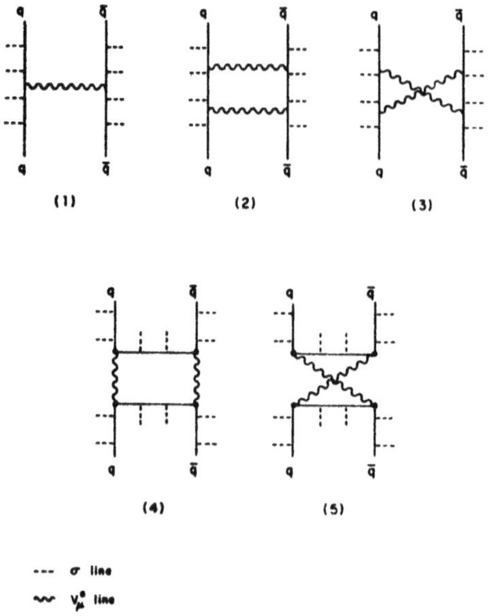

Fig. 8

Fig. 9

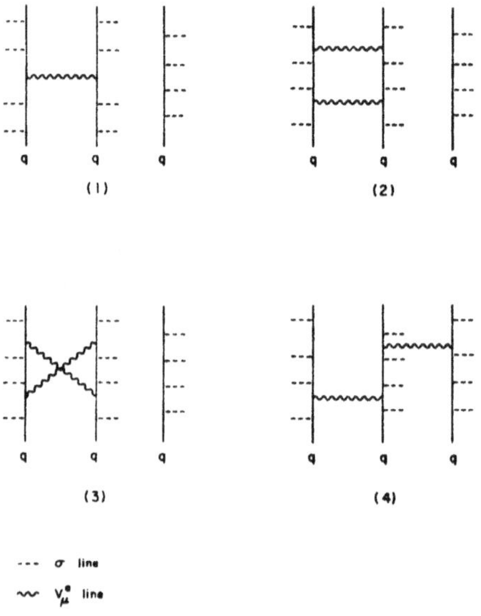

zero quark mass, the meson and the baryon masses, with the first-order diagrams included, can always be written in the form

$$M = \frac{N\xi}{R} + \frac{4\pi}{3} R^3 p \qquad (7.1)$$

where $N = 2$ for mesons and 3 for baryons. The parameter ξ is given by

$$\xi = \rho_0 - \alpha(I_{el} - \lambda I_{mag}) \tag{7.2}$$

where ρ_0 is the zeroth-order parameter given by (5.8)-(5.9)

$$\rho_0 = \begin{cases} 2.0428 & \text{MIT Bag} \\ 1 & \text{SLAC Bag} \end{cases} ,$$

$$\lambda = \begin{cases} -3 & \text{for } \pi \\ -1 & \text{for N} \\ 1 & \text{for } \Delta \text{ and } \rho \end{cases} ; \tag{7.3}$$

I_{el} and I_{mag} are dimensionless numbers, which can be computed from Figures 8 and 9. The radius R is then determined by minimizing M.

Let us assume I_{el} and I_{mag} have been calculated. Furthermore we have chosen a specific bag model so that ρ_0 is known. By using (7.1)-(7.2), we can express the masses of these four hadrons π, ρ, N and Δ in terms of two parameters α and ρ. By eliminating α and ρ, we can derive two mass formulas:

$$\left(\frac{2}{3}\right)^{3/4} \frac{M_\Delta}{M_\rho} = 1 , \tag{7.4}$$

$$\frac{3(M_\rho^{4/3} - M_\pi^{4/3})}{4(M_\Delta^{4/3} - M_N^{4/3})} = 1 . \tag{7.5}$$

As we shall see, exactly the same two mass formulas emerge when I_{el}, I_{mag} and ρ_0 are assigned different numbers. This is quite fortunate since the calculations of I_{el} and I_{mag} are complicated and (7.4)-(7.5) are the maximum information that can be gotten from (7.1)-(7.2) for these four hadron masses. The experimental values of the lefthand sides of (7.4) and (7.5) are 1.180 and 1.187 respectively, which agrees reasonably well with the theoretical predictions.

To see how these two remarkable mass formulas can be derived without first determining I_{el}, I_{mag} and ρ_0, we note that according to (7.3) the λ values for Δ and ρ are the same. Hence, because of (7.2), the ξ values for Δ and ρ must also be the same. From (7.1), we see that the only difference is $N = 2$ for ρ and $N = 3$ for Δ. By minimizing M with respect to R, we find M to be $\propto N^{3/4}$, and that leads to (7.4).

In (7.2) ρ_0 and αI_{el} only appear in the combination $(\rho_0 - \alpha I_{el})$. Thus, even if we treat I_{el}, I_{mag}, ρ_0, α and p all as unknown parameters, (7.1) actually depends only on three combinations $(\rho_0 - \alpha I_{el})$, αI_{mag} and p. By eliminating these three from the four hadron masses, we derive (7.5).

8. Remarks

Quark confinement is a large-scale phenomena. Therefore, at least on the phenomenological level, it should be understandable through a quasi-classical macroscopic theory, much like the London-Landau theory for superfluidity. In this talk, we suggest that the use of the color dielectric constant of the vacuum serves such a purpose. The hadrons are color-singlet composites of q and $\bar{q}$, describable by domain structures in the vacuum. In this way we are able to connect the bag models with QCD.

References

1. A. Chodos, R. L. Jaffe, K. Johnson, C. B. Thorn and V. F. Weisskopf, Phys. Rev. D9, 3471 (1974).
2. W. A. Bardeen, M. S. Chanowitz, S. D. Drell, M. Weinstein and T.-M. Yan, Phys.Rev. D11, 1094 (1975).
3. H. D. Politzer, Phys.Rev.Lett. 30, 1346 (1973); D. Gross and F. Wilczek, Phys. Rev.Lett. 30, 1343 (1973); G. 't Hooft, Phys.Lett. B61, 455, B62, 444 (1973), first presented in some brief remarks at the Marseilles meeting, 1972 (unpublished).
4. R. Friedberg, T. D. Lee and A. Sirlin, Phys.Rev. D13, 2739 (1976), Nucl.Phys. B115, 1, 32 (1976).
5. R. Friedberg and T. D. Lee, Phys.Rev. D15, 1694, D16, 1096 (1977), D18, 2623 (1978).
6. T. D. Lee, Phys.Rev. D19, 1802 (1979).

IX

Historical Papers

Reprinted From
Elementary Processes at High Energy

History of Weak Interactions (*)

T. D. LEE
Columbia University - New York, N. Y.

While the phenomenon of β-decay was discovered near the end of the last century, the notion that the weak interaction forms a separate field of physical forces evolved rather gradually. This became clear only after the experimental discoveries of other weak reactions such as μ-decay, μ-capture, etc., and the theoretical observation that all these reactions can be described by approximately the same coupling constant, thus giving rise to the notion of a universal weak interaction. Only then did one slowly recognize that the weak interaction force forms an independent field, perhaps on the same footing as the gravitational force, the electromagnetic force, and the strong nuclear and sub-nuclear forces.

In preparing this lecture, I was reminded, once more, of how relatively recent some of these « early developments » were. Indeed, most of the shaping from the single class of nuclear β-decay phenomena to the entire field of activities of what is now called « the weak interaction » took place after I become actively interested in physics, and I certainly consider myself (or rather, I would like to consider myself) to be one whose major scientific career still lies in the future.

1. Early history.

The difficulty with weak interactions is that, on the theoretical side, ever since in the early 1930's Pauli (1) introduced the notion of neutrino and Fermi (2) formulated the phenomenological Lagrangian of β-decay interaction, one

(*) This research was supported in part by the U.S. Atomic Energy Commission.

thinks one has already the complete theory. Almost all subsequent theoretical developments are, to a large extent, merely modifications and elaborations of the same old basic idea. The impressive success of Fermi's formulation enables us to describe, through a single parameter, the Fermi coupling constant G which is related to the nuclear mass m_N by ($\hbar = c = 1$)

$$G \simeq 10^{-5}/m_N^2\,, \tag{1}$$

the enormous complexities of all known weak reactions, ranging from the assorted allowed and forbidden β-decays of different nuclei to various other weak processes involving leptons, mesons and hyperons. Yet, as we shall see, Fermi's theory is only a phenomenological one; its success also implies that this large body of known weak processes is really quite insensitive to the underlying basic theory, and therefore it is difficult to make further theoretical progress.

On the experimental side, the difficulty has always been the presence of wrong but convincing experimental results. From the very beginning, experiments on β-decay have followed a tortuous path. At first, its spectrum appeared to consist of discrete lines; this was, of course, later proven to be incorrect. Nevertheless, in those early days of quantum jumps, it was much easier to understand, theoretically, discrete lines rather than a continuum spectrum. Even after Fermi's theory, for a very long time, the theoretical prediction did not agree with the experimental results. The correct β-energy spectrum was observed by Lawson and Cork ([3]) in 1940 and, in the low energy range, only since 1949 by C. S. Wu and her co-workers([4]), more than 50 years after the discovery of the β-ray. Thus, probably we should not feel too bad about the present uncertainties in the $K_{\ell 3}$ spectrum and the confusion over *CP* violations.

When I began my graduate study of physics at the University of Chicago in 1946, the pion was not known. Fermi and Teller ([5]) had just completed their theoretical analysis of the important experiment of Conversi, Piccioni and Pancini ([6]); they concluded that the mesotron, then discovered in cosmic radiation, could not possibly be the carrier of strong forces hypothesized by Yukawa. This made it somewhat more plausible, at least to me as a young graduate student, that perhaps the β-decay interaction, which becomes exceedingly singular according to the second-order perturbation formula, could be the strong nuclear force. At that time, the β-decay interaction was thought to be reasonably well understood. Fermi's original vector coupling form,

$$G(\psi_n^\dagger \gamma_4 \gamma_\lambda \psi_p)(\psi_e^\dagger \gamma_4 \gamma_\lambda \gamma_5 \psi_\nu) \tag{2}$$

was, after all, too simple; to conform to reality, it should be extended to include a Gamow-Teller term. [The presence of γ_5 has, of course, no physical significance. Nevertheless, it is curious why Fermi should choose this particular expression, which resembles the VA interaction, but with parity conservation.]

A year later, the discovery of the pion through its decay sequence π-μ-e by Powell and his group dramatically confirmed the original idea of Yukawa. The fact that the higher-order β-interaction is singular is not a good argument that it should simply become the strong interaction. Thus, it seems reasonable to regard β-decay as belonging to a new class of interactions. With this motivation, several graduate students at Chicago, Rosenbluth, Yang and myself (7), began in 1948 a more systematic investigation of whether there exist interactions, other than β-decay, that could belong to the same class.

After we found that if μ-decay and μ-capture were described by a similar four-Fermion interaction as β-decay, all their coupling constants appeared to be of the same magnitude, we naturally went to our teacher, Enrico Fermi. We told him of our discovery, including also our speculation that, in analogy with electromagnetic forces, the basic weak interaction should be represented by a universal interaction between a hypothetical heavy boson and a pair of Fermion fields. [Since then, this intermediate boson has acquired a definite name, $W^{\pm}$, though otherwise remaining hypothetical.] Fermi, with his usual deep insight, immediately recognized the full significance. He asked: if this is to be a universal interaction, then there must be reasons why some pairs of Fermions should have such interactions, and some pairs should not. For example, why is it that

$$\mu^- \nleftrightarrow e^+ + 2e^- ,$$

and

$$p \nleftrightarrow e^+ + 2\nu ,$$

and, in a similar manner, we may also ask why is it that

$$p \nleftrightarrow e^+ + \gamma ?$$

A few days later, he told us that he had found the answer; he then proceeded to assign various sets of numbers, + 1, — 1 and 0, to each of these particles. This was the first time to my knowledge that both the laws of baryon number conservation and of lepton number conservation were formulated. However, at that time (1948), my own reaction to such a scheme was totally

54

negative: surely it is not necessary to explain why $p \nrightarrow e + \gamma$, since everyone knows that the identity of a particle is never changed through the emission and absorption of a photon; as for the weak interaction, why should one bother to introduce a long list of mysterious numbers, when all one needs is to say that only three combinations ($\bar{n}p$), ($\bar{e}\nu$) and $\bar{\mu}\nu$) can have interactions with the intermediate boson.

Sometimes a discovery is made simply because the time is ripe; if one person does not make it, then surely another person will do it at about the same time. In looking back, what we did in finding the same weak-interaction coupling constant for different processes was precisely a discovery of such a nature. This is clear, since very similar observations were made independently by at least three other groups, Klein ([8]), Puppi ([9]), and Tiomno and Wheeler ([10]), at about the same time. Yet, Fermi's thinking was of a more profound nature. Unfortunately for physics, his proposal was never published. The full significance of these conservation laws was not realized until years later. While this might be the first time that I failed to recognize a great idea in physics when it was first presented to me, unfortunately this did not turn out to be the last time.

In the early fifties, extensive efforts were made on β-decay experiments. By then, the Konopinski-Uhlenbeck interaction was definitely ruled out. The absence of the Fierz interference term in the spectrum shows that the β-interaction must be either *V*, *A* or *S*, *T*. These two possibilities were further resolved by a series of β-ν angular correlation experiments. In an allowed transition, the distribution for the angle θ between β-ν is given by (neglecting Fierz term)

$$\left[1 + \lambda \left(\frac{P}{E}\right)_e \cos\theta \right] d\cos\theta\,, \tag{3}$$

where the subscript e refers to the momentum and energy of the electron. For a $\Delta J = 1$ transition,

$$\lambda = \begin{cases} +\frac{1}{3}, & \text{for } T\,, \\ -\frac{1}{3}, & \text{for } A\,. \end{cases}$$

The experiment on ^{6}He decay by Rustad and Ruby in 1953 gave ([11])

$$\lambda = +0.34 \pm 0.09\,, \tag{4}$$

which seemed to establish unquestionably that the β-decay interaction should be *S*, *T* with perhaps some unknown mixture of *P*.

With this new result, the theoretical idea of intermediate boson seemed to be definitely ruled out. It is bad enough to assume the possibility of two kinds of intermediate bosons of different spin-parity, one for the Fermi coupling and the other for the Gamow-Teller coupling. However, a tensor interaction with no derivative coupling is too much even for a liberal free-thinking theorist. It cannot even be transmitted by a spin-2 boson, since the former is described by an antisymmetric tensor and the latter by a symmetric one.

In the meantime, progress was also made in other domains of weak interactions, such as the electron spectrum in μ-decay. In such a decay the normalized electron spectrum $N(x)$ depends only on one parameter, called the Michel parameter ϱ:

$$N(x) = 6x^2[2(1-x) + \tfrac{4}{9}(-3 + 4x)\varrho], \tag{5}$$

where

$$\int_0^1 N(x)\,dx = 1$$

and x is related to the electron momentum p_e and the muon mass m_μ by

$$x = \frac{p_e}{\frac{1}{2}m_\mu}. \tag{6}$$

As one can see from the following graph, different ϱ-values give very different shapes of the electron spectrum.

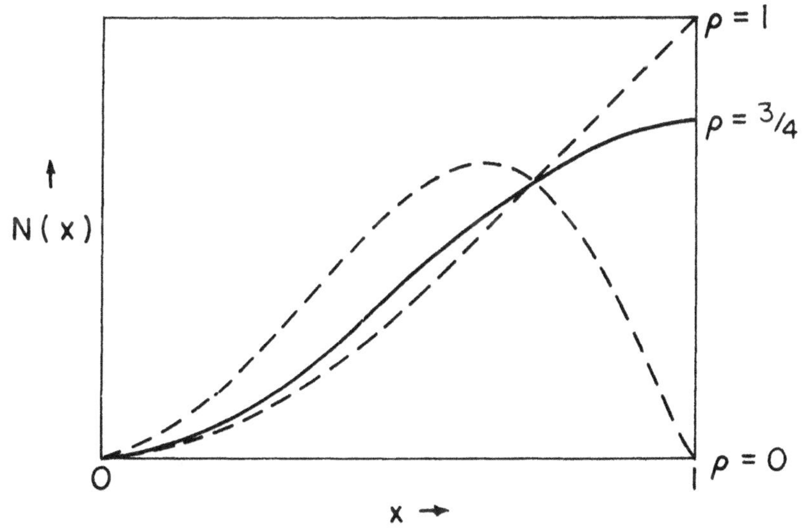

Fig. 1.

Yet, as shown by the next Figure, the experimental value in 1951 was

$$\varrho = 0.1 \pm 0.1\,,$$

which is to be compared with the presently known value of

$$\varrho = 0.752 \pm 0.003\,. \tag{7}$$

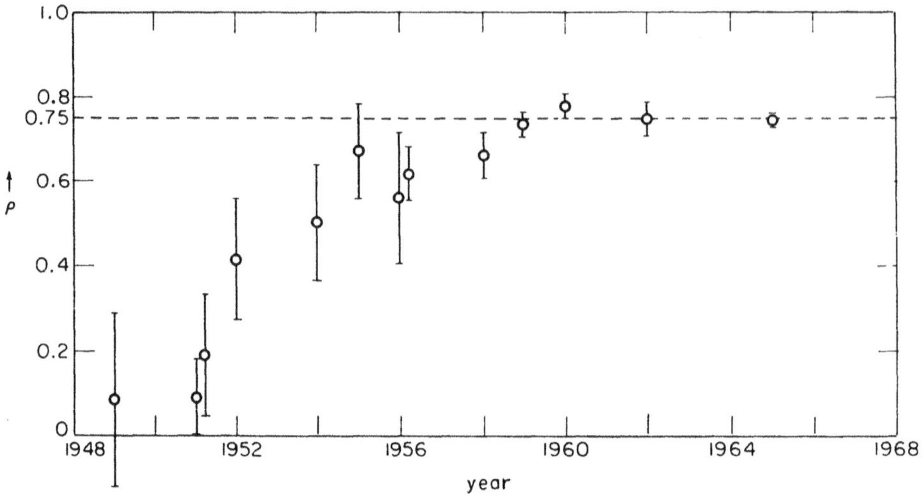

Fig. 2.

It is strange to observe the time dependence of the ϱ-value in Fig. 2, especially to note that at any given time, each experiment is *always* within the error of the preceding one.

Our understanding of weak interactions could have remained in such a misrepresented situation for a much longer period, had it not been for the appearance of an entirely different class of weak processes due to the then newly discovered strange particles.

2. Transition period.

The decays of the strange particles extended weak interactions to include, in addition to the already existing purely leptonic process μ-decay, and the strangeness conserving ($\Delta S = 0$) semi-leptonic processes μ-capture and β-decay, two entirely new classes of processes: the strangeness nonconserving semi-leptonic and nonleptonic transitions. Among these new processes, the most puzzling ones were the charged θ and τ mesons. Both mesons were defined according to their decay modes:

$$\theta^+ \to \pi^+\pi^0$$

and

$$\tau^+ \to \pi^+\pi^+\pi^-,$$

or

$$\tau'^+ \to \pi^+\pi^0\pi^0 .$$

The spin-parity of θ^+ is clearly $0+$, $1-$, $2+$, etc. As early as 1953, Dalitz ([12]) already pointed out that the spin-parity of τ^+ can be analysed through his Dalitz plot, and, by 1954, the then existing data was more consistent with the assignment $0-$ than $1-$. Although both mesons were known to have comparable masses (within ~ 20 MeV), there was, at that time, nothing extraordinary about this conclusion. However, by 1955, very accurate lifetime measurements became available. This, together with a statistically much more significant Dalitz plot of τ^+-decay, presented a very puzzling picture indeed. Except for very high spin assignments $J \geqslant 3$ [which was not preferred partly on esthetic grounds and partly because it would lead to angular correlations between the production and decay directions, while none was observed], the spin-parity of τ^+ was determined to be definitely $0-$; therefore, it appeared to be definitely a different particle from $0+$. Yet, within the experimental accuracy ($\sim$ a few percent) these two particles have exactly the same lifetime, and also comparable masses. This was known as the θ-τ puzzle.

My first reaction was to invent schemes within the conventional theories to explain this puzzle. Together with Orear ([13]), we proposed in the summer of 1955 a cascade mechanism: Accepting the preferred spin parity assignments, $0-$ for τ^+ and $0+$ for θ^+, it seems reasonable to assume that θ^+ should have a much shorter lifetime than τ^+ because of phase space considerations; now if one assumes τ^+ is, say, ~ 10 MeV heavier than θ^+, the 0-0 forbidden transition

$$\tau^+ \to \theta^+ + 2\gamma$$

can become comparable to the 3π decay mode of τ^+, resulting in the « apparent » θ lifetime being simply the lifetime of τ^+, provided that such lifetime measurements are (as they were) determined by the attenuation of the 2π decay mode. The approximate mass degeneracy was then incorporated into a parity-exchange symmetry.

However, when Alvarez's group failed to find such a 5 MeV γ in their bubble chamber, it finally became clear to me, in early 1956, that the solution to the θ-τ puzzle must lie in something much deeper; perhaps, parity is not conserved, and θ and τ are indeed the same particle. The immediate reaction to this simple proposal was « so what ». Unless it could be shown that parity conservation also extended to other processes, one would *never* know for *sure* whether in θ-τ decays parity conservation is, or is not, violated. Thus, it was necessary to investigate how possible parity nonconservation effects could be observed in other weak reactions.

At that time, the true meaning of parity operation P was not clear, at least not to me. Of course, I knew its mathematical property: P should be represented by a unitary operator in the Hilbert space, and under P, one has, *e.g.*, for a spin-$\frac{1}{2}$ Fermion field

$$P\psi(r, t)P^{-1} = e^{i\varphi}\gamma_4\psi(-r, t),$$

etc. I assumed that β-decay should be described by a more general Lagrangian consisting of ten coupling constants, the usual five C_i $(i = S, P, V, A, T)$ and the additional five parity violating constants C_i'. I then borrowed from C. S. Wu the authoritative book on β-decay edited by K. Siegbahn ([14]), and proceeded with C. N. Yang to calculate systematically all possible parity-violating effects. We started with the allowed spectrum, and found exactly the same expression after replacing the usual

$$C_i^* C_j \quad \text{by} \quad C_i^* C_j + C_i'^* C_j'. \tag{8}$$

Next came the forbidden spectrum, and the same thing happened. We then computed Coulomb effects, β-ν correlations, β-γ correlations and β-γ-γ correlations; while some of these calculations were rather complicated, in the end all interference terms between C_i and C_j' invariably were cancelled, and the same old expressions appeared, provided we made the same simple replacement (8). After we went through the entire Siegbahn book, rederived all these old formulas with this new interaction, it became obvious to us that not only was there, at that time, not a single evidence for parity conservation in β-decay, but that we must have been very stupid! There had to be an extremely simple reason why all those complicated interference terms $C_i^* C_j'$ cancelled each other. Once we stopped calculating and started to think, in a rather short time, it dawned on us that the reason for this lack of evidence was the simple fact that nobody had made any attempt to observe a pseudoscalar as distinct from an otherwise seemingly right-left symmetrical arrangement.

From then on, the theoretical problem came under control, and it resulted in, in 1956, the papers ([15]) on parity nonconservation, two-component theory and the general question of possible *C*, *P*, *T* noninvariance, including possible *CP* violations in neutral K-decays. It also led, in January 1957, to the important first experiment on *P* and *C* asymmetries in β-decay by Wu, Ambler, Hayward, Hoppes and Hudson ([16]), which was quickly followed by the observation of the same symmetry breaking in π and μ decays by Garwin, Lederman and Weinrich ([17]), and by Friedman and Telegdi ([18]). The experiment of Garwin *et al.* was done in a particularly elegant way.

An outburst of new experiments quickly followed, quite often with mutually contradictory results. The question of what should be the new β-decay interaction was a pressing one, but in the summer of 1957 it was clear that no theory could possibly agree with all then-existing experimental results. Some of these experiments had to be wrong. Privately, I thought, if anything, the old ^{6}He experiment must be right. First, it was done after years of patient experimental effort, not like one of those new quickies; secondly, it came from that great institution, Columbia University. Nevertheless, I kept my opinion reasonably quiet, and when Feynman passed through New York that summer asking me about my opinion of the experimental situation, I suggested that the best way was to flip a coin.

At the end of that summer, a manuscript by Gell-Mann and Feynman ([19]) arrived on my desk in Pupin. On its margin, Feynman had scribbled « I have flipped my coin, and this is the answer ». After I read their paper, I knew I should have borrowed his coin.

The fact that if the ^{6}He experiment were wrong, one could then have V-A interaction which would be theoretically more attractive, *e.g.*, one could have lepton conservation, small π_{e2} decay rate, etc., was known to me. The same reasoning also led, at about the same time, Marshak and Sudarshan ([20]), and Sakurai ([21]) to their independent suggestion of V-A interaction. However, I was particularly impressed by the added statement of Feynman and Gell-Mann that the strangeness conserving hadronic weak vector current V_μ^{wk} and the corresponding isovector part of the electromagnetic current eJ_μ could form an isotriplet

$$V_\mu^{wk}, \quad (J_\mu)_{I=1}, \quad V_\mu^{wk\dagger}, \tag{9}$$

which is very appealing and seems to carry a much deeper significance. From that, it follows

$$\frac{\partial V_\mu}{\partial x_\mu} = 0 \,. \tag{10}$$

Since there is no zero-mass particle with strong interactions, (10) implies that, neglecting radiative corrections

$$(G_V)_{obs} = (G_V)_{bare} \,, \tag{11}$$

which makes it possible to compare the weak coupling constant for hadrons with that for leptons. The V-A interaction was established mainly through the neutrino helicity measurement of Goldhaber, Grodzins and Sunyar ([22]). The later experiment on weak magnetism effects in $\beta^\pm$ decays of ^{12}B and ^{12}N by Wu and others ([23]), together with experiments on the lifetime of ([24])

$$\pi^+ \to \pi^0 + e^+ + \nu_e$$

give strong evidence for the correctness of the Gell-Mann–Feynman isotriplet hypothesis.

The following decade from 1957 to the present was a period of great clarification. For example, as shown in Fig. 2, the experimental ϱ-value is now in excellent agreement with the theory. Through the work of Gell-Mann, Cabibbo and others we now have a very simple interaction form, at least for the semi-leptonic processes. During the same period, one witnessed the innovation of high energy neutrino experiments which led to the discovery of two neutrinos

$$\nu_e \neq \nu_\mu \tag{12}$$

by Lederman, Schwartz, Steinberger and their coworkers ([25]). Except for *CP*([26]), one could have said that, finally, we have the correct phenomenological theory of weak interactions.

3. Present status.

Independently of the detailed dynamics, one expects at present the following general properties of the weak interaction to hold to all orders of weak interactions:

Lorentz Invariance
Charge Conservation
Baryon Number Conservation
Lepton Number Conservation
CPT Symmetry
Two Component Theory of Neutrino
μ-e Symmetry
$U_2 \times U_2$ Symmetry of Leptons.

Apart from these general properties, we know that the phenomenological theory of Fermi with V-A interaction holds at least at low 4-momentum transfer. Among the important unresolved problems, I may list:

1) Origin of *CP* Asymmetry.
2) Breakdown of Fermi Theory at High Energy (Violation of Unitarity Limit at Momentum $\sim$300 GeV/c).
3) Possible Existence of Intermediate Boson.

Besides these, there are also many questions connected with the detailed dynamics and selection rules such as the $|\Delta I| = \frac{1}{2}$ rule, validity of the local chiral $SU_3 \times SU_3$ current algebra, etc. The presence of these fundamental and crucial questions, and the prospect of using in the near future the ISR at CERN for possible production of $W^{\pm}$, and the very high energy protons, neutrinos and muons from the NAL 500 GeV machine to search for the answers to these questions, promise a most exciting future for the physicists working in this field.

REFERENCES

1) W. Pauli: *Proceedings of the Solvay Conference*, Brussels 1933, p. 324.
2) E. Fermi: *Zeits. Phys.*, **88**, 161 (1934).
3) J. L. Lawson and J. M. Cork: *Phys. Rev.*, **57**, 982 (1940).
4) C. S. Wu and R. D. Albert: *Phys. Rev.*, **75**, 315 (1949); C. S. Wu: *Rev. Mod. Phys.*, **22**, 386 (1950).
5) E. Fermi and E. Teller: *Phys. Rev.*, **72**, 399 (1947).
6) M. Conversi, E. Pancini and O. Piccioni: *Phys. Rev.*, **68**, 232 (1945).

7) T. D. LEE, M. ROSENBLUTH and C. N. YANG: *Phys. Rev.*, **75**, 905 (1949).
8) O. KLEIN: *Nature*, **161**, 897 (1948).
9) G. PUPPI: *Nuovo Cimento*, **6**, 194 (1949).
10) J. TIOMNO and J. A. WHEELER: *Rev. Mod. Phys.*, **21**, 153 (1949).
11) B. M. RUSTAD and S. L. RUBY: *Phys. Rev.*, **89**, 880 (1953).
12) R. H. DALITZ: *Phil. Mag.*, **44**, 1068 (1953); *Phys. Rev.*, **94**, 1046 (1954).
13) T. D. LEE and J. OREAR: *Phys. Rev.*, **100**, 932 (1955).
14) K. SIEGBAHN, editor: *Beta- and Gamma-Ray Spectroscopy*, North-Holland (1955).
15) T. D. LEE and C. N. YANG: *Phys. Rev.*, **104**, 254 (1956); *Phys. Rev.*, **105**, 1671 (1957); T. D. LEE, R. OEHME and C. N. YANG: *Phys. Rev.*, **106**, 340 (1957).
16) C. S. WU, E. AMBLER, R. W. HAYWARD, D. D. HOPPES and R. P. HUDSON: *Phys. Rev.*, **105**, 1413 (1957).
17) R. L. GARWIN, L. M. LEDERMAN and M. WEINRICH: *Phys. Rev.*, **105**, 1415 (1957).
18) J. J. FRIEDMAN and V. L. TELEGDI: *Phys. Rev.*, **105**, 1681 (1957).
19) R. FEYNMAN and M. GELL-MANN: *Phys. Rev.*, **109**, 193 (1958).
20) R. E. MARSHAK and E. C. G. SUDARSHAN: *Padua-Venice International Conference*, 1957; *Phys. Rev.*, **109**, 1860 (1958).
21) J. J. SAKURAI: *Nuovo Cimento*, **7**, 649 (1958).
22) M. GOLDHABER, L. GRODZINS and A. W. SUNYAR: *Phys. Rev.*, **109**, 1015 (1958).
23) Y. K. LEE, L. W. MO and C. S. WU: *Phys. Rev. Lett.*, **10**, 253 (1963).
24) R. B. BACASTOW, T. ELIOFF, R. R. LARSEN, C. E. WIEGAND and T. YPSILANTIS: *Phys. Rev. Lett.*, **9**, 400 (1962); P. DEPOMMIER, J. HEINTZE, C. RUBBIA and V. SOERGEL: *Phys. Lett.*, **5**, 61 (1963); A. F. DUNAITSEV, V. I. PETRUKHIN, YU. D. PROKOSHKIN and V. I. RYKALIN: *Proc. Intern. Conf. Fundamental Aspects of Weak Interactions*, Brookhaven National Laboratory, Upton, L. I., 1963.
25) G. DANBY, J. M. GAILLARD, K. GOULIANOS, L. M. LEDERMAN, N. MISTRY, M. SCHWARTZ and J. STEINBERGER: *Phys. Rev. Lett.*, **9**, 36 (1962).
26) J. H. CHRISTENSON, J. W. CRONIN, V. L. FITCH and R. TURLAY: *Phys. Rev. Lett.*, **13**, 138 (1964).

BROKEN PARITY

T. D. Lee

1. Prologue

On a dark and foggy day, two children played on the beach. One of them said, "Look, do you see a flickering light?" The other replied, "Yes, let's take a closer look." Both being enormously curious, they ran towards it, neck to neck. Sometimes one was ahead and sometimes the other. The challenge made them exert themselves more, racing faster and faster. The effort and the speed created a unique sensation for each of them, oblivious even of themselves.

"Here it is!", said the child who got there first, and he opened the door. The other rushed in. Dazzled by the immense beauty of the interior, he exclaimed, "How wonderful! How splendid!"

It turned out that they had discovered the hidden treasure of the Yellow Emperor. For their feat, these two were richly decorated and greatly admired. Their fame spread over four seas. Years and years later, both grew old and quarrelsome. Memories became dim and life on the whole dull. One of them decided to inscribe his own epitaph in gold: "Here lies the one who first saw the treasure." The other then said, "But I opened the door."

My first collaboration with Frank (C.N.) Yang on physics was in the fall of 1948, the next one in 1951-52, and then continuously from 1955-62. Recently, a friend showed me a book entitled Selected Papers 1945-80 with Commentary by Yang.* Its account of events concerning our collaboration astonished me.

Reluctantly and sadly I have to retrace the memory of a broken friendship.

---

* Chen Ning Yang, Selected Papers 1945-1980 with Commentary (San Francisco, W. H. Freeman and Company, 1983).

2. Early Collaborations

I entered the graduate school of the University of Chicago in the autumn quarter of 1946 at the age of nineteen. Frank Yang, who was a few years older than I and came from the same university in China, was already at Chicago when I arrived. Yang was extremely bright and especially talented in mathematical physics. I met him once in Kunming in 1945. At Chicago, we soon became close friends.

In the spring of 1948, after passing the basic examination, I started my thesis under Professor E. Fermi. At that time another student in Fermi's laboratory, Jack Steinberger, was doing experimental work on the e-spectrum in the decay of $\mu \rightarrow e + \cdots$. His measurements confirmed that mu decay, like beta decay, involves four fermions. Naturally my attention, as well as that of several others at Chicago, was attracted to problems in that area. The result was my first collaboration with Yang.

In the fall of 1948 Marshall Rosenbluth, Frank Yang and I began a systematic study of mu decay and mu capture. We discovered that these reactions could be described by a similar four fermion Fermi interaction with approximately the same coupling constant as in the beta decay. We then speculated that, analogous to Coulomb interactions, these reactions are in fact transmitted through a universally coupled intermediate boson field with respect to which all fermion pairs have the same "charge." In order to generate the short-range interaction, the quanta of such a field must be massive, and in order to escape detection it must have a very short lifetime. However, since both photons and gravitons are massless, the massiveness of the intermediate boson greatly troubled us. In addition, because of parity conservation it was difficult at that time to understand why there were both Fermi and

Gamow-Teller interactions in β-decay. We discussed these problems with Professor Fermi. It was only at his insistence that we overcame our reluctance and wrote up our work* around the New Year 1949.

In those Chicago days, I discussed a great deal of physics and other matters with Frank. His taste tended to be more mathematical, complementing my own. We had an open approach to all subjects and these discussions, often heated, had an important influence on my development, especially in those formative years; they greatly enhanced my appreciation of intellectual power different from mine. The impact of Professor Fermi was, of course, much more profound.** Even now, sometimes when I encounter difficulties, I try to imagine how Fermi might react under similar circumstances. The relationship between Fermi and his students was quite personal. I would see him one whole afternoon each week. Often he would mention some topics and ask me if I could think and read about them and then "give him a lecture." Of course, I obliged and usually felt very good afterward. Only much later did I realize that this was an excellent method of guiding the student to be independent.

I left Chicago at the end of 1949. Yang had left a few months earlier for the Institute for Advanced Study in Princeton.

When I arrived at the Institute in the fall of 1951, among the subjects most actively under discussion there was the Ising model. At that time Yang had just

---

* T. D. Lee, M. Rosenbluth and C. N. Yang, Phys. Rev. 75, 905 (1949).

** Besides Professor Fermi, I also learned a great deal of physics from Professor Chandrasekhar, the Mayers and others in Chicago. Yang's account of my training in physics differs from mine. See page 7 of his Selected Papers 1945-1980 with Commentary.

completed his magnetization computation of the two-dimensional Ising model, and wished to continue working in that field. Since I had taken my statistical mechanics course at Chicago from Joe and Maria Mayer, the Mayer theory of condensation had always held a fascination for me. Naturally our discussions quickly centered on the extension of these Ising results to gas-liquid phase transitions. Very soon Yang and I were able to make substantial progress in proving several rigorous theorems related to phase transitions, later published in two papers.*

Paper I contained two theorems, of which I proved the major part. After we finished that paper, Yang asked whether I minded his having his name before mine, because he was a few years older. I was surprised by his request. Because of the Chinese tradition of respect for seniority in age, I agreed. Later, I looked into the literature and realized that this was unfair. When we were writing the second paper, I showed him a few examples of other published papers to imply that seniority in age was usually not a consideration in the ordering of names. Therefore the order was reversed on paper II even though a decisive step of the unit circle theorem in that paper was made by Yang.**

As a result of this seemingly slight, but disconcerting, incident I felt that it might be better for me not to collaborate further with Yang. This is why we did not

---

* C. N. Yang and T. D. Lee, Phys.Rev. 87, 404 (1952); T. D. Lee and C. N. Yang, Phys.Rev. 87, 410 (1952).

** It took us more than a month of very intensive and arduous joint effort to go from the first conjecture of the unit circle theorem, through many partially successful and unsuccessful routes, to the final completion of the proof. In Yang's account to M. Kac in 1969 (for publication), he emphasized only his own contribution in that very last step of the collaborative work. I was unaware of this personal version by Yang, reading it for the first time in his Selected Papers 1945-1980 with Commentary, p. 15.

work together again until 1955, even though I remained at the Institute for a year and a half more (to 1953) and after that moved to Columbia University which is not far away.

Among the giants at the Institute, Einstein towered over everyone. The young members felt awed just by seeing him, often walking with K. Gödel on his way to and from his office. We were almost all much too shy to talk to him. One day in 1952 Einstein asked his assistant, Bruria Kaufman, to see if he could talk with Yang and me.* We immediately said "of course." I thought of bringing my copy of The Meaning of Relativity to Einstein to autograph, but could not bring myself to do it. I have regretted that ever since.

We went to Einstein's office. He said that he had read our two papers on statistical mechanics and was impressed by them. He first asked us about the foundation of the grand canonical ensemble. Apparently he was not at all familiar with that approach. This surprised me because I had always been under the impression that the whole method was invented for the derivation of Bose-Einstein condensation. His questions then turned to the physical applicability of lattice gas and details of the root distribution of the partition function. Our answers pleased him. The whole conversation was quite extensive and lasted a long time. At the end he stood up, shook our hands and said "Best wishes for your future success in physics." I remember his hand being larger and warmer than other people's. Altogether this was a most memorable experience.

---

* Yang's recollection of the event differs from mine. See page 15 of Selected Papers 1945-1980 with Commentary.

3. Symmetry Violation

For the next few years, although Yang and I did not collaborate, our families often visited each other.

In 1954 Yang and Mills published their paper on isospin gauge theory.* This was the first appearance of the celebrated "Yang-Mills equation," which was destined to play a most important role in modern physics. However, at that time its ultimate significance was unclear to me, and my role in this direction of research was a minor one. Nevertheless, the result was to bring Yang and myself closer, leading gradually to our very fruitful collaboration of the next seven years.

In early 1955, I expressed doubts about the validity of the "local gauge invariance" assumption in the paper by Yang and Mills. After all, since isospin is not globally conserved why should one require it to be locally gauge-invariant? To illustrate this point, one could take the best-established conservation law, the conservation of baryon number. By applying the same local gauge-invariance assumption, one would reach the conclusion that there must be a new long-range force, which however had no experimental support. Yang and I had several discussions. The result was our 1955 paper.**

From 1956-1962 Yang and I worked together on a total of 32 papers, ranging from particle physics to statistical mechanics. We were lucky to be in that extraordinary period of physics. In spite of the political tension of the cold-war period, international

---

* C. N. Yang and R. L. Mills, Phys.Rev. 96, 191 (1954).

** T. D. Lee and C. N. Yang, Phys.Rev. 98, 1501 (1955).

cooperation in physics was perhaps at its zenith, exemplified by the famous Rochester conferences initiated by R. E. Marshak and the formation of CERN through the joint effort of many nations. The collaboration between Yang and myself matched and reflected the spirit of that time. It was close and productive, competitive yet harmonious. Our working together brought out the best of our abilities. The result was much more than the sum of what each could do separately.

Among our papers the most important were on symmetry violation; I shall now concentrate on these, giving my personal account of that remarkable time, as accurately as I can recall. I would like to emphasize the very private nature of this account, which as it is drawn from my own memory includes only situations where I was present. Furthermore, these are not the sort of memories that it is standard to share. I do so only in response to Yang's quite different version of the events which took place.

The discovery of parity nonconservation was the fruit of a generation of physics. Yang and I were two of the many participants. The drama began with the theta-tau puzzle, which by 1955 was already a "hot" topic among physicists. Why should theta and tau, two particles of different spin parity, have the same lifetime and also comparable masses?

My first effort was on the wrong track. In the summer of 1955 J. Orear and I proposed a scheme within the bounds of conventional theory. We suggested a cascade mechanism:* accepting the preferred spin parity assignments, 0- for tau and 0+ for theta, it seemed reasonable to assume that theta ($\rightarrow 2\pi$) should have a much shorter lifetime than tau ($\rightarrow 3\pi$) because of phase space considerations. If one assumed tau

---

* T. D. Lee and J. Orear, Phys. Rev. 100, 932 (1955).

to be, say, about 10 MeV heavier than theta, the 0-0 forbidden transition

$$\tau \rightarrow \theta + 2\gamma$$

could become comparable to the 3π decay mode of tau, resulting in the "apparent" theta lifetime being simply the lifetime of tau, provided that the theta lifetime was determined by the attenuation method of the 2π decay mode.

However, in early 1956 when Alvarez's group failed to find such a 5 MeV gamma in their bubble chamber,* it was clear that our conventional scheme was incorrect. I then realized that the solution to the theta-tau puzzle must lie in something much deeper: perhaps parity is not conserved, and theta and tau are the same particle. I discussed this possibility with Yang, but we were not able to make any progress.**

At that point the hypothesis of parity nonconservation in theta-tau decay led, unfortunately, to a tautology. For if one assumed that parity nonconservation occurred only in the weak decay of theta-tau, then the maximal measurable effect would have been the same zero-spin particle decaying into 2π (theta mode) and 3π (tau mode), but that was just the original and already-observed phenomenon.

---

* L. Alvarez, in High Energy Nuclear Physics (Proceedings of the Sixth Annual Rochester Conference, April 3-7, 1956), (New York, Interscience Publishers, 1956), p. V-28.

** Our feeling was expressed by Yang in his response to Feynman's question at the Rochester Conference (High Energy Nuclear Physics, op.cit., p. VIII-27):

> " ... Pursuing the open mind approach, Feynman brought up a question of Block's: Could it be that the θ and τ are different parity states of the same particle which has no definite parity, i.e., that parity is not conserved. ... Yang stated that he and Lee looked into this matter without arriving at any definite conclusions."

After the 1956 Rochester Conference, I realized even more acutely that unless it could be shown that parity nonconservation also extended to other processes, one would never know for sure whether in theta-tau decays parity conservation was, or was not, violated. The crucial step came about a week later, around the middle of April, from my discussion with Jack Steinberger who was also at Columbia.

At that time, Steinberger and others were conducting extensive experiments on the production and decay of the hyperons Λ^o and Σ^-:

$$\pi^- + p \rightarrow \begin{cases} \Lambda^o + K^o \\ \Sigma^- + K^+ \end{cases} \qquad (1)$$

and

$$\left.\begin{matrix} \Lambda^o \\ \Sigma^- \end{matrix}\right\} \rightarrow \pi + N \ . \qquad (2)$$

The dihedral angle θ between the production plane and the decay plane is of importance for the determination of the hyperon spin. Jack and I had a discussion on the range of θ. As a dihedral angle, θ varies from 0 to π. In the course of the discussion, it emerged that if θ is measured as an angle of rotation around the Λ momentum-vector which is the intersection of these two planes, then the range of θ can be extended from 0 to 2π. I pointed out that if parity is not conserved, there could be an asymmetry * between events with θ from 0 to π and those from π to 2π.

---

* Immediately after the discussion Steinberger and his collaborators re-analyzed their data to look for such an asymmetry. Mel Schwartz recalls that their analysis following my suggestions was done in mid-April 1956; recently Steinberger also reminded me that the result of this first experimental attempt to test parity nonconservation was in fact published before the theoretical paper by Yang and myself (see R. Budde, M. Chretien, J. Leitner, N. P. Samios, M. Schwartz and J. Steinberger, Phys.Rev. 103, 1827 (1956)). I wish to thank Schwartz and Steinberger for sharing their recollections with me.

This might well be the missing key. Later, I realized that a simpler way would be to change the definition of θ to be the angle between the normal to the production plane and the final pion momentum. Parity nonconservation would imply an asymmetry between the "up" events with $\cos\theta$ from 0 to 1 and the "down" events from 0 to -1.

I then made the assumption that while parity conservation holds for strong interactions, at least in the weak decays of strange particles there would be parity violation. This would explain the theta-tau puzzle and give rise to an up-down asymmetry in Steinberger's hyperon experiment. At that time the total number of such hyperon events was quite limited, only about 40. But in another year or so, with more events one would be able to examine the up-down asymmetry in hyperon decay in a statistically meaningful way, and that would give a decisive test to the parity nonconservation idea.* Parity nonconservation obviously had many other and further consequences, which I began to investigate. I felt quite excited by my progress.

Around the beginning of May Steinberger came to my office, saying he had just given a seminar at Brookhaven on his experiment. Yang had been in the audience and had strongly objected to my idea that there could be an up-down asymmetry in Steinberger's hyperon experiment due to parity nonconservation. I then called Yang and asked him not to express his objections in public until we could talk it over.

The next morning Yang drove from Brookhaven to Columbia and we had a very intensive session. I told him my idea and analysis. He was quickly convinced of its importance and wanted to work on it with me.

---

* Indeed, this was observed by the Steinberger group in 1957 (F. Eisler et al., Phys. Rev. 108, 1353).

Yang is endowed with a highly critical mind. It always made me feel more confident if I could overcome his objections with substantive arguments. Furthermore, he is an excellent physicist. The implications of parity nonconservation extended over all facets of physics. I thought Yang's participation would undoubtedly enrich the final product. Therefore I expressed my welcome.

On that day, because of the parking regulations, Yang had to move his car at 11. Our discussion was carried over to the White Rose Cafe at the corner of 125th Street and Broadway, then to the Tien Tsin Restaurant a few doors away, and after lunch back to my office at Columbia. We covered an enormous variety of physical processes, arguing, debating, sometimes even shouting; occasionally we would switch our points of view, just to make sure. The tremendous intensity of our working together in opposition and in harmony, the immense open feeling that the whole world was lying in front of us, the fearlessness of youth, these are what made life meaningful!

Both of us, of course, knew there existed very good experimental evidence for parity conservation in the strong reaction involving pions and nucleons. But how good was the evidence for parity conservation in the strong interaction of other particles? By the end of the day, we were convinced that for strong interactions parity is indeed conserved even for strange particles. In addition, there was good proof for parity conservation in the electromagnetic interaction.

My original thesis, that all weak decays of strange particles could be parity violating, stood firm. However, for the weak interaction of non-strange particles further investigation was needed, and both of us agreed we should look into that.

Before Yang joined me on that day, I had been thinking of writing up the work I had already done on parity nonconservation in strange particle decays. However, Yang convinced me to hold off on this since it was a better idea to enlarge the scope to encompass the whole weak interaction.* We separated and each of us continued the examination of parity questions in beta decay.

As these are my personal recollections, I can describe only what happened at Columbia and not what took place at Brookhaven where Yang was at that time.

Beta decay is a field with a long history. A very large body of knowledge was available at that time. Since parity conservation had been an implicit assumption in all analyses, and the notion of parity was commonly used, it required a very careful examination of all the available experimental facts to see whether parity nonconservation could also be extended to beta decay. One of the world's great experts on beta decay is C. S. Wu, whose office was a few floors above mine at Columbia. I visited her and told her of these ideas. She was extremely interested in them and lent me the authoritative book on beta decay edited by K. Siegbahn.**

At that time the true meaning of the parity operation P was not clear to Yang and me. Of course, we knew its mathematical property: P should be represented by a unitary operator in the Hilbert space, and under P one has, e.g., for a spin-$\frac{1}{2}$ fermion field

$$P\,\psi(r,t)\,P^{-1} = e^{i\phi}\,\gamma_4\,\psi(-r,t)$$

---

* Yang's recollection of the events of that day differs from mine. See pages 26-28 of Selected Papers 1945-1980 with Commentary.

** K. Siegbahn, editor, Beta- and Gamma-ray Spectroscopy (Amsterdam, North-Holland Publishing Co., 1955).

etc. Without parity conservation, beta decay should be described by a general Lagrangian consisting of ten coupling constants, the usual five C_i ($i = S, P, V, A, T$) and the additional five parity-violating constants C_i'.

Yang and I started a systematic investigation of all known beta decay phenomena by using the general parity nonconserving interaction. We raced through the Siegbahn book and kept in touch often by telephone. It took us about two weeks to finish the entire β-decay analysis. In computational skill, Yang and I were evenly matched. This race was extremely intense, but in the best spirit of collaboration and competition, with both of us as winners.* Any attempt to try to place which one was ahead in such intensive and extensive calculations is meaningless.

We began with the allowed spectrum and we found, after some calculation, exactly the same β-distribution with a simple replacement of the usual

$$C_i^* C_j \qquad \text{by} \qquad C_i^* C_j + C_i'^* C_j' \quad . \tag{3}$$

Next came the forbidden spectrum, and the same thing happened. We then computed the Coulomb effect, the $\beta-\nu$ correlation, the $\beta-\gamma$ correlations and the (then-existing) β and γ results from polarized nuclei; while some of these calculations were rather complicated, in the end all interference terms between C_i and C_j' invariably were cancelled, and the same old expressions appeared, provided we made the same simple replacement.

---

* Yang's recollection of our joint endeavor in the search for parity nonconservation differs almost entirely from mine, right down to the peripheral details. For example, he even recalls differently the manner in which our results were written up. In contrast to his account (page 29 of Selected Papers 1945–1980 with Commentary), I remember that the writing was a joint effort; indeed, as usual in our collaborations, we debated almost continuously the wording and nuances of our presentation. There were several versions of the paper, revised over a period of time.

Finally we reached the end of the Siegbahn book, rederiving all the old formulas with this new interaction. It then became clear that there was not a single evidence for parity conservation in beta decay. We got together to compare notes. As we talked we realized that we must have been very stupid, and that there had to be an extremely simple reason why all those complicated interference terms $C_i^* C_j'$ cancelled each other. Once we stopped calculating and started to think, within a very short time it was clear that the reason for this lack of evidence was the simple fact that nobody had made any attempt to observe a physical pseudo-scalar from an otherwise seemingly right-left symmetrical arrangement!

From then on, the theoretical analysis could be made independently of any detailed calculations.* The next scene of the parity drama was set, with many more participants.

There remained the experimental question of how to detect possible parity violation in beta decay. Again I discussed with C. S. Wu the best way to measure asymmetry in spin-momentum correlation in β-decay. She suggested the possible use of a strongly polarized ^{60}Co β-source, for which however a low-temperature set-up was needed. Subsequently, she approached the Bureau of Standards in Washington, which led to the celebrated experiment by Wu, Ambler, Hayward, Hoppes and Hudson.**

The summer of 1956 I spent at Brookhaven. Yang and I talked a great deal to many experimental physicists there, S. DeBenedetti, M. Goldhaber, and others, encouraging them to do various parity experiments. The reception was mixed. We spent

---

* T. D. Lee and C. N. Yang, Phys.Rev. 104, 254 (1956).

** C. S. Wu et al., Phys.Rev. 105, 1413 (1957).

most of our collaborative effort during that summer on statistical mechanics and many-body problems. In 1955 Huang, Luttinger and Yang had been working on the properties of a dilute system of Bose hard spheres. They encountered some difficulties and Yang described these troubles to me. In our discussions, we were able to look at the problem from a different angle. This enabled us to make further progress. The result was the formulation of the binary collision method * and the collaboration with Huang on the energy levels of the hard sphere system. ** There were quite a few twists and turns in these pursuits. Apart from their physical interest, these calculations provided a needed alleviation (at least for me) of the anxiety of waiting for the experimental results on parity.

One day that summer, Yang showed me an extremely interesting letter written by R. Oehme in which he pointed out that under certain restrictions the requirements of charge conjugation (C) and time reversal (T) symmetries would imply that parity (P) must be conserved. Later I thought about the question raised by Oehme. In a short time I realized that in any relativistic local field theory the product CPT is always conserved. This seemed to me such a remarkable theorem that I immediately explained it to T. Kinoshita and then to R. Feynman, who happened to be near my summer office at Brookhaven. As I was explaining it I had the eerie feeling that I had heard the words I was speaking before, under different circumstances. Afterwards, I tried hard to remember where, and then there came a faint recollection that it had something to do with W. Pauli. Indeed, in 1954 Pauli had been at Columbia to give the Pupin

---

* T. D. Lee and C. N. Yang, Phys. Rev. 105, 1119 (1957).

** K. Huang, T. D. Lee and C. N. Yang, Phys. Rev. 106, 1135 (1957).

Lecture and spoke on the CPT theorem. At that time I, as well as others, firmly believed that C, P and T were separately conserved. The fact that the joint symmetry was a necessary consequence of relativistic local field theory held very little interest. Nevertheless, some deep impression must have remained in my subconscious, so that it could resurface years later and enable me to derive the CPT theorem in record time. Ever since, I have often made myself sit through lectures I do not fully comprehend, hoping that at some later time it might come in handy.

In 1956 Lederman and his group had just made the important discovery * of the long-lived K^0 meson (called θ_2^0 then), which was regarded as the rigorous proof of C invariance in strange particle decay, in accordance with the elegant theoretical prediction of Gell-Mann and Pais. ** Because of the CPT theorem, our work on parity nonconservation had to be extended to possible violation of C, CP and T. We found that the existence of the long-lived θ_2^0 was in fact totally independent of either C or CP invariance, and that there was no evidence whatsoever to assume any of these symmetries were valid in the weak interaction.

The outcome of this work, our joint paper with Oehme, *** discussed the possibility of CP and T violation in addition to suggesting several experimental methods to detect such violation. Interestingly, Yang's evaluation of this paper, then as now, **** differs from mine.

---

* K. Lande, E. T. Booth, W. Chinowsky, J. Impeduglia and L. M. Lederman, Phys. Rev. 103, 1901 (1956).

** M. Gell-Mann and A. Pais, Phys. Rev. 97, 1387 (1955).

*** T. D. Lee, R. Oehme and C. N. Yang, Phys. Rev. 106, 340 (1957).

**** C. N. Yang, Selected Papers 1945-1980 with Commentary, pages 32 and 58.

The final impetus on symmetry violation came almost entirely from the experimental side. As I recall, on Christmas that year I was just giving the finishing touch to the appendix of the paper with Oehme and Yang. As it was near the end of a year, I naturally pondered over what I had done in physics that year. I thought of the paper with Yang on G parity written at the beginning, * and then of the solution of the Bose hard sphere system obtained in the fall with Huang and Yang. Both of these appeared to be definitive pieces, and I thought that they should be of value to physics, even in the future. As for discrete symmetry violations, Wu's β-decay and Steinberger's hyperon-asymmetry experiments were still in progress, and here the theoretical speculation was being extended not only to P and C violations, but also to T and CP violations. I then looked at the last formula that I had just written; it was on the possible deviation of the asymmetry parameter **

$$r \equiv \frac{\text{Rate}\,(K_+^0 \rightarrow \pi^- e^+ \nu)}{\text{Rate}\,(K_+^0 \rightarrow \pi^+ e^- \bar{\nu})}$$

from unity as a proof of CP violation. (K_+^0 is now called K_L^0.) Since, from unitarity considerations, and assuming the optimum condition, the deviation from unity was already known then to be very small, it was estimated that at least 10^4- 10^5 long-lived K^0 decays would be needed. At that time, altogether only some 23 events of the long-lived K^0 had been observed. I felt that, whatever might turn up in the

---

* T. D. Lee and C. N. Yang, Nuovo Cimento 3, 749 (1956).

** Equation (A7) of Lee, Oehme and Yang, loc. cit.

future, I must be moving along a tangent, discussing equations that surely had no chance of being verified during my lifetime.

Events, however, moved much faster than I had anticipated. A few days later, on Saturday, December 29th, I received a call from C. S. Wu. During that half year, she had spent most of her time in the low-temperature laboratory at the Bureau of Standards in Washington. But we were in touch with each other. She told me that her ^{60}Co result indeed showed parity nonconservation, but that further checks would still be necessary to ascertain the precise magnitude of the violation. On January 3rd, Thursday of the following week, Wu came to my office and said that she had checked all corrections, and that the violation effect was very large. I assured her that this was excellent, since it agreed completely with the two-component theory. I then immediately called Leon Lederman at Nevis and Yang in Princeton.

During that year, Yang and I had had many discussions with Lederman urging him to do the $\pi - \mu - e$ experiment. Lederman was reluctant, however, partly because he was busy (as always) doing other experiments, and partly because he was afraid that the parity violation effects in both π-decay and μ-decay might not be maximal; the observable effect in the $\pi - \mu - e$ sequence depended on the product and therefore it might be quite small. Over the telephone, I tried to impress on him that from Wu's result it appeared that the two-component theory was definitely correct, that the effect in either π- or μ-decay should be maximal, and that the product of a 100% effect times a 100% effect should remain a 100% effect. Therefore, he had to do the $\pi - \mu - e$ experiment. Leon said he would come to Pupin the next day anyway, since it would be Friday. On Friday, then as now, we always have Chinese lunch.

The next day, January 4th, Leon came to the Shanghai Cafe with J. Orear and many other colleagues. After I reiterated Wu's result on parity violation, two-component theory, etc., our discussion quickly centered on how to do the $\pi - \mu - e$ experiment. Lederman said he had a student, Marcel Weinrich, who was doing the classical Conversi-Pancini-Piccioni experiment on μ-capture and μ-decay. To do the new experiment, he only needed to slowly rotate the counter around to observe if there were any angular asymmetries in the final electron direction. Another method was to use emulsion, and Orear said he would try that. As it turned out, V. Telegdi and J. Friedman had already started such an experiment at Chicago.

Of course, the question of possible depolarization of the μ-spin during its stopping confused me, as well as, perhaps, other participants in that discussion. Nevertheless, it appeared that the best thing was to try it out, and Lederman said he should know the answer in about two week's time. During the weekend R. Garwin joined the experiment. The pace became much swifter than anticipated. To my great surprise, three days later Lederman called me a little after six in the morning to say that parity was dead and they had a "22 standard deviation" effect!

I was in Princeton when the news of the 1957 Nobel physics prize came. I felt exhilarated and grateful, and wrote to my teacher, Professor Ta-you Wu, expressing my deep gratitude for his personal guidance when I was in China and for awarding me a fellowship in 1946, without which my life would have been totally different. During the month of November my wife, Jeannette, and I prepared for the trip to Sweden.

Yang and I also had to write our lectures and speeches. In the course of discussing what to do, Yang asked if our actual accepting of the prize during the ceremony could be done in age order. I was surprised, but acquiesced to his request.

After 1957, I returned to New York. Yang and I continued collaborating for several more years. At that time, we met fairly regularly about twice a week, a pattern that was developed soon after our joint work on parity nonconservation. Usually Yang would come to New York on Thursday and I went to Princeton on Tuesday.

At Columbia I had frequent contact with experimental colleagues; at the Institute for Advanced Study Yang did not have the same advantage. Since much of the progress in particle physics depends on close interaction between experimental and theoretical physicists, this created an asymmetry between Yang and myself, as illustrated by the following example.

In the fall of 1959, I was interested in the high energy behavior of the weak interaction, and had a number of discussions with Leon Lederman, Mel Schwartz and Jack Steinberger concerning the experimental possibilities. At one time, we thought of developing a high energy muon beam and using reactions such as $\mu + p \rightarrow \nu + \cdots$ as probes. Then one day Mel thought of using, instead, the high energy neutrinos. This is his account of that event:*

> "The Columbia neutrino experiment originated at one of the daily coffee hours on the 8th floor of Pupin Lab. where faculty and graduate students met and spent a pleasant half hour relaxing from their daily chores. ...

---

* In Adventures in Experimental Physics, Vol. 1, ed. B. Maglich (Princeton, World Science Communications, 1972), p. 82.

One Tuesday afternoon in November, 1959, I happened to arrive late at coffee to find a lively group around T. D. (which is what we called T. D. Lee) with a conversation ensuing as to the best way of measuring weak interactions at high energies. A large number of possible reactions were on the board making use of all the hitherto known beam particles—electrons, protons, neutrons. None of them seemed at all reasonable. ...

That night it came to me. It was incredibly simple. All one had to do was use neutrinos. ... The next morning I ran into T. D. Lee's office ..."

This turned out to be the essential step toward further progress. Yang and I then wrote on the various theoretical implications,* which together with Mel's paper** became the basis of the famous two-neutrino experiment of Lederman, Schwartz, Steinberger and others in 1962.***

Because of my proximity to these three colleagues during the course of their experiment, when they had theoretical questions they usually turned to me rather than to Yang, whereby the delicate symmetry between Yang and myself was disturbed. However, at that time (1959) I had no inkling that such a slight asymmetry would over a period of time seriously affect the parity between us and eventually play an instrumental role in breaking up our collaboration and our friendship.

---

* T. D. Lee and C. N. Yang, Phys.Rev.Lett. 4, 307 (1960).

** M. Schwartz, Phys.Rev.Lett. 4, 306 (1960).

*** G. Danby, J. M. Gaillard, K. Goulianos, L. M. Lederman, N Mistry, M. Schwartz and J. Steinberger, Phys.Rev.Lett. 9, 36 (1962).

4. Broken Friendship

In 1962 Jeremy Bernstein wrote a profile of Yang and me which appeared in the May 12th New Yorker magazine. We received, in April, copies of Jeremy's galley proof. After correcting some minor points, Yang said there was "something painful" that he wanted to discuss. In several places in the article he wished his name to precede mine: 1) in the title, 2) when the Nobel Prize was announced, and 3) when we received the prize. In addition the name of Chih Li, his wife, should precede Jeannette's because she was a year older. The next day he came to me and said that where the article mentioned "Lee and Yang wrote ...", there should be a statement that this was being done alphabetically in accordance with custom. I told him he was being silly. That evening he called me and said that maybe there shouldn't be such a statement, but that the article should always say "Yang and Lee wrote ...". I was dumbfounded.

Shortly after, on April 18th according to Yang, he came to my office* saying that he was even more deeply troubled by the name sequence, and the problem had spread to all our collaborations: "Lee and Yang" in alphabetical order did not make him happy, "Yang and Lee" would make him look ridiculous, and a random ordering would also look peculiar. This was indeed a "no go" situation, so I suggested that perhaps we should not collaborate in the future. He then became very emotional and started to cry, saying he wanted very much to work with me. I felt embarrassed and helpless, and talked to him gently for a long time. In the end we agreed we would stop collaborating at least for a limited time, and then decide.

---

* Yang's version of that session differs from mine. See page 53 of his Selected Papers 1945-1980 with Commentary. His statement that the meeting on April 18, 1962 functioned as a joint review of and agreement on the events of our collaboration is entirely at variance with my own recollection.

That June, when the two-neutrino experimental results of Lederman, Schwartz and Steinberger were ready for publication, Yang again became highly agitated, worrying over the ordering of our names in the acknowledgments in their paper. Quite unexpectedly, there followed a series of letters from Yang to me which was most unpleasant and full of hostility. I was greatly saddened by all these events and realized our friendship was no more.

My collaboration with Yang ended more than two decades ago. Its value, as represented by our scientific papers, has passed the test of time. There is no need for additional comment. I am truly distressed that the recent publication of Selected Papers 1945–1980 with Commentary by Yang has made this account necessary.

X

Gravity (Continuum Theory)

CU - TP - 305

ARE BLACK HOLES BLACK BODIES?

T. D. Lee

Columbia University, New York, N. Y. 10027

This research was supported in part by the U.S. Department of Energy.

ABSTRACT

We give a new derivation for the properties of radiation appearing to an observer bounded in a space-time domain that has a horizon; this includes the Rindler case (constant accelerating frame) and the Schwarzschild solution (black hole). Because of the global nature of the quantum state, which inevitably extends to space beyond the horizon, it is possible for the observer to gather information concerning the physical state beyond his horizon, similar to the Einstein-Podolsky-Rosen experiments. Depending on the quantum states, the radiation can appear to be either black-body or something quite different.

1. INTRODUCTION

Outside a black hole, or in a constant accelerating frame, one deals with space-time manifolds that possess horizons. It is well-known [1-10] that observers bounded within such a manifold can detect radiation whose characteristics bear a remarkable resemblance to that from a black body. This is puzzling, because black-body radiation has non-zero entropy; its origin lies in the ensemble average necessary in statistical mechanics. Yet, the use of a constant accelerating frame seems to be unrelated to any statistical averages. Likewise, a satisfactory quantum field theory must include gravity and therefore black holes; why should its state vector necessarily lose its coherence? Is this "merge" of field theory and statistical mechanics a necessity derivable from the accepted rules of quantum mechanics, or is it the result of some additional assumptions? In what follows we shall analyze these questions.

Let us concentrate first on the physics of a quantum field theory in an accelerating frame (Rindler space). To make the problem even simpler, we begin with a free zero-mass scalar field ϕ in a flat two-dimensional minkowski space X and T. The Lagrangian $\mathcal{L}$ and Hamiltonian $\mathcal{H}$ are given by (in units $\hbar = c = 1$)

$$\mathcal{L} = \frac{1}{2} \int_{-\infty}^{\infty} \left[\left(\frac{\partial\phi}{\partial T}\right)^2 - \left(\frac{\partial\phi}{\partial X}\right)^2 \right] dX \tag{1.1}$$

and

$$\mathcal{H} = \frac{1}{2} \int_{-\infty}^{\infty} \left[\mathcal{P}^2 + \left(\frac{\partial\phi}{\partial X}\right)^2 \right] dX \tag{1.2}$$

where $\mathcal{P} = \partial\phi/\partial T$. At equal time T, we have the usual equal-time commutation relation

$$[\mathcal{P}(X, T), \ \phi(X', T)] = -i\delta(X - X') . \tag{1.3}$$

The coordinate system X and T will be referred to as Σ .

Next, we introduce the familiar coordinates x and t in a constant accelerating frame Σ_{acc} :

$$X = \frac{1}{g} e^{gx} \cosh gt \tag{1.4}$$

and

$$T = \frac{1}{g} e^{gx} \sinh gt \tag{1.5}$$

where g denotes the constant acceleration. In Σ_{acc} , the entire domain $-\infty \leqslant x \leqslant \infty$ and $-\infty \leqslant t \leqslant \infty$ covers only the quarter

$$X \geqslant |T| \geqslant 0 , \quad \text{called } I \tag{1.6}$$

in Σ , as shown in Figure 1. The horizon is defined by $X = |T|$, and I refers to its inside. The corresponding Lagrangian and Hamiltonian in the accelerating frame Σ_{acc} are

$$L = \frac{1}{2} \int_{-\infty}^{\infty} \left[\left(\frac{\partial \phi}{\partial t}\right)^2 - \left(\frac{\partial \phi}{\partial x}\right)^2 \right] dx \tag{1.7}$$

and

$$H = \frac{1}{2} \int_{-\infty}^{\infty} \left[P^2 + \left(\frac{\partial \phi}{\partial x}\right)^2 \right] dx \tag{1.8}$$

where $P = \partial\phi/\partial t$ and, instead of (1.3),

$$[P(x, t), \ \phi(x', t)] = -i\,\delta(x - x') . \tag{1.9}$$

The field equations in both Σ and Σ_{acc} take on identical forms:

$$\frac{\partial^2\phi}{\partial T^2} - \frac{\partial^2\phi}{\partial X^2} = 0 \qquad (1.10)$$

and

$$\frac{\partial^2\phi}{\partial t^2} - \frac{\partial^2\phi}{\partial x^2} = 0 \ . \qquad (1.11)$$

In the Heisenberg picture, the usual expansion in terms of the annihilation and creation operators (A_K and $A_K^\dagger$ in Σ, and a_k and $a_k^\dagger$ in Σ_{acc}) assumes the expression

$$\phi(X, T) = \int_{-\infty}^{\infty} \frac{dK}{2\pi} (2\Omega)^{-\frac{1}{2}} (A_K e^{iKX-i\Omega T} + A_K^\dagger e^{-iKX+i\Omega T}) \qquad (1.12)$$

in Σ with $\Omega = |K|$, and

$$\phi(x, t) = \int_{-\infty}^{\infty} \frac{dk}{2\pi} (2\omega)^{-\frac{1}{2}} (a_k e^{ikx-i\omega t} + a_k^\dagger e^{-ikx+i\omega t}) \qquad (1.13)$$

in Σ_{acc}, where $\omega = |k|$. From (1.3) and (1.9) we see that

$$[A_K, A_{K'}^\dagger] = 2\pi\delta(K - K') \ , \qquad (1.14)$$

$$[a_k, a_{k'}^\dagger] = 2\pi\delta(k - k') \qquad (1.15)$$

and

$$[A_K, A_{K'}] = [a_k, a_{k'}] = 0 \ . \qquad (1.16)$$

The crucial point is that, viewed in Σ, while the first expansion (1.12) is valid in the entire X, T space, the second expansion (1.13) holds only in the region I. In the larger frame Σ , the ground state $|\,VAC>$ of the Hamiltonian $\mathcal{H}$ is determined by

$$A_K \,|\, VAC> \;=\; 0 \tag{1.17}$$

for all K. Assuming that the quantum state of the system is $|\,VAC>$, we may evaluate the expectation value of the occupation number operator $a_k^\dagger a_k$ in the accelerating frame Σ_{acc}. The result is the well-known black-body radiation formula [4-5]

$$< VAC \,|\, a_k^\dagger a_k \,|\, VAC> \;=\; \text{volume} \cdot (e^{2\pi\omega/g} - 1)^{-1} \quad . \tag{1.18}$$

Although the usual derivation is quite straightforward (and is given in Appendix A), the emergence of a Bose-Einstein distribution often comes as a surprise.

In sect. 2, we give an alternative derivation which makes it transparent why such a distribution should appear in the final expression. This derivation is based on a new theorem proved for a quantum field theory in curved space. It also extends the validity of (1.18) to a general class of problems with horizons, such as the Schwarzschild case. The central idea stems from the simple observation that the field equation in the minkowski space is hyperbolic; its characteristics are closely connected with the horizon. On the other hand, in the euclidean space the field equation is elliptic; hence there is no horizon. A much better perspective on the problem can therefore be obtained. Detailed rules of Feynman diagrams for the euclidean quantum field theory in polar coordinates are given in sect. 3. We show that, unlike the familiar cartesian case, there are additional diagrams* which may be interpreted as due to an external field, one that is coherent.

---

* While in the literature there exist many papers on this subject [5, 11, 12], the necessity for such additional diagrams seems not to have been recognized.

The result is a full-fledged quantum field theory with a unitary S-matrix, but without any statistical averages. The method developed is readily applicable to other problems, including the black hole. These will be discussed in sects. 4 and 5.

In these problems, because the dynamical variables outside the horizon are assumed to be undetected, there is always a loss of information whenever one performs a final measurement. However, such entropy increase is common to any inclusive measurement in physics.* This does not mean that the state vector itself (before the final measurement) has to be incoherent. It is in this sense we conclude that radiations resulting from horizons are inherently coherent, different from the black-body radiation in statistical mechanics.

Consider again the above example of an observer restricted to the accelerating frame Σ_{acc}. Let O be any observable in Σ_{acc} and H the corresponding Hamiltonian in the same frame. By assumption, both O and H are functions only of dynamical variables, denoted by q, available in Σ_{acc}. Let $| >$ be the state vector of the entire system in Σ. There are two important questions related to measurements that we shall examine in this paper:

(1) What is the condition on the state vector that will enable us to write the expectation value of O as a statistical ensemble average

$$< | O | > \;=\; \frac{\mathrm{tr}\,(e^{-2\pi H/g}\, O)}{\mathrm{tr}\; e^{-2\pi H/g}} \quad ? \tag{1.19}$$

As we shall see, if $| > = |\, \mathrm{VAC} >$ defined by (1.17), then the above formula holds for

---

* An inclusive experiment is one in which not all (mutually commuting) dynamical variables in the final state are measured. This covers, of course, most of the experiments in physics.

any observable O in Σ_{acc}. While $| \mathrm{VAC} >$ is a pure state, its coherence is between the variables q in Σ_{acc} (within the horizon) and other variables, say q', outside Σ_{acc} (beyond the horizon). Since the variables q' are not measured and $O = O(q)$, this coherence does not show up in $< \mathrm{VAC} \mid O(q) \mid \mathrm{VAC} >$.

(2) For what kinds of state vectors does (1.19) fail to hold? It will be shown that for a large class of excited states, there exist expectation values $< \mid O(q) \mid >$ that cannot be written in the form of an ensemble average (1.19). For example, assume the state vector $| >$ consists of $| \mathrm{VAC} >$, plus an additional coherent excitation outside Σ_{acc}, say a wave packet which lies completely outside Σ_{acc} and is described by the appropriate q'. Although this wave packet is beyond the reach of an observer in the accelerating frame, through measurements such as $< \mid O(q) \mid >$ the same observer can in fact detect its existence. Furthermore, he can also measure the long-range correlations between q and q' that are contained in $| \mathrm{VAC} >$. [See (2.38)-(2.39) below.] In this case, (1.19) no longer holds. Perhaps we can understand this situation better if we recall the resolution of the well-known question raised by Einstein, Podolsky and Rosen [13-16]. A pure quantum state is defined globally; its coherence may extend over field variables located at well-separated points on a space-like surface, e.g., like the above q and q'. To detect such correlations the simplest method is to employ two detectors, A and B, with A measuring q and B measuring q'. Assume that B is sensitive to the field amplitude which, because of its quantum nature, can create or annihilate a quantum. So far as A is concerned, the action of B is equivalent to altering the state by the additional amplitude of a particle, or antiparticle, located near B and therefore outside the light cone of A. Conversely, the presence of

such a coherent additional particle amplitude (described by q') in the state vector $| >$ can replace the necessity of having an actual detector B. In this case a single detector A is sufficient to measure the long-range correlation between q and q'. Hence, in our problem, the single observer bounded within Σ_{acc} may also, through $< | O(q) | >$, detect such a correlation, and with that the existence of a wave packet outside Σ_{acc}. This is why he can gather information concerning the state beyond the horizon.

These conclusions are quite general; they can be extended to the Schwarzschild problem, and therefore to an eternal black hole. For a realistic black hole, its non-static complexity makes a complete analysis difficult. In Appendix C, we examine in some detail a model calculation. We find that, at least in principle, it is possible for an observer within the Schwarzschild frame to detect physics inside the black hole.

2. A THEOREM

In quantum mechanics, a state vector $|\,>$ must be associated with a space-like surface S. As S varies, the evolution of $|\,>$ is determined by the familiar Schroedinger equation or its generalization [17]. For the simple problem (1.1)-(1.3) formulated in Σ, the usual rule of quantum field theory requires the surface S to cut the entire space-time manifold (X, T) into two disconnected parts. Thus, S cannot be contained within the accelerating frame Σ_{acc}, and that gives rise to the complication.

For definiteness, choose S to be the flat surface

$$T = 0 . \tag{2.1}$$

Let $< Q \,|\, >$ be the coordinate representation of the state vector $|\,>$ where the set

$$Q = \{Q_K\} \tag{2.2}$$

consists of all field-coordinates Q_K defined by the Fourier integral

$$\phi(X) = \int_{-\infty}^{\infty} \frac{dK}{2\pi} Q_K e^{iKX} . \tag{2.3}$$

In this section, we shall adopt the Schroedinger picture and regard the operator ϕ as T-independent. The $X > 0$ portion of the surface $T = 0$ coincides with the entire $t = 0$ surface in Σ_{acc}. In that region, the same operator ϕ may also be expanded as

$$\phi(X) = \int_{-\infty}^{\infty} \frac{dk}{2\pi} q_k e^{ikx} \quad \text{with} \quad X = \frac{1}{g} e^{gx} > 0 . \tag{2.4}$$

For X negative, we must introduce a new set of variables q_k' defined by

$$\phi(X) = \int_{-\infty}^{\infty} \frac{dk}{2\pi} q_k' e^{ikx'} \quad \text{when} \quad X = -\frac{1}{g} e^{gx'} < 0 . \qquad (2.5)$$

The ranges of x and x' are both from $-\infty$ to ∞ ; together they cover the entire X-region. Similar to (2.2), we define the sets q and q' to be

$$q = \{q_k\} \quad \text{and} \quad q' = \{q_k'\} . \qquad (2.6)$$

Together, q and q' form a complete set of field-coordinates, equivalent to Q. From (2.3)-(2.5) we see that q and q' are functions of Q,

$$q = q(Q) \quad \text{and} \quad q' = q'(Q) ; \qquad (2.7)$$

conversely, we can express Q as a function of both q and q',

$$Q = Q(q, q') . \qquad (2.8)$$

For example, in Σ the vacuum state $| \text{VAC} \rangle$, defined by (1.17), can be expressed either as a function of Q_K ; i.e., written as $\langle Q | \text{VAC} \rangle$ in the Q-representation, or in terms of q_k and q_k' as

$$\langle q, q' | \text{VAC} \rangle . \qquad (2.9)$$

The set q denotes the field-coordinates in Σ_{acc}. If one wishes, one may think of q' as those in another accelerating frame Σ'_{acc} which, when viewed from Σ, has an acceleration of the same magnitude as g but in the opposite direction. (We note that in the Heisenberg picture, (2.4) can be extended only to the region within I in Figure 1, and the other expansion (2.5) to the region within II, which is outside the horizon of Σ_{acc} but inside the horizon of Σ'_{acc} .)

For an observer confined to the accelerating frame Σ_{acc} , he may "pretend" that q forms a complete set. In that case, he would use the accelerating system Hamiltonian H , given by (1.8) for his Schroedinger equation

$$-\frac{1}{i}\frac{\partial}{\partial t}\;|t> \;=\; H\,|t> \;,$$

where the quantum state $|t>$ is associated with the space-like surface

$$t \;=\; \text{constant}$$

contained entirely in Σ_{acc} . Viewed from Σ , the physical meaning of such a state vector $|t>$ is quite unclear. Operationally one also notes that something is defective. The Green's function e^{-itH} can bring the state vector only up to the future horizon $X = T > 0$ when t varies towards ∞ , and up to the past horizon $X = -T > 0$ when t moves towards $-\infty$. By introducing interactions, it is not difficult to see that such a theory, which is formulated solely in the accelerating frame, would at least be considered incomplete.

On the other hand, a totally different vista is opened if instead of e^{-itH} we consider the corresponding euclidean Green's function $e^{-\tau H}$ with τ real. Under the analytical continuation

$$T \;=\; -iY \qquad \text{and} \qquad t \;=\; -i\tau \tag{2.10}$$

the coordinate transformation (1.4)-(1.5) becomes simply

$$X = \rho \cos\theta \qquad \text{and} \qquad Y = \rho \sin\theta \tag{2.11}$$

where

$$\rho = \frac{1}{g} e^{gx} \qquad \text{and} \qquad \theta = g\tau \ .$$

When θ varies from 0 to π, the Green's function $e^{-\tau H}$ takes a state vector from $X > 0$ in region I to $X < 0$ in region II. (See Figures 1 and 2.) There is <u>no</u> horizon in the euclidean space. Consequently, as we shall see, in contrast to the minkowski operator e^{-itH}, a skillful manipulation of the euclidean operator $e^{-\tau H}$ can give the complete solution of the quantum field theory. In particular, the state vector $|\,VAC\rangle$ is given by the matrix elements of $e^{-\pi H/g}$:

<u>Theorem</u>

$$\langle VAC\,|\,q, q'\rangle = \langle q'\,|\,e^{-\pi H/g}\,|\,q\rangle / (\text{trace}\ e^{-2\pi H/g})^{\frac{1}{2}} \tag{2.12}$$

where, as in (2.9), $|\,VAC\rangle$ is the ground state of the total Hamiltonian $\mathcal{H}$ in Σ. The numerator on the right-hand side of (2.12) refers to the matrix elements of $e^{-\tau H}$ with $\tau = \pi/g$ and H the Hamiltonian operator in Σ_{acc},

$$H = \tfrac{1}{2}\int_{-\infty}^{\infty} \left[P(x)^2 + \left(\frac{d\phi}{dx}\right)^2\right] dx \tag{2.13}$$

given before by (1.8). Here, because we are in the Schroedinger picture, both $\phi(x)$ and its conjugate momentum $P(x)$ are independent of τ.

<u>Proof</u> Express the matrix elements $\langle Q' | e^{-Y\mathcal{H}} | Q \rangle$ and $\langle q' | e^{-\theta H/g} | q \rangle$ in terms of Feynman's path integrals. In both cases, the integrand is exp (action). In the former,

$$\text{action} = \mathcal{A}(\phi) \equiv \int_0^Y dY \int_{-\infty}^{\infty} dX \, \tfrac{1}{2} \left[\left(\frac{\partial \phi}{\partial X}\right)^2 + \left(\frac{\partial \phi}{\partial Y}\right)^2 \right] \qquad (2.14)$$

with the boundary conditions that at $Y = 0$, the initial field configuration is given by $Q = \{Q_K\}$ through (2.3), and at the final Y, ϕ is given by the same expression but with Q replaced by $Q' = \{Q'_K\}$; in the latter,

$$\text{action} = A(\phi) \equiv \int_0^\theta d\theta \int_0^\infty \rho d\rho \, \tfrac{1}{2} \left[\left(\frac{\partial \phi}{\partial \rho}\right)^2 + \frac{1}{\rho^2} \left(\frac{\partial \phi}{\partial \theta}\right)^2 \right] \qquad (2.15)$$

with the boundary condition that at $\theta = 0$ the initial field configuration ϕ is given by $q = \{q_k\}$ through (2.4) and at the final θ, q is replaced by $q' = \{q_k\}$.

Because the action is a positive quadratic functional of ϕ , it is equal to its minimum (the classical action) plus a quadratic function depending only on fluctuations. The result of integration over these fluctuations is independent of the boundary values Q , Q' or q, q' . Thus, the matrix elements of the quantum operators $e^{-Y\mathcal{H}}$ and $e^{-\theta H/g}$ are related to their corresponding classical actions $\mathcal{A}_{cl}$ and A_{cl} :

$$\begin{aligned} \langle Q' | e^{-Y\mathcal{H}} | Q \rangle &= \langle 0 | e^{-Y\mathcal{H}} | 0 \rangle \, e^{-\mathcal{A}_{cl}} \quad , \\ \langle q' | e^{-\theta H/g} | q \rangle &= \langle 0 | e^{-\theta H/g} | 0 \rangle \, e^{-A_{cl}} \end{aligned} \qquad (2.16)$$

where $|0\rangle = |Q\rangle$ or $|q\rangle$, when all $Q_K = 0$ or all $q_k = 0$.

The classical action is obtained by finding the classical solution ϕ_{cl}. Here, ϕ_{cl} satisfies the Laplace equation with the boundary conditions specified by Q and Q' in the cartesian problem, or q and q' in the polar problem. (For definiteness, we take $\phi_{cl} = 0$ at $X = \pm\infty$ in the cartesian problem, and at $\rho = \infty$ in the polar problem.) Set

$$Y = \infty , \qquad \theta = \pi , \qquad Q' = 0$$

(i.e., $Q'_K = 0$ for all K). Since Q is related to q and q' by (2.8), the boundary condition of ϕ_{cl} described by Q at $Y = 0$ is identical to that described by q at $\theta = 0$ and q' at $\theta = \pi$. Thus the classical solutions of these two problems are the same, and so are their classical actions; i.e., $\mathcal{A}_{cl} = A_{cl}$. Consequently,

$$\frac{\langle q' \mid e^{-\pi H/g} \mid q \rangle}{\langle 0 \mid e^{-\pi H/g} \mid 0 \rangle} = \lim_{Y\to\infty} \frac{\langle 0 \mid e^{-Y\mathcal{H}} \mid Q \rangle}{\langle 0 \mid e^{-Y\mathcal{H}} \mid 0 \rangle} .$$

As $Y \to \infty$,

$$\langle Q' \mid e^{-Y\mathcal{H}} \mid Q \rangle \to e^{-E_0 Y} \langle Q' \mid VAC \rangle \langle VAC \mid Q \rangle$$

where E_0 is the vacuum energy (eigenvalue of $\mathcal{H}$ in the ground state $\mid VAC \rangle$). Combining these formulas, we find

$$\langle q' \mid e^{-\pi H/g} \mid q \rangle = \text{constant} \langle VAC \mid Q \rangle .$$

Since $Q = Q(q, q')$, we derive (2.12) and complete the proof. The denominator in (2.12) insures $\langle VAC \mid VAC \rangle = 1$.

The ground state wave function can always be set to be real. In that case, $< q, q' \mid VAC > = < VAC \mid q, q' >$, and therefore (2.12) can also be written as

$$< q, q' \mid VAC > \;=\; < q' \mid e^{-\pi H/g} \mid q > / \;(\text{trace } e^{-2\pi H/g})^{\frac{1}{2}} \;. \qquad (2.12')$$

Remarks

1. From this proof, it is clear that the theorem can be applied without change to any theory whose H is a quadratic function of ϕ ; this includes massive or massless fields in a Rindler or Schwarzschild space. Extension to interacting fields is straightforward; the details will be given in a separate paper.

2. At first sight, (2.12) may appear unusual since both the field-coordinates q and q' occur in the ket-vector on its left side, while on the right q' is in the bra-vector and only q in the ket-vector. A moment of reflection may reassure one that this is not so surprising. In the minkowski formulation, there is a clear distinction between a space-like surface and a time-like surface. No such difference exists in the euclidean formulation. In the minkowski space, the field equations are hyperbolic; hence, e.g., in the classical theory because of the characteristics of these differential equations, special care is required whenever there is a closed surface boundary. On the other hand, in the euclidean space the field equations are elliptic; the appropriate boundary conditions in such classical theories are most conveniently expressed in terms of closed surfaces. The existence of the horizon is closely connected with the characteristics of the hyperbolic equation; therefore it is absent in the euclidean space. The closed surface boundary condition for a classical euclidean field, when transposed to the quantum theory,

makes it possible to associate the initial state with part of the closed surface, and the final state with the remaining part. In the classical euclidean field theory (e.g., the Laplace equation in electrostatics) sometimes the closed surface boundary may consist of an open surface plus boundaries at infinity. In this case in its quantum version we can associate part of the open surface (which is also a part of the entire closed surface boundary) with the initial condition and the other part with the final state, and that is what occurs in the matrix element $\langle q' \mid \exp(-\pi H/g) \mid q \rangle$ in (2.12).

3. Any observable in the accelerating frame Σ_{acc} can be represented by an operator $O(q)$ which depends only on q_k and $-i\,\partial/\partial q_k$. In the q-representation, $O(q)$ takes on the matrix form $\langle q_f \mid O(q) \mid q_{in} \rangle$, where q_f and q_{in} refer to the final and initial values of q. Hence

$$\langle VAC \mid O(q) \mid VAC \rangle = \sum_{q'} \sum_{q_f, q_{in}} \langle VAC \mid q_f, q' \rangle \langle q_f \mid O(q) \mid q_{in} \rangle \cdot \langle q_{in}, q' \mid VAC \rangle .$$

Substituting (2.12) into the above expression and noting that $O(q)$ is independent of q', we have, after carrying out the sum over q',

$$\langle VAC \mid O(q) \mid VAC \rangle = \frac{\text{trace}\left[e^{-2\pi H/g}\, O(q)\right]}{\text{trace}\ e^{-2\pi H/g}} . \tag{2.17}$$

The right-hand side is precisely the canonical ensemble average in statistical mechanics, with a Hamiltonian given by H in Σ_{acc} and a temperature $= (2\pi)^{-1} g/$Boltzmann constant. Thus, when the state vector is $\mid VAC \rangle$, the ground state of the total Hamiltonian

$\mathcal{H}$ in Σ, observations in Σ_{acc} give the same results as if the system were in thermal equilibrium. However, as we shall see in remarks 5 and 6 below, the situation can be quite different, when the state vector is not $|\,VAC>$.

Setting $O(q) = a_k^\dagger a_k$, defined by (1.13), we find

$$< VAC \,|\, a_k^\dagger a_k \,|\, VAC > \;=\; \frac{\text{trace}\,(e^{-2\pi H/g}\, a_k^\dagger a_k)}{\text{trace}\; e^{-2\pi H/g}} \;. \qquad (2.18)$$

Since in accordance with (2.13)

$$H \;=\; \int_{-\infty}^{\infty} \frac{dk}{2\pi}\, (a_k^\dagger a_k + \tfrac{1}{2})\, \omega \;, \qquad (2.19)$$

we derive the Bose-Einstein distribution (1.18). Note that on account of (1.15), the commutator of a_k and $a_{k'}^\dagger$ is a δ-function, which gives rise to the factor "volume" in (1.18). In order to remove this factor, it is customary to convert the δ-function in the commutator into a Kronecker δ-symbol by introducing a large linear box of size V and defining

$$a_k \;=\; V^{-\frac{1}{2}}\, a_k \qquad (2.20)$$

so that when $V \to \infty$, H becomes

$$H \;=\; \sum_k \,(a_k^\dagger\, a_k \;+\; \tfrac{1}{2})\;\omega \qquad (2.21)$$

with a_k satisfying

$$[a_k\,,\, a_{k'}^\dagger\,] \;=\; \delta_{kk'} \qquad (2.22)$$

and $[a_k\,, a_{k'}\,] = 0$. The replacement (2.20) is convenient because the eigenvalues

n_k of $a_k^\dagger a_k$ are now all positive integers 0, 1, 2, $\cdots$. Consequently, the eigenvectors of H can be written as

$$|\, n > \qquad \text{with} \qquad n = \{n_k\} \quad . \tag{2.23}$$

It follows from (2.18)-(2.23),

$$< \mathrm{VAC} \,|\, a_k^\dagger \, a_k \,|\, \mathrm{VAC} > \;=\; (e^{2\pi\omega/g} - 1)^{-1} \quad , \tag{2.24}$$

without the "volume" factor in (1.18).

4. Equation (2.12) of the theorem can also be stated in the representation of the occupation number space. By using (2.21)-(2.23), we can write the numerator on the right-hand side of (2.12) as

$$< q' \,|\, e^{-\pi H/g} \,|\, q > \;=\; \sum_n < n \,|\, q' >^* \, < n \,|\, q > \, e^{-\sum_k (n_k + \frac{1}{2})\, \pi\omega/g} \quad . \tag{2.25}$$

Because the hermitian conjugate of the operator q_k satisfies the hermiticity condition

$$q_k^\dagger = q_{-k} \quad , \tag{2.26}$$

the q - representation of the state vector $|\, n >$ under a complex conjugation (which is like a time reversal in the present case) becomes

$$< n \,|\, q' >^* \;=\; < n' \,|\, q' > \tag{2.27}$$

where $n' = \{n_k'\}$ is related to the original set $n = \{n_k\}$ by

$$n_{-k}' = n_k \quad . \tag{2.28}$$

Combining (2.12'),(2.27) and (2.28), the result of the Theorem can also be stated as

$$< q, q' \mid VAC > \; = \; \text{constant} \cdot \sum_n < n \mid q >< n' \mid q' > e^{-\sum_k (n_k + \frac{1}{2}) \pi\omega/g} \; . \qquad (2.29)$$

Returning to the expansions (2.4) and (2.5), written for the surface $T = 0$ in the minkowski space (which is identical to $Y = 0$ in the euclidean space, since $T = -iY$), in the q_k and q_k' representation of the Hilbert space in Σ, q_k and $p_k = -i\,\partial/\partial q_k$ can be expressed in terms of a_k and $a_k^\dagger$:

$$q_k = (2\omega)^{-\frac{1}{2}} (a_k + a_{-k}^\dagger)$$

and (2.30)

$$-i \frac{\partial}{\partial q_k} = p_k = -i\omega (2\omega)^{-\frac{1}{2}} (a_{-k} - a_k^\dagger) \; .$$

Likewise, we introduce

$$q_k' = (2\omega)^{-\frac{1}{2}} (a_k' + a_{-k}'^\dagger)$$

and (2.31)

$$-i \frac{\partial}{\partial q_k'} = -i\omega (2\omega)^{-\frac{1}{2}} (a_{-k}' - a_k'^\dagger) \; .$$

Hence a_k', $a_k'^\dagger$ satisfy the same commutation relation (1.15), as a_k and $a_k^\dagger$; furthermore a_k' commutes with a_k and $a_k^\dagger$. As in (2.20), we define

$$\alpha_k' = V^{-\frac{1}{2}} a_k' \; . \qquad (2.32)$$

By identifying the eigenvalue of $\alpha_k'^\dagger \alpha_k'$ with n_k' in (2.29), we derive

$$\mid VAC > \; = \; \text{constant} \cdot e^F \mid vac > \qquad (2.33)$$

where

$$F = \int_0^\infty \frac{d\omega}{2\pi} e^{-\pi\omega/g} (a_\omega^\dagger a'^\dagger_{-\omega} + a'^\dagger_\omega a^\dagger_{-\omega}) \tag{2.34}$$

and the state $|\,vac>$ is defined by

$$a_k \,|\, vac> = a'_k \,|\, vac> = 0 \ . \tag{2.35}$$

Of course, the same result can also be derived directly in the minkowski space, as is well-known [4, 5]. The above euclidean derivation shows more explicitly the relevance of the multiplicative term $e^{-\pi\omega/g}$, which resembles the statistical factor. In this particular case, because we are dealing with the ground state of a non-interacting system, although $|\,VAC>$ is a coherent state the correlation is entirely between the quantum in Σ_{acc} and that in Σ'_{acc}. Since an observer restricted to Σ_{acc} cannot measure the quantum in Σ'_{acc}, he would see an incoherent black-body distribution of quanta in Σ_{acc}, in accordance with (2.18). However, the situation can be radically different if the system is in a different state, as will be illustrated by the examples below.

5. Consider now the hypothetical problem that a "mirror" is placed [18] at $X = 0$ in the minkowski space $\Sigma(X, T)$. Depending on the nature of the mirror, we may impose the Dirichlet condition $\phi = 0$ at $X = 0$ or the Neumann condition $\partial\phi/\partial X = 0$ at $X = 0$. If we stay in the Schroedinger picture, then $\phi = \phi(X)$; the former condition implies $\phi(X) = -\phi(-X)$ while the latter $\phi(X) = \phi(-X)$. By using (2.4)-(2.5) and (2.30)-(2.31), we see that the corresponding constraints on their Fourier components are $a_k = -a'_k$, or $a_k = a'_k$. Thus, the vacuum state becomes, as shown in Appendix B,

$$|\,VAC> = \text{constant} \cdot e^F \,|\, vac> ,$$

but instead of (2.34)

$$F = -\int_0^\infty \frac{d\omega}{2\pi}\, e^{-\pi\omega/g}\, a_\omega^\dagger a_{-\omega}^\dagger \qquad \text{(Dirichlet)}$$

or (2.36)

$$F = \int_0^\infty \frac{d\omega}{2\pi}\, e^{-\pi\omega/g}\, a_\omega^\dagger a_{-\omega}^\dagger \qquad \text{(Neumann).}$$

In either case, even an observer in Σ_{acc} would detect a pair-correlated B.C.S. distribution [19], which is coherent, has zero probability amplitude for any odd number of quanta and is totally different from a "black-body" distribution.

In Appendix B, we analyze the case when the mirror is placed at $X = -\ell < 0$, beyond the horizon of Σ_{acc}. We find

$$\langle VAC | a_\omega a_{-\omega'} | VAC \rangle = \begin{cases} \mp\ \pi(\text{csch}\ \pi\omega/g)\ \delta(\omega-\omega') & \text{if } \ell = 0 \\ \mp\ \dfrac{(\omega\omega')^{\frac{1}{2}}}{2\pi g^2}\,(g\ell)^{i(\omega'-\omega)/g}\ \Gamma(-i\frac{\omega}{g})\ \Gamma(i\frac{\omega'}{g})\ \Gamma(i\frac{\omega-\omega'}{g}) & \text{if } \ell > 0 \end{cases} \qquad (2.37)$$

where the upper sign is for the Dirichlet mirror and the lower sign for the Neumann mirror. Note that $\ell^{i(\omega'-\omega)/g} = \exp\left[i\,\frac{\omega'-\omega}{g}\ \ln \ell\right]$ which in the limit $\ell \to 0$, but at $\omega \neq \omega'$, carries an infinite phase; this together with $\Gamma(i\,\frac{\omega-\omega'}{g})$ leads to the δ-function when $\ell = 0$. Therefore, when the mirror is at $X = -\ell < 0$, although an observer in Σ_{acc} can never reach the mirror, yet by measuring correlation functions within the accelerating frame, such as $\langle VAC | a_\omega a_{-\omega'} | VAC \rangle$, the same observer can determine the existence, the location and the nature of the mirror!

6. As a further illustration, let us consider the case <u>without</u> a mirror. The ground state of $\mathcal{H}$ in Σ is then $|\,\mathrm{VAC}>$, given by (2.33)-(2.35). Assume that the state of interest is *

$$|> \;=\; \left[C_0 + \int \frac{dk}{2\pi}\, C_1(k)\, a_k'^{\dagger}\right] \;|\,\mathrm{VAC}> . \qquad (2.38)$$

Such a state can be readily prepared if there is an interaction. Note that the "primed" quantum generated by $a_k'^{\dagger}$ is beyond the horizon of Σ_{acc} and therefore outside the direct reach of an observer in Σ_{acc}. Through $C_1(k)$, the "primed" quantum may be viewed as a wave packet confined to the region II in Figure 1. Nevertheless, the observer in Σ_{acc} can infer its existence through an observation within the region I by

---

* This is representative of the physical situation when the state is known to be of low excitation, as observed in Σ. In such a case, $|>$ differs from $|\,\mathrm{VAC}>$ only by an amplitude consisting of a limited number of excited quanta. Without such a mild and (in many cases) quite reasonable assumption, by measuring only the variables q in Σ_{acc}, one cannot draw conclusions about the variables q', which are outside Σ_{acc} and commute with q. Note that, because of (2.33)-(2.34), we have $a_k'\,|\,\mathrm{VAC}> = e^{-\pi\omega/g}\, a_{-k}^{\dagger}\,|\,\mathrm{VAC}>$ and $a_k'^{\dagger}\,|\,\mathrm{VAC}> = e^{\pi\omega/g}\, a_{-k}\,|\,\mathrm{VAC}>$. Hence for any operator $v = v(q')$ outside Σ_{acc}, there exists an operator $u(q)$ inside Σ_{acc}, so that when operating on $|\,\mathrm{VAC}>$ their effects become identical: $v(q')\,|\,\mathrm{VAC}> = u(q)\,|\,\mathrm{VAC}>$. This gives the underlying reason why an observer in Σ_{acc}, through measurements of observables $O(q)$ within his horizon, can distinguish between the state vector $|> = v(q')\,|\,\mathrm{VAC}>$ and $|\,\mathrm{VAC}>$, provided that $v(q')$ is not unitary.

measuring, say,

$$< | a_k | > \; = \; \tfrac{1}{2} C_1(-k) * C_0 \operatorname{csch} \pi\omega/g \quad . \tag{2.39}$$

If there is a genuine thermal bath of black-body radiation, then the same expectation value would necessarily be zero. A nonzero value of (2.39) can tell the observer in Σ_{acc} that the state is a coherent mixture of $| \text{VAC} >$ plus an additional amplitude of a quantum beyond the horizon; in addition, from (2.39) he can infer the B.C.S. nature of $| \text{VAC} >$, which pair-correlates the $a_k^\dagger$ -quanta in Σ_{acc} with the $a'^\dagger_{-k}$ - quanta in Σ'_{acc}.

From these examples, we see that the existence of a horizon is simply due to an artificial restriction of the space-time domain influenced by a bad choice of coordinate system. The quantum mechanical state vector $| >$ is of a global character; its preparation is not in the control of an observer restricted to staying within the horizon. Yet, by careful experimentation the same observer can extract (from the state vector) information beyond his horizon. As mentioned in the Introduction, the present situation is analogous to the resolution of the Einstein-Podolsky-Rosen paradox. However, the experiments that can verify the global nature of a pure quantum state are usually in atomic or subatomic physics, such as the decay of positronium [16] or the K-meson [20]. Here, it happens on a macroscopic scale, and that accentuates the unusual circumstances.

3. EUCLIDEAN QUANTUM THEORY IN POLAR COORDINATES *

In this section, we wish to show that the quantum field theory in Σ_{acc} actually contains all the information of the complete theory in the larger frame Σ, provided we properly take into account its euclidean extension. In the standard euclidean quantum field theory, one starts from the minkowski space and makes the usual transformation $T = -iY$ in (2.10). The boundary-surfaces $T = T_0$ and $T = T_f$ that define the initial and final quantum states in the minkowski space become simply $Y = Y_0$ and $Y = Y_f$. However, once in the euclidean space, because of the absence of horizons, one does not have to stay with the cartesian coordinates; it is equally convenient to formulate the quantum field theory in terms of curved coordinates, such as the polar coordinates ρ and θ of (2.11).

Let us introduce interactions to the simple problem discussed above. In the cartesian frame, called Σ, the coordinates are X and Y. The Hamiltonian operator in the Schroedinger picture is now given by

$$\mathcal{H} \equiv \int_{-\infty}^{\infty} \left[\tfrac{1}{2}\wp^2 + \tfrac{1}{2}\left(\frac{d\phi}{dX}\right)^2 + V(\phi)\right] dX \qquad (3.1)$$

in which $V(\phi)$ can be an arbitrary function of ϕ, and the operators $\wp(X)$ and $\phi(X)$ depend only on X, not on Y. As before, they satisfy

$$[\wp(X), \phi(X')] = -i\delta(X - X') . \qquad (3.2)$$

---

* Readers who are not interested in the technical details of Feynman diagrams may wish to skip this section.

When there is an external current $j(X, Y)$, we modify the Hamiltonian operator to

$$\mathcal{H}_j(Y) = \mathcal{H} + \int_{-\infty}^{\infty} j(X, Y)\, \phi(X)\, dX \quad . \tag{3.3}$$

The Green's functions of (3.1) and (3.3) are

$$U(Y_f, Y_0) \equiv e^{-(Y_f - Y_0)\mathcal{H}} \tag{3.4}$$

and

$$U_j(Y_f, Y_0) \equiv \lim_{\epsilon \to 0} e^{-\epsilon \mathcal{H}_j(Y_N)} \cdot e^{-\epsilon \mathcal{H}_j(Y_{N-1})} \cdots e^{-\epsilon \mathcal{H}_j(Y_0)} \tag{3.5}$$

where

$$Y_f - Y_0 = (N+1)\,\epsilon \geqslant 0 \,, \tag{3.6}$$

$$Y_n = Y_0 + n\epsilon$$

and $n = 0, 1, \cdots, N+1$ with $Y_{N+1} = Y_f$. The generating function $W(j)$ is given by

$$W(j) = \lim_{\substack{Y_f \to \infty \\ Y_0 \to -\infty}} \frac{\langle VAC \mid U_j(Y_f, Y_0) \mid VAC \rangle}{\langle VAC \mid U(Y_f, Y_0) \mid VAC \rangle} \tag{3.7}$$

where $\mid VAC \rangle$ is the ground state of $\mathcal{H}$. Throughout our discussion, we adhere to the well-known fact that the full content of the field theory is completely contained in $W(j)$. In what follows, we show that the same $W(j)$ can also be computed in polar coordinates.

In terms of polar coordinates $X = \rho \cos\theta$ and $Y = \rho \sin\theta$, the Hamiltonians $\mathcal{H}$ and $\mathcal{H}_j$ are replaced by

$$H = \int_{-\infty}^{\infty} \left[\tfrac{1}{2} P^2(x) + \tfrac{1}{2} \left(\frac{d\phi}{dx}\right)^2 + e^{2gx} V(\phi)\right] dx \tag{3.8}$$

and

$$H_j(\theta) = H + \int_{-\infty}^{\infty} e^{2gx} j(x, \theta)\ \phi(x)\ dx \tag{3.9}$$

where, as before,

$$\rho = \frac{1}{g} e^{gx} \quad . \tag{3.10}$$

In the (polar coordinate) Schroedinger picture, $\phi(x)$ and its conjugate momentum $P(x)$ are operator functions of x, independent of θ. They satisfy

$$[P(x),\ \phi(x')] = -i\,\delta(x - x') \quad . \tag{3.11}$$

The factor e^{2gx} in (3.8) and (3.9) is due to the jacobian in

$$dX\,dY = \rho\,d\rho\,d\theta = e^{2gx}\,dx\,d\tau \tag{3.12}$$

where, as in (2.11),

$$\tau = \theta/g \quad . \tag{3.13}$$

Similar to (3.4)-(3.6), the Green's functions are

$$u(\theta) = e^{-\theta H/g} \tag{3.14}$$

and

$$u_j(\theta) \equiv \lim_{\epsilon \to 0} e^{-\epsilon H_j(\theta_N)} \cdot e^{-\epsilon H_j(\theta_{N-1})} \cdots e^{-\epsilon H_j(\theta_0)} \tag{3.15}$$

where

$$\theta = (N+1)\epsilon g \geqslant 0 \ , \tag{3.16}$$

$$\theta_n = n \epsilon g$$

and $n = 0, 1, \cdots, N+1$ with $\theta_0 = 0$ and $\theta_{N+1} = \theta$. The corresponding generating function in polar coordinates is defined to be

$$w(j) \equiv \frac{\text{trace } u_j(2\pi)}{\text{trace } u(2\pi)} \ . \tag{3.17}$$

In what follows, we shall examine the precise relationship between $W(j)$ and $w(j)$. As we shall see, although different, they are closely related to each other. For example, from $w(j)$ we can derive $W(j)$. Therefore, $w(j)$ also contains all the information of the original quantum theory in cartesian coordinates.

That they are different can be seen by considering the special case $V = 0$. In the cartesian coordinates, from (3.1)-(3.7) we derive the familiar expressions:

$$W(j) = \exp\left[\tfrac{1}{2}\int j(1) \langle 1 \,|\, \mathcal{D} \,|\, 2 \rangle\, j(2) \prod_{p=1}^{2} dX_p\, dY_p\right] \tag{3.18}$$

and

$$\langle 1 \,|\, \mathcal{D} \,|\, 2 \rangle = \mathcal{D}(X,Y) \equiv \int_{-\infty}^{\infty} \frac{dK_x}{2\pi} \int_{-\infty}^{\infty} \frac{dK_y}{2\pi} (K_x^2 + K_y^2)^{-1} \exp i(K_x X + K_y Y) \tag{3.19}$$

where $j(p)$ refers to $j(X, Y)$ at $X = X_p$ and $Y = Y_p$, the coordinates of point $p = 1$ or 2. The propagator $\mathcal{D}(X, Y)$ depends only on the relative coordinates

$$X = X_2 - X_1 \qquad \text{and} \qquad Y = Y_2 - Y_1 \ ; \tag{3.20}$$

it satisfies the Poisson equation

$$\left(\frac{\partial^2}{\partial X^2} + \frac{\partial^2}{\partial Y^2} \right) \mathcal{D} = - \delta(X)\, \delta(Y) \quad . \tag{3.21}$$

Similarly, for the same problem, (3.17) can be written as

$$w(j) = \exp \left[\tfrac{1}{2} \int j(1) < 1 \mid D \mid 2 > j(2) \prod_{p=1}^{2} \rho_p \, d\rho_p \, d\theta_p \right] \tag{3.22}$$

where ρ_p and θ_p refer to the polar coordinates of p. As will be shown, these two propagators $\mathcal{D}$ and D are not the same. To evaluate $< 1 \mid D \mid 2 >$, we expand the right-hand side of (3.17) in powers of j and extract the term quadratic in j. Equating that with the corresponding term in (3.22), we derive when $\theta_2 > \theta_1$

$$< 1 \mid D \mid 2 > = \frac{\operatorname{trace} \left[u(2\pi - \theta_2)\; \phi(\rho_2)\; u(\theta_2 - \theta_1)\; \phi(\rho_1)\; u(\theta_1) \right]}{\operatorname{trace}\; u(2\pi)} \tag{3.23}$$

and when $\theta_1 > \theta_2$

$$< 1 \mid D \mid 2 > = < 2 \mid D \mid 1 > \quad . \tag{3.24}$$

For $V = 0$, H is the same one given by (2.13). Hence,

$$\operatorname{trace}\; u(2\pi) = \prod_k \tfrac{1}{2} \operatorname{csch} \omega \pi / g \tag{3.25}$$

and the numerator in (3.23) is, when $\theta = \theta_2 - \theta_1 > 0$,

$$\sum_n < n \mid e^{-(2\pi - \theta) H/g} \mid n > < n \mid \phi(\rho_2)\, e^{-\theta H/g}\, \phi(\rho_1) \mid n > \tag{3.26}$$

where, as in (2.23), $n = \{n_k\}$. By using (2.4) and (2.30) we find that (3.26) equals

$$\int_{-\infty}^{\infty} \frac{dk}{2\pi} \Big[\prod_{k' \neq k} \tfrac{1}{2} \operatorname{csch} \omega' \pi/g \Big] \sum_{n_k=0}^{\infty} \Big[(n_{-k}+1)\, e^{-\theta\omega/g} + n_k e^{\theta\omega/g} \Big] (2\omega)^{-1}$$

$$\cdot\ e^{-2\pi(n_k+\frac{1}{2})\,\omega/g}\ e^{ikx} \tag{3.27}$$

where

$$\omega = |\,k\,| \ , \quad \omega' = |\,k'\,| \quad \text{and} \quad x = x_2 - x_1 = \ln(\rho_2/\rho_1) \ . \tag{3.28}$$

Combining (3.23)-(3.28) we obtain

$$<1\,|\,D\,|\,2> \;=\; D(x\,,\,\theta) \;=\; \int_{-\infty}^{\infty} \frac{dk}{2\pi}\, f_k(\theta)\, e^{ikx} \tag{3.29}$$

where for $\theta = \theta_2 - \theta_1 > 0$

$$f_k(\theta) \;=\; \frac{1}{2\omega} \left[\frac{e^{-\theta\omega/g}}{1 - e^{-2\pi\omega/g}} + \frac{e^{\theta\omega/g}}{e^{2\pi\omega/g} - 1} \right] , \tag{3.30}$$

and for $\theta < 0$, $f_k(\theta) \equiv f_k(-\theta)$ so that

$$D(x,\,\theta) \;=\; D(x,\,-\,\theta) \tag{3.31}$$

in accordance with (3.24). From (3.30) we see that $f_k(\pi + a) = f_k(\pi - a)$, hence $f_k(0) = f_k(2\pi)$ where

$$\omega f_k(0) \;=\; (e^{2\pi\omega/g} - 1)^{-1} + \tfrac{1}{2} \ . \tag{3.32}$$

The function $f_k(\theta)$ can be extended to arbitrary θ through the periodic condition

$$f(\theta) \;=\; f_k(\theta + 2\pi) \ , \tag{3.33}$$

While $f_k(\theta)$ is continuous, its derivative $f_k'(\theta) \equiv df_k(\theta)/d\theta$ is not:

$$f_k'(\theta) = \begin{cases} -\dfrac{1}{2g} & \text{at} \quad \theta = 0+ \\[2ex] \dfrac{1}{2g} & \text{at} \quad \theta = 0- \end{cases} \quad . \tag{3.34}$$

Setting $\theta = g\tau$, we see that $D(x, \theta)$ satisfies the Poisson equation

$$\left(\frac{\partial^2}{\partial x^2} + \frac{\partial^2}{\partial \tau^2} \right) D = -\delta(x)\, \delta(\tau) \quad . \tag{3.35}$$

Since x and τ are related to X_p and Y_p coordinates of $p = 1$ and 2 by

$$\begin{aligned} x &= x_2 - x_1 \quad , \qquad \tau = \theta/g = \tau_2 - \tau_1 \quad , \\ X_p &= \frac{1}{g} e^{gx_p} \cos g\tau_p \quad \text{and} \quad Y_p = \frac{1}{g} e^{gx_p} \sin g\tau_p \end{aligned} \tag{3.36}$$

we have, keeping X_1 and Y_1 fixed (and therefore also x_1 and τ_1 fixed),

$$\frac{\partial^2}{\partial X^2} + \frac{\partial^2}{\partial Y^2} = e^{-2gx_2} \left(\frac{\partial^2}{\partial x^2} + \frac{\partial^2}{\partial \tau^2} \right)$$

and

$$\delta(X)\, \delta(Y) = e^{-2gx_2}\, \delta(x)\, \delta(\tau) \tag{3.37}$$

where, as before, $X = X_2 - X_1$ and $Y = Y_2 - Y_1$. Thus, both $D(x, \theta)$ and $\mathcal{D}(X, Y)$ satisfy the Poisson equation (3.21), or (3.35). Their difference is a solution of the Laplace equation, as we shall see. (Note that x and θ alone do not uniquely determine X and Y , and vice versa. Hence, there must be a difference between these two propagators.)

By combining the Fourier series

$$e^{\pm\theta\omega} = \sum_{m=-\infty}^{\infty} \frac{1 - e^{\pm 2\pi\omega}}{2\pi(im \mp \omega)} e^{im\theta} \tag{3.38}$$

with (3.30) and carrying out the k-integration in (3.29), we can write $D(x, \theta)$ as

$$D(x, \theta) = D_0(x) + \frac{1}{2\pi} \sum_{m=1}^{\infty} \frac{1}{m} e^{\mp mx} \cos m\theta \tag{3.39}$$

where the upper sign is for $x > 0$, the lower sign for $x < 0$, and

$$D_0(x) = \int_0^{\infty} \frac{d\omega}{2(\pi\omega)^2} \cos \omega x \; . \tag{3.40}$$

The derivative of $D_0(x)$ is

$$-\int_0^{\infty} \frac{d\omega}{2\pi^2\omega} \sin \omega x = \begin{cases} -(4\pi)^{-1} & \text{if} \quad x > 0 \\ (4\pi)^{-1} & \text{if} \quad x < 0 \; . \end{cases} \tag{3.41}$$

Integrating (3.41) and using (3.28), we find

$$D_0(x) = \text{constant} + \begin{cases} -\frac{1}{4\pi} \ln \rho_2/\rho_1 & \text{if} \quad \rho_2 > \rho_1 \\ \frac{1}{4\pi} \ln \rho_2/\rho_1 & \text{if} \quad \rho_2 < \rho_1 \; . \end{cases} \tag{3.42}$$

From (3.36) and

$$\rho_2 = \frac{1}{g} e^{gx_2} \; , \qquad \rho_1 = \frac{1}{g} e^{gx_1} \; ,$$

it follows that the distance r between the two points 1 and 2 is

$$r = (X^2 + Y^2)^{\frac{1}{2}} = (\rho_1^{\,2} + \rho_2^{\,2} - 2\rho_1\rho_2 \cos\theta)^{\frac{1}{2}} \; ,$$

and therefore

$$-\frac{1}{2\pi}\ln r = \begin{cases} -\frac{1}{2\pi}\ln \rho_2 + \frac{1}{2\pi}\sum_{m=1}^{\infty}\frac{1}{m}e^{-mx}\cos m\theta & \text{if} \quad x > 0 \\ -\frac{1}{2\pi}\ln \rho_1 + \frac{1}{2\pi}\sum_{m=1}^{\infty}\frac{1}{m}e^{mx}\cos m\theta & \text{if} \quad x < 0 \ . \end{cases}$$

Hence, the euclidean propagator in polar coordinates is given by

$$\langle 1 \mid D \mid 2 \rangle = D(x, \theta) = \text{constant} - \frac{1}{2\pi}\ln r + \frac{1}{4\pi}\ln \rho_1 \rho_2 \ . \qquad (3.43)$$

On the other hand, from (3.19) we see that the euclidean propagator in cartesian coordinates is

$$\langle 1 \mid \mathcal{D} \mid 2 \rangle = \mathcal{D}(X, Y) = \text{constant} - \frac{1}{2\pi}\ln r \ . \qquad (3.44)$$

Hence

$$\langle 1 \mid \mathcal{D} \mid 2 \rangle = \langle 1 \mid D \mid 2 \rangle + C(1) + C(2) \qquad (3.45)$$

where

$$C(p) = -(4\pi)^{-1}\ln \rho_p + \text{constant} = -(4\pi)^{-1} g x_p + \text{constant} \qquad (3.46)$$

which is independent of $\tau_p = \theta_p / g$ and satisfies the Laplace equation

$$\left(\frac{\partial^2}{\partial x_p^2} + \frac{\partial^2}{\partial \tau_p^2}\right) C(p) = 0 \ . \qquad (3.47)$$

Viewed in Σ_{acc}, introduced in the previous section, $C(p)$ corresponds to an "external" field generated by a source located at its boundary: $x_p = \pm\infty$.

In Figure 3, we represent the cartesian propagator $\langle 1 \mid \mathcal{D} \mid 2 \rangle$ by a solid line connecting 1 and 2, and the polar propagator $\langle 1 \mid D \mid 2 \rangle$ by a dashed line.

Equation (3.45) can be pictured as the decomposition of the solid line into the sum of a dashed line plus two graphs each consisting of two separate pieces, one with a line ending on a circle representing $C(p)$ and the other with only an open line, representing the factor unity. The inverse of (3.45)

$$< 1 \mid D \mid 2 > \;=\; < 1 \mid \mathscr{D} \mid 2 > - \; C(1) - C(2) \quad , \tag{3.48}$$

may be viewed as the decomposition of the dashed line into the sum of a solid line plus two similarly disconnected graphs.

The inclusion of an arbitrary $V(\phi)$ is completely straightforward. By following the standard procedures in deriving Feynman diagrams, we can expand the logarithm of either $W(j)$ of (3.7) or $w(j)$ of (3.17) as a double power series in V and j. In terms of diagrams, these two series, one for $\ln W(j)$ and the other for $\ln w(j)$, are otherwise identical, except that in the former we use the cartesian propagator and in the latter the polar propagator. Substituting the decomposition (3.48) into the expansion of $\ln w(j)$, we see that the diagrams in polar coordinates can be regrouped as

$$\ln w(j) \;=\; \ln W(j) + S_1(j) + S_2(j) + \cdots \tag{3.49}$$

where $S_1(j)$ is linear in the external field $C(p)$, $S_2(j)$ is quadratic, etc.

Thus, by starting from the minkowski space, adopting the Schroedinger picture in the accelerating frame Σ_{acc} of coordinates x and t and taking its Hamiltonian operator H into the euclidean space through the analytic continuation $t = -i\tau$, we arrive at $w(j)$ by using polar coordinates. Next, through (3.49), we can extract the part of $\ln w(j)$ that is independent of the "external" field $C(p)$; the result is $\ln W(j)$. Reverting back to the minkowski space, we succeed in deriving the complete S-matrix of $\mathcal{H}$. An alternative way is to compute $\ln W(j)$ directly by using the polar propagator. This can be done as follows:

Use (3.45) and expand $\ln W(j)$ around $\ln w(j)$ in powers of $C(p)$:

$$\ln W(j) = \ln w(j) + s_1(j) + s_2(j) + \cdots \qquad (3.50)$$

where $s_1(j)$ is linear in $C(p)$, $s_2(j)$ quadratic, etc. The amplitude of any scattering process involving n quanta in Σ is determined by the part in $\ln W(j)$ that is proportional to the nth power of j . By using (3.50), we can also calculate the same amplitude as a sum of terms, given by its right-hand side: the first refers to diagrams without $C(p)$, the second to those linear in $C(p)$, etc. All these diagrams now employ the polar propagator $< 1 \mid D \mid 2 >$ expressed in terms of the variables x and τ in Σ_{acc} , as illustrated in Figure 4.

Remarks

1. The "external" field $C(p)$ is used only in the calculation of the euclidean scattering amplitudes via the recipe (3.50). To obtain the physical S-matrix, we must perform the analytic continuation back to the real time T on $W(j)$. The resulting S-matrix connecting the in- and out-states in the frame $\Sigma(X, T)$ is, of course, unitary.

2. The presence of the "external" field $C(p)$ in (3.45) and (3.50) is somewhat reminiscent of the situation when there is a black-body radiation. Let $B(p)$ represent the amplitude of such a radiation at point p . In that case, any scattering matrix element can

also be written as a power series in B(p) . There are however several major differences; among them we may mention:

In the black-body radiation, after squaring the amplitude there is an ensemble average, which makes any physical probability a function of $| B(p) |^2$. Here, there is no statistical average; the probability depends on odd as well as even powers of C(p) . Furthermore, according to (3.46), C(p) is linear in the space coordinate x_p in Σ_{acc} ; therefore it corresponds to a "constant gradient field", not a black-body distribution. The so-called black-body radiation information is contained in the entire polar propagator; e.g., when points 1 and 2 coincide $< 1 | D | 2 >$ gives (3.32), which is the familiar Bose-Einstein distribution formula for the average of $n_k + \frac{1}{2}$.

Therefore, we conclude that the apparent radiation in the accelerating system Σ_{acc} should not be viewed as a genuine black-body radiation.

3. Consider now the question of measurements in the accelerating frame Σ_{acc} . Returning to the minkowski space, we assume that the complete system in Σ is described by a coherent initial wave function, say at time $T = -\infty$. Through the Schroedinger equation, we determine the final wave function $< q , q' | S >$ on any space-like surface $S = S_I + S_{II}$ in Σ , where S_I is the part of the surface and q part of the field coordinates inside Σ_{acc} , while S_{II} and q' are outside. To an observer restricted to Σ_{acc} , the outcome of his measurement on S_I can be predicted through the density matrix, obtained by squaring the final wave function and summing over q' . This gives rise to an entropy increase that is typical of any inclusive measurement. However, the system immediately before the final measurement may well be described as a pure state; its formation is determined by the initial condition, which in turn depends on information

outside Σ_{acc} , as illustrated by the examples given at the end of the previous section. The inability of the observer to measure q' does not mean the state vector itself has to be incoherent.

4. GENERALIZATIONS

We shall now extend the above method to any quantum field theory in which the metric $g_{\mu\nu}$ of the space is known to have a horizon. For example, the square of the invariant length-element can be

$$ds^2 = e^{2gx}(-dt^2 + dx^2) + dy^2 + dz^2 \tag{4.1}$$

as in the constant accelerating frame Σ_{acc} of the previous sections (but now in a three space- and one time-dimensional world), or it may be the Schwarzschild solution

$$ds^2 = -(1 - \frac{2GM}{r})\, dt^2 + (1 - \frac{2GM}{r})^{-1} dr^2 + r^2(d\alpha^2 + \sin^2\alpha\, d\beta^2) \tag{4.2}$$

where G is Newton's constant, M the mass, α the polar angle and β the azimuthal angle. (In this paper, effects of quantum gravity will not be included.) In either case the horizon implied by the metric is known to be connected with a bad choice of coordinates, unrelated to any genuine physical singularity. Label such a coordinate frame with the horizon as Σ_H, and the region within the horizon as I. Starting from I, there are infinitely many particle-trajectories which can cross the horizon within a finite proper time of the particle. By following these trajectories, we can easily extend I and arrive at a larger frame. When a particle crosses the horizon, if it is leaving from I, let F be the region that it is going into; if it is coming into I, then call P the region that it comes from. By construction, F lies in the future light cones of I, and P in its past light cones.

In order to set up a quantum field theory in a space with a minkowski signature, we must associate each state vector $| >$ with a space-like surface S. The important question is whether such an S can be contained entirely within Σ_H. To put it another way, write

$$S = S_I + S_{II} \tag{4.3}$$

where S_I is in I and S_{II} is not. The question is: must S_{II} be nonzero? Let the region spanned by all S_{II} be II. By definition, II is outside I, and therefore beyond the horizon of Σ_H. Since S is space-like, II is also outside F and P. Consequently, unlike F and P, the existence of II cannot be ascertained by following particles crossing the horizon.

Actually, this question is a universal one in any space with a minkowski signature. To underline the issue involved, consider an arbitrary finite simply-connected three-dimensional space-like surface S_I. At each point on the boundary set up its light cone. Let I be the corresponding simply-connected four-dimensional region, which contains S_I and is bounded by these light cones. Call the boundary of I the "horizon". Label F and P the two regions outside the horizon, with F lying in the future light cones of I, and P in the past. As before, the region outside I, F and P is called II. In a classical field theory, by restricting our interest only to the region within I we can simply ignore the other regions F, P and II. In that case, II is irrelevant since it is space-like with respect to I. If the retarded Green's function is used for the integration of the field equation, then the effect of P can always be cast into the appropriate initial conditions within I; likewise for F through time reversal. Quantum mechanically, the situation is quite different. The specification of a state vector requires a complete set of mutually commuting operators (representing observables); this is why we need S_{II} as well as S_I, since the field variables on S_{II} commute with those on S_I.

To determine whether II exists or not, we may consider the trajectories of signals radiated from any point in P. If they do not all reach I, then II must exist. Likewise, think of signals received in F. If these do not all emanate from I, again II must exist.

Then, there is still the admittedly slippery course of common sense. For example, from (4.1) one can easily see that the space is flat, although the coordinates are curved. The restriction to $X > 0$ (in our previous notation) seems artificial. Therefore, it may appear self-evident that the region $X < 0$ should also exist, and that is II. In the case of a black hole, consider the situation just before its creation. Concentrate first on the finite space-like region from $r = 0$ to $R > 2GM$. By excluding $r > R$, it is not difficult to set up a new time coordinate T in which the black hole can be formed within a finite T. At the instant $T = 0-$, assume that inside $r = 2GM$ there is only a mass $M - \epsilon$ where $\epsilon = 0+$. At $T = 0+$, the entire mass M is inside, thus forming a black hole. Let S_I be the space-like surface defined by $r > 2GM$ at $T = 0$. At the instant before, we know that the region $r < 2GM$, denoted by $S_{II} \equiv S - S_I$, existed; it was nonzero and sizable. It seems unreasonable that S_{II} should disappear altogether at $T = 0+$.

In what follows, we assume that II exists as a physical reality. The problem is how to construct a complete quantum field theory by knowing <u>only</u> the Hamiltonian operator H in I. Because of the horizon, this is difficult if we stay in the space with a minkowski signature. However, by changing into the euclidean signature we can bypass this obstacle. The method developed in the previous sections can be readily generalized, and will be summarized below:

(i) Start from I and construct the Hamiltonian H in Σ_H in a space with a minkowski signature.

(ii) Go to the euclidean signature by setting time $t = -i\tau$ with τ real. The field equation becomes elliptic and there is no horizon.

(iii) The theorem proved in sect. 2 holds in general for a free field, and (2.12) gives the physical vacuum state.

(iv) For interacting fields, use (3.17) to define the generating function $w(j)$ (constructed entirely from information available in I). Following the discussion given in sect. 3, we can derive $W(j)$ which contains the full information of the quantum field theory in the entire space I, II, F and P.

As an illustration, we may consider, say, a free massive, or massless, scalar field in the four-dimensional constant accelerating frame Σ_{acc} described by (4.1). The Hamiltonian H in Σ_{acc} can be written in terms of its normal modes as

$$H = \sum_n \tfrac{1}{2}(p_n^2 + \omega_n^2 q_n^2) = \sum_n (a_n^\dagger a_n + \tfrac{1}{2})\, \omega_n \tag{4.4}$$

where q_n and p_n are the canonical variables, ω_n their frequency and a_n, $a_n^\dagger$ the corresponding annihilation and creation operators. Using (2.12), one sees that the vacuum wave function in the larger frame (I, II, F and P) is given by

$$< VAC \mid q, q' > = < q' \mid e^{-\pi H/g} \mid q > / (\text{trace } e^{-2\pi H/g})^{\frac{1}{2}} \tag{4.5}$$

and therefore, as in (2.24),

$$\begin{aligned} < VAC \mid a_n^\dagger a_n \mid VAC > &= \text{trace } (e^{-2\pi H/g} a_n^\dagger a_n)/\text{trace } e^{-2\pi H/g} \\ &= (e^{2\pi\omega_n/g} - 1)^{-1} \quad . \end{aligned} \tag{4.6}$$

In contrast, a conventional proof of the same formula along the line given in Appendix A can become somewhat complicated in this case.

5. BLACK HOLE

As a further illustration, we apply the above procedure to a scalar field ϕ in a Schwarzschild metric given by (4.2). (Because we do not discuss the evolution of the black hole nor the quantum gravitational effect, this serves only as a mathematical model, sometimes referred to as an eternal black hole.) The coordinates are

$$x^0 = t \quad , \quad x^1 = r \quad , \quad x^2 = \alpha \quad \text{and} \quad x^3 = \beta \; . \tag{5.1}$$

The non-vanishing elements of the metric $g_{\mu\nu}$ are

$$g_{00} = -e^{2\lambda} = -(1 - \frac{2GM}{r}) \; , \quad g_{11} = e^{-2\lambda} = (1 - \frac{2GM}{r})^{-1} \; ,$$

$$g_{22} = r^2 \quad \text{and} \quad g_{33} = r^2 \sin^2 \alpha \; . \tag{5.2}$$

The region within the horizon is referred to as

$$I: \quad r > 2GM \quad \text{and} \quad -\infty \leqslant t \leqslant \infty \; . \tag{5.3}$$

Let S_I be the space-like surface within I given by

$$S_I: \quad r > 2GM \quad \text{and} \quad t = \text{constant} \; . \tag{5.4}$$

The Lagrangian of a scalar field in I is

$$L = \int_{S_I} d^3\vec{r} \left[-\tfrac{1}{2} \phi^\mu \phi_\mu - V(\phi) \right] \tag{5.5}$$

where

$$\phi_\mu = \partial\phi / \partial x^\mu \; , \qquad \phi^\mu = g^{\mu\nu} \phi_\nu \; ,$$

$$d^3\vec{r} = \sqrt{-|g|}\; dx^1\, dx^2\, dx^3 = r^2 \sin\alpha \; dr\, d\alpha\, d\beta \tag{5.6}$$

and $V(\phi)$ is a polynomial in ϕ. The conjugate momentum π and the Hamiltonian H are given by

$$\pi = \frac{\delta L}{\delta \phi_0} = e^{-2\lambda} \frac{\partial \phi}{\partial t} \tag{5.7}$$

and

$$H = \int_{S_I} d^3\vec{r} \left[\tfrac{1}{2} e^{2\lambda} \pi^2 + \tfrac{1}{2} \phi^i \phi_i + V(\phi)\right] . \tag{5.8}$$

As usual, the repeated greek indices are summed over from 0 to 3, and the roman indices from 1 to 3.

Adopting the Schroedinger picture, ϕ and π depend only on the 3-vector $\vec{r}$; their commutation relations are

$$[\pi(\vec{r}), \phi(\vec{r}\,')] = -i\,\delta^3(\vec{r}\,' - \vec{r}) . \tag{5.9}$$

Following the step (ii) outlined in sect. 4, we set $t = -i\tau$. Introduce (in analogy to the Kruskal-Szekeres coordinates in the space with a minkowski signature)

$$\rho \equiv \left(\frac{r}{2GM} - 1\right)^{\frac{1}{2}} e^{r/4GM} \tag{5.10}$$

and

$$\theta \equiv \tau / 4GM . \tag{5.11}$$

The euclidean coordinates can be either

$$\theta, \rho, \alpha \text{ and } \beta$$

or (5.12)

$$\tau, r, \alpha \text{ and } \beta .$$

Their ranges are

$$0 \leqslant \theta \leqslant 2\pi \ , \qquad 0 \leqslant \rho \leqslant \infty \ ,$$
$$0 \leqslant \alpha \leqslant \pi \quad \text{and} \quad 0 \leqslant \beta \leqslant 2\pi \ ; \tag{5.13}$$

correspondingly,

$$2GM \leqslant r \leqslant \infty \quad \text{and} \quad 0 \leqslant \tau \leqslant 8\pi GM \ . \tag{5.14}$$

To understand these ranges, we may return briefly to the minkowski case.

In the minkowski case, the Kruskal-Szekeres coordinates are u, v, α and β where [21, 22]

$$\begin{cases} u = \rho \cosh t/4GM \\ v = \rho \sinh t/4GM \end{cases} \quad \text{in I} \ , \tag{5.15}$$

$$\begin{cases} u = -\rho \cosh t/4GM \\ v = -\rho \sinh t/4GM \end{cases} \quad \text{in II} \tag{5.16}$$

and ρ remains given by (5.10). The future and past regions F and P are related to t and r by

$$\begin{cases} u = \sigma \sinh t/4GM \\ v = \sigma \cosh t/4GM \end{cases} \quad \text{in F} \tag{5.17}$$

$$\begin{cases} u = -\sigma \sinh t/4GM \\ v = -\sigma \cosh t/4GM \end{cases} \quad \text{in P} \tag{5.18}$$

where

$$\sigma = \left(1 - \frac{r}{2GM}\right)^{\frac{1}{2}} e^{r/4GM} \ . \tag{5.19}$$

At $t = 0$, the minkowski space-like surface S_I defined by "$0 \leqslant u \leqslant \infty$ and $v = 0$" coincides with "$0 \leqslant \rho \leqslant \infty$ and $\theta = 0$" in the euclidean case. Next, we stay euclidean, and change the surface by varying θ. When $\theta = \pi$, the euclidean surface $0 \leqslant \rho \leqslant \infty$ meets the minkowski space-like surface S_{II} "$-\infty \leqslant u \leqslant 0$ and $v = 0$". When $\theta = 2\pi$, one returns to S_I. This explains the range of parameters given by (5.13).

[The customary designation of F and P by $r < 2GM$ may produce some unnecessary confusion. Ordinarily without a black hole, when one refers to $r > R$ as the region outside a sphere of radius R and $r < R$ as the inside region, one is thinking of the situation in which the union of these two regions forms a space-like surface; e.g., t = constant and r can be arbitrary. In (5.17) and (5.18), r acts like a "time" coordinate; it is II of (5.16) that plays the role of what would ordinarily be called the (space-like) region, not included in $r > R$.]

To carry out steps (iii) and (iv) given in sect. 4, we note that in the euclidean region (5.13) there is no singularity. Both (2.12) and (3.17) are well-defined. The details will be given elsewhere [23]. Here, we discuss only the simple example $V = \frac{1}{2} m^2 \phi^2$, i.e., a free scalar field of mass m (which may or may not be zero). Define the Laplace operator to be

$$\Delta \equiv \frac{1}{r^2} \frac{\partial}{\partial r} (r^2 e^{2\lambda} \frac{\partial}{\partial r}) + \frac{1}{r^2 \sin\alpha} \frac{\partial}{\partial \alpha} (\sin\alpha \frac{\partial}{\partial \alpha}) + \frac{1}{r^2 \sin^2\alpha} \frac{\partial^2}{\partial \beta^2} \tag{5.20}$$

where, as before, $e^{2\lambda} = 1 - (2GM/r)$. Let $f_n(r, \alpha, \beta)$ be the solution of

$$(-\Delta + m^2) f_n = \omega_n^2 e^{-2\lambda} f_n \tag{5.21}$$

in the region $r > 2GM$. It can be shown that the eigenvalue ω_n^2 is real and positive. Hence, we can always choose ω_n to be also real and positive. The set of all these eigenfunctions f_n is complete. From (5.21), one can readily verify that

$$\int_{r>2GM} f_n^* f_m e^{-2\lambda} d^3\vec{r} = 0 \qquad \text{if} \qquad \omega_n^2 \neq \omega_m^2 . \tag{5.22}$$

Choose f_n to be all real and properly normalized so that

$$\int_{r>2GM} f_n f_m e^{-2\lambda} d^3 r = \delta_{nm} . \tag{5.23}$$

Expand

$$\phi(r,\alpha,\beta) = \sum_n f_n(r,\alpha,\beta)\, q_n$$

and

$$\pi(r,\alpha,\beta) = \sum_n f_n(r,\alpha,\beta)\, e^{-2\lambda} p_n . \tag{5.24}$$

From (5.8)-(5.9) it follows that

$$[p_n, q_m] = -i\delta_{nm} \tag{5.25}$$

and

$$H = \sum_n \tfrac{1}{2}(p_n^2 + \omega_n^2 q_n^2) . \tag{5.26}$$

Substituting H into (2.12) and replacing g by $(4GM)^{-1}$ because of (5.11), we see that the field-coordinate representation of the vacuum state $|VAC>$ in the larger space, defined on the space-like surface $S_I + S_{II}$, is given by

$$\frac{<q' | e^{-4\pi GMH} | q>}{(\text{trace } e^{-8\pi GMH})^{\frac{1}{2}}} . \tag{5.27}$$

Therefore, the vacuum average of the occupation number operator (defined in I only) is

$$< VAC \mid a_n^\dagger a_n \mid VAC > \; = \; \frac{\text{trace } (e^{-8\pi GMH} a_n^\dagger a_n)}{\text{trace } e^{-8\pi GMH}}$$

$$= \; (e^{8\pi GM\omega_n} - 1)^{-1} \qquad (5.28)$$

where a_n and $a_n^\dagger$ are the standard annihilation and creation operators of the nth oscillator in the sum (5.26). This distribution corresponds to a Bose-Einstein distribution at the Hawking temperature $(8\pi GM)^{-1}$. Because the Schwarzschild metric is assumed to hold in the region I at all time from $t = -\infty$ to $t = +\infty$, the "radiation" is a static one, similar to the situation in the accelerating frame discussed before. As emphasized in the previous sections, this radiation should not a priori be treated as incoherent, a situation quite different from the black-body radiation in statistical mechanics.

A realistic black hole is nonstatic and therefore more complex. In Appendix C we study a soluble two-dimensional model in which the metric is a given time-dependent function, simulating the formation of a black hole. From this model calculation and from the above discussion of the static case, we are led to the conclusion that a black hole is not a black body. At least in principle, an observer outside the black hole can retrieve information about the physical state inside the black hole, beyond his horizon.

In these discussions, we are concerned with the application of quantum field theory to a limited space-time domain bounded by a horizon. The main question is how to properly take into account the global nature of a quantum state, which inevitably extends over to the space outside the horizon. As in the well-known Einstein-Podolsky-Rosen experiments, the same global properties of the quantum state can give correlations between particles that are well separated on a space-like surface, therefore seemingly unrelated in any causal way. Here, because of the macroscopic scale of the objects involved, the situation could be more striking. We hope the central issue raised in this paper may lead to a much deeper understanding of the relation between quantum mechanics and general relativity.

ACKNOWLEDGMENTS

I wish to thank James Anderson and James Hartle for discussions, and Gary Horowitz and Malcolm Perry for conversations which stimulated my interest in this subject.

APPENDIX A

CONVENTIONAL PROOF OF THE RADIATION FORMULA

A straightforward derivation of (1.18) is to express the operators a_k and $a_k^\dagger$ in the accelerating frame Σ_{acc} in terms of A_K and $A_K^\dagger$ in the larger frame Σ. We note that the inverse of (1.13) is

$$(2\omega)^{-\frac{1}{2}}(a_k e^{-i\omega t} + a_{-k}^\dagger e^{i\omega t}) = \int_{-\infty}^{\infty} \phi(x, t)\, e^{-ikx}\, dx \quad . \tag{A1}$$

Since the expansion (1.12) is valid everywhere in the (X, T) plane (including the region I shown in Figure 1), it may be substituted into the above expression. Comparing both sides of the new (A1) at $t = 0$, and then their t-derivatives at the same instant, we find

$$\begin{aligned} a_k = \tfrac{1}{2}\int_{-\infty}^{\infty} \frac{dK}{2\pi} \Big\{ & A_K \left[\left(\frac{\omega}{\Omega}\right)^{\frac{1}{2}} \langle k \mid K \rangle + \left(\frac{\Omega}{\omega}\right)^{\frac{1}{2}} (k \mid K) \right] \\ & + A_K^\dagger \left[\left(\frac{\omega}{\Omega}\right)^{\frac{1}{2}} \langle k \mid -K \rangle - \left(\frac{\Omega}{\omega}\right)^{\frac{1}{2}} (k \mid -K) \right] \Big\} \end{aligned} \tag{A2}$$

where

$$\langle k \mid K \rangle = \int_{-\infty}^{\infty} dx \ \exp i \left(\frac{K}{g} e^{gx} - kx\right) \ , \tag{A3}$$

$$(k \mid K) = \int_{-\infty}^{\infty} e^{gx}\, dx \ \exp i \left(\frac{K}{g} e^{gx} - kx\right) \ , \tag{A4}$$

$\omega = |k|$ and $\Omega = |K|$. It can be verified that

$$\langle \omega \mid \pm\Omega \rangle = \langle -\omega \mid \mp\Omega \rangle^* = f_\pm(\omega)\, \Omega^{i\omega/g} \tag{A5}$$

where $f_{\pm}(\omega)$ can be expressed in terms of the gamma function:

$$f_{\pm}(\omega) = g^{-i\frac{\omega}{g} - 1} \Gamma(-i\frac{\omega}{g}) e^{\pm\pi\omega/2g} \quad . \tag{A6}$$

Consequently,

$$(k \mid K) = \frac{k}{K} < k \mid K > \ , \tag{A7}$$

and (A2) becomes

$$a_{\pm\omega} = \int_0^\infty \frac{d\Omega}{2\pi} \left(\frac{\omega}{\Omega}\right)^{\frac{1}{2}} \left[A_{\pm\Omega} < \pm\omega \mid \pm\Omega > + A^{\dagger}_{\pm\Omega} < \pm\omega \mid \mp\Omega > \right] , \tag{A8}$$

where the upper signs are for $k = \omega > 0$ and the lower ones for $k = -\omega < 0$. Using (A8) and its hermitian conjugate, we can readily evaluate the vacuum expectation value of $a_k^{\dagger} a_k$. The result is (1.18):

$$< \mathrm{VAC} \mid a_k^{\dagger} a_k \mid \mathrm{VAC} > = \mathrm{volume} \cdot (e^{2\pi\omega/g} - 1)^{-1} \quad . \tag{A9}$$

This derivation is quite simple. It relies on the specific form of the transformation matrix elements (A5)-(A7). However, in a general case, say a four-dimensional massive field, the corresponding matrix elements can be more complicated. The alternative derivation given in the text does not depend on any such explicit calculations; it extends in a simple way the validity of (A9) to a much wider class of problems: the quantum may be massive, the space dimension can be arbitrary, the underlying metric may have a non-vanishing Riemann tensor as in the Schwarzschild case, etc.

APPENDIX B

MIRROR BEYOND THE RINDLER FRAME

1. In the coordinate system $\Sigma(X, T)$ introduced in sects. 1 and 2, let us impose the boundary condition

$$\phi = 0 \qquad \text{at} \qquad X = -\ell \quad ; \tag{B1}$$

this corresponds to a (Dirichlet) mirror at $X = -\ell$, which separates the region $X > -\ell$ from $X < -\ell$. The Lagrangian in the region $X > -\ell$ is

$$\mathcal{L} = \int_{-\ell}^{\infty} \frac{1}{2}\left[\left(\frac{\partial\phi}{\partial T}\right)^2 - \left(\frac{\partial\phi}{\partial X}\right)^2\right] dX \quad . \tag{B2}$$

At a given T, expand ϕ and its conjugate momentum in the same region as

$$\phi = \int_0^{\infty} \frac{d\Omega}{\pi} \; \mathcal{Q}_\Omega(T) \; \sin\Omega(X+\ell)$$

and (B3)

$$\mathcal{P} = \int_0^{\infty} \frac{d\Omega}{\pi} \; \mathcal{P}_\Omega(T) \; \sin\Omega(X+\ell)$$

where $\mathcal{P} = \partial\phi/\partial T$. The Hamiltonian in terms of these Fourier components can be written as

$$H = \int_0^{\infty} \frac{d\Omega}{2\pi} \; \frac{1}{2}\left[\mathcal{P}_\Omega^2 + \Omega^2 \mathcal{Q}_\Omega^2\right] \quad . \tag{B4}$$

From (B3), it follows that

$$\mathcal{Q}_\Omega = \int_{-\ell}^{\infty} dX \; 2\phi \; \sin\Omega(X+\ell)$$

and (B5)

$$\mathcal{P}_\Omega = \int_{-\ell}^{\infty} dX \; 2 \; \frac{\partial\phi}{\partial T} \; \sin\Omega(X+\ell) \quad .$$

Substituting (B5) back into (B3), we may extend ϕ from $X > -\ell$ to $X < -\ell$.

Set $T = 0$. Rewrite (B3), over the entire range $-\infty \leqslant X \leqslant \infty$, as

$$\phi(X) = \int_{-\infty}^{\infty} \frac{dK}{2\pi} e^{iKX} Q_K \tag{B6}$$

and

$$\mathcal{P}(X) = \int_{-\infty}^{\infty} \frac{dK}{2\pi} e^{-iKX} \Pi_K \tag{B7}$$

where

$$\begin{aligned} Q_\Omega &= -i\,\mathcal{Q}_\Omega e^{i\Omega\ell} , & Q_{-\Omega} &= i\,\mathcal{Q}_\Omega e^{-i\Omega\ell} , \\ \Pi_\Omega &= i\,\mathcal{P}_\Omega e^{-i\Omega\ell} , & \Pi_{-\Omega} &= -i\,\mathcal{P}_\Omega e^{i\Omega\ell} \end{aligned} \tag{B8}$$

and $\Omega = |K|$ as before. Use (2.4)-(2.5) to introduce q_k and q_k':

$$\phi(X) = \int_{-\infty}^{\infty} \frac{dk}{2\pi} e^{ikx} q_k \quad \text{when} \quad X = \frac{1}{g} e^{gx} > 0 \tag{B9}$$

and

$$\phi(X) = \int_{-\infty}^{\infty} \frac{dk}{2\pi} e^{ikx'} q_k' \quad \text{when} \quad X = -\frac{1}{g} e^{gx'} < 0 . \tag{B10}$$

Likewise, define p_k and p_k' by

$$P(X) = \int_{-\infty}^{\infty} \frac{dk}{2\pi} e^{-ikx} p_k \quad \text{when} \quad X = \frac{1}{g} e^{gx} > 0 \tag{B11}$$

and

$$P(X) = \int_{-\infty}^{\infty} \frac{dk}{2\pi} e^{-ikx'} p_k' \quad \text{when} \quad X = -\frac{1}{g} e^{gx'} < 0 \tag{B12}$$

where P is related to $\mathcal{P}$ in (B3) by

$$P(X) = gX\mathcal{P}(X) . \tag{B13}$$

[Since, according to (1.4)-(1.5), $\partial\phi/\partial t = gX(\partial\phi/\partial T) + gT(\partial\phi/\partial X)$ which, at $T = 0$, reduces to (B13); i.e., $P = \partial\phi/\partial t = gX(\partial\phi/\partial T) = gX\,\mathcal{P}$.] From (B9) and (B11), we can express q_k and p_k in terms of $\phi(X)$ and $P(X)$:

$$q_k = \int_{-\infty}^{\infty} dx \; e^{-ikx} \, \phi(X) \tag{B14}$$

and

$$p_k = \int_{-\infty}^{\infty} dx \; e^{ikx} \, P(X) \; . \tag{B15}$$

Substituting (B6) into (B14), (B7) into (B15), and using (B13), we obtain

$$q_k = \int_{-\infty}^{\infty} \frac{dK}{2\pi} < k \mid K > Q_K \tag{B16}$$

and

$$p_k = \int_{-\infty}^{\infty} \frac{dK}{2\pi} \, (k \mid K)^* \, \Pi_K \tag{B17}$$

where, as in (A3)-(A4),

$$< k \mid K > \; \equiv \int_{-\infty}^{\infty} e^{iKX - ikx} \, dx \; , \tag{B18}$$

$$(k \mid K) = \int_{0}^{\infty} e^{iKX - ikx} \, dX \tag{B19}$$

with $X = g^{-1} e^{gx}$. It can be verified that in addition to (A5)-(A7),

$$\int_{-\infty}^{\infty} \frac{dK}{2\pi} < k \mid K > (k' \mid K)^* = 2\pi\delta(k - k') \tag{B20}$$

and

$$\int_{-\infty}^{\infty} \frac{dK}{2\pi} < k \mid -K > (k' \mid K)^* \, e^{-2iK\ell} = 0 \quad \text{if} \quad \ell \geqslant 0 \; . \tag{B21}$$

From (B8) and (B16)-(B17), the variables q_k and p_k can be written as functions of the original canonical variables $\mathcal{Q}_\Omega$ and $\mathcal{P}_\Omega$:

$$q_k = \int_0^\infty \frac{d\Omega}{2\pi} \left[- i < k \mid \Omega > e^{i\Omega\ell} + i < k \mid -\Omega > e^{-i\Omega\ell} \right] \mathcal{Q}_\Omega$$

and (B22)

$$p_k = \int_0^\infty \frac{d\Omega}{2\pi} \left[i (k \mid \Omega)^* e^{-i\Omega\ell} - i (k \mid -\Omega)^* e^{i\Omega\ell} \right] \mathcal{P}_\Omega \quad .$$

The quantization condition is

$$\left[\mathcal{P}_\Omega , \mathcal{Q}_{\Omega'} \right] = - i \, 2\pi \delta (\Omega - \Omega') \quad ; \tag{B23}$$

therefore, by using (B20)-(B23) we derive

$$\left[p_k , q_{k'} \right] = - i \, 2\pi \delta (k - k') \qquad \text{if} \qquad \ell \geqslant 0 \quad . \tag{B24}$$

Thus, when the mirror is placed at $X = - \ell \leqslant 0$, beyond the horizon of Σ_{acc} , q_k and p_k are also canonical variables. [In contrast, if $X = - \ell > 0$ the mirror is inside Σ_{acc} , then q_k and p_k , defined by (B9) and (B11), do not satisfy (B24) and therefore are not canonical.] Define

$$\mathcal{Q}_\Omega \equiv (2\Omega)^{-\frac{1}{2}} (A_\Omega + A_\Omega^\dagger) \quad , \tag{B25}$$

$$\mathcal{P}_\Omega \equiv - i (\Omega/2)^{\frac{1}{2}} (A_\Omega - A_\Omega^\dagger) \quad , \tag{B26}$$

$$q_k \equiv (2\omega)^{-\frac{1}{2}} (a_k + a_{-k}^\dagger) \tag{B27}$$

and

$$p_k \equiv - i (\omega/2)^{\frac{1}{2}} (a_{-k} - a_k^\dagger) \quad . \tag{B28}$$

In terms of the annihilation and creation operators A_Ω and $A_\Omega^\dagger$ in Σ, the corresponding annihilation operators in Σ_{acc} become

$$a_\omega = \int_0^\infty \frac{d\Omega}{2\pi}\left(\frac{\omega}{\Omega}\right)^{\frac{1}{2}} \left[-i<\omega\,|\,\Omega> e^{i\Omega\ell} A_\Omega + i<\omega\,|-\Omega> e^{-i\Omega\ell} A_\Omega^\dagger\right], \tag{B29}$$

$$a_{-\omega} = \int_0^\infty \frac{d\Omega}{2\pi}\left(\frac{\omega}{\Omega}\right)^{\frac{1}{2}} \left[i<\omega\,|\,\Omega>^* e^{-i\Omega\ell} A_\Omega - i<\omega\,|-\Omega>^* e^{i\Omega\ell} A_\Omega^\dagger\right]. \tag{B30}$$

As before, let $|\,VAC>$ be the ground state in Σ, determined by

$$A_\Omega\,|\,VAC> \;=\; 0 \qquad \text{for all } \Omega\,. \tag{B31}$$

When the (Dirichlet) mirror is placed at $X = 0$ (i.e., $\ell = 0$), the vacuum expectation value of $a_k a_{k'}$ observed in the accelerating frame Σ_{acc} is

$$<VAC\,|\,a_k a_{k'}\,|\,VAC> \;=\; -\,\pi\,(\operatorname{csch}\pi\omega/g)\;\delta(k+k') \tag{B32}$$

where $\omega = |\,k\,|$. On the other hand, if the mirror is at $X = -\ell < 0$ then the corresponding expectation value of, say, $a_\omega a_{-\omega'}$ (i.e., $a_k a_{k'}$ with $k = \omega > 0$ and $k' = -\omega' < 0$) can be shown to be given by (2.37):

$$<VAC\,|\,a_\omega a_{-\omega'}\,|\,VAC> \;=\; -\,\frac{(\omega\omega')^{\frac{1}{2}}}{2\pi g^2}\,(g\ell)^{i(\omega'-\omega)/g}\;\Gamma(-i\frac{\omega}{g})\;\Gamma(i\frac{\omega'}{g})\;\Gamma(i\frac{\omega-\omega'}{g})\,. \tag{B33}$$

As emphasized before, although the mirror is beyond the horizon, by staying within the accelerating frame Σ_{acc}, an observer can nevertheless determine the existence and the position of the mirror.

2. To understand the significance of (B32)-(B33), let us examine the special case when the (Dirichlet) mirror is at $X = -\ell = 0$. In this case, (B29)-(B30) can readily be inverted by using (B20)-(B21) and (A5)-(A7):

$$A_\Omega = i\int_0^\infty \frac{d\omega}{2\pi}\left(\frac{\omega}{\Omega}\right)^{\frac{1}{2}}\Big[<\omega\,|\,\Omega>^*\, a_\omega - <\omega\,|\,\Omega>\, a_{-\omega} - <\omega\,|\,-\Omega>\, a_\omega^\dagger + <\omega\,|\,-\Omega>^*\, a_{-\omega}^\dagger\Big]$$

$$= i\int_0^\infty \frac{d\omega}{2\pi}\left(\frac{\omega}{\Omega}\right)^{\frac{1}{2}}\Big[-\Omega^{i\omega/g}\,(f_-(\omega)\, a_\omega^\dagger + f_+(\omega)\, a_{-\omega}) + \Omega^{-i\omega/g}\,(f_+^*(\omega)\, a_\omega + f_-^*(\omega)\, a_{-\omega}^\dagger)\Big]\,. \qquad \text{(B34)}$$

Multiply $A_\Omega\,|\,VAC>$ by $\Omega^{-\frac{1}{2}\mp i(\omega/g)}\,d\Omega$ and integrate from $\Omega = 0$ to ∞, we derive, on account of (B34) and $\int_0^\infty \Omega^{i(\omega' - \omega)/g}\,d\Omega/\Omega = 2\pi g\,\delta(\omega' - \omega)$,

$$(f_-(\omega)\, a_\omega^\dagger + f_+(\omega)\, a_{-\omega})\,|\,VAC> \;=\; 0$$

and $\qquad$ (B35)

$$(f_+^*(\omega)\, a_\omega + f_-^*(\omega)\, a_{-\omega}^\dagger)\,|\,VAC> \;=\; 0\,.$$

Let $|\,vac>$ be the vacuum state in Σ_{acc}; i.e.,

$$a_k\,|\,vac> \;=\; 0 \qquad \text{for all } k\;. \qquad \text{(B36)}$$

Defining

$$F \equiv -\int_0^\infty \frac{d\omega}{2\pi}\, e^{-\pi\omega/g}\, a_\omega^\dagger\, a_{-\omega}^\dagger\,, \qquad \text{(B37)}$$

we have

$$[a_k\,,\,F] \;=\; -\,e^{-\pi\omega/g}\, a_{-k}^\dagger \qquad \text{(B38)}$$

where $\omega = |k|$. Because of (A6), it follows that

$$\frac{f_-(\omega)}{f_+(\omega)} = e^{-\pi\omega/g} \tag{B39}$$

and therefore

$$(f_-(\omega)\, a_\omega^\dagger + f_+(\omega)\, a_{-\omega})\, e^F\, |\,\mathrm{vac}>$$

$$= e^F(f_-(\omega) - f_+(\omega)\, e^{-\pi\omega/g})\, a_\omega^\dagger\, |\,\mathrm{vac}> = 0 \ .$$

Likewise

$$(f_+^*(\omega)\, a_\omega + f_-^*(\omega)\, a_{-\omega}^\dagger)\, e^F\, |\,\mathrm{vac}>$$

$$= e^F(- f_+^*(\omega)\, e^{-\pi\omega/g} + f_-^*(\omega))\, a_{-\omega}^\dagger\, |\,\mathrm{vac}> = 0 \ .$$

In terms of $|\,\mathrm{vac}>$ in Σ_{acc}, the vacuum in Σ is given by the first equation in (2.36):

$$|\,\mathrm{VAC}> = \text{constant} \cdot e^F\, |\,\mathrm{vac}> \ . \tag{B40}$$

This expression gives directly the correlation function (B32).

3. By following the same argument, but replacing the Dirichlet condition "$\phi = 0$ at $X = 0$" by the Neumann condition "$\partial\phi/\partial X = 0$ at $X = 0$", we derive the second equation in (2.36):

$$|\,\mathrm{VAC}> = \text{constant} \cdot e^F\, |\,\mathrm{vac}>$$

where instead of (B37)

$$F = \int_0^\infty \frac{d\omega}{2\pi} e^{-\pi\omega/g} a_\omega^\dagger a_{-\omega}^\dagger \quad . \tag{B41}$$

In both cases, Dirichlet or Neumann, the quantum distribution in Σ_{acc} is described by a pair-correlated B.C.S. formula, completely different from a black-body radiation.

APPENDIX C

A TWO-DIMENSIONAL MODEL OF GRAVITATIONAL COLLAPSE

In gravitational collapse, one deals with a nonstatic metric. In the literature, a two-dimensional model [4, 5] has often been used to illustrate the kind of physical consequences that might be present in the formation of a black hole, especially if the collapsing matter is distributed mainly on a shell. The same model will be studied in this Appendix, but with some minor variations. As we shall see, while an observer outside the shell can detect "Hawking" radiation, it is also possible for him, through measurements of various functions of field variables, all restricted to the same outside region, to ascertain that the state may be quite <u>different</u> from an incoherent mixture of black-body radiation quanta.

1. The Model

In this model the coordinates are $X^0 = T$ and $X^1 = X$, with the trajectory of the shell along a given function $X = L(T)$. The metric is assumed to be

$$g_{01} = g_{10} = 0 \qquad \text{everywhere,} \tag{C1}$$

$$\begin{cases} g_{00} = -\left(1 - \frac{2M}{X}\right)\left(\frac{dt}{dT}\right)^2 \\ g_{11} = \left(1 - \frac{2M}{X}\right)^{-1} \end{cases} \qquad \text{when } X > L(T) , \tag{C2}$$

and

$$\begin{cases} g_{00} = -1 \\ g_{11} = 1 \end{cases} \qquad \text{when } X < L(T) \tag{C3}$$

where M is a constant and t is a function of T. The continuity requirement of ds^2 across the dividing line $X = L(T)$ relates $t(T)$ to $L(T)$:

$$\left(\frac{dt}{dT}\right)^2 = \frac{L}{L - 2M}\left[1 + \left(\frac{dL}{dT}\right)^2 \frac{2M}{L - 2M}\right] . \tag{C4}$$

We assume

$$L(T) = \begin{cases} L_0 \equiv 2M + \Delta & \text{when} \quad T \leqslant 0 \\ L_1 \equiv 2M + \epsilon & \text{when} \quad T \geqslant T_1 > 0 \end{cases} \tag{C5}$$

with $\Delta > \epsilon > 0$. Physically, this simulates a collapsing shell moving from L_0 at $T = 0$ to L_1 at $t = T_1$. Much of the analysis can be carried out without specifying the details of how the shell changes from L_0 to L_1. [An example is shown in Figure 5.] Equation (C4) gives

$$t(T) = \begin{cases} (L_0/\Delta)^{\frac{1}{2}}\, T & \text{when} \quad T \leqslant 0 \\ (L_1/\epsilon)^{\frac{1}{2}}\,(T - T_1) + t_1 & \text{when} \quad T \geqslant T_1 \end{cases} \tag{C6}$$

where t_1 is a constant, so that $t = t_1$ when $T = T_1$. The case of a very small ϵ is especially interesting, since when $\epsilon \to 0$ we have $L_1 \to 2M$ and $t_1 \to \infty$, resembling in a crude way the formation of a black hole. However, this limit is somewhat delicate, as we shall see. In (C5) we assign $L(T) = 2M + \epsilon$ for all $T \geqslant T_1$. When $\epsilon \to 0$, this has the effect of stretching out the interval ϵ near zero, and thereby allows more flexibility in our examination of the limit.

Let ϕ be a complex massless scalar field required to satisfy the boundary condition that at all T,

$$\phi(X, T) = 0 \qquad \text{at} \qquad X = 0 \tag{C7}$$

(which simulates the usual radial boundary condition "$r\phi = 0$" when the radius $r = 0$ in a three-space-dimensional problem). The Lagrangian and Hamiltonian for a free field are

$$\mathcal{L} = -\int_0^\infty \sqrt{-|g|}\ \phi_\mu^\dagger \phi^\mu \, dX$$

and (C8)

$$\mathcal{H} = \int_0^\infty \left[-(-|g|)^{-\frac{1}{2}} g_{00} \mathcal{P}^\dagger \mathcal{P} + (-|g|)^{\frac{1}{2}} g^{11} \frac{\partial \phi^\dagger}{\partial X} \frac{\partial \phi}{\partial X} \right] dX$$

where $\dagger$ denotes the hermitian conjugate as before, $\phi_\mu = \partial\phi/\partial X^\mu$, $\phi^\mu = g^{\mu\nu} \phi_\nu$, and

$$\mathcal{P} = -\sqrt{-|g|}\ g^{00} \frac{\partial \phi^\dagger}{\partial T} , \qquad \mathcal{P}^\dagger = -\sqrt{-|g|}\ g^{00} \frac{\partial \phi}{\partial T} \tag{C9}$$

are the conjugate momenta of ϕ and $\phi^\dagger$. The quantization condition is the equal-time commutator

$$\left[\mathcal{P}(X, T),\ \phi(X', T) \right] = -i\,\delta(X - X') . \tag{C10}$$

In the Schroedinger picture, ϕ and $\mathcal{P}$ are T-independent, and the state vector $|T>$ satisfies

$$\mathcal{H}\,|T> = i \frac{\partial}{\partial T} |T> .$$

The Hamiltonian $\mathcal{H}$ although T-dependent is a hermitian, and that insures the unitarity of the theory.

In the following, we shall stay with the Heisenberg picture. The state vector $|>$ is now T-independent and the variables $\phi(X,T)$ and $\mathcal{P}(X,T)$ satisfy

$$\frac{\partial \phi}{\partial T} = i\left[\mathcal{H}, \phi\right]$$

and (C11)

$$\frac{\partial \mathcal{P}}{\partial T} = i\left[\mathcal{H}, \mathcal{P}\right] .$$

For $\mathcal{H}$ given by (C8), the Heisenberg equations (C11) yield the familiar wave equation

$$\frac{\partial}{\partial X^{\mu}} \left(\sqrt{-|g|}\ \phi^{\mu} \right) = 0 \tag{C12}$$

where $|g|$ is the determinant of the metric tensor. By using (C1)-(C3) we see that ϕ satisfies

$$- \frac{\partial^2 \phi}{\partial t^2} + \frac{\partial^2 \phi}{\partial x^2} = 0 \qquad \text{when} \quad X > L(T)$$

and (C13)

$$- \frac{\partial^2 \phi}{\partial T^2} + \frac{\partial^2 \phi}{\partial X^2} = 0 \qquad \text{when} \quad X < L(T)$$

where

$$x = X + 2M \ln (X - 2M) \ , \tag{C14}$$

so that $dx = dX / [1 - (2M/X)]$.

2. S-matrix

When T is less than 0, $\mathcal{H}$ becomes T-independent on account of (C6). It is convenient to expand the operator $\phi(X, T)$ in terms of the plane wave solutions:

$$\phi(X, T) = \int_0^{\infty} \frac{d\omega}{2\pi} \frac{1}{\sqrt{2\omega}} \left[A_{\omega}^{in}\ \phi_{\omega}^{in}(X, T) + B_{\omega}^{in\dagger}\ \phi_{\omega}^{in}(X, T)^* \right] \tag{C15}$$

where for $T \leqslant 0$

$$\phi_{\omega}^{in}(X, T) \equiv 2e^{-i\omega t} \cdot \begin{cases} \sin(\omega x + \delta_0) & \text{when} \quad X > L_0 \\ \sin(\omega\sqrt{L_0/\Delta}\ X & \text{when} \quad X < L_0 \ , \end{cases} \tag{C16}$$

with the phase shift

$$\delta_0(\omega) = \omega\,(L_0 \sqrt{L_0/\Delta} - \ell_0) \tag{C17}$$

in which ℓ_0 is the value of x when $X = L_0$; i.e.,

$$\ell_0 = L_0 + 2M \ln \Delta \ . \tag{C18}$$

Likewise, for T greater than T_1 , $\mathcal{H}$ is also independent of T . The corresponding expansion is

$$\phi(X,T) = \int_0^\infty \frac{d\omega}{2\pi} \frac{1}{\sqrt{2\omega}} \left[A_\omega^{out}\, \phi_\omega^{out}(X,T) + B_\omega^{out\dagger}\, \phi_\omega^{out}(X,T)^* \right] \tag{C19}$$

where for $T \geqslant T_1$

$$\phi_\omega^{out}(X,T) \equiv 2e^{-i\omega t} \cdot \begin{cases} \sin(\omega x + \delta_1) & \text{when} \quad X > L_1 \\ \sin(\omega \sqrt{L_1/\epsilon}\; X) & \text{when} \quad X < L_1 \end{cases} \tag{C20}$$

with the phase shift δ_1 given by

$$\delta_1(\omega) = \omega\,(L_1 \sqrt{L_1/\epsilon} - \ell_1) \ , \tag{C21}$$

and

$$\ell_1 = L_1 + 2M \ln \epsilon \ . \tag{C22}$$

These two sets of c. number solutions $\{\phi_\omega^{in}(X,T)\}$ and $\{\phi_\omega^{out}(X,T)\}$ satisfy the following orthonormality relations:

when $T \leqslant 0$,

$$-\int_0^\infty \phi_\omega^{in}(X,T)^*\, \phi_{\omega'}^{in}(X,T)\, \sqrt{-|g|}\; g^{00} \frac{dt}{dT}\, dX = 2\pi\delta(\omega - \omega') \ , \tag{C23}$$

and when $T \geqslant T_1$,

$$- \int_0^{\infty} \phi_\omega^{out}(X, T)^* \; \phi^{out}(X, T) \sqrt{-|g|} \; g^{00} \frac{dt}{dT} \, dX \;=\; 2\pi\delta(\omega - \omega') \; . \tag{C24}$$

By construction, the operators A_ω^α and B_ω^α (α = in or out) are independent of X and T ; because of (C 10), they satisfy

$$[A_\omega^\alpha \, , \; A_{\omega'}^{\alpha\dagger}] \;=\; [B_\omega^\alpha \, , \; B_{\omega'}^{\alpha\dagger}] \;=\; 2\pi\delta(\omega - \omega') \tag{C25}$$

and within each set, in or out, other commutators are zero. The transformation matrix connecting these two sets of operators is the S-matrix.

By using the wave equation (C12), we can extend the solutions ϕ_ω^{in} and ϕ_ω^{out} to all time T in a standard way: Introduce the light-cone coordinates

$$z_\pm \;\equiv\; x \pm t$$

and

$$Z_\pm \;\equiv\; X \pm T \; . \tag{C26}$$

According to (C13), any c. number solution of the wave equation can be written as

$$f_+(z_+) + f_-(z_-) \qquad \text{when} \quad X > L(T)$$

and

$$F_+(Z_+) + F_-(Z_-) \qquad \text{when} \quad X < L(T) \; . \tag{C27}$$

Along the dividing line $X = L(T)$ occupied by the shell, we may regard, say, Z_+ as the independent variable (instead of T), thus through (C 4)-(C 6) we can express z_+ as a function of Z_+ :

$$z_+ \;=\; \hat{z}_+(Z_+) \tag{C28}$$

in which the left-hand side is the value of the dependent variable z_+ and, following Unruh, $\hat{z}_+(Z_+)$ denotes the $\hat{z}_+$-function of the independent variable Z_+. Conversely, we may regard z_+ as the independent variable and write along $X = L(T)$

$$Z_+ = \hat{Z}_+(z_+) \quad . \tag{C29}$$

Likewise, we define the functions $\hat{z}_-$ and $\hat{Z}_-$ by regarding along $X = L(T)$, either Z_- or z_- as the independent variable and writing

$$z_- = \hat{z}_-(Z_-) \qquad \text{or} \qquad Z_- = \hat{Z}_-(z_-) \quad . \tag{C30}$$

Clearly, the functions $\hat{z}_\pm$ and $\hat{Z}_\pm$ are the inverse of each other; i.e., for an arbitrary variable y,

$$\begin{aligned} y &= \hat{z}_+(\hat{Z}_+(y)) = \hat{Z}_+(\hat{z}_+(y)) \\ &= \hat{z}_-(\hat{Z}_-(y)) = \hat{Z}_-(\hat{z}_-(y)) \quad . \end{aligned}$$

Next, we apply the wave equation (C12) to derive the continuity condition that relates the two solutions in (C27). On the dividing line $X = L(T)$, let

$$dS_0 \equiv \frac{dL}{dT}\, dT \qquad \text{and} \qquad dS_1 \equiv -\, dT$$

be a line element normal to $X = L(T)$; from (C12), it follows that

$$\sqrt{-|g|}\;\; \phi^\mu dS_\mu \tag{C31}$$

is continuous across $X = L(T)$. Of course, ϕ must also be continuous. These conditions lead to

$$F_{\pm}(Z_{+}) = f_{\pm}(\hat{z}_{\pm}(Z_{\pm})) \tag{C32}$$

which is identical to the converse relation

$$f_{\pm}(z_{\pm}) = F_{\pm}(\hat{Z}_{\pm}(z_{\pm})) \ . \tag{C33}$$

The boundary condition (C7) at $X = 0$ requires

$$F_{-}(Z_{-}) = -F_{+}(Z_{+}) \qquad \text{when} \quad Z_{-} = -Z_{+} \ . \tag{C34}$$

Thus, among the four functions in (C.27), only one is independent. For example, if we choose the independent function to be $f_{+}(z_{+}) = e^{-i\omega z_{+}}$, then the corresponding $F_{\pm}(Z_{\pm})$ and $f_{-}(z_{-})$ are $\pm\exp\left[-i\omega\hat{z}_{+}(\pm Z_{\pm})\right]$ and $-\exp\left[-i\omega\hat{z}_{+}(-\hat{Z}_{-}(z_{-}))\right]$. For T negative, these functions $\hat{z}_{+}$ and $\hat{Z}_{-}$ can be determined by considering only the initial straight portion of the dividing line: $X = L_0$. Thus, $\phi_{\omega}^{in}(X, T)$ defined by (C16) can be written as

$$\phi_{\omega}^{in}(X,T) = i\,e^{-i\delta_0} \cdot \begin{cases} e^{-i\omega z_{+}} - e^{-i\omega\hat{z}_{+}\left[-\hat{Z}_{-}(z_{-})\right]} & \text{when } X > L(T) \\ e^{-i\omega\hat{z}_{+}(Z_{+})} - e^{-i\omega\hat{z}_{+}(-Z_{-})} & \text{when } X < L(T)\,. \end{cases} \tag{C35}$$

Likewise, $\phi_{\omega}^{out}(X, T)$ defined by (C20) becomes

$$\phi_{\omega}^{out}(X,T) = -i\,e^{i\delta_1} \cdot \begin{cases} e^{i\omega z_{-}} - e^{i\omega\hat{z}_{-}\left[-\hat{Z}_{+}(z_{+})\right]} & \text{when } X > L(T) \\ e^{i\omega\hat{z}_{-}(Z_{-})} - e^{i\omega\hat{z}_{-}(-Z_{+})} & \text{when } X < L(T)\,. \end{cases} \tag{C36}$$

Both (C35) and (C36) are valid at all T; so are, then, the expansions (C15) and (C19). Through (C19) and (C24) A_{ω}^{out} and B_{ω}^{out} can be expressed in terms of $\phi(X, T)$ and its derivative at any $T > T_1$. Substituting (C15) into these expressions, we find

$$A_\omega^{out} = \int_0^\infty \frac{d\omega'}{2\pi} \left([\omega \mid \omega']_+ A_{\omega'}^{in} + [\omega \mid \omega']_-^* B_{\omega'}^{in\dagger} \right)$$

and (C37)

$$B_\omega^{out} = \int_0^\infty \frac{d\omega'}{2\pi} \left([\omega \mid \omega']_-^* A_{\omega'}^{in\dagger} + [\omega \mid \omega']_+ B_{\omega'}^{in} \right)$$

where

$$[\omega \mid \omega']_\pm = -\tfrac{1}{2} \left(\frac{\omega}{\omega'}\right)^{\frac{1}{2}} \int_0^\infty dX \sqrt{-|g|} \; g^{00} \frac{dt}{dT} \phi_\omega^{out*} \left(\phi_{\omega'}^{in} \pm \frac{i}{\omega} \frac{\partial \phi_{\omega'}^{in}}{\partial t} \right)$$

(C38)

in which the integration over X is to be evaluated at any fixed time $T \geqslant T_1$. Since A_ω^a and B_ω^a (a = in or out) are T-independent, the transformation matrices $[\omega \mid \omega']_\pm$ are also independent of T.

To compute $[\omega \mid \omega']_\pm$, it is convenient to take a very large T, in which case in most of the X-integration in (C38) the relevant $\hat{z}_\pm$ and $\hat{Z}_\pm$ functions are determined only by the final position of the shell, $X = L_1$ (which is assumed to exist from $T = T_1$ to ∞) ; it is quite easy to see that as $T \to \infty$, that portion of the integration by itself would give

$$2\pi \delta(\omega - \omega')\, e^{i(\delta_1 - \delta_0)} \qquad \text{(C39)}$$

to $[\omega \mid \omega']_+$ and zero to $[\omega \mid \omega']_-$. The remaining part depends on the details of $X = L(T)$ for $T \leqslant T_1$ which, in the integral (C38), refers to an X-region near infinity when $T \to \infty$ (e.g., the region $z_- > 2\ell_1 - \ell_0$ in Figure 5). Of course, this part cannot be neglected even in the limit, since it contains all the information about Hawking radiation, as we shall see. However, the existence of (C39) shows that one is not going to arrive at a pure Hawking radiation by using the S-matrix. This is hardly surprising, since ϕ_ω^{out} carries information of the whole physical domain, both inside and outside the shell.

3. Hawking Radiation

To derive the Hawking radiation, it is necessary to concentrate on measurements dealing with fields only in the outside region, which for $T \geqslant T_1$ is simply $X > L_1$.

As mentioned before, when the parameter ϵ in (C5) approaches 0 , the shell collapses from $L_0 = 2M + \Delta$ at $T = 0$ to $L_1 = 2M$ at T_1 , simulating the formation of a black hole. In this limit, on account of (C6) and (C22), ℓ_1 approaches $-\infty$ and $t_1 \geqslant \ell_0 - \ell_1$ becomes ∞. Correspondingly, the allowed range of x extends from $\ell_1 \to -\infty$ to ∞. Thus, an observer restricted to the outside of the shell might think that his outside world in x and t could be the entire physical space-time domain. Natural basis vectors for such an observer to adopt (in this region $X > L_1$) would be the plane wave solution of the first equation in (C13), $e^{i(kx-\omega t)}$, but without any $X < L_1$ portion, in contrast to ϕ_ω^{out} in (C20). To enhance the sensitivity of our analysis, we assume the positive parameter ϵ to be nonzero but extremely small, $\ll 2M$.

In the following, we shall be interested only in $T = T_1$ and its immediate neighborhood. From (C35), to construct ϕ_ω^{in} at $T = T_1$, we need to know $\hat{Z}_-(z_-)$, especially over the range of z_- for which the collapsing part of the shell-trajectory $L(T)$ is important. This corresponds to the region between the two dashed curves in Figure 5:

$$\ell_1 - t_1 \leqslant z_- \leqslant \ell_0$$

i.e., at $t = t_1$

$$\ell_1 \leqslant x \leqslant t_1 + \ell_0 . \tag{C40}$$

In this region, consider the expansion

$$\phi(X,T) = \int_{-\infty}^{\infty} \frac{dk}{2\pi} \frac{1}{\sqrt{2\omega}} \left[a_k e^{ikx-i\omega(t-t_1)} + b_k^\dagger e^{-ikx+i\omega(t-t_1)}\right] \qquad (C41)$$

where $\omega = |k|$. The inverse of (C41) is

$$\left.\begin{matrix} a_k \\ b_{-k}^\dagger \end{matrix}\right\} \equiv \left(\frac{\omega}{2}\right)^{\frac{1}{2}} \int_{\ell_1}^{\ell_0+t_1} dx \left[\phi(x,t) \pm \frac{i}{\omega} \frac{\partial\phi(x,t)}{\partial t}\right] e^{-ikx\pm i\omega(t-t_1)} \qquad (C42)$$

in which $\phi(x,t)$ refers to the same operator $\phi(X,T)$. If one wishes, one may regard (C42) as the definitions of a_k and $b_{-k}^\dagger$. Equation (C41) is then valid only inside the region (C40); outside, its right-hand side becomes zero. The parameter $\epsilon > 0$ can be arbitrary. Note that if $\epsilon \to 0$, we have $\ell_1 \to -\infty$ and $\ell_0 + t_1 \to \infty$ and therefore the integral in (C42) extends over the entire range of x.

Substituting the expansion (C15), which is valid everywhere, into (C42), we find, setting $t = t_1$,

$$\begin{aligned} a_k &= \int_0^\infty \frac{d\omega'}{2\pi} \left(\{k|\omega'\}_+ A_{\omega'}^{in} + \{k|\omega'\}_-^* B_{\omega'}^{in\dagger}\right), \\ b_k &= \int_0^\infty \frac{d\omega'}{2\pi} \left(\{k|\omega'\}_-^* A_{\omega'}^{in\dagger} + \{k|\omega'\}_+ B_{\omega'}^{in}\right) \end{aligned} \qquad (C43)$$

where, with $\omega = |k|$,

$$\{k|\omega'\}_\pm = \tfrac{1}{2}\left(\frac{\omega}{\omega'}\right)^{\frac{1}{2}} \int_{\ell_1}^{\ell_0+t_1} dx\, e^{-ikx}\left(\phi_{\omega'}^{in} \pm \frac{i}{\omega} \frac{\partial\phi_{\omega'}^{in}}{\partial t}\right). \qquad (C44)$$

So far, all these formulas are exact without approximations. While (C43)-(C44) are similar to (C37)-(C38), unlike A_ω^{out}, $A_\omega^{out\dagger}$ and B_ω^{out}, $B_\omega^{out\dagger}$, these operators a_k, $a_k^\dagger$ and b_k, $b_k^\dagger$ do not form a complete set of independent bona fide annihilation and creation operators.

The function ϕ_ω^{in} at $T = T_1$ depends on the details of $X = L(T)$ between $T = 0$ and T_1. A particularly simple case is when the shell collapses from $L_0 = 2M + \Delta$ to $L_1 = 2M + \epsilon$ at the light-velocity; i.e.,

$$L(T) = L_0 - T \qquad \text{for } T_1 \geqslant T \geqslant 0 \quad . \tag{C45}$$

When $T = T_1 = L_0 - L_1$, t is

$$t_1 = \ell_0 - \ell_1 \tag{C46}$$

and therefore (C40) can be written as

$$2\ell_1 - \ell_0 \leqslant z_- \leqslant \ell_0 \quad , \tag{C47}$$

as illustrated in Figure 5. When z_- is $> \ell_0$, we need only the initial shell position $X = L_0$ to determine the relevant $\hat{z}_+$ and $\hat{Z}_-$ functions in (C35); in the region (C47) the collapsing trajectory (C45) plays an important role. The result is, for $X > L_1$ and at $T = T_1$ (plus its immediate neighborhood),

$$\phi_\omega^{in} = i\, e^{-i\delta_0}\left(e^{-i\omega z_+} - e^{-i\omega \hat{z}_+[-\hat{Z}_-(z_-)]}\right) \tag{C48}$$

where

$$-\hat{z}_+[-\hat{Z}_-(z_-)] = \begin{cases} z_- + (2\delta_0/\omega) & \text{when } z_- > \ell_0 \\ 2(L_0/\Delta)^{\frac{1}{2}}\, \hat{X}\left[\tfrac{1}{2}(\ell_0 + z_-)\right] - \ell_0 & \text{otherwise.} \end{cases} \tag{C49}$$

The function $\hat{X}$ of a variable x is determined by

$$x = X + 2M \ln (X - 2M) \qquad \text{if } X = \hat{X}(x) \quad . \tag{C50}$$

As mentioned before, when $\epsilon \to 0$, we have $\ell_1 \to 2M \ln \epsilon \to -\infty$ and $t_1 \to -2M \ln \epsilon \to \infty$. Since $\frac{1}{2}(\ell_0 + z_-) = \frac{1}{2}(x + \ell_1)$ at $t = t_1$, we have, because of (C50)

$$\hat{X}\left[\tfrac{1}{2}(\ell_0 + z_-)\right] = 2M + \frac{1}{e} \exp\left[\frac{1}{4M}(\ell_0 + z_-)\right] + O(\epsilon) \qquad \text{(C51)}$$

where the estimate $O(\epsilon)$ is valid for $-\infty \leqslant x \leqslant O(1)$. By using (C48)-(C51), the integration (C44) can be readily evaluated. We find, neglecting the $O(\epsilon)$ term in (C51),

$$a_\omega = \int_0^\infty \frac{d\omega'}{2\pi} \left(\frac{\omega}{\omega'}\right)^{\frac{1}{2}} \left[< \omega \mid \Omega > e^{i\theta} A_{\omega'}^{in} + < \omega \mid -\Omega > e^{-i\theta} B_{\omega'}^{in\dagger} \right] ,$$

$$b_\omega = \int_0^\infty \frac{d\omega'}{2\pi} \left(\frac{\omega}{\omega'}\right)^{\frac{1}{2}} \left[< \omega \mid -\Omega > e^{-i\theta} A_{\omega'}^{in\dagger} + < \omega \mid \Omega > e^{i\theta} B_{\omega'}^{in} \right], \qquad \text{(C52)}$$

$$a_{-\omega} = i A_\omega^{in} e^{i\omega t_1} \qquad \text{and} \qquad b_{-\omega} = i B_\omega^{in} e^{i\omega t_1}$$

where

$$\Omega = \frac{1}{2eM} \left(\frac{L_0}{\Delta}\right)^{\frac{1}{2}} e^{\ell_1/4M} \, \omega' ,$$

$$\theta = \left[4M \left(\frac{L_0}{\Delta}\right)^{\frac{1}{2}} - \ell_0\right] \omega' - \frac{\pi}{2} \qquad \text{(C53)}$$

and

$$< \omega \mid \pm\Omega > = 4M (4M\Omega)^{i4M\omega} \, \Gamma(-i4M\omega) \, e^{\pm 2M\pi\omega} .$$

The last equation is identical to (A5)-(A6), with $g = (4M)^{-1}$. Using the commutation relation (C25) and the orthogonality relations between $< \omega \mid \pm\Omega >$, established in (B20)-(B21), we find that (C52) gives

$$[a_\omega , a^\dagger_{\omega'}] = [b_\omega , b^\dagger_{\omega'}] = 2\pi\delta(\omega - \omega') ,$$

$$[a_\omega , a_{\omega'}] = [b_\omega , b_{\omega'}] = 0 , \tag{C54}$$

$$[a_\omega , b_{\omega'}] = [a_\omega , b^\dagger_{\omega'}] = 0$$

and, in addition,

$$[a_{-\omega} , a^\dagger_{-\omega'}] = [b_{-\omega} , b^\dagger_{-\omega'}] = 2\pi\delta(\omega - \omega') ,$$

$$[a_{-\omega} , a_{-\omega'}] = [b_{-\omega} , b_{-\omega'}] = 0 , \tag{C55}$$

$$[a_{-\omega} , b_{-\omega'}] = [a_{-\omega} , b^\dagger_{-\omega'}] = 0 .$$

However, the commutator between $a_{-\omega}$ and b_ω , or $a^\dagger_\omega$, is proportional to $e^{i\omega t_1}$. In the limit when $\epsilon \to 0+$, $e^{i\omega t_1}$ oscillates infinitely fast and, therefore, quite often can be neglected. A word of caution may be necessary. For a free field, each a_k and b_k can undergo a gauge transformation $a_k \to a_k \exp i\theta_a$ and $b_k \to b_k \exp i\theta_b$. Consequently, it is dangerous to approximate $e^{i\omega t_1}$ as zero indiscriminately. Only for physical observations (which are gauge independent) can we neglect such rapid-oscillation terms.

In the Heisenberg picture, the state vector is independent of T . Let $|\,VAC>$ be the vacuum state in the in-representation; i.e.,

$$A^{in}_\omega |\,VAC> = B^{in}_\omega |\,VAC> = 0 . \tag{C56}$$

By using (C52) and (C53), one can readily verify that

$$< VAC \mid a^{\dagger}_{\omega'} \, a_{\omega} \mid VAC > \;=\; < VAC \mid b^{\dagger}_{\omega'} \, b_{\omega} \mid VAC > \;=\; \frac{2\pi\delta(\omega' - \omega)}{e^{8\pi\omega M} - 1} \quad ,$$

but (C57)

$$< VAC \mid a^{\dagger}_{-\omega'} \, a_{-\omega} \mid VAC > \;=\; < VAC \mid b^{\dagger}_{-\omega'} \, b_{-\omega} \mid VAC > \;=\; 0 \quad .$$

For experiments restricted to measuring only quadratic functions of these operators, the situation corresponds exactly to the well-known phenomenon of Hawking radiation. There is a black-body-like radiation of quanta with momentum $k = \omega > 0$, but none for $k = -\omega < 0$.

4. Deviation from Black-body Radiation

(i) There are other experimental possibilities if one deals with higher-order correlation functions. Note that

$$a^{\dagger}_{-\omega} \, a_{-\omega} \; , \quad b^{\dagger}_{-\omega} \, b_{-\omega} \; , \quad a^{\dagger}_{-\omega} \, b_{-\omega} \quad \text{and} \quad b^{\dagger}_{-\omega} \, a_{-\omega}$$

do not carry any rapidly oscillating factors $e^{\pm i\omega t_1}$. Their product with a_{ω} , b_{ω} and $a^{\dagger}_{\omega}$, $b^{\dagger}_{\omega}$ can produce some interesting non-vanishing expectation values, even when the state vector is $\mid VAC >$. For example,

$$< VAC \mid b^{\dagger}_{\omega} \, a^{\dagger}_{-\omega'} \, b_{-\omega'} \, a_{\omega} \mid VAC > = < VAC \mid a^{\dagger}_{\omega} \, b^{\dagger}_{-\omega'} \, a_{-\omega'} \, b_{\omega} \mid VAC >$$

$$= \frac{8\pi M}{\omega'} \; \frac{1}{e^{8\pi M\omega} - 1} \qquad \text{(C58)}$$

and

$$< VAC \mid b^{\dagger}_{\omega} \, a^{\dagger}_{-\omega'} \, a_{-\omega'} \, b_{\omega} \mid VAC > = < VAC \mid a^{\dagger}_{\omega} \, b^{\dagger}_{-\omega'} \, b_{-\omega'} \, a_{\omega} \mid VAC >$$

$$= \frac{8\pi M}{\omega'} \; \frac{1}{e^{8\pi M\omega} - 1} \quad . \qquad \text{(C59)}$$

Both would be zero if the radiation were genuinely black-body. By using (C42), the above operator products can be expressed entirely in terms of ϕ in the outside region $X > L_1$. Thus, it is possible for an observer outside the collapsing shell to ascertain that the state under consideration may be quite different from a black-body radiation.

Because of the boundary condition (C7), the present model somewhat resembles the situation when there is a mirror in the Rindler case that was discussed in sect. 2. This is why even for the $|\,VAC>$ state, one can observe non-black-body correlations between quanta. [Here, (C58)-(C59) play a role analogous to the previous (2.37).]

(ii) As shown in (2.38)-(2.39), it is possible for an observer in the accelerating frame to detect the existence of a wave packet beyond his horizon. In the present case, although (strictly speaking) there is no horizon, one can pose a similar problem: for an observer restricted to the region outside the shell (say, $X > L_1$ and $T \geqslant T_1$), is it possible for him through field measurements to gather information about the physical world inside the shell, $X < L_1$? As we shall see, the answer is yes.

From (C37)-(C38), one notes that the $|\,VAC>$ state defined by (C56) contains excitations when expressed in terms of quanta generated by the $A_\omega^{out\dagger}$ - operators. These excitations are described by ϕ_ω^{out} of (C20), which correlates a wave of frequency (or wave number) ω in the outside region with that of a much higher frequency

$$\nu = (L_1/\epsilon)^{\frac{1}{2}}\,\omega \tag{C60}$$

in the inside region. [Note that when $X \to \infty$, $x \to X$ on account of (C14); hence asymptotically wave numbers in x and X become the same.] Therefore, if the state vector $| >$ at T near T_1 is not $| VAC >$, but $| VAC >$ plus an additional coherent wave packet amplitude of frequency ν inside the shell [analogous to $\int dk\, C_1(k)\, a_k'^{\dagger} | VAC >$ in (2.38)], it would be possible for the observer, still restricted to $X > L_1$, to detect the deviation of $| >$ from $| VAC >$. In other words, just as in (2.39), he would be able to determine the existence of such a wave packet, even though it was beyond his physical reach; in addition, he should be able to measure the long-range correlations inherent in the ϕ_ω^{out}-excitations that are contained in $| VAC >$.

Next, we turn to the question of how such a state $| >$ can be prepared. For convenience, adopt the Schroedinger picture. Consider the example that initially at $T \ll 0$ the state $| >$ consists of $| VAC >$ plus an additional wave packet amplitude of frequency ω in the outside region, very far from the collapsing shell. This wave packet, traveling at light-velocity, is assumed to move faster than the collapsing matter. Eventually, say at T near T_1, it passes the collapsing shell and goes behind $X = L(T)$, as indicated by the shaded region in Figure 5. By then, its frequency is shifted from ω to near ν, providing the kind of state vector $| >$ being discussed. The final state is a coherent one, if the initial state is.

REFERENCES

[1]. S. W. Hawking, Nature 248, 30 (1974); Comm.Math.Phys. 43, 199 (1975); Phys.Rev. D14, 2460 (1976).

[2]. W. Rindler, Am.J.Phys. 34, 1174 (1966).

[3]. S. Fulling, Phys.Rev. D7, 2850 (1973).

[4]. W. G. Unruh, Phys.Rev. D14, 870 (1976).

[5]. N. D. Birrell and P. C. W. Davies, Quantum Fields in Curved Space, Cambridge University Press, 1982.

[6]. R. Wald, Comm.Math.Phys. 45, 9 (1975); Phys.Rev. D13, 3176 (1976).

[7]. L. Parker, Phys.Rev. D12, 1519 (1976).

[8]. D. G. Boulware, Phys.Rev. D11, 1404 (1975); Phys.Rev. D13, 2169 (1976).

[9]. W. Israel, Phys.Lett. 57A, 107 (1976).

[10]. W. Unruh and R. Wald, Phys.Rev. D29, 1047 (1984).

[11]. J. B. Hartle and S. W. Hawking, Phys.Rev. D13, 2188 (1976).

[12]. G. Gibbons and M. Perry, Proc.Roy.Soc. (London) A358, 467 (1978).

[13]. A. Einstein, B. Podolsky and N. Rosen, Phys.Rev. 47, 777 (1935).

[14]. D. Bohm and Y. Aharonov, Phys.Rev. 108, 1070 (1957); Nuovo Cimento 17, 964 (1960).

[15]. J. S. Bell, Physics 1, 195 (1964).

[16]. L. R. Kasday, J. D. Ullman and C. S. Wu, Nuovo Cimento 25B, 633 (1975).

[17]. S. Tomonaga, Prog.Theor.Phys. 1, 27 (1946).

[18]. K. Freese, C. T. Hill and M. Mueller, "Covariant Functional Schroedinger Formalism and Application to the Hawking Effect" (Fermilab preprint).

[19]. J. Bardeen, L. N. Cooper and J. R. Schrieffer, Phys.Rev. 108, 1175 (1957).

[20]. M. Goldhaber, T. D. Lee and C. N. Yang, Phys.Rev. 112, 1796 (1958).

[21]. M. D. Kruskal, Phys.Rev. 119, 1743 (1960).

[22]. G. Szekeres, Publ.Mat.Debreen 7, 285 (1960).

[23]. T. D. Lee and Y. Pang, in preparation.

FIGURE CAPTIONS

Figure 1. An observer in a constant accelerating frame Σ_{acc} "sees" only a quarter of the space $(I : X \geqslant |T|)$ in Σ.

Figure 2. The (X, T) plane defines the minkowski space in Σ, with $\overline{OH}$ along one of the horizons in Σ_{acc}. The corresponding euclidean space is the (X, Y) plane. When θ varies from 0 to π, the point p moves from I in Figure 1 to II.

Figure 3. The decomposition of the cartesian propagator $< 1 \mid \mathcal{D} \mid 2 >$ into the polar propagator $< 1 \mid D \mid 2 >$ plus two other diagrams, which can be interpreted as due to an "external" field coming from the horizon. [See Eq. (3.45).]

Figure 4. Example of a one-loop correction to the scattering amplitude $1 + 2 \rightleftarrows 3 + 4$. The left-hand side is in cartesian coordinates, and the right side in polar coordinates. [See Eq. (3.50).]

Figure 5. An example of a shell collapsing from $(X, T) = (L_0, 0)$ to (L_1, T_1) along $Z_+ = L_0$, i.e., $z_+ = \ell_0$ [$Z_\pm = X \pm T$ and $z_\pm = x \pm t$ with t and x given by (C6), (C14) and (C46); when $T = T_1 = L_0 - L_1$, $t = t_1 = \ell_0 - \ell_1$].

FIGURE 1

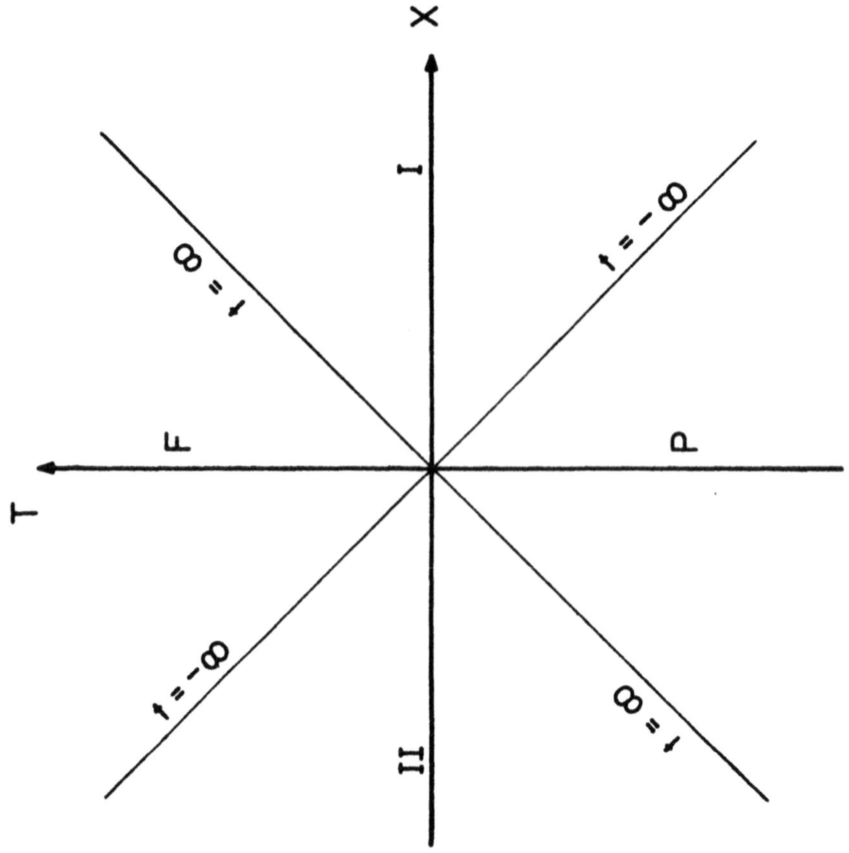

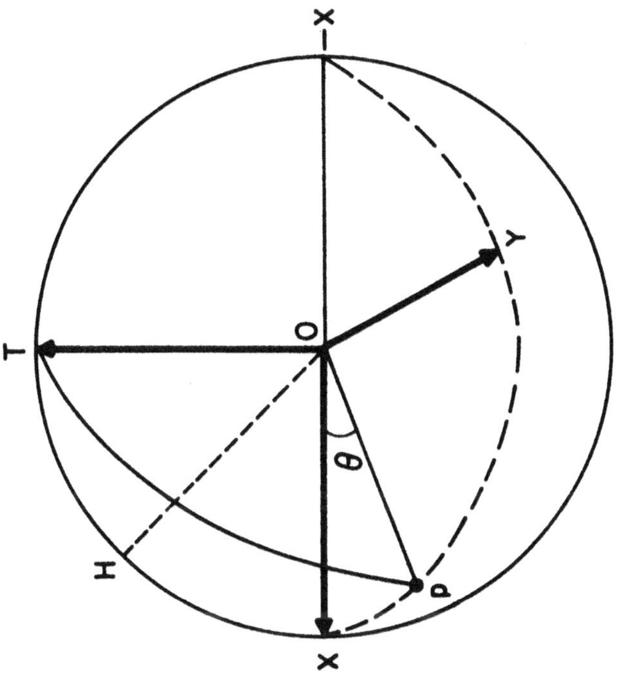

FIG. 2

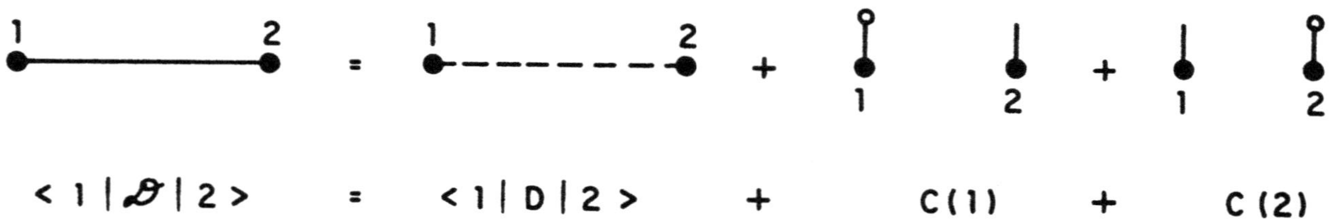

FIG. 3

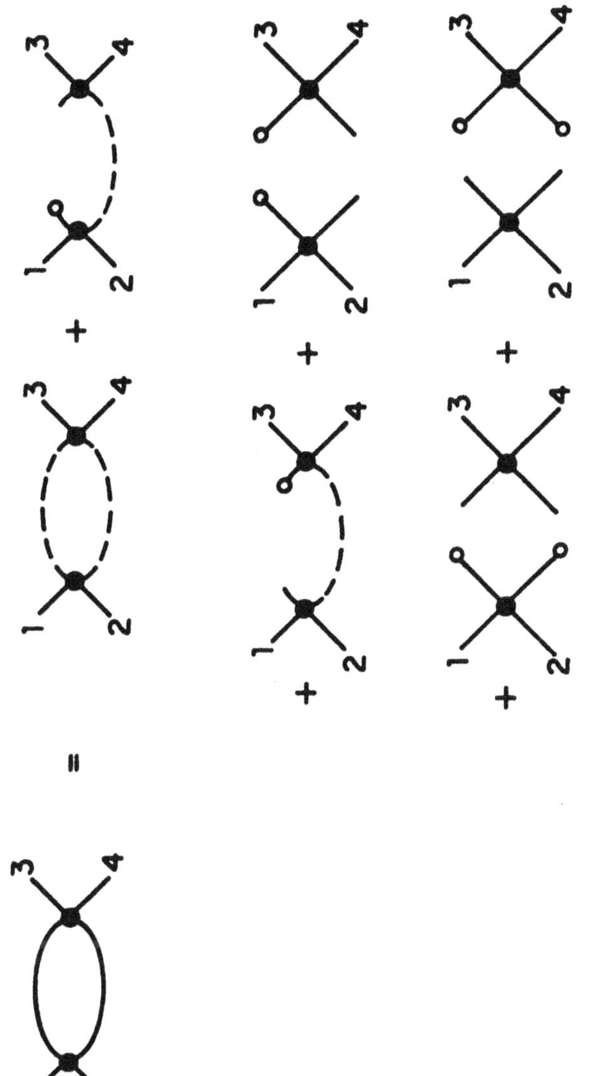

FIG. 4

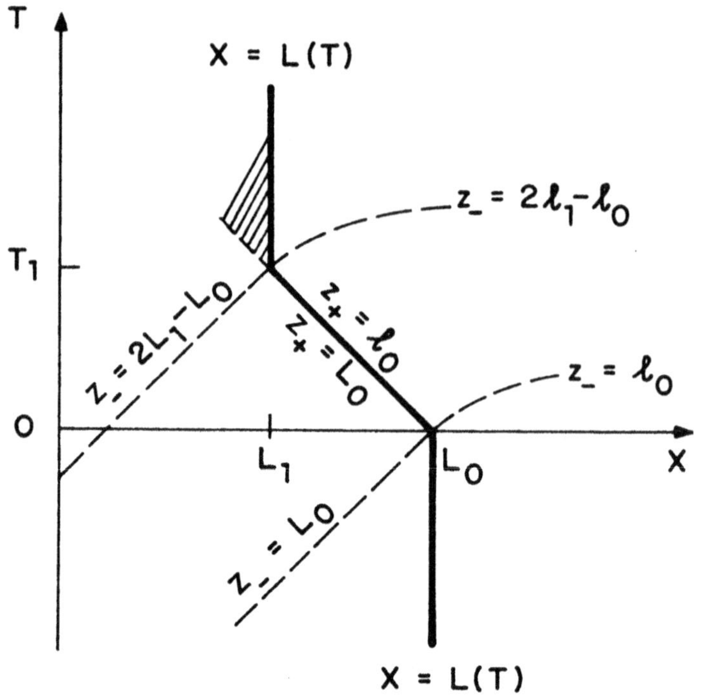

FIGURE 5.

Reprinted from
Nucl. Phys. **B264,** 437 (1986)

Permissions

Birkhäuser Boston would like to thank the original publishers of the papers of T. D. Lee for granting permission to reprint specific papers in this collection:

VII. Discrete Physics

[179] Reprinted from *Nucl. Phys.* **B202,** © 1982 by North-Holland Physics Publishing.
[180] Reprinted from *Nucl. Phys.* **B210** (FS6), © 1982 by North-Holland Physics Publishing.
[181] Reprinted from *Nucl. Phys.* **B210** (FS6), © 1982 by North-Holland Physics Publishing.
[182] Reprinted from *Phys. Letters* **122B**, © 1983 by North-Holland Physics Publishing.
[183] Reprinted from *Nucl. Phys.* **B225** (FS9), © 1983 by North-Holland Physics Publishing.
[185] Reprinted from *Proceedings of the International School of Subnuclear Physics,* © 1986 by Plenum Press.
[188] Reprinted from *Nucl. Phys.* **B242,** © 1984 by North-Holland Physics Publishing.
[189] Reprinted from *Nucl. Phys.* **B245,** © 1984 by North-Holland Physics Publishing.
[195] Reprinted from *The Lesson of Quantum Theory,* edited by J. de Boer, E. Dal and O. Ulfbeck, © 1986 by North-Holland Physics Publishing.

VIII. Strong Interaction Models

[72] Reprinted from *Phys. Rev.* **135,** © 1964 by the American Physical Society.
[143] Reprinted from *Phys. Rev.* **D5,** © 1972 by the American Physical Society.
[144] Reprinted from *Phys. Rev.* **D6,** © 1972 by the American Physical Society.
[145] Reprinted from *Phys. Rev.* **D6,** © 1972 by the American Physical Society.
[152] Reprinted from *Revs. Mod. Phys.* **47,** © 1975 by the American Physical Society.
[155] Reprinted from *Report of the Workshop on BeV/Nucleon Collision of Heavy Ions—How and Why,* BNL Report No. 50445, © 1975 by Brookhaven National Laboratory.
[156] Reprinted from *Phys. Rev.* **D11,** © 1975 by the American Physical Society.
[156a] Reprinted from *Phys. Rev.* **D12,** © 1976 by the American Physical Society.
[170] Reprinted from *Mesons in Nuclei, Vol. III,* edited by M. Rho and D. Wilkinson, © 1979 by North-Holland Physics Publishing.
[176] Reprinted from *Proceedings of the 1980 Guangzhou Conference on Theoretical Particle Physics,* edited by Ning Hu and Hung-yuan Tzu, © 1980 by Science Press, Beijing and Van Nostrand Reinhold Company Inc.

IX. Historical Papers

[136] Reprinted from *Elementary Processes at High Energy,* edited by A. Zichichi, © 1971 by Academic Press.

X. Gravity (Continuum Theory)

[192] Reprinted from *Nucl. Phys.* **B264,** © 1986 by North-Holland Physics Publishing.

Contemporary Physicists

James Glimm and Arthur Jaffe: Collected Papers
0-8176-3273-5 (complete set)
0-8176-3271-9 (Volume 1)
0-8176-3272-7 (Volume 2)

T.D. Lee: Selected Works
0-8176-3344-8 (complete set)
0-8176-3341-3 (Volume 1)
0-8176-3342-1 (Volume 2)
0-8176-3343-X (Volume 3)